ICANN '94

Proceedings of the International Conference on Artificial Neural Networks
Sorrento, Italy
26–29 May 1994

Volume 1, Parts 1 and 2

Edited by
Maria Marinaro and Pietro G. Morasso

Springer-Verlag
London Berlin Heidelberg New York
Paris Tokyo Hong Kong
Barcelona Budapest

Maria Marinaro
University of Salerno
Dipartimento di Fisica Teorica e S.M.S.A
Via S. Allende
84081 Barnossi, Salerno
Italy

Pietro G. Morasso
University of Genova
Dipartimento di Informatica,
Sistemistica, Telematica
Via Opera Pia 11A
16145 Genova
Italy

ISBN-13:978-3-540-19887-1   e-ISBN-13:978-1-4471-2097-1
DOI: 10.1007/978-1-4471-2097-1

Typesetting: Camera ready by contributors

34/3830-543210 Printed on acid-free paper

Preface

This book contains the proceedings of the International Conference on Artificial Neural Networks which was held between May 26 and 29 1994 in Sorrento, organized by the Department of Theoretical Physics of the University of Salerno, IIASS and the Istituto Italiano per gli Studi Filosofici (Napoli), and cosponsored by ENNS, JNNS, IEEE and SIREN. It is the fourth in a series after Helsinki (1991), Brighton (1992) and Amsterdam (1993). The conference is the main event of the European Neural Network Society, which is fostering the growth of the interdisciplinary community of European researchers and supporting the interaction and cooperation with other international societies in the same scientific area.

Neural network is a living proof that the re-birth in science is possible. The first life was driven by a handful of pioneers who, in the fifties and early sixties, discovered the analogies between machines and biological systems as regards communication, control, computing. In a sense, this was not completely new, and we are sure that the mechanical toys of the eighteenth century are a beautiful although naive witness of its illuministic roots. However, the idea was a daring one, and the technological/methodological constraints of the time did not allow it to flourish and trigger a virtuous growth cycle through significant successes in the real world.

We are now in the second life. Computation is cheap; software tools are powerful; mathematical tools, developed by computer-aware people, accumulated steadily. Moreover, the brain is less of a black box and can teach us some things. Researchers in many fields have demonstrated in the last decades that they are willing to try again: mathematicians, physicists, psychologists, neuroscientists, and, last but not least engineers who have the crucial task to prove the feasibility of the field and make things work. Of course, there is hype, not all promises will be maintained, but even if only 10% of them succeed, as is happening now in many application areas, this will allow the second wave to expand and consolidate. We can forsee the limits, mainly due to technology, and we can already guess the real break-throughs, which will make the third wave possible and make neural networks as pervasive as are microcomputers now, will happen when new materials, self-organizing, adaptable wet-ware become available. For all these reasons it is mostly appropriate to dedicate this conference to Eduardo R. Caianiello, one of the pioneers of the first wave and an eminent researcher of the second one. Eduardo R. Caianiello, who was also one of the founding fathers of ENNS, who put the machine of this conference in motion, died on October 22nd 1993.

We would like to thank all organizers and volunteers for their invaluable help that made the conference possible and the distinguished referees who assured the necessary high standard.

Sorrento, May 1994   Maria Marinaro
Pietro G. Morasso

The financial supports of

      Università di Salerno
      Università di Napoli
      Dipartimento di Scienze Fisiche, Università di Napoli
      Istituto di Cibernetica, Arco Felice, Napoli
      Consiglio Nazionale delle Ricerche
      Elsag Bailey s.p.a
      SGS – Thompson
      DIST, Università di Genova
      Regione Campania

are gratefully acknowledged.

Referee Committee

Contents, Volume 1

Part 2 • Mathematical Model

PART 1

Neurobiology

Why Bright Kanizsa Squares Look Closer: Consistency of Segmentations and Surfaces in 3-D Vision

Stephen Grossberg†
Department of Cognitive and Neural Systems
Boston University
Boston, Massachusetts 02215 USA

1. Introduction: The Need for Boundary-Surface Consistency

When a human observer views a Kanizsa square under appropriate viewing conditions, the bright square appears to be closer than its inducing pac man wedges (Figure 1). Much experimental evidence suggests that the square's apparent brightness and depth covary relative to those of the picture background (Bradley and Dumais, 1984; Kanizsa, 1955, 1974; Purghé and Coren, 1992). This interaction between the illusory contours that frame the square, the brightness percept that fills it in, and the depthful pop-out of the square from its background illustrate in a dramatic way how fundamentally different are biological vision processes from those of traditional machine vision algorithms. The present article sketches an explanation of this percept as part of a larger theory of biological vision that develops a solution of the classical figure-ground problem (Grossberg, 1993, 1994).

A key property of the theory, called *boundary-surface consistency*, suggests how only those boundary segmentations that are capable of supporting filled-in surface representations survive in the final 3-D percept. Feedback signals between boundary and surface representations are needed to ensure boundary-surface consistency. These feedback signals help to pop-out the brighter Kanizsa square so that it appears in front of its background.

2. Interscale and Interstream Interactions

The theory is called FACADE Theory because it suggests how representations of **F**orm-**A**nd-**C**olor-**A**nd-**DE**pth are generated in extrastriate cortex, notably area V4. FACADE theory describes the neural architecture of two parallel subsystems, the Boundary Contour System (BCS) and the Feature Contour System (FCS). The BCS generates an emergent 3-D boundary segmentation of edges, texture, shading, and stereo information at multiple spatial scales, whereas the FCS compensates for variable illumination conditions and fills-in surface properties of brightness, color, and depth among multiple spatial scales. See Grossberg (1994) for a self-contained exposition of the theory and its explanations of other data.

In its original form (Grossberg, 1987a, 1987b), FACADE Theory did not posit interactions between the different spatial scales of the BCS and the FCS,

† This research was supported in part by ARPA (ONR N00014-92-J-4015) and the Office of Naval Research (ONR N00014-91-J-4100). The author wishes to thank Cynthia E. Bradford for her valuable assistance in the preparation of the manuscript.

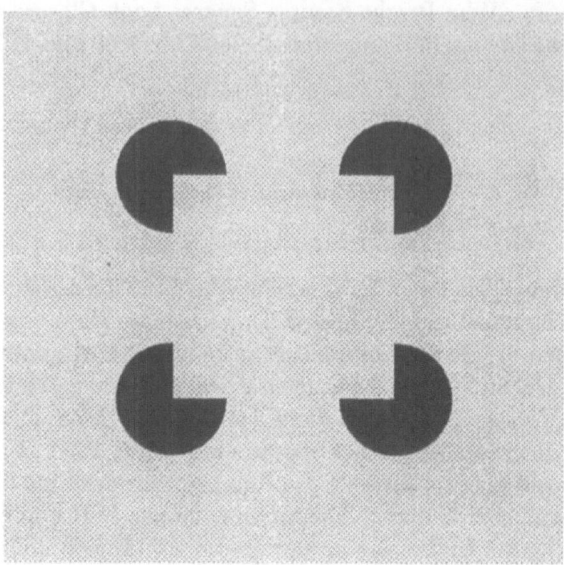

Figure 1. A Kanizsa square.

or from the FCS to the BCS. Such interactions were not needed to explain the data analysed in previous articles. Grossberg (1994) shows how interactions within and between BCS and FCS scales lead to explanations of a much wider body of data about 3-D visual perception than could be handled before.

The theory posits the existence of seven types of interactions that complement, and are consistent with, previously defined BCS and FCS mechanisms (Figure 2). These interactions clarify how the visual system can generate globally unambiguous 3-D surface representations from image data which contain several different types of local ambiguities. In these interactions, larger scales tend to influence smaller scales, and larger disparities tend to influence smaller disparities. Thus the new interactions tend to be *partially ordered* across scale and disparity.

The first new interaction takes place among the complex cells of the BCS. Model complex cells with large receptive fields can binocularly fuse more disparities than cells with small fields. Inhibitory competitive interactions occur between complex cells that code different disparities at the same position and size scale. These interactions are called *BB Intrascales*. Typically, active BCS complex cells that code larger disparities inhibit complex cells that code smaller disparities—an example of partial ordering. This competition sharpens the disparity tuning curves of the BCS complex cells, and selects those complex cells whose disparity tuning best matches the binocular disparities derived from an image.

Interactions called *BB Interscales* are excitatory cooperative interactions from bipole cells to hypercomplex cells that code the same disparity and position, across all scales. Each such CC Loop network is called a *BCS copy*. Each BCS copy generates its own emergent boundary segmentations corresponding to a prescribed disparity range, or relative depth from the observer. Each segmentation forms the best spatial compromise between all the scales that are sensitive to its disparity range. Due to the combined effect of these cooperative interactions and of competitive interactions among BCS hypercomplex cells, the larger scales tend to inhibit the smaller scales within each BCS copy in the manner reported in psychophysical data (Tolhurst, 1972; Watt, 1987; Wilson, Blake, and Halpern, 1991). These interactions are predicted to include the Interstripes in cortical area V2.

Each disparity-sensitive 3-D boundary segmentation in a BCS copy interacts with a Monocular FIDO, or Filling-In-DOmain, of the FCS. These BCS signals select those monocular brightness and color signals that are consistent with the binocular BCS segmentation, and suppress the rest. These BCS → FCS interactions are called *BF Intracopies* because each BCS copy selects binocularly consistent monocular data from a corresponding FCS copy.

Thus the illuminant-discounted monocular FCS representation is transformed into multiple FCS copies, or Monocular FIDOs, one for each BCS copy. This one-to-many transformation carries out two functions. First, it maps the monocular positions of FCS signals into the binocular positions of the corresponding BCS copy. It is hypothesized that the BF Intracopy signals act as teaching signals to realign the FCS → FCS pathways based on their mutual correlation during visual experience. This adaptive process was used to help explain monocular McCollough effect data in Grossberg (1987b). Second, this one-to-many transformation enables monocular FCS signals that do not positionally match binocular BF Intracopy signals in a given FCS copy to be suppressed. The same monocular FCS signals may be selected for further processing in a different FCS copy where they do positionally match the corresponding BF Intracopy signals. This one-to-many transformation is called *Monocular FF Intercopies*.

In addition, reciprocal interactions exist from the FCS to the BCS. They are called *FB Intercopies*, and ensure boundary-surface consistency. These FCS output signals are derived from those filled-in FCS regions at the monocular FIDOs that are surrounded by connected boundaries. These filled-in connected domains, which represent those monocular surface representations that are binocularly consistent, are used to build up the final 3-D surface representation at the Binocular FIDOs. In particular, the filled-in connected FCS regions activate contrast-sensitive FCS → BCS pathways that generate FCS output signals at the edges of the filled-in connected regions. These outputs excite BCS cells corresponding to the same disparity and position at its BCS copy while inhibiting BCS cells corresponding to smaller disparities at that position. The FB Intercopy signals hereby inhibit the BCS boundaries of any occluded region that occur at the same positions as the boundaries of an occluding region.

Possible neural loci for these BF Intracopies and FB Intercopies are suggested by the neural interpretation of the BCS in terms of the Interblob corti-

6

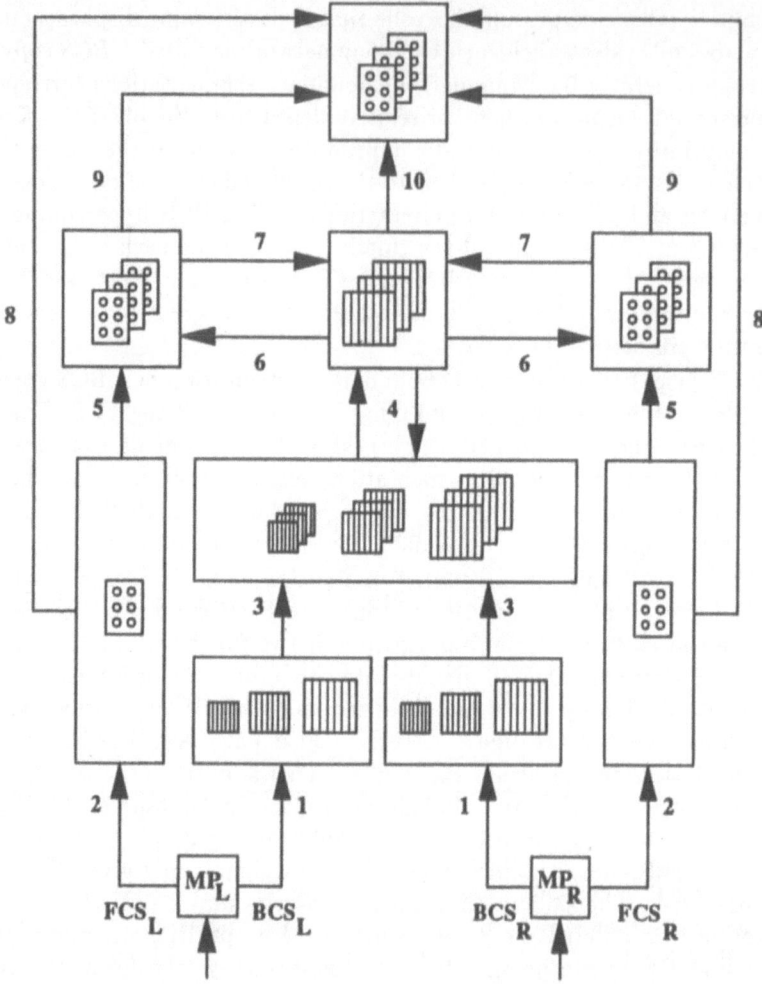

Figure 2. Macrocircuit of monocular and binocular interactions of the BCS and FCS: Left eye and right eye monocular preprocessing stages (MP_L and MP_R) send parallel pathways to the BCS (boxes with vertical lines, designating oriented responses) and the FCS (boxes with three pairs of circles, designating opponent colors). Output signals from MP_L and MP_R activate BCS simple cells with multiple receptive field sizes via pathways 1. MP_L and MP_R outputs are also transformed into opponent FCS signals via pathways 2. Pathways 3 generate multiple cell pools that are sensitive to multiple disparities and scales. BB Intrascales are at work among the resultant cells. Pathways 4 combine the multiple scales corresponding to the same depth range into a single BCS copy via BB Interscales. Multiple copies exist corresponding to different (but possibly overlapping) depth ranges. Pathways 5 are the Monocular FF Intercopies. Pathways 6 are the BF Intracopies. Pathways 7 are the FB Intercopies. Pathways 8 are the excitatory Binocular FF Intercopies. Pathways 9 are the inhibitory Binocular FF Intercopies. Pathways 10 are the BF Intercopies. See the text for further details.

cal stream and of the FCS in terms of the Blob parvocellular stream. Within cortical area V2, Thin Stripes should be investigated as possible Monocular FIDOs, with Interstripe-to-Thin Stripe pathways as the BF Intracopies, Blob-to-Thin Stripe pathways as the Monocular FF Intercopies, and Thin Stripe-to-Interstripe pathways as the FB Intercopies.

In addition to these FF, BF, and FB interactions, *Binocular FF Intercopies* are predicted to occur. Both excitatory and inhibitory output signals are generated, as in the case of FB Intercopies. The excitatory signals from each eye activate Binocular FIDOs that correspond to the same disparity and position. The inhibitory signals suppress Binocular FIDOs corresponding to smaller disparities at the same position. These interactions obliterate the brightness and color signals that could otherwise erroneously fill-in surface representations of occluded objects in the regions where they are occluded. The surviving excitatory signals from both eyes are binocularly matched to trigger the filling-in of the 3-D surface representation. These Binocular FF Intercopies occur within the Blob cortical stream. The *excitatory* Binocular FF Intercopies arise from the same source of illuminant-discounted FCS signals as the Monocular FF Intercopies. In contrast, the *inhibitory* Binocular FF Intercopies arise from the edges of the filled-in connected regions within the Monocular FIDOs, as do the FB Intercopies. The excitatory Binocular FF Intercopies form a one-to-many map to the Binocular FIDOs. They are positionally aligned among the Binocular FIDOs using BCS $\rightarrow$ FCS boundary signals as teaching signals. These are the BF Intercopies that were used in Grossberg (1987b) to help explain data about binocular transfer of the McCollough effect. The positions of the inhibitory Binocular FF Intercopies are defined by the binocularly shifted BF Intracopies that define the filled-in domains whose edges activate them. The inhibitory FF Intercopies also converge upon the Binocular FIDOs, where they suppress FCS signals that would otherwise trigger filling-in of occluded regions.

The final interactions are called *BF Intercopies*. These are BCS $\rightarrow$ FCS boundary signals from a given disparity and position that add to the BCS boundaries of all smaller disparities at that position in order to prevent all nearer occluding surfaces from appearing transparent due to filling-in of their positions by the brightness and colors of farther occluded surfaces.

3. Boundary-Brightness-Depth Interactions in Kanizsa Square Percepts

The percept of closer depth and enhanced brightness induced by a Kanizsa display may be explained by these mechanisms. Consider the percept under conditions of binocular viewing. BB Interscales of the CC Loop form an illusory contour around the Kanizsa square. This square boundary encloses a connected region. Within each copy of the BCS, there is a largest disparity at which the complex cells can induce the formation of such a connected square. BB Interscales hereby form multiple copies of the square boundary within multiple BCS copies, where each copy is capable of binocularly fusing a different range of non-zero disparities. These illusory boundaries are no stronger than the boundaries that are formed around the pac man inducers themselves by a similar process.

The connected. BCS boundaries use BF Intracopies to form filling-in domains within the monocular FIDOs via the pathways labelled 6 in Figure 2. The discounted feature contour signals from the monocular preprocessing stages then trigger filling-in of the connected regions. The interior of the Kanizsa square has a higher level of filled-in activity than the background due to the spatial distribution of these feature contour signals. As a result, the contrast between the filled-in activity of the Kanizsa square and the pac man figures is greater than the contrast between the filled-in activity of the background and the pac man figures. This contrast difference is one of the key properties in the explanation.

A second key property concerns the way FB Intercopies respond to this contrast difference. In particular, contrast-sensitive FB Intercopy signals via the pathways labelled 7 in Figure 2 excite BCS boundaries corresponding to the same disparity and position, but inhibit BCS boundaries corresponding to smaller disparities at that position. Due to this excitatory FB feedback, the illusory BCS boundaries of the Kanizsa square become stronger than the (remaining) BCS boundaries of the pac man figures. Using competitive BCS interactions at the hypercomplex cells, these strengthened Kanizsa square boundaries can now cause gaps in the boundaries, called end cuts, to form where the pac man boundaries join the square boundary. The boundaries around the pac man regions are no longer connected. Consequently they cannot contain the filling-in process within the corresponding monocular FIDO. Activity hereby diffuses out of the pac man figures.

This excitatory FB feedback and end cutting take place in the BCS copy corresponding to the largest disparity that can respond to the image. The escape of activity from the corresponding pac man FIDO eliminates FB Intercopy signals from the pac man boundaries, both excitatory and inhibitory. Removing the inhibitory FB Intercopy signals from the pac man boundaries enables the BCS copies that are sensitive to smaller disparities to form pac man boundaries. Pac man boundaries do not, however, form at the common edge of any pac man with the square, because FB Intercopy signals from the square region of the largest disparity copy inhibit the BCS square boundaries at all smaller disparities. With the square boundaries out of the way, the pac man boundaries can complete an (almost) circular boundary within the BCS copies of the smaller disparities. In all, a square boundary is completed at a BCS copy corresponding to a nearer depth, while (almost) circular boundaries are completed at a BCS copy corresponding to a slightly farther depth. When these segmentations are input to the Object Recognition System in temporal cortex, they lead to recognition of a Kanizsa square in front of circular regions.

Why do we see a square surface in front of the unoccluded surfaces of the pac men? Why are the boundaries that are completed behind the square recognized but not seen? BF Intercopies add larger disparity boundaries to smaller disparity boundaries. FF Intercopies transmit the filling-in generators of the high contrast square to the largest disparity binocular FIDO, while they inhibit the filling-in generators of the square at smaller disparity binocular FIDOs. The filling-in generators of the background and of the pac man figures are not inhibited at these smaller disparity FIDOs. The resultant FACADE representation fills-in a brighter surface representation of the square at a larger disparity binocular FIDO. The pac man figures and their background fill-in at a smaller disparity binocular FIDO, but cannot fill-in behind the square, due to the action of BF Intercopies. Consequently, the brighter Kanizsa square looks closer than the background. In addition, the pac man figures are recognized, but not seen, behind the occluding square surface.

At bottom, this interaction between boundary, brightness, and depth percepts is a consequence of the surface filling-in that compensates for variable illumination, the use of connected monocular surfaces to reinforce consistent boundary segmentations which thereupon control the formation of binocular surface representations, and the use of end cuts to detach the boundaries of nearer surfaces from those of farther surfaces to facilitate figure-ground pop-out.

References

Bradley, D.R. and Dumais, S.T. (1984). The effects of illumination level and retinal size on the depth stratification of subjective contour figures. *Perception*, **13**, 155–164.

Grossberg, S. (1987a). Cortical dynamics of three-dimensional form, color, and brightness perception, I: Monocular theory. *Perception and Psychophysics*, **41**, 87–116.

Grossberg, S. (1987b). Cortical dynamics of three-dimensional form, color, and brightness perception, II: Binocular theory. *Perception and Psychophysics*, **41**, 117–158.

Grossberg, S. (1993). A solution of the figure-ground problem for biological vision. *Neural Networks*, **6**, 463–483.

Grossberg, S. (1994). 3-D vision and figure-ground separation by visual cortex. *Perception and Psychophysics*, in press.

Kanizsa, G. (1955). Margini quasi-percettivi in campi con stimolazione omogénea. *Rivista di Psicologia*, **49**, 7–30.

Kanizsa, G. (1974). Contours without gradients or cognitive contours. *Italian Journal of Psychology*, **1**, 93–113.

Purghé, F. and Coren, S. (1992). A modal completion, depth stratification, and illusory figures: A test of Kanizsa's explanation. *Perception*, **21**, 325–335.

Tolhurst, D.J. (1972). Adaptation to square-wave gratings: Inhibition between spatial frequency channels in the human visual system. *Journal of Physiology*, **226**, 231–248.

Watt, R.J. (1987). Scanning from coarse to fine spatial scales in the human visual system after the onset of a stimulus. *Journal of the Optical Society of America*, **4**, 2006–2021.

Wilson, H.R., Blake, R., and Halpern, D.L. (1991). Coarse spatial scales constrain the range of binocular fusion on fine scales. *Journal of the Optical Society of America*, **8**, 229–236.

Spatial Pooling and Perceptual Framing by Synchronizing Cortical Dynamics

Stephen Grossberg* and Alexander Grunewald[t]

Center for Adaptive Systems and Department of Cognitive and Neural Systems

Boston University, Boston, USA

1 Introduction

The primate visual system performs the complex task of analyzing the visual environment in several stages. At the first stage, the retina, the incoming image is transduced into neural signals. These signals are then transmitted to the lateral geniculate nucleus (LGN) and from there to the striate cortex (V1). Cells in all these stages have comparatively small receptive fields, with the biggest being in V1. The maximum size of striate receptive fields have a diameter of about one degree (Hubel & Wiesel, 1968). Unlike the receptive fields of cells in the retina and the LGN, receptive fields of striate neurons tend to have a preferred orientation. These cells fire optimally when a bar of their preferred orientation is in their receptive fields. Since the receptive fields are rather small, it can be said that striate neurons respond to local features, and hence they decompose the retinal image into its main local orientations.

It has been possible to build computers that perform the same operation. However, what has thus far eluded engineering approaches is a mechanism that integrates the information across the entire internal cortical representation. This is necessary since the interpretation of an image, which includes the recognition and the localization of objects in the image, requires global information. How can the visual system achieve this task? In the present study it is shown that horizontal integration within the visual cortex can improve performance of single cortical cells, and thus can form a starting point for the understanding of visual images.

Another issue that has to be solved by the visual system is that the processing of different parts of an image may happen at different rates, so that the cortical representation of the image may be desynchronized. As long as the retinal image is constant this does not cause serious problems. However, when there is motion in the retinal image, the visual system needs to ensure that all the parts corresponding to the same retinal image are processed together, to avoid wrong correspondences. The process that ensures this is called perceptual framing (Varela, Toro, John, & Schwartz, 1981). The network model used in the present study was also used to show that perceptual framing can be implemented with the same type of horizontal connections that have been postulated in a model of form perception and perceptual grouping (Grossberg & Mingolla, 1985).

*Supported in part by AFOSR F49620-92-J-0499, ONR N00014-92-J-4015, and ONR N00014-91-J-4100.

[t]Supported in part by AFOSR F49620-92-J-0225 and AFOSR F49620-92-J-0334.

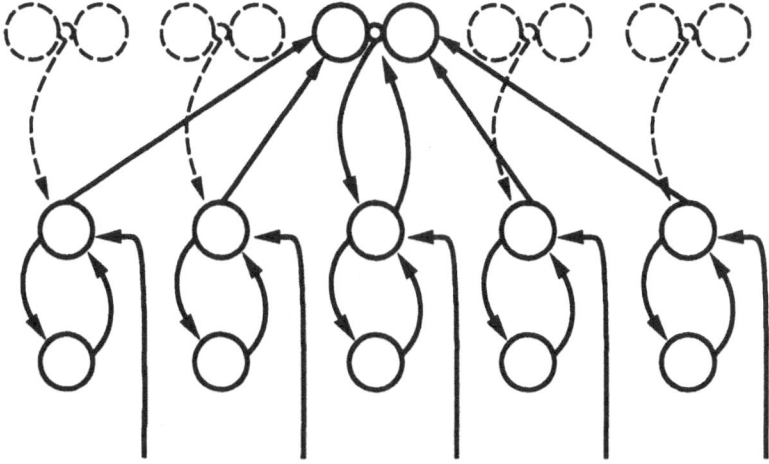

Figure 1: The architecture of the model proposed. A layer of fast–slow oscillators is coupled via a layer of bipole cells. In contrast to previous versions of the model, there is a direct signal from each oscillator to its corresponding bipole cell which facilitates boundary completion.

2 Spatial Pooling

Spatial summation is the effect that when stimuli are very small (typically smaller than the size of a striate receptive field), then an increase of stimulus size leads to a reduction of threshold contrast (Thomas, 1978). These experiments were conducted under the assumption that spatial summation only occurs within the receptive field, and hence little data are available that indicate cooperative interactions over sizes that go significantly beyond a single receptive field. One exception is a study by Essock (1990), which showed a reduction of threshold contrast up to line lengths of 5 degrees. This length is too long to allow an explanation within a striate receptive field, thus suggesting some kind of horizontal cortical cooperation. In the present study we call this effect spatial pooling to distinguish it from spatial summation proper.

3 Perceptual Framing

Perceptual framing is the process of binding together parts of neural representations corresponding to the same image that may have come temporally out of register due to early processing. A possible mechanism for this would be some clocking device. Here we model how synchronization of distributed cortical activities can temporally realign out-of-phase image parts. It has been found that cortical activities synchronize in the cat and in the monkey when a stimulus is present in the visual field (Eckhorn, Bauer, Jordan, Brosch, Kruse, Munk, & Reitboeck, 1988; Gray & Singer, 1989), even when the receptive fields of the units recorded do not overlap.

A way to test this notion of perceptual framing has been to link it to temporal order judgments of two separate visual stimuli. Perceptual framing suggests that there is a definite nonzero lower bound for the time between two such stimuli that is necessary to allow reliable performance. Hirsch and Sherrick (1961) have found such a lower bound with highly trained subjects.

4 Description of the Model

Grossberg and Mingolla (1985) developed a model called the Boundary Contour System (BCS) for the generation of emergent boundaries by the visual cortex. This model was later adapted to show that cortical synchronization of neural activities does not require the presence of a central clocking mechanism (Grossberg & Somers, 1991). In the present study, we further develop and modify this model. There are two layers, one consisting of fast-slow neural oscillators (Ellias & Grossberg, 1975), and the other of bipole cells, that receive input to two separate lobes, in addition to receiving direct bottom-up input. In the present simulations, bipole cells fire if at least two of its three receptive zones are activated. The model is shown in Figure 1.

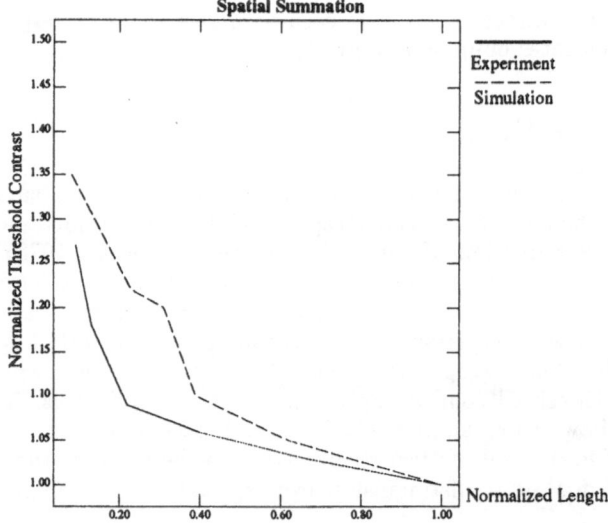

Figure 2: Comparison between psychophysical data and computer simulations. Normalized threshold contrast as a function of normalized stimulus size. Solid line: results from experimental study. Dashed line: results from simulations.

We tested the model against psychophysical data on spatial pooling (Essock, 1990) by finding the minimal inputs necessary to yield oscillations (and therefore activities beyond 0.5) for different input sizes. The results from the simulations and from the experiment were normalized. Normalization was performed by dividing the input (or the contrast) by the value at which it asymptotes for large stimulus sizes. Since smaller stimuli require more contrast to be detected, the normalized contrast for these stimuli is greater than one. Size was also normalized with respect to the asymptoting size. The normalized experimental data and computer simulations are shown in figure 2.

For different stimulus onset asynchronies (SOA) between two stimuli we found the internal time difference Δt for the corresponding neural signals in our model. The probability that two neural events that are separated by Δt ms is then given by $P = \Phi\left(\frac{\Delta t}{2\sigma}\right)$ where Φ is the normal distribution function. Since identical stimuli were used in the simulation as in Hirsch and Sherrick (1961) the comparison between the two is most appropriate, and this is shown in Figure 3.

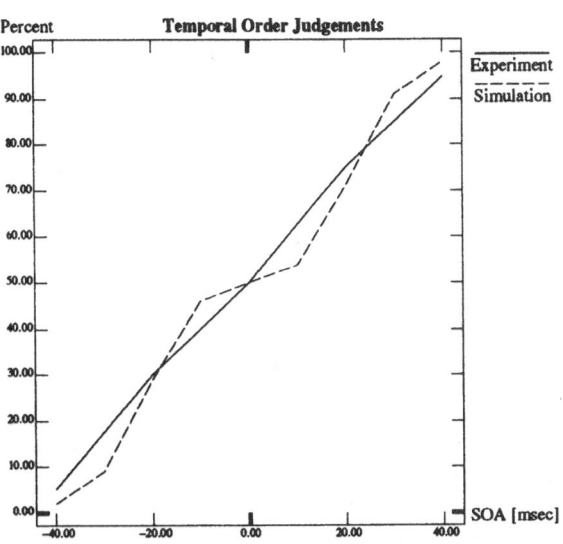

Figure 3: Accuracy of temporal order judgement as a function of SOA. Comparison between experimental results and the model proposed. SOA indicates the time by which stimulus one (e.g. the "right stimulus") leads the other stimulus in a two stimulus presentation task. The ordinate gives the percent responses that stimulus one appeared first. Solid line: results from experimental study. Dashed line: results from simulation of the model.

5 Simulations

In the simulations of the model there were 64 oscillators arranged along a ring. Each oscillator consisted of two nodes each, one fast and one slow. The activity of the fast node is denoted by x_i, of the corresponding slow node by y_i. The index i denotes the position of the oscillator, and ranges from 1 to 64. Oscillators with indices differing by one are neighbors. Since the oscillators are arranged as a ring, units indexed by 1 and 64 respectively are also neighbors. This structure was chosen to avoid edge effects. Care was taken to ensure that input was sufficiently far removed from the wrap around position to avoid undesirable side effects. The input to the network is denoted by I_i and it is position specific. Associated with every oscillator there is a bipole cell, whose activity is denoted by z_i. The equations governing the oscillators are:

$$\frac{dx_i}{dt} = -Ax_i + (B - x_i)(Cf_o(x_i) + C\alpha f_o(z_i) + I_i) - Dx_i f_o(y_i) \quad (1)$$

$$\frac{dy_i}{dt} = E(x_i - y_i) \quad (2)$$

where the signal function f_o is given by

$$f_o(x) = \frac{x^{n_o}}{Q_o^{n_o} + x^{n_o}} \quad (3)$$

and A, B, C, D, E and α are parameters of the network. The parameters n_o and Q_o determine the signal function of the oscillator. The equation governing the bipole cells is:

$$z_i = [Pf_b(L_i) + Pf_b(R_i) + P^* f_b(C_i) - \Gamma_{cpl}]^+ \quad (4)$$

where

$$[x]^+ = \max(x, 0) \quad (5)$$

and the bipole signal function is

$$f_b(x) = \frac{x^{n_b}}{Q_b^{n_b} + x^{n_b}}. \quad (6)$$

Where P, P^* and Γ_{cpl} are parameters. The parameters n_b and Q_b determine the signal function of the bipole cell. The kernels are given by

$$L_i = \frac{1}{w} \sum_{j=1}^{w} f_o(x_{i-j}) \quad (7)$$

$$R_i = \frac{1}{w} \sum_{j=1}^{w} f_o(x_{i+j}) \quad (8)$$

$$C_i = f_o(x_i) \quad (9)$$

where w is the halfwidth of the kernel. The initial conditions of the network where chosen to be $x_i = 0.2$, $y_i = 0.4$, and $z_i = 0$ for all i. The initial value of the slow variable is maintained by tonic input, which is quenched when an input comes on. Scaling of time was done by taking into account that the period of oscillations should be about 20 ms. Thus it was found that putting a timestep of 1 unit in the model equal to 1 ms yields good results. Thus the integration stepsize used was $H = 0.1$ ms. The parameters used throughout this report are $A = 1, B = 1, C = 20, D = 33.3, E = 0.05, \alpha = 0.05, n_o = 4, Q_o = 0.9, n_b = 5, Q_b = 0.001$ or $Q_b = 0.006, P = 1, P^* = 0.5, \Gamma_{cpl} = 1, w = 6$.

In the spatial pooling simulations the background activity was set to zero, to avoid unwanted lateral interactions. For each stimulus size all units that received input received the same value. Threshold input was the lowest value (up to 0.01) that led to oscillations. In the temporal order judgment simulations each node received a constant level of background activity ($I_i = 0.2$). Two nodes received an input ($I_i = 0.6$). The first input ($i = 30$) comes at simulation onset, the second input ($i = 34$) comes on later by an amount specified with SOA.

6 Discussion

In this study we have shown that spatial pooling and perceptual framing can be explained through the same process of cortical cooperation across space and time in a neural network. This process synchronizes activities across the network, and can set up a resonant state that drives learning processes, as in Adaptive Resonance Theory (Grossberg, 1976). Moreover, this process leads to long-range cooperation, which can provide the basis for global integration of visual information.

Reference

Eckhorn, R., Bauer, R., Jordan, W., Brosch, M., Kruse, W., Munk, M., & Reitboeck, H. J. (1988). Coherent oscillations: a mechanism of feature linking in the visual cortex?. *Biological Cybernetics, 60*, 121–130.

Ellias, S. A., & Grossberg, S. (1975). Pattern formation, contrast control, and oscillations in the short term memory of shunting on-center off-surround networks. *Biological Cybernetics, 20*, 69–98.

Essock, E. A. (1990). The influence of stimulus length on the oblique effect of contrast sensitivity. *Vision Research, 30*(8), 1243–1246.

Gray, C. M., & Singer, W. (1989). Stimulus-specific neuronal oscillations in orientation columns of cat visual cortex. *Proceedings of the National Academy of Sciences USA, 86*, 1698–1702.

Grossberg, S. (1976). Adaptive pattern classification and universal recoding, II: feedback, expectation, olfaction, illusions. *Biological Cybernetics, 23*, 187–202.

Grossberg, S., & Mingolla, E. (1985). Neural dynamics of perceptual grouping: textures, boundaries, and emergent segmentations. *Perception & Psychophysics, 38*(2), 141–171.

Grossberg, S., & Somers, D. (1991). Synchronized oscillations during cooperative feature linking in a cortical model of visual perception. *Neural Networks, 4*, 453–466.

Hirsch, I. J., & Sherrick, C. E. (1961). Perceived order in different sense modalities. *Journal of Experimental Psychology, 62*(5), 423–432.

Hubel, D. H., & Wiesel, T. N. (1968). Receptive fields and functional architecture of monkey striate cortex. *Journal of Physiology (London), 195*, 215–243.

Thomas, J. P. (1978). Spatial summation in the fovea: asymmetrical effects of longer and shorter dimensions. *Vision Research, 18*, 1023–1029.

Varela, F. J., Toro, A., John, E. R., & Schwartz, E. L. (1981). Perceptual framing and cortical alpha rhythm. *Neuropsychologia, 19*(5), 675–686.

Vertebrate retina: sub-sampling and aliasing effects can explain colour-opponent and colour constancy phenomena.

J. HERAULT

NEURONICS, INPG-TIRF, 46 Ave. Félix Viallet, F-38031 Grenoble Cedex, France

1. Introduction

The spatial distribution of elementary colour receptors in the mammalian retina [2, 3], their spatial coupling and a simple model of the neural circuits of outer and inner plexiforme layers [4, 7, 9, 10, 11] can provide an interesting hypothese of the spatio-temporal processing of colour and luminance signals. Considering the spatial sampling scheme of colours in fovea and parafovea, and applying the spatio-temporal filters of retinal circuitry [1, 5], we derive the colour-opponancy phenomenon and its relation to the spatio-temporal properties of X ganglion cells as shown in [3, 6], and also the achromatic and spatio-temporal properties of Y cells [8]. All these properties are explained by means of signal processing in the spatio-temporal frequency domain, in conjunction with spatial sub-sampling of colour signals. As a consequence of colour-opponancy, it is easy to postulate a scheme for the colour-constancy phenomenon at a higher level.

2. Model of retinal sampling of colour

Let us consider the foveal and parafoveal areas of the retina. Histological data suggest that the colour receptors (cones) are regularly spaced (in a first approximation) on an hexagonal centred grid of side Δl, which will be taken as unity: $\Delta l=1$. In the fovea, one find only red and green receptors the latter ones being twice as numerous as the former ones. In parafovea, the three kinds of (red - green - blue) receptors are supposed to be equally numerous and uniformly scattered, the presence of rods being neglected. These assumptions therefore are well suited to hexagonal sampling.

In order to model this sampling scheme, let us first consider a one-dimensional space where the receptors, in the parafoveal area are regularly placed at every sampling step ($\Delta l=1$): r - g - b - r - g ... Let us then imagine a fictive retina with three colour receptors at every sampling step k: r(k), g(k), b(k), r(k+1), ... The frequency spectra of the sampled colour signals are periodic with periodicity $1/\Delta l=1$. The real retina appears as a sub sampling of these three signals, each colour sample being shifted by one sample from the preceding one. The functions which allow to transform the fictive retina signals into those of the real one are the following modulation functions:

for red samples: $\quad r'(k) = r(k) \cdot \dfrac{1}{3}\left[1 + \cos(2\pi\,\dfrac{k}{3})\right] \quad \neq 0$ only for k=3p,

for green samples: $\quad g'(k) = g(k)\cdot\dfrac{1}{3}\left[1 + \cos(2\pi\dfrac{k-1}{3})\right] \quad \neq 0$ only for k=3p+1,

for blue ones: $\quad b'(k) = b(k)\cdot\dfrac{1}{3}\left[1 + \cos(2\pi\dfrac{k+1}{3})\right] \quad \neq 0$ only for k=3p+2,

Then, the corresponding frequency spectra are :

$$R'(f) = R(f) * \frac{1}{3}\left[\delta(f) + \delta(f\text{-}\tfrac{1}{3}) + \delta(f\text{+}\tfrac{1}{3})\right], \; G'(f) = G(f) * \frac{1}{3}\left[\delta(f) + \left\{\delta(f\text{-}\tfrac{1}{3}) + \delta(f\text{+}\tfrac{1}{3})\right\} e^{-j2\pi f}\right]$$

and:
$$B'(f) = B(f) * \frac{1}{3}\left[\delta(f) + \left\{\delta(f\text{-}\tfrac{1}{3}) + \delta(f\text{+}\tfrac{1}{3})\right\} e^{+j2\pi f}\right]$$

This leads to spectral components of the three colours centred around frequencies f=n/3. If we consider the signal at the outputs of cones as a unique one s(k), sampled at every step, it can be obtained by merely summing r', g' and b' :

$$s(k) = \frac{r(k)+g(k)+b(k)}{3} + \frac{2}{3}\left[r(k)\cos(2\pi\tfrac{k}{3}) + g(k)\cos(2\pi\tfrac{k\text{-}1}{3}) + b(k)\cos(2\pi\tfrac{k\text{+}1}{3})\right]$$

The first term represents the luminance signal sampled every step, which spectrum has a period f=1 i.e. low-pass. The second term represents modulated colour signals with spectra centred at f = n±1/3, i.e. high-pass signals. It is already apparent that *colour and luminance signals can be separated by a simple frequency filtering.*

3. Basic neural circuits of the retina

Following [4, 12, 13], the architecture of the mammalian retina is rather well known. The main characters that will be used hereafter are those of the outer plexiforme layer in the foveal and parafoveal areas.

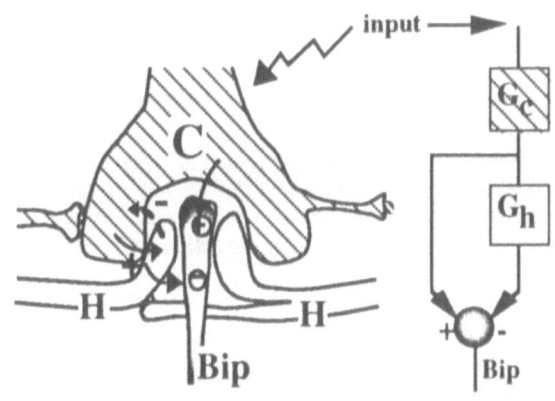

Fig. 1. *Left: Synaptic organisation of cone, bipolar and horizontal cells. Right: Functional block-diagram. G_c: cone-coupling low-pass filter. G_h: horizontal cells coupling low-pass filter. a: feed-back gain of horizontal cells to cone loop.*

Considering figure 1, the colour receptors (C) of same colour are spatially coupled through special synapses called "gap junctions", which behave mainly as pure resistors. Hence, the electrical potential of a cone can diffuse towards its neighbours, leading, because of membrane transconductance and capacitance, to a spatio-temporal smoothing (*low-pass filtering* of transfer function G_c) of the input signal r, g or b(k).

Then, in a structure called "triadic synapse", the resulting signal s(k) is delivered simultaneously to bipolar (Bip) and horizontal (H) cells, with the same polarity, either excitatory or inhibitory, according to the ON or OFF nature of the concerned bipolar cell. In the mean time, the horizontal cell acts on the bipolar one with reverse polarity. The horizontal cells are also coupled each other through gap junctions, producing for the same reasons as above, a low-pass spatio-temporal filtering of cone signals with a transfer function G_h. Therefore, the bipolar cells compute the difference between a signal and its low-pass version: this results in a *high-pass filtering*. We will see now how to model these interactions.

4. Model of retinal filtering

4. 1. Low-pass filter through the coupling of cones

As said before, cones of the same colours are known to be coupled through gap junctions. A simple model of such electrically coupled cells [1] can be represented by a R-C circuit (Figure 2a), which spatio-temporal transfer function is (if resistors R are placed every 3 sampling steps) :

$$G_c(f_x,f_t) = \frac{1}{1 + \beta_c + 2\alpha_c \left(1 - \cos(2\pi f_x/3)\right) + j\, 2\pi\, \tau_c\, f_t}, \text{ with } \alpha_c = r/R,\ \beta_c = r/r_f,\ \tau_c = rC.$$

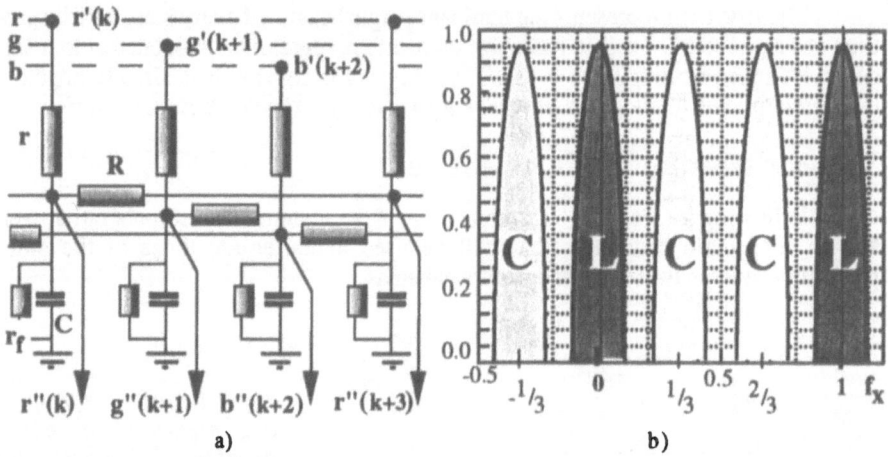

Fig. 2. a) Electrical circuit equivalent to cone coupling. The r, g, b(k) signals sampled every two steps give r', g', b'(k) which are filtered. b): the resulting spectra in the spatial frequency f_x domain. L: spectral components of the luminance part of s(k), C: spectral components of the colour part.

The spatial aspect of the transfer function (at temporal frequency $f_t=0$) has a periodicity of 1/3; thus shaping the same way the *low-pass* (luminance) and the *high-pass* (colour) components of signal s(k), see figure 2b.

4.2. High-pass filter through the Synaptic triad

The signal s'(k) delivered to the horizontal cells will be low-pass filtered through an R-C circuit (figure 3) of the same type than the above filter, excepted that resistors R are now at every sample step k. This produces the signal h(k) which is a *low-pass version of s'(k)* obtained after passing through the horizontal spatio-temporal transfer function $G_h(f_x, f_t)$:

$$G_h(f_x,f_t) = \frac{1}{1 + \beta_h + 2\alpha_h \left(1 - \cos(2\pi f_x) \right) + j\, 2\pi\, \tau_h\, f_t}, \text{ with } \beta_h = r/r_f,\ \alpha_h = r/R,\ \tau_h = rC.$$

Then, the outputs of bipolar cells Bip(k) will be given by the difference between s'(k) and its low-pass version h(k), i. e. they appear as the *high-pass filtering of s'(k)* through a filter which transfer function is: $G_{hb}(f_x, f_t) = [1-G_h(f_x, f_t)]$, shown as a dashed line on figure 3b. This model of filtering has first been derived in [7], then widely used and extended as for example in [1, 5].

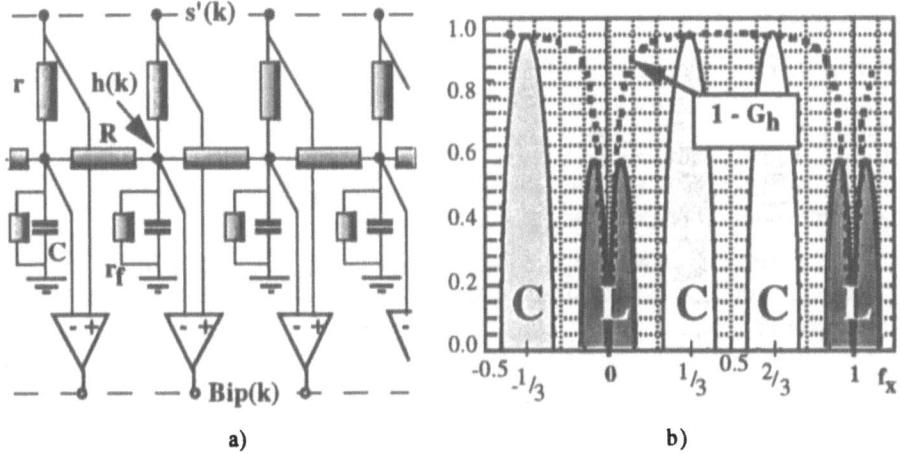

Fig. 3. *a) Equivalent circuit of the outer plexiforme layer of the retina. b) The resulting filtering in the domain of spatial frequencies.*

The total transfer function between input signal s(k) and bipolar cells is now:

$$G_{tot}(f_x, f_t) = G_c(f_x, f_t) \left[1 - G_h(f_x, f_t)\right]$$

The first term, $G_c(f_x, f_t)$, acts as an anti-aliasing filter for the sub-sampled colour signals r', g', b'(k) as shown in figure 2b. The second one strongly modifies the low-frequency part of s'(k), i.e. the luminance part of the signal: the $[1-G_h(f_x, f_t)]$ high-pass filter is responsible for *spatio-temporal contrast enhancement*. However, the high frequency part of s'(k) is not modified because, in this region the high-pass filter gain is unity.

Then, by further low-pass filtering the output of bipolar cells (easily done at the Ganglion cells level), one can get a spatio-temporal high-pass filtered luminance signal (achromatic) of *transient type just as Y ganglion cells are*.

4.3. Colour-opponancy and X ganglion cells

How to retrieve colours? We have said (§ 2) that the high frequency part of the spectrum of s(k) corresponds to colour signals modulating cosine functions. By a series of three synchronous demodulations of these signals, it is possible to retrieve pure colour signals: the first function 1/3(1+2 cos 2π k/3), will select one sample every 3rd (1/3 sub-sampling) in front of the red signals r"(k):

$$\frac{2}{3}\left[r(k)\cos(2\pi\frac{k}{3}) + g(k)\cos(2\pi\frac{k-1}{3}) + b(k)\cos(2\pi\frac{k+1}{3})\right] \cdot \frac{1}{3}\left[1 + 2\cos(2\pi\frac{k}{3})\right] =$$

$$\frac{2}{9}\left[r(k) - \frac{g(k) + b(k)}{2}\right], \text{ i. e. (Red - Cyan)}$$

the two other functions 1/3[1+2 cos 2π (k-1)/3] and 1/3[1+2 cos 2π (k+1)/3] will do the same in front of green and blue signals. We then obtain three kinds of colour signals which show colour-opponant properties: $r(k) - \frac{g(k) + b(k)}{2}$, $g(k) - \frac{b(k) + r(k)}{2}$, and $b(k) - \dfrac{r(k) + g(k)}{2}$, i. e. (Red-Cyan), (Green-Magenta) and (Blue-Yellow). Hence we retrieve the original colour signals only filtered by the cone-coupling spatio-temporal low-pass function (*no transient behaviour* either spatially or temporally) with *colour-opponent characteristics*. This property is found in X ganglion cells as well as in parvocellular region of the Lateral Geniculate Nucleus (figure 4). In the foveal region, we can replace the Blue signal by a green one, this results in two colour-opponent signals: (Red-Green) and (Green-Red).

It is amazing to remark that this property is due to the sampling of a high-pass signal without respect of shannon theorem: in this case, aliasing contains useful informations!

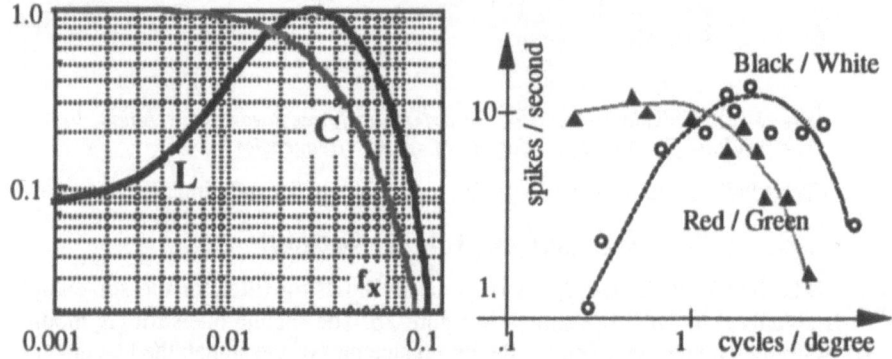

Fig. 4. Spatio-temporal behaviour of the luminance (L) and colour-opponant (C) signals. Left: spatial behaviour according to our calculations. Right: Spatial behaviour for biological data (after [3]).

5. Colour constancy

Let us now consider the border region between two colours C_1 and C_2 in the input image. The spatial sequence is then as displayed in first line of table 1. The colour-opponant signals obtained at the ganglion cell level are of the form: 2R'-(G'+B'), 2G'(B'+R'), 2B'-(R'+G').

	Color C_1				Color C_2			
colour components	$..G_1$	B_1	R_1	G_1	B_2	R_2	G_2	$B_2..$
col.- opp. signals	$\dfrac{G_1^2}{R_1B_1}$	$\dfrac{B_1^2}{G_1R_1}$	$\dfrac{R_1^2}{G_1B_1}$	$\dfrac{G_1^2}{R_1B_2}$	$\dfrac{B_2^2}{G_1R_2}$	$\dfrac{R_2^2}{B_2G_2}$	$\dfrac{G_2^2}{R_2B_2}$	$\dfrac{B_2^2}{G_2R_2}$
after low-pass filter	$...1$	$\dfrac{B_1}{B_2}$	$\dfrac{R_1}{R_2}$	$\dfrac{G_1}{G_2}$	$\dfrac{B_2}{B_1}$	$\dfrac{R_2}{R_1}$	$\dfrac{G_2}{G_1}$	$1...$

Table 1. Signals around a colour border between two colours C_1 and C_2.

Because the receptors have a response of logarithmic type, we can consider that these signals are obtained from the logarithms of R-G-B signals, and hence correspond to forms like $Log\{R^2(k)/G(k-1).B(k+1)\}$... as shown at the second line of table 1, after omission of the "Log" symbol. Let us suppose that the colour-opponent signals are then further low-pass filtered, for example with kernels (1-2-3-2-1): this results in the third line of table 1. Here, the signals appear as the sequence of the *ratios* of the same colour components from each side of the border (R_1/R_2, ...).

Now, let us remember that each colour component C_i results from the product of the illuminating light component L_i by the corresponding reflection coefficient ρ_i of the illuminated surface: $C_i = L_i \, \rho_i$. Hence, our signal depends *only* on the ratios between the reflection coefficients of the two surfaces: ρ_i/ρ_j, and *not on the colour components of the illuminating light*. This phenomenon is called by psychophysiologists "Colour constancy". According to our model, we can postulate that it is made possible because 1) of early vision signal processing which provides colour-opponent properties and 2) of higher level processing (Lateral Geniculate Nucleus and/or V1 cortical area) of simple low-pass type.

6. References

[1] **Beaudot W, Palagi P, Hérault J** (1993) Realistic Simulation Tool for Early Visual Processing including Space, Time and Colour Data. International Workshop on Artificial Neural Networks 93, Barcelona, June 1993

[2] **Derrico J.B., Bushbaum G.**, (1991), A computational model of spatio chromatic Image coding in early vision. J. of Visual Communication and Image Representation. Vol. 2, 1 , pp. 31-37.

[3] **De Valois R., De Valois K.** (1990), Spatial Vision Oxford Psychology Series N° 14. Oxford University Press.

[4] **Dowling JE** (1992) Synapses in the Retina. In: Neurones and Networks: An Introduction to Neuroscience. The Belknap Press of Harvard University Press, Cambridge, Massachusetts, pp 318-321

[5] **Hérault J, Beaudot W** (1993) Motion Processing in the Retina: About a Velocity Matched Filter. European Symposium on Artificial Neural Networks 93, Brussels, April 1993

[6] **Kremers J., Lee B.B., Kaiser P. K.**, (1992) Sensitivity of the macaque retinal ganglion cells and human observers to combined luminance and chromatic temporal modulation. J. Opt. Soc. Am. A, Vol 9, N° 9.

[7] **Mead C, Mahowald M** (1988) A silicon model for early visual processing. Neural Networks, 1:1, 91-97

[8] **Richter J, Ullman S** (1982) A model for the temporal organization of X- and Y-type receptive fields in the primate retina. Biological Cybernetics 43, 127-145

[9] **Siminoff R**, (1984) Electronic simulation of cones, horizontal cells and bipolar cells of generalized vertebrate cone retina. Biological Cybernetics, Vol. 50, n° 3, pp 173-192

[10] **Siminoff R** (1991) Simulated bipolar cells in fovea of human retina: I. Computer Simulation. Biological Cybernetics, vol. 64, n° 6, pp 497-510

[11] **Usui S, Kamiyama Y, Sakakibara M** (1988) Physiological engineering model of the retinal horizontal cell layer. ICNN IEEE, vol. II, pp 87-93, San Diego 1988

[12] **Wässle H, Boycott BB** (1991) Functional architecture of the mammalian retina. Physiological Reviews, Vol. 71, n°2, pp 447-480

[13] **Witkovsky P, Stone S, Tranchina D** (1989) Photoreceptor to Horizontal Cell Synaptic Transfer in the Xenopus Retina: Modulation by Dopamine Ligands and a Circuit Model for Interactions of Rod and Cone Inputs. Journal of Neurophysiology, vol. 62, n°4, pp 864-881

[14] **Zaidi Q., Shapiro A.**, (1993) Adaptive orthogonalization of opponent color signals. Biol. Cybern. 69, 415-428.

RETINA: a Model of Visual Information Processing in the Retinal Neural Network

Faure A.[1], Rybak I.[2], Golovan A.[3], Cachard O.[1]
Shevtsova N.A.[3], Podladchikova L.N.[3]

[1] Laboratory LACOS, University of Le Havre
Le Havre, France
[2] Department of Chemical Engineering, University of Pennsylvania
Philadelphia, USA
[3] A.B.Kogan Research Institute for Neurocybernetics, Rostov State University
Rostov on Don, Russia

1 Introduction

In computer vision, there exist many problems of visual information representation and recognition that have not yet found their effective treatment (Faure A., 1985, Pugh A., 1983). The most of these problems are connected with bulky information processed in real time and with image distortion in the real world. It seems us that the second problem will be adequately treated on the base of the imitation of the biological visual perception mechanisms.

One of the points of the development in this field is a new class of the foveal system (Burt P.J,1988; Hecht-Nielsen R. et al,1992; Zeevi Y.Y,1990). The aims of such artificial systems is to simulate respectively the non uniformity of the distribution of the receptor on the retina , and the eye movements. The expected result is to fix the artificial fovea on the most important fragments of the scene that should be processed with a higher spatial resolution.

With the purposes of both simulation of the biological visual perception and implementation of developed earlier algorithm of high level vision (Rybak I. et al, 1991), we are carrying out works in order to develop an artificial vision system using related data about structural and functional organization of the real scenes.

In this paper, we present our basic model: RETINA, with some results related to the study of the responses to moving stimuli of various shapes.

2 Basic model of RETINA

2.1 Primary transform

The model RETINA is described by a neural network consisting of excitatory and inhibitory elements. The primary transform of the initial image $I=\{X_{ij}\}$ simulates the decrease of resolution from the center of the artificial retina to the periphery, by forming two input images: $I^*=\{X^*_{ij}\}$ and $I^\circ=\{X^\circ_{ij}\}$. The first one of which is the input to the excitatory elements of RETINA, and the second one is the same to the inhibitory elements.

Both I* and I° depend on the adjusted position of the retinal center and on the adjusted level of resolution l_0 in the vicinity of the center. The central point is considered to be the center of a serie of concentric circles whose radii are:

$$R_k(l_0) = 3 \cdot 2^{(l_0-1+k)} \quad \text{with } k=\{0,...,5\} \quad \text{and} \quad l_0 \in \{1,2,3\} \qquad (1)$$

Thus, the input images I* and I° are subdivided into the set of areas with different resolutions: the central one is the area of the circle with the radius $R_0(l_0)$, the next ones are the areas of successive rings surrounding the central circle. Within the central area of I* the image is represented with resolution l_0, the resolution within the first surrounding ring is at l_0+1 level, and so on. Beginning with the level of resolution l_0+1, we have a similar spreading out of the levels of resolution applied to the image I°. To represent the image I=$\{X_{ij}\}$, at resolution level l, the recurrent computation of the Gaussian convolution at each point is used. So, we obtain:

$$X_{ij}^l = \sum_{pq} G_{pq} \cdot X_{i-2^{l-1}p, \, j-2^{l-1}q} \qquad (2)$$

where G_{pq} is the sampled Gaussian distribution, with $p,q \in \{-2;-1;1;2\}$

2.2 Neural network

The schema of the model of the used neural network is shown in Fig.1

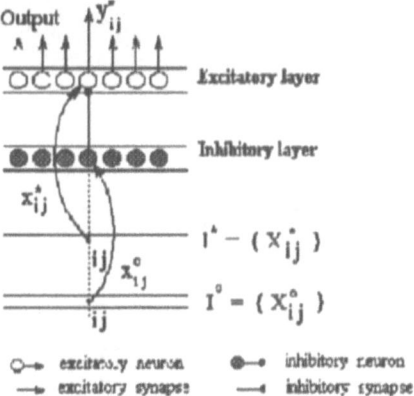

Fig 1: schema of the used model

The network is described by the following differential equations related to the membrane potential of respectively the excitatory and inhibitory artificial neurons.

$$\tau^* \cdot \frac{dU_{ij}^*}{dt} = -U_{ij}^* - Y_{ij}^0 + X_{ij}^* - Q^*(l) \quad \text{and} \quad \tau^0 \cdot \frac{dU_{ij}^0}{dt} = -U_{ij}^0 + X_{ij}^0 - Q^0(l) \qquad (3)$$

U_{ij} is the membrane potential of the ij neuron, Y_{ij} is the output activity of the same neuron, $Q(l)$ is the threshold (that depends on resolution), τ^* and τ^o are respectively the time constant of excitatory and inhibitory neuron, t is the time. Digital solution of (3) was carried out by using the method of exponential approximation (Mac Gregor R.J., 1987);

$$\text{if } \frac{dE(t)}{dt} = -A(t) \cdot E(t) + B(t), \text{ then}$$
$$E(n+1)=E(n).\exp(-A.h)+(1-\exp(-A.h)).B/A \qquad (4)$$

where h is the step of calculation. The system (3) was been solved with $\tau^*=4.h$ and $\tau^o=16.h$.

3 Computer simulation

During the computer simulation, the responses of the described above neural network model RETINA to different visual stimuli were examined. The squares, edges and lines having different sizes, intensities and moving with different velocities were used as a visual stimulus in the most of experiments. Stationary stimuli were applied and the responses of model neurons without primary transform were tested in some cases.

If the all other stimulus parameters are fixed, the maximum of response depends on the stimulus velocity. The relation of the maximum of the response of single excitatory neuron, located in different resolution areas, to a moving square is shown in fig 2.

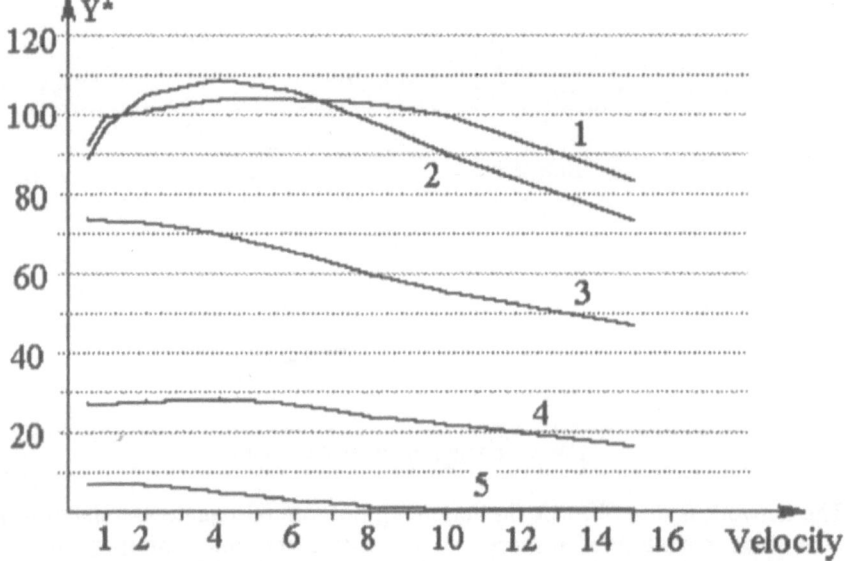

Fig 2: Variations of the maximum of the response (Y*) of a neuron of the excitatory layer vs time.(from N° 1 for the center to N° 6 for the periphery) The stimulus is a moving square

This fact indicate that RETINA neurons can be tuned on different optimal velocity in dependence of stimulus shape, i.e. they can detect the stimulus shape. The results of performed qualitative computer simulation allow us to suppose that change of each stimulus parameters reflected in specific dynamics of response parameters of single neuron and network as a whole (excitation pattern of RETINA)

One of the most interest phenomenon among many different ones revealed during the computer simulation is the complex response dynamics of the peripheral RETINA neurons. In some cases, an oscillatory activity has been detected, and these wavelets can be used fro pretuning of central RETINA neurons.

4 Conclusion

As showed the first results of the qualitative simulation in the developed neural network structure RETINA, there exist several possibilities for the detection of visual stimulus parameters. These possibilities are based on: the location dependence of the response dynamics of the RETINA neurons, the specific response dynamics of the same neuron and the specific excitation pattern of the RETINA, as a whole.

In present software version of the RETINA model, all network parameters were fixed, and the detectory properties of the model should be depend on these parameters. So, such dependence is now investigated either by computer simulation than by analytical approach.

5 References

Burt P.J. (1988) "Smart sensing within a pyramid vision machine", Proc of the IEEE, 76, 1006-1015

Faure A.(1985) "Perception et reconnaissance des formes",EdiTests edit., Paris *id°* (1989) Machinostroenie edit.,Moscou.

Hecht-Nielsen R. and Zhow Y.T. (1992) "A low cost foveal vision system", Report on Government Conference on Neural Networks, Ohio, USA

Mac Gregor R.J.(1987) "Neural and brain modeling", Acad. Press, San Diego, USA

Pugh A. (1983) "Robot vision", Springer-Verlag, Berlin

Rybak I.A, Golovan A.V, Gusakova V.I, Shevstova N.A, Podladchikova L.N. (1990), "A neural network system for active visual perception and recognition", Neural Network World , 4, 245-250

Zeevi Y.Y., Ginosar R. (1990) "Neural computers for foveating vision systems", *in* "Advanced neural computers", R.Eckmiller editor, Elsevier Science Publishers B.V., North-Holland, 323-330

The influence of the inhomogeneous dendritic field size of the retinal ganglion cells on the fixation

T. Yagi †, K. Gouhara ††, Y. Uchikawa †††

† Dept. of Electronic-Mechanical Engineering, Nagoya University, Nagoya, Japan
†† Dept. of Electrical Engineering, Nagoya University, Kasugai, Japan
††† Dept. of Electronic-Informatic Engineering, Nagoya University, Nagoya, Japan

1   Introduction

This paper describes how the variation with eccentricity in the dendritic field size of the retinal ganglion cells affects the fixation. Although the visual information processing in the retina does not directly cause the fixation but gives the input to the oculomotor system [2], this morphological feature is one of causes to make the spatial resolution inhomogeneous. Therefore it must have some relations with the fixation, which is necessary activity in seeing [4][6].

2   Inhomogeneous model of the retina

According to the neurological studies, the dendritic field size of the retinal ganglion cells increases with eccentricity *linearly* in a macaque monkey [5] and gold fish [3]. In this paper, we hypothesize that this type of the variation of the dendritic field size plays a great role in the determination of the fixation points. To confirm this hypothesis, we evaluate several types of the variation in a mathematical model of the retina. Define D as the diameter of the dendritic field size of the retinal ganglion cells and E as the eccentricity from the center of the retina. D increases with the following equation;

$$D = cE^n \quad \text{where} \quad c = D|_{max}(E) \tag{1}$$

Parameter n determines the type of the variation, and parameter c the size of the dendritic field at 40 degrees away from the center of the retina (Fig. 1). For the simplification of the retinal structure, we make some assumptions. Each retinal ganglion cell radially extends its dendrites and directly connects the photoreceptor cells (rods) with the equal effective strength. Moreover the distribution density of the cells is uniform.

Define $C(x,y)$ as the total input signal to each retinal ganglion cell located at the position (x,y), $I(x,y)$ as the input signal from the displayed image, and the effective strength between the retinal ganglion cell and the photoreceptor cells as $F_{x,y}(\zeta,\eta)$, and the integration area as $A_{x,y}$. Then $C(x,y)$ is expressed in convolution with $F_{x,y}(\zeta,\eta)$ and $I(x,y)$ as follows;

$$c(x,y) = \iint_{A_{x,y}} Fx,y(\xi,\eta) \cdot I(x+\xi, y+\eta) d\xi d\eta \quad \text{where} \quad \xi^2 + \eta^2 \leq \left(\frac{D}{2}\right)^2 \tag{2}$$

Assume each retinal ganglion cell outputs the mean value of all input signals. In this case, output signal O(x,y) is denoted as;

$$O(x,y) = \frac{C(x,y)}{\iint_{Ax,y} Fx,y(\xi,\eta)d\xi d\eta} \qquad [3]$$

Using this model, we process the displayed images, and detect the position of the maximum signal level which we assume as the fixation point.

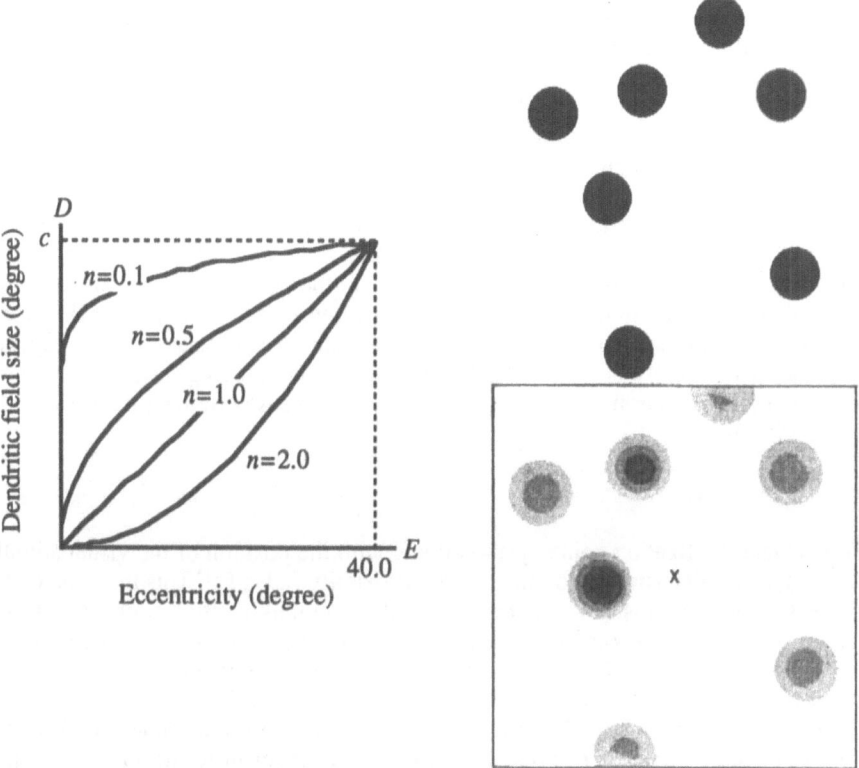

Fig. 1 Various types of the variation of the dendritic field size with eccentricity .

Fig. 2 a. One example of the displayed images (above). b. The output signal levels of each ganglion cell in a model (below).

3 Computed and human fixations

3.1 Displayed images

According to the previous psycho-physical studies on the fixation, there are various factors which shift eye movement for the fixation [1][4][6]. Among various factors, the factor, "the proximity of the stimuli", must be mainly related to the inhomogeneous feature mentioned above. Therefore focusing on this factor and trying to reduce others in order to make more quantitative analysis of the fixation, we use the simple displayed images in a computer simulation of the model and a psychophysical experiment on the fixation [Fig. 2a]. The details are described in [7].

3.2 *Output of each ganglion cell, computed fixations and human fixations*

Fig. 2b indicates the output signal levels of each ganglion cell in a model which has one of types of the variation of the dendritic field size. Each pixel (300 x 300) corresponds to each retinal ganglion cell which is assumed to be distributed in the retina two-dimensionally. The darkness in the figure is proportional to the output signal level of the retinal ganglion cells, and the center of the figure (denoted by a symbol "x") corresponds to the center of the retina. It is obvious that the inhomogenity of the morphological feature makes the difference between the signal level of the central ganglion cells and the peripheral. The darkest position is assumed to be the next fixation point. To obtain the sequential fixation, we relocate the center of the retina to that position, process the image again, and repeat the same procedure above [7]. The results depended on the parameters, c and n. Some resulted in the repetitive fixations of the same visual stimuli, and the other in the random fixations. Some model seemed to emulate the fixations caused by the proximity of the visual stimuli. On the other hand, human eye movements were observed in the psychophysical experiment. As we expected, human eyes were moving towards the closest visual stimuli. Human fixations seemed to be mainly caused by the proximity of the visual stimuli [7].

3.3 *Analysis*

To evaluate the fixation points quantitatively with the position of the visual stimuli, we analyze results using the defined index, Proximity Index [7]. This index gives 1.0 if the fixated stimulus is located closer to the previous fixation point, and 0.0 if farther from the fixation point. All computed and human fixations were analyzed with this procedure, then the frequencies of Proximity Index in every 0.1 were graphed as a histogram in Fig. 3. The percentage of human fixations which occurred towards the closest visual stimuli was 46.9%, while the fixations by the model differ in the respect of the parameter, c and n. For example, a model in $(c, n)=(13.3, 0.5)$ had similar distribution of the Proximity Index to the human; however, in $c=2.7$ and $n=2.0$ they were quite different.

4 Conclusion

According to the analysis above, a certain model where the variation of the dendritic field size increases almost linearly ($n=0.5\sim1.0$) emulated the human fixation. This type of variation coincides the previous neurological finding. Therefore the inhomogenity of the morphology in the retina seems to play a great role to determine the fixation points if the displayed visual stimuli have the same features in the size, color, intensity and shape. Our present study, however, contains various assumptions in a model and the displayed images. Hence other morphological features and more complicated displayed images will be taken into account in our future study. The achievement of this study is expected to contribute to the machine vision research. We will also develop the applicability and feasibility of our study to the industries.

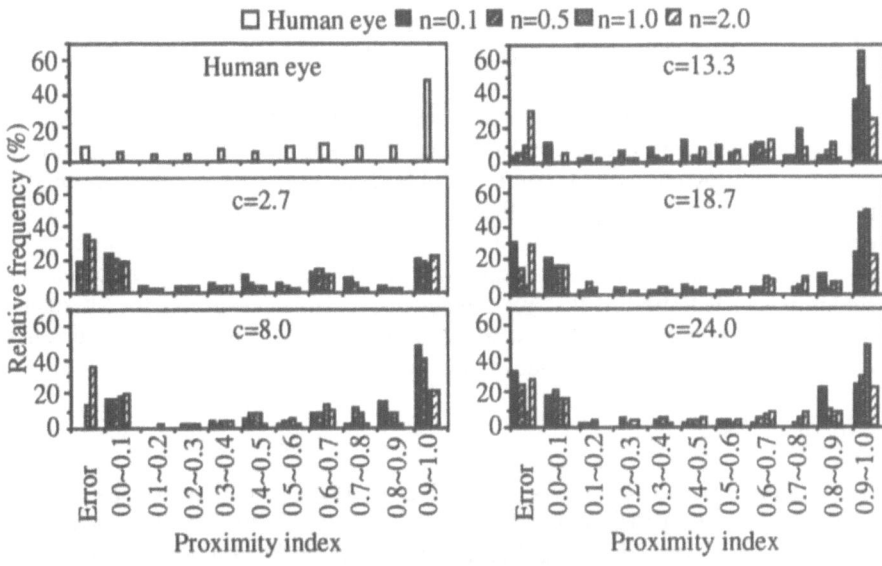

Fig. 3 A histogram of the Proximity of Index

5   Reference

[1] Abbott, A.L. (1992) A survey of selective fixation control for machine vision, IEEE Control Systems, Vol. 12, No. 4, 25-31

[2] Eckmiller, R. (1983) Neural control of foveal pursuit versus saccadic eye movements in primates - single unit data and models, IEEE Transaction on Systems, Man and Cybernetics, SMC-13, 980-989

[3] Hitchcock, P. F. and Easter, S. S. Jr. (1986) Retinal ganglion cells in goldfish, Journal of Neuroscience, Vol. 6, No. 4, 1037-1050

[4] Noton, D. and Stark, L. W. (1971) Scanpaths in eye movements during pattern perception", Science, Vol. 171, 308-311

[5] Perry, V. H. and Cowey, A. (1984) Retinal ganglion cells that project to the superior colliculus and pretectum in the macaque monkey, Neuroscience, Vol. 12, 1125-1137

[6] Yabus, A. L. (1967) Eye movement and vision, Plenum Press

[7] Yagi, T., Gouhara, K., and Uchikawa, Y. (1993) Retinal neural network model for selective fixation in machine vision, Proc. IJCNN '93 Nagoya, Vol. 2, 1199-1202

Top-Down Interference in Visual Perception

C. Taddei Ferretti°. C. Musio°. R.F. Colucci^

° Istituto di Cibernetica. CNR. 80072 Arco Felice (Na). Italy
^ Datitalia Processing S.p.A.. 80100 Napoli. Italy

Abstract - Both bottom-up sensory information and top-down expectations contribute to the the perception processes. It is known that attention. memory. imagery and possibly will interfere on ambiguous visual pattern perception. We investigated sistematically the effect of the top-down volitional factor on the perceptual alternations of a multistable ambiguous visual pattern. as well as the effect of the combination of such top-down factor with the effect of the bottom-up subliminal visual stimuli. that we already proved to interfere on the alternance mechanism. In both cases the will interference is effective. The possible non linear interactions between top-down and bottom-up influences are discussed.

1 Introduction

It is known that two factors contribute to the perception: i) the sensory information and ii) the expectations elaborated by hypothesis-testing processes based on past experience and knowledge. While Gibson (1950; 1966; 1979) stressed the importance of the bottom-up component. Bruner (1957). Gregory (1970; 1972) and Neisser (1967) pointed out the interaction of bottom-up/top-down influences. Thus we were interested on the possible influence of another top-down factor on visual perception. the influence of will. The will has been defined by Minsky (1989. p. 139) as an old and vague idea. Despite this judgement. a lot of experimental work has been accumulated. concerning the physiology and pathology of will in relation with action. *i.e.*. with voluntary movements. However. the discussions on the nature of will are not at end (*e.g.*. Brown 1988. pp. 315-320; Dennett 1984; Honderich 1973; Lucas 1970; Van Inwagen 1983; Watson 1982); James (1950. pp. 562-575) considered the essential feature of willing as a mental event consisting in an effort of attention (to be distinguished from muscular effort) to an idea that offers resistance to being attended to; according to Frankfurt (1971). free will is carachterized by the possibility to generate second order volitions (willing of willing).

Leaving apart the theoretical aspects of the will problem. we focused on the observed effects of attention. memory and imagery on ambiguous pattern perception (Horlitz & O'Leary 1993) and especially on the *en passant* observation of Ditzinger & Haken (1989). concerning the influence of will on the same type of perception. Experimental and theoretical researches on the perception of bistable reversible figures are well known (Borsellino *et al.* 1972; Caglioti & Caianiello 1978; Radilova' & Pöppel 1990; Radilova' & Radil Weiss 1984; Riani *et al.* 1984; 1986). On the other hand. we have already studied the perceptual alternations of a multistable ambiguous pattern. giving rise either to a solid figure oriented to left (L). or to right (R). or to a flat figure: we produced a bottom-up interference in the perception (P) alternance mechanism by means of single subliminal either seconding (S). or contrasting (C) stimuli; the effect of the stimulus depends on its time relation with the preceding perceptual interpretation (Colucci *et al.* 1992; Taddei Ferretti *et al.* 1993). At this point we planned to investigate the possible top-down influence of a volitional factor (W) on such multistable pattern's P.

2 Material, Methods and Subjects

A special purpose. specifically designed and implemented programming language was used for the ambiguous pattern presentation according to different complex protocols. for the subliminal stimuli presentation. for the monitoring of the subject reactions and for the organization of the output data (Colucci *et al.* 1993). We tested 5 subjects during a total of 9 experimental sessions. each one composed by a variable number of phases of 2 min duration. The ambiguous pattern was a modification of the Necker cube. with the two more distant vertices horizontally aligned. We used 2 different experimental protocols. i) either without. ii) or with the presentation of subliminal (14 ms duration) S or C stimuli (LS. RS. LC or RC = respectively. to second the perception of the pattern as a solid oriented to the left. to second the perception to the right. to contrast the perception to the left or to second the perception to the right) 1 s after each P event (LP or RP = perception of the pattern as a solid oriented respectively to the left or to the right). in addition to the W effort in both cases (LW or RW = sustained will to perceive the pattern as a solid oriented respectively to the left or to the right). See Taddei Ferretti *et al.* (1993) for more details. The phases of experimental protocol i) were 4 (3 for subject 5) and those of protocol ii) were 1 (0 for subject 5) for each type of imposed interference to P or combination of interferences (thereafter indicated in brackets). We calculated the total time of LP and RP of all the experimental phases for each subject and for each type of interference to P or combination of intreferences.

3 Results

A) Experimental protocol i):
the values of the ratio LP(LW)/LP(RW) and of the ratio RP(RW)/RP(LW) for the 5 subjects were respectively: 2.286 and 1.993: 46.376 and 66.633: 2.06 and 2.199: 1.169 and 1.163: 1.959 and 1.878.
B) Experimental protocol ii):
when using S subliminal stimuli. the values of the ratio LP(LW.LS.RS)/LP(RW.LS.RS) and of the ratio RP(RW.LS.RS)/RP(LW.LS.RS) for the first 4 subjects were respectively: 0.932° and 1.158: 5.998 and no RP(LW)^: 1.223 and 1.054: 0.915° and 1.269;
when using C subliminal stimuli. the values of the ratio LP(LW.LC.RC)/LP(RW.LC.RC) and of the ratio RP(RW.LC.RC)/RP(LW.LC.RC) for the first 4 subjects were respectively: 2.039 and 1.312: 76.876 and no RP(LW)^: 2.072 and 1.975: 1.118 and 3.699.
In all cases at the end of each experiment all subjects spontaneously reported a great effort in order to maintain the sustained W.

32

4 Discussion

The positive values of the ratios reported in all the results (except in 2 cases, indicated by °) clearly indicate that the reversal mechanism of ambiguous pattern's P is not incapsulated with respect to W.

The performance varied with the subjects: with subject 2, the influence exerted by W was much greater than with the other subjects and in 2 cases (indicated by ^) the non wanted P was completely abolished.

The results of experimental protocol ii) indicate that the top-down influence of W is effective also in the cases of administration of those subliminal S or C stimuli that were already proved to exert a bottom-up interference (Colucci et al. 1992; Taddei Ferretti et al. 1993) on P.

It has to be noted that in the case of S stimuli, which seconde both types of P of solid figures from the ambiguous pattern (LP and RP), the top-down and bottom-up influences have the same sign with respect to the P which is the goal of W, while they have the opposite sign with respect to the P inhibited by W. On the contrary, in the case of C stimuli, which contrast both types of P, the top-down and bottom-up influences have the opposite sign with respect to the P which is the goal of W, while they have the same sign with respect to the P inhibited by W. The combinations (LW.LS.RC) and (RW.LC.RS) on one side, and (LW.LC.RS) and (RW.LS.RC) on the other side will also be used in the near future.

The fact that better results were obtained with C subliminal stimuli than with S ones could be an indication of non linear interactions among top-down and bottom-up influences of the same and different sign and of the possible different weight of the influences of the same type and different sign. Further experiments are planned to elucidate the problem.

Also the effect of W on the occurrence of the ambiguous pattern's P as a flat figure will be investigated.

Acknowledgements

We are indebted to Prof. E.R. Caianiello for fruitful comments and suggestions at the previous phase presentation (Taddei Ferretti et al. 1993) of this research: the present work is dedicated to his memory.

We wish to thank Dr S. Santillo and Mr A. Cotugno for the valid help in the data processing. The work was supported in part by a grant from the CNR Special Project "Non Linear Dynamics in Biological Systems".

References

- Borsellino, A., A. De Marco, A. Allazetta, S. Rinesi & S. Bartolini (1972). Reversal time distribution in the perception of visual ambiguous stimuli. *Kybernetik* 10:139-144.
- Brown, J.W. (1988). *The life of the mind. Selected papers*. LEA. Hillsdale. NJ.
- Bruner, J.S. (1957). On perceptual readiness. *Psycholog. Review* 64:123-152.
- Caglioti, G. & E.R. Caianiello (1978). A model for non-resolvable ambiguities. *Biol. Cybern.* 31:205-208.

- Colucci. R.. C. Musio & C. Taddei Ferretti (1993). EXPLAN - A programming language for complex visual stimuli presentation. *Int. J. Bio-Medical Computing.* in press.

- Colucci. R.F.. A. Cotugno. M. Martino. C. Musio & C. Taddei Ferretti (1992). Influence of subliminal stimulation on multistable visual pattern perception. *Neurosci. Lett.* 43:S28.

- Dennett. D. (1984). *Elbow room: The varieties of free will worth wanting.* MIT Press. Cambridge. MA.

- Ditzinger. T. & H. Haken (1989). Oscillations in the perception of ambiguous patterns. A model based on synergetics. *Biol. Cybern.* 61:279-287.

- Frankfurt. H. (1971). Freedom of the will and the concept of a person. *J. Phil.* 68:5-20.

- Gibson. J.J. (1950). *The perception of the visual world.* Houghton Mifflin. Boston.

- Gibson. J.J.(1966). *The senses considered as perceptual systems.* Houghton Mifflin. Boston.

- Gibson. J.J. (1979). *The ecological approach to visual perception.* Houghton Mifflin. Boston.

- Gregory. R.L. (1970). *The intelligent eye.* McGraw-Hill. New York.

- Gregory. R.L. (1972). Seeing as thinking. *Times Literary Suppl.*, June 23.

- Honderich. T. ed. (1973). *Essays on freedom of action.* Routledge & Kegan Paul. London.

- Horlitz. K.L. & A. O'Leary (1993). Satiation or availability? Effects of attention. memory and imagery on the perception of ambiguous figures. *Percept. & Psychophys* 53(6):668-681.

- James. W. (1950; 1st ed. 1890). *The principles of psychology.* Vol. II. Dover. New York.

- Lucas. J.R. (1970). *The freedom of the will.* Clarendon. Oxford.

- Minsky. M. (1989). *La società della mente.* It. transl.. Adelphi. Milano.

- Neisser. U. (1967). *Cognitive psychology.* Appleton-Century-Crofts. New York.

- Radilova'. J. & E. Pöppel (1990). The perception of figure reversal as a function of contrast reversal exemplified with the Schröder staircase. *Acta Neurobiol. Exp.* 50:37-40.

- Radilova'. J. & T. Radil-Weiss (1984). Subjective figure reversal in two- and three-dimensional perceptual space. *Int. J. Psychophysiol.* 2:59-62.

- Riani. M.. G.A. Oliva. G. Selis G.. Ciurlo & P. Rossi (1984). Effect of luminance on perceptual alternation. *Percept. & Mot. Skills* 58:267-274.

- Riani. M.. M.T. Tuccio. A. Borsellino. J. Radilova' & T. Radil (1986). Perceptual ambiguity and stability of reversible figures. *Percept. & Mot. Skills* 63:191-205.

- Taddei Ferretti. C.. C. Musio. R.F. Colucci & M. Martino (1993). Multistable ambiguous pattern perception. *Proc. World Congress on Neural Networks, INNS Annual Meeting.* Erlbaum. Hillsdale. NJ. II: 155-158.

- Van Inwagen. P. (1983). *An essay on free will.* Clarendon. Oxford.

- Watson. G. ed. (1982). *Free will.* Oxford University Press. New York.

Dynamic Vision System: Modeling the prey recognition of common toads *Bufo bufo*

E. Stolte, E. Littmann[1], and H. Ritter

Computer Science Department, Bielefeld University, D-33501 Bielefeld, FRG
email: etzard,littmann,helge@techfak.uni-bielefeld.de

Abstract

The tectal structures underlying the prey-catching of toads are one of the best-known mechanisms in neurobiology. In our paper we present a simplified model of the prey recognition of the common toad *Bufo bufo*. We show that the recognition performance can be explained by coupling two neuron layers that exhibit locally excitatory interaction while there is a topographic inhibition between the layers. The neurons are modeled as nodes with gaussian lateral excitation and time-dependent exponential decrease in activity. We derive a method to learn the adjustable parameters by supervised training.

1 Introduction

The development of artificial neural networks was elicited and in the following often inspired by discoveries in neurobiology of mechanisms how biological neurons and neural assemblies function. The mainstream of today's networks, however, hardly reminds of biological networks. There are only few attempts to develop models based on existing neurobiological data, since suitable data is often rare. One opportunity are the neural mechanisms underlying the prey-catching of toads. The visual system of the toad is a robust tool to distinguish worm-like moving stimuli (elongated objects moving lengthwise) from opposite stimuli (elongated objects moving vertical to their length axis), and functions for a wide range of different visual backgrounds. Therefore, the development of a model of this system with similar properties would provide a powerful tool for dynamic object segmentation and recognition.

In longterm studies several neural cell types and assemblies were reliably identified that contribute to the neural and behavioral response of the toad [1,2]. Approaches have been made earlier to model certain properties of these structures [3,5]. In our paper we present a model of the visual prey recognition system of the common toad *Bufo bufo* that attempts to use a minimal number of adjustable parameters. We show that the recognition performance can be achieved by coupling two neuron layers that exhibit locally excitatory interaction while there is a topographic inhibition between the layers. The neurons are modeled as nodes with gaussian-distributed, excitatory interaction with the surround and time-dependent exponential decrease of their activity. The data on the neural response of the toad is used to define a function for the desired network output. Based on this target function, we derive a method that allows to learn the adjustable parameters by supervised training and discuss ways to evaluate the similarity of the model with the biological original.

2 Vision System Model

Our current model of the toad's vision system [Fig.1] is based on several assumptions to simplify the structure and to minimize the number of adjustable parameters. The input is the time- and space-variant activation of the retina resp. of the ganglion cells.

---

[1]to whom correspondence should be sent

This activation is mapped topographically onto two neuron layers. For our current simulations we constrained the stimuli to binary, rectangular signals of length l and width w moving lengthwise with velocity v. The retinal activity $A(x, y, t)$ at time t and location (x, y) can then be written as

$$A(x, y, t) = \left[\Theta((x - vt) + \frac{l}{2}) - \Theta((x - vt) - \frac{l}{2}) \right] \left[\Theta(y + \frac{w}{2}) - \Theta(y - \frac{w}{2}) \right] \quad (1)$$

The first neuron layer ("Tectum") models the $T5.1$ neurons in the tectal region. The other one ("Pretectum") corresponds to the pretectal neuron layer. Within the layers the interaction between the neurons is modeled as radial symmetric, lateral excitation that is gaussian-distributed and thus characterized by the variances σ_T and σ_P. The nodes are assumed to be "leaky integrators" that show a time-dependent exponential increase and decay of their activity with time-constants τ_T and τ_P. These parameters are equal for all nodes within one layer. Thus, instead of using large weight *matrices*, we describe the connectivity of the system by a single weight *function* $V(x, y, t)$ that is common to all nodes and that requires only four adjustable parameters σ_T, σ_P, τ_T, and τ_P for its full specification:

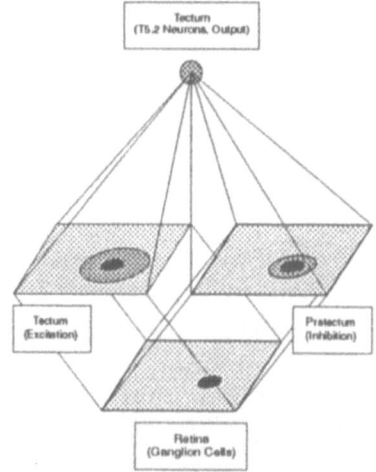

Fig. 1: Dynamic vision system

$$V_{T,P}(x, y, t) = \frac{1}{2\pi\sigma_{T,P}^2} \frac{1}{\tau_{T,P}} \times \quad (2)$$

$$\times \underbrace{\int_0^t \int_{-\infty}^{\infty} \int_{-\infty}^{\infty} exp\left[-\frac{(x - x')^2 + (y - y')^2}{2\sigma_{T,P}^2} \right] exp\left[-\frac{t - t'}{\tau_{T,P}} \right] A(x', y', t')\, dx' dy' dt'.}_{f_{T,P}(\sigma_{T,P}, \tau_{T,P})}$$

3 Training Algorithm

The evaluation of the neural activity in the tectal/pretectal system is supposed to be achieved by $T5.2$ neurons in the tectum. We model these neurons as static classifiers for prey and non-prey (predator) resp. These neurons receive weighted input from both layers. The weights are different for both layers but identical within one layer. Thus, we define two input weights w_T, w_P for the $T5.2$ neuron. Its output is then given by $O = w_T V_T + w_P V_P$. The inhibition of the tectal layer by the pretectal layer can be taken into account if we constrain w_T to be positive and w_P to be negative. If we regard the $T5.2$ neurons as simple classifiers, we can define the desired output (target function) as $T \in \{ 0\ ("predator"), 1\ ("prey") \}$. A more complex approach will try to model the output function as measured in [3], thus yielding a continuous-valued target function T based on the measurements.

Given these target functions, gradient descent can be applied to minimize the quadratic error and thus to adapt the input weights of the $T5.2$ neurons. Furthermore, the error signal can be back-propagated and adaptation signals can be calculated for the variance of the neurons as well as for the relaxation time constant.

4 Results

In order to evaluate this approach, simplified simulations have been regarded so far. The training data consist of rectangular stimuli traversing a 40×40 receptive area at a constant velocity of 2 units per time step. The signals are "worm-like" rectangles of width 1 and lengths $l = 2, 4, 6, 8, 10$ (e.g. Fig.2 (left)), the predator stimuli are rectangles of length 1 and widths $w = 2, 4, 6, 8, 10$. A third set of stimuli consists of squares of edge lengths $l = 2, 4, 6, 8, 10$.

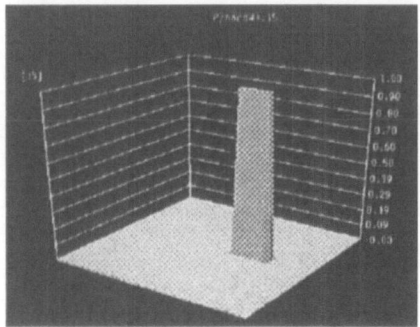 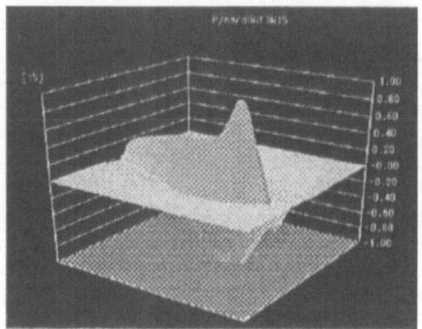

Fig. 2: Left: Worm-like signal (6×1). Right: system response

Starting with initial parameters $\{(w_E = 1.0, w_I = -10.0), (\sigma_E = 1.0, \sigma_I = 4.0), (\tau_E = 10.0, \tau_I = 1.0)\}$, the system was trained to yield an output value of 1 for worm-like stimuli, and 0 otherwise. Thresholds for worm and antiworm were defined to 0.7 and 0.3, resp. Training only one parameter at a time, the system converged to a parameter constellation $\{(8.51, -19.1), (1.0, 4.0), (5.25, 3.31)\}$. This constellation yields output activity functions as shown in the right image of Fig.2 for the worm-like stimulus. The output of the system is shown in Fig.3. Compared to the biological data in Fig.4 [3] we find similar curves for worm and antiworm stimuli. For quadratic stimuli the curves differ considerably. This is due to the fact that in the current training process squares and antiworms are treated equally.

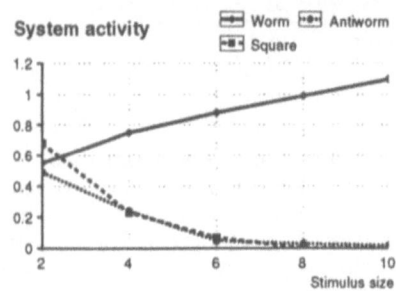 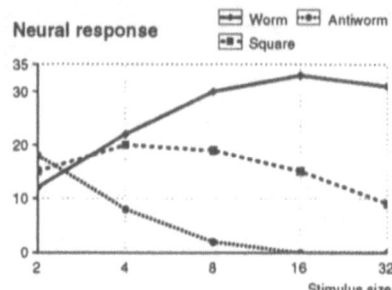

Fig. 3: System response   Fig. 4: Neuronal response

In order to improve our understanding of the system, the stability of this solution was investigated. The stability intervals if varying only one parameter are listed in Tab. 1. Within these ranges the system converges reliably to the original constellation. The interaction between parameter pairs can be illustrated by 3D error surface plots. For the pair (τ_E, τ_I) the error surface is plotted in Fig.5 on a logarithmic scale covering a range of $[0.1, 100]$. During the training, the system parameters follow the gradient along these surfaces.

Stability intervals		
Parameter	Minimum	Maximum
w_E	2.1	50
w_I	-100	100
σ_E	0.8	2
σ_I	1	10
τ_E	0.5	50
τ_I	0.1	100

Tab.1: Stability intervals for parameters

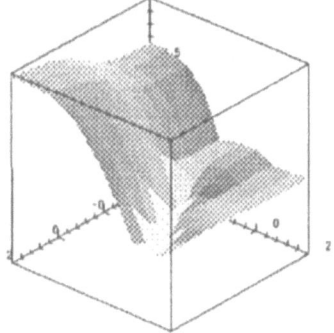

Fig. 5: Error surface (τ_E, τ_I)

5 Discussion

Previous work on modeling the prey-catching behavior of the toad has focused on modeling the behavior of the *existing* system. Our approach investigates the *evolution* of such systems. The results above show that the model system can be *trained* to correctly classify the stimuli. This is especially remarkable as it includes the adaptation of time characteristics. The adaptation of the time decay constants (as well as the variances of the lateral excitation) is driven by the gradient provided by the error function.

One goal of this work is to develop a model system that shows a behavior similar to the original system. Since the biological experiments are very difficult the number of measurements are limited. Given a similar model system, a variety of experiments concerning the statistics of the stimuli can be performed. The most informative results could then be verified (or falsified) by the biological experiment. This procedure would facilitate to improve our knowledge about both systems.

Furthermore, we hope to exploit such systems for practical image recognition tasks. As the recognition process heavily relies on the *dynamic* behavior of the objects, it seems to be well suited for online recognition of moving objects in front of noisy background. Due to the radial symmetry of the approach, the recognition is independent of the movement direction. The proposed model uses properties exhibited by existing physical systems (artificial retina [4], leaky integrators) and thus could be easily implemented in hardware, as the adaptation process only changes the global parameters of the variances and time decay constants.

References

[1] Ewert J.-P. (1970), Neural mechanisms of prey-catching and avoidance behaviors in the toad (*Bufo bufo* L.), in *Brain Behav. Evol.* **3**, pp. 36–57.

[2] Ewert J.-P. (1984), Tectal mechanisms that underlie prey-catching and avoidance behaviors in toads, in *Comparative neurology of the optic tectum*, ed. H. Vanegas, Plenum Press, New York, London, pp. 247–416.

[3] Ewert J.-P., and v. Seelen, W. (1974), "Neurobiologie und System-Theorie eines visuellen Mustererkennungsmechanismus bei Kröten", in *Kybernetik* **14**, pp. 167–183.

[4] Koch, C. (1989), Seeing Chips: Analog VLSI circuits for computer vision, in *Neural Computation* **1**, pp. 184–200.

[5] Wang, D.L., and Ewert, J.-P. (1992), "Configurational pattern discrimination responsible for dishabituation in common toads *Bufo bufo* (L.): Behavioral tests of the predictions of a neural model", in *J. Comp. Physiol. A* **170**, pp. 317–325.

Emergence of Long Range Order in Maps of Orientation Preference

F. Wolf, K. Pawelzik, T. Geisel

Institut für Theoretische Physik, Universität Frankfurt

60054 Frankfurt/M., Germany

e-mail: fred@chaos.uni-frankfurt.d400.de

Introduction

The formation of feature-maps in the developing visual cortex has become a central system for the study of cooperative phenomena in large neural networks, both theoretically[1, 2] and experimentally [3, 4]. With advanced optical imaging techniques [5, 6]it has now become possible to monitor the activity of neural populations synchronously with a high spatial resolution. This kind of measurement should finaly enable a comparison of experiment and mathematical theory in quantitative detail.

In the adult visual cortex the pattern of preferred orientations exhibits a particular complex spatial organization, characterized by a large number of point defects (pinwheels) [7]. Because the orientation preference map in adult cats is close to optimally smooth, the structur of the map can be predicted from the knowledge of position and chirality of its defects [8]. Moreover in this system optimal smoothness results in a particular kind of long range order, that allows to predict preferred orientation across cortical distance [9]. In this contribution we show that this global spatial coherence cannot be achieved easily during the initial phase of development, in which the pattern of preferred orientations arises via an instability mechanism. Therefore we propose a two stage model for the process of map formation which explains the establishment of long range order in the visual cortex. We predict the occurence of a second phase in the process of map formation, during which cells cooperate effectively over large cortical distances. Furthermore we show how this process could be observed directly in optical imaging experiments [1].

Long Range Order in Visual Cortex

The spatial pattern of orientation preference is given by a scalar function $\theta(\mathbf{x})$ assigning the preferred orientation θ to any cortical location $\mathbf{x}$. In this representation the presence of a pinwheel at $\mathbf{x}_i$ on the cortical plane means that the surface $\theta(\mathbf{x})$ forms a staircase centered on $\mathbf{x}_i$, which runs clockwise or counterclockwise depending on its chirality $q_i = \pm 1/2$. The assumption that the surface $\theta(\mathbf{x})$ connecting the staircases at $\mathbf{x}_i$ is optimally smooth in the sense

---

[1] Corresponding experiments are currently performed by D.-S. Kim and T. Bonhoeffer (priv.com.).

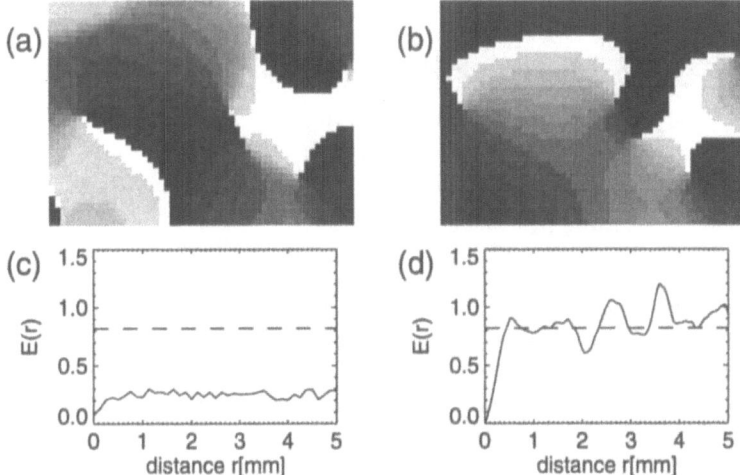

Figure 1: (a) orientation preference map generated according to (2) and (b) the optimally smooth map according to Eq.(1), containing an identical configuration of pinwheels. (c) predictability function $E(r)$ of a measured map from area 17 of a cat with respect to the field analogy model and (d) predictability of a map generated according to Eq.(2). While the actual measurement shows a significant degree of spatial predictability, the synthetic map does not.

that it minimizes the functional

$$S[\theta(\mathbf{x})] = \int_{area} d^2\mathbf{x} \mid \nabla\theta(\mathbf{x}) \mid^2$$

determines a unique map structure. This principle can be tured into an explicit model of the orientation preference map[9],

$$\theta^{th}(\mathbf{x}) - \theta_0^{th} = \left(\oint_{\mathbf{x}_0}^{\mathbf{x}} d\mathbf{x} \times \sum_i q_i \frac{\mathbf{x} - \mathbf{x}_i}{|\mathbf{x} - \mathbf{x}_i|^2} \right) \bmod \pi, \qquad (1)$$

which gives the optimally smooth map $\theta^{th}(\mathbf{x})$ as a function of the singularity configuration $\{\mathbf{x}_i, q_i\}$ and the preferred orientation θ_0^{th} at a single location $\mathbf{x}_0$. Individual measurements or model solutions can be compared directly with this field analogy model by extracting $\{\mathbf{x}_i, q_i\}$ from the map and calculating the theoretical prediction for that particular pinwheel configuration.

The field analogy model implies infinite long range order. This becomes particularly clear from the fact that for a given singularity distribution specifying the preferred orientation θ_0^{th} at a single and arbitrary point $\mathbf{x}_0$ fixes the whole map. Fitting $\theta^{th}(\mathbf{x})$ to an observation $\theta^{exp}(\mathbf{x})$ at a single location $\mathbf{x}_0$ by the choice $\theta_0^{th} = \theta^{exp}(\mathbf{x}_0)$ yealds a prediction of $\theta_{exp}(\mathbf{x})$ all over the observed area. To investigate the quality of this spatial prediction for a map from A17 of a cat, we estimated the mean squared error $E(\theta_0^{th}, r) = \langle | \theta^{th}(\mathbf{x}) - \theta^{exp}(\mathbf{x}) |^2 \rangle_{|\mathbf{x}-\mathbf{x}_0|=r}$ between predicted and measured orientation preference at a distance r from $\mathbf{x}_0$. $E(\theta_0^{th}, r)$ approaches repibly a vlue of about 0.2 which is significantly below the expectation value for unrelated maps $\pi^2/12 \sim 0.82$. Can this kind of global coherence arise spontaneously in cortical development?

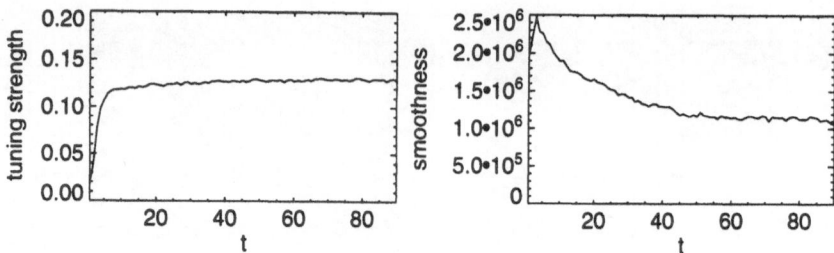

Figure 2: Time course of orientation tuning strength (right) and smoothness functional $S[\theta(\mathbf{x})]$ (left) in the developmental model proposed by Durbin and Mitchison. Saturation of the orientation tuning strength indicates the end of the instability phase of the temporal evolution. After the instability phase is finished the smoothness functional of the map decreases towards the minimum defined by the field analogy model.

A large number of models have been proposed, to explain the formation of the orientation preference map [10, 11, 12, 2]. In any of them the spatial pattern of orientation preference arises via the instablity of a band of fourier modes, that are selected by the range of horizontal interactions within the layer [11, 2]. The effect of this instabilities is to amplify random fluctuations within the unstable band of fourier modes (cf. [13]), which leads the establishment of a pattern that resembles bandpass filteren random noise [14, 15]. We first test the hypothesis, that the pattern of OP observed in the visual cortex arises by this mechanism without further refinement. To enable a direct comparson with the measured maps we construct maps exhibiting the identical powerspectrum $P_{ex}(\mathbf{k}) = \left| \int_{area} d^2\mathbf{x} \exp(-i\mathbf{k}\mathbf{x}) \exp(i2\theta(\mathbf{x})) \right|^2$ as the actual measurement

$$\theta^{inst}(\mathbf{x}) = 0.5 \arctan \int d^2\mathbf{k} \exp(i\mathbf{k}\mathbf{x}) \sqrt{P_{ex}(\mathbf{k})}\, \eta(\mathbf{k}), \qquad (2)$$

where $\eta(\mathbf{k})$ is a field of gausian white noise. Fig.1 shows an example compared to the optimaly smooth map containing the same configuration of singularities. We now calculate the spatial predictability function of the map with respect to the FAM (Fig.1d). While the correlation function is by construction identical to the measurement, the predictability function saturates rapidly to the expectation value for totally unrelated maps. This demonstrates that the primary instability mechanism is not sufficent to produce the global coherence observed in the actual measurement.

Development of Optimal Smoothness

In principle there are two ways how global coherence could come about. The developing orientation preference map could be patterned by an unknown mechanism, that preestabishes long range order. More plausibly global coherence could arise dynamically in a self organization process following the primary instability. This is indeed what happens in a biologicaly plausible dynamical model for the formation of the orientation preference map. Fig.2 shows the time course of orientation tuning strength and the smoothness functional in a simulation of the elatic net model[12]. While the orientation tuning strength saturates at the end of the instability phase, the smoothness functional decreases afterwards for a much longer time finally reaching the minimum value of the optimaly smooth solution.

Because it is unclear how a global prepatterning of the orientation preference map could be brought about in the real brain, we propose that during development of the visual cortex, the primary establishment of the orientation preference map is followed by a comparable optimization of global smoothness. This hypothesis can be tested experimentaly by calculating the time course of S from optical imaging mesurements at successive times during development in a particular individual.

Conclusions

Up to now the role of cooperation in the formation of neural maps, has been resticted to wavelenght selection during a primary instability. We have shown that this mechanism is not sufficent to explain the global coherence exhibited by the biological system. Instead the dynamics far beyond the instability is important for a global optimisation of smoothness during development and consequently for the emergence of long range order. It will be fascinating to see whether these processes are actually operating in the real brain.

Acknowledgement: We acknowledge fruitful discussions with D.-S. Kim, T. Bonhoeffer and S. Löwel. Experimental data from cat area 17 was provided by D.-S. Kim and T. Bonhoeffer. This work has been supported by the Deutsche Forschungsgemeinschaft (LA 441/5-1 and SFB 185).

References

[1] Miller, K.D., Keller, J.B. & Stryker, M.P., Science **245**,605-615 (1989).

[2] Obermayer, K., Blasdel, G.G., & Schulten, K., Phys. Rev. A **45**,7568-7589 (1992).

[3] Löwel, S., Singer, W., Science **255**,209-212 (1992).

[4] Yuste, R., Alejandro, P., Katz, L.C., Science **257**,665-669 (1992).

[5] Grinvald, A., Lieke, E., Frostig, R.D., Gilbert, C.D. & Wiesel, T.N., Nature **324**,361-364 (1986).

[6] Blasdel, G.G., Salama, G., Nature **321**,579-585 (1986).

[7] Bonhoeffer, T., Grinvald, A., Nature **343**,429-431 (1991).

[8] Wolf, F., Pawelzik, P., Geisel, T., Kim, D.-S., Bonhoeffer, T., in *ICANN'93* , eds. Gielen, S. & Kappen, B., (Springer, 1993).

[9] Wolf, F., Pawelzik, P., Geisel, T., Kim, D.-S., Bonhoeffer, T., in *Computation and Neural Systems II*, eds. Marder, E., Rinzel, J., Eeckman, F., (Kluwer, 1993).

[10] von der Malsburg, Ch., Biol. Cybern. **14**,85-100 (1973).

[11] Swindale, N.V., Proc. R. Soc. London **215**,211-230 (1982).

[12] Durbin, R., Mitchison, G., Nature **343**,644-647, (1990).

[13] Manneville, P., *Dissipative Structures and Weak Turbulence*, (Academic Press, 1990).

[14] Rojer, A.S., Schwartz, E.L., Biol. Cybern. **62**,381-391 (1990).

[15] Wörgötter, F., Niebur, E., Biol. Cybern. in press (1993).

Oriented Ocular Dominance Bands in the Self-Organizing Feature Map

H.-U. Bauer

Institut für Theoretische Physik and SFB "Nichtlineare Dynamik",
Universität Frankfurt, Robert-Mayer-Str. 8-10, 60054 Frankfurt, Germany

1 Introduction

The formation of ocular dominance (OD) columns is an often investigated example for activity dependent emergence of ordered maps in the brain. Many models have been developed to explain this self-organization process (von der Malsburg, 1979; Swindale, 1980; Miller et al, 1989; Goodhill, 1993). Two more recent models are based on Kohonen's Self-Organizing Feature Map algorithm (SOFM, Obermayer et al. 1992), and on the elastic net algorithm (Goodhill and Willshaw 1990).

All of these models reproduce the band structure of OD-bands quite well. In order to gain further insights into the underlying mechanisms, and possibly to differentiate between the different models, additional features of the OD-band structure should be taken into account. One example for such additional features is the different appearance of OD-bands in cat and monkey. In cat the OD-stripes appear irregularly branched, and do not seem to have a preferred direction (Anderson et al., 1988). In monkey the OD-bands appear like a series of parallel stripes, which run perpendicular to the representation of the horizontal meridian (LeVay et al., 1985).

LeVay et al. noted that the projection from the LGN to the cortex in monkey involves magnification factors which are twice as large in the elongated direction of area V1 than in the perpendicular direction. They hypothesized that this anisotropy is the reason for the zebra-like (Swindale, 1980) appearence of the OD-bands in monkey. Jones et al. complemented this hypothesis with the observation, that the geniculocortical projection in cat has isotropic magnification factors (Jones et al., 1991). Using a non-developmental model, they were able to reproduce the different OD-band structures in cat and monkey as a consequence of the different projection geometry only.

Since these author's model does not contain elements which could be identified with ontogenetic processes, the question remains open, whether developmental models can also generate oriented OD-bands as a consequence of projection geometry. I describe in this contribution results for anisotropic projections in the SOFM, a well established map formation algorithm which has already been applied to the generation of OD-bands (Obermayer et al., 1990, 1992). The

SOFM is interesting in this regard not only, because it has been successfully applied to many mapping problems in various sensory modalities, but also because a framework for the analytical description of instabilities in the map has been developed (Ritter and Schulten, 1988). Results for oriented OD-bands in the elastic net (Durbin and Willshaw, 1987), a second general purpose map formation algorithm, can be found in (Goodhill et al., 1992; Bauer, 1993).

2 Analysis of Ocular Dominance Band Structure in the Self Organizing Feature Map

Due to lack of space in a four-page contribution, we skip a detailed description of the SOFM and refer the reader to other publications for a more thorough treatment (Kohonen 1989; Ritter et al. 1990). The map projects stimuli $\mathbf{v}$ from some input space V onto neurons located at $\mathbf{r}$ in some output space A. These neurons are characterized by receptive field centers $\mathbf{w_r} \in V$. A stimulus $\mathbf{v}$ is mapped onto that neuron $\mathbf{r}$, whose receptive field $\mathbf{w_r}$ lies closest to $\mathbf{v}$. The learning rule enforces that neighboring neurons in A have neighboring receptive field centers in V. A crucial element of the learning rule is a cortical neighborhood function $h_{\mathbf{r},\mathbf{r}'}$ which is characterized by a length scale σ,

The self-organizing feature map has been applied to the modeling of ODC-bands by mapping stimuli from an input space of dimensions $1 \times 1 \times 2s$ ($s << 1$) onto an output space of dimensions $N \times N$. As long as s remains small, the receptive field center positions all remain centered in the additional dimension. By deriving a Fokker-Planck-equation for fluctuations about this equlibrium solution, and by computing the stability of Fourier modes of the fluctuations in terms of Eigenvalues of some geometry-dependent matrices, Ritter et al. were able to analytically derive a critical thickness s^* for structure along the third dimension to occur,

$$s^* = \frac{\sigma}{N}\sqrt{3e/2} \approx 2.02\frac{\sigma}{N}. \qquad (2.1)$$

In order to test LeVay et al.'s and Jones et al.'s geometry hypothesis, one can now adapt the above described procedure to maps from an input space of dimensions $1 \times 1 \times 2s$ onto an output space of dimensions $2N \times N$. The geometry dependent matrices have to be modified, and the the crucial eigenvalue equation now reads

$$\lambda_3^B(\mathbf{k}) = \left(1 - \frac{s^2}{3}(4k_x^2 + k_y^2)\exp(-k^2\sigma^2/2)\right) = 0, \qquad (2.2)$$

(k_x, k_y: wavevectors of the relevant modes in x and y-direction). Using this condition, one can further see, that the first instability in the system occurs purely along the x-direction, at a critical width

$$s_1^* = \frac{\sigma}{N}\sqrt{3e/8} = \frac{\sigma}{2N}\sqrt{3e/2} \approx 1.01\frac{\sigma}{N}. \qquad (2.3)$$

44

With increasing values of s, also modes with k_y-component become unstable. Modes which are oriented purely in the y-direction become unstable only at values of s above a value $s_2^* = \frac{\sigma}{N}\sqrt{3e/2} \approx 2.02\frac{\sigma}{N}$, which is identical to the critical thickness of the $N \times N$-system.

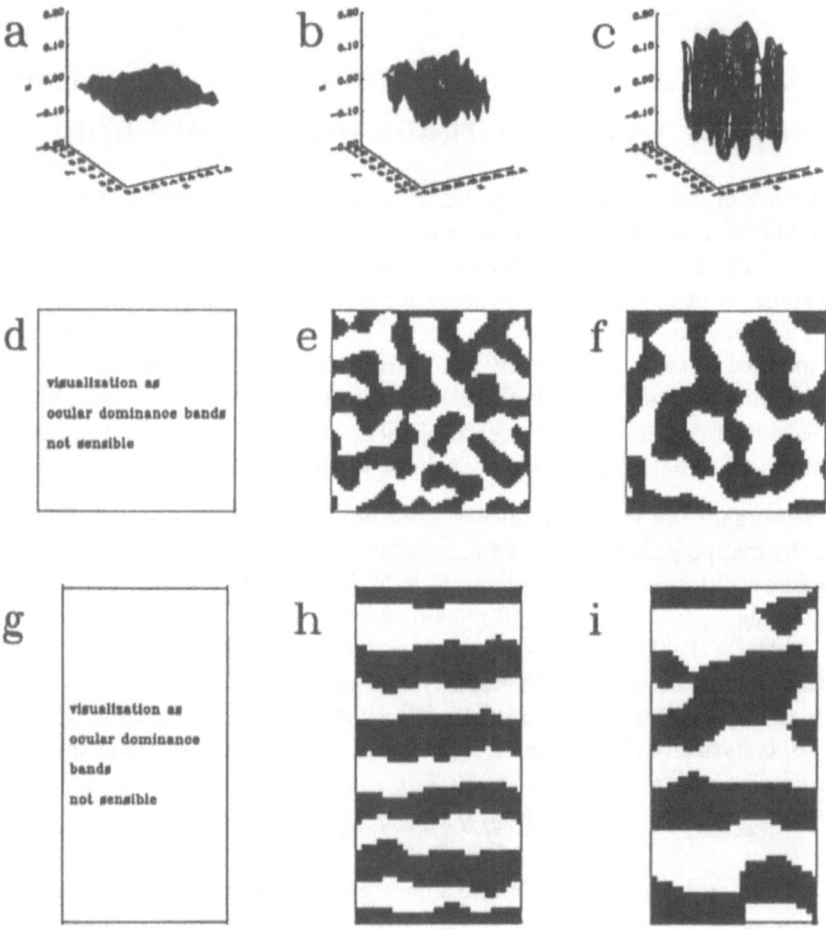

Fig. 1: OD-bands generated by the SOFM for the case of isotropic elongations (64×64, $s = 0.05$ (subcritical, a,d), $s = 0.1$ (critical, b,e), $s = 0.2$ (critical, c,f)), and anisotropic elongations (32×64, $s = 0.05$ (both directions subcritical, d), $s = 0.1$ (one direction critical, h), $s = 0.2$ (both directions critical, i)). The stimuli (x,y,z) were evenly distributed in $0 < x, y < 1, -s < z < s$; neurons with $w_z > 0$ were displayed in black, neurons with $w_z < 0$ in white in the above figures. Simulations parameters were $\sigma = 2$, $s = 0.1$, $\epsilon = 0.2$, 10000 steps, initialization retinotopic in x, y-direction, random in z-direction, periodic boundaries.

In Fig. 1 the results of simulations are displayed which illustrate the above considerations. Identifying Figs. 1d-f with the cat geometry, and Figs. 1g-i with the monkey geometry we conclude that the SOFM is able to reproduce the geometry effect described by (Jones et al., 1991).

3 Acknowledgements

Helpful discussions with Klaus Pawelzik, Fred Wolf and Ken Miller are gratefully acknowledged. This work has been supported by the Deutsche Forschungsgemeinschaft (Sonderforschungsbereich 185 "Nichtlineare Dynamik", TP E3).

4 References

Anderson, P.A., Olavarria, J., Van Sluyters, R.C. (1988) The Overall Pattern of Ocular Dominance Bands in Cat Visual Cortex. J. Neurosci. **8**, 2183-2200.

Bauer, H.-U. (1993) Development of Oriented Ocular Dominance Bands as as Consequence of Areal Geometry, submitted to Neur. Comp..

Durbin, R., and Mitchison, G., 1990. A Dimension Reduction Framework for Understanding Cortical Maps. Nature **343**, 644-647.

Durbin, R., and Willshaw, D., 1987. An Analogue Approach to the Travelling Salesman Problem Using an Elastic Net Method. Nature **326**, 689-691.

Goodhill, G.J., and Willshaw, D.J. (1990). Application of the Elastic Net Algorithm to the Formation of Ocular Dominance Stripes. Network **1**, 41-59.

Goodhill, G., (1992) Correlations, Competition and Optimality: Modelling the Devel. of Topography and Ocular Dominance. CSRP 226, U. of Sussex, GB.

Goodhill, G., (1993) Topography and Ocular Dominance: A Model Exploring Posiive Correlations. Biol. Cyb. **69**, 109-118.

Jones, D.G., Van Slyuters, R.C., Murphy, K.M. (1991) A Computational Model for the Overall Pattern of Ocular Dominance. J. Neurosci. **11**, 3794-3808.

Kohonen, T., 1989. Self-Organization and Associative Memory, Springer.

LeVay, S., Connolly, M., Houde, J., Van Essen, D.C. (1985) The Complete Pattern of Ocular Dominance Stripes in the Striate Cortex and Visual Field of the Macaque Monkey. J. Neurosci. **5**, 486-501.

Miller, K.D., Keller, J.B., Stryker, M.P. (1989) Ocular Dominance Column Development: Analysis and Computation. Science **245**, 605-615.

Obermayer, K., Ritter, H., Schulten, K., (1990) A Principle for the Formation of the Spatial Structure of Cortical Feature Maps. PNAS USA **87**, 8345-8349.

Obermayer, K., Blasdel, G.G., Schulten, K. (1992) Statistical-Mechanical Analysis of Self-Organization and Pattern Formation during the Development of Visual Maps. Phys. Rev A **45**, 7568-7589.

Ritter, H., and Schulten, K. (1988) Convergence Prop. of Kohonen's Topology Cons. Maps: Fluct., Stability and Dimension Selection. Biol. Cyb. **60**, 59-71.

Ritter, H., Martinetz, T., Schulten, K. (1990) Neuronale Netze, Add. Wesley.

Swindale, N.V. (1980) A Model for the Formation of Ocular Dominance Stripes. Proc. R. Soc. London **B 208**, 243-264.

von der Malsburg, C. 1979. Development of Ocularity Domains and Growth Behaviour of axon Terminals. Biol. Cyb. **32**, 49-62.

How To Use Non–Visual Information for Optic Flow Processing in Monkey Visual Cortical Area MSTd

M. Lappe†, F. Bremmer†, and K.-P. Hoffmann†

† Dept. Zoology & Neurobiology, Ruhr University Bochum
D-44780 Bochum, Germany

1   Visual and non–visual information in MSTd

Area MSTd, part of the visual motion pathway in monkey cortex, contains cells that respond selectively to various large–field, random–dot, optic flow patterns [2] [9] [12]. It receives major input from area MT which contains cells directionally selective for local motions. An earlier network model proposes a way to achieve the visual response properties of MSTd neurons from the output of MT–like neurons [5] and links these properties to the psychophysics of human heading detection from optic flow [6] [7]. However, human heading detection has been shown to sometimes depend on non–visual eye movement information [11] [13] [14], and area MSTd has been found to contain extraretinal eye movement [8] as well as eye position [1] information. For instance, pursuit neurons in MSTd fire during smooth pursuit even in the absence of visual stimulation. Many other MSTd cells show a different behavior possibly related to non–visual input: They distinguish between active, self–induced visual motion resulting from active eye movements in a stationary environment and passive, externally–induced visual motion resulting from movements in the outside world, a finding not present in area MT [3].

In this paper we incorporate extraretinal information into the network model in a biologically plausible way by introducing an independent population of pursuit neurons, enabling the model to account for important psychophysical data. We show that the model neurons then attain the active/passive properties described for MSTd. However, we also show that these neuronal properties can be achieved visually without the need for extraretinal input, thereby posing the question, whether the active/passive distinction in MSTd is necessarily a reflection of some non–visual input or whether it may be due to complex optic flow processing allowing for such a differentiation.

2   Detecting egomotion visually

The network model is a biologically plausible implementation of a least-square algorithm of heading detection from optic flow. This algorithm computes the most likely heading direction $\mathbf{T}$ by minimizing a residual function

$$R(\mathbf{T}) = \|\Theta^t \mathbf{C}^\perp(\mathbf{T})\|^2, \tag{1}$$

in which Θ is a collection of several optic flow vectors θ_i, and $\mathbf{C}^\perp(\mathbf{T})$ is a matrix depending on the visual field locations of these flow vectors [4]. The network implements this minimization in a two–layer scheme. The first layer represents the optic flow input by direction selective neurons modelled after cells in area MT. Several neurons with different preferred directions e_{ik} and tuning functions s_{ik} are assumed to form a population encoding of an optic flow vector

$$\theta_i = \sum_{k=1}^{n} s_{ik} e_{ik}. \tag{2}$$

Their outputs feed into a second layer which contains cell populations that each represents one direction $\mathbf{T}_j$ and becomes maximally excited when $R(\mathbf{T}_j) = 0$, so that the peak of activity in this layer signals the best matching direction of heading. However, a single neuron l within population j evaluates only part of the argument of $R(\mathbf{T}_j)$. Its output activity is described by

$$u_{jl} = g(\sum_{i=1}^{m} \sum_{k=1}^{n} J_{ijkl} s_{ik} - \mu), \tag{3}$$

where $g(x)$ is a sigmoid function, μ a threshold, and J_{ijkl} the synaptic connection from a first layer cells ik. The synaptic strengths are chosen to satisfy

$$\sum_{i=1}^{m} \sum_{k=1}^{n} J_{ijkl} s_{ik} = \Theta^t C_l^{\perp}(\mathbf{T}_j), \tag{4}$$

and the threshold μ is chosen such that the population activity $U_j = \sum_l u_{jl}$ peaks when $R(\mathbf{T}_j) = 0$ [6].

3 Combining visual and non–visual input

This visual–only heading detection system is complemented by a separate population of pursuit neurons with preferred directions e_k and directional tuning functions p_k. They provide the second layer cells with the direction and — up to a scalar gain factor γ — the speed of a pursuit eye movement. Thus, the eye rotation is encoded as

$$\Omega = \gamma \sum_{k=1}^{n} p_k e_k. \tag{5}$$

The second layer neurons use this information to subtract the eye movement related visual motion from the input flow field. The component of the optic flow that is a result of the pursuit eye movement can be written as

$$\theta_i^p = \gamma \sum_{q=1}^{3} \sum_{k=1}^{n} w_{iq} p_k e_k \tag{6}$$

with suitable coefficients w_{iq}. Taking the input from the pursuit neurons into account the activity of a single second layer neuron becomes

$$u_{jl} = g(\sum_{i=1}^{m} \sum_{k=1}^{n} J_{ijkl} s_{ik} + \sum_{k=1}^{n} J_{jkl}^p p_k - \mu). \tag{7}$$

Instead of eq. 4 the calculation of the synaptic connections then proceeds with

$$\sum_{i=1}^{m} \sum_{k=1}^{n} J_{ijkl} s_{ik} + \sum_{k=1}^{n} J_{jkl}^p p_k = (\Theta - \gamma \Theta^p)^t C_l^{\perp}(\mathbf{T}_j). \tag{8}$$

48

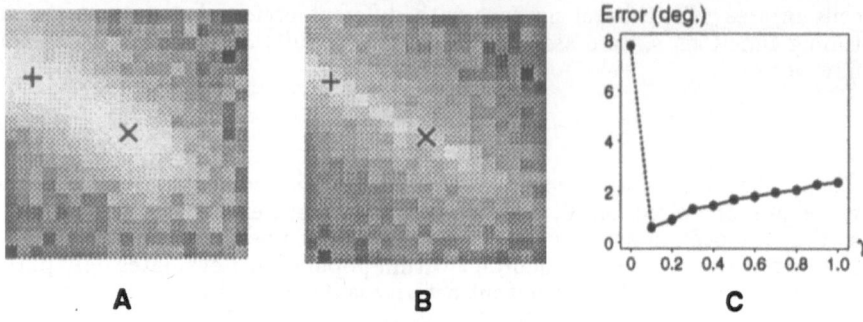

Figure 1:
A: Output activities of the second layer populations in response to an optic flow input when approaching a frontoparallel plane while performing an eye movement. Non–visual eye movement information was not available to the network ($\gamma = 0$). B: Same situation with the non–visual information switched on ($\gamma = 1$). C: Mean error of 100 simulation runs as a function of gain γ.

4 Using non–visual information: system level

Humans cannot accurately determine their direction of heading when an optic flow field corresponding to approaching a frontoparallel plane is confounded by the visual effects of a simulated eye movement [10]. When real eye movements are performed instead the task is solved easily [14]. The model was tested with such flow stimuli (Fig. 1). Without the input from the pursuit neurons ($\gamma = 0$) the network erroneously computes a direction towards the visual field center ($\times$) instead of the correct direction ($+$). When the pursuit information is present ($\gamma > 0$) the network identifies the direction of heading ($+$) correctly. This behavior is consistent with the data reported for humans [10] [14]. However, accurate information about the speed of the eye movement is not required, since the network works best with $\gamma < 1$ (Fig. 1C). The sharp drop between $\gamma = 0$ and $\gamma = 0.1$ in Fig. 1C shows that the eye movement information is only necessary to resolve the ambiguity present in the flow field. Similar to humans, in many cases other than this restricted situation the network functions with visual input only, and does not depend on extraretinal information at all [6].

In conclusion the model reproduces a characteristic human deficiency in processing certain retinal optic flow fields as well as its disappearance when non–visual eye movement information is available.

5 Using non–visual information: neuron level

Similar to cells in MSTd the neurons in the second layer of the model respond selectively to various optic flow patterns [6]. In the model different degrees of selectivity which have been described in MSTd [2] can be achieved by introducing biologically plausible constraints on the eye movements [5]. Fig. 2 shows that by using extraretinal eye movement information the second layer neurons can also distinguish between active, self–induced stimulation (as happens during smooth pursuit over a lit background) and passive, externally induced stimulation (when a visual pattern is moved across the receptive field). However, the dependance of this distinction on the non–visual input is also influenced by said eye movement constraints. The neuron in Fig. 2A–B looses its selectivity

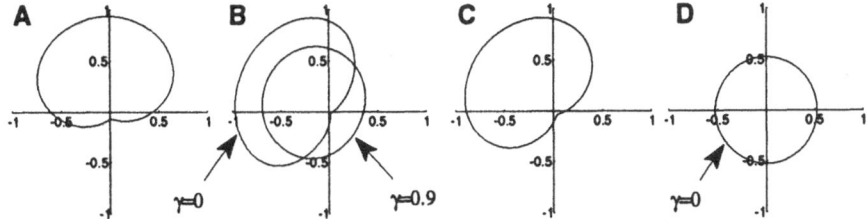

Figure 2:
A: Polar plot of passive direction selectivity of a second layer neuron preferring upward motion. B: Responses induced by an optic flow pattern experienced during active pursuit without ($\gamma = 0$) and with ($\gamma = 0.9$) non–visual input. C–D: Passive (C) and active (D) responses of a different neuron.

for the direction of an active pursuit only if $\gamma \gg 0$, whereas the neuron in Fig. 2C–D is not selective for the direction of an active pursuit even when no eye movement information is present ($\gamma = 0$). The synaptic weights of the latter neuron were calculated using eq. (8), while in constructing the former neuron eq. (8) was modified, as described in [5], by assuming that only smooth pursuit eye movements to fixate a environmental target were to be performed.

Thus we conclude that in addition to several visual properties of MSTd cells [5] [6] the model can also generate cells with the ability to discriminate self–induced from externally–induced visual motion. However, the model shows that this ability must not necessarily depend on non–visual input, leaving this question to future experimental work.

References

[1] F. Bremmer and K.-P. Hoffmann. *Soc.Neurosci.Abstr.*, 19, 1993.

[2] C. J. Duffy and R. H. Wurtz, *J.Neurophysiol.*, 65(6):1329–1345, 1991.

[3] R. G. Erickson and P. Thier, *Exp.Brain Res.*, 86:608–616, 1991.

[4] D. J. Heeger and A. Jepson, *Int.J.Comp.Vis.*, 7(2):95–117, 1992.

[5] M. Lappe and J. P. Rauschecker, In C. L. Giles *et al.*, ed., *Advances in Neural Information Processing Systems 5*. Morgan Kaufmann, 1993.

[6] M. Lappe and J. P. Rauschecker, *Neural Comp.*, 5:374–391, 1993.

[7] M. Lappe and J. P. Rauschecker, *Perception*, 22 (supp.), 1993.

[8] W. T. Newsome *et al.*, *J.Neurophysiol.*, 60(2):604–620, 1988.

[9] G. A. Orban *et al.*, *Proc.Nat.Acad.Sci.*, 89:2595–2599, 1992.

[10] D. Regan and K. I. Beverly. *Science*, 215:194–196, 1982.

[11] C. S. Royden *et al.*, *Nature*, 360:583–585, 1992.

[12] K. Tanaka *et al.*, *J.Neurosci.*, 6(1):134–144, 1986.

[13] A. V. van den Berg, *Nature*, 365:497–498, 1993.

[14] W. H. Warren and D. J. Hannon, *Nature*, 336:162–163, 1988.

A Learning Rule for Self-organization of The Velocity Selectivity of Directionally Selective Cells

Ken-ichiro Miura, Koji Kurata* and Takashi Nagano

College of Engineering, Hosei University
3-7-2, Kajino-cho, Koganei, Tokyo, 184, JAPAN
e-mail ken@keiei.hosei.ac.jp

Department of Biophysical Engineering, Faculty of Engineering Science, Osaka University
Machikaneyama-cho 1-1, Toyonaka, Osaka, 560, JAPAN

Abstract

We first present mathematical analysis about the relation between the parameters and the behavior of the basic module in the neural network model for visual motion detection proposed by one of the authors[1]. Based on the analytical results, a learning rule is proposed that can develop the velocity selectivity of directionally selective cells. The proposed learning rule is simple and plausible in the actual nervous system in that it is described only with local information. Numerical simulation results showed that the basic module learned self-organizingly to acquire the selectivity for velocity of an input stimulus.

1.Introduction

Many cells in, for example, middle temporal area have selectivity for direction and velocity of moving stimuli[2]. These two selectivities aren't determined by nature but can be acquired by post-natal learning[3]. One of the authors proposed a model, called "basic module" hereafter, that can explain the mechanism for direction and velocity selectivity[1]. We proposed a learning rule for the basic module to acquire the selectivity for velocity and direction[4]. This learning rule, however, has a problem that acquired selectivity is not necessarily tuned exactly to input velocity.

In this paper, we first analyze the relation between the synaptic weights and the optimum velocity of the basic module. By using the analytical results we propose a new learning rule that enables the basic module to respond maximally to a stimulus moving with the same velocity as the training stimulus. Numerical simulations are conducted to confirm the self-organizing process of the basic module.

2. Basic module
2.1 Structure and behavior of the basic module

For simplicity, we consider one basic module which has selectivity only to one direction. The structure of the basic module is shown in Figure.1. It is composed of three layers Y, P, and D. In the following, a cell in layer Y is denoted by Y_i, and the output of Y_i is denoted by $y_i(t)$, where i and t denote the cell's position and time respectively. The same notation is used for cells in layer P and D. Y_i responds transiently if a stimulus moves across or flickers on its receptive field and outputs an impulse. P_i receives input signals from Y_i, P_{i-1} and P_i via synapses with positive weights W_{yp}, L_i and S_i respectively and outputs $p_i(t)$ given by the sum of the weighted inputs. D_i has a threshold θ_d and gives output 1 when its input exceeds θ_d, and gives output 0 otherwise. The potentials of cells are

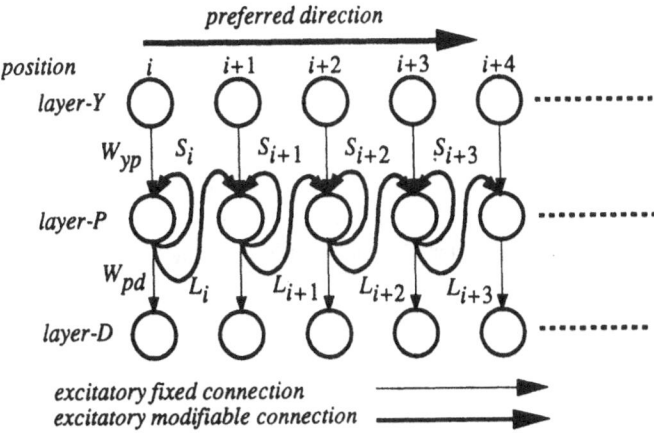

Figure.1 *Structure of the basic module*

supposed to be updated at every unit time Δt. The weights W_{yp} and W_{pd} are fixed and assumed to be 1 without loss of generality. L_i and S_i are modifiable.

A basic module responds selectively to specific direction and velocity. Its preferred direction is decided by the direction of lateral connections in layer P. The optimum velocity is decided by the weights L_i and S_i under the restriction $S_i+L_i=\alpha$ $(0<\alpha<1)$[1]. It is required that the values of L_i and S_i are almost constant at any position in the basic module in order that all the output cells in the basic module may have the same optimum velocity. Lateral connections in layer P cause the gradual increase of the activity of cells in the layer as a stimulus moves to the preferred direction of the module. When the activity exceeds the threshold, the module detects the motion of an input stimulus. The larger the ratio of L_i to S_i becomes, the larger the optimum velocity becomes.

2.2. Analysis of the basic module

It is clear that the behavior of the basic module is determined by the connections in layer P. So, the relation between the weights L_i and S_i and the optimum velocity is analyzed mathematically in order to clarify the behavior of the basic module in detail. $p_i(t)$ is described by the following differential equation if the module is assumed to work continuously.

$$T_i \cdot \frac{dp_i(t)}{dt} = -p_i(t) + x_i(t) \tag{1}$$

, where

$$x_i(t) = L_{i-1} \cdot p_{i-1}(t) + y_i(t) \tag{2}$$

$$T_i = \frac{1}{(1-S_i)} \tag{3}$$

So, the transfer function of P_i is given by

$$P_i(s) / X_i(s) = \frac{1}{1+T_i s} \tag{4}$$

, where $P_i(s)$ and $X_i(s)$ are the Laplace transform of $p_i(t)$ and $x_i(t)$ respectively. Here, the values of L_i and S_i are assumed to be constant at every position $(L_i=L, S_i=S, T_i=T)$ and $y_i(t)$ is assumed to be the impulse described by δ-function. Consider the response property of P_i

to an impulse input given to P_1 (the leftmost cell in Figure.1). The cell's response in the s-domain to the impulse is expressed by $1/(1+Ts)$. The response of the second cell is $L/(1+Ts)^2$ because its input is $L/(1+Ts)$. In such a way, the response of the n-th cell P_n to the impulse given to the first cell P_1 is generally described by: $L^{n-1}/(1+Ts)^n$. That is, the n-th cell has the response property of the n-th order lag to the impulse given to P_1. $r_n(t)$, the response of P_n to the input impulse to P_1 in the t-domain has the maximum value at $t_{peak}=(n-1)T$. This result shows that the time interval between the peak of $r_i(t)$ and that of $r_{i+1}(t)$ for the impulse is constant for all i and is given by T. A stimulus moving to the preferred direction causes each Y_i to give an impulse output and the potential of P_n before receiving an impulse from Y_n is expressed by sum of the responses to the impulses given at positions from 1 to m $(m \leq n-1)$. This means that, the basic module responds most strongly when the input interval between Y_i and Y_{i+1} is equal to the peak interval T.

3. Learning rule

A learning rule that enables the basic module to be organized to respond maximally to the stimulus moving at the same velocity as the training stimulus is described in the following. The result in 2.2. shows that the optimum velocity of the basic module is determined by the time constant T_i of the cells in layer P. So, we propose a learning rule that modifies the time constant T_i based on the state of the basic module so as the basic module to be tuned to input velocity. If a P cell receives an input from Y before its potential reaches the maximum, that is, input velocity is larger than the optimum one, the time constant T_i must be decreased in order to make the time of the peak of the cell's potential equal with the time when an input from Y arrives at the cell. On the contrary, when a cell receives an input from Y after the cell gives the maximum potential, that is, the input stimulus velocity is smaller than the optimum one, the time constant T_i must be increased. In the former case, the value of $(dp_i(t)/dt)$ is positive at the time when an input from Y arrives at. In latter case, $(dp_i(t)/dt)$ is negative at that time. Therefore, if the value of T_i is changed proportional to $-(dp_i(t)/dt)$ the basic module can be organized to responds maximally to the input stimulus.

T_i is updated by changing the weight S_i on the module shown in Figure.1. ΔS_i, the change of S_i in the basic module, can be given by the following equations. Here, all the weights are assumed to be changed at every time when an input from Y arrives at the cell.

$$\Delta S_i = -\eta \cdot \{p_i(t_{input} - 0) - p_i(t_{input} - \Delta t)\} \tag{5}$$

, where η denotes the learning rate, t_{input} denotes the time when a P cell receives an input from Y and $p_i(t_{input}-0)$ denotes the potential of P_i just before an impulse from Y_i arrives at. Also L_i is changed according to the relation $L_i = \alpha \cdot S_i$. Initial values of S_i and L_i are assumed to be the same at every position. When all the S_i's converge, the learning process is finished.

4. Computer simulation

In order to confirm the self-organizing behavior of the basic module to be the expected one, we conducted some numerical simulations. One of the results is shown here. In this simulation, a basic module was exposed iteratively to a stimulus moving with constant velocity. Here, velocity was defined by $V=1/t_{interval}$, where $t_{interval}$ shows the input interval between Y_i and Y_{i+1}. Figure.2 shows the values of S_i's after learning in the cases of $V=0.50$, 0.33, 0.25 and 0.20. The horizontal and vertical axes in Figure.2 show the position of cells and S_i respectively. The values of L_i's are easily obtained by using the relation $L_i = \alpha - S_i (\alpha = 0.99)$.

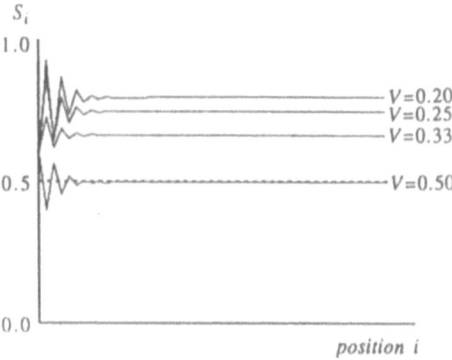

Figure.2 *Values of S_i's after learning*

5. Results and Discussion

It can be seen in these results that the variations of S_i in the basic module were very small except at the initial part (the leftmost part in Figure.2). This means that most of the cells in the basic module were tuned to the same velocity, that is, the basic module had the same selectivity regardless of cell's position. Figure.2 shows that the values of $(1 - S_i)$ after learning were equal to the velocity of each training stimulus. This means that the basic module was organized so that the model had the optimum selectivity for the velocity of the input stimulus. The values of S_i's after learning at the initial part greatly differed from those of the other in the basic module. This is caused by the fact that learning never occurs at the first cell. This makes the following few cells to converge to a little deviated value from the expected one. But the deviation rapidly decreases as cell's positions move on to the right. This initial part does not have much influence on the total behavior of the basic module because the part is short enough to be neglected. Consequently, the basic module can be optimally organized by using the proposed learning rule in that the turned velocity is always equal to the stimulus velocity used in the training period. In addition, the proposed rule is described only with local information: the pre- and post-potentials of a cell. So it is natural to assume the existence of the rule in the actual nervous system. We also introduced the learning rule into the mass model[1] which is constructed by many basic modules and that has inhibitory connections between them, and it was confirmed that the learning rule can be successfully applied to the mass model.

This research was supposed by Grant-in-Aid #04246107 for Scientific Research on Priority Areas on "Higher-Order Brain Functions", the ministry of Education, Science, and Culture of Japan.

Reference

[1] M.Hirahara, T.Nagano [1993] A neural network model for visual motion detection that can explain psychophysical and neurophysiological phenomena. *Biol.Cybern.*, 68, pp.247-252

[2] J.H.R.Maunsell, D.C.VanEssen [1983] Functional properties of neurons in middle temporal visual area of the macaque monkey.I.Selectivity for stimulus direction, speed and orientation. *J. Neurophysiol.*,49[5],pp.1127-1147

[3] W.Singer [1976] Modification of orientation and direction selectivity of cortical cells in kittens with monocular vision. *Brain Res.*, 118, pp.460-468

[4] K.Miura, T.Nagano [1993] Self-organization of the velocity selectivity of directionally selective cells. *Proc. IJCNN-Nagoya '93*, Vol.1, pp.49-52

Motion analysis with recurrent neural nets

A. Psarrou†,H. Buxton‡

† School of Computer Science, Uni. of Westminster, London, U.K.
‡ COGS, Uni. of Sussex, Brighton, U.K.

1 Introduction

218zVisual tasks such as the interpretation of cell images (Psarrou and Buxton, 1993) and the recognition of moving vehicles require to track objects along their trajectory and to predict their future position in their environment. It was noted that objects move purposely in an environment and effective prediction on their trajectories can be achieved by modelling the spatio-temporal regularities associated with their moving purposes with visually augmented hidden Markov Models (Gong and Buxton, 1992). Temporal prediction and recognition require (a) a short-term memory that retains aspects of the input sequence relevant to prediction and recognition, (b) the specification of a function that combines the current memory and the current input in order to form a new temporal context (Mozer, 1993), (c) to identify and learn the regularities from the temporal sequence. Feedforward neural networks (Figure 1a) can be trained with the backpropagation algorithm to represent, predict and recognise temporally ordered events since (a) they are capable of extracting "common features" from a temporal sequence, (b) can encode such features in the hidden units of the network and (c) these features encode information that relates past events with future input values. However, this approach requires the spatial representation of events by parallelising time which involves several drawbacks: (a) only allows fixed window size on the event representation, (b) large memory consumption and (c) the network cannot easily distinguish the relative temporal position of an element in a sequence from its absolute temporal position (Elman, 1990). In this paper we explore the work of (Elman, 1990), (Cleeremans et al, 1989) and (Mozer, 1993) to show how recurrent neural networks can be used to address the problem of temporal prediction in computer vision applications.

2 Why Recurrent Neural Networks

A popular way to recognise and predict sequences is to use partially recurrent networks. In these architectures the connections are mainly feedforward, but include a carefully chosen set of feedback connections either from the hidden layer or the output layer (Figure 1b). The feedback or context units remember some aspects of the recent past, and so the state of the whole network at time t depends on the aggregate of previous states as well as on the current state. In both cases, feedback is easily implemented by extending the input field with an additional feedback vector containing the hidden or output unit values generated by the preceding input. In most cases the feedback connections are fixed, so standard back-propagation may easily be used for training. Elman suggested the architecture shown in Figure 1b in which the context units hold a copy of the activations of the hidden units from the previous time step. As it is pointed out, the hidden unit patterns of activation represent an "encoding" of the features of the input patterns that are relevant to the task. The context units patterns of activation represent an "encoding" of the relevant features of the past input elements (Elman, 1990). Thus, a hidden layer pattern now can encode information about the relevant features of two consecutive input elements. Furthermore, (a) the event can now be processed sequentially without the need of a buffer and (b) there is not an absolute temporal position of an element. Such a network is able to recognise and produce short continuations of known sequences. Cleeremans (Cleeremans et al, 1989) has shown that when this network is trained with strings from a particular finite-state grammar, it can learn to be a perfect finite-state recognizer for the grammar (Figure 2b).

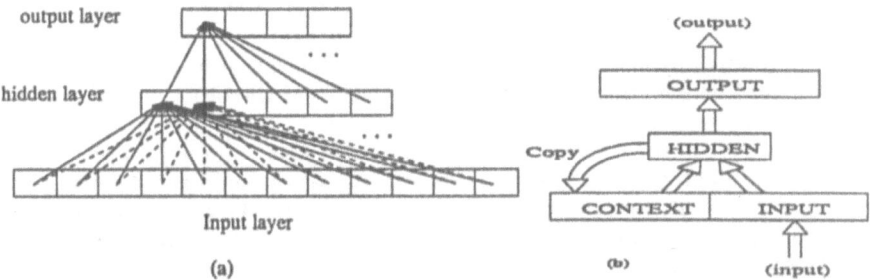

Figure 1: (a) Feedforward network (b) Recurrent network

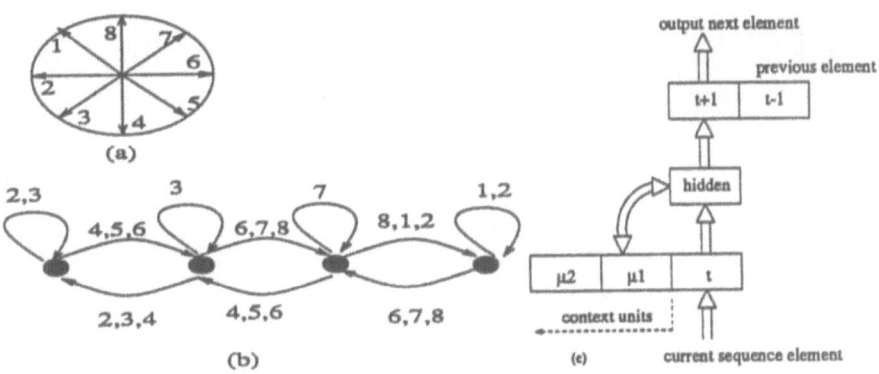

Figure 2: (a) Qualitative directional states. (b,c) A finite-state machine modeled by a recurrent neural network is used to model and predict the trajectory of purposively moving vehicles.

Prediction using Recurrent Neural Networks: Trajectories that include hidden regularities can be represented by a finite state machine where the location of these regularities correspond to the states of the machine and the orientation and dispacement vectors correspond to the transition arcs. Such a representation can be modelled on a neural network based on Elman's architecture that learns the significant locations of the trajectory of a moving object and encodes such information in its hidden units (Figure 2c). Figure 2b shows a finite state machine representation of a circular trajectory. The states of the machine correspond to the locations in a cyclic trajectory where the orientation changes suddenly, namely at 0, 90, 180 and 270 degrees from the horizontal. The symbols on the transition arcs correspond to the set of possible displacement the object may exhibit. The object may either tranverse to the next state or loop on the same state. Their course of direction can be described by eight qualitative directional states that denote the next possible movement of the object and are shown in Figure 2a.

3   Architecture and Experimental Results

When mapping object trajectories on recurrent neural networks the following issues should be addressed: (a) the representation of the object's trajectory in terms of data elements and (b) the representation of the trajectory in the short-term memory. In our experiments the object trajctory was represented using only displacement vectors. The displacements were presented as normalised real values in the input units. Trajectories are mapped in the short-term memory using an exponential function (Mozer, 1993) of the past hidden unit activations. The architecture used is shown in Figure 2c. In comprises of: (1) *input layer*

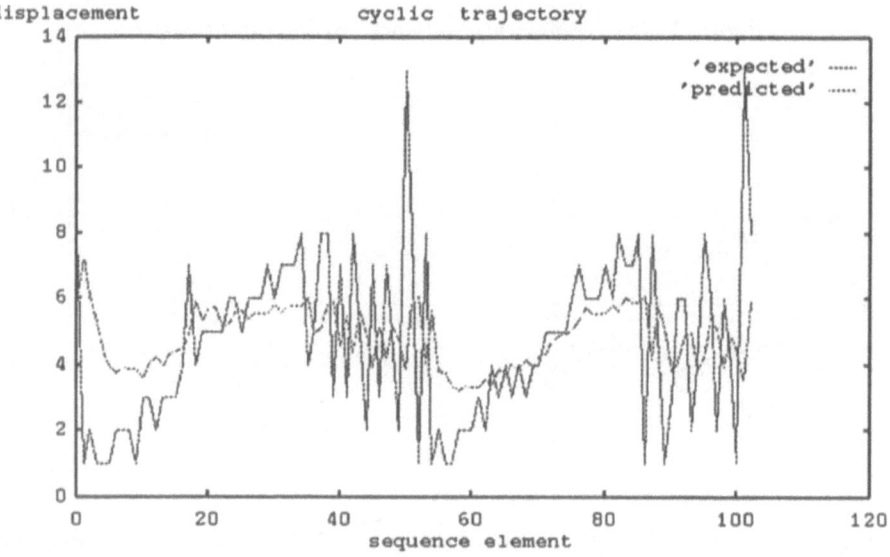

Figure 3: Expected and predicted trajectories

of one unit and represents the element of the sequence at time t; (2) *hidden layer* of four units; (3) *output layer* of two units representing the element of the sequence at time $t+1$ and $t-1$; (4) *context layer* of eight units. The value of the context units is an exponential encoding of the hidden unit activation values.

They maintain moving averages of past hidden value activations according to the equation $x_i(t) = (1 - \mu_i)x(t) + \mu_i x_i(t - 1)$, where μ_i lies in the interval $[1, -1]$ and allows for the representation of averages spanning various intervals of time, $x(t)$ represents the vector of the hidden units activation values at time t and, $x_i(t)$ represents the memory vector i at time t. An exponential trace memory is formed using the kernel function $c_i(t) = (1 - \mu_i)\mu_i^t$. Figure 3 shows the expected and predicted cyclic trajectory of a moving vehicle. As it is shown the trajectory predicted from the network follows the pattern of the expected orientation along the trajectory. The results show that (a) there may be a scaling factor between the expected and predicted displacement value (b) a different data representation where the displacement values are encoded in the gradient of the input sequence may be prefered.

4   Discussion

In this paper we described how observed trajectories can be mapped on recurrent neural networks. During our experiments we showed that the network was able to learn the sequence that was presented to it and produce continuations of the observed sequence given the initial element. Investigation of the hidden unit activation values showed that they exhibit characteristic patterns associated with the orientation pattern of the trajectory, however a cluster analysis on the hidden unit values did not yet provide any conclusive evidence of state representation. This is mainly attributed to the absolute displacement value predicted by the network. Past experiments have shown that the (a) the number of memory elements in the context units and (b) the data representation of the input sequence may affect considerably the recovery of quantitative results. In our current work we (a) exploit further the parameter setting of our network architecture and especially the structure of the context layer (b) exploit alternative data representation of our input sequence as suggested in the previous section and finally we apply this network to a real world scenario.

5   References

Cleeremans, A. et. al. (1989), "Finite State Automata and Simple Recurrent Networks", *Neural Computation*, 1, 372-381.

Elman, J. (1990), "Finding structure in time", *Cognitive Science*, 14, 179-211.

Gong, S., Buxton, H. (1992), "On the Expectations of Moving Objects", ECAI-92, Vienna, Austria.

Mozer, M. (1993) "Neural net architectures for temporal sequence processing", *Predicting the future and Understanding the past* A. Weigend & N. Gershenfeld (eds), Redwood City, CA: Addison-Wesley.

Psarrou, A., Buxton, H. (1993)' "Hybrid architecture for understanding motion sequences", *Neurocomputing*, 5, 221-241.

Self-Organizing a Behaviour-Oriented Interpretation of Objects in Active-Vision

H.-M. Gross, H.-J. Boehme, D. Heinke, T. Pomierski, R. Moeller

Department of Neuroinformatics, Technical University of Ilmenau

D-98684 Ilmenau, Germany

1 Introduction

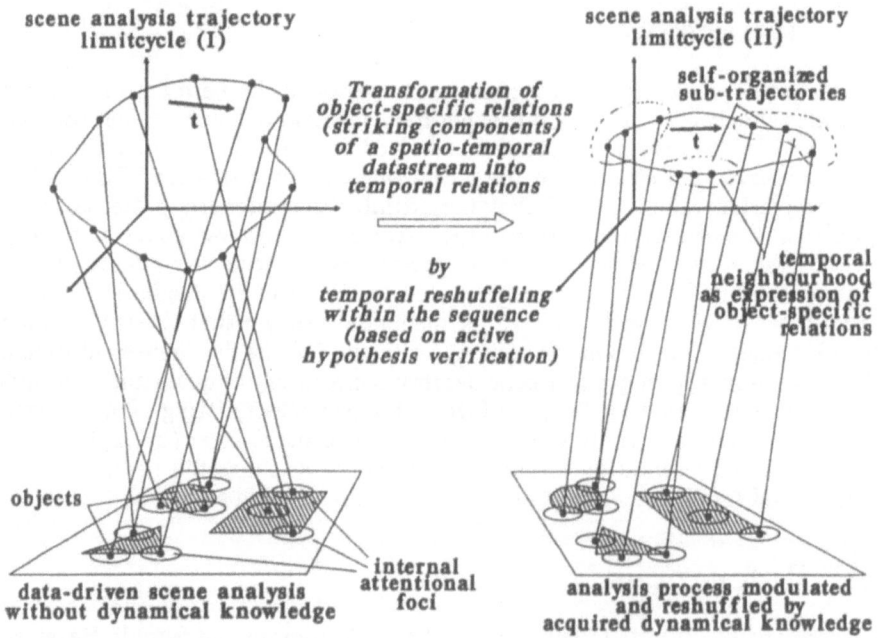

Figure 1: *Evolution of stable temporal relations within sub-trajectories*

This paper deals with a new approach towards self-organization of an intrinsic (behaviour oriented) spatio-temporal object-understanding in the context of an active-vision process in the widest sense. The focus is on the functional architecture and the dynamical principles suited for self-organization of knowledge about complex visual structures, and for a *behaviour-oriented interpretation* of objects in real-world scenes. In our mind, such a self-organization of intrinsic dynamical knowledge is a prerequisite for active, autonomous learning under real-world conditions. In this context an approach is used in our model, whose sub-sequential states of learning are related to the development of the systems behaviour in visual scene analysis. It is well known, that the visual input to an analyzing system dealing with real-world problems is not a set of preselected, figure-ground segregated objects which are to be properly arranged, but the system itself has to select reliably detectable features or input components out of the massively parallel input (v.d. Malsburg 1992). Selective attention processes (both active-vision and internal scanning not related to eye movement) and dynamical processing at different organizational levels are widely accepted mechanisms explaining the decomposition of a complex visual scene into components and the subsequent reassembling the recalled internal representations towards an unitary decision (Crick et al., 1990). Therefore, an important aspect of our approach is the data and/or hypotheses driven dissolution of the highly parallel visual input into meaningful components, which can be reassembled freely to new complex visual structures. Only in this case, the analyzing system is able to handle the present input on the base of the knowledge already acquired at any time. In addition, only by such a continuous interaction a self-organization is possible, the aim of which is to bring the actual perception into maximal consistency with the acquired knowledge. The goal of our approach is to find useful ways of exploiting the wealth of dynamic behaviour for such aspects like active, autonomous learning of internal representations, which are assumed to be fundamentally for a behaviour-oriented understanding of visual objects during active-vision. Of our particular interest are such concepts like generation and active verification of dynamical hypotheses about the input in a feedback coupled process of *Sensory Controlled Internal Simulation.* In the context of the systems behaviour during active vision that means the generation and testing of hypothesis about *what* components are to be expected *when* and *where* in the visual field – this is an internal anticipation of a real spatio-temporal selective attention or active-vision process. Therefore our model concept proposed later should be able to map the temporal and spatial characteristics of the data driven serial processing within the intervals between the eye movements shown strongly simplified in Figure 1 into characteristic and dynamically stable internal representations coding stable striking feature relations within the the objects. Without any internal knowledge our functional architecture is not able to establish suitable hypotheses about objects to anticipate an internal scan process. Self-organization of an object understanding means, that typical, reliably detectable striking visual components and their object-specific relations detected in preceding active vision processes more frequently, gradually can be coupled or linked in the temporal domain as *temporal adjacent*. The evolution of stable temporal relations (temporal neighborhood) within subtrajectories is an expression of stable object-specific relations within input data stream and is to understand as a behaviour-oriented internal understanding which components of a visual structure (object) belong together. In our opinion, *temporal neighborhood* in selective attention processes could be a good, possibly the only criterion for an unsupervised segmentation and learning of objects arranged in highly structured visual scenes. Figure 1 (on the right) shows the final state of such a behaviour-oriented transformation. By knowledge-based reshuffling the input sequence the development of such a sequence (limit cycle) of object-specific sub-trajectories is forced which organizes best all relevant input components into a globally consistent decision.

2 Functional architecture

For dealing with autonomous learning and self-organization of such a behaviour-oriented object-understanding under real world conditions we developed a neurobiologically inspired functional architecture. Our concept presented here is

60

an improved but still very simple computational model of a modular processing hierarchy compared to an earlier approach of us (Koerner et al., 1991). The architecture to be developed in the NAMOS-project is to decompose a complex visual input into a reverberating sequence of reliably detectable fragments (components) ranked by its visual conspicuousness (complexity) and controlled by the self-organized knowledge. The organizational levels of our model schematized in Figure 2 define the following basic abilities and information processing tasks: The **Saliency System** proposed in (Gross et al., 1992) has

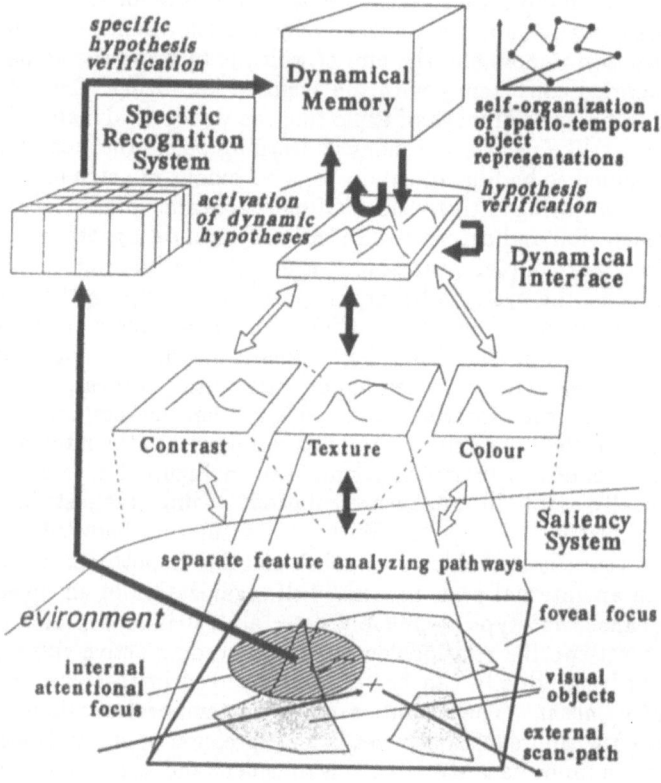

Figure 2: *Sensory Controlled Internal Simulation*

been influenced essentially by the neurophysiological concepts of primary visual processing (DeYoe et al., 1988). The parallel representations at different feature maps yield a measure of the conspicuity of a location in the scene. This is prerequisite for a data-driven decomposition of the visual input into striking and reliably detectable components that can classified as known or unknown by the following levels and can be reshuffled and reassembled freely.

Based on its inherent dynamics the **Dynamical Interface** carries out a sequential search for the most striking components within the visual field by shifting its internal attentional focus. The Dynamical Interface is modulated both bottom-up by the sensory input from *Saliency System* and top-down by spatio-temporal recall from *Dynamical Memory* in context of an unspecific hypotheses verification. In this way the Interface and with that the sensory data stream can be controlled by activated hypotheses about the input so that the interesting components can be reshuffled in time according to the state of internal hypthesis activation (Gross et al. 1992, Heinke et al., 1993). The **Specific Recognition System** operates on the internal attentional focus controlled by the *Dynamical Interface* within the *Saliency System*. In cooperation with the *Dynamical Memory* it determines in a *specific hypothesis verification* the similarity of the actual internal focus feature set to the feature sets extracted and learned autonomously in previous cycles (see Pomierski et al., 1993). The **Dynamical Memory - (DM)** is the highest organizational level of our architecture. DM is activated and driven by the established spatio-temporal sequence of striking input components and can act as a guide in attentional control and input decomposition based on the knowledge already accumulated within the system. Therefore DM is interacting reciprocally with the *Dynamical Interface* in so-called 'hypothesize-verification-cycles'. All activated hypotheses interfere back to the *Dynamical Interface* and try to control the course of data-driven search within this system. Via this feed-back DM can search for that input components which would support one of the activated hypotheses. In this sense, DM uses its internal self-organized knowledge for flexible activation and continuous verification of hypotheses about the input (see (Boehme et al., 1994)). This forces the development of such a sequence of decisions which organizes best all selected input components into a globally consistent decision. If it is impossible to activate internal hypotheses by a data-driven input sequence this input has to be accepted by the DM as a new sensory situation. On this background a model was developed and presented in (Boehme et al., 1992) as a first very simple attempt to extract the inherent structure of a spatio-temporal data stream in active-vision or selective attention processes. Based on this first model the DM tries to map each sensory input sequence into a characteristically memory trace. So, the sequence of decisions on certain striking input components is transferred into a spatio-temporal representation within DM. For the case, that the input is partially unknown, DM can generate sub-hypotheses on the base of the already accumulated knowledge. If no interpretation of the input pattern sequence is possible, this input sequence will be learned. In first simulations (see Heinke et al., 1993) our system was able to change its behaviour in scanning unknown complex scenes, that is a result of the transformation of detectable object-specific relations uncoupled with respect to time at the beginning into more and more stable temporal relations by autonomous learning. In next time we will couple all different organizational levels of our model in a comprehensive simulation.

References

Boehme, H.-J. et al. (1992); Proc. of ICANN'92, 1381-84

Boehme, H.-J. et al. (1994); submitted paper to ICANN'94

Crick, F. & Koch, Ch. (1990); Sem. in Neurosc. 2 (1990) 263-275

DeYoe, E.A., Van Essen, D.C. (1988) TINS, 11 (1988) 5, 219-227

Gross, H.-M., Koerner, E. (1991); Proc. ICANN'92, 825-828

Heinke, D., Gross, H.-M. (1993); Proc. ICANN'93, 63-66

Koerner, E., Boehme, H.-J. (1991) Proc. of ICANN'91, 873-78

v.d.Malsburg, Ch., Buhmann, J. (1992) Biol. Cybern., 67 (1992) 233-242

Pomierski, T., Gross, H.M. (1993) Proc. of ICANN'93, 142-47

Hybrid Methods
for Robust Irradiance Analysis
and 3-D Shape Reconstruction
from Images

F. Callari [†], U. Maniscalco, P. Storniolo

DIE- Dipartimento di Ingegneria Elettrica, University of Palermo
Viale delle Scienze, I-90128 Palermo, Italy

† DIE and Computer Science Dept., University of Colorado at Boulder,
CB 430, Boulder, CO, 80309/0430, USA
e-mail: callari@cs.colorado.edu

1. Introduction

The analysis of the differential structure of images is an interesting task in machine vision, among other reasons because it can provide relevant featural representation of images, suited for higher level information processing task like geometry reconstruction and object recognition. The importance of invariants of the field of isophotae on lambertian surfaces in shape perception by means of chiaroscuro is discussed in (Koenderink and Van Doorn, 1980). In their approach to shape from shading, (Breton et al , 1992) represent the shading of the image by means of its shading flow field, i.e. by the first order differential structure of the image expressed as the isoluminance direction and gradient magnitude. The (Grossberg and Mingolla, 1985) model of low level visual processes uses input sensors approximately sensitive to brightness gradient, providing a map of segments oriented along constant brightness lines. This last approach inspired our investigation of connectionist architectures aimed to robust first and second order intensity analysis of gray scale images, whose study we pursued according to a previously proposed visual perception model (Ardizzone et al, 1992). The proposed architectures compute estimates, optimal in the least squares sense, of the shading flow and of the shading flow's variational structure by mask-filtering through purposely designed receptive fields. The feasibility of 3-D reconstruction using a featural image representation based on the quantized shading flow field has been investigated in (Breton et al , 1992), (Ardizzone et al, 1991) and (Maniscalco et al, 1993). The last approach is followed herein to approximate single objects in images by 3-D geometric primitives (superquadrics, (Barr, 1980)). The approximation is achieved through an optimization algorithm matching the shading flow map of the object image with the (analytically computed) isophotae of approximating superquadrics. Promising experimental results thus achieved are reported.

2 First order (shading flow) analysis

We begin by approximating the irradiance density distribution (intensity per unit area) $i(\bullet)$ in a circular neighborhood of radius R of a given a point on the image by its second-order truncated Taylor series, i.e. by a quadratic form such as

$$a_{11} - a_{22} = \frac{8}{\pi R^4} \int\limits_{\alpha=0}^{\pi/2} m_\varepsilon(\alpha) \sin 2\alpha d\alpha \quad , \quad a_{12} = -\frac{4}{\pi R^4} \int\limits_{\alpha=0}^{\pi/2} m_\varepsilon(\alpha) \cos 2\alpha d\alpha$$

In this case only two of the three stationariety equations are independent, consistently with our previous geometric interpretation: we cannot recover all three hessian parameters from the measure of two independent quantities, namely the two components of the vectorial composition of mask outputs.

4   The connectionist architectures

The results of the first order vectorial compositions can be thought as a discretized field of vectors aligned according to a robust estimate of direction of the isophotae, and whose magnitude is proportional to the local value of the brightness gradient. All the computations previously outlined can be performed by linear parallel-distributed neural networks whose input is the set of image pixels (fig. 2). A uniform square grid is superimposed on the image and two linear units, C_{ij} and S_{ij} are connected to each grid cell. For the (first order) shading flow analysis case, the activations of each unit pair represent the components of a vector tangential to the isophota in the cell center. The network weights w_{Cij} and w_{Sij} impinging on the pixels within a region are arranged so that for the net inputs to each pair of neurons, equal to their outputs, the following approximations apply:

$$NetC_{ij} = \sum_{l,k} w_{Cij} I(\rho_1, \varphi_k) \cong \int\limits_{\alpha=0}^{\pi} m_\varepsilon(\alpha) \cos \alpha d\alpha \quad , \quad NetS_{ij} = \sum_{l,k} w_{Sij} I(\rho_1, \varphi_k) \cong \int\limits_{\alpha=0}^{\pi} m_\varepsilon(\alpha) \sin \alpha d\alpha$$

where the sums run over the pixel matrix within the cell, $I(\cdot)$ being the image brightness, whose density we previously indicated as $i(\cdot)$. A closed-form expression for the weights is obtained when the previous integrals are approximated by composing the outputs of N masks with equally spaced orientations. Analogous considerations apply for the second order case.

5   The recovery of parametric surfaces from the shading flow field

We follow the approach detailed in (Maniscalco *et al*, 1993) to reconstruct the shapes of synthetic and real objects from their above estimated shading flow fields using superquadrics as approximation primitives. The approximation algorithm,searches an optimal match between the shading flow extracted from the image by the neural net and the analytically generated isophotae of approximating superquadrics. In the implementation described herein only single superquadrics in canonical position are searched, their surface reflection properties are defined according to the lambertian model and the imaging projection is assumed to be orthographic. These assumptions are consistent with those used in current literature on this topic.(Breton *et al* , 1992), (Horn, 1986), (Pentland, 1982). One sample of the results of the experiments insofar performed is shown in fig 3, while a detailed report of system characteristics and performances can be found in (Callari *et al*, 1993).

$$i(x, y) = a_{11}x^2 + 2a_{12}xy + a_{22}y^2 + 2a_{13}x + 2a_{23}y + a_{33}$$

where the origin is in the given point, see fig. 1. In polar coordinates,

$$i(\rho, \varphi) = \rho^2\left(a_{11}\cos^2\varphi + a_{22}\sin^2\varphi\right) + 2\rho\left(a_{12}\sin\varphi\cos\varphi + a_{13}\cos\varphi + a_{23}\sin\varphi\right) + a_{33}$$

Now the result m(•) of masking the neighborhood by an α-oriented first order analyzer (see fig. 1) of radius R, defined by

$$D_1(\varphi, \alpha) \equiv \begin{cases} -1 & , \quad \alpha \le \varphi < \alpha + \pi \\ +1 & , \quad \alpha + \pi \le \varphi < \alpha + 2\pi \end{cases}$$

is evaluated as

$$m_1(\alpha, a_{13}, a_{23}) \equiv \int_{\rho=0}^{R} \int_{\varphi=0}^{2\pi} i(\rho, \varphi)D_1(\varphi, \alpha)\rho d\rho d\varphi = \frac{8}{3}R^3\left(a_{13}\sin\alpha - a_{23}\cos\alpha\right)$$

where the dependency of m_1(•) on the gradient of i(•), i.e. on the linear coefficients of the Taylor expansion is made explicit. Due to model inaccuracies, to numeric error, to noise, what is actually measured by a detector is an erroneous function $m_\varepsilon(\alpha) = m_1(\circ) + \varepsilon$, which may be considered depending only on the detector orientation α, since α and R are the only free parameters (they identify one particular detector). The aim is then to minimize in the least squares sense the error term ε over all possible detector orientations, i.e. the integral squared error.

$$E_1(a_{13}, a_{23}) \equiv \int_{\alpha=0}^{\pi} \left[m_\varepsilon(\alpha) - m_1(\alpha, a_{13}, a_{23})\right]^2 d\alpha$$

where the upper limit of integration is π, since the detector output m_1 is obviously an odd function of α, periodic with period π. By formally solving the equations for the stationary points it follows that, when E_1 is stationary,

$$a_{13} = \frac{3}{5\pi R^3} \int_{\alpha=0}^{\pi} m_\varepsilon(\alpha)\sin\alpha d\alpha \quad , \quad a_{23} = -\frac{3}{5\pi R^3} \int_{\alpha=0}^{\pi} m_\varepsilon(\alpha)\cos\alpha d\alpha \quad (1)$$

Equations (1) can be explained as follows (see fig. 1): if the region is masked with a set of α-oriented detectors, each considered affected by a vector oriented as α and with magnitude equal to the detector's output, then the smaller is the summed squared error, the more the composition of such vectors is orthogonal to the brightness gradient. Hence, if it can be assumed that the third-order brightness variations within the receptive field are small , then an optimal linear estimate for the shading flow field vector s in the given point is $s = [-a_{23}, a_{13}]^T$, where the expression for the estimates of a_{13} and a_{23} are the above (1).

3   Second order analysis

The methods used in the above analysis can be repeated with regard to the second order detectors depicted in fig. 3. An analogous expression for the detector output $m_2(\alpha, a_{11}, a_{12}, a_{22})$ under the second order Taylor approximation is computed, an integral square error measure $E_2(a_{11}, a_{12}, a_{22})$ is defined as above, and when it'is stationary the following relations hold between the compositions of mask outputs and the components of the hessian matrix of the brightness density:

65

References

Ardizzone, E.,Chella, A., Pirrone, R., Sorbello, F.: A System Based on Neural Architectures for the Reconstruction of 3-D Shapes from Images. In: Ardizzone, E., Gaglio, S., Sorbello, F. (ed.s), *Trends in Artificial Intelligence*, Springer-Verlag, Berlin, 1991.

Barr, A. H.: Superquadrics and AnglePreserving Transformations, *IEEE Computer Graphics and Applications*, 1, 11-23, 1981

Breton, P., Iverson, L. A., Langer, M. S., Zucker, S. W.: A New Approach to Shape from Shading. In: Carpenter, G., Grossberg, S. (eds.): *Neural Networks for Vision and Image Processing*, Bradford Book, MIT Press, Cambridge, 1992

Callari, F. G., Maniscalco, U., Storniolo, P.: Robust Shading Flow Estimation and Applications to Parameric Surface Recovery, CS & AI Lab. Tech. Report. 9/1/1993, Elec. Eng. Dept., Univ. of Palermo, Italy.

Grossberg, S., Mingolla, E.: Computer Simulations of Neural Networks for Perceptual Psychology. *Behavior Research Methods, Instruments, and Computers*, 18 (6), 601-607, 1986.

Horn, B.: *Robot Vision*, MIT Press, Cambridge, Mass, 1986.

Koenderink, J. J., Van Doorn, A.:Photometric Invariances Related to Solid Shape, Optica Acta, 27 (7), 981-996, 1980.

Maniscalco, U., Pirrone, R., Sorbello, F., Storniolo, P: A Shape From Shading Hybrid Approach to Estimate Superquadric Parameters. In: Caianiello, E. R. (ed.), Proceedings of Sixth Italian Workshop on Parallel Architectures and Neural Networks , World Scientific Publ. Co, Singapore, 1993.

Pentland, A.: Finding the Illuminant Direction, J. Opt.Soc. Amer., 72, 448-455, 1982

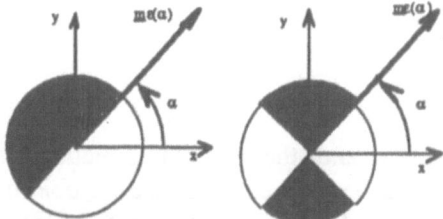

Fig. 1 - A pictorial illustration of first (left) and second (right) order masks (shading flow and hessian components detectors).

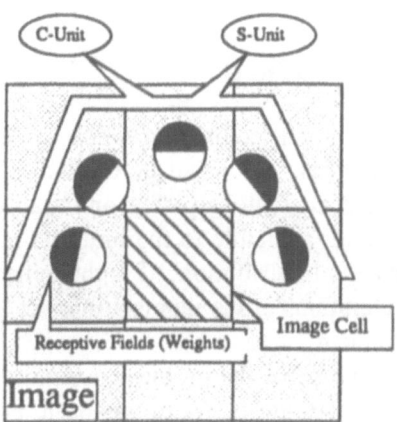

Fig. 2 - A schematic diagram of the connectionist architecture for the evaluation of the shading flow field.

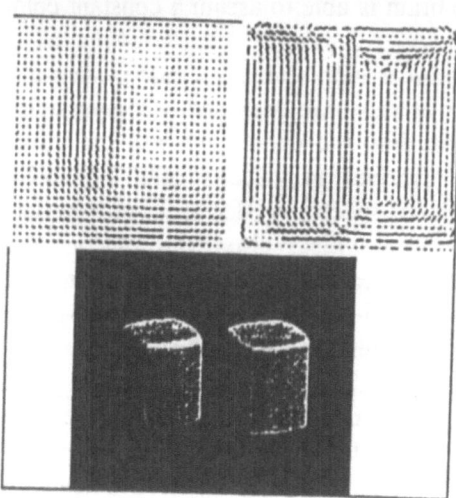

Fig. 3 - A sample 3-D reconstruction.

A Parallel Algorithm for Simulating Color Perception

Linmi Tao[1]   Yunzu Chen[2]   Guozhen Yao[3]

[1] International Institute for Advanced Scientific Studies (IIASS)
Via S.Pellegrino,19-84019 Vietri sul Mare [SA]-Italy

[2] Institute of Intelligent Machines, Academia Sinica
Hefei 230031, P.R.China

[3] Institute of Biophysics, Academia Sinica
Beijing 100101, P.R.China

1. Introduction

Human color perception is constancy and the color information is important for visual perception. The visual system can describe an objects color attribute rapidly and stably in natural environment, but the visual stimuli available to the brain do not offer a stable code of information. The energy distribution of light reflected from surfaces changes along with alterations in the illumination, yet the brain is able to assign a constant color to them.

To solve the problem, we must consider both animal's visual system and its environment. With this basic point of view, we proposed a computaional theory for human color vision [1]. At first, the underlying task for visual information processing is detecting invariant in visual world. For color information processing, the invariant is spectral reflectance of physical surfaces in visual environment. This is a typical ill-posed problem. And second, we introduced two constraints, named Full-spectrum Constraint and Full-color Constraint, to transform the ill-posed problem to well-posed problem. The Full-spectrum Constraint relates to incident light and the Full-color Constraint does to the physical surfaces in visual world. After the proof of algorithm existance, we construct a neural representation, relative reflectance, for human color vision. Finally, the third constraint, Objectivity Constraint, is introduced to connect the constructed neural representation with spectral reflectance directly.

In this paper, we present a parallel algorithm to construct neural representation of color information processing. The algorithm is performed by neural network and the results of computer simulations match well with that of psychophysical experiments.

2. Parallel algorithm and network structure

The algorithm maps input, radiance energy of reflective light, to output, relative reflectance. The centeral problems are if the algorithm has the attribute of color constancy and if the limitation of the relative reflectance is the spectral refletance of the testing physical surfaces.

Color information is processed in a hierachical and parallel system. On the first level, retina, there are three kinds of cons which sample in different spectrum region independently and simultaneously. The input to retina is the light arrays reflected by physical surfaces and carrying the information of visual world. And on the succeeding levels, from retina to visual area of brain, the input is mapped to inner representation, relative reflectance. That is color perception with the property of color constancy. On the cooperation of advanced functions of brain, such as learning and memory, visual cognition on the color property of visual environment is built.

Parallel algorithm is predicate on above principles and some basic attributes of the model are as follows.

- *The structure of neural network*

The parallel algorithm is performed by a neural network containning three layers, input layer, hidden layer and output layer. There are one hundred units which are counterpart with one hundred points distributed on a testing physical surface randomly. Every such unit consists of three basic elements. In the same kind of elements, connections are complete between two layers and there is no connection among different kinds of elements. The learning rule is Generalized δ-rule with back propogation algorithm [2].

- *The input and output of the algorithm*

The input and output are 3-D vector, relating to red, green and blue. In input layer, the three elements samples in long, middle and short wavelength of visible light. The input value is a function of radiance energy received by computer system [3]. The outputs are the relative reflectances of testing points.

3. Results of psychophysical experiment and computer simulation

A psychophysical experiment is designed to detect the color constancy of human visual system [4] and compare the results with that of computer simulation directly.

Under Full-color Constraint, Mondrian collage [5] is adopt as a scene of the experiment. Light source fits well with Full-spectrum Constraint and its spectral energy distribution can be adjusted freely (Fig. 1). Five persons with normal color sensation are chosen as subjects. They match respectively colored blocks, selected randomly from the scene, with that in Munsell Book. This procedure is repeated with different incident light.

The computer simulation includes two steps. At first step, the computer system receives reflected light from the scene, which is illuminated by standard light source B, C and D65 (Fig.1b). The spectral reflectance of the selected point are inputed to train the neural network. At second step, the computer system performs same psychophysical experiments, the scene is illuminated by the light in test set (e.g. Fig. 1a), and output relative reflectance, which can be mapped to the color block of Munsell Book.

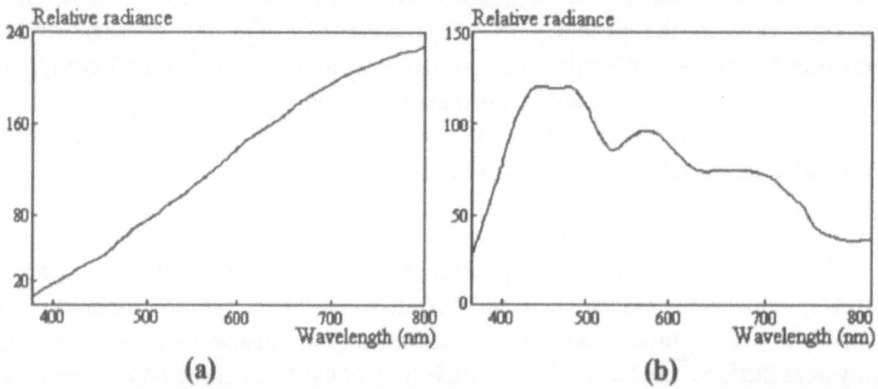

Fig. 1 - Light source illumilating the test scene in the psychophysical experiments and the computer simulation. (a) is Light A in test set, (b) is Ligth C in training set.

We define the mean irradiance of the standard light source as

$$I_{(\lambda)}^{m} = \frac{1}{3}(I_{(\lambda)}^{B} + I_{(\lambda)}^{C} + I_{(\lambda)}^{D65}) .$$

The change rate of the test light is defined as

$$CR^{A} = \frac{1}{\Delta\lambda}\int_{\Delta\lambda} \frac{\left|I_{(\lambda)}^{A} - I_{(\lambda)}^{m}\right|}{I_{(\lambda)}^{m}} d\lambda \times 100\% ,$$

in which $\Delta\lambda$ is the visible area in spectrum and $I_{(\lambda)}^{A}$ is the irradiance of test light A. Digital calculation result shows that the mean change rate is 67.75% in our experiment, and the change rate of light A (Fig. 1a) is 95.87%.

Constancy rate has been defined by R. H. Thouless in 1931, that is

$$Thouless\ rate = \frac{\log R - \log S}{\log A - \log S},$$

in which, A is the reflectance of standard stimulus, S is the reflectance matched with stimulus and R is the reflectance matched by subject.

Under the illumination of test light, the result of psychophysical experiments is that mean *Thouless rate* equals 85.20% and that for computer simulation is 92.07%. The mean error of our experiment and simulation is estimated as 12%.

The conclution, inferring from the results, is that the parallel algorithm can calculate the relative reflectance of physical surface in natural environment. It has the property of color constancy and fit well with the Objectivity Constraint.

4. Discussion

Algorithms are bridge between brain researches and artificial intelligence. It is necessary for the computational theory to verify its properties on computational capability, completeness and objectivity. It is always an important goal for machine vision to recognize color in natural environment. The parallel algorithm is the first step to approach this aim.

References

[1] Tao L.M., Yao G.Z., Wang Y.J., A computational theory for human color vision, Acta Psychologica Sinica, 25(3), pp. 233-240 (1993).

[2] McClelland J.L., Rumelhart D.E., Explorations in Parallel Distributed Processing, The MIT Press, pp. 121 - 160, Cambridge (1988).

[3] Vos J.J., Walraven D.L., On the derivation of the foveal receptor primates, Vision Res., 11, pp. 799 - 818 (1971).

[4] Hao B.Y., Zhang H.C., et al, Experimental Psychology, Beijing, University Press, pp. 548 - 554, Beijing (1983).

[5] Land E.H., Resent advances in retinex theory, Vision Res.,26(1), pp. 7-21 (1986).

Positional Competition in the BCS*

Lothar Wieske

Department of Computer Science, University of Hamburg

Hamburg, Germany

1 Introduction and Motivation

A large part of human and mammal brains is dedicated to visual perception and the combination of locally ambiguous visual information into a globally consistent and unambiguous representation of the visual environment. So, one of the key ideas for the development of the FACADE architecture by Stephen Grossberg and coworkers is the conviction that modular approaches with special purpose procedures for texture, stereo, and motion are not very useful for an understanding of real-world vision.

The FACADE architecture aims at an explanation of how the visual system is able to detect relatively invariant surface colours under variable illumination conditions, to detect relatively invariant object boundaries under occlusion conditions, and to recognize familar objects or events in the environment. The Feature Contour System (FCS), Boundary Contour System (BCS), and the Object Recognition System (ORS) are designed in correspondence to the above processing goals of the overall architecture ([Grossberg90]). Processing of the visual input in retina and LGN relies on local measurements which introduce some amount of uncertainty into the representation of the visual environment. The FACADE architecture sets up parallel and hierarchical interaction schemes that can resolve these uncertainties.

The MP (monocular preprocessing) stage defines oriented receptive fields for each perceptual location which are sensitive to local contrast in the intensity function in accordance to the hypercolumn model of the primary visual cortex by Hubel and Wiesel. The local contrast detectors feed into the competitive-cooperative loop of the Boundary Contour System which generates an emergent segmentation of boundaries in the scene by means of spatially short-range competitive interactions and spatially long-range cooperative interactions. Preattentive processing in the Boundary Contour System is purely data-driven and does not rely on memorized templates or expectancies. Emergent segmentation provides a way to generate boundaries (*illusory contours*) which have no direct physical correlate in the intensity function but are perceived by human subjects in psychophysical experiments.

In the first competitive layer a cell of prescribed orientation excites like oriented cells at the same location and inhibits like-oriented cells at nearby positions. This on-center off-surround organization of like-oriented cells exists around every perceptual location. In the second competitive layer cells compete that represent different orientations at the same perceptual location. The competitive layers feed into the cooperative stage which defines spatially long-range interactions for boundary completion. The output of the cooperative stage is enhanced by a further competitive layer and fed back into the first competitive layer. This feedback allows for the discontinuous completion of continuous boundaries. (For details see [GrossbergMingolla85],[GrossbergMingolla87].)

*This work was funded by the German Federal Ministry of Science and Technology (NAMOS/413-4001-01 IN 101 C 1)

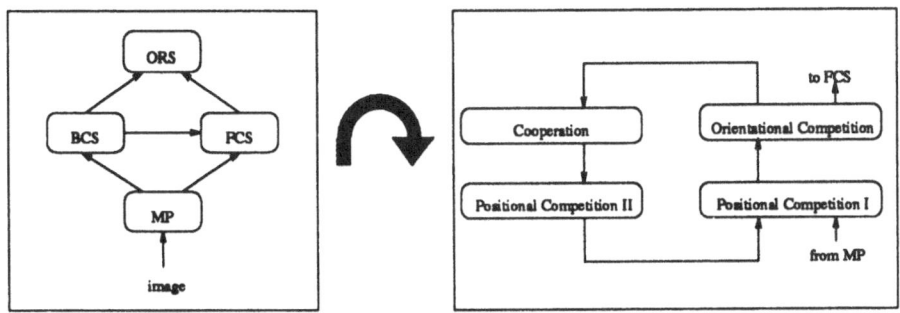

FACADE   Boundary Contour System

2 Positional Sharpening and Positional Competition

In [GrossbergMingolla85] and [GrossbergMingolla87] the authors define the following process for the second positional competition layer to realize the postulate of positional sharpening

$$v_{ijk} = \frac{h(z_{ijk})}{1 + \sum_{(p,q)} h(z_{pqk}) \cdot W_{pqij}}$$

with a circular inhibitory weighting kernel and a threshold-linear signal function.

$$W_{pqij} = \begin{cases} W & \text{if} \quad (p-i)^2 + (q-j)^2 \le W_0^2 \\ 0 & \text{otherwise} \end{cases} \qquad \text{and} \qquad h(z) = L \cdot [z - M]^+$$

The variables z and v denote the output of the cooperative stage and the subsequent competitive stage. The pair $< i, j >$ indexes a perceptual location whereas k is an index to an orientation band. Since there are no interactions between different orientation bands, we will leave out the index k in our further considerations. The definition of the first positional competition layer is quite similar and uses the same competitive mechanism, but is not so well suited for an analysis of positional sharpening since it superposes two inputs.

3 Analysis of Positional Competition

The above transformation may be rewritten as the concatenation of two transformations where a point operation (thresholding) is followed by a positional competition transformation, i.e. $T(I) = T_2(T_1(I))$.

$$\{T_1(I)\}(x, y) = h(I(x, y)) \qquad \text{and} \qquad \{T_2(I)\}(x, y) = \frac{I(x, y)}{1 + \{I \star W\}(x, y)}$$

We want to relate two one-dimensional stimuli I_1 and I_2 where I_2 is a transformed version of stimulus I_1. We assume that I_1 and I_2 are positive unimodal functions having their maximum in $t = 0$. Defining the meaning of I_2 is *sharper* than I_1 is a debatable issue. We start with the definition of the opposite. The intuition of this definition is that the function I_2 is something like a cheese cover of the same height for the function I_1, if I_2 is *broader* than I_1. (See the illustration at the end.) We will show that the sharpening-effect of the overall transformation T is due to the threshold operation T_1 for a large class of inputs and inhibitory kernels, since the transformation T_2 *broadens* its input.

Definition 1: Given two positive unimodal functions I_1 und I_2 having their maximum in $t = 0$, we say that I_2 **is broader** than I_1, if

$$\frac{I_2(t)}{I_2(0)} = \overline{I_2(t)} > \overline{I_1(t)} = \frac{I_1(t)}{I_1(0)} \qquad \text{for} \qquad t \neq 0$$

●

Lemma 1: Let f and g be positive, even functions which are montone decreasing with distance from the origin. we require f to be strictly decreasing such that

$$
\begin{array}{lll}
f & \in & C^\infty \\
f(t) & > & 0 \\
f(t) & = & f(-t) \\
f(t_1) & > & f(t_2)
\end{array}
\qquad
\begin{array}{lll}
g & \in & W^{1,1} \\
g(t) & \geq & 0 \\
g(t) & = & g(-t) \\
g(t_1) & \geq & g(t_2) \qquad 0 \leq t_1 < t_2
\end{array}
$$

$$\int_0^\infty \Phi(t) \cdot g'(t)\, dt \quad < \quad 0 \qquad \forall \Phi \in C^\infty, \Phi(t) > 0$$

Then their convolution product

$$\{f \star g\}(t) = \int_{-\infty}^{\infty} f(u) \cdot g(t - u) du$$

takes its maximum in $t = 0$. ●

Proof of Lemma 1: We show that $\{f \star g\}(0) - \{f \star g\}(t)$ takes a global minimum in $t = 0$. Now

$$\frac{\partial}{\partial t} \left(\{f \star g\}(0) - \{f \star g\}(t) \right) = \int_0^\infty (f(t - u) - f(t + u)) \cdot g'(u)\, du$$

The sign of this derivative directly depends on the sign of $f(t-u) - f(t+u)$. For $t > 0$ we have $f(t-u) - f(t+u) > 0$ since f is even and strict decreasing with distance from the origin. If $t = 0$ the difference is 0. For negative t we conclude that the difference is negative because of the identity $f(t - u) - f(t + u) = -(f(|t| - u) - f(|t| + u))$. So the function $\{f \star g\}(0) - \{f \star g\}(t)$ has an extremum in $t = 0$, and is strictly monotone decreasing with distance from the origin, which finishes the proof. ✓

Proposition 1: Let the stimulus $I(t)$ fulfill the conditions of the function f in lemma 1, and the inhibitory weighting kernel $K(t)$ fulfill the conditions of the function g. Then the positional competition transformation T_K broadens I, i.e. $T_K(I)$ is *broader* than I.

$$\{T_K(f)\}(t) = \frac{f(t)}{1 + \{f \star K\}(t)}$$

Proof of Proposition 1: We have to show, that

$$\frac{\overline{\{T_K(I)\}(t)}}{\overline{I(t)}} = \frac{1 + \{I \star K\}(0)}{1 + \{I \star K\}(t)} > 1 \quad \text{if } |t| > 0.$$

and we see that $\{I \star K\}(0) > \{I \star K\}(t)$ has to be shown for $|t| > 0$. We verify the inequality $\{I \star K\}(0) > \{I \star K\}(t)$ for $t \neq 0$ by invoking lemma 1. ✓

Proposition 2: Let the stimulus $I(x, y) = I(x)$ fulfill the conditions of the function f in lemma 1, and the circular inhibitory weighting kernel $(K(x, y) = K(\sqrt{x^2 + y^2}))$ be such that $g(u) = \int_{-\infty}^{\infty} K(u, v)dv$ fullfills the properties of the function g. Then the positional competition transformation $\mathcal{T}_K$ broadens I, i.e. $\{\mathcal{T}_K(I)\}(x, y) = \{\mathcal{T}_K(I)\}(x)$ is *broader* than $I(x, y) = I(x)$.

$$\{\mathcal{T}_K(f)\}(x, y) = \frac{f(x, y)}{1 + \{f \star K\}(x, y)}$$

Proof of Proposition 2: Transformation of the convolution integral

$$\{I \star K\}(x, y) = \int_{\infty}^{\infty} I(x - u) \cdot \left[\int_{\infty}^{\infty} K(u, v)\right] du$$

shows that we may see the bracketed integral as an one-dimensional kernel and invoke proposition 1. ✓

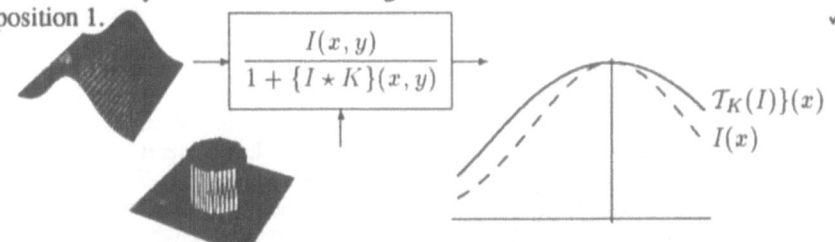

4 Conclusion

We have analyzed the positional sharpening capabilities of the Boundary Contour System. Positional sharpening as set up by Grossberg and Mingolla may be seen as a two-stage process, where a threshold-linear signal functions transforms the input and drives the subsequent positional competition stage. We come to the conclusion that the sharpening effect of the positional sharpening process is due to the transformation by the threshold-linear signal function. A similar analysis has been given in [ElliasGrossberg75] for other types of feedforward shunting equations.

Acknowledgement. I would like to thank O. Ludwig and R. Sprengel for fruitful and stimulating discussions.

References

[ElliasGrossberg75]   S.A. Ellias and S. Grossberg. Pattern formation, contrast control, and oscillations in the short term memory of shunting on-center off-surround networks. *Biological Cybernetics*, 20:69–98, 1975.

[Grossberg90]   S. Grossberg. Neural facades: Visual representations of static and moving form-and-color-and-depth. *Mind & Language*, 5:411–456, 1990.

[GrossbergMingolla85]   S. Grossberg and E. Mingolla. Neural dynamics of perceptual grouping: Textures, boundaries, and emergent segmentation. *Perception and Psychophysics*, 38:141–171, 1985.

[GrossbergMingolla87]   S. Grossberg and E. Mingolla. Neural dynamics of surface perception: Boundary webs, illuminants, and shape-from-shading. *Computer Vision, Graphics, and Image Processing*, 37:116–165, 1987.

A Computational Model for Texton-based Preattentive Texture Segmentation

Mehdi N. Shirazi, Mitsuo Hida and Yoshikazu Nishikawa

Elec. Eng. Dept., Kyoto University Yoshida-Honmachi, Sakyoku, Kyoto, Japan

Abstract- A hierarchical Markov random field with two layers is proposed as a computational model for texton-based preattentive texture segmentation. Different textures are assumed to be different arrangements of bars (textons) with different orientations.

1 Introduction

A fundamental property of human visual system is its ability to discriminate between textures. A systematic approach to texture discrimination was pioneered and pursued by Julesz [3]. Julesz has made it clear that human visual system operates in two distinct modes called "preattentive vision" and "attentive vision." In preattentive vision, texture differences are perceived by an observer almost instantaneously and effortlessly whereas in the attentive vision, they are perceived by a time-consuming serial search and scrutiny. Julesz and his colleagues have hypothesized that a preattentive texture discrimination is done instantaneously and effortlessly based on a few local conspicuous features, which they called textons, [4]. On the other hand, there have, in parallel, been efforts to describe the preattentive texture discrimination in terms of linear filter models and their nonlinear extensions (see the references given in [8].)

The texton theory is not a computational theory. The texton theory has been proposed to explain, as closely as possible, the psychophysical findings regarding the ability of the human preattentive visual system. On the contrary to the texton theory, the linear spatial filtering theory is a computational theory constructed in such a way to be in consistent with the known physiological findings regarding the early vision of primates. The linear filtering theory cannot fully explain the psychophysical findings and there have been some efforts to to extend it by adding some nonlinear elements in order to explain quantitatively the psychophysical data regarding the human preattentive vision. Nevertheless, the filtering theory is not still capable of explaining some psychophysical findings which can be explained by the texton theory [8].

Our main goal is to construct a computational model for preattentive texture discrimination (segmentation) based on texton theory. In this paper, we propose a hierarchical Markov random field with two layers as a preliminary computational model for texton-based preattentive texture segmentation.

2 Test Images

We assume that test images (images displayed to an observer for a brief time for a preattentive segmentation or to the preattentive segmentation system which we are going to make)

consist of $N \times N$ texton subimages. Each subimage might consists of $n \times n$ pixels, where n is an integer which its actual value is not important. We assume that the subimages are located on an $N \times N$ rectangular lattice $\mathcal{L} = \{(i,j)\}, 1 \leq i,j \leq N$. Each subimage is assumed to be either empty or depicts a texton. In this paper, we only consider textures consisting of bars (texton) with 0°, 45°, 90° and 135° angles and encode them by 2, 3, 4, and 5, respectively and keep 1 as a code for empty subimages. We associate a random variable Y_{ij} to the (i,j)-subimage and treat test images as realizations of the random field $Y_{\mathcal{L}} = \{Y_{ij}\}$, $(i,j) \in \mathcal{L}$, where Y_{ij} takes a value from the set $\{1, 2, 3, 4, 5\}$. In Fig. 1, we display the possible states of the (i,j)-subimage and their codes.

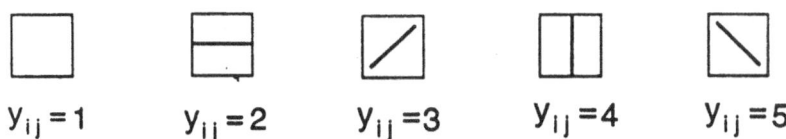

Figure 1: The subimages' possible states and their associated codes.

3   Modeling of the Test Images

We model test images consisting of regions of different textures (texton arrangements) by a hierarchical Markov random field (HMRF) with two layers. The first layer consists of an MRF [2] which is considered as a model for the underlying unobservable regions while the second layer incorporates MRFs which are assumed to generate the observable distinct textures.

3.1   A Hierarchical Markov Random Field with Two Layers

We assume that a test image consisting of regions of different textures is a realization of a collection of interacting random variables $(X_{\mathcal{L}}, Y_{\mathcal{L}})$. The image process $Y_{\mathcal{L}} = \{Y_{ij}\}$, $(i,j) \in \mathcal{L}$ is assumed to be a function of the underlying region process $X_{\mathcal{L}} = \{X_{ij}\}$, $(i,j) \in \mathcal{L}$. The interacting processes $(X_{\mathcal{L}}, Y_{\mathcal{L}})$ can be characterized completely by a joint probability density function $P(x_{\mathcal{L}}, y_{\mathcal{L}})$ or equivalently, according to Bayes' rule, by $P(x_{\mathcal{L}})$ and $P(y_{\mathcal{L}} \mid x_{\mathcal{L}})$. In the following, we precisely describe $(X_{\mathcal{L}}, Y_{\mathcal{L}})$ in terms of $P(x_{\mathcal{L}})$ and $P(y_{\mathcal{L}} \mid x_{\mathcal{L}})$.

We consider test images consisting of M different textures (textons with different arrangements), i.e., we assume that X_{ij} is a discrete-valued random variable taking a value from the set $Q_X = \{1, \ldots, M\}$. We further assume that $X_{\mathcal{L}}$ is an MRF characterized by local conditional distributions $P(x_{ij} \mid x_{\eta'_{ij}})$, where η'_{ij} denotes η_{ij} with the (i,j)-subimage deleted. For details see [6].

The conditional joint distribution $P(y_{\mathcal{L}} \mid x_{\mathcal{L}})$ is given by

$$P(y_{\mathcal{L}} \mid x_{\mathcal{L}}) = (Z^{Y|X})^{-1} e^{-\mathcal{E}(y_{\mathcal{L}}|x_{\mathcal{L}})} \tag{1}$$

where the global conditional energy function is

$$\mathcal{E}(y_{\mathcal{L}} \mid x_{\mathcal{L}}) = \sum_{(i,j) \in \mathcal{L}} \sum_{C \in C_{ij}^Y(x_{\eta_{ij}^Y(x_{ij})})} \mathcal{E}(y_C \mid x_{ij}) \tag{2}$$

and the conditional partition function is given by

$$Z^{Y|X} = \sum_{y_c \in \Omega_Y} e^{-\mathcal{E}(y_c|x_c)}. \tag{3}$$

In (3), $\Omega_Y(x_c) = \prod_{(i,j) \in \mathcal{L}} Q_Y$, where $Q_Y = \{1,2,3,4,5\}$. For details see [6].

4 Preattentive Texture Segmentation

We formulate the problem of preattentive texture segmentation as an optimization problem. Let $P(x_{\mathcal{L}})$ and $P(y_{\mathcal{L}} \mid x_{\mathcal{L}})$ be given. We assume that the preattentive segmentation is carried out by calculating the Maximum A Posteriori (MAP) estimate of the region process $X_{\mathcal{L}}$, when a target image $y_{\mathcal{L}}$ is given as a realization of $Y_{\mathcal{L}}$. By definition, the MAP estimate of $X_{\mathcal{L}}$ is

$$\hat{x}_{\mathcal{L}} = \arg \max_{x_{\mathcal{L}} \in \Omega_X} P(x_{\mathcal{L}} \mid y_{\mathcal{L}}). \tag{4}$$

We have already proposed a parallel deterministic relaxation algorithm based on the mean field approximations to solve the above large scale optimization problem [7]. The local deterministic updating rule of the algorithm is given by

$$\hat{x}_{ij}^{(p+1)} = \arg \max_{x_{ij} \in Q_X} P(y_{ij} \mid x_{ij}, x_{\eta_{ij}'Y}^{(p)}, y_{\eta_{ij}'Y}) P(x_{ij} \mid x_{\eta_{ij}'X}^{(p)}) \tag{5}$$

at the $(p+1)$-th iteration.

5 Simulation Result

To test our computational model, we applied it to the test image shown in Fig. 2(a). It is well known that the human preatentive vision segments the test image into two regions of tilted Ts and upright Ts and Ls. As it is shown in Fig. 2, our computational model shows the same behaviour. For the region process, we used a 2nd-order MRF model with the following local conditional distribution [1]

$$p(x_{ij} \mid x_{\eta_{ij}'X}) = \frac{e^{\beta \mathcal{E}_{\eta_{ij}'X}(x_{ij})}}{\sum_{k=1}^{M} e^{\beta \mathcal{E}_{\eta_{ij}'X}(k)}}$$

where $\mathcal{E}_{\eta_{ij}'X}$ denotes the number of x_{ij}-valued pixels in $\eta_{ij}'^X$, $M = 3$ and $\beta = 0.8$. For the texture processes, we used 2nd-order MRF models with singleton and doubleton (pairwise) cliques only, given respectively as follows

$$\mathcal{E}(y_{ij} \mid x_{ij}) = \begin{cases} -\alpha_0(x_{ij}) & \text{when } y_{ij} = 0 \\ 0 & \text{otherwise} \end{cases}$$

and

$$\mathcal{E}(y_C \mid x_{ij}) = \begin{cases} -\alpha_C(y_{ij}, x_{ij}) & \text{if pixels in C have the same value} \\ \alpha_C(y_{ij}, x_{ij}) & \text{otherwise} \end{cases}$$

where C denotes a pair-wise clique. If we assume that the MRF models are homogenous and isotropic then it is easy to see that our model has sixty four parameters in total. In our simulation study, we set the horizontal and vertical bonding parameters to 1.0 for the textures consisting of upright T's and L's, the diagonal bonding parameters to 1.5 for the texture consisting of tilted T's, the singleton parameters to 10.0 and the rest of parameters to 0.0.

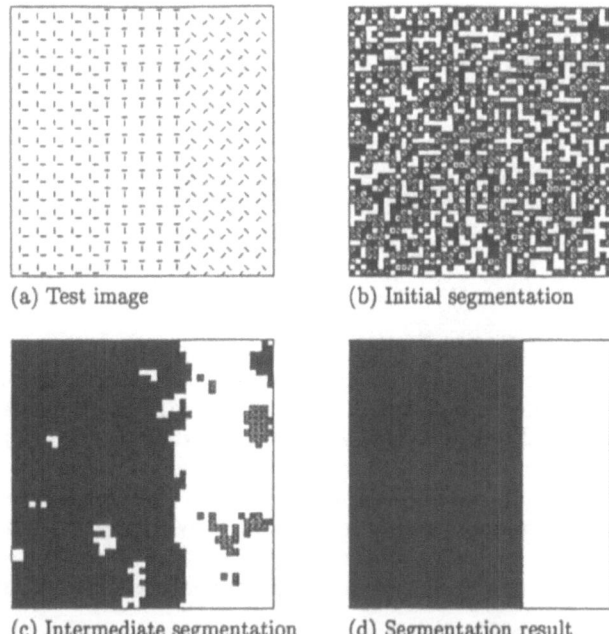

(a) Test image   (b) Initial segmentation

(c) Intermediate segmentation   (d) Segmentation result

Figure 2: Segmentation result after 20 iterations

References

[1] J. E. Besag, "On the Statistical Analysis of Dirty Pictures", J. Roy. Statis., Soc. B, 48, No. 3, pp.259-302, 1986.

[2] S. Geman and D. Geman, "Stochastic relaxation, Gibbs distributions, and the Bayesian restoration of images," IEEE Trans. on Pattern Analysis and Machine Intelligence, PAMI-6, pp. 721-741, 1984.

[3] B. Julesz, "Visual Pattern Discrimination," IRE Trans. Info. Theory, IT-8, pp. 84 92, 1962.

[4] B. Julesz and J. R. Bergen, "Textons, the Fundamental Elements in Preattentive Vision and Perception of Textures," Bell System Technical Journal 62(6), pp. 1619-1645, 1983.

[5] J. Malik and P. Person, "Preattentive Texture Discrimination with Early Vision Mechanisms," J. Opt. Soc. Am. A, Vol., No., pp.923 932, 1990.

[6] Mehdi N. Shirazi and H. Noda, "Textured Image Segmentation: Hierarchical Markov Random Fields and Mean Field Approximations," submitted to IEEE Transactions on Image Processing.

[7] Mehdi N. Shirazi and H. Noda, "A Deterministic Iterative Algorithm for HMRF-Textured Image Segmentation," Proceedings of IJCNN'93-NAGOYA, Japan, October 1993.

[8] D. Williams and B. Julesz, "Filters Versus Textons in Human and Machine Texture Discrimination," in Neural Networks for Perception Vol. 1, Human and Machine Perception edited by H. Wechster, pp. 145 175, 1992.

Hopfield Neural Network for Motion Estimation and Interpretation

G. Convertino†, M. Brattoli†, A. Distante†

† IESI-CNR, Bari - Italy

1 Introduction

We propose an algorithm for the estimation and interpretation of *optical flow
(OF)* field (the field of apparent velocities associated to the image brightness (I)
variation of a TV image sequence). The algorithm takes as input the brightness
spatial and temporal gradient (I^x, I^y, I^t) associated to two frames of a TV
sequence and returns a map of OF vectors in which areas subject to the same
motion are isolated and classified. The algorithm is based on the geometric theory of differential equations and is implemented by a Hopfield neural network
(HNN). In the paper we first introduce the phases of estimation and interpretation, then their interaction and cooperation is presented. Finally some results
are shown.

2 Optical Flow Estimation

A linear version of OF is considered. If (u, v) are the components of the OF,
taking the first Taylor series expansion of OF about a point P_0, we have:

$$\begin{pmatrix} u \\ v \end{pmatrix} = \mathbf{b} + \mathbf{B}X = \begin{pmatrix} u_0 \\ v_0 \end{pmatrix} + \begin{pmatrix} Exp + Shear1 & Shear2 - Rot \\ Shear2 + Rot & Exp - Shear1 \end{pmatrix} X \quad (1)$$

where:

$$\begin{array}{llll} u_0 = u_{p_0} & Shear1 = \frac{1}{2}(\frac{\partial u}{\partial x} - \frac{\partial v}{\partial y}) & Exp = \frac{1}{2}(\frac{\partial u}{\partial x} + \frac{\partial v}{\partial y}) \\ v_0 = v_{p_0} & Shear2 = \frac{1}{2}(\frac{\partial u}{\partial y} + \frac{\partial v}{\partial x}) & Rot = \frac{1}{2}(\frac{\partial v}{\partial x} - \frac{\partial u}{\partial y}) \end{array} \quad (2)$$

By means of (1) the OF field is approximated by a linear combination of *elementary motions (translation, expansion, rotation, deformation)*. An estimation of
OF vectors (1) is obtained using the *image brightness constancy equation*:

$$(I^x u + I^y v + I^t) = 0 \quad (3)$$

The (3) by itself does not allow to estimate the six motion components of OF.
As additional constraint we assume that in a given region (window) the linear
coefficients are constant. To map the OF problem onto an Hopfield network
we divide the image in $N \times M$ windows of dimension $w \times w$. Six neural units
(V_k, $k = 0..5$) representing (2) are associated to each window. If we identify
each region with ij, we write the following network energy function is written:

$$E(V_{ijk}) = \sum_{ij} \sum_{r=-w}^{w} \sum_{s=-w}^{w} (I^x_{i+r,j+s}(V_{ij0} + (V_{ij2} + V_{ij1})s + (V_{ij4} - V_{ij5})r) + \quad (4)$$

$$I^y_{i+r,j+s}(V_{ij3} + (V_{ij4} + V_{ij5})s + (V_{ij2} - V_{ij1})r) + I^t_{i+r,j+s})^2$$

I^x, I^y and I^t are computed as finite differences. The input/output characteristic of the units is a linear function. The V_{ijk} are synchronously updated with a gradient descent rule, as follows: $\frac{dV_{ijk}}{dt} = -\tau \frac{\partial E}{\partial V_{ijk}}$.

3 Optical Flow Interpretation

By means of the geometric theory of differential equation a set of symbolic descriptors of OF patterns is derived (Arrowsmith, 1982) (Rao, 1990). We are interested with 2D linear systems of differential equations, expressed as:

$$\dot{x}(t) = \mathbf{X}(x) = \mathbf{A}(x) \tag{5}$$

where x is a vector in $\mathbf{R}^2$, $\mathbf{X} : \mathbf{R}^2 \to \mathbf{R}^2$ is a linear mapping, and $\mathbf{A}$ (*coefficients matrix*) is a 2D matrix. x(t) is a solution of the system if satisfies (5). The qualitative behaviour of the (5) solutions is represented in the plane $x_1 - x_2$ as a family of curves directed with increasing t and called *phase portrait* (PP). When two systems have the same *dominant qualitative features* (same global behaviour) they can be considered equivalent, even if their PPs are not identical. Two systems with similar coefficients matrices are qualitatively equivalent. Similarity is an equivalence relation on the 2D non-singular matrices, that defines four equivalence classes. The classes are usually identified by the so-called *Jordan forms*, that are:

$$(a) = \begin{pmatrix} \lambda_1 & 0 \\ 0 & \lambda_2 \end{pmatrix} (b) = \begin{pmatrix} \lambda_0 & 0 \\ 0 & \lambda_0 \end{pmatrix} (c) = \begin{pmatrix} \lambda_0 & 1 \\ 0 & \lambda_0 \end{pmatrix} (d) = \begin{pmatrix} \alpha & -\beta \\ \beta & \alpha \end{pmatrix} \tag{6}$$

where λ_0, λ_1, λ_2, α, β are real numbers. The Jordan form of a matrix is identified by the nature of its eigenvalues. Therefore all the systems having coefficient matrices belonging to the same class are equivalent. The PPs associated to the Jordan forms can be considered as representative of the systems behaviour of the corresponding class. Therefore the PPs can be considered as *symbolic descriptors* of 2D systems. The PPs associated to (6.b), (6.a), and (6.d) are shown in fig [1.a], [1.b], [1.c]. When A is singular the PPs associated to the system can be of two kinds, depending on the rank of A (0 or 1 fig [1.d]). The OF equations (1) can be regarded as a 2D linear system of differential equation, therefore the PPs can be used as descriptors of OF patterns. The PP that best matches a given OF pattern is easily detected identifying the Jordan form associated to the matrix B that defines the pattern.

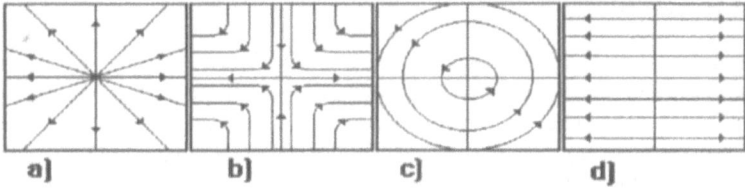

Figure 1: PPs examples

4 A Neural Algorithm for Motion Analysis

The integration of the estimation-phase and the interpretation-phase in a unique algorithm is presented in this section. Each of the equivalence classes defined in the previous section identifies a motion class. In the following table the classes are listed. For each one the Jordan form corresponding to **B** of (1) is also given:

Class	S	R_z	RT_z	T_x	T_y	T_z	T_{xy}
JF	9.a	9.d	9.d	sin B	sin B	9.b	sin B
M. Par.	Sh1, Sh2	Rot	Rot+Exp	u_0	v_0	Exp	$T_x + T_y$

Table 1: motion classes

Each class is characterized by one or more dominant elementary motion parameters as underlined in the third row of the table. To represent the motion classes we associate a neural unit (act_{ijk}) to each unit V_{ijk} of the HNN defined in section 2. The values of act are in the range $[0,1]$. The units indicate the *activation level* of the corresponding V unit. When a pixel is classified as belonging to a class, we disactivate $(act = 0)$ the units representing parameters not dominant in that motion class and activate the others $(act = 1)$. The final classification is given by the values of the units act_{ijk}. With the introduction of the units act_{ijk} the energy function of the network is modified as follows:

$$E(V_{ijk}) = \sum_{i,j,r,s} (I^x_{i+r,j+s}(U_{ij0} + (U_{ij2} + U_{ij1})s + (U_{ij4} - U_{ij5})r) + \quad (7)$$

$$I^y_{i+r,j+s}(U_{ij3} + (U_{ij4} + U_{ij5})s + (U_{ij2} - U_{ij1})r) + I^t_{i+r,j+s})^2 + \lambda E_1(V_{ijk})$$

where $U_{ijk} = V_{ijk} \times act_{ijk}$. The term $E_1(V_{ijk})$ is defined as:

$$E_1 = \sum_{ijk} (V_{i+1jk} - V_{ijk})^2 + (V_{ijk} - V_{i-1jk})^2 + (V_{ij+1k} - V_{ijk})^2 + (V_{ijk} - V_{ij-1k})^2 \quad (8)$$

E_1 is introduced to have a more homogeneous segmentation, constraining neighbouring windows to have similar parameter values. The following updating schema is used: firstly n (usually $n = 100$) cycles of updating of units V_{ijk} are performed as fixed in sect. 2, then the unit act are updated. To update the act first the motion class of each window is determined then the units act_{ijk} are modified as previously described. The disactivation (activation) of the units is performed gradually to recover possible wrong classification.

5 Results

Some preliminary results are presented. The tests have the purpose to evaluate the goodness the classification and to evaluate the influence that the parameters involved in the algorithm (τ, λ, w) have on the final results. In this paper we present the results relative to the two frames (64×64) of a synthetic sequence (fig [2.a]), in which the sphere is translating towards the observer (class T_z in tab. 1) and the rectangle is translating parallel to the image plane (class T_x). The parameters have been set to $\lambda = 10000$, $\tau = 10^{-7}$, $w = 7$. In fig [2.b] the linear OF map is shown. In fig [3.a] the final segmentation is represented: the darkest area correspond to the class T_z, the brightest to the class T_x. Another area is wrongly detected, the region involved is the one on the boarder on the sphere, (areas in which the OF is not linear). In any case the region is classified as uniform translation (T_x, or T_y, or T_{xy}), that is not so wrong because far away from the focus of expansion (FOE) the flow becomes similar to a translation. Finally in fig. [3.b] the map of the *fixed points* for T_z (points in which the OF vanishes, the FOE for the class T_z) is shown. The darkest points correspond to the most "voted" points. The most voted point is the real fixed point (pixel 33, 29).

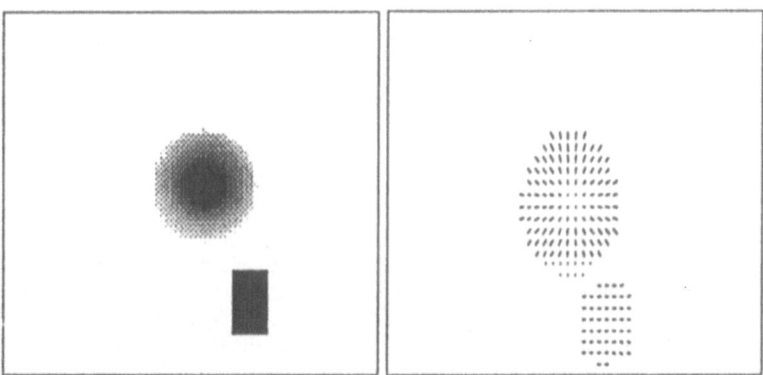

Figure 2: Original image. Linear OF

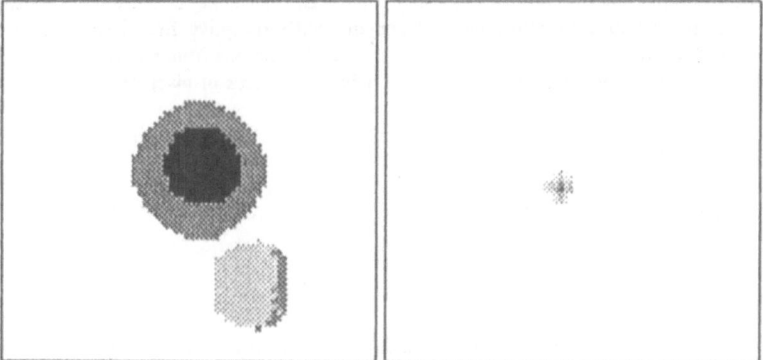

Figure 3: Segmentation map. Fixed point map

References

[1] D.K. Arrowsmith, C.M. Place *Ordinary Differential Equations* Chapman and Hall (1982)

[2] A.R. Rao, *A Taxonomy for a Textured Description and Identification*, Springer-Verlag (1990)

Phase Interactions Between Place Cells During Movement

J. G. Taylor and L. P. Michalis
King's College, London

Abstract

A model is described of phase interactions between "place" cells during movement. While mean firing phases - during walking- are "narrowly" distributed near the positive peak of the dentate theta [Fox et al, 1986], individual cells shift their phase as a function of place. [Burges et al, 1994]. Phase shifts in place cells have been seen as having their origin in similar shifts made at an earlier time and stage in the firing of entorhinal cells [Burges et al 1994]. Phase shifts might instead arise from associative interactions between place cells in what might be seen as a system of coupled oscillators. A gating function has been associated with the theta rhythm. Entorhinal inputs can only excite place (and granule) cells during a critical period which is centred around the transition from the positive to the negative peak of theta. Place cells activated during this period will tend to "pull" the activity of neighbouring cells towards a phase close to the middle of theta causing a phase shift that matches the shift of activity across CA3.

1. Introduction

Phase has been interpreted as a way of encoding the position of a cell's place field relative to the rat [Burges et al, 1994] - place cells with firing fields ahead of the rat fire late in the theta cycle, while cells with place fields behind the rat fire early in the (theta) cycle [Burges et al 1994]. This is a property which in their model is shared by all cells "throughout the model", and its earlier manifestation is seen in entorhinal cells [Burges et al., 1994].

Phase shifts in place cells reflect shifts made at an earlier time and stage in the firing of entorhinal cells. Like shifts in place cells , phase shifts in entorhinal cells encode the position of a cell's firing field relative to the rat, and like place cells, entorhinal cells have firing, or receptive fields. Only two "sensory cells" - each encoding distance from a distinct cue - may project to each entorhinal cell and the cell's receptive field is peaked at the centroid of the two cues [Burges et al, 1994]. Direct projections from entorhinal cells onto a merged CA3-CA1 field determine the firing fields and phases of place cells.

Phase shifts in place cells need not be generated by (phase) shifts in entorhinal cells, or rely upon receptive field properties which entorhinal cells do not seem to have. Neither do they have to be - or arise from some intrinsic properties of place cells. Instead, they can arise from interactions between the phases of distinct inputs to CA3.

2. Phase Modulation of Place Cells

Oscillations in the firing of principal (pyramidal and granule) cells and interneurons during theta (are thought to) have their origin in the rhythmic activity of medial septal cells. Distinct populations of septal cells are active during, and generate each component of theta. The atropine sensitive component of theta is so (thought to be) mediated by a cholinergic projection and the atropine resistant component by a serotonergic projection having terminals in pyramidal (and granule) cells as well as on interneurons. There is also an inhibitory (GABAergic) projection which is active during both the components of theta and terminates onto inhibitory interneurons only (Figure 1).

The frequency and phase of firing of both principal cells and interneurons is initially set by rhythmic excitatory (and inhibitory) septal inputs. Interactions within, or across populations may also influence the phase and frequency of what otherwise might have been independent oscillators. The firing of place cells is so coupled to that of inhibitory interneurons - by means of reciprocal connections between principal cells and interneurons, and to the firing of both granule and other place cells by means of mossy fibre and associative projections. The effects that interactions between place cells and inhibitory interneurons have upon the firing phase of place cells might be described by an equation of the form:

$$\frac{d\theta_i}{dt} = \acute{\omega}(I) + w_{ij} \, h(\theta_j - \theta_i), \quad i = 1,....N \quad (1)$$

where θ_i, θ_j are the phases of the ith place cell the jth interneurons and $h(\theta)$ is a periodic function of its argument. Similar is the effect - and might be modelled by an equation of the form of (1) - that granule cells have upon the phase of place cells.

CA3 and the network of associative projections in CA3, in its simplest form resembles - and might be modelled as - a system of independent oscillators in which only nearest neighbours are coupled. The effects that associative interactions have upon the phase of each oscillator might be modelled by equations of the form:

$$\frac{d\theta_j}{dt} = \acute{\omega}(I) + w_{ij-1} \, h(\theta_{j-1}, \theta_j) + w_{j+1j} \, h(\theta_{j+1}, \theta_j), \quad j = 1...., N \quad (2)$$

where $\acute{\omega}(I)$ reflects the effect that an entorhinal input encoding spatial position has upon the (firing) phase of the jth place cell. $\acute{\omega}(I)$ might be chosen as a Gaussian function of distance centred at τo:

$$\acute{\omega}(\tau) = c \, \exp \, (-(\tau - \tau o)^2 / 2d^2) \quad (3)$$

For phase shifts dependent upon external (entorhinal) inputs, and for a constant velocity trajectory $\tau = \tau o + vt$:

$$d\theta_j / dt = c \, \exp \, (-v^2 \, t^2 / 2 \, d^2) \quad (4)$$

For $c = (\sqrt{2}\pi) \, v/d$, $\theta(+\infty) = 2\pi$ making a shift of 360° throught the theta cycle.

Stimulation of the perforant path, or entorhinal input along the perforant path has the effect of an external perturbation upon the system's oscillating activity. In the absence of sensory stimulation, say under urethane anaesthesia, individual place (and granule) cells are "strongly phase-locked to the phase of theta" [Fox et al., 1986], regardless of whether mean firing phases are "narrowly" or widely" distributed within theta. Activation of the perforant path causes the phase of place (and granule) cells to shift to an angle near the "middle" (phase) of theta.

The shift in the phase of place cells might be interpreted in terms of a critical period in the phase of theta. Entorhinal inputs can only influence the firing of granule and place cells only during this period which is centred around the transition from the positive to the negative peak of theta. Stimulation applied (to the perforant path) at any other time (or phase of) theta can have no effect upon the firing of place or granule cells.

Similar to the "gating" of entorhinal inputs by the theta rhythm are interpretations of theta as a process serving the quantization and timing of sensory information around the "basic four-synapse hippocampal circuit" [O'Keefe and Nadel, 1978], or alternatively, around a larger system that also includes the septal area and (what Gray calls) the "subicular loop" [Gray, 1982]. The time for information to flow around this larger system seems closer to the period of theta than the time around the hippocampal formation which is an order of magnitude too short of even the fast theta rhythm [Gray, 1982].

Entorhinal inputs which can only excite place (and granule) cells during the transition from the positive to the negative peak of theta may explain the firing of these cells at a phase near the "middle" of theta. Neither the shift to the "middle" phase of theta nor the return to what

84

might be seen as an equilibrium phase have the form of discontinuous transitions between discrete states of cell firing. The phase of place cells gradually shifts away from, and returns to an equilibrium phase as the rat approaches to, and moves away from a firing field. Gradual shifts in the phase of place cells do not seem to have their origin in similar shifts at the entorhinal level. Instead, they might arise from the (phase) coupling of place cells in CA3. Place cells activated by entorhinal inputs at an earlier phase of theta may so tend to "pull" the phase of neighbouring cells towards a value close to their own phase causing a phase shift that match the shift of activity across CA3.

Phase interactions, like interactions between firing rates depend upon the synaptic strengths of associative connections between cells. They also depend upon phase differences. For a fixed (initial) phase difference, phase interactions have a maximal effect upon cells with strong associative connections between them. Since associative projections are thought of as connecting cells with adjacent firing fields, larger phase shifts result among cells with proximal firing fields.

Phase shifts also depend upon (initial) phase differences between cells. If $h(\theta)$ in (1) is chosen as a sinusoidal function of its argument, the effect that the jth cell has upon the ith cell is maximum when the two cells are $\pi/2$ out of phase. There is no effect when the two cells are π out of phase so that cells firing near the "middle" phase of theta can have no effect upon the phase of those firing "late" in the theta cycle.

Neither can cells firing near the "middle" phase of theta have any effect upon the (firing) rate of cells that are approximately π out of phase locked to the "late" (or "early") part of theta. Only cells activated during the same theta cycle and by the same entorhinal input can influence each other's activity by their associative connections. Even then, single bursts fired by simultaneously active cells may not be - and are not sufficient to induce LTP. While a temporary memory trace might be formed during this state, its potentiation does not occur until a later stage (which is) at the end of theta and which has been identified with SPWs [Buzsaki, 1989]. Concurrent population bursts from the release of subcortical inhibition during this stage might function to potentiate the synaptic strengths (of the projections) between cells that were active during the theta state. [Buzsaki, 1989]

References

1.   Burges N., O'Keffe J, Recce M (1994), A model of hippocampal function, submitted to Neural Networks -issue on Neurodynamics and behaviour.

2.   Fox S, Wolfson S., Ranck J.B. (1986), Hippocampal Theta Rhythm and the firing of neurons in walking and urethane anaesthetized rats, Exp. Brain Res. 62:495-508

3.   Buzsaki G. (1989), Two stage model of memory trace formation: a role for "noisy" brain states, Neuroscience vol.31, No 3, 551-570.

4.   O'Keefe J. and Nadel. (1978), The Hippocampus as a cognitive map, Oxford: Clarendon Press.

5.   Steward M. and Fox S (1990), Do septal neurons pace the hippocampal theta rhythm?, TINS, Vol.13, No 5, 1990.

6.   Traub R, Miles R, Muller R, Gulya's 4 (1992), Functional organisation of the hipporampal (43 region, Network 3, 465-488

7.   Gray, J (1982) The Neurosychology of Fear and Anxiety, Oxford, Clarendon Press.

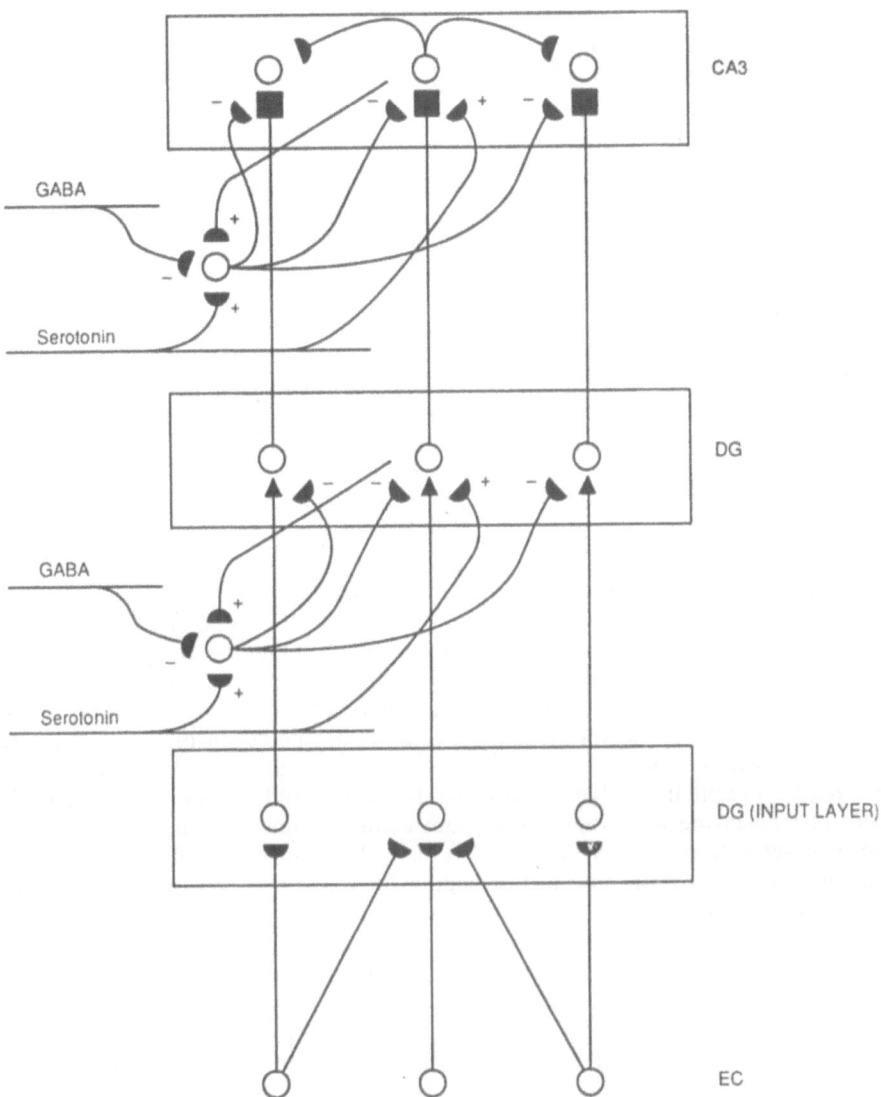

Figure 1: The frequency and phase of firing of both principal cells and interneurons is (initially) set by rhythmic septal inputs. Entorhinal inputs have the effect of an external perturbation upon the system's oscillatory activity. CA3 and the network of associational projections in CA3 might be seen as a system of coupled oscillators each acting to "pull" the phase of neighbouring cells towards its own.

Self-Organization of an Equilibrium-Point motor controller[*]

V. Sanguineti and P. Morasso

Department of Informatics, Systems and Telecommunications (DIST),
University of Genova, Genova, Italy

1   Equilibrium-Point models for motor control

One of the main problems in motor control is how the Central Nervous System (CNS) can deal with the multiplicity of muscles and the complex geometry of the human body. The observation that muscles behave as *elastic elements* (that is, as springs), thus being able to store elastic potential energy, implies that a *posture* corresponds to a minimum potential energy configuration of the whole musculoskeletal system; moreover, there is some experimental evidence that even during movements the intermediate body configurations are also equilibrium postures. These findings led to the hypothesis (known as *Equilibrium Point (EP)* hypothesis, see (Feldman, Adamovich, Ostry, & Flanagan, 1990) for a review) that the generation of movements is made by the CNS in terms of the equilibrium or *virtual* trajectory, which is only influenced by the "static" components of the involved mechanical structures (muscles, ligaments, joints, bones), whereas viscous and inertial effects act as "perturbations", so that real trajectories may differ from virtual ones; The main advantage is that there is no need to explicitly solve the inverse dynamic problem; however, the problem of translating a desired equilibrium configuration into the corresponding muscle activations is still ill-posed because the number of muscles is usually much greater than the number of degrees of freedom. To this purpose, several computational models were proposed, originating from slightly different formulations of the *EP* hypothesis; one interesting variant, known as λ-model, is due to (Feldman et al., 1990); it reduces the problem of generating the set of muscle activations to a coordinate transformation from configuration space to muscle activation space. The most interesting feature of this approach is that it allows to deal simultaneously with the muscle-joint configurations and the *level of coactivation*. In fact, a number of neurophisiological studies (Humphrey & Reed, 1983) hypothesize distinct channels and cortical areas, responsible for separately specifying these quantities; moreover, the FLETE (Factorization of LEngth and TEnsion) model (Bullock & Grossberg, 1989), though not an *EP* model, is also based on integrating two different kinds of central commands, one specifying the equilibrium posture and the other related to force magnitudes.

After a review of the computational aspects of an extended formulation of the λ-model, this paper proposes a new learning schema for training a feedforward controller in order to compute the muscle activations corresponding to a desired equilibrium configuration of the body.

[*]This work was partly supported by the Esprit Basic Research Action SPEECH-MAPS.

2   Generation of muscle activations

Let us consider the general case of a musculoskeletal system with p degrees of freedom and m muscular actuators, with (in general) $m > p$. For such a system, the equilibrium configuration is given by:

$$\tau = J_M^T(\mathbf{q}) \cdot \mathbf{f}_M(\mathbf{l}, \mathbf{u}) + \tau_g(\mathbf{q}) = 0 \tag{1}$$

where $\mathbf{q} \in \mathcal{Q} \subset R^p$ is the body configuration in joint coordinates; the matrix $J_M(\mathbf{q})$ is the jacobian of the transformation between body configuration and muscle lengths, $\mathbf{l} = \mathbf{l}(\mathbf{q})$ (in particular, the element J_{Mij} corresponds to the moment arm of the i-th muscle with respect to the j-th joint); $f_{Mi}(l_i, u_i)$ is the force exerted by the i-th muscle. The term $\tau_g(\mathbf{q})$ accounts for 'passive loads (gravity, ligaments, etc), non depending on muscle activation; without loss of generality, this term may be neglected (it can be hypothesized that gravity and other passive loads are compensated by peripheral reflexes, non involving an explicit central control). In the EP modelling framework, the motor control problem consists of solving Eq. 1 with respect to $\mathbf{u}$ (the vector of muscle activations). We now show that such a solution has an attractively simple form if we make a few (biologically motivated) assumptions on the family of length-tension curves $f_{Mi} = f_{Mi}(l_i, u_i)$.

Hypothesis 1: *For each muscle i, for each muscle length l_i and muscle force $f_{Mi} \in \mathcal{F} \subset (-\infty, 0]$, there is a single value of muscle activation, $u_i \in U$, so that $f_{Mi} = f_{Mi}(l_i, u_i)$.* An important consequence of this assumption is that u_i is uniquely determined if the desired length and force can be separately specified.

Hypothesis 2: *For each muscle i, given an activation level u_i, there is a unique rest length $l_{0i} = l_{0i}(u_i)$: $f_{Mi}(l_i, u_i) = 0$ for $l_i < l_{0i}$). Moreover, the mapping $l_{0i} = l_{0i}(u_i)$ is invertible.* A consequence is that the muscle length-tension curve may be written in the equivalent form $f_{Mi} = f_{Mi}(l_i, l_{0i})$, where the rest length is the controlled variable.

The λ-model was inspired by a set of experimental *Invariant Characteristics (ICs)* (i.e. length-tension curves of muscles measured in-vivo), that were found to satisfy the above assumptions; moreover, the controlled variable was identified by Feldman with the *stretch reflex threshold*, λ (corresponding to the rest length of each *IC*. In the case of the λ-model, the motor control problem reduces to determining the vector of muscle forces $\mathbf{f}_M$ that is solution of the homogeneous linear system $J_M^T(\mathbf{q}) \cdot \mathbf{f}_M = 0$, with the constraint $f_{Mi} \geq 0$ for each i (which accounts for the monodirectional behavior of each muscle, i.e. the fact that it can only exert a contractile force); a particular value of muscle force uniquely determines the activation value $\lambda_i = \lambda_i(l_i(\mathbf{q}), f_{Mi}(\mathbf{q}))$. An additional hypothesis, also introduced by Feldman from observations on the *ICs*, is that muscles are described by $f_{Mi}(\lambda_i, l_i) = f_{Mi}(A_i)$, where $A_i = l_i - \lambda_i$, with $A_i \geq 0$; in this case, λ_i is given by $\lambda_i = l_i(\mathbf{q}) - A_i(f_{Mi})$. As regards the computation of f_M, the above homogeneous linear system has ∞^n solutions, with $n \leq m - p$, whose general form is

$$\mathbf{f}_M = \sum_{i=1}^{n} \phi_i(\mathbf{q}) c_i = \Phi(\mathbf{q}) \cdot \mathbf{c} \tag{2}$$

where $\{\phi_i(\mathbf{q})\}$ is a set of linearly independent (normalized) solutions of the homogeneous system, or *coordinative modes*, i.e. a *basis* for *ker* $J_M^T(\mathbf{q})$.

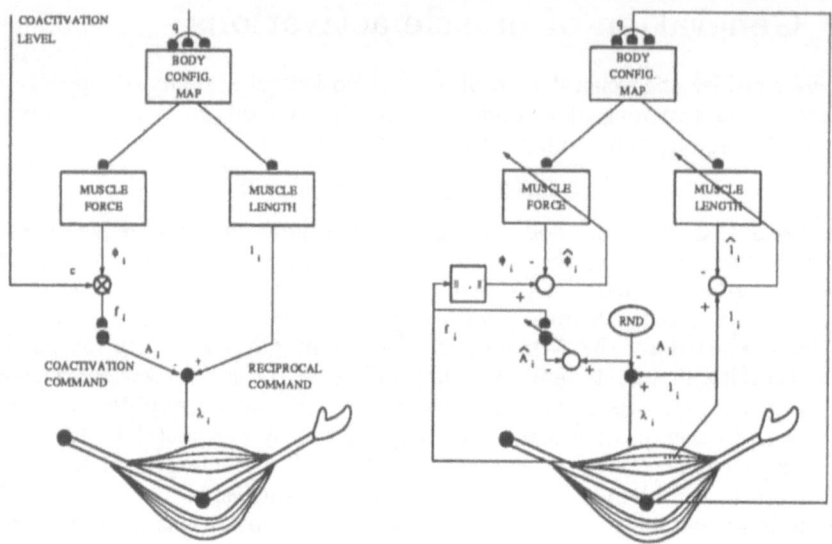

Figure 1: Block diagram of the computational model (left) and the relaxation learning scheme (right)

Let us consider the case $n = 1$ first; in this case, there is only one coordinative mode $\phi(\mathbf{q})$, and the scalar quantity c is the *level of coactivation*; the expression for λ_i can be re-written as follows:

$$\lambda_i = l_i(\mathbf{q}) - A_i(\phi_i(\mathbf{q})\,c) \qquad (3)$$

This formulation suggests that muscle activation is given by the sum of two terms: a *reciprocal command*, responsible for establishing the activation threshold of each muscle corresponding to the desired equilibrium configuration $\mathbf{q}$, and a *coactivation command* that is responsible (via the coactivation level c) for the magnitude of muscle forces and, indirectly, of the global body stiffness. However, different from the Feldman's formulation (Feldman et al., 1990), both terms are non-linearly dependent on $\mathbf{q}$ and c.

3 Self-organization of the controller

The computational architecture emerging from Eq. 3 involves a number of static mappings: (i) $\mathbf{l} = \mathbf{l}(\mathbf{q})$, i.e. muscle lengths given the body configuration; (ii) $\phi = \phi(\mathbf{q})$, i.e. coordinative mode given the body configuration; (iii) $A_i = A_i(f_{Mi})$, i.e. coactivation command given force (for each muscle i). The whole architecture may be neurally implemented as in Fig. 1 (left) where, for simplicity, only the connections related to the i-th muscle are displayed; the *body configuration map* is a neural field or *cortical map* of processing elements, where a particular body configuration $\mathbf{q}$ is population-coded by the peak of activation. The *force* and *length* layers approximate, respectively, the (normalized) force ϕ and the vector of muscle lengths $\mathbf{l}$ from the pattern of activation in the map; for the i-th muscle the quantity ϕ_i, multiplied by the coactivation level, is then translated into the coactivation command A_i.

Regarding the learning scheme, we propose the Piagetian concept of *circular reaction*, i.e. the ability of the body to measure the sensory consequences of a given pseudo-random, exploratory movement (Morasso & Sanguineti, 1993). A simple action-directed scheme, based on randomly exploring the space of muscle activations, is not suitable because there is no guarantee that a given λ would correspond to any stable and non-neutral equilibrium configuration. The proposed solution is indeed based on exploring the space of coactivation commands (see Fig. 1, right); at each learning step, a pseudo-random coactivation command **A** is generated and subtracted to the present length vector, measured by proprioceptive sensors; the resulting muscle activation modifies the equilibrium configuration of the musculoskeletal system, which *relaxes* until an equilibrium configuration is reached for which **A** is a 'legal' coactivation command. At that time, the pairs $\{q, l\}$, $\{q, \|f_M\|\}$ and (for each muscle) $\{f_{Mi}, A_i\}$ may serve as teaching signals for the supervised learning of the involved mappings. We propose for such a learning scheme, based on a sort of feedback mechanism and on the elastic behavior of the musculoskeletal system, and somehow related to the *feedback-error learning* scheme (Kawato, 1990), the name of *relaxation learning*.

In the most general case, when $n > 1$, there are more coordinative modes; however, many of them may be disadvantageous in energetic terms, and/or may correspond to similar postural behaviors. In general, the hypothesis of the motor system that is able to arbitrarily set all feasible patterns of muscular force seems to be unrealistic, so that we can simplify the problem with additional constraints. A biologically plausible criterion consists into selecting the particular coordinative mode corresponding to minimizing the potential energy at equilibrium. It is easy to show that such an additional constraint may be taken into account by adding a correction term in the standard supervised learning rule for the force layer. Promising simulations have been performed.

Reference

Bullock, D., & Grossberg, S. (1989). VITE and FLETE: Neural modules for trajectory formation and postural control. In Hershberger, W. (Ed.), *Volitional Action*, pp. 253–297. North-Holland/Elsevier, Amsterdam.

Feldman, A. G., Adamovich, S. V., Ostry, D. J., & Flanagan, J. R. (1990). The origins of electromyograms – Explanations based on the equilibrium point hypothesis. In Winters, J. M., & Woo, S. L.-Y. (Eds.), *Multiple Muscle Systems – Biomechanics and Movement Organization*. Springer-Verlag, New York.

Humphrey, D. R., & Reed, D. J. (1983). Separate cortical systems for control of joint movement and joint stiffness: reciprocal activation and coactivation of antagonist muscles. In Desmedt, J. E. (Ed.), *Motor control mechanisms in health and disease*, pp. 347–372. Raven Press, New York.

Kawato, M. (1990). Feedback-Error-Learning Neural Network for Supervised Motor Learning. In Eckmiller, R. (Ed.), *Advanced Neural Computers*, pp. 365–372. North-Holland/Elsevier, Amsterdam.

Morasso, P., & Sanguineti, V. (1993). Self-organizing body-schema for motor planning. *Journal of Motor Behavior*. in press.

Study of a Purkinje unit as a basic Oscillator of the Cerebellar Cortex

P. Chauvet[*], G.A. Chauvet[*,**]

[*]Institut de Biologie Théorique, Université d'Angers, 49100 Angers (France)
[**]Department of Biomedical Engineering and Program for Neuroscience, University of Southern California, Los Angeles, CA 90089 (USA)

1. Introduction

Different models have been proposed to describe cerebellar function (Tyrrel and Willshaw, 1992), but the implication of its neuro-anatomical structure in its capacity to learn motor coordination is not again well understood. In previous papers, we introduced a basic element of the cerebellar neural network called a Purkinje unit (Chauvet G.A., 1986; Chauvet P. and Chauvet G.A., 1993). This basic circuit is a neural network defined with biological constraints, i.e.: (i) real connectivity, (ii) specific activating or inhibiting synaptic property, (iii) anatomical hierarchical structure. Then, the global cerebellar neural network is a network of neural networks.

We study here the stability and behaviour in retrieval phase of a single Purkinje unit, introducing propagation delays of neural signals in its components. We show that this circuit can associate to a constant input an output varying in time (Chapeau-Blondeau and Chauvet G.A., 1991), and that this unit has oscillatory capabilities under biological plausible conditions. Then, motor coordination could be the exchange of synchronized informations between all Purkinje units, as in the case of coupled oscillators (Müller-Wilm, 1993).

2. The Purkinje unit

A Purkinje unit consists of one Purkinje cell, the single output system of the cerebellar cortex, with some of the neurons connected to it (Chauvet G.A.,1986). Figures 1 and 2 show the two components of a Purkinje unit.

The granule subsystem includes a layer of g granule cells and the Golgi cell. The subsystem's outputs are: a pattern $\underline{U}$ of m elements 1 or 0 that represent information propagated along the m mossy fibers; the external context X_e representing

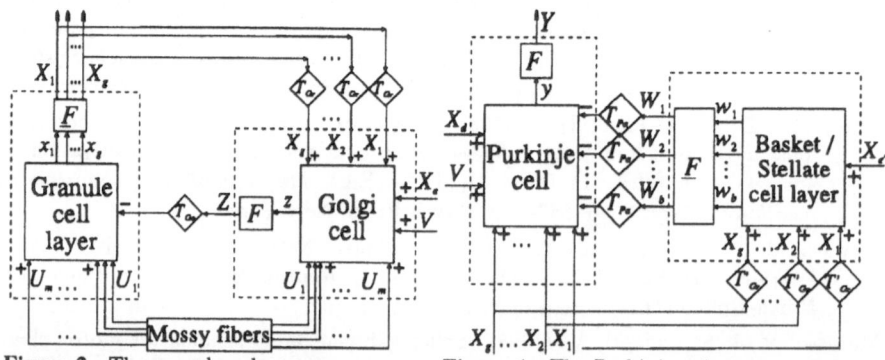

Figure 2 : The granule subsystem.   Figure 1 : The Purkinje subsystem.

activities propagated along the parallel fibers connected with the Golgi cell but that do not issue form this unit; the activity V on the climbing fiber. The Purkinje subsystem includes a layer of b basket or stellate cells, and one Purkinje cell. It receives as inputs: the vector $\underline{X}$ of activities X_i , which is the output of the granule subsystem; the external context X_e and X_d , which are activities propagated along parallel fibers connected, respectively, with the basket cells by sysnapses $\underline{\gamma}_e$, and with the Purkinje cell by the synapses μ_d ; and the climbing fiber V. The output activity along the Purkinje cell axon is the main output of the unit, and it is denoted as Y. All the synaptics weights are positive because the excitatory or inhibitory effects of synapses are included in signs between the elements of the following equations.

We introduce delays of propagation and transformation between the unit components. We denote as T_{Go} , T_{Gr} , T_{Gr}' and T_{Pa} the delays between, respectively: the Golgi cell and the granule cells, the granule cells and the Golgi cell, the granule cells and the basket cells, the baskets cell and the Purkinje cell.

The output S of a neuron, a mean frequency, is equal to $F[s_0 + {}'\underline{w}.\underline{E}]$ where there are n synaptic weights w_i and inputs E_i , s_0 a basic activity and ${}'\underline{w}.\underline{E}$ the scalar product between $\underline{w}$ and $\underline{E}$. F is a firing function equal to the identity or: $F(s) = (1 + e^{-as})^{-1}$ with $a>0$. Then, the output $\underline{S}$ of a layer is $\underline{F}[\underline{s}_0 + W\underline{E}]$ where W is the matrix of the synaptic weights.

Equations that give Y as a function of time, inputs and parameters are now established. The granule cells layer receives excitatory impulses from mossy fibers $\underline{U}$ by the way of synaptic matrix S_m and inhibitory impulses from the Golgi cell, denoting as Z its output, by synaptic vector $\underline{\sigma}_G$. The Golgi cell is excited by parallel fibers $\underline{X}$ via the synaptic vector $\underline{\eta}_p$, by X_e , by mossy fibers $\underline{U}$ via the synaptic vector $\underline{\eta}_m$ and by climbing fiber. Then:

$$Z(t) = F[z_0 + {}'\underline{\eta}_p.\underline{X}(t-T_{Gr}) + {}'\underline{\eta}_m.\underline{U} + \eta_e X_e(t) + \eta_c V(t)]$$
$$\underline{X}(t) = \underline{F}[\underline{\sigma}_0 + S_m\underline{U} - \underline{\sigma}_G Z(t-T_{Go})]$$

(1)

The basket cells layer, denoting as $\underline{W}$ its output, is excited by parallel fibers via the synaptic matrix G_p , and the Purkinje cell is inhibited by basket cells and excited by parallel fibers and climbing fiber:

$$\underline{W}(t) = \underline{F}[\underline{\gamma}_0 + G_p\underline{X}(t-T_{Gr}') + \underline{\gamma}_e X_e(t)]$$
$$Y(t) = F[\mu_0 + {}'\underline{\mu}_p.\underline{X}(t) - {}'\underline{\mu}_p.\underline{W}(t-T_{Pa}) + \mu_c V(t) + \mu_d X_d(t)]$$

(2)

We then obtain in the linear case:

$$\underline{X}(t) = -G\underline{X}(t-T_G) + \underline{X}_0 + S\underline{U} - H_0(t)\underline{\sigma}_G$$

(3)

$$Y(t) = Y_0 + {}'\underline{\mu}_p.\underline{X}(t) - {}'\underline{\gamma}_p.\underline{X}(t-T_P) + H(t)$$

(4)

where $\quad T_G = T_{Gr} + T_{Go}$, $\quad T_P = T_{Gr}' + T_{Pa}$,

$$G = \underline{\sigma}_G \cdot \underline{\eta}_p \ , \ \underline{X}_0 = \underline{\sigma}_0 - \underline{\sigma}_G z_0 \ , \ S = S_m - \underline{\sigma}_G \cdot \underline{\eta}_m \ , \ Y_0 = \mu_0 - {}^t\underline{\mu}_b \cdot \underline{Y}_0 \ , \ {}^t\underline{Y}_p = {}^t\underline{\mu}_p \cdot G_p \ ,$$

and $H_0(t) = \eta_e X_e(t) + \eta_c V(t)$, $H(t) = \mu_d(t) X_d(t) - {}^t\underline{\mu}_b \cdot \underline{Y}_e X_{e'}(t - T_{Pa}) + \mu_c V(t)$.

3. The study of a Purkinje unit in retrieval phase

3.1 Stability

We study now the stability of variables $\underline{X}$ and Y during the retrieval phase when H_0 and H are constant. Because of the delays, the unit has internal dynamics, that associate to a constant input $\underline{U}$ time-varying outputs $\underline{X}$ and Y. It is demonstrated in the linear case (Chauvet P., 1993) that, for t in $[nT_G, (n+1)T_G]$:

$$\underline{X}(t) = (-1)^{n+1} G^{n+1} \underline{X}^0(t - (n+1)T_G) + \sum_{i=0}^{n} (-1)^i G^i [\underline{X}_0 + S\underline{U}] - \sum_{i=0}^{n} (-1)^i H_0(t - iT_G) G^i \underline{\sigma}_G$$

where $\underline{X}^0$ is the initialization of $\underline{X}$ for $t<0$. After some calculus with a norm on $\mathbf{R}^g$, it comes that the solution of equation (3) is asymptotically stable if $\|G\| < 1$ and simply stable if $\|G\| = 1$. If $\|G\| > 1$ then the output $\underline{X}$ is unstable and diverge.

In the non-linear case, the granule subsystem is stable if the second term of equation (1) is contractant in respect to $\underline{X}$: it is an application of the fixed-point theorem. Then, $\underline{X}$ is asymptotically stable if $a^2 \|G\| < 16$ and simply stable if $a^2 \|G\| = 16$, where a is the maximum of the derivative of F. If $a^2 \|G\| > 16$ the output is unstable but cannot diverge. The conditions of stability of the Purkinje subsystem depend only on the conditions of stability of those for the granule subsystem: if the output of the granule subsystem is stable (respectively unstable) then the output of the Purkinje subsystem is stable (respectively unstable). This analysis show that the stability of one unit depends on the {granule layer - Golgi cell} feedback loop, which is not modifiable because it seems that synaptic weights in G are unmodifiable (Ito, 1984).

3.2 Variations in time of the outputs $\underline{X}$ and Y

G is a square matrix, and if the eigenvectors of G form a base it is possible to write in the linear case, denoting as $\lambda_1, \lambda_2, \ldots, \lambda_g$ the eigenvalues of G :

$$\Psi_i(t) = -\lambda_i \Psi_i(t - T_G) + c_i \ , \ i = 1, \ldots, g \ , \text{ with } \underline{X}(t) = P \, \underline{\Psi}(t) \tag{5}$$

If λ_i is strictly positive then Ψ_i oscillates, with a pseudo-period equal to T_G and with discontinuities at instants (nT_G), n integer. In our case, the synaptic weights $\underline{\sigma}_G$ and $\underline{\eta}_p$ are randomly chosen with a Gaussian distribution around means equal to σ_G and η_p . If the standard deviation is not too large with respect to the mean (for example, 10^{-2} for means that equal to 10^{-1} on figure 3), eigen values are:

$$\lambda_i = \varepsilon_i (\sigma_G \eta_p)^g \ , \ i = 1, \ldots, g \ , \ \lambda_g = (g(1+\varepsilon) + \varepsilon_g)(\sigma_G \eta_p)^g$$

where ε is the standard deviation and ε_i an error factor (with the same example, ε_i is in the order of 10^{-4}). λ_g is greater than zero and much greater than the other eigen values. Then, oscillations of Ψ_g are much greater than the possible oscillations of Ψ_i for i different from g. Each output X_i is a linear combination of Ψ_i, but it depends strongly on Ψ_g. The output of the granule subsystem oscillates due to the {granule layer - Golgi cell} feedback loop and the regularity of connections in this loop. In the non-linear case the same behaviour is observed. The only difference is that, with the conditions of unstability, the output oscillates after a finite time between zero and the maximum of F, whatever the values of parameters and $\underline{U}$. Then the unit has no ability of distinction.

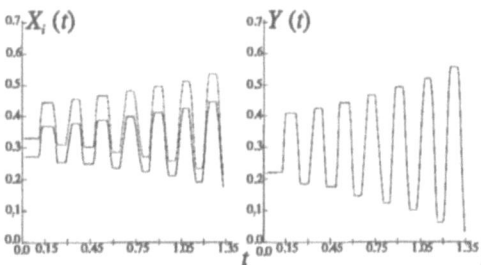

Figure 3 : Outputs of a linear Purkinje unit ($g=2$, $b=2$). The norm of the matrix G is greater than one.

4. Conclusion

In this paper a Purkinje unit with internal dynamics has been defined using the real connectivity around a Purkinje cell and distance between cells. Each neuron is formal and its transfer function is a sigmoïd or the identity. Mathematical conditions of stability have been stated in linear and non-linear cases during retrieval phase. The variation in time of outputs has been studied and conditions of oscillations have been found.

Necessary conditions of stability and convergence for a network of Purkinje units without delays between units have been determined (Chauvet P., 1993). In a network with delays between units, the same condition of stability has been found for one unit plus a set of sufficient conditions on those parameters that are included in the global connectivity. The study of the global net, using this hierarchical approach, allows us to demonstrate and to anticipate certain of its behaviours at the higher level from the study of the lower levels, specifically from interactions between Purkinje circuits and their own individual properties.

References

Chapeau-Blondeau, F., Chauvet, G.A. (1991). A neural network model of the cerebellar cortex performing dynamic associations, *Biological Cybernetics*, **65** 267-279.

Chauvet, G.A. (1986). Habituation rules for a theory of the cerebellar cortex, *Biological Cybernetics*, **55** 1-9.

Chauvet, P., Chauvet, G.A. (1993). On the ability of cerebellar Purkinje units to constitute a neural network, *Proceedings of the World Congress on Neural Networks, Portland, July 11-15, 1993, WCNN'93*.

Chauvet P. (1993). Etude d'un réseau de neurones hiérarchique à propos de la coordination du mouvement, *PhD Thesis, Université d'Angers*.

Ito, M. (1984). The cerebellum and neural control, *Raven press, New York*.

Müller-Wilm, U. (1993). A neuron-like network with the ability to learn coordinated movement patterns, *Biol. Cybern.*, **68**, 519-526.

Tyrrell, T., Willshaw, D. (1992). Cerebellar cortex: its simulation and the relevance of Marr's theory, *Phil. Trans. R. Soc. Lond.*, **B 336** 239-257.

Compartmental Interaction in the Granular Layer of the Cerebellum

L.N.Kalia*

Neural Systems Engineering, Dept of Electrical Engineering
Imperial College, London SW7 2BT

1   Introduction

The cerebellum is a layered structure, each layer distinguished by its own neural architecture and hence its own functional role. Consequently each layer constitutes a network in its own right. The granular layer is examined here. It is responsible for providing a representational surface for the principal cerebellar afferent system, the mossy fibres. This paper is concerned with the mechanisms which underlie the construction of this representation and with the functional properties it possesses. As in previous work [1, 2] the granular layer network is conceived as having a compartmental organisation. Whereas some modellers [2] assumed independent action, in [1] it was suggested that compartmental *interaction* could be a functionally-relevant mechanism operating within the granular layer. In this paper it will be demonstrated that this is indeed the case; this is shown by using the example of the XOR problem.

In order to do so, the anatomy of the granular layer is described and the form of the granular layer compartment outlined. It is then shown how pathways within the network can give rise to compartmental interaction. The functional implications associated with the representation resulting from compartmental interaction are then briefly discussed.

2   Architecture of the granular layer network

The principal neuronal constituents of the granular layer are the tiny and very numerous granule cells and the much larger but less prevalent Golgi cells. Both of these cell types receive excitatory input from the so-called mossy fibres which arise from various cells throughout the CNS [4].

Granule cells receive 2–7 mossy fibre inputs, all from different fibres. Co-localised with these is an inhibitory input from a Golgi cell. Each granule cell gives rise to a T-shaped axon which ascends perpendicularly to the granular layer. Each branch of the axon, referred to as a parallel fibre, makes synaptic contact with the dendritic trees of Golgi and Purkinje cells positioned at intervals along its length (Purkinje cells are associated with the molecular layer network and are the output stage of the cerebellar cortex). In total, each parallel fibre makes contact with a number of Golgi cells and with many more Purkinje cells.

Since the axonal ramification of the Golgi cell is bounded by an approximately hemispherical volume, the population of granule cells inhibited by a

*I should like to thank the U.K. SERC for their financial assistance and my colleagues in the NSE group for their help in the preparation of this paper.

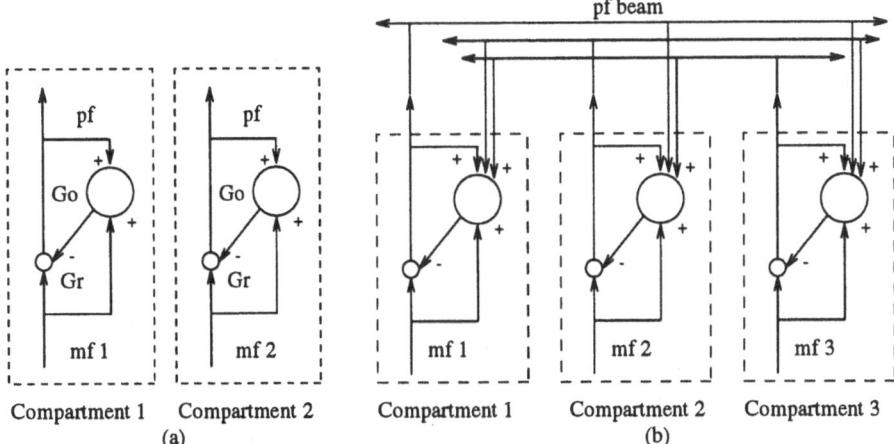

Figure 1: (a)Previous models assumed compartments operated independently. (b) Proposed model for granular organisation: parallel fibre input to a Golgi cell from extra-compartmental granule cells (i.e cells from a different compartment) enables mutual interaction to occur. Note: Gr = *population* of granule cells.

particular Golgi cell is similarly spatially confined. This, together with similar spatial confinement in the connectivity with mossy fibres (which lie well within the volume circumscribed by the axonal ramification) has led to the notion of the *granular layer compartment*, considered to be the elemental structure from which the entire granular layer is made up. A compartment is defined as a Golgi cell and the population of granule cells it inhibits.

3   Origins of compartmental interaction

Some modellers [2] have depicted the granular layer as a series of separate and disjoint compartments which are assumed to operate in isolation from one another (Fig 1a). However, following [1] it is argued here that a truer depiction would be that shown in Fig 1b which incorporates compartmental mutual interaction. This interaction is achieved through parallel fibre input to Golgi cells from extra-compartmental granule cells (i.e cells from a different compartment). It is the fact that a parallel fibre is longer than the axonal span of the Golgi cell which makes this possible.

Having therefore demonstrated the existence of interaction, it is possible to define an *interacting set* of compartments on the basis that they each lie along the length of a parallel fibre beam (a collection of similarly oriented parallel fibres) to which each contributes and from which each samples.

In order to demonstrate the importance of compartmental interaction in the construction of the overall representation produced by the granule cells (i.e across the interacting set) one can use a version of the XOR problem to show the advantages of interaction when it is included and the shortcomings of the resulting representation when it is not.

4 Illustration of the effects of interaction using the XOR problem

For the sake of simplicity the network under consideration is composed of just two compartments. Each compartment is assumed to receive input from its own specific mossy fibre source (the finding of fractured somatotopy supports this assumption [6]). Both send their output (the representations distributed across their constituent granule cells) to the same Purkinje cell, the overall output unit of the network.

The behaviour of the network in response to some event A carried by the set of mossy fibres ({mf 1}) associated with compartment-1 (C1) and some event B presented, via {mf 2}, to compartment-2 (C2) is to be examined. In what follows it is assumed that the Purkinje cell is required to fire ('1') if event A or B is present, at their appropriate input sites, but should not fire if both are present. If neither is present, the cell is again assumed to be quiescent. In other words the network is required to solve a distributed version of the XOR problem.

If no interaction is assumed to occur then the representation formed by C1 in response to A will be the same irrespective of the input occurring at C2. Similarly, the representations formed by C2 are invariant to the activity of C1.

If A alone is to cause the Purkinje cell to fire, C1 must produce an appropriate representation. Similarly for B alone and C2. Thus, if A alone (at C1) or B alone (at C2) can cause the Purkinje cell to fire, when A and B occur together the Purkinje cell will erroneously fire since it is the same representations which are formed, at their appropriate sites, as when each event was present by itself.

However, if compartmental interaction is permitted, then the representations formed by each compartment are made sensitive to the activity of the other compartment and, ultimately, to the input of the other compartment. Consequently, the representations formed by C1 in response to A when B is present at C2's input will be different from that formed when B is not present. Similarly for the representations formed by C2 in response to B. Since each of the locally-formed representations are different appropriate Purkinje cell behaviour is made possible.

Therefore the model incorporating mutual interaction is able to solve the XOR problem whereas the non-interacting model cannot. This example demonstrates that interaction between compartments leads to enhanced functional capabilities.

5 Context-sensitive encoding: an emergent property

The difference between the two schemes considered above can be represented thus:

Let x_i = mossy fibre input to the ith compartment of the interacting set, i=1, 2, ..., n.

Let y_i = granular representation formed by i^{th} compartment in response to x_i.

If interaction is *not* present then

$$y_i = y_i(x_i)$$

If interaction *is* present then

$$y_i = y_i(x_1, \ldots, x_i, \ldots, x_n)$$

Through mutual interaction the local representations formed by each compartment are made a function of the overall context (the total input to the interacting set) and not just of their local mossy fibre input, i.e each compartment constructs a *context-sensitive* representation of its local mossy fibre input. It is able to do this since interaction leads to the combination of information from different functional modalities, which is particularly pertinent to the cerebellum given the localisation of mossy fibre afferents implied by fractured somatotopy [6]. This augmentation of the capabilities of an individual compartment represents an emergent property of the granular layer network. As was hypothesised in [1] such properties could assist in cerebellar control of coordination, a role assigned to it by a number of authors [3, 5].

6   Summary

This paper has described how compartmental interaction can arise within the granular layer through extra-compartmental parallel fibre input to Golgi cells. The XOR problem was used as an example to demonstrate that mutual interaction leads to functionally-relevant properties, thereby adding support to the proposals made by Chapeau-Blondeau and Chauvet [1]. Given the fractured somatotopic distribution of input signals within the granular layer [6], the presence of interaction is a vital component in the process by which representations are constructed since it enables information, received at different sites within the layer, to be combined. Thus, future models of the cerebellum should incorporate this feature if they are to accurately depict the functionality inherent in this structure.

References

[1] F.Chapeau-Blondeau and G.Chauvet.(1991) A neural network model of the cerebellum performing dynamic associations. *Biol. Cybern.* 65:267–279.

[2] M.Fujita. (1982) Adaptive filter model of the cerebellum. *Biol. Cybern.* 45:195–206.

[3] P.F.C.Gilbert.(1974) A theory of memory that explains the function and structure of the cerebellum. *Brain Res.* 70:1–18.

[4] M.Ito. (1984) *The Cerebellum and Neural Control.* New York. Raven Press.

[5] W.T.Thach, H.P.Goodkin and J.G.Keating. (1992) The cerebellum and the adaptive coordination of movement. *Annu. Rev. Neurosci.* 15:403–442.

[6] W.I.Welker. Cerebellar representations: the importance of micromapping and natural stimuli. In *The Cerebellum and Neuronal Plasticity* ed. M.Glickstein, J.F.Stein and C.Yeo. New York: Plenum 1988 pp 109–118

Modeling Biologically Relevant Temporal Patterns

W. Zander, B. Brueckner, T. Behnisch, T. Wesarg

Institute of Neurobiology

Magdeburg, Germany

1. Introduction

On the basis of the observations of coherent oscillations and thereby caused synchronization in the visual cortex (Gray et al., 1987; Engel et al. 1991; Singer et al., 1993) the temporal output pattern (code?) of a spiking neuron model is getting increasingly interesting (Gerstner et al., 1993; Kirillov et al., 1993).

In the submitted paper we show that it is possible to create biologically relevant patterns of neural activity without expansive numerical effort. Starting-point of our modeling was the fact that the spike train of a biological neuron is describable only with few parameters, regardless of the circumstance that the neuron itself has 46 electrophysiologically measurable properties (Bullok, 1976). We are presenting a neuron model beeing able to generate biologically relevant temporal patterns with small numerical effort.

2. Model

The neuron model is realized on the abstraction level of the membrane potential (Fig. 1). The membrane potential strived for by the neuron is represented in the model by the introduction of the target function $f_a(t)$. The membrane potential of the neuron $f_m(t)$ approaches that target function in an exponential way and in dependence on the time constant τ. The target function $f_a(t)$ is determined by the input function $f_i(t)$ and by the afterhyperpolarization $f_{ahp}(t)$. The input function $f_i(t)$ is computed by the sum of impulses from the synapses $p_{ij}(t)$ (ith synapse, jth action potential on the transmitter neuron) arriving at the axon hillock of the neuron. In the resting-state of the neuron $f_a(t)$ is equal to the resting membrane potential p_{rest}. If $f_m(t)$ exceeds the threshold th, an action potential followed by an afterhyperpolarization $f_{ahp}(t)$, is generated. The variables p_{old} and t_{old} are put new at each alteration of $f_a(t)$.

The neuron model was implemented in the simulation system NESSI (Zander et al., 1993; Brueckner et al., 1993). A part of the graphic-interactiv user interface of NESSI is presented in Figure 2.

2..1 Parameters

synapse i	amplitude (with electrotonic decrement)	a_i
	duration	t_i
	delay	d_i
neuron n	length constant	λ
	time constant	τ
	threshold	th
	afterhyperpolarization - amplitude	a_{ahp}
	afterhyperpolarization - duration	d_{ahp}

2..2 Equations

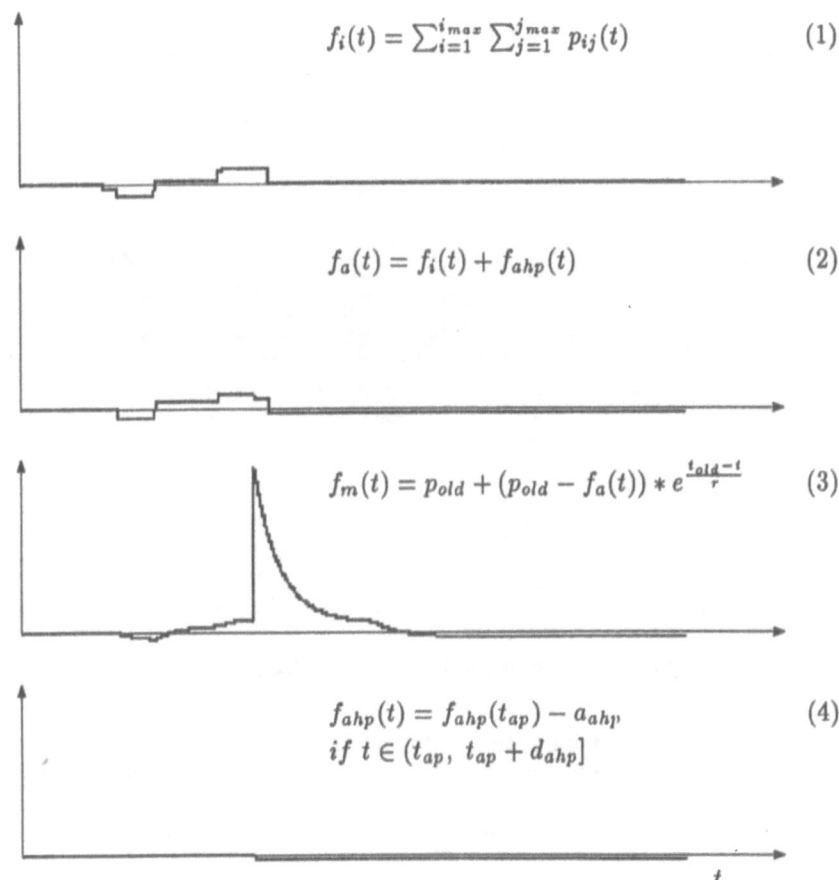

$$f_i(t) = \sum_{i=1}^{i_{max}} \sum_{j=1}^{j_{max}} p_{ij}(t) \qquad (1)$$

$$f_a(t) = f_i(t) + f_{ahp}(t) \qquad (2)$$

$$f_m(t) = p_{old} + (p_{old} - f_a(t)) * e^{\frac{t_{old}-t}{\tau}} \qquad (3)$$

$$f_{ahp}(t) = f_{ahp}(t_{ap}) - a_{ahp} \qquad (4)$$
$$if \ t \in (t_{ap}, \ t_{ap} + d_{ahp}]$$

t

Figure 1: Neuron model

100

3. Results

Comparisons between experiments from neurophysiology and our computer simulations show that our model neurons generate biologically relevant spike trains, corresponding to their parameters and to the given impulses. The neurons behave adaptively, they are able to create bursts and to change their fire frequency depending on their depolarization.

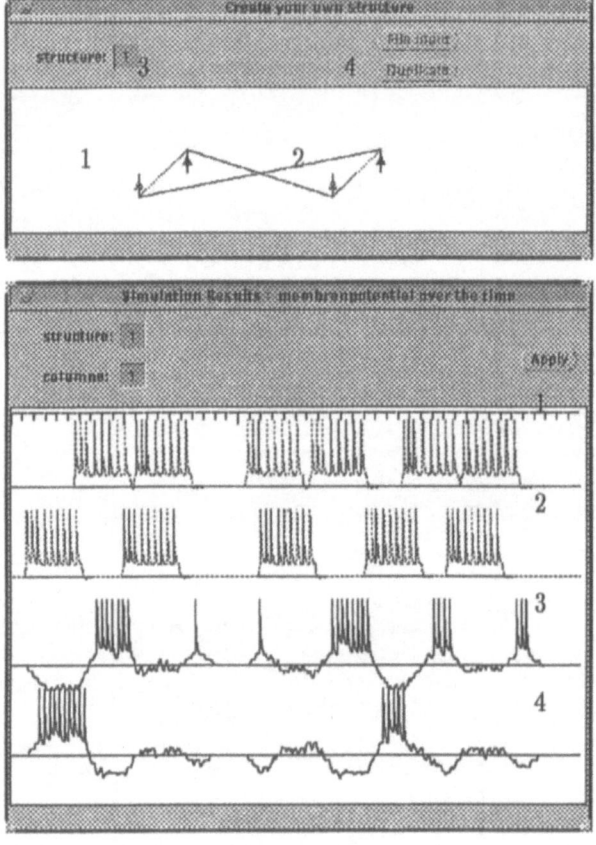

Figure 2: Circuit of an inhibition between antagonists and simulation results

In the example of inhibition between antagonists (Fig. 2) we show the behaviour of our neurons in the simulation. The circuit is simple, neuron 1 excitates neuron 3 and inhibits neuron 4, neuron 2 realizes the contrasting effect. The results of a simulation with a simulated time of 5 seconds are presented in the lower picture of Figure 2. Neurons 1 and 2 were stimulated by rectangular impulses in accordance with a random pattern and respond to the arriving impulses in an adaptive way. Their excitations inhibit the related antagonists and excitate the related agonists. Neurons 3 and 4 are firing in dependence on their depolarizations and with an adequate frequency.

4. Discussion

The specific dynamics of the model neurons follows from the time constant τ, a measure for the exponential approach, as well as from the target function $f_a(t)$. The adaptive behaviour of the model neurons and the generation of bursts only result from the afterhyperpolarization $f_{ahp}(t)$, because the threshold th is constant and thus a relative refractory period was not realized.

This effect corresponds to the observation of biological neurons. The medium afterhyperpolarization with a duration of a few hundred milliseconds is the reason for adaptation and bursts (Lorenzon et al., 1992). The role of afterhyperpolarization can be explained by its duration which is greater then the duration of the refractory period. The medium afterhyperpolarization causes summation processes (slow afterhyperpolarization) on the neuron and therefore realizes something like a "memory" of the neuron.

With the neurophysiological observations and the aid of our simulation results we postulate that afterhyperpolarization is a decisive factor in modeling of biologically relevant temporal output patterns of neurons.

Acknowledgements
This work is supported by the DFG grant Br 1289/1-1.

References
Brueckner, B. and Zander, W. (1992) Neurobiological Modelling and Structured Neural Networks. In: ICANN'92. Gielen, S., Kappen, B. (eds.), Springer, London, 43-46.

Bullock, T.H. (1976) In search of principles in neural integration. In: Simpler Networks and Behavior. Fentress, J. (ed.) Sinauer, Sunderland, 52-60.

Engel, A.K., Koenig, P., Kreiter, A.K. and Singer, W. (1991) Interhemispheric synchronisation of oscillatory neuronal responses in cat visual cortex. Science, 252, 1177-1179.

Gerstner, W. and van Hemmen, J.L. (1993) Spikes or Rates? - Stationary, Oscillatory, and Spatio-temporal States in an Associative Network of Spiking Neurons. In: ICANN'93, Gilen, S., Kappen, B., (eds.), Amsterdam, 633-637.

Gray, C.M., Koenig, P., Engel, A.K. and Singer, W. (1987) Stimulus-specific neuronal oscillations in the cat visual cortex: A cortical functional unit. Soc. Neurosci. Abstr., 13, 404.3.

Kirillov, A.B. and Woodward, D.J. (1993) Synchronisation of Spiking Neurons: Transmission Delays, Noise and NMDA Receptors. In: WCNN'93, Portland, II-594-II-597.

Lorenzon, N.M. and Foehring, R.C. (1992) Relationship Between Repetitive Firing and afterhyperpolarizations in Human Neocortical Neurons. J. Neurophys., Vol.67, No.2, 350-363.

Singer, W., Artola, A., Engel, A.K., Koenig, P., Kreiter, A.K., Loewel, S. and Schillen, T.B. (1993): Neuronal Representations and Temporal Codes. In: Exploring Brain Functions: Models in Neuroscience, Poggio, T.A., Glaser, D.A., eds., Jonhn Wily and Sons Ltd.

Zander, W., Brueckner, B., Brankatschk, G. (1993) Simulation of biological relevant neural networks. In: 21th Goettingen Neurobiology Conference, Thieme, Goettingen, 876.

A Model of the Baroreceptor Reflex Neural Network

J.S. Schwaber, I.A. Rybak and R.F. Rogers

Neural Computation Group, The Experimental Station, E.I. DuPont & Co.
Wilmington, DE 19880-0323, USA

Introduction

The baroreceptor vagal reflex is an important part of the cardiovascular control system. It may be defined as the biological neural control system responsible for the short-term blood pressure regulation. A simplified scheme of the baroreceptor vagal reflex circuit is the following. Baroreceptors located in the great arteries provide the sensory information to second-order neurons located in the nucleus tractus solitarii (NTS) in the lower brainstem [5]. Via a network of interneurons, the second-order neurons affect motor neurons which in turn control heart rate and total peripheral resistance and thus blood pressure [3]. The first-order neurons (baroreceptors) encode each pressure pulse with a frequency-adapting train of spikes [1]. One of the mysteries of the identified baroreflex related second-order neurons in NTS is that in spite of receiving direct monosynaptic inputs from the first-order neurons, they do not show any pulse-rhythmic activity that contains the frequency component corresponding to cardiac frequency. They do not respond to each heart beat and probably perform a low pass filtration of their input signals [6]. Inhibitory interactions in the NTS network of second-order neurons probably play the important role in this process as well as in the functioning of the baroreflex neural network in general. This is confirmed by the findings of inhibitory postsynaptic potentials (IPSP) in the barosensitive NTS neurons [5] and by the experiments in which the behavior of these neurons was investigated in the conditions of blocking GABAergic synaptic inhibition [7]. Recent investigation [6] revealed a group of barosensitive presumed second-order NTS neurons called "active" neurons. A typical response of a neuron from this group to induced blood pressure challenges is shown in Fig. 1. Analysis of neuronal responses allows us to assume and underline some behavioral properties of these neurons that were employed in the development of a neural model of early stages of the baroreflex:

(i) These barosensitive second-order NTS neurons respond to blood pressure changes with an expressed burst of activity whose frequency is much lower than the frequency of cardiac cycle.

(ii) The character of their responses allows to assume that they are inhibited just before and just after the bursts.

(iii) The bursts of these neurons are the sources of regulatory signals to the object of regulation (the heart), which in turn provide compensatory changes of blood pressure by changing cardiac output.

(iv) Possibly, each of these neurons responds to pressure changes and provides the above regulation in a definite static and dynamic range of these changes.

Modeling Results

The developed baroreflex model was based on two main hypotheses. The first one is the hypothesis of "barotopic" organization. It has been offered earlier [8] and consists in the supposition that individual baroreceptors' pressure thresholds are topically distributed in a working area of the pressure space, and that each second-order barosensitive neuron receives inputs from the first-order ones, whose thresholds lie in a definite small area of pressure. There are some anatomical [2] and physiological [6] data which support this supposition. The second hypothesis assumes that projections of the first-order neurons onto second-order ones are organized like ON-center-OFF-surround receptive fields in the visual and most other sensory systems. It means that each group of the second-order neurons gets "lateral" inhibition from neighboring neural structures tuned to higher and lower levels of blood pressure. This supposition is derived from the above statement (ii) and corresponds to general principles of organization of sensory systems.

The developed single neuron model is a modification of some spiking neuron models described earlier [4]. The model of the baroreflex neural net consists of the first- and second-order neurons (Fig. 2). The network is incorporated into a generalized model of the system for blood pressure control. The first order neurons get the excitatory input signal that is proportional to the blood pressure. These neurons are arranged sequentially with respect to increasing pressure threshold. The second-order neurons get synaptic excitation and inhibition from the first-order ones as it is shown in the Fig. 2. Parallel outputs of the second-order neurons are integrated, and via an intermediate subsystem (INT), provide negative feedback to a subsystem for blood pressure generation. This subsystem simulates the heart and vessels. It operates under the influence of an uncontrolled input signal that causes a deviation of the mean pressure from the setpoint.

Fig. 3 shows the responses of four first-order neurons (baroreceptors) with different blood pressure thresholds (the four upper rows; thresholds increase from the bottom to the top) to the increasing mean blood pressure (the bottom row). These neurons demonstrate the adaptive type of responses and respond with the spike to each pulse of the pressure. Fig. 4 and Fig. 5 show the responses of four first-order neurons (the 2nd-5th rows) and one second-order one (the upper row) to the pressure dynamics (the bottom row). Because of barotopical distribution of thresholds the first-order neurons start to respond to increasing mean blood pressure sequentially. In Fig. 4 the feedback control loop is broken off. In the Fig. 5 the feedback control loop is closed, and the second-order neuron performs the regulation of the pressure. The second-order neuron demonstrates low frequency spike bursts like its real prototype (Fig. 1). The behavior of the second-order neuron in the model corresponds in the first approximation to the behavior of real neurons and answers to the above statements (i)-(iv). It supports the hypotheses which the described model is based on.

One of the most interesting features of the considered model is that each second-order neuron responds to changes of the mean blood pressure and provides its regulation only in a definite static and dynamic range of these changes. The set of the second-order neurons controls the pressure via some sequence of control actions of individual second-order neurons that is formed automatically and adaptive to dynamics of pressure changes. Thus, the set of second-order neurons may be considered to be a set of interacting controllers, each of which dominates and provides control in a definite range of changes of the controlled parameter.

1. Abboud, F.M., and Chapleau, M.W. *J. Physiol.*, 1988, 401, 295-308.
3. Donoghue,S., Garcia,M., Jordan,D., and Spyer, K.M. *J. Physiol.*,1982, 322, 337-352.
3. Karemaker, J.M. In *The Beat-By-Beat Investigation of Cardiovascular Function* (R.I.Kitney and C. Rompelman, Eds.) Claredon Press, Oxford, 1987, 27-49.
4. Koch, C, and Segev, I. (Eds.) *Methods in Neuronal Modeling.* , MIT Press, 1989.
5. Miffin, S.W., and Felder, R.B. *Am. J. Physiol.*, 1990, 259, H653-H661.
6. Rogers, R.F., Paton, J.F.R., and Schwaber, J.S. *Am. J. Physiol.*(in press).
7. Suzuki, M., Kuramochi, T., and Suga, T. *J. Aut. Nervous Syst.*, 1993, 43, 27-35.
8. Schwaber, J.S., Paton, J.F.R., Rogers, R.F., Spyer,K.M., and Graves, E.B. In Computation and Neural Systems (F. Eekman and J. Bower, Eds.)Kluwer Acad.,1993,86-89

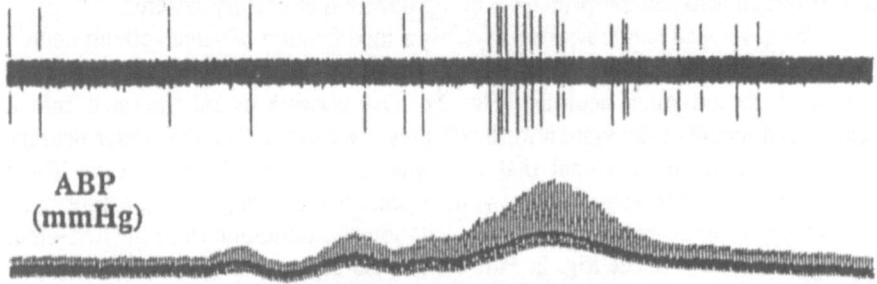

Fig. 1. Typical response of an "active" second-order neuron to the increasing mean arterial blood pressure (ABR).

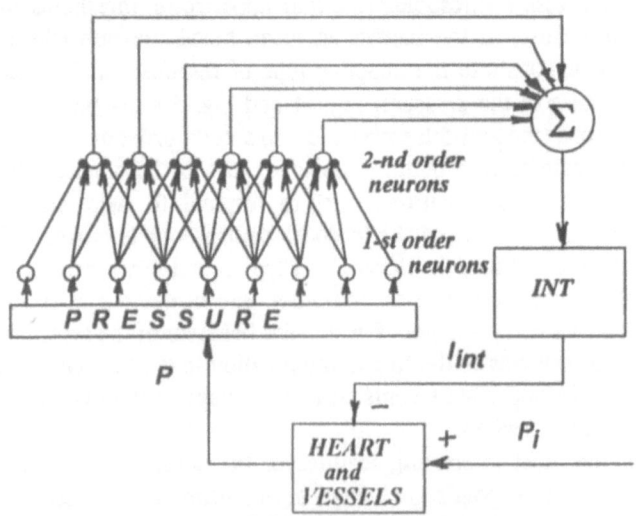

Fig. 2. Scheme of the model.

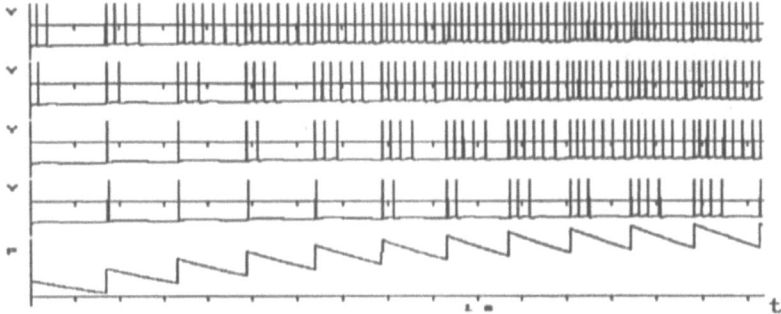

Fig. 3. Responses of four first-order neurons with different pressure thresholds
to the increasing mean blood pressure

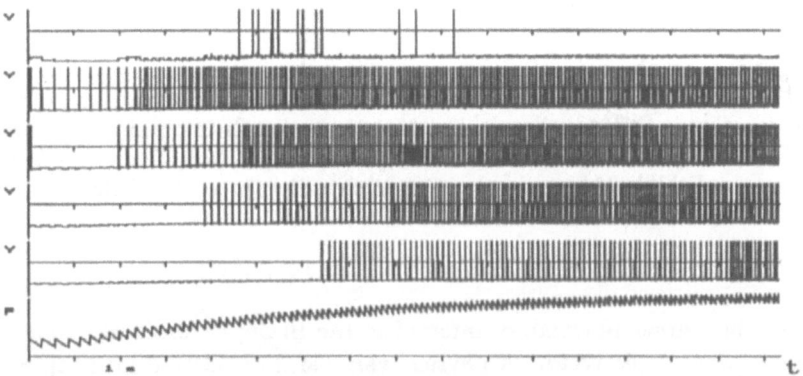

Fig. 4. Responses of four first-order and one second-order (at the top) neurons
(the feedback control loop is broken).

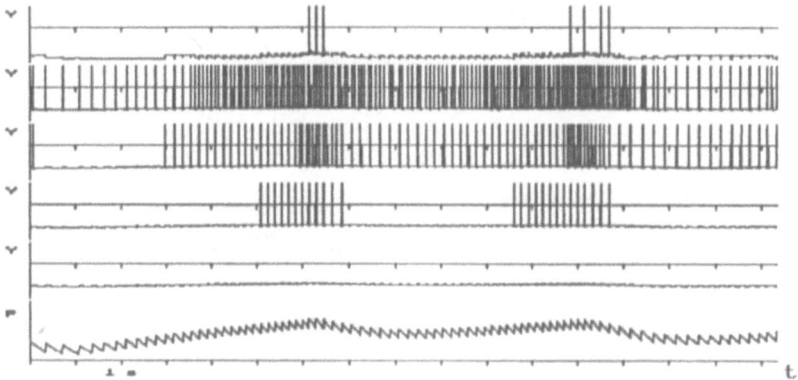

Fig. 5. Responses of four first-order and one second-order (at the top) neurons
(the feedback control loop is closed).

Modelization of Vestibulo-Ocular Reflex (VOR) and Motion Sickness Prediction

L. Zupan†, J. Droulez ‡, C. Darlot†*, P. Denise* & A. Maruani†

† Télécom Paris, Dept Image/PIAO

46, rue Barrault 75635 Paris Cedex 13, France

‡Labo. de Physio. de la Perception et de l'Action, Collège de France, Paris

*Laboratoire de Physiologie, Université de Caen, France

1   Introduction

Investigations on the role of vision and vestibular organs in eye movement control have revealed the importance of interactions between sensory signals. A central question is therefore how to combine those signals to obtain an accurate estimate of the physical variables. A comprehensive model of combination of the visual, otolithic and semicircular canals signals is presented below.

2   Presentation of the VOR-Model

Each variable, either physical, or internal to the Brain, is considered as a 3D-coordinate craniotopic vector. A physical variable $\vec{X}_j$ is filtered by a captor of transmittance $T_j(s)$ (s-Laplace notation), considered here as a first-order high-pass (HP) or low-pass (LP) filter. The output of the T_j-filter is the sensory input to the Brain.

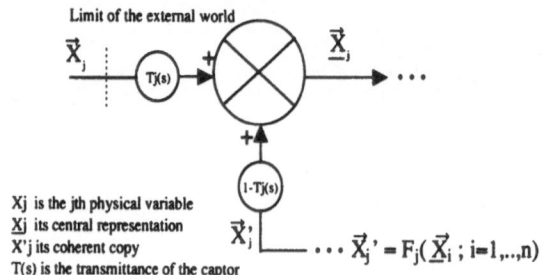

Figure 1: Principle of Sensory Integration

Each external physical variable is represented by a set of two internal variables : its central representation $\vec{X}_j$ and its coherent copy $\vec{X}'_j$. The former is the most accurate representation computed by the Brain; its computation is the core of the problem addressed in this paper. The latter is computed from the representations of a set of non-independent variables, linked to $\vec{X}_j$ by a definite physical relationship. This relationship between the internal variables is assumed to be similar to that existing between the physical variables in the external world. The principle of combination between the different internal representations of a variable [Droulez et al., 1989] is to complement the frequency content of the sensory signal, by the frequency content of the coherent copy, filtered through $[1 - T_j(s)]$, to obtain the central representation. This combination scheme is applied to all sensory variables (cf table).

PHYSICAL VAR.	SOURCE OF INFORMATION	NOT.
Head Angular Velocity	Semicircular Canals (HP)	$\vec{H}$
Gravitational Acceler.	Otolithic Regular Units (LP)	$\vec{G}$
Inertial Acceleration	Otolithic Irregular Units (HP)	$\vec{A}$
Gaze Angular Velocity	Retinal Slip (LP)	$\vec{R}$
Head Linear Velocity	Image Deformation (LP)	$\vec{V}$

In order to perform, from the total acceleration $\vec{Acc}$, the separation between gravitational $\vec{G}$ and inertial $\vec{A}$ acceleration, the source input to $\underline{\vec{G}}$ (resp. $\underline{\vec{A}}$) is $\vec{Acc}$ filtered through a LP-filter (resp. HP-filter); these two signals are completed by $-\vec{A}'$ (resp. $-\vec{G}$) filtered in accordance [Mayne et al., 1974]. The difference between the two ways $\vec{Acc}$ is completed is due to two necessities :
1/ stability since the use of $\vec{Acc} - \vec{A}$ instead of $\vec{Acc} - \vec{A}'$ would lead to instability in the range of usual VOR frequences.
2/ accurate fitting of experimental results on human beings.

The relationships between the variables set the structure of the computing circuits. Let $\underline{\vec{E}} = -a \cdot \underline{\vec{H}} - b \cdot \underline{\vec{R}}$ be the eye velocity in the orbit : a is the gain of the direct vestibulo-ocular pathway and b that of the direct optokinetic pathway.

- The relationship between Head and Gaze Angular Velocities reads :

$$\vec{R}' = \underline{\vec{H}} + \underline{\vec{E}} \quad \text{and} \quad \vec{H}' = c \cdot \left[\underline{\vec{R}} - \underline{\vec{E}}\right] + (1 - c) \cdot \vec{H}'_G \tag{1}$$

c is a weighting coefficient between the two coherent copies of $\vec{H}$: $\underline{\vec{R}} - \underline{\vec{E}}$ and $\vec{H}'_G$, as defined in (4).

- The relationships between Head Orientation and Angular Velocity and between Head Linear Velocity and Acceleration : to take into account the angular velocity $\vec{H}_{\mathcal{R}/\mathcal{W}}$ of the frame of reference $(\mathcal{R})$ with respect to the external world $(\mathcal{W})$, the temporal derivative of a vector $\vec{U}$ is computed as follows :

$$\frac{d\vec{U}_{\mathcal{W}}}{dt} = \frac{d\vec{U}_{\mathcal{R}}}{dt} + \vec{H}_{\mathcal{R}/\mathcal{W}} \wedge \vec{U}_{\mathcal{R}} \tag{2}$$

So, to build the coherent copy $\vec{G}'$ (resp. $\vec{V}'$), $\vec{U}$ is replaced by $\vec{G}$ (resp. $\vec{V}$) in (2), taking into account that $d\vec{G}_W/dt = \vec{0}$. Thus :

$$\vec{G}' = \int \left[\underline{\vec{G}} \wedge \underline{\vec{H}}\right] dt \quad \text{and} \quad \vec{V}' = \int \left[\underline{\vec{A}} - \underline{\vec{H}} \wedge \underline{\vec{V}}\right] dt \qquad (3)$$

In addition, as $d\vec{G}/dt + \left(\vec{H} \wedge \vec{G}\right) = \vec{0}$, by applying the double cross-product law to that equation, the following relation between Head Angular Velocity and Gravity Acceleration is obtained:

$$\vec{H}'_G = 1/g^2 \cdot \left[\frac{d\vec{G}}{dt} \wedge \underline{\vec{G}} + \langle \underline{\vec{H}}, \underline{\vec{G}} \rangle \cdot \underline{\vec{G}}\right] \ with \ g = 9.81 \ m \cdot s^{-2} \qquad (4)$$

3 Off-Vertical Axis Rotation (OVAR)

The simulation was implemented on MATLAB 4.1 and SIMULAB 1.2 and addressed OVAR in darkness for $\theta = 30°$, $\omega = 1 \ rad \cdot s^{-1}$ and $d0 = 5 \ m$:

$$\vec{H} = \omega \cdot \vec{k}, \vec{Acc} = \begin{pmatrix} -g \cdot \sin \theta \cos \omega t \\ g \cdot \sin \theta \sin \omega t \\ g \cdot \cos \theta \end{pmatrix}, \vec{P} = d0 \cdot \vec{i} \ \text{[Buizza et al., 1980]}$$

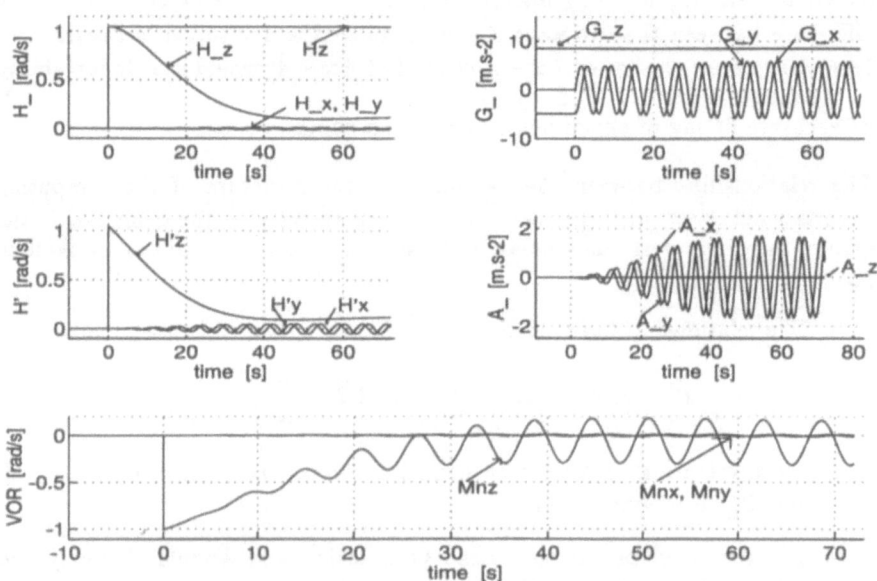

Figure 2: OVAR simulation for $\theta = 30°$ and $\omega = 1 \ rad \cdot s^{-1}$

In darkness, $\vec{R}$ and $\vec{V}$ are equal to $\vec{0}$. $\vec{Mn}$ (see fig.2) represents the order sent to the motoneurons commanding eye muscles. It reproduces well the experimental results obtained on human beings.

4 Motion Sickness Prediction

The incoherence between $\vec{X}_j$ and $\vec{X}'_j$ is assumed to be correlated to motion sickness. Ninety OVAR experiments were simulated for different values of θ and ω. For each variable $\vec{X}_j$, an incoherence I_X_j is calculated by :

$$I\_X_j(t) = \int_0^t \|\underline{\vec{X}_j} - \vec{X}'_j\| \, dt \quad \text{where} \quad \|\cdot\| \text{ is the euclidian norm} \qquad (5)$$

Human beings are assumed to get sick the quicker, the greater the asymptotic value of the temporal derivative of I_X_j (means the limit slope). The surface obtained for the different values of θ and ω (see fig.3) fits the experimental results obtained on human beings [Miller et al., 1973], with a resonance phenomenon between $\omega = 2 \; rad \cdot s^{-1}$ and $\omega = 2.5 \; rad \cdot s^{-1}$, depending on θ.

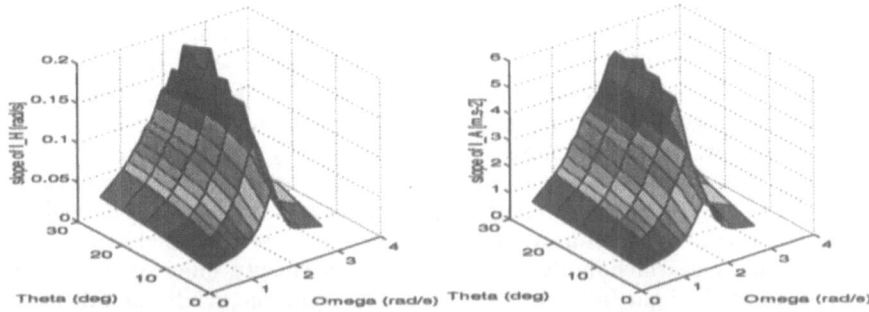

Figure 3: Limit slope of I_H and I_A for different values of θ and ω

Conclusion : Experimental results can be reproduced by a VOR model whose structure is similar to identified anatomical connections.

Acknowledgements :
This research was supported by DRET contract 91/1291A/DRET.

5 References

Buizza, A., Leger, A., Droules, J., Berthoz, A., and Schmid, R. (1980) Influence of otolithic stimulation by horizontal linear acceleration on optokinetic nystagmus, *Experimental Brain Research*, 39, 165-176.

Droules, J., and Darlot, C. (1989) The Geometric and Dynamic Implications of the Coherence Constraints, *Performance and Attention*, Vol. XIII (Ed. Jeannerod), 495-523.

Mayne, R. (1974) A system concept of the vestibular organs, *Handbook of sensory physiology*, Vol VI/2, 493-580.

Miller, E., and Graybiel, A. (1973) Susceptibility to motion sickness, *NASA report L-43518*.

Kernel Correlations of Movements in Neural Network

Naohiro Ishii

Department of Intelligence and Computer Science,
Nagoya Institute of Technology
Gokiso-cho, Showa-ku, Nagoya 466, JAPAN

1 Intoduction

An enormous amount of information is processed in the biological visual systems. Motion detection of objects is one of problems in the visual perception. Arbib(1991) presented models based on biological neural networks of rabbit, frog and toad, about the detection of the movement of objects. However, many of details of the network connections among cells, are unknown in them. Then, their studies assume some functions and connections in the networks.

In the past, Naka, Marmarelis et al. extensively developed their studies of catfish retinal networks to clarify the function of cells in the network, by applying Winners white noise analysis (Sakuranaga and Naka, 1987; Korenberg, Sakai and Naka, 1988). From Winner nonlinear analysis developed here, we can derive equations, which imply the directional movements; one is in case of the movement of the light from the left side to the right one, while the other is in case of that right side to the left one. Further, it is shown that temporal pattern discrimination can be done in this asymmetrical networks. The ganglion cell, which is the output cell in the retina, is assumed to behave like a multiplication functional unit. From the functional operation, the output of the ganglion cell, becomes the auto-correlation of the first order kernel and the cross-correlation of the second order kernel.

2 Spatial and Temporal Interaction in Neural Network

From studies of cell functions(Naka, Sakai and Ishii, 1988), Naka presented a simplified, but essential network of catfish inner retina. To verify the proposed network, we have decomposed into two subcircuits; one is **C**-circuit and the other is **N**-circuit(Ishii, 1991). The **N**-circuit, an asymmetric network, might process spatial interactive information between bipolar cells, while the **C**-circuit, also an asymmetric network, might process temporal interaction. The fundamental **N**-circuit is shown in Fig.1, where **B** shows a bipolar cell, **C** and **N** show amacrine cells.

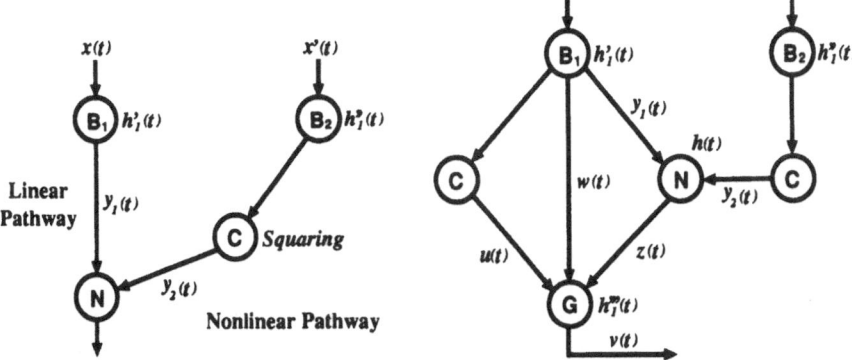

Figure 1: Asymmetric Neural Network for Spatial Interaction.

Figure 2: Extended Neural Network.

On the linear pathway in Fig.1 the first order kernel is derived as follows,

$$C_{11}(\lambda) = h_1'(\lambda)$$

The second order kernel becomes

$$C_{21}(\lambda_1, \lambda_2) = \alpha^2 h_1''(\lambda_1) h_1''(\lambda_2)$$

, where the ratio α is a mixed coefficient of $x(t)$ and $x'(t)$ and α in the above equation shows the second order Wiener kernel. Similarly, on the nonlinear pathway in Fig.1, we can derive the first order kernel $C_{12}(\lambda)$ and the second order kernel $C_{22}(\lambda_1, \lambda_2)$, respectively.

Under the condition that power spectrums, p and p'' of inputs $x(t)$ and $x''(t)$, are known in advance. Then, the ratio k between p and p', is computed easily. From the second order kernels C_{21} and C_{22}, which are abbreviated in the representation,

$$(C_{21}/C_{22}) = \alpha^2$$

holds. Then,

$$\alpha = \sqrt{\frac{C_{21}}{C_{22}}}.$$

From the first order kernels C_{11} and C_{12}, which are abbreviated, the following equation holds,

$$\frac{C_{12}}{C_{11}} = \frac{\sqrt{\frac{C_{21}}{C_{22}}}}{\frac{C_{21}}{C_{22}} + k\left(1 - \sqrt{\frac{C_{21}}{C_{22}}}\right)^2}$$

, which shows the movement of stimulus light from the left side to the right one.

3 Kernel Correlations in Extended Neural Network

In order to perceive the ratio α in the movement, we discuss the cell function in the extended neural network. The asymmetric network in Fig.1 is extended to the network as shown in Fig.2. In the previous chapter 2, The first order kernel on the linear pathway $C_{11}(\lambda)$ and the second order kernel $C_{22}(\lambda_1, \lambda_2)$ on the nonlinear pathway, do not change their values according to the ratio α, which is the moving parameter, since the values $h'(\lambda)$ and $h_1''(\lambda_1)h_1''(\lambda_2)$, are cell's characteristic functions, respectively. But, the second order kernel on the linear pathway and the first order kernel on the nonlinear pathway, change their values according to the ratio α. What functions are needed in the ganglion cell in Fig.2, to perceive the moving ratio α. First, we introduce a multiplicative operation in the ganglion cell. A multiplicative operator is used in Rechardt model for the perception of motion. Second, since the ganglion cell behaves as a linear filter(Korenberg, 1989), the ganglion cell is assumed to have two functions; multiplication and linear filtering. Further, we assume two kinds of multiplication between $w(t)$ and $z(t)$, while the other is a multiplication between $u(t)$ and $z(t)$ in Fig.2. Thus the output of the ganglion cell in the former multiplication, becomes

$$v_1(t, \delta) = \int_0^\infty h'''(\eta) \cdot w(t - \eta) \cdot z(t - \eta - \delta) d\eta$$

, where

$$w(t) = \int_0^\infty h_1'(\tau) \cdot x(t - \tau) d\tau$$

$$z(t) = \int_0^\infty h(\tau')\{y_1(t - \tau') + y_2(t - \tau')\} d\tau'$$

Then

$$v_1(t, \delta) = \int\int h'''(\eta)h_1'(\lambda)x(t - \lambda - \eta)d\lambda \cdot \int\int h(\eta)h_1'(\tau)x(t - \tau - \eta - \delta)d\tau d\eta$$

The impulse response function of **N** cell; $h(t)$, was clarified to be differential(Ishii, 1992), and the ganglion cell is considered to be **N**-type, i.e., $h'''(t) \simeq h(t)$. Then the above equation becomes

$$v_1(t, \delta) = \{h_1'(t) \cdot h_1'(t - \delta)\}$$

This equation does not change to the α. similarly, the output of ganglion cell in the latter multiplication, becomes

$$
\begin{aligned}
v_2(t, \delta) &= \int_0^\infty h'''(\eta) \cdot u(t - \eta) \cdot z(t - \eta - \delta) d\eta \\
&= \{h_1'(t) \cdot h_1'''(t - \delta)\}^2(\alpha^2 + k(1 - \alpha)^2) \cdot p^2
\end{aligned}
$$

Since responses of bipolar cells are almost same in the absolute values, the equation

$$\mid h_1'(t) \mid \simeq \mid h_1''(t) \mid$$

is assumed here. Then we define

$$V(t) = \frac{v_2(t, \delta)}{\{v_1(t, \delta)\}^2} \simeq \alpha^2 + k(1 - \alpha)^2$$

In case of $k \geq 1$, $V(t)$ shows monotonic increasing function according to the increasing of α $(0 < \alpha \leq 1)$. Then the movement is perceived from the left to the right by the increasing value of $V(t)$. We call here the first order kernel correlation function $v_1(t, \delta)$, and the second kernel correlation function $v_2(t, \delta)$.

4 Conclusion

We discussed an asymmetric neural network suggested from retinal network in catfish. Since the asymmetric network contains a cell with quadratic nature, we applied Wiener nonlinear analysis to them. To detect the movement in the neural network, we introduced kernel correlations, which will be made in the higher neural network.

References

[1] Barlow, H.B. and Levick, R.W.(1965) *The mechanism of directional selectivity in the rabbit's retina.* J. physiol. **173:** pp.377–407.

[2] Ishii, N.(1991) *Differentiation processing of linear and nonlinear information in retinal neural network.* Proc. IJCNN. **2**, Seattle, WA. pp.663–666.

[3] Ishii, N.,(1992a) *Motion detection by biological asymmetrical neural network.* Proc. IJCNN. **3**, Boltimore, MD. pp.390–395.

[4] Ishii, N.,(1992b) *Modified differentiation and nonlinear function in motion detection of neural network.* Proc. Int. Fuzzy Logic & Neural Networks, **2**, Iizuka, pp.867–870.

[5] Korenberg, M.J., Sakai, H.M. and Naka, K.-I.(1989) *Dissection of the neuron network in the catfish inner retina.* J. Neurophysiol. **61:** pp.1110–1120.

[6] Liaw, J.S. and Arbib, M.A.(1991) *A biologically inspired neural network model for 3-D motion detection.* Proc. IJCNN. **1:**, Seattle, WA. pp 661-665.

[7] Naka,K.-I., Sakai, H.M. and Ishii, N.(1988) *Generation and transformation of second order nonlinearity in catfish retina*, Annals of Biomedical Engineering. **16:** pp. 53–64.

[8] Reichardt, W.(1961) *Autocorrelation, a princeple for the evaluation of sensory information by the central nervous system.* Rosenblith, WA., Wiley, New York.

[9] Sakuranaga, M., Ando, Y., and Naka, K.-I.(1987) *Dynamics of the ganglion cell response in the cat fish and frog retinas*, J. Gen. Physiol. **90**; pp.229–259.

Analysis of the Golf Swing from Weight-Shift using Neural Networks

Ho Sub Yoon, Chang Seok Bae, Byung Woo Min
AI Division, Systems Engineering Research Institute / KIST
P.O.Box 1 YouSung Gu, Taejon 305-600 KOREA

Abstract

Weight-Shift of human body motion means the continuous change of weights which is loaded on left and right feet respectively. In this study, a neural network method is employed to identify golf swing from a continuous weight-shift wave form. We defined eight input features which can classify various shapes of swing pattern. The adopted network is a three-layered error back-propagation model consisting of eight input ten hidden and two output nodes. According to experimental results, the identifying success rate is 97.75% using 10 hidden nodes and 5000 epochs.

1. Introduction

Image analysis and pattern matching in AI technology have been studied to analyse motion. Various methods have been proposed to analyse human body's motion[1]. In this paper, we describe a method for identifying golf swing from weight-shift which arises from human body's motion. If we succeed in finding swing pattern, we can analyse and correct golf swing pose automatically. In golf swing motion, weight-shift loaded on left and right feet respectively and used as an input signal. Figure 1 shows hardware configuration of weight shift analyzing system.

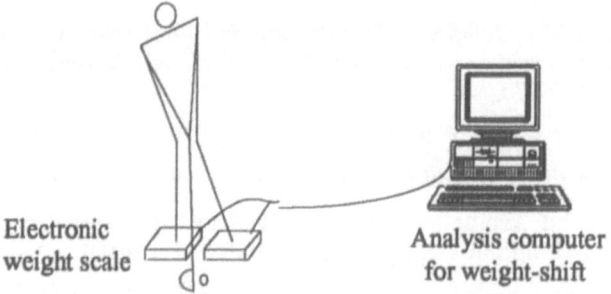

Electronic
weight scale

Analysis computer
for weight-shift

Figure 1. Hardware configuration of weight-shift analyzing system

By analyzing weight-shift, it is possible to analyse overall human body's motion. Input signal is composed of various patterns depending on personal traits. It is difficult to classify these various input patterns using rule-based approaches. Because the general conditions to classify various shapes of swing patterns do not exist and the input signal includes noise patterns which are similar to swing pattern. In order to remedy these problems, we describe the methods how to classify input patterns using neural network which is known as powerful means for solving pattern classification [2] and how to choose the input features which can classify input patterns efficiently.

2. Weight-Shift Signal

Weight-shift analysis contributes to correct swing pose of golfers or baseball players, etc. At first, weight-shift analysis was processed manually. It is a tedious operation. Therefore, this paper describes automatic analysis of weight-shift. We tried to test weight-shift of golf swing and detect swing pattern automatically. Figure 2 shows an input weight-shift signal representing weights loaded on the left and right feet, and measured by two electronic scale respectively. The signal is separately transferred through a serial port ofthe host computer at 12 data points per sec.

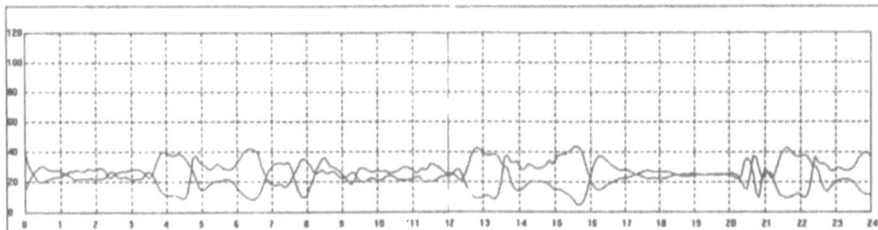

Figure 2. Input weight-shift signal

In general, golf swing is composed of nine steps. They are Address(Ad), Back Swing 1(B1), Back Swing 2(B2), Top(Tp), Down Swing 1(D1), Down Swing 2(D2), Impact(Im), Follow Thru(Ft) and Finish(Fn). Each step is mapped onto one point in the weight-shift signal. Weight-shift signal of one cycle standard golf swing is shown in Figure 3 (a).

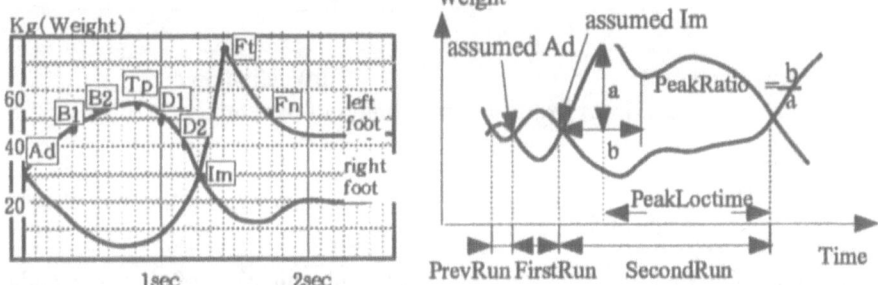

Figure 3. (a) Weight-shift signal of standard golf swing (b) Definitions for feature extraction

3. Selection of input features

This section illustrates our approach to the extracting and the selecting input features which describes a special property of pattern. The input vector is the most important factor for obtaining higher recognition rate. We analyse a lot of input signals, and find common properties which are occurred in majority people. Through many experiments, eight input features are selected. For the sake of input features selection, we divide the continuous input signal into candidate swing signal. In the first step, we find the cross point of weights where the left weight is increasing and the right weights id decreasing. It is assumed to be the impact point. Next, the cross point in front of impact point is found. This point is assumed to be the address point. If two cross points are detected, we are able to extract swing signal or noise signal with two cross points. In order to search input features, we define 'PrevRun', 'FirstRun',

116

'Second Run', 'PeakLoctime', 'PeakRatio' as in Figure 3 (b).   We defines function V
() that extract 8 input features, each item having a real value from 0.0 to 1.0.

Definition 1: V(PrevRun)
  Function V(PrevRun) means the time of a range from cross point before
  assumed address point to address point as in figure 4 (a).
Definition 2 : V(FirstRun)
  Function V(FirstRun) means the time of a range from assumed address point
  to assumed impact point as in figure 4 (b).
Definition 3 : V(SecondRun)
  Function V(FirstRun) means the time of a range from assumed impact point to
  next cross point as in figure 4 (c).
Definition 4 : V(PeakRatio)
  Function V(PeakRatio) means the ratio of width to height of the first peak point
  in the SecondRun as in figure 4 (d).

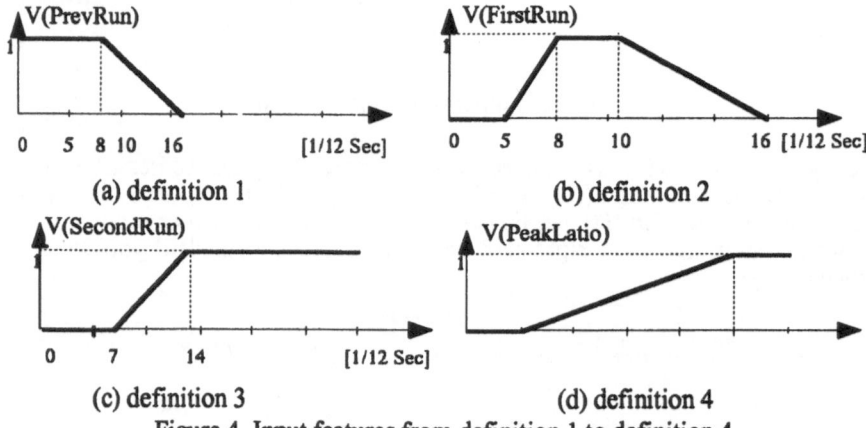

(a) definition 1   (b) definition 2

(c) definition 3   (d) definition 4
Figure 4. Input features from definition 1 to definition 4

Definition 5 : V(PrevRatio)
  Function V(PrevRatio) means the ratio of the minimum value of the right weight
  to the maximum value of the left weight.
Definition 6 : V(FirstRatio)
  Function V(FirstRatio) means the ratio of the minimum value of the left weight
  to the maximum value of the right weight.
Definition 7 : V(SecondRatio)
  Function V(SecondRatio) means the ratio of the minimum value of the right
  weight to the maximum value of the left weight in SecondRun.
Definition 8 : V(PeakLoc)
  Function V(PeakLoc) means the ratio of PeakLoctime to the length of SecondRun.

4. Application using neural network

An error back-propagation neural network is employed to solve this problem because
of its better ability to classify pattern than the others[3,4,5]. The three-layered error
back-propagation neural network has an input layer, a hidden layer, and an output
layer. We constitute 8 input nodes and 2 output nodes. In order to the find optimal
hidden nodes count and learning epochs count, many tests are performed. The
learning rate and moment were set to 0.1 and 0.9. Initial weights which connect node
to node were set to 0.0 ~ 1.0 by random generator. Training data which have 178

swing regions and 422 non swing regions were obtained from 60 golfers. Test data containing 135 swing regions and 386 non swing regions were extracted from 50 golfers. Table 1 and table 2 show recognition rate according to various hidden node counts and learning epochs. It shows good result of 99.44% when hidden node is set to 11 and epoch is set to 6000 in table 1, but it shows lower result of 91.85% under the same environment in table 2. These results mean that learned weight fall into memorization. The other way, it has good result of 100% when hidden nodes is set to 6 and epochs is set to 7000 in table 2, but it is shows not lower result of 92.13% under the same environment in table 1. These results mean that learned weight is fall into generization. In optimal case of 10 hidden nodes and 5000 epochs, the recognition rate of 97.75% is shown in Table 1 and Table 2.

Table 1. Recognition result from training data (a unit : %)

	3000	4000	5000	6000	7000
4	96.62	94.94	94.94	97.19	97.19
5	97.75	97.75	97.75	97.75	97.75
6	98.31	96.63	96.63	96.63	**92.13**
7	97.75	96.63	96.63	93.82	92.13
8	97.75	96.63	96.63	97.19	95.51
9	97.19	96.63	98.31	98.31	97.19
10	97.19	97.75	97.75	97.75	97.75
11	94.94	94.94	98.44	**99.44**	98.31
12	98.31	98.31	97.75	96.06	96.06

Table 2. Recognition result from test data (a unit : %)

	3000	4000	5000	6000	7000
4	91.85	94.81	95.59	97.59	91.11
5	92.59	95.55	95.55	92.59	92.59
6	92.59	96.29	97.03	97.03	**100.0**
7	95.55	95.55	95.55	96.29	97.03
8	95.55	95.55	94.07	94.07	96.29
9	95.55	94.81	93.33	93.33	96.29
10	96.29	94.81	97.77	97.77	97.77
11	94.81	94.81	92.59	**91.85**	93.33
12	94.81	94.81	95.55	94.07	94.07

5. Conclusion

Weight-shift analysis contributes to correct swing pose of golfer or baseball player. At first, weight-shift analysis was processed manually. We tried to test weight-shift of golf swing and this paper describes automatic analysis of weight-shift. We presented identifying method of golf swing from weight shift signal using neural networks. We defined 8 input features and implemented error back-propagation 3 layered neural networks. As results of repeating experiments, we achieved the identifying success rate of 97.75 % by using 10 hidden nodes and 5000 epochs. According to the experimental results, it is ascertained that weight-shift analysis problem which is difficult to solve by using rule-based approach can be solved easily by using neural network. As future works, continuous researches for improving learning speed and recognition rate are needed. Moreover, a syntactic analysis including image information as well as weight-shift is needed also.

References

[1] E. Charniak and D. McDermott, Introduction to Artificial Intelligence, Addison-Wesley, 1985.
[2] Jacek M. Zurada, Artificial Neural Systems, West Publishing Company, 1992.
[3] Jones, William P. and Josian Hoskins. "Back-Propagation." Byte, vol.12, no. 11, October 1987):155-62.
[4] King, Todd. "Using Neural Networks for Pattern Recognition." Dr.Dobbs's Journal, vol. 14, no. 1, 1989.
[5] Marilyn McCord Nelson and W.T.Illigworth, A Practical Guide to Neural Nets, Addison-Wesley, 1992.

Dry Electrophysiology:
An approach to the internal representation of brain functions through artificial neural networks

Shiro USUI and Shigeki NAKAUCHI

Department of Information and Computer Sciences,
Toyohashi University of Technology
1-1 Hibarigaoka Tempaku Toyohashi 441, Japan

1 Introduction

In this article, we attempt to develop an approach called *Dry Electrophysiology* (DEP) that aims to understand the internal representation of brain functions. DEP adopts a combined strategy of top-down and bottom-up analysis via neural networks; a neural network is trained to perform a certain computational task of particular part of brain according to an optimality criterion, and then, components of the neural network are probed and the emerging internal representation is compared with the *wet* experimental data from real neurons or brain functions.

This article reviews several considerations of the computational principles in the neural computation and the hardware constraints in the neural systems, and investigates the internal representation acquired by supervised and unsupervised learning models. We demonstrate that neural networks can be expected to do more than perform given tasks. They can also provide an insight into *what* real neural systems are doing and *why* they do so, by analyzing *how* they create the internal representation to realize a particular computational task.

2 What is Dry Electrophysiology ?

Two strategies exist for understanding brain function; top-down and bottom-up. Both of these approaches take the same stance on basic concepts concerning brain, that is, individual neurons are regarded as functional elements of brain and the brain function is considered to be a realization of the assembly of neurons. Nevertheless, they stand on different levels of description. The bottom-up approach takes the sigle-unit recordings as a starting point to characterize individual neurons. The top-down approach, on the other hand, starts by putting an entire system into one black box model rather than focusing on behavior of individual neurons.

Due to the differences of their view points, it seems that these two approaches conflict. The bottom-up side points out that top-down approach, using only black box models, does not have the ability to describe "how real neurons behave". Conversely, the complaint about the bottom-up approach is its inability to know "how the information processing is taken place" using single-unit recordings. Both of these criticisms are partly correct. However, the most crucial but difficult issue concerning brain function is the relation between properties of individual neurons and overall brain functions, in other words, the problem of how the individual neurons interact and constitute a neural circuit which realizes a certain function.

We present an alternative approach called *Dry Electrophysiology* that attempts to investigate the emerging internal representation in the *artificial* neural networks trained to realize a certain computational task according to an optimality criterion. To do this, DEP focuses on several virtues of the neural networks; we can embed a certain computational principle or task, inspired by the top-down view, into supervised or unsupervised learning procedures. Moreover, behavior of *real* neurons or hardware constraints, derived by the bottom-up view, can be represented as the assembly of functional elements connected in parallel structure of the neural network. These virtues allow the neural networks to not only realize a certain function but also *disassemble* it; analysing the trained artificial neural networks and probing the emerging internal representation can form a bridge between brain function and behavior of individual neurons and may tell us how the brain functions, and how the computational principles and the hardware constraints relate to each other.

3   Crucial clues about the brain function

3.1   Computational principles guiding neural computation in the brain

Several studies have made attempts to ascertain the computational principles in neural information processing. One of these attempts is concerned with the *efficiency of the information representation* as the optimality criterion. Since natural stimuli from the animal's environment are typically redundant, one can argue that the neural system is well adapted to processing highly redundant signals and is geared to building efficient information representations based on the expectation of this redundancy, this clearly would has several evolutional advantages (Field, 1987).

4 Neural networks, learning procedures and examples of DEP

As mentioned in section 2, neither computational principles nor experimental data suffice for understanding brain functions. We thus were seeking the computational models with features of (1) a parallel architecture like the brain's and (2) the ability to learn according to a certain optimality criterion, rather than the black box models or the single-neuron models. Artificial neural networks on which DEP foucuses posses such features. In this section, we review various approaches to the understanding of brain functions through neural networks with supervised and unsupervised leaning abilities.

4.1 Supervised learning and DEP

The most widely used supervised learning method for multilayer neural networks is the back-propagation (BP) algorithm. BP works according to the criterion of minimizing the error between the model outputs and the external (teaching) signals. BP networks can provide a powerful tool for generating a nonlinear mapping and have been successfully applied to what are essentially engineering problems, such as an speech recognition or handwritten character recognition. Rather than such engineering goals, we here emphasize that this ability of BP networks allow us to model the system function of particular part of brain. A more interesting feature of the BP network is the activation patterns of hidden units; hidden units which, in some functional respects, behave like real neurons, arise in the hidden layer.

Several authors have reported on the internal representations in the hidden layers of the neural networks trained to perform certain computational tasks; shape-from-shading (Lehky and Sejnowski, 1988), caliculating the target position from the retinal image and eye position (Zipser and Anderson, 1988), binocular depth perception (Pouget and Sejnowski, 1990) and color discrimination (Usui, Nakauchi and Nakano, 1992; Usui and Nakauchi, 1993).

Although it is unlikely that the brain employs a back-propagation learning mechanism, the learning rule itself is not important; the BP learning procedure is just one of many methods for minimizing the error. An essential point of DEP with a supervised leaning procedure is that the internal representation which reflects characteristics of the given computational task arises automatically rather than being designed by hand. Thus, such learning procedures can provide valuable insights into the relation between the computational task and possibly provide clues to the internal representation of brain function.

4.2 Unsupervised learning and DEP

It has been shown that some of the optimal criteria mentioned above can be embeded in a Hebbian unsupervised leaning procedure. Barlow and Földiák (1989) showed that minimum redundancy principle can be instantiated by a decorrelating network with anti-Hebbian learning. Various Hebbian learning algorithms for PCA have also been proposed (Oja, 1989).

Recently progress has been made in constructing unsupervised learning models of self-organization in the early stage of the brain, based on information-theoretical considerations. Linsker (1988) proposed a neural network model based on the Infomax principle for self-organization of receptive fields and the cortical maps. Usui et al. (1993) showed that certain kinds of color receptive fields found in the visual system are closely related to improving redundacy of spatio-chromatic signals from natural environment from the Infomax perspective. Another application of unsupervised learning, shown by Usui, Nakauchi and Miyamoto (1992), is the application of a neural network with anti-Hebbian learning, based on the minimum redundancy principle, which successfully explains a possible mechanism of color constancy.

Unlike the supervised learning procedure, unsupervised learning does not requires external signals. Therefore, DEP with unsupervised learning may provide an effective method for investigating the relation between the information representation and the task-independent criteria with respect to the statistical properties of stimulus environment.

5 Conclusions

In this paper, a novel approach, *Dry Electrophysiology*, to the understanding brain function has been addressed. DEP has one foot in computational principles/tasks guiding neural information processing, and an another foot in considerations of the hardware constraints in the neural system. That is, DEP regards supervised and unsupervised learning procedures as powerful tools that instantiate computational tasks/principles, and artificial neural networks with parallel functional elements as appropriate computational models, based on considerations of hardware constraints in the neural system. We argue that probing the trained neural network to display its internal representation is a potent approach to obtaining clues that may elucidate brain functions.

References

Atick, J.J. and Redlich, A.N. (1990) Towords a theory of early visual processing, Neural Computation, 2, 308–320

Barlow, H.B. (1989) Unsupervised learning, Neural Computation, 1, 245–311

Barlow, H.B. and Földiák, P.(1989) Adaptation and decorrelation in the cortex, In: The Computing Neuron, chap.4, 54–72, Addison-Wesley Pub.

Barlow, H.B., Kaushal, T.P., Hawken, M. and Parker, A.J. (1987) Human contrast discrimination and the threshold of cortical neurons, J.Opt.Soc.Am.A, 4, 12, 2366–2371

Churchland, P.S. and Sejnowski, T.J. (1992) The Computational Brain, MIT Press

Field, D.J. (1987) Relations between the statistics of natural images and the response properties of cortical cells, J.Opt.Soc.Am.A, 4, 12, 2379–2394

Lehky, S.R. and Sejnowski, T.J. (1988) Network model of shape-from-shading: neural function arises from both receptive and projective fields, Nature, 332, 2, 452–454

Linsker, R. (1988) Self-organization in a perceptual network, Computer, 21, 105–117

Oja, E. (1989) Neural Networks, principal components and subspaces, Int.J. Neural Systems, 1, 1, 61–68

Pouget, A. and Sejnowski, T.J. (1990) Neural models of binocular depth perception, Cold Spring Harbar Symposia on Quantitative Biology, 55, 765–777

Usui, S. and Nakauchi, S. (1993) Color opponency as the internal representation acquired by a three-layered neural network model, Proc. of ICNN (San Francisco), IEEE, 3, 1327–1332

Usui, S., Nakauchi, S. and Miyamoto, Y. (1992) A neural network model for color constancy based on the minimally redundant color representation, Proc. of IJCNN (Beijin), 1, 696–701

Usui, S., Nakauchi, S. and Nakano, M. (1992) Reconstruction of Munsell color space by a five-layered neural network, J.Opt.Soc.Am.A, 9, 4, 516–520

Usui, S., Nakauchi, S. and Takahashi, K. (1993) Self-organization of color receptive fields using random color noise images, Proc. of WCNN (Portland), 1, 72–75

Zipser, D. and Anderson, R.A. (1988) A back-propagation network that simulates response properties of a subset of posterior parietal neurons, Nature, 331, 25, 679–684

ANNs and MAMFs: Transparency or Opacity?

Lawrence W. Stark

Neurology and Telerobotic Units

Bioengineering Graduate Group

University of California at Berkeley

481 Minor Hall

Berkeley, California 94720

1 Introduction

The goals of artificial neural net, ANN, research are several. The very name bespeaks

i) A biological aim. There are a number of examples of straight-forward models of biological systems generated from knowledge about a system's neural elements and the behavior controlled by that neural system (or speculation regarding such behavior). Outstanding neurophysiological research is the Lettvin-McCulloch (1959) paper "What the frog's eye tells the frog's brain." This was, of course, based upon the McCulloch-Pitts formal neuron theory that initiated the neural network concept (Fig. 1, left)(McCulloch, 1945; McCulloch and Pitts, 1943). John von Neumann, after meeting McCulloch (on the train station at Princeton Junction), used such neurons to depict the logical design of ALU operators of the EDVAC (Fig. 1, right)(Von Neumann, 1945).

ii) Bionics or biomimetics suggest that we can learn about a biological system that works (and biologists suppose that these systems work optimally) in order to make an engineering copy of this machine. Perhaps, we may even develop better machines (e.g., from silicon) that will carry out the principles of the biological machine (from carbon).

Reverse engineering is related to the above, and in this context has a biological science goal. We wish to understand the workings of a natural biomachine designed by evolution to carry out a particular function, such as color vision. We first study an artificial network that has been trained to carry out this same function. We discover the actions of its artificial neural elements, especially the 'hidden units.' Thus, we may obtain insights as to how the natural net performs its role. Indeed, Usui and his group at Toyohasi University, studying color vision and opponens cells, and Anderson and Zipser studying ANN models for parietal lobe neurons, have carried out pioneering work in these areas.

iii) An engineering goal for ANNs is to make machines that work, and work well — that is: accurately, predictably, and robustly under changing conditions. A main advantage of the neural net as a machine is that it programs itself, and thus is designed by experience. The Perceptron of Rosenblatt is the outstanding and pioneering example of trainable neural nets functioning as pattern recognition devices (Minsky, 1954; Rosenblatt, 1962; Widrow, 1962). Note that early neural nets were supervised and trained; they were also opaque and seemed to have a mysterious intelligence.

124

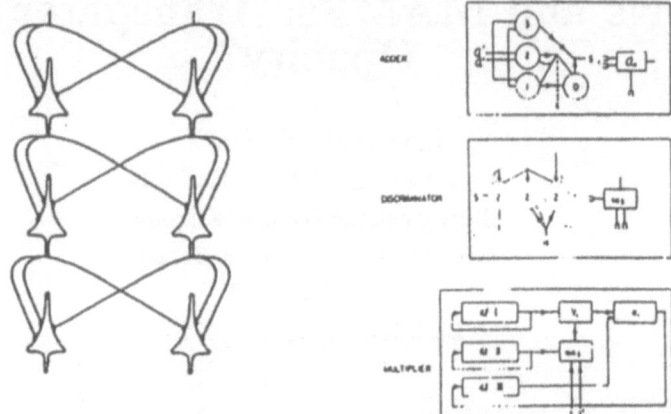

Figure 1 McCulloch-Pitts Formal Neurons
   A small net of formal neurons (left) created by McCulloch and Pitts (1943); these were as "poverty stricken" so as to clearly illustrate the encompassing ability of a brain composed of such neurons. John Von Neumann (1945), after meeting McCulloch (on the train station at Princeton Junction), used such neurons to depict the logical design of ALU operators of the EDVAC (right).

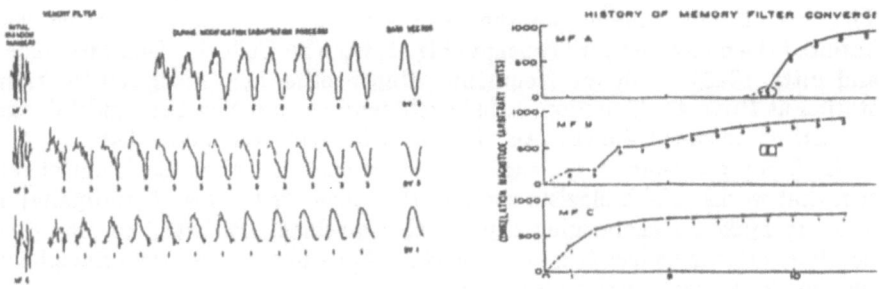

Figure 2 Multiple Adaptive Matched Filters; MAMFs
   Above, MFs as they adapt; below, history of MF convergence. Original noisy library matched filters, MF A, B, C... (left upper); data vectors, DV 1, 2, 3... (right upper) produce a random sequence of events from successive data vector presentations. Adaptation or matching of original noisy library MFs at times proceeds smoothly (lower panels, both above and below) until with modification of coefficients to data events MF C approximate DV 1. Ordinate below is cross-correlation amplitude of MFs to DVs, abscissae are successive instances of filter modification.
   At times, several DVs are accepted by a MF especially early in adaptation process (middle panels, both above and below); finally, although DV 2 is maximally matched to MF B (see box) and according to first order winner-take-all algorithm should be incorporated into MF B, its cross-correlation coefficient is below a rising threshold (not shown). A new library filter MF C is pressed into service (upper panel both above and below) [in another experiment, this could have been a daughter filter (see references to genetic algorithms in text)]. Now MF B and MF C each develop smoothly, showing convergent adaptation and thus separate DV 2 and DV 3. (From Stark, Okajima and Whipple, 1962).

Another endeavor, also occurring in the '50s and '60s, employed multiple adaptive matched filters, MAMFs, to set up a scheme of adaptive coefficients for pattern classification. Contrary to the ANNs, the MAMF approach was an example of unsupervised learning. Changes in the coefficient patterns were observable and understandable; indeed, the transparency of MAMFs disguised their equivalence to ANNs for several decades.

iv) The aim of this paper is to document how the transparency of MAMFs can illustrate their learning behavior. Since MAMFs are mathematically equivalent to ANNs, an important question is raised as to why we have been content with the opacity and mysterious intelligence of ANNs.

2 MAMF, Multiple Adaptive Matched Filter Scheme

An early attempt in this direction was carried out by a 1960 MIT group (Stark, 1961; Stark et al, 1962 and 1965; Yasui et al, 1964). The MAMF approach to ECG diagnosis was an example of unsupervised learning. This MAMF scheme, equivalent to a neural net, is described in the figure legends explaining the remaining figures.

The MAMF scheme is of Eliashberg-Chomsky, E-C, Order 0 (Chomsky, 1956; Eliashberg, 1981, 1989, and 1990). (Order 0 in the E-C system is equivalent to a Turing machine with random access memory. E-C 1 is equivalent to a context-free grammar and does not have an analog in computational devices or algorithms. E-C 2 is a device with only stack memory. E-C 3 is a finite state machine that has associative feedback. E-C 4 is a combinatorial machine with only a lookup table memory, and is realized in a programmable logic array, PLA.)

The MAMF illustrates self-organization or learning without a teacher. In its adaptive phase it develops (a) changes of synaptic weights or equivalent filter coefficients (winner take all tactics), (b) changes of thresholds that protect against destructive modification, (c) spontaneous generation of new filters, either with random coefficients or as daughter filters (related to later development of genetic algorithms) and (d) initial preprocessing, such as time and amplitude warping. Adaptive filters for signal extraction from noise are to be contrasted with this scheme for signal extraction from a background of many similar signals. It was tested for feasibility on ECG and earlier on EEG data. According to Nilsson (Nilsson, 1965), it was the first piece-wise linear categorizer with the mode-seeking training rule. Grossberg and his school have redeveloped this scheme and used it to justify ART theories (Grossberg, 1982).

Note how with the MAMF scheme the convergence of the weights can be documented and understood. Also, since there are only two layers, the winner-take-all algorithm operates without the complexity of back propagation from layer to layer to layer. Feature space can be mapped metaphorically onto two dimensions, as in the figure. (See Kohonen for recent developments in this area.)

126

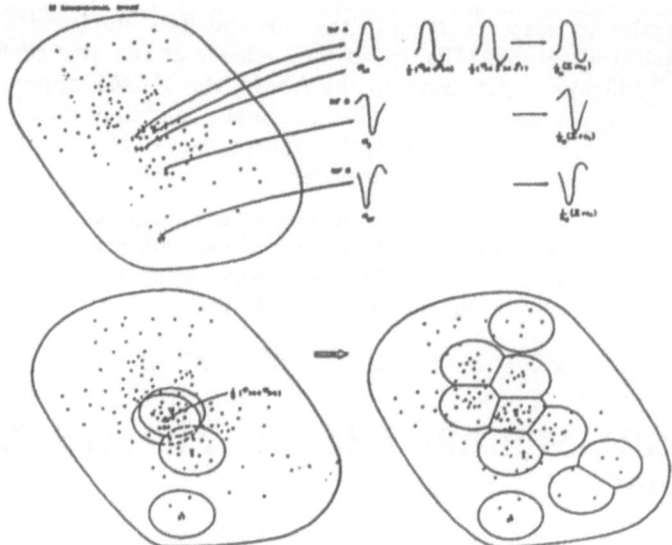

Figure 3 Feature Space View of Multiple Adaptive Matched Filters, MAMFs
As the matched filters adapt (upper right), they come to represent clusters of similar events in feature space (upper left and middle left). After adaptation they fill event space and are separated by hyperplanes (middle right).

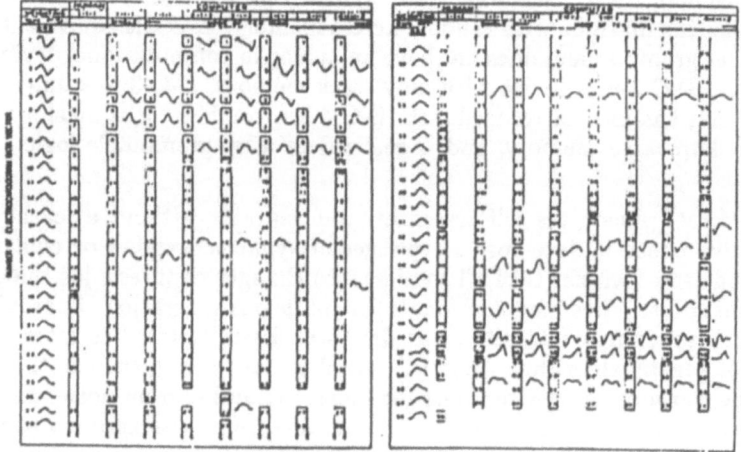

Figure 4 Classification by Humans and Computer
Data vectors are 55 ECG QRS patterns presented in random and repetitive sequences. After self-organization by MAMF about a dozen classes were obtained (upper panels). Amplitude weighting (see column labels) influenced the classification obtained as did random order of presentation. For example, divergences (D, ordinate) from the modal classification of ten runs (T, abscissa) varied from 20 to 3 (lower left). Divergences in classifying a pattern were almost always to a neighboring cluster, so MAMF errors were similar to human ones.

3 Engineering Goals To Reverse Engineering

The MAMF scheme was driven by an engineering goal, pattern classification by a self-organizing, unsupervised set of engineering coefficients. It is transparent and thus easily open to reverse engineering approaches. Since it is identical with ANNs, this poses a challenge for us to use reverse engineering on ANNs in order to study them and perhaps make them more effective. Especially now that ANNs are used to define membership functions and also premise-consequence rules in a number of fuzzy control systems, Takagi, working in Zadeh's group, has initiated such studies.

4 Appendix A: A Mechanical Metaphor

Within a simple engineering goal framework, neural nets might be considered as an engineering black box whose unknown design is considered satisfactory if its performance is satisfactory. An exaggerated mechanical metaphor may be appropriate: — suppose we have a network of mechanical elements under the hood of an automobile. This mechanical net could be rewarded for forward motion when the accelerator pedal is depressed, and punished for backward motion or absence of forward motion. With some scheme to connect all these mechanical elements in all possible configurations and then to solidify connections that are active in a successful trial, and to reduce connections during an unsuccessful trial, this mechanical net "should" eventually make a satisfactory car engine. What a nice automatic way to accomplish mechanical design! An example of this in the neural net field is Perceptron — either the old perceptron from the 1950's or 1960's, or new perceptrons from the 1980's or 1990's. These now perform successful pattern recognition, and are being further developed to provide membership functions for fuzzy controllers.

For the reverse engineering paradigm, however, we must also LOOK UNDER THE HOOD and examine the actual mechanical design of the machine created. "Aha! It is a heat engine of type xyz." Reverse engineering to understand specific brain functions has at least two solid, recent and successful contributions by Andersen and by Usui. Here, nets have been created as models of specific parts of the nervous system. These nets were rewarded and punished for correct and incorrect behavior until they performed satisfactorily, thus achieving engineering goals. These two researchers then proceeded further with the reverse engineering paradigm. They explored the new particular mechanisms in the neural nets — neurons, synapses, weights and connections — to obtain clues to how neurons of particular parts of the nervous system do in fact function. There remains a logical problem because the uniqueness of the design solution is not guaranteed; still, it is a potent approach.

5 References

- R.A. Anderson, G.K. Essick, and R.M. Siegel (1985), "Encoding of spatial location by posterior parietal neurons," *Science*, vol. 230, pp. 456-458.

- N. Chomsky (1956), "Three models for the description of language," *I.R.E. Transactions on Information Theory*, vol. IT-2 pp. 113-124.

- V. Eliashberg (1981), "The concept of E-machine: On brain hardware and the algorithms of thinking," *Proceedings of the Third Annual Meetings of the Cognitive Science Society*, pp. 289-291.

- V. Eliashberg (1989), "Context-sensitive associative memory: 'Residual excitation' in neural networks as the mechanism of STM and mental set," *Proceedings of IJCNN-89*, Washington, D.C.

- V. Eliashberg (1990), "Universal Learning Neurocomputers," *Proceedings of the Fourth Annual Parallel Processing Symposium*, April 1990.

- S. Goodman, and R. Andersen (1989), "Microstimulation of a neural-network model for visually guided saccades," *Journal of Cognitive Neuroscience*, vol. 1, No. 4, pp. 317-326.

- S. Grossberg (1982), *Studies of Mind and Brain*, Boston: Reidel Press.

- J.Y. Lettvin and H. Maturana, W. McCulloch, and W.H. Pitts (1959), "What the frog's eye tells the frog's brain," *Proceedings of the I.R.E.*, vol. 47, No. 11, pp. 1940-1959.

- W. McCulloch (1945), "A Heterarchy of values determined by the topology of nervous nets," *Bulletin of Mathematical Biophysics*, vol. 7 pp. 89-93.

- W. McCulloch, and W. Pitts (1943), "A logical calculus of the ideas immanent in nervous activity," *Bulletin of Mathematical Biophysics*, vol. 5 pp. 115-133.

- M. Minsky (1954), "Neural nets and the brain-model problem," Unpublished doctoral dissertation, Princeton University.

- S. Nakauchi, S. Usui, and S. Miyake (1990), "A three-layered neural network model which simulates color-opponent processing." *Proceedings of the International Conference on Fuzzy Logic and Neural Networks*, pp. 481-484.

- N.J. Nilsson (1965), *Learning Machines: Foundations of trainable pattern-classifying systems*, pp. 125-126, California: McGraw-Hill Inc.

- M. Okajima, L. Stark, G. Whipple, and S. Yasui (1963), "Computer recognition techniques; Some results with real electrocardiographic data," *IEEE Transactions on Biomedical Electronics, BME*, vol. 10 pp. 106-114.

- F. Rosenblatt (1962), *Principles of Neurodynamics*, Washington, D.C.: Spartan Books.

- D. Sankoff and J.B. Kruskal, editors (1983) *Time Warps, String Edits, and Macromolecules: The Theory and Practice of Sequence Comparison*, Massachusetts: Addison-Wesley Inc.

- L. Stark (1961), "Pattern recognition for electroencephalographic diagnosis," *Quart. Prog. Report No. 61, Research lab. of Electronics, M.I.T.*, pp. 215-219.

- L. Stark, M. Okajima, and G.H. Whipple (1962), "Computer pattern recognition techniques: Electrocardiographic diagnosis," *Communications of the Association for Computing Machinery*, vol. 5 pp. 527-532.

- L. Stark, and J. Dickson (1965), "Remote computerized medical diagnostic systems," *Computers and Automation*, vol. 14 pp. 18-21.

- S. Usui, S. Nakauchi, and S. Miyake (1993), "Acquisition of the color-opponent representation by a three-layered neural network model," *Proceedings of IEEE Conference.*

- J. Von Neumann (1945), "First draft of a report on the EDVAC," *Contract Report, Moore School of Electrical Engineering, University of Pennsylvania*, June 30, 1945.

- B. Widrow (1962), "Generalization and information storage in networks of Adaline neurons," in *Self-organizing systems, Washington, D.C.: Spartan Books*; Cambridge, MA: MIT Press.

- S. Yasui, G. Whipple, and L. Stark (1964), "Comparison of human and computer electrocardiographic waveform classification and identification," *American Heart Journal*, vol. 68 pp. 236-242.

- D. Zipser, and R.A. Andersen (1988), "A back-propogation programmed network that simulates response properties of a subset of posterior parietal neurons," *Nature*, vol. 331 pp. 697-684, 1988.

COLLECTIVE BRAIN AS DYNAMICAL SYSTEM

Michail Zak

Center for Microelectronics Technology
Jet Propulsion Laboratory
California Institute of Technology
Pasadena, CA 91109

The concept of the collective brain has appeared recently as a subject of intensive scientific discussions from theological, biological, ecological, social, and mathematical viewpoints. It can be introduced as a set of simple units of intelligence (say, neurons) which can communicate by exchange of information without explicit global control. The objectives of each unit may be partly compatible and partly contradictory, i.e., the units can cooperate or compete. The exchanging information may be at times inconsistent, often imperfect, non-deterministic, and delayed. Nevertheless, observations of working insect colonies, social systems, and scientific communities suggest that such collectives of single units appear to be very successful in achieving global objectives, as well as in learning, memorizing, generalizing and predicting, due to their flexibility, adaptability to environmental changes, and creativity.

In this note collective activities of a set of units of intelligence will be represented by a dynamical system which imposes upon its variables different types of non-rigid constraints such as probabilistic correlations via the joint density. It is reasonable to assume that these probabilistic correlations are learned during a long-term period of performing collective tasks. Due to such correlations, each unit can predict (at least, in terms of expectations) the values of parameters characterizing the activities of its neighbors if the direct exchange of information is not available. Therefore, a set of units of intelligence possessing a "knowledge base" in the form of joint density function, is capable of performing collective purposeful tasks in the course of which the lack of information about current states of units is compensated by the predicted values characterizing these states. This means that actually in the collective brain global control is replaced by the probabilistic correlations between the units stored in the joint density functions.

Since classical dynamics can offer only fully deterministic constraints between the variables. we will turn to its terminal version introduced by Zak, M., (1989a, b, 1990a, b. 1991a. b. 1992). The main departure from classical dynamics here is in violations of the Lipschitz condition at equilibrium points. Because of that, terminal neurodynamics can "generate" randomness without any random inputs. As shown by Zak, M. (1991a, 1993), random motions can be prescribed and stored in a form of a new type of attractor - stochastic attractor - represented by a stationary stochastic process with prescribed probability distribution. Based upon this phenomenon as a paradigm, we will develop a dynamical system whose solutions are stochastic processes with prescribed joint density. Such a dynamical system introduces more sophisticated relationships between its variables which resemble those in biological or social systems, and it can represent a mathematical model for the knowledge base of the collective brain. One of the most remarkable properties of this model is that the joint density evolution is described by the Fokker-Planck equation whose diffusion and drift coefficients are uniquely defined by fully deterministic synaptic interconnections of the original dynamical system. In addition to that, one can introduce a learning device which compares the predicted and the available values of the corresponding variables, and, based upon this comparison, readjust the synaptic interconnections of the dynamical system, and therefore, the parameters of the associated Fokker-Planck equation.

A mathematical framework for such a model of collective brain is the following.

Based upon terminal version of Newtonian dynamics [10], one can introduce the following neural net:

$$\dot{x}_i = \gamma_i \sin^{1/3} \frac{\sqrt{\omega}}{\alpha} \phi_i(y_i)] \sin \omega t, i = 1, 2, \cdots n \tag{1}$$

where γ_i, α, ω are constant,

$$y_i = \sum_{j=1}^{n} T_{ij} x_j, \ T_{ij} = \text{const} \quad \frac{d\phi_i}{dy_i} = \begin{cases} > 0 & \text{for } |y_1| < N_i \\ = 0 & \text{for } |y_i| > N_i \end{cases} \ N_i < \infty \tag{2}$$

and T_{ij} form a symmetric positive-definite matrix.

The solution to Eq. (1) is non-deterministic: it splits into two branches at each (terminal) equilibrium point representing an n-dimensional random walk. The joint probability density f of this solution (for $\omega \to \infty$) satisfies the n-dimensional Fokker-Planck equation, and it relaxes to a stochastic attractor:

$$f(x_1, \cdots x_n) = \Pi_{i=1}^{n} p_i'(y_i) \cdot \det |T_{ij}|, p' = \frac{dp}{dy} \tag{3}$$

where y_i is expressed via x_i by Eq. (2). As follows from (3), the probabilistic property of the solution to (1) is uniquely defined by fully deterministic synaptic interconnections T_{ij} which can be adapted by the neural net (1) by learning in the course of performing a certain class of collective tasks. Based upon the solution (3), each neuron can predict values of the rest neurons if the direct exchange of information between the neurons is not available. Due to this property, the terminal neural nets represent a dynamical paradigm for decentralized control which can be applied to biological, economical, and social systems.

Since the main property of the model of collective brain is its ability to perform in a more "human" way when rigid rules are replaced by multi-choice ones, the problems with fuzzy objectives constitute the best "match" for this model whose loose coupling is compensated by a high degree of universality. Fussy objectives given in the form of inequalities can be introduced in Eq. (1) by assuming that

$$\omega = \begin{cases} 0 & \text{if } \theta < 0 \\ \omega_0 & \text{if } \geq 0 \end{cases} \qquad \omega_0 > 0 \tag{4}$$

in which thr function $\theta(x_i, \cdots x_n)$ plays the role of a discrimination surface and implements the simplest fuzzy objective in terms of "yes" or "no".

Thus, the dynamical model of collective brain is represented by a system of ordinary differential equations (1) with terminal attractors and repellers, and it does not contain and "man-made" digital devices. That is why this model can be implemented only be analog elements. The last property allows us to assume that (at least, phenomenologically) the proposed dynamical architecture can simulate not only ecological and social systems, but also a single brain as a set of neurons performing collective tasks where the global coordination is combined with learned correlations between neurons.

REFERENCES

1. Huberman, B., (1989), "The Collective Brain", Int. J. of Neural Systems, Vol. 1, No. 1, 41-45.

2. Seeley, T., and Levien, R., "A Colony of Mind", The Sciences, July, 1988, 39-42.

3. Zak, M., (1989a), "Terminal Attractors in Neural Networks", Neural Network, Vol 2, No. 3.

4. Zak, M., (1989b), "Spontaneously Activated Systems in Neurodynamics", Complex

Systems No. 3, pp. 471-492.

5. Zak. M., (1990a), "Weakly Connected Neural Nets", Appl. Math. Letters, Vol. 3, No. 3.

6. Zak. M., (1990b), "Creative Dynamics Approach to Neural Intelligence", Biological Cybernetics, Vol 64, No. 1, pp. 15-23.

7. Zak. M., (1991a), "Terminal Chaos for Information Processing in Neurodynamics", Biological Cybernetics, 64, pp. 343- 351.

8. Zak. M., (1991b), "An Unpredictable Dynamics Approach to Neural Intelligence", IEEE, Expert, August, pp. 4-10.

9. Zak, M., (1992), " To the Problem of Irreversibility in Newtonian Dynamics", Int. J. of Theoretical Physics, No. 2.

10. Zak, M., (1993), "Terminal Model of Newtonian Dynamics", In. J. of Theoretical Physics, No. 1.

Temporal Pattern Dependent Spatial-Distribution of LTP in the Hippocampal CA1 Area Studied by an Optical Imaging Method

Minoru Tsukada, Takeshi Aihara and Makoto Mizuno

Department of Information-Communication Engineering, Tamagawa University, Machida, Tokyo, 194 Japan

Abstract

Long-term potentiation (LTP) in the CA1 area of hippocampus are highly sensitive to the higher order statistical characteristics of the stimulus (correlation between successive pairs of inter-stimulus intervals) (Tsukada 1991, 1992, 1993, Aihara 1991). The temporal-pattern sensitivity in LTP was identified in a slice preparation by using the optical imaging method. In this experiment, we found that spatial pattern of LTP in CA1 area differed depending on the temporal pattern of stimulus; positively correlated sequences were much more effective (large area) in producing LTP, while negatively correlated sequences were ineffective (small area). In addition, the spatial pattern of LTP was closely related to the position where the repetitive firing of population spikes, involving dynamic formation of LTP through the activities of NMDA channels, was evoked during the period of temporal-pattern stimuli. These results suggest that there is a transformation from the temporal pattern into the spatial pattern in coding process of the hippocampal learning system, and that hippocampus uses a temporal code as an index.

Introduction

Since LTP was first reported when the electrical high frequency stimulation "tetanus" of nerve fibres was applied (Bliss, 1973) many investigator used high frequency stimulation to induce and estimate LTP (Teyler, 1984). However, such high frequency stimulation are well beyond the physiological firing range of hippocampal neurons (Rank, 1973). Recently some investigators have addressed the issue of what stimulus patterns optimally generate LTP under physiological conditions (Rose and Dunwiddie, 1978). In their report, a stimulus consisting of 200 msec intervals, corresponding to the frequency of the theta rhythm in the hippocampus, generates maximal LTP. In general, we have reported that in the CA1 area of the hippocampus the magnitude of LTP was highly sensitive to the higher order statistical characteristics of temporal modulation of Schaffer collateral stimulation (Tsukada et al. 1991, 1992, 1993, Aihara et al. 1991, 1992)

While Grinvald et al. (1982) detected the spread evoked electrical activity at the CA1 region of rat hippocampal slices by using the optical of imaging method of a matrix of 100 photodetectors. Since each photodetector receives light from several neural element, the observed optical signal probably reflects the intracellular activity of a population. Recentry the spatial firing pattern of the hippocampal slice was estimated in detail by using 128×128 channels. The stimuli used in these experiments were high frequency stimulation of "tetanus" which is well beyond the normal physiological firing range of hippocampal neurons. In this paper, we use the temporal pattern stimuli with a low mean rate and show temporal pattern-dependent LTP by using the optical imaging method.

Material and Method

The slice was stained for 15 minutes with 0.2 mg / ml RH155 in normal medium   and then was washed away and recovered for an additional 10 minutes. Stimulation was delivered to the Schaffer commissural-collateral (SC) fibers linking CA1 with CA3 using a monopolar tungsten electrode and the area of optical recording covered CA1 area.

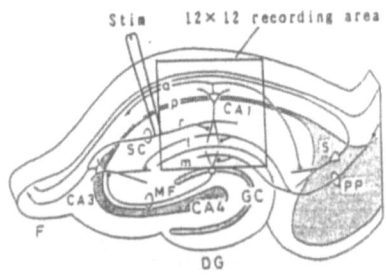

Fig.1 Stimulation and recording area

Fig.1 Stimulation and recording area

The position of the stimulating electrode and the optical recording area are shown Fig.1. The magnitude of induced LTP was estimated by the mean percentage changes in the amplitude of population spike before and after each type of MS for at least 20-30 min after a Markov Stimulus. A naive slice was used for each stimulus sequence of test-Markov-test.

The optical apparatus:

Slices were viewed with $10 \times$ objective. The voltage-sensitive dye signals were recorded with 700 ± 30nm interference filter. The transmitted light was detected by a 12×12 square array of photodiodes, each 1.3×1.3mm. Each photodiode received light from a $130 \times 130 \mu$m area of the microscope objective field, and was coupled to a current-to-voltage converter and amplifier; r.c.-filtering time constant 0.1 msec, a.c.-coupling time constant 0.05 msec $1000 \times$ amplified.   The output of the amplifiers was multiplexed and digitized by four multiplexes and 8-bit A/D converters. The object field was viewed with CCD camera and superimposed to the spatial-temporal pattern of population spike.

The temporal pattern of stimuli (Markov Stimuli):

The electrical stimulation consisted of mixed short and long stimulation intervals generated by a Markovian stochastic process. This generation process is defined by the correlation coefficient q, which is the probability that succeeding stimulus intervals have the same interval length. For given pair of short and long intervals (s and l), we used different stimulus sets, with two types of correlation coefficients; the negative correlation q=0.1, type N and the positive correlation q=0.9, type P). These stimuli has different second order statistics (correlation between successive ISIs) but identical first order statistics (mean intervals and histogram) In Fig.2, we show examples of the Markovian generated stimuli (Markov Stimuli). We tested two different short and long stimulus interval pairs ((s, l) = (100msec, 400msec), (100msec, 900msec)), with two correlation coefficients (q = 0.1, 0.9).

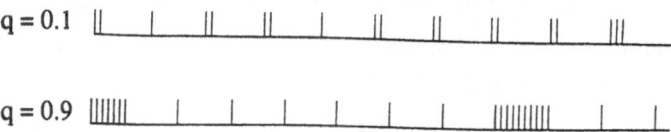

Fig.2 An example of Markov Stimuli

Results and Discussion

Fig.3 respectively shows two spatial distribution of the magnitude of the LTP induced by the two types of MS with the different components (S, L) = (100msec, 900msec), (100msec, 400msec). These results show that the LTP induced by type P was distributed spatially in the large area of CA1 region while that by type N was in the extremely small area. This corresponds to the result of temporal pattern dependent LTP estimated by the % change of the amplitude of population spikes obtained by the single extradellular recording; type P >> type N (Aihara et.al., 1992). By this optical method, we can estimate the spacific position of LTP induction depended on temporal pattern stimuli.

Fig.4 shows an example of the spatial distribution of repetitive firing at 75s after the initiation of type P stimulus, of which the LTP distribution corresponds to Fig.3b. This repetitive firing of population spikes was found only during stimulation which involes LTP formation through the activities of NMDA channnels (Tsukada, 1993). From this result; the position of induced LTP corresponds to that of the repetive firing; there is a strong correlation between the two spatial distribution.

These results obtained contain spatio-temporally the more detail information than those obtained by the single electrode method, and are interesting in comparision with the anatomical structure of hippocampal networks.

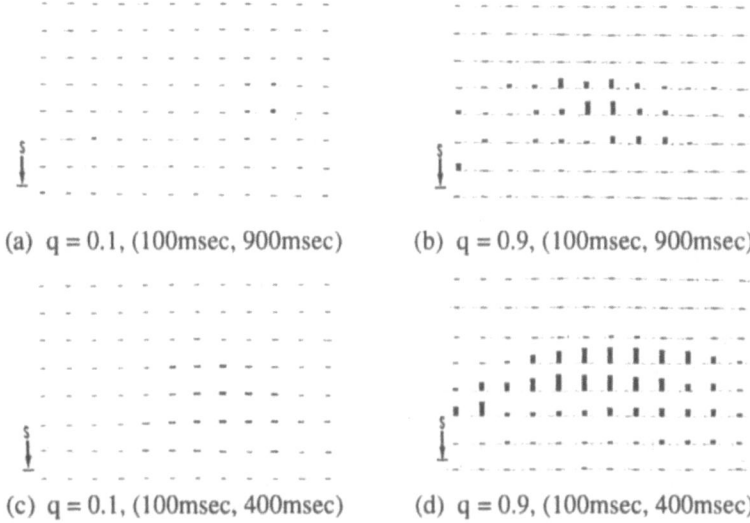

(a) q = 0.1, (100msec, 900msec)   (b) q = 0.9, (100msec, 900msec)

(c) q = 0.1, (100msec, 400msec)   (d) q = 0.9, (100msec, 400msec)

Fig.3 LTP (the % change of the peak amplitude of population spike)

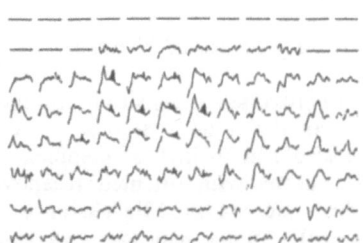

Fig.4 An typical example of the spatial distribution of repetitive firing
at 75s after type P Markov stimulus

References

Aihara T et al. (1992) IBRO Workshop, in Greek, 47.

Aihara T et al. (1991) Artifical Neural Network Vol.1, 1405

Bliss T. V. P. and Lφmo T. (1973) J.Physiol. 232, 331-356.

Grinvald A et al. (1982) J.Physiol. 333, 269-291. Neuroscience Res.

Ito K, Miyakawa H, Igumi, Kato H (1989) Biomed. Res.,10: 111-124.

Rank Jr., J. B. (1973) Exp. Neurol. 41: 462-531.

Rose G. M. and Dunwiddie T. V. (1986). Neurosci. Lett. 69: 244-248.

Teyler, T. J. and Discenna, P. (1984) Brain Res. Rev. 7: 15-28.

Tsukada M. (1992) Concepts in Neurosciences Vol.3, No.2, 213-224

Tsukada M. et al. (1991) IJCNN'91 Singapore, 3, 2177-2182

Tsukada M. et al. (1993) Biological Cybernetics (in press)

Synchronization-Based Complex Model Neurons[1]

G. Hartmann, S. Drüe
Fachbereich 14 Elektrotechnik, Universität-GH Paderborn
Warburger Straße 100, D 33098 Paderborn, Germany

Introduction

There is a lot of evidence that cortical neurons show synchronized activity (Eckhorn et al., 1989; Gray, Singer, 1987). There is also a variety of synchronization-based models and simulations, but there are almost no arguments excluding the possibility that the investigated phenomena could also be explained without synchronization. Our results from comparison of two alternative models of complex neurons show quantitatively the superiority of a synchronization-based mechanism.

We look at the problem from a vision system designer's point of view: Our robot vision system is based on a holistic approach (Hartmann, 1991) assuming that objects up to a relativly high degree of complexity are recognized by matching of suitable neural representations. Region based representations including colour representations are very robust against minor errors in foveation, and against errors due to parametric mappings (Hartmann et al., 1993) providing invariances, but they are not very sensitive to shapes.

Contour Matching by Complex Model Neurons

We could significantly improve shape sensitivity by including contour representations. Spatial tolerance could not be achieved, however, by only changing the similarity measure, we also had to include complex model neurons for contour representation. Model neurons with oriented receptive field (Hartmann, 1983) similar to simple cortical neurons provide chains of acitivity in response to a contour structure (fig. 1a). Model neurons with enlarged, highly overlapping receptive fields like complex cortical cells (fig. 1b), however, respond to contours with a cloud of activation. If we learn the representation e. g. of a rectangle (fig. 2), and present the image in a diagonally shifted position, there is decreased overlap in the case of complex neurons, but no overlap in the case of simple neurons.

At the moment we have investigated two types of complex model neurons with receptive fields of double and quadruple size compared to our simple neurons, but in principle all sizes are possible. There is an equal number of simple and complex neurons per unit area and so the large sized fields have high overlap, necessary for cloud representation. The shapes of the receptive fields are similar to those of the simple cells (Hartmann, 1983) and allow response to contours with orientations of $\pm 15°$ related to the field axis. Field axes are oriented at $\varphi = 0°$, $15°$, $30°$ ... $180°$ related to a hexagonal pixel matrix.

All these features are very similar to those of biological complex cells, but they are also indispensable for our system. They provide spatial tolerance and orientation selectivity necessary for representation of contours by orientation clouds. Insensitiveness against contrast phase helps to cope with changing contrasts due to ilumination. Finally, response to only continuous uninterrupted contours provides a clean up effect, showing only essential parts of contour structures in the oriented cloud. By verification of continuity recognition becomes more reliable, as noise and

[1] Supported by the German Minister of Research and Technology (BMFT), ITN 910 506

oriented textures are not learnt together with the contour structure of an object. So continuity is an essential point in our model of a complex neuron.

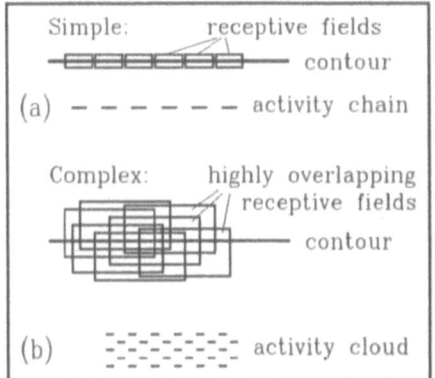

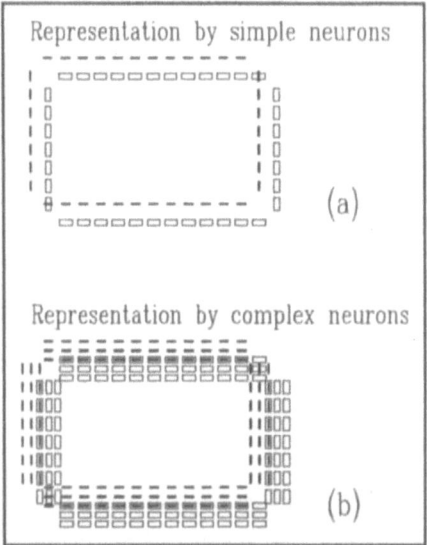

Fig. 1: Simple model neurons (a) with small oriented receptive fields provide chains of activity in response to a contour, while complex model neurons (b) respond with a cloud of activity.

Fig. 2: Activity pattern of simple (a) and complex (b) model neurons in response to a rectangular contour. Overlap between activity pattern (-) and learnt pattern (□) of simple neurons is completely lost (a) due to a diagonal shift of the image by only one pixel. In a representation by complex neurons (b) overlap is only reduced.

Alternatives for Implementation of Complex Model Neurons

There are actually two alternatives for implementing all the features of our complex model neuron: one using synchronization, the other one using only an appropriate interconnection scheme in a straightforward way. In both cases we define size and position of the complex receptive field, and within this area we select all the combinations of simple cells which may be simultaneously activated by any contour, matching to the receptive field of the complex cell. So the condition for activation of the complex neurons is the simultaneous acitivity of one of these combinations of simple neurons. In a straightforward model we could use linking neurons for verification of simultaneous activity. A linking neuron l_1 only becomes supraliminal if all the neurons a, b, c, d of combination c_1 are active (fig. 3a) and another linking neuron l_n is acitvated by neurons s, t, u, v of combination c_n. The complex neuron itself is activated by only one out of the linking neurons due to a very low threshold adjustment. The interconnection scheme is very simple (Fig. 4a) and no synchronization is necessary.

The real problem with this architecture is the tremendous number of linking neurons. In order to allow arbitrary contours to be encoded, small parts of the visual field are encoded by complete sets of simple neurons with receptive fields of different position, orientation, and shape. There is a detailed description of this set in (Hartmann, 1983) and fig. 3b shows only a simplified scheme. But the graph in fig. 3b will clearly show that a contour entering at point p will have 5^4 possibilities to cross the complex field. So already in our small sized example there are more than 5000 combinations and a corresponding number of linking neurons l_n if all possible entering points are considered . An architecture, however, which needs 5000 useless interneurons in order to connect about 100 simple neurons to one complex neuron is evidently unaceptable expensive, and so we did not really realize it.

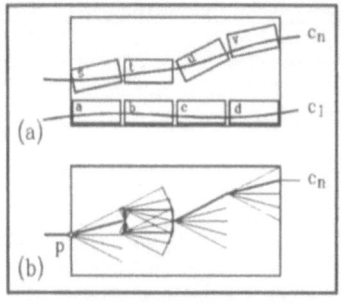

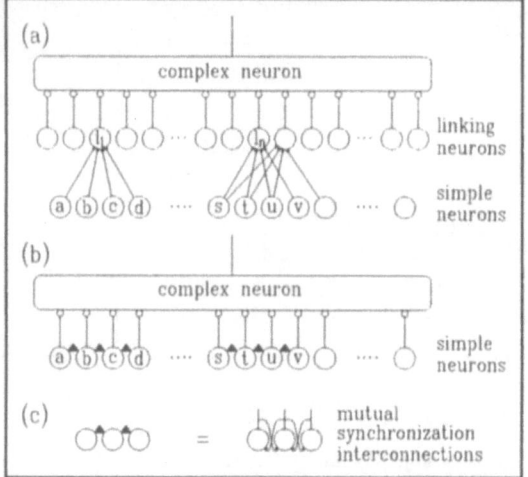

Fig. 3: The receptive fields of the contributing simple neurons form sequences (a) and the high number of possible combinations can be estimated on this base of these (b).

Fig. 4: Due to the hight number of linking neurons the straightforward architecture (a) is very expensive compared with the synchronization-based architecture (b). The symbols for synchronizing interconnections (b) are explained in (c).

The alternative architecture is based on synchronization between pulse coded model neurons (French&Stein, 1970). All pairs of simple neurons in the visual field with adjacent in-line receptive fields, like (a, b) or (t, u) in fig. 3a, are mutually interconnected via excitatory synapses (fig. 4c). We could previously show (Hartmann, 1990) that all neurons responding to an arbitrary continuous contour are synchronized by this interconnection scheme. Consequently, the combinations c_n (Fig. 3a), the building blocks of our complex neuron, are also synchronized by continuous contours. Vice versa, continuity of a contour can easily be verified by sensing synchronized activity.

This leads to the very simple and inexpensive architecture in fig. 4b. All those simple neurons contributing to valid combinations c_n of a complex neuron are interconnected to it by excitatory synapses. These synapses have a short time constant in the range of 1 ... 3 ms, and so the membrane potential can only become supraliminal, if more than three spikes pile up within a short time slot. This occurs as soon as a combination c_n of simple neurons is simultaneously activated and synchronized. This simple architecture obviously provides all the features of our complex model neuron.

Biological Aspects and Conclusion

Our actual research is in the field of robot vision systems, and our model neurons sucessfully provide clouds of activation (Fig. 5). The above results, however, could also support the assumption that sychnronization provides superior solutions in biology. Especially, we would like to discuss two points. Hubel and Wiesel have already proposed a model of complex neurons receiving input from simple neurons. The spikes of the simple neurons are not synchronized in this model and so the synapses of the complex neuron must have longer time constants and smaller weights in order to integrate the uncorrelated rates. This leads to a delayed activity of the complex model neuron which is not observed in biology. This delay was one of the strongest objections against this model, but this delay does not occur in our synchronization based model. The first set of synchroneous spikes will immediatelly cause a spike of the complex neuron, and there is almost no delay. Moreover, the complex model neuron is synchronized to the simple neurons in accordance with recent experimental results.

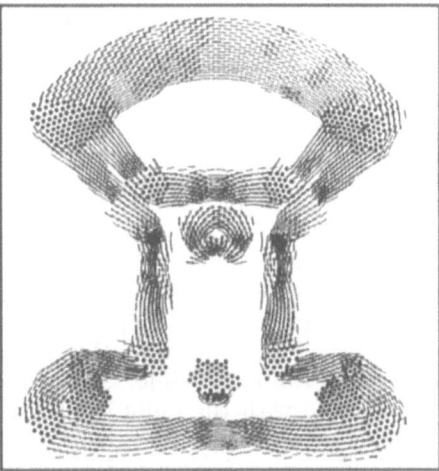

Fig. 5: Contour representation of workpiece (b) by "activity" cloud of complex neurons (a).

The second point which should be included into the discussion deals with the fast direct inputs to biological complex neurons. These additional inputs are not considered in our simulations. However, it would be compatible with our model, if fast inputs would prepare a complex neuron by rising the membrane potential. Sychronized spikes from simple cells could then more easily add the missing part of the membrane potential.

References

Eckhorn, R. et al.: Feature linking via stimulus-evoked oscillations: Experimental results from cat visual cortex and functional implications from a network model. Proc. IJCNN89, IEEE, 1.723-1.730 (1989)
Gray, C. M., Singer, W.: Stimulus specific neuronal oscillations in the cat visual cortex: a cortical functional unit. Soc. Neurosc. abstr. 404.3 (1987)
Hartmann, G.: Hierarchical Neural Representation by Synchronized Activity: A Concept for Visual Pattern Recognition. In: Taylor, J. G. et al. (Hg.): Neural Network Dynamics. London et al. (Springer-Verlag) 1991, S. 356-370
Hartmann, G.; Drüe, S.; Kräuter, K. O.; Seidenberg, E.: Simulations with an artificial retina. In: Proceedings of the World Congress on Neural Networks. WCNN 1993. S. III-689-III-694
Hartmann, G.: Processing of Continuous Lines and Edges by the Visual System. In: Biological Cybernetics 47 1983, S. 43-50
French, A. S., Stein, R. B.: A flexible neural analog using integrated circuits. IEEE Trans. Biomed. Eng., 17, 248-253 (1970)
Hartmann, G.; Drüe, S.: Verification of Continuity, Using Temporal Code. In: Proc. of the International Joint Conference on Neural Networks (IJCNN), II, San Diego 1990, S. 459-464

Synchronization of Integrate-and-fire Neurons with Delayed Inhibitory Lateral Connections.

Leslie S. Smith†, David E. Cairns†[1], Alfred Nischwitz‡

†CCCN, Department of Computing Science, University of Stirling
Stirling FK9 4LA, Scotland, UK
‡Lehrstuhl für Nachrichtentechnik, Technische Universität München
D–80333 München, Federal Republic of Germany

1   Introduction

Integrate-and-fire (leaky integrator) neurons are both mathematically tractable and have a degree of biological plausibility. Systems of two neurons, interacting via symmetric pulsatile coupling with zero delay and zero absolute refractory period have been studied by (Mirolla and Strogatz 90). For positive coupling, they found two fixed points, an unstable one with the units out of phase, and a stable one with the units in phase. For negative coupling, the stable and unstable fixed points are reversed, if a refractory period is assumed.

We show that for delayed symmetric pulsatile inhibitory coupling, a pair of integrate-and-fire neurons can synchronise. (Glünder and Nischwitz 93) and (Nischwitz et al 92) observed this in simulations of ring-shaped networks of such units. Delayed inhibitory connections have been used by (König and Schillen 91), but using a non-spiking unit, and (Schuster and Wagner 90) use an activation dependent (rather than pulsatile) coupling.

2   The Network

The network examined here consists of two symmetrically connected integrate-and-fire neurons, with equation

$$\frac{dx_i}{dt} = T - \gamma x_i + \varepsilon Y_{|i-1|}(t - \tau) \tag{1}$$

where x_i is the voltage-like state variable of the neuron ($i \in \{0, 1\}$), T is the constant tonic input, γ is the dissipation, ε the coupling strength, and τ the coupling delay. On the potential $x_i(t)$ reaching threshold, θ, the output, $Y_i(t)$ becomes instantaneously $\delta(t)$, resulting in a sudden change in potential of the other unit by ε a (short) time τ later. x_i is then reset to 0. The unit will oscillate with period $\frac{\log(T/(T-\theta))}{\gamma}$ for $\theta < T$ and $\varepsilon = 0$.

3   Analysis and Results

---

[1]David Cairns was partially supported by SERC.

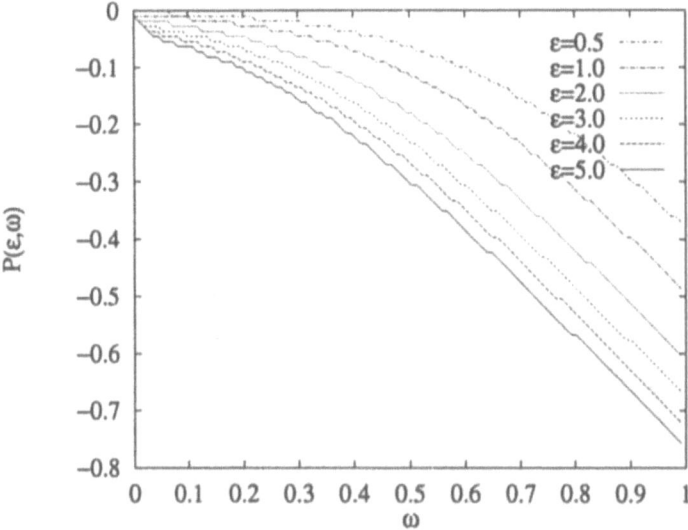

Figure 1: Phase response graph for the neuron, with an inhibitory input of varying strength (ε). ω is the point in the cycle of the arrival of the (inhibitory) input where 0 and 1 are identified with each other, and are the point at which the spike is produced. The Y axis shows the resultant change in phase, $P(\varepsilon, \omega)$. The effect is always to extend the cycle, and the strength of the effect is smallest just after spiking, and largest just before spiking. $T = 20$, $\theta = 19.96$ and $\gamma = 0.95$.

One key to the behaviour of the system is the phase response function (Rinzel and Ermintrout 89), $P(\varepsilon, x)$, the change in phase at one unit caused by an impulse from the other arriving at phase point x. Figure 3 shows this for a single neuron for a number of varying negative coupling strengths, ε. The graph shape depends on whether the unit can have a negative potential: if so then $P(\varepsilon, 0) < 0$, and if not the graph starts with a line of gradient -1 from the origin as shown. $P(\varepsilon, x)$ is strictly decreasing in x on $[0, 1)$, and $P(\varepsilon, x) = P(\varepsilon, x + k)$ for integer k. An informal analysis of the main behaviour is presented: a detailed analysis can be found in (Nischwitz 94) and is the subject of ongoing work.

Figure 2a shows the behaviour of the system with a delay (τ) of 0. If the units start off out of phase, they become antiphase with the period extended due to the inhibition. If B leads A by $\Delta\omega_i$, then the phase difference at A's next spike is

$$\Delta\omega_{i+1} = \Delta\omega_i + P(\varepsilon, \Delta\omega_i) - P(\varepsilon, 1 - \Delta\omega_i - P(\varepsilon, \Delta\omega_i)) \tag{2}$$

For $\Delta\omega_i < 0.5$, the effect on B is larger than that on A, so that $\Delta\omega_{i+1} > \Delta\omega_i$. As (Mirolla and Strogatz 90) suggest, and is proven in (Nischwitz 94) for units with a refractory period, two fixed points exist, in phase, and antiphase, and the basin of attraction of the antiphase one is all initial $\Delta\omega_i$ except 0.

Figure 2b illustrates one possible result of a delay. The effect of the first spike from B on A (1) is larger, since it arrives nearer the end of the period

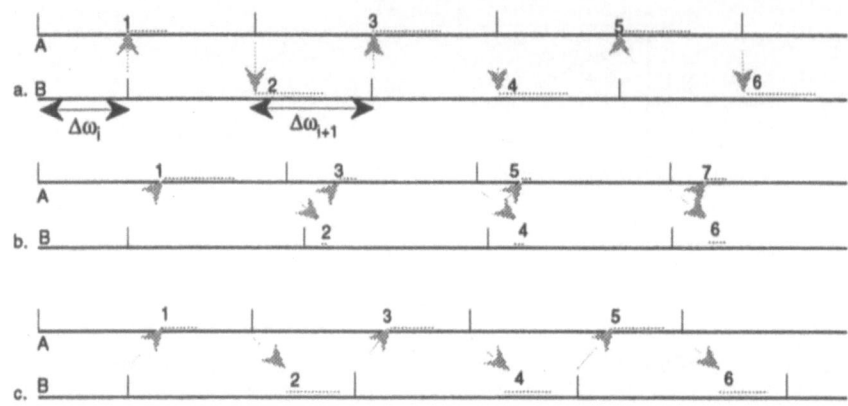

Figure 2: a. Behaviour of 2-unit system with inhibitory connections with 0 delay. The short horizontal lines show the delay caused by the arrival of an inhibitory spike. Each inhibitory spike arrival is numbered. b. Behaviour of a 2-unit system with delay (of 0.2 × period). c. Behaviour of 2 unit system, with the effect of the inhibitory connection halved.

(due to the delay). This causes the next spike at A to be delayed so that its inhibitory effect on B does not occur until after B spikes (2). It therefore has only a small effect. We call this crossover. The effect of the next spike at B on A (3) is also small, since it arrives a little way through A's cycle. Thus both spikes have their effect near the start of the other unit's cycle, making both effects small. This relies on the delay being nonzero, but small relative to the period. If A spikes again before B (as shown) the effect of B on A (3) will be bigger than the previous effect of A on B (2), so that the units become closer in phase. If B spikes again before A, the opposite happens, and the units again become closer in phase. This continues until the units are in phase, at which point the effect of each on the other is identical, and the synchronisation is stable.

Figure 2c shows the effect of a delay, but with a less strong inhibitory connection. In this case, the effect of the first spike at B on A is large (1), but small enough to permit the next spike at A to have its effect at B (2) before B spikes. This spike's inhibitory effect will be larger than the previous effect of B on A, since it arrives nearer the end of the cycle. However, its effect is not so large as to make the inhibitory effect of B's next spike arrive after A spikes (3), and this continues, forcing the units to remain out of phase.

If we assume that the potential is not allowed to become negative, it is possible to derive the basins of attraction of these fixed points (see (Nischwitz 94)). The derivation rests on the strictly decreasing nature of $P(\varepsilon, x)$, and on finding exactly when crossover can occur. For small ε, and large θ/T, so that the initial gradient of -1 lasts for a short time, the basin of attraction of the in-phase fixed point is

$$\omega \in [0, \tau - P(\varepsilon, \omega + \tau)] \cup [1 - (\tau - P(\varepsilon, \omega + \tau)), 1)$$

and the rest of the $[0, 1)$ interval for the out-of-phase fixed point.

4 Discussion

This brief paper shows how delayed inhibitory interconnections can permit either in-phase and out-of-phase oscillations, depending on the precise connection strengths and delays. We have investigated only the simplest possible system, that of two neurons: a detailed analytical discussion is in (Nischwitz 94). Further work is required to investigate these effects in more complex systems. Multiple stable phase effects in larger systems are described in (Nischwitz et al 92), using varying T, and in (Nischwitz and Glünder 94) for weakly coupled systems. Because the system's synchronization capabilities are limited, such systems with weak inhibitory interconnections may be able to support multiple simultaneous synchronous sets of oscillators, as suggested by (Shastri 89), without falling into a single phase, as found in (Cairns et al 93).

References

Cairns D.E., Baddeley R.J., Smith L.S. (1993), Constraints on synchronising oscillator networks, Neural Computation, 5, 260-266.

Glünder H. and Nischwitz A. (1993), On spike synchronization, in Brain Theory, 251-258, ed Aertsen A., Elsevier.

König P. Schillen T. (1991), Stimulus-dependent formation of oscillatory responses, Neural Computation, 3, 155-166.

Nischwitz A. (1994) Impuls-Synchronisation in neuronalen Netzwerken, Dissertation, T.U. Muenchen, in preparation.

Nischwitz A., Glünder H. (1994), Local lateral inhibition: a key to spike synchronisation, Biol. Cybern, submitted.

Nischwitz A, Glünder H, von Oertzen A., Klausner P. (1992), Synchronization and label-switching in networks of laterally coupled model neurons, in Artificial neural networks 2, 852-854, ed Aleksander I and Taylor J, Elsevier.

Mirollo R.E., Strogatz S.H. (1990), Synchronization of pulse-coupled biological oscillators, SIAM J. Applied Mathematics, 50, 6.

Rinzel J and Ermentrout G. B. (1989), Analysis of neural excitability and oscillations, in Methods in neuronal modelling - from synapses to networks, ed Koch C and Segev I, MIT Press,.

Schuster H.G., Wagner P. (1990), A model for neuronal oscillations in the visual cortex, Biological Cybernetics, 64, 77-82.

Shastri L. (1989), From simple associations to systematic reasoning: a connectionist representation of rules, variables, and dynamic bindings, Tech report, University of Pennsylvania.

Complex Patterns of Oscillations in a Neural Network Model with Activity-Dependent Outgrowth

A. van Ooyen and J. van Pelt

Netherlands Institute for Brain Research,

Meibergdreef 33, 1105 AZ Amsterdam, The Netherlands

1 Introduction

Many processes that play a role in shaping the structure of the nervous system are modulated by electrical activity. For example, electrical activity can affect neurite outgrowth: high levels of activity, resulting in high intracellular calcium concentrations, cause neurites to retract, whereas low levels of activity, and consequently low calcium concentrations, allow further outgrowth [1]. As a result of this and other activity-dependent processes, a reciprocal influence exits between the formation of connectivity ("slow dynamics") and activity ("fast dynamics"). We have made a start at unravelling the implications of activity-dependent neurite outgrowth [2, 3], and have been able to show that several interesting properties arise as the result of interactions among outgrowth, excitation and inhibition: (i) a transient overproduction ('overshoot') during development with respect to connectivity; (ii) the neuritic fields of inhibitory cells tend to become smaller than those of excitatory cells; (iii) the spatial distribution of inhibitory cells becomes important in determining the level of inhibition; (iv) pruning of connections can no longer take place if the network has grown without activity for longer than a certain time ('critical period'). The results show many similarities with findings in cultures of dissociated cells.

Previously, we studied networks in which ϵ, the level of activity for which the neurites of a cell neither grow out nor retract [see eqn(3)], is the same for all cells. Here, we show that excitatory networks in which ϵ is distributed over a range of values can display complex patterns of oscillations in electrical activity and outgrowth. Oscillations in neurite outgrowth have indeed been observed in tissue cultures of hippocampal cells (S. B. Kater, personal communication).

2 The Model

We use a distributed, excitatory network, with neuron dynamics governed by

$$\frac{dX_i}{dT} = -X_i + (1 - X_i) \sum_{j}^{N} W_{ij} F(X_j) \qquad (1)$$

where X_i is the membrane potential, N is the total number of cells, W_{ij} represents the connection strength ($W > 0$), and $F()$ is the firing rate:

$$F(u) = \frac{1}{1 + e^{(\theta - u)/\alpha}} \qquad (2)$$

where α determines the steepness and θ represents the firing threshold. Growing cells are modelled as expanding circular neuritic fields, and neurons become connected when their fields overlap. The outgrowth of each cell depends upon its own level of electrical activity:

$$\frac{dR_i}{dT} = \rho\left[1 - \frac{2}{1 + e^{(\epsilon_i - F(X_i))/\beta}}\right] \tag{3}$$

where R_i is the radius of the field, ρ is the rate of outgrowth, ϵ_i is the firing-rate at which $\frac{dR_i}{dT} = 0$, and β determines the non-linearity. Eqn(3) is just a description of Kater's hypothesis [1] that the depolarization level of the neuron influences its outgrowth. In the simulations we took $\theta = 0.5, \alpha = \beta = 0.5$, and ρ very small so that connectivity is quasi-stationary relative to membrane potential dynamics ($\rho = 0.0001$ in full model, and 0.005 in eqn(5)).

3 Results

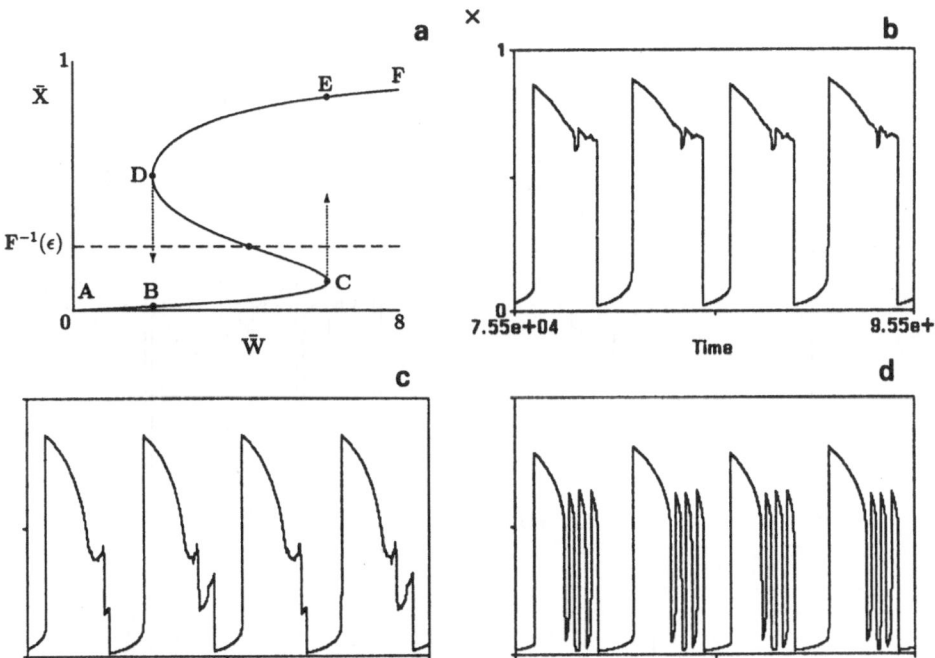

Fig.1 a Equilibrium manifold of $\bar{X}$ ($\frac{d\bar{X}}{dT} = 0$). **b,c,d** The membrane potential, X, of three different cells in a network in which ϵ is uniformly distributed between 0.3 and 0.8.

To understand the ocurrence of complex oscillations, first consider a network in which all the cells have the same ϵ. If the variations in X_i are small (relative to $\bar{X}$, the average membrane potential of the network), we find that for a given connectivity $\mathbf{W}$ the equilibrium points are solutions of:

$$0 \simeq -\bar{X} + (1 - \bar{X})\bar{W}F(\bar{X}) \tag{4}$$

148

This gives us the equilibrium manifold of $\bar{X}$ ($\frac{d\bar{X}}{dT} = 0$) as depending on the average connectivity $\bar{W}$ (Fig. 1a). The equilibria are stable on the branches ABC and DEF, and unstable on CD. The size of a neuritic field remains constant if $X_i = F^{-1}(\epsilon)$, where F^{-1} is the inverse of F; thus, since all cells have indentical ϵ, $\bar{W}$ remains constant if $\tilde{X} = F^{-1}(\epsilon)$. An intersection point of this line on ABC (quiescent state) or DEF (activated state) of the manifold results in a stable point of the whole system, whereas an intersection point on DC results in oscillations following the path $ABCEDBCEDBC$.... These oscillations, in connectivity and activity, are similar for all cells. On the other hand, a network in which the cells have different ϵ values exhibits a complex pattern of oscillations: cells can differ in frequency, phase and amplitude, while other cells show no oscillations at all (Fig. 1b, c ,d).

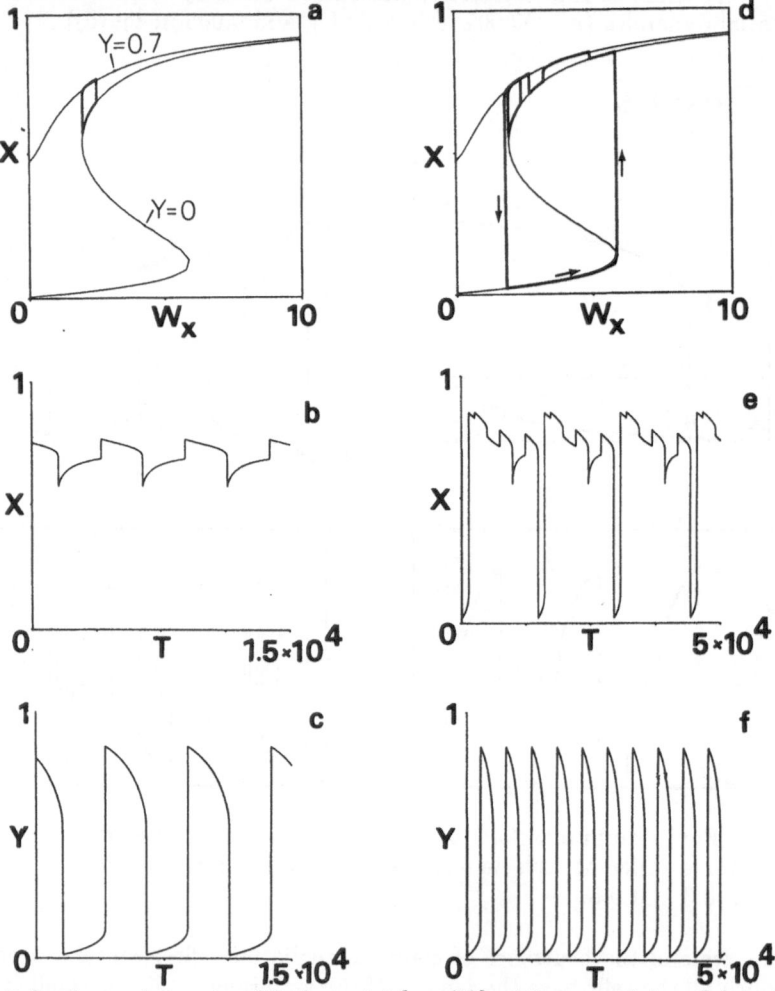

Fig.2 Behaviour of the simplified model [eqn(5)] with $\epsilon_Y = 0.4$ and in **a,b,c** $\epsilon_X = 0.7$ and in **d,e,f** $\epsilon_X = 0.685$. In **a,d** the manifolds of X for $Y = 0$ and $Y = 0.7$ are drawn together with the trajectory of X (thick line) (skipping the initial transients). In **b,c,d,f** X and Y are plotted against time.

Simplified models can be used to explain how these patterns can arise:

$$\frac{dX}{dT} = -X + (1-X)(W_X F(X) + F(Y))$$

$$\frac{dY}{dT} = -Y + (1-Y)W_Y F(Y)$$

$$\frac{dW_X}{dT} = \rho(\epsilon_X - X)$$

$$\frac{dW_Y}{dT} = \rho(\epsilon_Y - Y)$$

(5)

where X (W_X) and Y (W_Y) are the average membrane potential (connectivity) of two different cell populations which differ in their ϵ value. We assume that the influence of X on Y can be neglected and that the connection strength between X and Y is constant. The connectivity in both populations depends on their activity according to a simplified form of eqn(3). In the examples (Fig. 2) ϵ_Y is such that Y oscillates, and ϵ_X is such that, without the effect of Y, X goes to a stable point. Thus, Y oscillates independently of X between activated and quiescent state, and X is "forced" by Y. Depending on ϵ_X, X oscillates also between quiescent and activated state (possibly with a different frequency: Fig. 2e) or remains in the activated state (Fig. 2b). In Fig. 2 the equilibrium manifold of X as depending on W_X is drawn for Y at the quiescent state ($Y = 0$) and the activated state ($Y = 0.7$). As can be seen, X jumps between these two manifolds.

4 Conclusions

We have shown that complex patterns of oscillations in connectivity, and consequently in activity, can arise when cells grow depending on their own level of activity. The emergence of these oscillations (on the timescale of growth) hinges upon the presence of a hysteresis loop between activity and connectivity (Fig. 1a). Simplified models can help to understand the occurrence of oscillations of different forms and frequencies, in terms of trajectories on manifolds. Extensions of the simplified model presented here are readily possible.

5 References

[1] Kater SB, Guthrie PB, Mills LR (1990) Integration by the neuronal growth cone: a continuum from neuroplasticity to neuropathology. Progress in Brain Research 86: 117-128.
[2] Van Ooyen A, Van Pelt J (1993) Activity-dependant outgrowth of neurons and overshoot phenomena in developing neural networks. Journal of Theoretical Biology, in press.
[3] Van Ooyen A, Van Pelt J (1993) Implications of activity-dependent neurite outgrowth for developing neural networks. In: S. Gielen and B. Kappen (eds.), Proceedings of ICANN'93, pp. 177-182.

Learning and the Thalamic–NRT–Cortex System

J. G. Taylor, F. N. Alavi

Centre for Neural Networks, Department of Mathematics
King's College London, The Strand
London WC2R 2LS, United Kingdom

1 Introduction

Learning in the brain can be divided into implicit and explicit modes, where the former corresponds to skill learning or priming (or other types) and the latter to semantic or to episodic coding in which awareness of the memory is possible to the subject. The former type of learning is thought to take place in non–cortical regions (cerebellum, thalamus, etc.) and posterior associative cortex, the latter in associative cortex and more specifically in hippocampus (if only temporarily).

In this paper, we will not attempt to distinguish between those forms of learning, but consider instead a number of types of biologically plausible learning mechanisms which may be involved in the above processes, although they may have their underlying basis in only one of them, namely Hebbian learning (Hebb 1949). We present certain modified forms of this rule, and also simulations performed with this rule on the thalamus–NRT[1]–cortex complex in the brain.

2 Hebbian Learning

Let us consider firstly simple Hebbian learning. This latter process is well known to require both pre–synaptic and post–synaptic activity, so that the connection weight W_{ij} of a synapse from neuron j to neuron i changes under experience as

$$\dot{W}_{ij} = \eta \text{ACT}_i \text{OUT}_j + \text{decay term,} \tag{1}$$

where ACT_i, the activity impinging on neuron i, need not be identical to its output OUT_i; η is the learning rate. We shall take the decay term to be of the following simple form[2]:

$$- \beta W_{ij}, \qquad (\beta > 0). \tag{2}$$

What is of concern to us is the modular construction apparently enjoyed by the cortical sheet. Mini– and hyper–columns in visual cortex (Hubel & Wiesel 1968), columnar structures in somato–sensory cortex (Mountcastle 1957) and similar parcellations in numerous other cortical areas are well known. The modules may be distinguishable by anatomical means, or only have a functional

---

[1] Nucleus Reticularis Thalamus.
[2] Other forms are considered in Taylor & Alavi (1993).

character, and moreover can appear in terms of the afferent or the lateral connections in the cortex. The mini–columns appear to have a well–defined afferent structure. There is also strong lateral connectivity between iso–orientation columns, so corresponding to a well–defined "lateral modularity". Encoding of faces in inferotemporal cortex appears to possess a repeatability of $\sim 3-4$ mm, each patch being about 3 mm across (Harries & Perret 1991).

Two different approaches have been taken to derive these modular structures, using Hebbian adaptive weights obeying (1). The first (Miller, Keller & Stryker 1989) uses correlations in the inputs to drive afferent excitatory connection weights into a wave–like dependence on cortical co–ordinates. The other (Chernjowsky & Moody 1990) uses spatially collective modes, arising from laterally inhibited activity driven by random inputs, to sculpt the excitatory weights to follow the periodicity of the major collective mode. The lateral inhibition needs a difference-of-Gaussians form, and although there is evidence for this from neuroanatomy (Lund 1993), it is doubtful that the inhibition will extend much beyond half a hyper–column.

The TH–NRT–C complex that is of interest to us possesses long–range excitations, arising even from localised inputs. It is therefore possible to expect that Hebbian learning on cortical afferents or lateral connections would lead to a modular structure from the effects of the NRT waves brought about by uncorrelated inputs. One would then add the effects arising from the other two models at whatever level they are accepted or confirmed. Here, we will discuss the possibility of modular structure arising in cortex solely in the former manner.

Let us start our analysis with the development of afferent connections to the cortex from the thalamus. We use a similar approach to that of Miller, Keller and Stryker (1989), with an afferent function $A(x - y)$ denoting the number of synapses from the thalamic cell located at y to the cortical cell at x. The synaptic strength at time t for the thalamo–cortical afferent from the cell at y to that at x is denoted by $w(x, y; t)$. Then the Hebbian learning rule (1) becomes

$$\dot{w}(x,y) = \eta A(x - y)I(y)\text{ACT}(x) - \lambda w(x,y), \qquad (3)$$

where $I(y)$ is the thalamic output at y and $\text{ACT}(x)$ is the activity on the cortical cell at x. This latter we take to have the standard form

$$\int dy' w(x,y')A(x - y')I(y'), \qquad (4)$$

where no account is being taken of lateral cortical effects. Then, averaging (3) and (4) over inputs I, we obtain

$$\dot{w}(x,y) = \eta A(x - y) \int dy' w(x,y')A(x - y')C(y,y') - \lambda w(x,y), \qquad (5)$$

where $C(y,y') = \langle I(y)I(y') \rangle$ denotes the correlation function of the thalamic inputs. The simplest way to analyse (5) is to consider a simplified model of the TH–NRT–C complex in which parameters of the system are chosen so that semi–autonomous waves on NRT are created whose wavelength are roughly input independent. The set of inputs $I(y)$ over which averaging must be performed are therefore the set $\sin[k(y - y_0)]$, as y_0 varies, with fixed wavelength

$l = 2\pi/k$. In that case, the analysis of Taylor & Alavi (1993) shows that the Hebbian learning rule (1) implies that

$$\dot{w}_l(x,y) = \eta\text{ACT}(x) \cdot \text{ACT}(y) - \lambda w_l(x,y). \tag{6}$$

This is the form for the learning rule that we shall consider in our simulations.

3 Simulations

The simulation model we investigated is essentially one–dimensional, which corresponds to *lines* of thalamic, NRT and cortical neurons.

The x–axes in the simulation plots represent the spatial positions of a line of 100 neurons. The y–axes represent $\text{OUT}(u_T)$ and $\text{OUT}(v_T)$ for the cases of the thalamic excitatory and inhibitory neurons respectively, and user–scaled raw voltage output from the excitatory neurons in the NRT. Time delays for signal propagation between neurons were set to zero. The uppermost graphs in each set show the changes in time of the value of $W_{j,50}$, for $j = 1, \ldots, 100$, as the system ages. This corresponds to an exploration of the weight centred on the 50'th cortical neuron, which is connected to 100 thalamic neurons.

Figure 1 is intended to demonstrate the behaviour predicted in the previous section, where we endowed the TH–NRT–C system with Hebbian learning capability on the afferent connections to the cortex directly from the thalamus. The Figure shows the evolution of the weights (with constant input being fed into the thalamus) according to (6). The learning rate η was set to 1.0 and the decay term λ to 0.125. The weights are initially set to zero. There is linear (as opposed to Gaussian) fan–in of the connections from the thalamus, with a user–definable spread, into the cortex. The spread was set to 30. It is clear from the plots that the system learns its connections in a manner directly governed by the activity on the NRT. This lateral modularity is precisely what was predicted in §2.

4 References

Chernjowsky, A. and Moody, J. (1990), *Spontaneous Development of Modularity in Simple Cortical Models*, Neural Computations, **2**, 334–354.

Harries, M. H. and Perret, D. I. (1991). Visual Processing of Faces in Temporal Cortex: Physiological Evidence for a Modular Organisation and Possible Anatomical Correlates, J. Cog. Neurosci., **3**, 9–23.

Hebb, D. O. (1949), *The Organization of Behaviour*, Wiley.

Hubel, D. and Wiesel, T. (1968), *Receptive Fields and Functional Architecture of Monkey Striate Cortex*, J. Neurophysiol., **195**, 215–243.

Taylor, J. G. and Alavi, F. N. (1993). A Global Competitive Neural Network (submitted to Biol. Cyb.).

Lund, J. (1993), *Comparison of Intrinsic Connectivity in Different Areas of Macaque Monkey Cerebral Cortex*, Cerebral Cortex (to appear).

Miller, K. D., Keller, J. B. and Stryker, M.P. (1989), *Ocular Dominance Column Development: Analysis and Simulations*, Science, **245**, 605–615.

Mountcastle, V. (1957), *Modality and Topographic Properties of Single Neurons of Cat's Somatic Sensory Cortex*, J. Neurophysiol., **20**, 408–434.

Figure 1. Simulation run showing a TH–NRT–C system endowed with Hebbian learning capability on the afferent connections from the thalamus to the cortex. The x-axes in these plots represents the positions of 100 neurons. (a) and (b) represent the state of the system at the second and sixteenth iteration of (6), respectively. The initial weights were set to zero. It is clear that the system learns its connections in a manner directly governed by the wave-like activity on the NRT.

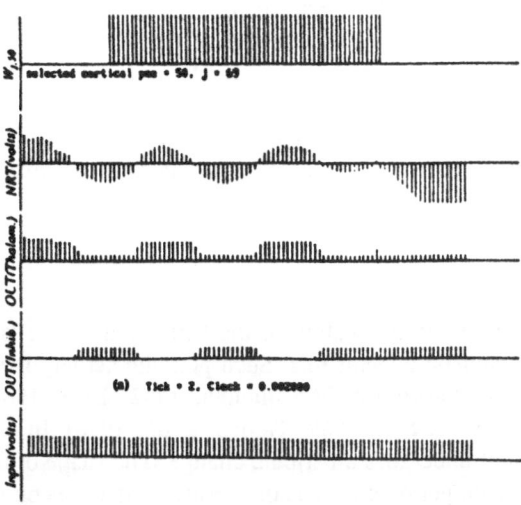

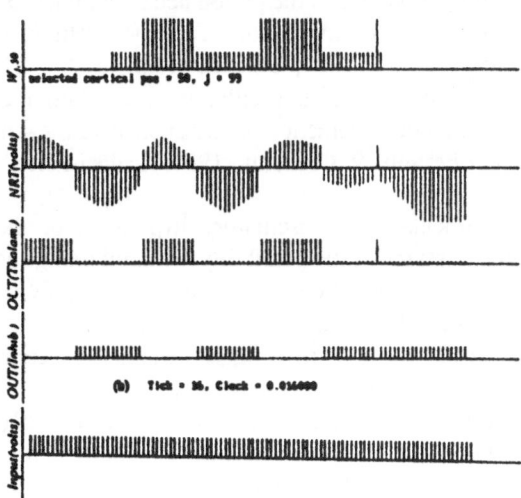

Resetting the Periodic Activity of *Hydra* at a Fixed Phase

C. Taddei-Ferretti[*], C. Musio[*] & S. Chillemi[^]

[*]Istituto di Cibernetica, CNR, 80072 Arco Felice (NA), Italy
[^]Istituto di Biofisica, CNR, 56100, Pisa, Italy

Abstract - It is hypothesize that the mechanism responsible for the periodic shortening-elongation behaviour of *Hydra* has an oscillatory nature, which is reflected by the periodical variation of the *Hydra* sensitivity to photic stimuli. We tested such hypothesis by searching the possibility of mantaining the system at a fixed phase by repetitive light pulse stimulation of suitable frequency. The role of the intinsic noise of the system on the degree of congruence of the experimental results with the expected ones is then discussed.

1. Introduction

The periodic shortening and elongation of the Cnidarian *Hydra* is brought by the activity of two antagonistic effector sets. Such periodic activity is variously affected by light stimulation (Passano & McCullough, 1962, 1963, 1964, 1965; Rushforth *et al.*, 1963; Tardent & Frei, 1969; Tardent *et al.*, 1976). In particular, the sensitivity to light stimuli undergoes a periodic change. The shape of the phase response curve expressing such periodic change of sensituvity depends on the pulse duration, relative intensity, wavelength and polarity (Taddei-Ferretti & Cordella, 1976; Taddei-Ferretti *et al.*, 1988); entrainment of the actity is obteinable by an external Zeitgeber of different frequency (Taddei-Ferretti & Cordella, 1976); the level of steady background stimulation affects the period behavioural period both in undisturbed condition and during light stimulation (Taddei-Ferretti & Cotugno, 1988; Taddei-Ferretti, Di Maio *et al.*, 1987; Tardent *et al.*, 1976). The above phenomena are justified by the hypothesis of an oscillatory pacemaking mechanism carried out by the functional inhibitory interaction of two subnets of the structurally isotropic nervous n*et* (Taddei-Ferretti & Chillemi, 1987; Taddei-Ferretti *et al.*, 1988).

In order to test experimentally the oscillatory hypothesis of the *Hydra* behaviour triggering system, we searched the possibility of maintaining the system at a fixed phase, by suppressing the occurrence of all other phases (Ypey *et al.*, 1982; Kawato & Suzuki, 1978). This situation, representing a limit case of entrainment,should be obtained by repetitive pulse stimulation continuously resetting the system, i.e. of repetitive rephasing.

The suppression could be obtained and maintained provided that (Ypey *et al.*, 1982): at least at some phases of the system cycle, the effect of a defined single stimulus is an immediate phase shift of the overall cycle; the period of repetition between the stimuli is not higher than the maximum delay of the overt reference event obtainable by a single pulse stimulation; the application phase of each stimulus of the repetitive stimulation, used to obtain the suppression of the overt reference point, is such that the above delay corresponding to that application phase is equal to the inter-stimulus interval; the system is not affected by intrinsic noise; the slope of the phase response curve expressed in the format "New phase at which the system is brought by a stimulus versus Old phase at which the stimulus is applied" is included between +1 and -1.

2. Materials and Methods

A Swiss strain of *Hydra attenuata* Pall. was used. For the culture conditions see Taddei-Ferreti & Cordella, 1975. For the monitoring and recording of the bioelectric pulse (organized in a train) correlated to the periodic shortening activity see Taddei-Ferretti & Cordella, 1976.

We applied either white light pulses, 10 s duration, 3000 lx intensity at the *Hydra*, at repetition frequencies included between 1.5 s-1 and 45.5 s-1, or darkness pulses, with the same duration and frequency values, interrupting the 3000 lx illumination. For the photic stimulation details see Taddei-Ferretti *et al.*, 1988.

3. Results

1) With repetitive white light pulses the suppression of the contraction pulse (and the continuous resetting of the elongation phase) is obtained for more or less long periods (up to 23 times the natural cycle length) interrupted by a single escaping contraction pulse train.

2) With repetitive darkness pulses, the natural cycle length is shortened, but the events are abortive contraction pulse trains consisting of single contraction pulses.

3) The range of the inter-stimulus intervals useful for the suppression is more limited on the side of the higher values, especially in the case of the darkness pulses.

4) The magnitude of the above described effects decreases with a decrease of the intensity both of the light pulses and of the starting level of the darkness pulses.

4. Discussion

At this point, taking in mind the conditions necessary to obtain and maintain the suppression (Ypey *et al.*, 1982) and the characteristics of the phase response curves experimentally considered (both that obtained with single light pulses and that obtained with darkness pulses) is clear that:

a) The possibility of suppression is a direct consequence of the existence of parts of the phase response curve corresponding to delays.

b) The not complete efficiency of the used stimulations is explainable both by the existence of parts of the phase response curves corresponding to advances and to the noise which is evident by the spread of experimental points that is observed in the phase response curves.

c) The minor range of inter-stimulus intervals which produce the suppression than expected is another consequence of the noise.

d) The efficiency decrease with the pulse level decrease is comprehensible by taking into account the fact that the modifications of the phase response curve slope produced by an intensity decrease of a light or darkness stimulus occur in the opposite direction than the slope modification giving rise to an increase of the entrainment stability.

e) The minor efficiency of the darkness pulses compared to light pulses is due to the quantitative and qualitative differences of the phase response curve related to single light pulse stimuli compared to that related to darkness stimuli (the range of values of the Old phase to which a delayed New phase corresponds is higher in the case of light pulses than in the case of darkness pulses; the range of values of the Old phase to which a delayed New phase corresponds is higher than the range to which an advanced New phase corresponds in the case of light pulses, but it is lower in the case of darkness pulses; the parts of the phase response curve where a slope included between +1 and -1 are higher in the case of light pulses than in the case of darkness pulses; such parts with slope included between +1 and -1 correspond mainly to delays in the case of light pulses, to advances in the case of darkness pulses; if considering the delay zones, the spread of experimental points is higher in the case of darkness pulses).

The noise could be due either to variations of the cycle period length (the same phase, corresponding to the same responsiveness to a given stimulus, should be reached at times different from the forecasted one), or to variations of the cycle shape (the same phase, corresponding to sensitivities different from the forecasted one, should be reached at the same time). In both cases, after the same amount of time different sensitivities should be obtained.

It has been already observed that in *Hydra* the length of each cycle varies slightly (Taddei-Ferretti & Cordella, 1976) and that the shape of a bioelectric event reflecting the system state (Taddei-Ferretti *et al.*, 1976), varies slightly in subsequent cycles: for the above reasons it seems that in *Hydra* the noise could be due to a combination of both causes.

Acknowledgements - We are indebted to Mr. A. Cotugno for continuous valuable assistance.

References

Kawato, M. and R. Suzuki (1978) "Biological oscillators can be stopped. Topological study of phase response curve", Biol. Cybern., 30, 241-248.

Passano, L.M. and C.B. McCullough (1962) "The light response and the rhythmic potentials in *Hydra*", Proc. Natl. Acad. Sci. USA, 48, 1376-1382.

Passano, L.M. and C.B. McCullough (1963) "Pacemaker hierarchies controlling the behaviour of hydras", Nature, 199, 1174-1175.

Passano, L.M. and C.B. McCullough (1964) "Co-ordinating systems and behaviour in *Hydra*. I. Pacemaker system of the periodic contractions", J. Exp. Biol., 41, 643-664.

Passano, L.M. and C.B. McCullough (1965) "Co-ordinating systems and behaviour in *Hydra*. II. The rhythmic potential system", J. Exp. Biol., 42, 205-31.

Rushforth, N.B., A.L. Burnett, and R. Maynard (1963) "Behavior of *Hydra*: contraction responses of *Hydra* pirardi to mechanical and light stimuli", Science, 139, 760-761.

Taddei-Ferretti, C. and S. Chillemi (1987) "Modulation of *Hydra attenuata* rhythmic activity. V. A revised interpretation", Biol. Cybern., 56, 225-235.

Taddei-Ferretti, C. and L. Cordella (1975) "Modulation of *Hydra attenuata* rhythmic activity: Photic stimulation", Arch. Ital. Biol., 113, 107-121.

Taddei-Ferretti, C. and L. Cordella (1976) "Modulation of *Hydra attenuata* rhythmic activity. II. Phase response curve", J. Exp. Biol. 65, 737-751.

Taddei-Ferretti, C., L. Cordella, and S. Chillemi (1976) Analysis of *Hydra* contraction behaviour. In: Coelenterate ecology and behavior (G.O. Mackie, ed.) New York.: Plenum, pp. 685-694.

Taddei-Ferretti, C. and A. Cotugno (1988) "Wavelength effect of background illumination upon pulse stimulation effect in *Hydra*", Proc. IX National Congress of Gruppo Nazionale Cibernetica Biofisica del CNR, Trento, I, p. 102.

Taddei-Ferretti, C., V. Di Maio, A. Cotugno, and M. Durante (1987) "Photoresponses of symbiotic and aposymbiotic *Hydra*", Proc. VIII National Congress of Società Italiana Biofisica Pura e Applicata, Viareggio, I, p. 62.

Taddei-Ferretti, C., V. Di Maio, S. Ferraro, and A. Cotugno (1988) *Hydra* Photoresponses to different wavelengths. In: Light in Biology and Medicine (R.H. Douglas, J. Moan and F. Dall'Acqua, eds.) New York.: Plenum, Vol. 1, pp. 411-416.

Tardent, P. and E. Frei (1969) "Reaction patterns of dark- and light-adapted *Hydra* to light stimuli", Experientia, 25, 265-267.

Tardent, P., E. Frei, and M. Borner, (1976) The reaction of *Hydra attenuata* Pall. to various photic stimuli. In: Coelenterate ecology and behavior (G.O. Mackie, ed.) New York: Plenum, pp. 671-683.

Ypey, D.L., W.P.N. Van Meerwijk, and G. De Bruin, (1982) "Suppression of pacemaker activity by rapid repetitive phase delay", Biol. Cybern., 45, 187-194.

Integral equations in compartmental model neurodynamics.

Paul C. Bressloff
Department of Mathematical Sciences, Loughborough University of Technology, Loughborough, Leicestershire LE11 3TU

1.   Introduction.

In this paper we consider the dynamics of a recurrent analogue neural network consisting of identical compartmental model neurons (Rall, 1964). We show how the associated set of ordinary differential equations can be reduced to a much smaller set of Volterra integro-differential equations in which the state of each neuron is represented by a single scalar variable. The kernel of the Volterra equations is of the convolution type, and is determined by the single neuron response function. Each neuron in the network effectively performs a temporal summation over all previous inputs to that neuron as determined by the convolution integral. Thus one can consider the network as having infinite or continuously distributed delays. The reduction presented here provides a compact and analytically tractable way of incorporating dendritic structure into neural network models. It should be noted that Poggio and Torre (1977) have also considered the representation of dendritic structure in terms of functional equations, however their analysis is purely at the single neuron level.

2.   The model.

Consider a fully-connected network of compartmental model neurons each with identical dendritic structure. Let $V_{i\alpha}$ denote the membrane potential of compartment α belonging to the ith neuron of the network, where $i = 1,...,N$, $\alpha \in \hat{\Gamma}$, and $\hat{\Gamma}$ specifies the topology of the dendritic tree (i-independent). We shall assume that the soma is a terminal node of the tree labelled by $\alpha = \alpha_0$ and shall take $\alpha = \alpha_1$ to be the unique dendritic compartment adjoining the soma. For convenience, we shall denote the membrane potential $V_{i\alpha_0}$ of the ith neuron's soma by ϕ_i. Let J_{ij}^{α} be the synaptic weight of the connection from neuron j impinging on the αth compartment of neuron i. The associated input is taken to be of the form $J_{ij}^{\alpha} f(\phi_j)$ where f is a sigmoidal output function

Under the above assumptions, an application of Kirchoff's law leads to the set of differential equations

$$C_{\alpha_0} \frac{d\phi_i}{dt} = -\frac{\phi_i}{R_{\alpha_0}} + \frac{V_{i\alpha_1} - \phi_i}{R_{\alpha_0\alpha_1}} + \sum_{j \neq i} J_{ij}^{\alpha_0} f(\phi_j) \qquad (1a)$$

$$C_\alpha \frac{dV_{i\alpha}}{dt} = -\frac{V_{i\alpha}}{R_\alpha} + \sum_{<\beta\in\Gamma;\alpha>} \frac{V_{i\beta} - V_{i\alpha}}{R_{\alpha\beta}} + \frac{\phi_i - V_{i\alpha_1}}{R_{\alpha_0\alpha_1}}\delta_{\alpha,\alpha_1} + \sum_{j\neq i} J_{ij}^\alpha f(\phi_j),$$
(1b)

for $\alpha \in \Gamma$, where Γ is the tree obtained by removing the somatic node from $\hat{\Gamma}$, i.e. $\hat{\Gamma}\backslash\alpha_0$, and $<\beta;\alpha>$ indicates that the summation over β is restricted to nearest neighbours of α. Note that the membrane leakage resistance R_α, capacitance C_α, and the junctional resistance $R_{\alpha\beta}$ are all assumed to be i-independent. One could now proceed to analyse the dynamics of the recurrent network by solving the $N \times |\hat{\Gamma}|$ coupled equations, where $|\hat{\Gamma}|$ is the number of nodes of the tree $\hat{\Gamma}$, that is, the number of compartments per neuron. It is clear, however that for a large number of compartments this is considerably more complicated than solving the standard analogue model network in which each neuron is taken to be a point processor (single compartment model $|\hat{\Gamma}| = 1$). Therefore, we shall proceed by eliminating the "auxiliary" variables $V_{i\alpha}$, $\alpha \in \Gamma$, $i = 1,...,N$ to obtain N equations for the variables ϕ_i.

3.   Integral equations.

Solving equation (1b) for $V_{i\alpha}$ in terms of ϕ_i under the initial conditions $V_{i\alpha}(t_0) = 0$ gives (on absorbing the factor $1/C_\alpha$ into each J_{ij}^α)

$$V_{i\alpha}(t) = \int_{t_0}^t dt' \sum_{\beta\in\Gamma} G_{\alpha\beta}(t - t')\left[\sum_{j\neq i} J_{ij}^\beta f(\phi_j(t')) + \frac{1}{\tau_{\alpha_1\alpha_0}}\phi_i(t')\delta_{\beta,\alpha_1}\right], \quad (2)$$

where $G_{\alpha\beta}(t)$ is of the form $[e^{t\mathbf{Q}}]_{\alpha\beta}$, $\alpha,\beta \in \Gamma$, with $\mathbf{Q}$ given by

$$Q_{\alpha\beta} = -\frac{\delta_{\alpha\beta}}{\tau_\alpha} + \sum_{<\beta'\in\Gamma;\alpha>} \frac{\delta_{\beta,\beta'}}{\tau_{\alpha\beta'}}, \quad \frac{1}{\tau_\alpha} = \frac{1}{C_\alpha}\left[\sum_{<\beta'\in\hat{\Gamma};\alpha>} \frac{1}{R_{\alpha\beta'}} + \frac{1}{R_\alpha}\right], \quad (3)$$

and $\tau_{\alpha\beta} = C_\alpha R_{\alpha\beta}$ for $\alpha,\beta \in \hat{\Gamma}$. We identify $G_{\alpha\beta}$ as the single neuron response function. That is, $G_{\alpha\beta}(s)$ determines the membrane potential response at time t of dendritic compartment $\alpha \in \Gamma$ induced by a unit impulse stimulation of compartment β at an earlier time $t-s$.

It is convenient to introduce some additional simplifications. First, assume that $J_{ij}^{\alpha_0} = 0$ for all i,j, that is, there is no direct stimulation of the soma. Second, J_{ij}^α, $\alpha \in \Gamma$, is taken to have the product form $J_{ij}^\alpha = w^\alpha J_{ij}$, $\sum_{\alpha\in\Gamma} w_\alpha = 1$, which means that the spatial distribution of the input from neuron j across the compartments of neuron i is independent of i and j. If equation (2) is now substituted into equation (1a) one obtains N coupled nonlinear Volterra integro-differential equations for the N variables ϕ_i, which are given by

$$\frac{d\phi_i}{dt} = -\frac{\phi_i}{\hat{\tau}} + \frac{1}{\hat{\gamma}} \int_{t_0}^{t} dt' \left[\sum_{j \neq i} J_{ij} H(t - t') f(\phi_j(t')) + \tilde{H}(t - t') \phi_i(t') \right] \quad (4)$$

with $\tilde{H}(t) = G_{\alpha_1 \alpha_1}(t) / \hat{\gamma}_0$, $\quad H(t) = \sum_{\alpha \in \Gamma'} w^\alpha G_{\alpha_1 \alpha}(t)$, $\quad \tau_{\alpha_0} = \hat{\tau}$, $\tau_{\alpha_1 \alpha_0} = \hat{\gamma}_0$, and $\tau_{\alpha_0 \alpha_1} = \hat{\gamma}$. Equation (4) is the functional extension of the familiar ordinary differential equation describing the dynamics of a standard (single compartment) analogue neural network to the case of neurons with dendritic structure. This dendritic structure is represented compactly in terms of the neuron response functions $G_{\alpha\beta}(t)$.

Recently, Bressloff and Taylor (1993) have developed a method for evaluating $\exp(t\mathbf{Q})$ which leads to simple analytical expressions for the response function $G_{\alpha\beta}(t)$ of an arbitrary dendritic tree provided that each branch of the tree is uniform and certain conditions are imposed on the membrane properties of compartments at the nodes and terminals of the tree. Under such assumptions, one can show that $e^{t\mathbf{Q}} = e^{-t/\tau} e^{t\mathbf{K}/\gamma}$ where τ and γ are global membrane time constants and junctional time constants of the system, and the matrix $\mathbf{K}$ generates paths along the tree. In particular, modulo additional constant factors arising from the boundary conditions at terminals and nodes, $[\mathbf{K}^m]_{\alpha\beta}$ is equal to the number of possible paths consisting of m steps between the compartments α and β on the tree, where a step is a single jump between neighbouring compartments. Thus the calculation of $G_{\alpha\beta}(t)$ reduces to (i) determining the sum over paths $[\mathbf{K}^m]_{\alpha\beta}$ and then (ii) evaluating the infinite series $\sum_{m \geq 0} (t / \gamma)^m \left[\mathbf{K}^m \right]_{\alpha\beta} / m!$ obtained by expanding $e^{t\mathbf{K}/\gamma}$ in powers of t. The simplest example is that of a uniform, infinite dendritic chain for which steps (i) and (ii) can be performed explicitly to yield $G_{\alpha\beta}^{(0)}(t) = e^{-t/\tau} I_{|\beta - \alpha|}(2t / \gamma)$ where I_n is a modified Bessel function of integer order n. Using arguments similar to Abbott et al (1991) one can then express the response function of an arbitrary tree as $G_{\alpha\beta}(t) = \sum_\mu c_\mu G_{\alpha + L_\mu, \beta}^{(0)}(t)$ where the right-hand side involves a summation over all possible trips μ (paths that can only reverse direction at terminal and branching nodes) and L_μ is the length (total number of steps) of a trip. The coefficients c_μ arise from additional factors picked up at terminal and branching nodes, and there are systematic rules for calculating these coefficients (Bressloff and Taylor, 1993).

4.   Discussion.

We end this paper with a brief discussion of one of the several issues that arises from our analysis. (A detailed study will be presented elsewhere). This concerns the onset of oscillations in compartmental networks due to the presence of continuous delays associated with the convolution integrals. One well known feature of delays is that they can lead to destabilization of a fixed point and the simultaneous creation of a stable limit cycle via a supercritical Hopf bifurcation (Cushing, 1977). This was studied previously by Marcus and Westervelt (1988) in the case of an analogue network with discrete delays. In order to study (local) destabilization in compartmental networks, one first needs to consider the linearization of equation (4) about an equilibrium ϕ^*. Setting $y_i = \phi_i - \phi^*_i$ this takes the form

$$\frac{dy_i}{dt} = -\frac{y_i}{\hat{\tau}} + \frac{1}{\hat{\gamma}} \int_0^t dt' \left[\sum_{j=1}^N C_{ij}(t-t') y_j(t') \right] \tag{5}$$

where $C_{ij}(t) = H(t) J_{ij} df(\phi^*)/d\phi_j$, $j \neq i$ and $C_{ii}(t) = \tilde{H}(t) df(\phi^*)/d\phi_i$. (We have assumed that $t_0 < 0$, and have taken the contribution of the linearized integral over the interval $[t_0, 0]$ to be a higher order correction to equation (7), cf. Cushing (1977)). Linear stability analysis then involves studying the roots of the characteristic equation $\det((z+\hat{\tau}^{-1})\mathbf{I} - \hat{\gamma}^{-1} \tilde{\mathbf{C}}(z))$ = 0 where $\tilde{\mathbf{C}}(z)$ is the Laplace transform of the kernel $C(t)$. One particularly useful feature of our analysis is that $\tilde{\mathbf{C}}(z)$ can be calculated explicitly for arbitrary dendritic topologies using the construction of the single neuron response function presented in Bressloff and Taylor (1993).

References.

Abbott L. F., Farhi E. and Gutmann S. (1991) The path integral for dendritic trees. Biol. Cybern. **66**: 61-70.

Bressloff P. C. and Taylor J. G., (1993) Compartmental response function for dendritic trees. Biol. Cybern. **70**: 199-207

Cushing J. M. (1977). Integrodifferential equations and delay models in population dynamics. Lecture notes in biomathematics **20**.

Marcus C. M. and Westervelt R. M. (1989). Stability of analog neural networks. Phys. Rev. A **39** 347-359.

Poggio T. and Torre V (1977) A new approach to synaptic interactions. In: Lecture notes in Biomathematics, 21: 89-115.

Rall W. (1964) Theoretical significance of dendritic trees for neuronal input-output relations. In: Reiss R. F. (ed) Neural theory and modeling. Stanford University Press, Stanford, pp 73-97.

Hysteresis in a Two Neuron-Network: Basic Characteristics and Physiological Implications

K. Pakdaman[†], A. van Ooyen[‡], A.R. Houweling[‡], J.-F. Vibert[†]

[†] B3E INSERM U 263/Faculté de Médecine Saint-Antoine

Paris France

[‡] Netherlands Institute for Brain Research

Amsterdam the Netherlands

1 Introduction

Physiological evidences show that electrical activity affects a neuron's neurite outgrowth by modulating the concentration of intra-cellular calcium (Kater *et al.*, 1990). The change of connectivity thus produced modifies in turn the electrical activity. Based upon such evidences, van Ooyen and van Pelt devised a model for activity dependant neurite growth in neural networks (van Ooyen and van Pelt, 1992; 1993). Their study shows that when the neuron model has a sigmoidal transfer function with a spontaneous activity, the network connectivity goes through an overshoot followed by elimination before stabilizing. This behavior was explained by the activity being a hysteresis function of connectivity. In this paper we further investigate the basic characteristics required for the network activity to display such a behavior.

2 The model

Van Ooyen and van Pelt show that a two neuron-network connected with identical excitatory connections captures the main features of phase transition and hysteresis in activity dependant neurite outgrowth. The model network is governed by the following equations.

$$\frac{dX}{dt} = -\gamma X + (A - X) W F(Y)$$

$$\frac{dY}{dt} = -\gamma Y + (A - Y) W F(X) \qquad (1)$$

$$\frac{dW}{dt} = R(X, Y)$$

Where X (resp. Y) is the membrane potential of neuron 1 (resp. 2), A is the saturation potential, γ determines the rate of decay, and W is the connection strength between cell 1 and 2 as well as between cell 2 and 1. Throughout this study the connection is excitatory and therefore W is a positive number. The firing rate of a neuron having a membrane potential X is determined by the function F, which is referred to as the neuron's transfer function. The basic assumptions on F are that it be a sigmoid function such that:

$$\begin{cases} F(0) = a, \ a > 0 \\ F'(x) > 0, \forall x \end{cases} \qquad (2)$$

During growth the connection weight changes and this evolution takes place according to the third part of equation (1). The function R satisfies the following constraints:

$$\begin{cases} R \text{ is symmetrical in } X \text{ and } Y \text{ that is } R(X,Y) = R(Y,X). \\ \text{There exists } X_0 > 0 \text{ such that: } \begin{cases} \forall X < X_0 & R(X,X) > 0 \\ & R(X_0, X_0) = 0 \\ \forall X > X_0 & R(X,X) < 0 \end{cases} \end{cases} \qquad (3)$$

This study is concerned with the asymptotic evolution of the equations (1).

3 Results

The dynamics of weight modification are on a much slower time scale than the electrophysiological activity of the neurons. This makes it possible to separate the system into slow and fast dynamics. Such methods were first applied by Zeeman in the modelling of neural behavior (Zeeman, 1972).

3.1 The slow manifold

The slow manifold corresponds to the set of equilibria of the fast dynamics with the weight considered as a fixed parameter. It represents therefore the bifurcation scheme of the fast dynamics for values of the parameter W. The slow manifold is determined by the two following equations.

$$\begin{cases} -\gamma X + (A - X) W F(Y) = 0 \\ -\gamma Y + (A - Y) W F(X) = 0 \end{cases} \qquad (4)$$

Taking into account the second constraint on F, it can be shown that system (4) is equivalent to the following system.

$$\begin{cases} X = Y \\ -\gamma X + (A - X) W F(X) = 0 \end{cases} \qquad (5)$$

The dynamics of equations (1) display a hysteresis when the slow manifold is S-shaped or equivalently the bifurcation diagram of the fast dynamics has exactly two turning points where $\frac{dW}{dX} = 0$. The following result can then be deduced:

When the transfer function F satisfies constraints (2), system (1) displays a hysteresis if and only if the function v defined below has exactly two zeros on the interval $]0, A[$, and changes sign at these zeros.

$$v(X) = F'(X) - \frac{AF(X)}{X(A - X)} \qquad (6)$$

From this point on we suppose v satisfies the condition described above and let X_1 and X_2 be the two zeros.

3.2 The isoclines

For $\frac{\gamma X_1}{(A-X_1)F(X_1)} < W < \frac{\gamma X_2}{(A-X_2)F(X_2)}$ the two isoclines (each isocline is defined by one of the equations in (4)) intersect in three points. Moreover the study of the slow manifold shows that these points are necessarily on the plane defined by $x = y$. Therefore for a weight given in this range each isocline crosses three times the line $x = y$.

3.3 `Asymptotic behavior

As the slow manifold lies in the $x = y$ plane, equilibria of the system (1) must satisfy the same constraint. In fact an equilibrium point of the system is a point (X, X) on the slow manifold for which $R(X, X) = 0$. According to (3), there is only one point $\mathcal{P} = (X_0, X_0, \frac{\gamma X_0}{(A-X_0)F(X_0)})$ for which this condition holds. The analysis of the stability of this equilibrium point leads to the following result.

There are two values $X_1' \equiv X_1$ and $X_2' \equiv X_2$ such that:
for $X_0 < X_1'$ or $X_0 > X_2'$ the system globally asymptotically converges to the equilibrium point $\mathcal{P}$.
At X_1' (resp. X_2') the system goes through a Hopf bifurcation, thus for $X_1' < X_0 < X_2'$ the system has a hyperbolic equilibrium point $\mathcal{P}$, and an attracting limit cycle lying in the plane $x = y$.

3.4 Application

In the case discussed by van Ooyen and van Pelt F is defined by:

$$F(X) = \frac{1}{1 + e^{(\theta - X)/\alpha}} \tag{7}$$

Applying equation (6) to this function shows that the system displays hysteresis for $\alpha < \alpha_0$, with α_0 satisfying:

$$A + (4\alpha_0 - A)e^{(\theta - 2\alpha_0)/\alpha_0} = 0 \tag{8}$$

4 Discussion

In this paper the results obtained by van Ooyen and van Pelt were complemented. Here it was shown analytically that any neuron model defined by similar equations would present the same hysteresis phenomenon as long as its transfer function satisfies the constraints (2 and (6). Moreover the geometrical interpretation of this constraint as the number of times the isocline (4) crosses the line $x = y$ for a given weight enables further generalization of this result to any neuron model. In fact the isocline curve represents the stationary potential X reached by a neuron when excited by another neuron at a fixed potential Y, through a synapse with strength W, and can therefore be evaluated for any neuron model. Let us note this function as $X = G(Y, W)$. Necessary and sufficient conditions for a neuron model to display hysteresis are then: 1) Spontaneous sporadic activity 2) Acceleration.

Spontaneous activity implies that $G(0, W) = a > 0$. This is a common living neuron property and is taken into account in many biologically plausible models. Acceleration refers to the fact that the isocline crosses the line $x = y$ for a range of weights, and that therefore for these values the post-synaptic neuron has a higher potential than the pre-synaptic neuron. In other words, under special circumstances, a neuron receiving a spike train with a rate f fires with $f' > f$. This kind of behavior is observed in living neurons. The accelerating power of a neuron depends on 1) post-synaptic potential characteristics such as its duration and amplitude 2) refractory period characteristics. The former depends on the neurotransmitter and the ionic channels involved as well as the nature of the connection (axo-somatic etc), and a detailed study of this is beyond the scope of this paper. The latter is mainly characterized by the post-spike potassium currents that depolarize the neuron. It should be noted that the duration of the refractory period is modulated by two such currents, referred to as I_C and I_{ahp}. I_C is a calcium-dependant, voltage-dependant potassium current, activated by μM concentrations of intra-cellular calcium at -50 mv, and it speeds up the post-spike depolarization thus shortening the refractory period and increasing the firing rate. I_{ahp}, on the other hand, is calcium-dependant voltage-independant potassium current, activated by mM concentrations of intra-cellular calcium. It slows down the post-spike depolarization thus increasing the refractory period and decreasing the firing rate. These currents illustrate well the fact that the intra-cellular calcium concentration which affects the neurite outgrowth may also be responsible for the changes in the firing rate observed in neurons.

5 Conclusion

Starting from a model developed by van Ooyen and van Pelt which accounted for phenomenon observed in neurite growth, methods for the analysis of dynamical systems were used to determine the geometrical characteristics required to reproduce the behavior. These features were then interpreted in terms of physiologically relevant observables. Further investigations in this direction are under way to improve the plausibility of the growth model.

6 References

Kater, S.B., Guthrie, P.B. and Mills, L.R. (1990) Integration by the neuronal growth cone: a continuum from neuroplasticity to neuropathology. Progress in Brain Research 86, 117–128.

van Ooyen, A. and van Pelt, J. (1992) Phase transitions, hysteresis and overshoot in developing neural networks. in Artificial Neural Networks, 2. I. Aleksander and J. Taylor (Eds). Elsevier Science Publishers B.V.

van Ooyen, A. and van Pelt, J. (1993) Activity-dependant outgrowth of neurons and overshoot phenomena in developing neural networks. Journal of Theoretical Biology, in press.

Zeeman E.C. (1972) Differential Equations for the Heartbeat and Nerve Impulse. Mathematics Institute University of Warwick Coventry, England.

Cooperation within networks of cortical automata based networks

Latifa Boutkhil[†], Frank Joublin[††], Sylvie Wacquant[††]
[†] Neurosciences et Modélisation,U.M.P.C- Paris VI, Paris, France
[††] L.C.I.A, Neurovision, I.N.S.A - Mont Saint-Aignan, France

1 Introduction

During the last few years, connectionist models obviously evolved. In particular, modular models involving several types of architectures or algorithms now replace the "one-block" original networks (Bechtel, 1993). When designing these architectures, it's possible to point out many kinds of inter-networks cooperations.
The cortical automata model (Burnod, 1988) is especially well suited to the implementation of such relationships. We present a vision system having invariance capabilities acquired through active exploration of the environment (Otto & al, 1992; Boutkhil & al., 1992; Marchal, 1992) that serves to illustrate the ways such networks may cooperate.

2 Short presentation of the model

2.1 The frame of the cortical automata model

The following biologically plausible model is described in (Burnod, 1988; Alexandre & al., 1991; Wacquant, 1993). Its basic unit does no more correspond to the formal neuron of McCulloch and Pitts or similar ones but to the cortical column, micro-circuit representing about a hundred of neurons and repeated all over the cortical surface. The so called "cortical automaton" is divided in layers[1] which locally sort information on origin or destination of links. Automaton outputs, weights and internal variables are locally computed through transfer functions which may be logical ones.
Automata are included in a structural hierarchy. The "low structures" or maxicolumns[2] are composed of groups of automata that share common inputs allowing their functional differentiation. Low structures make up "high structures" or cortical areas[1] which represent a network already characterized by a functional specialization. At the top, the "formal brain" results from the cooperation between those different structures, by mean of layers.

2.2 An example of "network of networks"

Experimental research on the visual system shows that invariant recognition of object is the result of many cooperative computations : local features extraction, multilayered processing, top-down interaction from the partially recognized features,

---

[1]Not to be confused with layers of perceptrons.
[2]Biological term.

use of ocular movements for the successive exploration of attractive points in the scene.

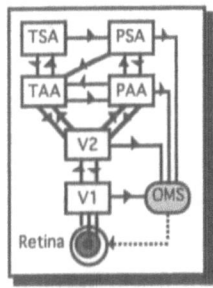

With the purpose of neuro-anatomical and physiological coherence, a cortical automata based network of invariant visual recognition (Otto & al., 1992) has been developed which join usual neural networks properties (generalization, noise immunity) but also architectural and functional properties of the primate visual system.

The multilayered architecture is composed of two branches that form a "Y" (cf fig 1). The temporal branch or "What" pathway enables the recognition of local features and

fig 1

configurations while the parietal branch or "Where" pathway extracts their locations. Invariant recognition results both from the difference in receptive fields increase within the two branches and from the temporo-parietal connections allowing the generation of an ocular movement command which resets the external pattern in the foveal zone.

The network architecture is composed of the following high structures (cf fig 1) :
- a retina, split into three zones (fovea, perifovea, far periphery).
- two primary areas (V1, V2) in charge, respectively, of simple and complex features extraction.
- above the primary areas, two associative areas manage learning: TAA (Temporal Associative Area) and PAA (Parietal Associative Area).
-at the top of the networks two output areas give the results: TSA (Temporal Semantic Area) and PSA (Parietal Semantic Area). These two outputs are determined by population coding. A peak of activity on TSA means the most likely pattern; the vector on PSA corresponds to the ocular movement needed by the OMS (ocular-motor system) to center the pattern on the fovea.

3 Different types of cooperations

The following classification describes different kinds of cooperations (ACTH, 1992) that could be found within large networks. For each type, we present the interest of such a relationship and comment what it brings in the particular example of the invariant visual recognition network. In each figure, the left icon symbolizes the cooperation type and the right one localizes it on the illustrated network architecture.

3.1 The "feedforward/feedbackward" cooperation

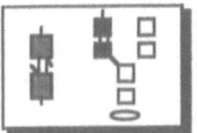

In this type of cooperation, the networks collaborate both in the feedforward direction by carrying out a progressively more elaborated information processing and in the feedbackward direction by allowing corrections on previous processings or more generally the realization of attentional processes by filtering feedforward activity.

In the current example, feedforward connections enable the filtering of incoming visual information (with directional masks) whereas feedback connections enable, at associative areas level, the matching of feedforward information with the one stored during the learning phase.

168

3.2 The "lateral" cooperation

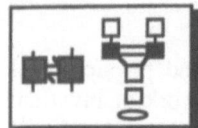

This type of cooperation happens between two parallel processing branches which can bring each other complementary information. An example of this cooperation is illustrated by the parieto-temporal interaction.

The processing of feedforward information arriving at TAA is controlled by the parietal branch, whose information of location drives a dynamical "internal shift" of the activity in TAA. This internal shift allows to get back the learned configuration, enabling recognition in TSA of the perifoveal target. This spares an ocular movement , thus increasing the invariant region on the perifovea.

3.3 The "competitive" cooperation

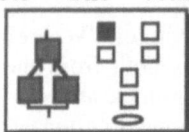

In this case, the same information is analyzed by different systems which specialize themselves on a specific processing.

In the recognition task, especially in TSA, there are extractions of several configurations. The cooperation occurs while automata compete to determine an answer —the most likely pattern— on a quality criterion. The underlying biological process is lateral inhibition ("winner-take-all" mechanism), which can be implemented through the local connections of the cortical automaton.

3.4 The "associative" cooperation

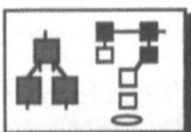

This cooperation results from the association of two or more informations of different types in order to compute a new one. By this way, the fusion of several information channels (visual and motor, for example) can be achieved and implemented by the inter-layer connectivity of automata.

In the illustration network, when the input pattern is outside the fovea, the non recognition in TSA provokes a call of PSA. This area locally combines this call with the feedforward activity coming from PAA (location of the target) to generate an appropriate command for OMS. As a result, the pattern is recentered on the retina: the recognition can then occurs.

3.5 The "sensori-motor" cooperation

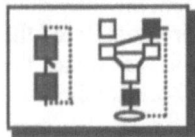

In the case of systems integrated in an environment on which they may act and from which they may take information — typically, the case of biological systems—, another structural cooperation type could be defined. This one is unidirectional within the system but bidirectional considering the external return loop. In fact, motor activity helps to structure sensory system by its action on the environment and, in reverse, sensory structuration sharpens behavioural motor activity.

In the illustration, such a sensory-motor loop is used by the associative areas to learn the internal shift mechanism occuring at the level of TAA : the ocular movements generated when the form is presented in the perifoveal zone are thus "internalized".

3.6 The "temporal" cooperation

Unlike previous structural cooperations which were bound with connections, this one takes into account temporal development of networks.

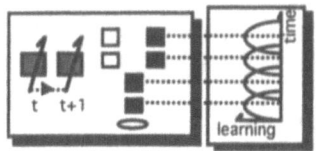

It concerns specifically the structuration of networks —i.e to sharpen or change, temporary or permanently, networks tasks— with time.

One illustration of this type of cooperation is in the studied model, the progressive setting of the high structures. Two factors are taken into account: a maturational factor and a learning factor. The first one refers to the critical period of plasticity of cortical areas while the second one refers to the learning rate during this period. Both factors have consequences on an architecture based upon known anatomical connections between cortical areas.

Within the studied network, the acquisition of perceptive invariance (Boutkhil & Burnod, 1992) is decomposed in a set of progressive learning stages. In this progressive structuration of the visual cortical areas, the active exploration of the environment by means of ocular movements is crucial (Marchal, 1992). These stages are in direct correspondence with psychological data concerning the visual exploration capabilities of the newborn.

Conclusion

This non-restricted cooperation classification may be generalized to others types of networks (ACTH, 1992). These cooperations lead to design a"network of networks" which include several interesting points, in particular, the increase of networks number develops cognitive functions level and a modular architecture allows incremental development of complex networks, thus acquiring new capabilities by adding new modules and cooperations.

References

ACTH: L. Boutkhil, H. Cardot, F. Joublin, V. Lorquet, JD. Muller, O. Sarzeaud& S. Wacquant (1992). Structures et algorithmes neuromimétiques coopératifs pour la résolution de problèmes complexes, Groupe de travail de l'ACTH, *5èmes Journées Internationales Neuro-Nîmes 92*, Nimes, France, Novembre 1992.

Alexandre, F., Guyot, F., Haton, J.P., Burnod, Y. (1991) The cortical column: a new procesing unit for multilayered networks. *Neural Networks*, 4, 15-25.

W. Bechtel (1993) Currents in connectionism. Minds and Machines : *Journal for Artificial Intelligence Philosophy and Cognitive Science*, 3 (2) : 125-153.

L., Boutkhil, Y., Burnod (1992) Modélisation par réseaux de neurones de l'acquistion de la constance perceptive chez le jeune enfant, *Association pour la recherche cognitive*, Nancy, 24-26 Mars 1992, 159-184.

Y. Burnod (1988). An adaptive neural network : the cerebral cortex, Prentice Hall.

Y. Burnod, P. Grandguillaume, I. Otto, S. Ferraina, P.B. Johnson & R. Caminiti (1992), Visuomotor transformations underlying Arm Movements toward Visual targets: A Neural Network Model of Cerebral Cortical Operations,. *Journal of Neuroscience*, 12 : 1435-1453.

P. Marchal (1992) Coopération entre la reconnaissance et l'exploration : modélisation fonctionnelle des aires corticales visuelles. Thèse de doctorat, Université Paris VI, France.

I. Otto, E. Guigon, P. Grandguillaume, L. Boutkhil & Y. Burnod (1992). Direct and Indirect Cooperation between Temporal and Parietal Networks for Invariant Visual Recognition. *Journal of Cognitive Neuroscience*, 4 (1) : 35-57.

Wacquant S. (1993) Contribution à l'étude d'un modèle de réseaux d'automates corticaux : principes et outils logiciels. Thèse de doctorat, Université de Rouen, France.

Anisotropic Correlation Properties in the Spatial Structure of Cortical Orientation Maps

S.P. Sabatini, R. Raffo, G.M. Bisio

Department of Biophysical and Electronic Engineering, University of Genoa

Via Opera Pia 11a - 16145 Genova - ITALY

1   Introduction

Experimental and theoretical studies on visual information processing have evidenced the advantages of a delocalized representation of information through *feature maps* which mix topological contiguity with proximity in the feature space. The peculiar dispositions of cortical orientation-selective cells can be understood in terms of dimension-reducing mappings which translate neighborhood relations in the orientation-subspace in spatial neighborhood relations on the cortical surface [Durbin and Mitchison, 1990]. Thus, cortical maps are not only a repository of information about the features present in the image, but are the substrate for coordinated interactions among features. The basic computational principles of these maps can be investigated with different approaches (heuristic, theoretical information, computational). In this paper, we'll provide information-theoretic insights about spatial arrangements of simple cell receptive fields in several biologically plausible orientation maps, by studying both autocorrelation and directional mutual information of cortical activity.

2   Autocorrelation of cortical activity

The retino-cortical pathway can be modeled by two layers, interconnected through feed-forward connections. The layers represent the retina ($\mathbf{x}$) and the cortical surface ($\mathbf{u}$). If $i(\mathbf{x})$ is the visual input, the activity $e(\mathbf{u})$ of a cortical cell can be written as $e(\mathbf{u}) = \int \int w_{\mathbf{u}}(\mathbf{x}) i(\mathbf{x}) \, d\mathbf{x}$ where the linear transfer function $w_{\mathbf{u}}(\mathbf{x})$ represents the receptive field. In this paper, we assume that cells have the same elongated Gaussian receptive field profile (aspect ratio 3.0), indepently of their position on the retinal plane, and that they are selective to a specific orientation θ that varies with cortical coordinates, according to a fixed orientation map. To estimate the capabilities of representation of visual information through a population of cells, we evaluate, in absence of noise, the spatial autocorrelation function of cortical activity $e(\mathbf{u})$. The autocorrelation function is strictly related to the statistics of input signal. However, to focus our attention only on the structural properties of orientation maps, we assumed that visual inputs are completely uncorrelated. Under this assumption, to evaluate the autocorrelation function of cortical activity it is sufficient to consider the deterministic autocorrelation function

[Papoulis, 1987] $Q(\mathbf{u};\mathbf{u}') = \int\int w_{\mathbf{u}}(\mathbf{x}) w_{\mathbf{u}'}(\mathbf{x}) d\mathbf{x}$. Given a cell located at position $\mathbf{u}^*$ on the cortical surface, the autocorrelation function centered around that cell is $Q(\mathbf{u}^*;\mathbf{u}^*+\mathbf{s})$. According to the characteristics of the orientation map, $Q(\mathbf{u}^*;\mathbf{u}^*+\mathbf{s})$ changes from point to point on the cortical surface. The values of the autocorrelation function can be used to define the ensemble of cells belonging to the hypercolumn centered around any given cell. In fact, $Q(\mathbf{u}^*;\mathbf{u}^*+\mathbf{s})$ is related to the maximum set of dependent channels (receptive fields) with respect to a reference cell, and it can be used to define the hypercolumn radius. In Fig. 1 three different maps generated through a model recently proposed by Wörgötter and Niebur [Wörgötter and Niebur, 1993] are shown. The shaded regions represent the hypercolumns referred to the cell located in the center of the map. The 2D profiles of the autocorrelation functions provide clues about the signal processing capabilities of the spatial arrangements of receptive fields within the hypercolumns, as well as about global properties of the whole map. The elongated shapes, observed in stripe-like maps as the one depicted in Fig. 1(left), point out that, in this type of map, there are preferential directions along which strongly correlated activities occur. It is worth noting that, from the point of view of coding theory, optimal mappings of input signals onto cortical surface, could be achieved, on the contrary, when autocorrelation functions are characterized by radial simmetry similar to the one observed for the more realistically-looking maps shown in Figs. 1(center) and 1(right).

Figure 1: Contour plots of the autocorrelation function referred to the cell located in the center of the map. Their shapes depend on the background orientation maps evaluated according to [Wörgötter and Niebur, 1993].

3 Directional mutual information

More information about the anisotropies of information properties of hypercolumns can be obtained considering the mutual information of cortical activity and evaluating it along various direction departing from a given cell in the orientation map. The information capacity of an assembly of cells, that act as a single information-carrying channel, can be measured in terms of the conditional entropy of their activity. For any given direction ϕ departing from a fixed cortical cell, let's consider the conditional entropy $H_\phi\left(e_0 \mid e_1^\phi, \ldots, e_d^\phi\right)$, i.e. the uncertainty about the activity e_0 of a fixed cell, when the activity of d cells along the ϕ direction have been observed. For distances greater than

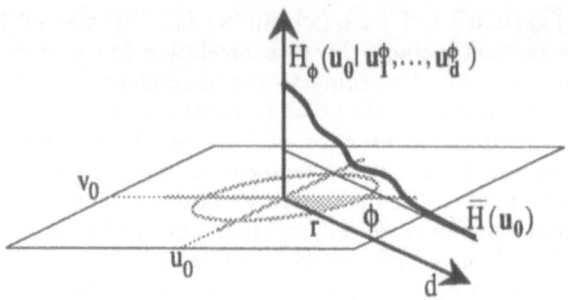

Figure 2: Illustration showing how directional entropy has been evaluated.

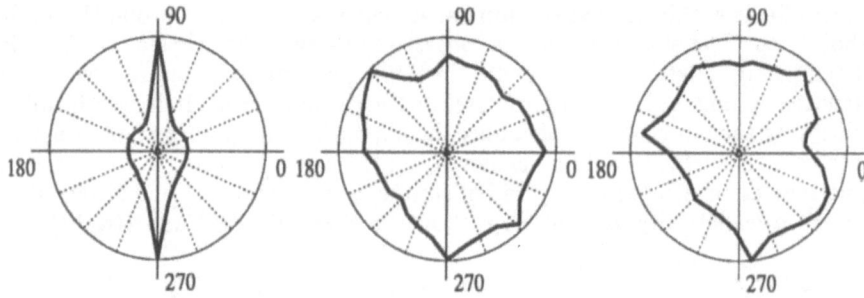

Figure 3: Polar plots of mutual information I_ϕ for the three hypercolumns shown in Fig. 1.

the hypercolumn radius the conditional entropy approaches its asymptotical value representing the *entropy rate* $\bar{H}_\phi(e_0)$ of the ensemble of cells along the direction ϕ(see Fig. 2). If the input signals are mutually spatially independent with normal distribution the entropy rate $\bar{H}_\phi(e_0)$ is related only to the spatial structure of the orientation map and the mutual information can be written as [Papoulis, 1987]

$$I_\phi\left(e_0, e_1^\phi, \dots, e_\infty^\phi\right) = H(e_0) - \bar{H}_\phi(e_0) = \frac{1}{2}\left(\ln(S_0) - \lim_{d\to\infty}\ln\frac{\Delta_{d+1}^\phi}{\Delta_d^\phi}\right)$$

where Δ_d^ϕ is the determinant of the correlation matrix R_d^ϕ of the activity of the first d cells along the direction ϕ, and S_0 is the energy of the receptive field profile. I_ϕ measures the information that the activity of the ensemble of cells disposed along the direction ϕ conveys about the input signal.

The evaluation of directional mutual information along several direction and for different centers on the map evidences anisotropies in the amount of information. In Fig. 3 are shown the polar plots of mutual information I_ϕ; from left to right the diagrams refer to the maps depicted in Fig. 1. The stripe-like map (Fig. 3 left) reveals highest asymmetries in mutual information while patch-like maps achieve good simmetry in mutual information.

4 Discussion and Conclusion

Information processing capabilities of visual cortex can be evaluated at different levels of observation. At single cell level, one deals with the properties of receptive fields considered as local feature extractors. At global level the whole orientation map should be characterized to point out how these structured assemblies of neurons cooperate and gain higher levels of description (e.g. feature enhancement and texture segregation). This can be pursued investigating how information about any small locus in visual space affects the activity of nearby cells on the basis of their overlap and mutual interactions. In this paper, we have focussed our analysis on the overlap. The way in which neighboring receptive fields overlap depends on their shape and extent, and on their relative disposition dictated by the orientation map. This overlap affects the local representation of visual information and can be quantified considering the autocorrelation function of cortical activity. In order to have the same representation and computation capabilities across the visual space, it is necessary to have an equal allocation of cells of different orientation preference for each point (cf. the uniform *coverage* criterium of [Swindale, 1991]). We should prefer those orientation maps that exhibit average invariance characteristics of the autocorrelation functions, thus maximizing spatial uncertainty. Comparable quantitative conclusions can be drawn by measuring directional mutual information at each cortical locations, and using it as a further measure of uniformity of coverage. Under this perspective, the (an-)isotropies in both the autocorrelation function and in mutual information become figures of merit of orientation maps or act as constraints for the design of artificial orientation maps.

Using these quantitative criteria we can characterize different orientation maps. We have observed (see Fig. 1(center), 1(right) and also Fig. 3(center), 3(right)) that patch-like arrangements of orientations lead to isotropic autocorrelation functions, thus ensuring good space invariant properties. Conversely, strong anisotropies have been observed (see Fig. 1(left) and Fig. 3(left)) for stripe-like maps.

References

[Durbin and Mitchison, 1990] R. Durbin and G. Mitchison. A dimension reduction framework for understanding cortical maps. *Nature*, 343:644–647, 1990.

[Papoulis, 1987] A. Papoulis. *Probability, random variables, and stochastic processes*. McGraw-Hill, 1987.

[Swindale, 1991] N.V. Swindale. Coverage and the design of striate cortex. *Biol. Cybern.*, 65:415–424, 1991.

[Wörgötter and Niebur, 1993] F. Wörgötter and E. Niebur. Cortical column design: A link between the maps of preferred orientation and orientation tuning strength? *Biol. Cybern.*, in press, 1993.

PART 2

Mathematical Model

Application of Neural Network and Fuzzy Logic in Modelling and Control of Fermentation Processes

N.A. Jalel, B. Zhang and J.R. Leigh

Industrail Control Centre,
University of Westminster,
London, UK

1 Introduction

The purpose of this paper is to investigate the use of neural networks and fuzzy logic in the modelling and control of an industrial fed-batch fermentation process. The reason why we choose neural networks and fuzzy logic for this application is because of the nature of the fermentation process and the features of the neural networks and fuzzy logic. Neural networks are composed of many processing elements and connections. From a system-theoretic point of view, a neural network can be viewed as a complex nonlinear function. The power of neural networks is that they are capable of representing various complex nonlinear functions. Neural networks have been trained to perform complex nonlinear functions in many fields of applications including pattern recognition, system identification, classification, speech recognition, image processing and control systems. For control applications in particular, neural networks can perform some tasks which are difficult for conventional approaches. Due to many uncertainties and non-linearities encountered, the control of fed-batch fermentation processes is a very difficult task. The mechanism of the processes is usually poorly understood. Some process variables are difficult to measure. Typically some variables are determined by a slow infrequent off-line laboratory analysis, which causes an undesirable time delay in using these variables for control. The practical situation we have encountered is that there is a massive amount of data of previous batches of an industrial fermentation process which have been recorded and available to us. Considering the difficulty in obtaining an analytical model of the process, and with such massive data available, we decided to use neural networks to model the process since neural networks can easily approximate any reasonable functions with no need to specify the structure of the functions (Bhat et al, 1990). Based on the previously recorded data, the neural networks are trained to predict, using on-line measurements, some variables which have to be obtained off-line by laboratory analysis. Hence these off-line measurements can be estimated on-line with the neural network model. The controller is then designed and activated according to the output of the neural networks. In this paper, the modelling of a fed-batch fermentation process using neural networks and the control of this process using self-organising fuzzy logic and neural network feedback controller are presented. Firstly the details of modelling the process using neural networks are described. After showing the success of the neural network model in representing the process being modelled, a self-organising fuzzy logic controller and the neural network controller are designed based on the neural network model. Then some simulation results of the control system based on the neural network and the fuzzy logic controller are shown. Finally some conclusions and discussion are made.

2 Modelling of the fermentation process

In a fermentation process some state variables are unmeasurable. In practice the values of these variables are obtained by infrequent laboratory analysis. From operation experience, it is known that the fermentation process can be much better performed if some of these variables can be well controlled. In order to control these variables, it is necessary to know the values of the these variables. When these variables cannot be measured on-line, it is natural to think the way of estimating these variables using some on-line measurements which are correlated with them. Different techniques for estimating the unmeasurable state variables in the fed-batch fermentation process have been investigated (Johnson, 1987). Among them neural networks have been found to be one of the most suitable approaches for performing the task (Jalel et al, 1993). In this paper, a multi-layer feedforward neural network is used for on-line estimate the residual nitrogen which is an important state variable, with the inputs to the network being some on-line measurements. According to the knowledge of experts of the process and based on the correlation analysis of the data, it has been found that six on-line measurement variables are strongly correlated or affect the residual nitrogen in the fermenter, namely, pH, change of pH, dissolved oxygen, power, oxygen uptake rate and nitrogen feed rate. Therefore it is decided to use these six variables as the inputs to the network. The output of the neural network is the estimated residual nitrogen in the fermenter. The neural network has an architecture 6-10-5-1, which means it consists of six input, one output and two hidden layers with ten and five processing elements respectively. Each processing element contains a bias input as well. The output from each processing element in the hidden layer is sigmoidal. The output from the element in the output layer is pure linear. The training of the network is based on the recent ten batches of data. The input data to the network are from recorded on-line measurements. The target output data are from the off-line laboratory analysis results. Comparing with the sampling frequency of the on-line measurements, the target output data are quite infrequent and sparse. Therefore we have to pre-process the raw data before training the network. The input training data are extracted from the recorded log files to pair with the off-line measurements. The back propagation learning algorithms with momentum is used to avoid trapping in local minima. The learning rate and the momentum is chosen to be 0.9 and 0.4 respectively. The initial weights and biases of the network are randomly generated in the range (-1, 1) with an uniform distribution.

2.1 Simulation results
The performance of the network was examined after the training and figure 1 shows the estimated value and the off-line measurement for the residual nitrogen. From the figure it is possible to see that the network was able to provide a good prediction of the residual nitrogen concentration.

3 Self organising fuzzy logic controller
The basic design of a self organizing fuzzy logic controller has been described in several papers (Procky et al, 1979). It consists of two levels, the first containing a simple fuzzy logic controller and the second containing the self organizing mechanism, figure 2. The input signals to the controller taken at each sampling instant are, the error signal calculated by subtracting the process output from the desired value, and the change in error calculated by subtracting the error of the previous sample from the present one. The resulting signal is then mapped to the corresponding discrete level by using the error GE(E) and change in the error GCE(E) scaling factors prior to rule evaluation. The control output signal is calculated using the input signals to the controller and the control rules inside the controller. The output signal is defuzzified and scaled to engineers' value using the output scaling factor and is fed to the process being controlled. The fuzzy sets are formed upon a discrete support universe of discourse of 13 elements for the error, change in the error and the output, defined in the range (-+ 6). The membership function is chosen between 0 and 1. The output rules are usually viewed as linguistic conditional statements in the form:

IF Error is Positive Big AND Change in Error is Negative Small THEN Output is Positive Medium

The basic function of the performance index is to calculate the deviation from the desired trajectory, issue appropriate control actions at the output of the controller based on evaluating the performance and to modify the control action. The performance index mechanism involves performance feedback where the performance index operates by assigning a credit or reward value to the individual control actions that contributed to the present performance. The performance index has a knowledge about the process being controlled which takes the form of a matrix indexed by E and CE with the entry being the reinforcement. The performance index is usually in the form of a look-up table. The reinforcement issued by the performance index is sent to the rules modifier which modifies the m samples in the past rule that contributed to the present state. If the controller is empty then an initial rules must be created. Further rules will be generated when required by using the modification procedure.

3.1 SOFLC results based on neural network
SOFLC has been used to control the residual nitrogen around a desired trajectory by controlling the amount of nitrogen fed inside the fermenter and using the neural network model which relates the nitrogen fed with residual nitrogen. The inputs for the network are the on-line measurements which includes the nitrogen fed rate, the control signal generated by the SOFLC while the output of the network is taken to be the residual nitrogen which needs to be estimated and controlled around a desired level. In the first run the controller starts initially with no rules in the data base thus the control performance is poor. In the second and third runs the SOFLC control the residual nitrogen using the existing rules, generated from the previous run, and at the same time modifying the existing rules. The system learns to control the process when the rules are not modified any more. Figure 3 illustrates the controlled residual nitrogen in the first, second and third runs.

4 Neural network controller

One of the important aspects in the industrial fed batch fermentation process is to control the state variables of the process around a desired trajectory by controlling the amount of nutrient feed. In this work the Multi-Layered Perceptron is designed to act as a feedback controller for the process with the aim of controlling the residual nitrogen around a desired level by controlling the amount of nitrogen feed. The controller network is fully connected, feed-forward network consists of two processing elements in the input layer with five in the first hidden layer and three in the second and learn the control action by standard back-propagation algorithm. The inputs to the network are given to be the error between the estimated value of the residual nitrogen concentration and the desired output (E) and the Time (T) (the age of the batch). The estimated value is derived from the network which is acting as an estimator of the process. The output of the network is the desired control signal which is the amount of nitrogen fed rate in hour (NFH). The control signal generated by the neural network is used as an input to the network which is acting as an estimator of the process. The close loop diagram of the two networks which are acting as a model and a controller of the process is illustrated in figure 4. The inputs and output of the controller network are scaled between zero and one. The output of each processing element in the hidden layer is passed through a sigmoid function while the output layer is taken to be a linear. The network weights are initialised to a random starting weights between -0.1 and +0.1. The learning rate of the network is chosen as 0.9 while the momentum term is equal to 0.4.

4.1 Network results

Neural network is tasted on simulation in order to examine the ability of the neural network to control the residual nitrogen concentration around a desired trajectory. Figure 5 illustrates the control signal of the neural network and the desired nitrogen fed of the process. The nitrogen fed generated from the neural network is fed as an input to the neural net which is acting as a model of the process. Figure 6 shows the model output, and the off-line measurement of the residual nitrogen concentration, where it is possible to see that the network is able to control the state along the desired trajectory.

5 Conclusion

The fermentation process is highly nonlinear and uncertain process. In this paper the neural network has been used to model the state variables of the fed batch fermentation process. SOFLC has been used to control the residual nitrogen around a defined trajectory by controlling the amount of nitrogen fed to the fermenter. Good control has been achieved in simulation using this approach. SOFLC approach has achieved good results due to their ability to handle the variations and disturbances in the model parameters. Controlling the process is also achieved using neural network and by training the network to follow the existing control strategy. The neural network controller is able to control the state around a desired trajectory and also to provide a smoother control signal. From the simulation results it is possible to model the nonlinear industrial fermentation process using artificial neural network and to control the states of the process through the use of SOFLC and neural network controller based on the derived neural network model. The feasibility of using these approaches in fermentation control is currently being further evaluated.

6 References

1. Bhat N.V., Minderman P.A., McAvoy T. and Wang N.S. (1990), Modelling Chemical Process Systems Via Neural Computation, IEEE Control Systems Magazine, April, 24-30.
2. Jalel N. and Leigh J.R. (1993), Modelling the Fed Batch Ferementation Process Using An Artificial Neural Network, ICANN, Amsterdam, Holland.
3. Johnson A. (1987), The Control of Fed Batch Ferementation Process- A Survey, Automatica, 23, 6, 691-705.
4. Procky T.J. and Mamdani E.H. (1979), A Linguistic Self-Organising Process Controller, Automatica, 15, 15-30.

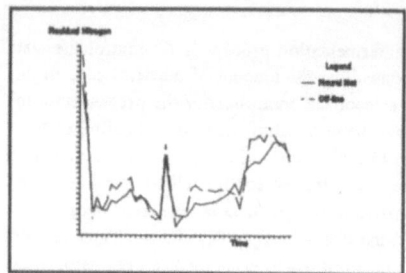

Figure 1 Neural Network Estimation

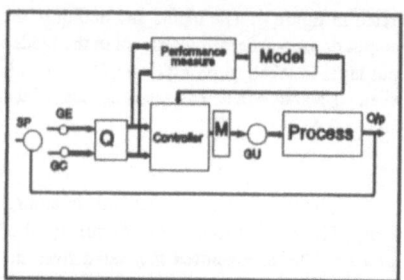

Figure 2 Self Organising Fuzzy Logic Controller

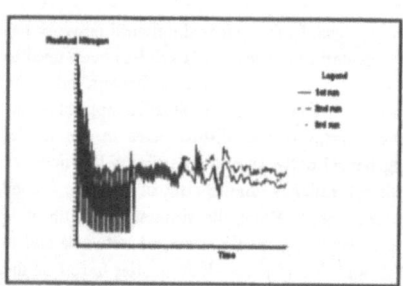

Figure 3 SOFLC Controlling Residual Nitrogen

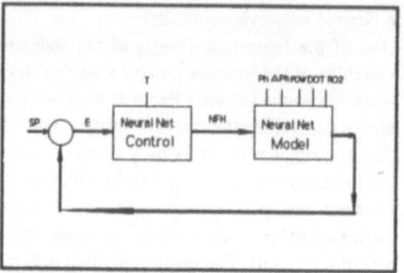

Figure 4 Neural Network as a Model and Controller

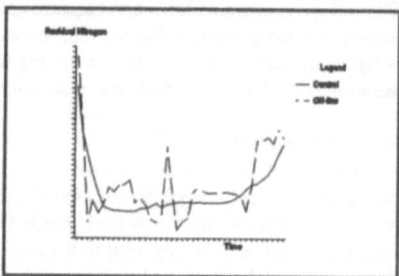

Figure 5 Neural Network Control

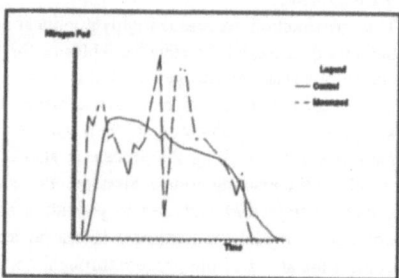

Figure 6 Neural Network Control Signal

Neural Networks
for the Processing of Fuzzy Sets

G. Bortolan
LADSEB-CNR
Padova, Italy

1   Introduction

In the last years the connectionist approaches have been successfully applied in several areas like image processing, speech recognition, pattern recognition and signal processing. Lately some attempts have been made in order to permit the neural networks to process and to manage uncertain information. A promising strategy is the use of the fuzzy logic, which has been successfully applied for the management of uncertainty and imprecision. Several authors have combined the fuzzy logic with the connectionist approach by different point of view (Hayashi et al., 1993, Ishibuchi et al., 1993, Keller and Tahani, 1992, Pedrycz, 1993). The main purpose of this study is to investigate an architecture in which each node is able to process fuzzy sets. The simplicity of the back-propagation algorithm has been preserved by the use of normalized trapezoidal fuzzy sets.

2   Fuzzy neural networks

The proposed approach adds to the feed-forward neural network architecture the particular property that all the single nodes are able to process and to transform normalized trapezoidal fuzzy sets (NTFS). A NTFS $\tilde{y}$ is completely defined by 4 terms, $y^a, y^b, y^\alpha, y^\beta$, with the following notation (Fig. 1):

$$\tilde{y} = \{y^a, y^b, y^\alpha, y^\beta\}.$$

The weights that characterize the connections between two nodes are defined by a quadruple $\underline{w}_{ji,l} = (w^a_{ji,l}, w^b_{ji,l}, w^\alpha_{ji,l}, w^\beta_{ji,l})$, to maintain the possibility to propagate independently the four components of the NTFS trough the entire neural network.

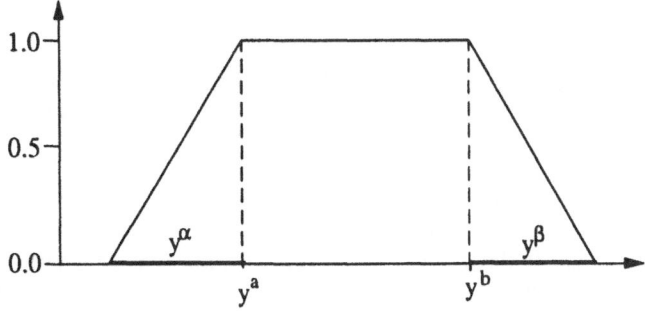

Fig. 1. A normalized trapezoidal fuzzy set $\tilde{y} = \{y^a, y^b, y^\alpha, y^\beta\}$

182

The state of a generic node $\widetilde{net}_{j,l}$ is given by the outer-product between the output of the connected nodes ($\widetilde{y}_{i,l-1}$) and the weights $\underline{w}_{ji,l} = (w_{ji,l}^a, w_{ji,l}^b, w_{ji,l}^\alpha, w_{ji,l}^\beta)$ that characterize such connections, that is:

$$\widetilde{net}_{j,l} = \sum_{i=1}^{\#(l-1)} \underline{w}_{ji,l} \cdot \widetilde{y}_{i,l-1} = \left\{ net_{j,l}^a, net_{j,l}^b, net_{j,l}^\alpha, net_{j,l}^\beta \right\}$$

$$= \left\{ \sum_{i=1}^{\#(l-1)} w_{ji,l}^a \cdot y_{i,l-1}^a, \sum_{i=1}^{\#(l-1)} w_{ji,l}^b \cdot y_{i,l-1}^b, \sum_{i=1}^{\#(l-1)} w_{ji,l}^\alpha \cdot y_{i,l-1}^\alpha, \sum_{i=1}^{\#(l-1)} w_{ji,l}^\beta \cdot y_{i,l-1}^\beta \right\}$$

In order that the resulting fuzzy set $\left\{ net_{j,l}^a, net_{j,l}^b, net_{j,l}^\alpha, net_{j,l}^\beta \right\}$ be a meaningful NTFS, some constraints or restrictions may be adopted such as:

$$net_{j,l}^a \le net_{j,l}^b \ , \ \ net_{j,l}^\alpha \ge 0 \ , \ \ net_{j,l}^\beta \ge 0.$$

The activation function f( ) used in this study is the usual sigmoidal function. Such activation function will operate on NTFS and will produce normalized fuzzy sets (Fig. 2), and this transformation can be interpreted as a compatibility measure (Zadeh, 1975). In order to work in a homogeneous environment, the fuzzy set y=f(net) will be approximated by a NTFS $\{y^a, y^b, y^\alpha, y^\beta\}$, that is:

$$\widetilde{y}_{j,l} = \widetilde{f}(\left\{ net_{j,l}^a, net_{j,l}^b, net_{j,l}^\alpha, net_{j,l}^\beta \right\}) \cong \left\{ y_{j,l}^a, \ y_{j,l}^b, \ y_{j,l}^\alpha, \ y_{j,l}^\beta \right\}$$

$$= \left\{ f(net_{j,l}^a), f(net_{j,l}^b), f(net_{j,l}^a) - f(net_{j,l}^a - net_{j,l}^\alpha), f(net_{j,l}^b + net_{j,l}^\beta) - f(net_{j,l}^b) \right\}$$

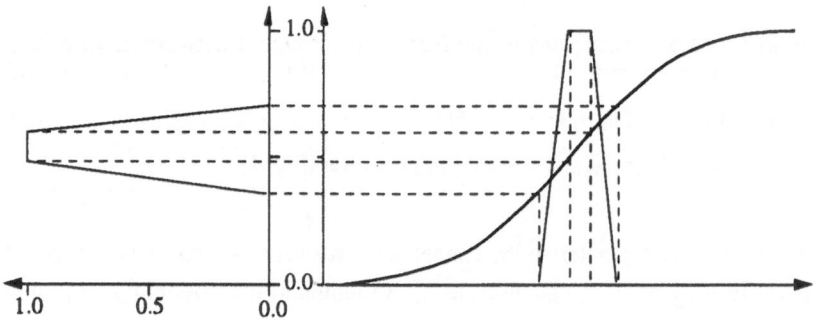

Fig. 2. Transformation produced by the activation function on a NTFS

3 Fuzzy generalized delta rule

In the learning phase the weights are upgraded in order to minimize an appropriate error function, which should consider the difference between the desired output $\widetilde{t} = \left\{ t^a, t^b, t^\alpha, t^\beta \right\}$ and the output of the neural network $\widetilde{y} = \left\{ y^a, y^b, y^\alpha, y^\beta \right\}$ The choice of this error function may cause some problems. In fact using the fuzzification of the usual error function, it may not reach the zero value also in case of perfect simulation, due to the presence of the fuzzy subtraction. This problem can be overcome using a distance measure between two fuzzy sets, the desired and the

computed output. Several distance measures proposed in literature (Degani and Bortolan, 1988, Zwick et al., 1987) may be used.
The alternative strategy used in this study is to minimize the errors in the four components of the trapezoidal fuzzy sets:

$$E = \left(E^a, E^b, E^\alpha, E^\beta\right) = \left(\sum (t^a - y^a)^2, \sum (t^b - y^b)^2, \sum (t^\alpha - y^\alpha)^2, \sum (t^\beta - y^\beta)^2\right)$$

In this way it is possible to define a strategy in the training phase which adjusts only the corresponding components in which the output of the neural network shows a deviation from the target.

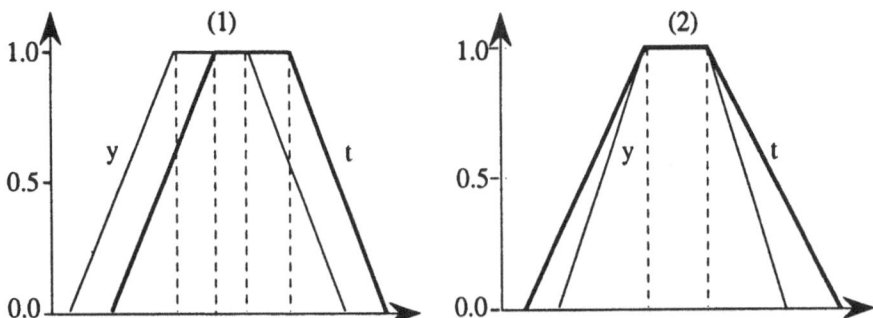

Fig. 3. Examples of target and output of a fuzzy neural network

For example in case 1) of Fig. 3 the target differs from the output of the NN by a constant, that is:

$$t^a = y^a + \delta \ , \ t^b = y^b + \delta \ , \ t^\alpha = y^\alpha \ , \ t^\beta = y^\beta$$
$$E^a = E^b > 0 \ , \ E^\alpha = E^\beta = 0$$

and consequently only E^a, E^b will have effect in the minimization process. In case 2) of Fig. 2 the target and the output of the NN differs only in the shape, that is:

$$t^a = y^a \ , \ t^b = y^b \ , \ t^\alpha = y^\alpha + \phi \ , \ t^\beta = y^\beta + \phi$$
$$E^a = E^b = 0 \ , \ E^\alpha = E^\beta > 0$$

and consequently only E^α, E^β will have effect upgrading the weights.
This approach has the advantage of simplifying the analysis of the generalized delta rule. In fact, for the computation of the delta rule:

$$\Delta \underline{w}_{ji,l} = -\eta \cdot \frac{\partial E}{\partial \underline{w}_{ji,l}} = \left(\Delta w^a_{ji,l}, \ \Delta w^b_{ji,l}, \ \Delta w^\alpha_{ji,l}, \ \Delta w^\beta_{ji,l}\right)$$

the four components of $\Delta \underline{w}_{ji,l}$ can be considered as independent terms, if the upgrade process is made in sequential order. In this way the fuzzy generalized delta rule can be derived as:

$$\Delta \underline{w}_{ji,l} = \left(-\eta \cdot \frac{\partial E^a}{\partial w^a_{ji,l}}, \ -\eta \cdot \frac{\partial E^b}{\partial w^b_{ji,l}}, \ -\eta \cdot \frac{\partial E^\alpha}{\partial w^\alpha_{ji,l}}, \ -\eta \cdot \frac{\partial E^\beta}{\partial w^\beta_{ji,l}}\right)$$

obtaining for example (Bortolan, 1993):

$$\Delta w^a_{ji,l} = \eta \cdot \delta^a_{j,l} \cdot y^a_{i,l-1}$$

where:

$$\delta_{j,L}^a = (t_{j,p}^a - y_{j,L}^a) \cdot f'(net_{j,L}^a)$$

$$\delta_{j,l}^a = \sum_{k=1}^{\#(l+1)} \delta_{k,l+1}^a \cdot w_{kj,l+1}^a \cdot f'(net_{j,l}^a) \quad for \; l = L - 1, \ldots, 1$$

This procedure may cause some inconsistencies during the process of weights updating, and it is necessary to add some particular constraints:

$$w_{ij,l}^a \neq w_{ij,l}^b \neq w_{ij,l}^\alpha \neq w_{ij,l}^\beta \qquad\qquad w_{ij,l}^a = w_{ij,l}^b \; ; \; w_{ij,l}^\alpha = w_{ij,l}^\beta$$

$$w_{ij,l}^a = w_{ij,l}^\alpha \; ; \; w_{ij,l}^b = w_{ij,l}^\beta \qquad\qquad w_{ij,l}^a = w_{ij,l}^b = w_{ij,l}^\alpha = w_{ij,l}^\beta$$

and each strategy will influence the behaviour of the back-propagation algorithm.

4 Back-propagation algorithm

The order of execution of the various phases of the back-propagation algorithm will determine the global behaviour of the learning process. Different back-propagation algorithms may be considered (Bortolan, 1993). In the simplest algorithm the four components are simultaneously upgraded, that is:

- compute $\overset{\approx}{y}_{*,p}$ and $\overset{\approx}{E}_p$

- compute $\Delta w_{ji,l}^a$, $\Delta w_{ji,l}^b$, $\Delta w_{ji,l}^\alpha$, $\Delta w_{ji,l}^\beta$

- upgrade $w_{ji,l}^a$, $w_{ji,l}^b$, $w_{ji,l}^\alpha$, $w_{ji,l}^\beta$

Some experiments have been performed with this algorithm. The weights have been initially set in a random way, with the constraint (4), and successively they have been upgraded independently. The bahaviour of the trained neural network has been satisfiable. This result has confirmed the potentiality of the proposed architecture of fuzzy neural networks, in which all the single nodes are able to process fuzzy sets.

5 References

Bortolan G. (1993) Fuzzy neural networks for linguistic processing. In M.N. McAllister (Ed.), Proceedings of the North-American Fuzzy Information processing Society NAFIPS (pp. 42-46). Allentown, PA.

Degani R., Bortolan G. (1988) The problem of linguistic approximation in clinical decisionmaking. Int. J. Approx. Reasoning, 2, 143-162.

Hayashi H., Buckley J. J., Czogala E. (1993) Fuzzy neural network with fuzzy signals and weights. Int. J. of Intelligent Systems, 8, 527-537.

Ishibuchi H., Tanaka H. (1993) An architecture of neural networks with interval weights and its application to fuzzy regression analysis. Fuzzy Sets and Systems, 57, 27-39.

Keller J. M., Tahani H. (1992) Implementation of Conjunctive and Disjunctive Fuzzy Logic Rules with Neural Networks. Int. J. Approx. Reasoning, 6, 221-240.

Pedrycz W. (1993) Fuzzy neural networks and neurocomputations. Fuzzy Sets and Systems, 56, 1-28.

Zadeh L.A. (1979) A theory of approximate reasoning. In J.E. Hayes, D. Michie and L.I. Mikulich (Eds.), Machine Intelligence (Vol. 9, 149-194). Chichester: Wiley.

Zwick R., Carlstein E., Budescu D. V. (1987) Measures of similarity between fuzzy concepts: a comparative analysis. Int. J. Approx. Reasoning, 1, 221-242.

Human Sign Recognition
Using Fuzzy Associative Inference System

Toru YAMAGUCHI†, Tomohiko SATO‡, Hirohide USHIDA‡, Atsushi IMURA‡

†Faculty of Engineering, Utsunomiya University,
2753 Ishii-cho, Utsunomiya, 321 JAPAN
‡Laboratory for International Fuzzy Engineering Research,
Siber Hegner Building. 3F, 89-1 Yamamashita-cho, Naka-ku, Yokohama, 231,
JAPAN

1. Introduction

The spotting recognition system is one system that recognizes human motion to a certain extent using moving images, it uses a dynamic programming method (Takahashi et al., 1993). This method, however, is limited because it is difficult to recognize the motions of unspecified people. This is because the system compares input patterns with standard patterns in its memory.

This paper proposes a human sign recognition method that can recognize the motion of unspecified people. Human signs are defined as significant human motions in this paper. The basic idea of this method is to transform a space-time pattern for a human motion into a state-transition pattern and extract characteristic states from the pattern, and then to recognize the human motion as a correct sign using their states by means of fuzzy associative inference(Yamaguchi et al., 1993a). Since the fuzzy associative inference approaches the input pattern to the global nearest pattern gradually by top-down and bottom-up processing, its inference result is not affected by the local error between the input pattern and standard patterns in its memory. Therefore, fuzzy associative inference is used in our method. As an example, this method is applied to instructions given to an autonomous robot (a micro-mouse) by human signs. The effectiveness of this method is shown by experiments. Finally, it is also compared with conventional fuzzy inference.

2. Fuzzy Associative Inference System

Fuzzy inference is effective for human sign recognition because it is more robust than other inference systems in recognizing motions of unspecified people. With conventional fuzzy inference (Mamdani, 1974), however, the degree of fuzziness increases as inference proceeds and the conclusion becomes more fuzzy. To solve this, a fuzzy associative inference system has been developed. It is driven by fuzzy associative memories combined with several bidirectional associative memories (BAMs)(Kosko, 1987), as shown in Fig. 1. It consists of three layers : the if-layer (x-layer), the if-then-rule-layer (r-layer), and the then-layer (y-layer). Fuzzy associative inference is performed as follows. As data is input into the system, activation values for nodes are propagated by reverberation. After the limited reverberation times, the activation values for each node in then-layer are integrated and become inference output (Yamaguchi et al., 1993b). The effectiveness of fuzzy associative inference can be described using a simple example. If the membership grade is 0 in conventional fuzzy inference, it cannot be performed. Human beings, however, can recognize objects even if some information is lacking. In the same way, fuzzy associative inference can be performed even if some information is lacking. This reason is that the inference result is not affected by the local error between the input pattern and standard patterns in its memory since the input pattern approaches to the global nearest pattern gradually and the degree of fuzziness decreases by top-down and bottom-up processing performed in the inference. At this point, it is superior to conventional fuzzy inference. Therefore, fuzzy associative inference is used in our method.

3. Human Sign Recognition System
A human sign recognition system that performs the proposed method is shown in Fig. 2. This system consists of a tracking module, a feature extraction module, and a fuzzy associative inference module. Let us now describe these modules.

3.1. Tracking module
First a person makes signs in front of a CCD camera. The tracking module detects the position of the moving object (hand) by extracting the color. And a time-series pattern of the detected position is created. Since this process is carried out (every 1/60 second) in some Japanese products, the algorithm is not described in this paper. The patterns are then transformed to a computer via interface devices.

3.2. Feature extraction module
In this module the space-time pattern is regarded as the state-transition pattern, and features are extracted from the pattern. We think that human beings recognize the motion of the object (hand) not by its position, but by the transition of its position in a space-time pattern. In this system, mountains and valleys for the data pattern can be regarded as the transition of the object (hand) position. Therefore, they are extracted as features from time-series data.

3.3. Fuzzy associative inference module
This module performs fuzzy associative inference using extracted features. The motion of the object (hand) is, consequently, recognized as a human sign. Here, membership functions are defined as suitable features for signs and are embedded within fuzzy inference rules. When the motion of the object (hand) is recognized as a human sign, the sign is transmitted to a robot (micro-mouse).

4. Experimental Method
The following five functions are used as human signs. (1)forward, (2)backward, (3)left, (4)right, (5)stop (Fig. 3). The position of hand gravity is tracked by color extraction in the tracking module. The transition of the position shown by time-series data is regarded as two wave graphs time-x, and time-y (Fig. 4). These wave graphs are different for each sign. The two mountains (or two valleys) in these graphs are extracted as features for each sign. The height of the mountains (or the depth of the valleys) are used as inputs into the membership functions in the fuzzy associative inference process. When outputs of fuzzy associative inference satisfy some conditions, a human sign represented by an output node having the highest activation value is recognized.

5. Experimental Results
Three people performed each sign three times and fuzzy rules and membership functions were made using these 45 sample data. The membership functions made for each sign data and the fuzzy rules implemented in a fuzzy associative inference system are shown in Fig. 1 and Fig. 5. When fuzzy associative inference was performed for the sample data used to make these membership functions, the average correct recognition ratio was 0.984. The ratio for four other people was 0.817. All together, 93 data were recognized as correct signs among 105 data

collected on seven people, and the ratio was 0.886 (Table 1). The number of data not recognized correctly was twelve. Among these, six data items could not be recognized at all, and the other six were recognized incorrectly. The former were for the 'left', and almost entirely for one specified person. The latter data items were incorrectly scattered across all signs almost uniformly except for the 'right', and all the data were for the same person.

6. Discussion

The correct recognition ratio ranged between 80 and 90% for all data. It can be said that data recognized entirely incorrectly and not recognized at all were for specified cases because they were for one specified person. On the whole, recognition error fell within the expected range. Effectiveness of the proposed method for human sign recognition is confirmed by the above results. We also compared the proposed method with conventional fuzzy inference. The average correct recognition ratio obtained by the method and the conventional method for same data on seven people was 0.942, and 0.876, respectively. Therefore it is also confirmed that the proposed method is superior to conventional fuzzy inference.

7. Conclusion

A human sign recognition method using a fuzzy associative inference system has been proposed. This method regards time-series data as state-transition patterns, and extracts features from these data. The human motion is recognized as a sign by means of fuzzy associative inference using fuzzy associative inference system. This method is superior to conventional fuzzy inference because the human motion can be recognized as a correct sign even if there are the local error between the input pattern and standard patterns in its memory. This reason is that the input pattern approaches the global nearest pattern gradually and the local error decreases by top-down and bottom-up processing performed in the system. The effectiveness of this method was confirmed by means of some experiments. We intend to combine this proposed method with a learning method to recognize more complex human signs.

Acknowledgments

We express our deepest gratitude to Mr. Sato and Mr. Yoshihara of Utsunomiya University for their useful advice and for providing us with important data.

References

Kosko, B. (1987). Adaptive Bidirectional Associative Memories, Applied Optics, Vol.26, No.23, pp.4947-4960

Mamdani, E.H. (1974). Applications of Fuzzy Algorithms for Control of Simple Dynamic Plant, Proc. IEE, Vol. 121, No. 12, pp.1585

Takahashi, K., Seki, S., and Oka, R. (1993). Spotting Recognition of Human Gestures from Motion Images, Technical Report of IEICE., IE92-134, PRU92-157, pp. 9-16 (in Japanese)

Yamaguchi, T., Goto, K., and Takagi, T. (1993a). Two-Degree-of-Freedom Fuzzy Model Using Associative Memories and Its Applications, Information Sciences, 71, pp.65-97.

Yamaguchi, T., Sekine, S., Montgomery, D., and Endo T. (1993b). Intelligent Interface Based on Fuzzy Associative Inference and its Application to Command Spelling Corrector, T.IEE Japan, Vol. 113-C, No.9, pp.709-718 (in Japanese)

188

R1: if x1 is FO, x2 is FO, y1 is FO, y2 is FO, then action is FORWARD.
R2: if x1 is BA, x2 is BA, y1 is BA, y2 is BA, then action is BACKWARD.
R3: if x1 is LE, x2 is LE, y1 is LE, y2 is LE, then action is LEFT.
R4: if x1 is RI, x2 is RI, y1 is RI, y2 is RI, then action is RIGHT.
R5: if x1 is ST, x2 is ST, y1 is ST, y2 is ST, then action is STOP.

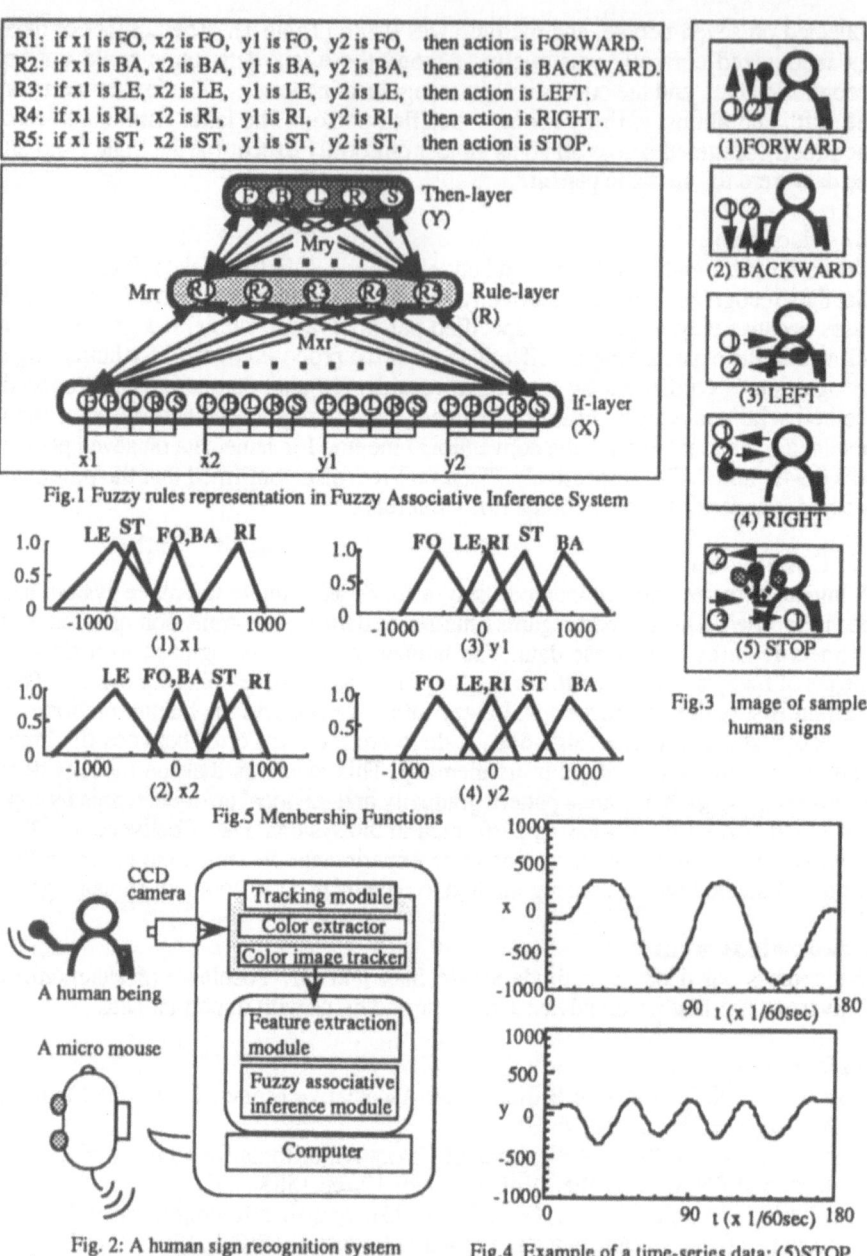

Fig.1 Fuzzy rules representation in Fuzzy Associative Inference System

Fig.5 Menbership Functions

Fig.3 Image of sample human signs

Fig. 2: A human sign recognition system

Fig.4 Example of a time-series data: (5)STOP

Table1 : The number of samples and correct recognition ratio

sign	sample	correct	incorrect	not recognize	recognition ratio
forward	21(12)	19(10)	2(2)	0(0)	0.905(0.833)
backward	21(12)	20(11)	1(1)	0(0)	0.952(0.917)
left	21(12)	14(6)	1(1)	6(5)	0.667(0.500)
right	21(12)	21(12)	0(0)	0(0)	1.000(1.000)
stop	21(12)	19(10)	2(2)	0(0)	0.905(0.833)
Total	105(60)	93(49)	6(6)	6(5)	0.886(0.817)

The number in ( ) shows the number of samples except for ones used to make membership functions.

Bayesian Properties and Performances of Adaptive Fuzzy Systems in Pattern Recognition Problems

F. Masulli[1,3], F. Casalino[2], and F. Vannucci[3]

(1) Department of Physics - University of Genoa
Via Dodecaneso 33 - 16146 Genova (Italy)

(2) DISI - Department of Computer and Information Sciences
University of Genoa - Via Benedetto XV, 3 - 16132 Genova (Italy)

(3) INFM Research Unit of Genoa
Via Dodecaneso 33 - 16146 Genova (Italy)

1 Introduction

Generally, Neural Networks are used to solve problems for which a-priori knowledge is provided, in an implicit way, through numerical relationships among variables (e.g., pattern recognition). Fuzzy Systems are successfully employed mainly to solve problems for which a-priori knowledge is available in linguistic form (e.g., process control).

As will be shown in this article, Adaptive Fuzzy Systems (AFS) [1] can be used to handle problems for which a-priori knowledge is available only in numerical form, or for which it could be too expensive to render knowledge explicit in linguistic form.

Multi-Layer Perceptrons (MLP) and Adaptive Fuzzy Systems can be trained, in a supervised way, to map variables without using explicit hypotheses about the analytical dependences among them. Such methods are usually referred to as function approximators (*model-free estimators*) [2, 3].

Therefore, both types of models can be applied in various domains (e.g., signal processing, process control, pattern recognition, etc.), and the criteria for choosing between them for a given application are not yet clear.

In this paper, we shall use an Adaptive Fuzzy System to solve a pattern recognition problem, i.e. off-line recognition of handwritten characters. We shall demonstrate that the AFS approximates a Bayesian discriminant function; moreover, we shall experimentally verify that, in the training phase, this system is some order of magnitude faster than a Multi-Layer Perceptron.

2 The Adaptive Fuzzy System

Fuzzy sets, proposed by Zadeh in 1965 [4], can be defined through a membership function m_F that maps the elements of the universal set in the unit range $[0, 1]$.

The form of the membership function is arbitrary. In this way, it is possible to model set of objects that fulfil a given property in different ways.

Fuzzy rules are expression of the type:

$$if\ A\ then\ B$$

where A and B are labels associated with fuzzy sets that are characterized by suitable membership functions.

An Adaptive Fuzzy System (AFS) [1] is a feedforward system that could be regarded as a Multi-Layer Perceptron with only one hidden layer; the units of the MLP correspond to fuzzy rules.

Let us describe our AFS implementation.

If there are K units in the input layer, J rules in the hidden layer and I units in the output layer, the activation of the j-th rule can be expressed as:

$$R_j = \prod_k \mu_{jk}(x_k),$$

where the quantity $\mu_{jk}(x_k)$ is the value of the membership function of the component x_k of the input vector for the j-th rule.

The membership function can be defined as:

$$\mu_{jk}(x_k) = \exp(-\frac{|x_k - m_{jk}|^2}{2\sigma_{jk}^2}).$$

Therefore, the receptive fields overlap with one another.

The values of the output units are obtained by means of defuzzification process based on the centroid rule [2] :

$$y_i = \frac{\sum_j R_j s_{ij}}{\sum_j R_j}.$$

The AFS is trained to work as a classifier by minimizing the Mean Square Error between the output of the net and the label vector $\vec{\mu}$, whose components are defined as follows:

$$\mu_j = \begin{cases} 1 & \text{if the example belongs to the class } j \\ 0 & \text{otherwise.} \end{cases}$$

The learning formulas for the the system parameters (i.e. m_{jk}, σ_{jk} and s_{ij}) are obtained by the Back-Propagation technique [1].

In our implementation, we make the learning rates adaptive, thus obtaining a sharp reduction in the convergence time, as compared with fixed learning rates [5].

It is worth noting that a classification approach using an Adaptive Fuzzy System allows us to insert available a-priori knowledge in the rules before the training phase and to interpret the learned values of parameters in terms of rules.

3   The AFS as an Approximation to a Bayes Optimal Discriminant Function

As stated in the Introduction, an AFS can approximate functions [2]. So it is possible to demonstrate that the AFS can approximate the Bayesian optimal discriminant function [6], if a large training set is used. This can be easily accomplished on the basis of the demonstration developed by Ruck et al for a Multi-Layer Perceptron [7].

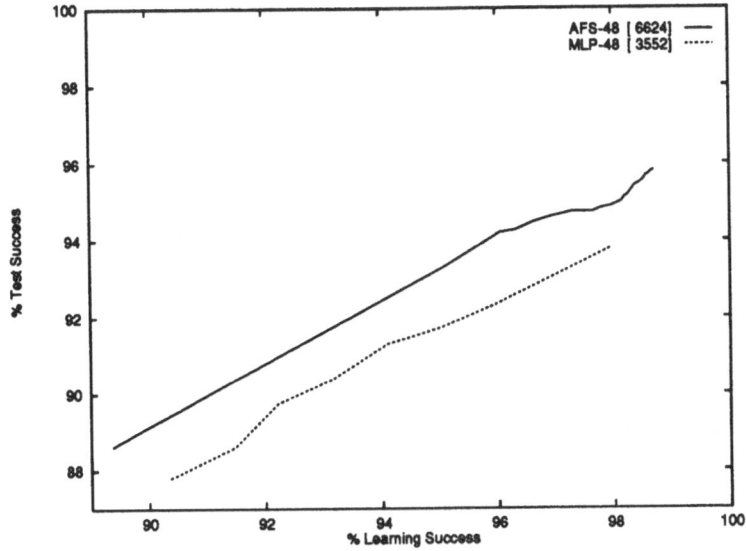

Figure 1

rules	parameters	(%L, %T)	epochs	(%L1, %T1)	Time (min)
24	3312	(97.55, 94.00)	30	(91.0, 88.0)	3.5
32	4416	(98.05, 95.00)	30	(89.0, 87.0)	4.0
48	6624	(98.65, 95.81)	30	(91.5, 88.5)	16.0
64	8832	(98.32, 96.02)	20	(93.5, 92.0)	23.0
128	17664	(98.97, 96.20)	30	(94.5, 93.5)	46.0

Table 1

4 Data Set and Preprocessing

The samples of the training and the test sets were extracted from the NIST-3 CD-ROM [8], which includes 313389 segmented characters in 128×128 binary image format and their corresponding labels in ASCII format.

During the preprocessing phase, the current character is normalized to a 32×32 format. A low-pass filter is used to fill small holes and remove some small noisy spots. Then a shear transformation is applied to the character. Finally, the dimensions of the input space are further reduced so that each character may be represented as a 64-component vector. Each component is associated with the number of black pixels in 4×4 disjoint image subsquares.

The resulting data-base contains a learning set and a test set, both made up of $10,000$ decimal numerals.

5 Results and Discussion

In Figure 1, a comparison between the performances of the Multilayer Perceptron (MLP) and of the Adaptive Fuzzy Systems (AFS) is reported. The numbers within brackets refer to the parameters used by the two networks.

The MLP and the AFS exhibit similar generalization accuracies. This fact can be theoretically explained by the two networks' capabilities for approximating the Bayesian optimal discriminant function, as previously discussed in Section 3.

Table 1 gives the number of rules, the number of adaptive parameters, the percentages of learning success (%L) and of test success (%T), the number of epochs required by the training phase, the percentages of learning success (%L1) and of test success (%T1), and the duration (in min) of each epoch. Some results are quite impressive: e.g., in the case of an AFS with 128 rules, a single epoch of 46 min, is enough to reach a test success %T1 = 93.5.

6 Acknowledgments

This work was supported by grants from CNR-Progetto Strategico Reti Neurali, GNCB-CNR, Consorzio INFM and MURST. Part of this work was carried out in Summer 1993 while F. Masulli was a Senior Visiting Scientist at the International Computer Science Institute in Berkeley (USA). We thank Alessandro Sperduti and Maurizio Martelli for helpful discussions.

References

[1] C.C. Jou, "On the mapping capabilities of fuzzy inference systems", *in IJCNN International Joint Conference on Neural Networks*, pp. 703–713, Baltimore, MD, USA, 7-11 June 1992, 1992. IEEE, New York, NY.

[2] B. Kosko, editor, *Neural networks and fuzzy systems : a dynamical systems approach to machine intelligence*, Englewood Ciffs Prentice Hall, NJ, 1992.

[3] K. Funahashi, "On the approximation realization of continuous mappings by neural networks", *Neural Networks*, vol. 2, pp. 183–192, 1989.

[4] L.A. Zadeh, "Fuzzy sets", *Information and Control*, vol. 8, pp. 338–352, 1965.

[5] F. Casalino, "Fuzzy systems for handwriting recognition (in italian)", Laurea thesis in computer science, University of Genoa, Genoa - Italy, 1993.

[6] R.O. Duda and P.E. Hart, *Pattern Classification and Scene Analysis*, Wiley, New York, 1973.

[7] D.W. Ruck, S.K. Rogers, M. Kabrisky, M.E. Oxley, and B.W. Suther, "The multilayer perceptron as an approximation to a bayes optimal discriminant function", *IEEE Transactions on Neural Networks*, vol. 1, pp. 296–298, 1990.

[8] M.D. Garris and R.A. Wilkinson, *NIST Special Database3 Handwritten Segmented Characters*, National Institute of Standard and Technology, Gaithesburg, MD, USA, 1992.

The Representation of Human Judgment by Using Fuzzy Techniques

A. Cannavacciuolo[*], G. Capaldo[**], A. Ventre[***], A. Volpe, G. Zollo[****]

[*] Fiat Research Center, Orbassano (Torino), Italy
[**] Faculty of Engineering, University of Naples, Italy
[***] Inst. of Mathematics, Fac. of Architecture, Univ. of Naples, Italy
[****] ODISSEO, Dept. of Computer Science and Systems, Univ. of Naples, Italy

1 The use of individual knowledge in the judgment

A rater R makes his judgment about the possibility of the candidate X to occupy the position P by using a predefined organizational procedure. Due to the rigidity of the formal procedure of evaluation, only a small part of the rater's knowledge is encoded in the organizational memory. Most of that knowledge is hidden in the mind of the evaluator or circulates within the organization as discourses (Capaldo, Zollo, 1993). The natural language is the most complete tool to describe organizational situations, because it allows the representation of meaning shades, ambiguities and conflicts, usually put away by formal methods for the sake of coherence and linearity. We can have an idea of the complex knowledge used by the evaluator when the rater R is called by one of his colleague to justify his overall judgment. In this case the rater R builds up a discourse D, where his knowledge is embedded in a complex way.

2 The elicitation of the categories from the discourse

According to the semiotic and pragmatic approaches to the discourse (Eco, 1976), to represent the complex messages encoded in each sentence, the following hypothesis should be made:

- the message the speaker wishes to send to a listener can be encoded in several different discourses, ranging from a discourse D_0, totally implicit, to a discourse D_∞, totally explicit. The length of the discourse D_0 is 0, while the length of D_∞ is infinite, like the biography of the famous extremely pedantic biographer. The real discourse D_i of the speaker is between these two poles;

- the speaker builds his discourse D_i choosing a definite degree of explicitation i, which maximizes the trade-off between the purpose of clarity of the message (deriving from the possibility of misunderstanding) and the cost of communication (time and speaker's abilities);

- in order to use the discourse as source of data for the final judgment, we must transform the discourse from the current form D_i to a more extended form D_j, where $j > i$, expliciting the frames and categories hidden in the sentences.

The most simple form of the justification discourse is D_i: "The rating of the candidate X for the position P is V because $S_1, S_2, S_3,..., S_n$". Using a special notation that is used in literature for defining the conditioning probability, the discourse D_i is written as

$$D_i : \mu P\,(X) = V |\, \{S_1, S_2, S_3,...\, S_n\} \tag{1}$$

where the vertical bar is employed to link the grade of membership of X in P to the explaining sentences.

Each sentence Sj evokes one or more situations, where facts concerning the candidate are evaluated against one or more tacit frames or categories belonging to the public memory. For example, the sentence S_1: "Usually the collaborators of X are satisfied with the decision of X", contains the fact F_1 : "A candidate to the position P is good if he is appreciated by his collaborators", and the implicit category A: "The cooperation between supervisor and subordinates is important". If the rater is asked to demonstrate the validity of the sentence S_1, he builds a new discourse. In symbolic terms:

$$S_1 : \mu_A\,(F_1(X)) = V_1 |\, \{S_{11}, S_{12}, S_{13},...\, S_{1n}\} \tag{2}$$

The explanation can continue if the questioner asks more questions about the validity of the sentences S_{ij}. Doing so, the explicitness degree of the discourse will become larger and larger (Schank, 1986). From the theoretical point of view the discourse can be enlarged indefinitely. In the reality the discourse will end in a finite point after that there is no utility to continue the explanation.

Field interviews confirm that point of view. The evaluators usually end the discourse when the following shared concepts of *Ideal candidate* are used:

- *Ideal Candidate A*: (Excellent): "The candidate is ideal when his profile is absolutely excellent";

- *Ideal Candidate B* (Fitting): "The candidate is ideal when his profile is coincident with the profile of the position requested by the company";

- *Ideal Candidate C* (Exceeding): "The candidate is ideal when his profile is equal to or exceeding the profile of the position requested by the company".

These concepts of ideal candidate have been defined as normal fuzzy sets A, B, C, characterized by the fuzzy membership functions: μ_A, μ_B, μ_C. Usually the membership functions are interpreted as operators transforming mathematical distance into perceived distance within different cognitive frames (Zimmermann, 1987).

3   An application

The evaluator, to explain his judgment of the candidate X, evokes 15 facts Fi which can be referred to the three categories A, B and C. The term set used is the following: very, very low (VVL); very low (VL); low (L); behind average (A-); average (A); above average (A+); high (H); very high (VH); very very high (VVH).

Several methods have been proposed in the fuzzy literature to link empirical evidences to categories by a fuzzy degree or a fuzzy membership function (Dubois and Prade, 1980). In our approach we tried to derive the fuzzy relationships between facts and categories from the justificative discourse, and managed the discourse itself as an experiment. For what concerns the methodology used to manage and analyze the justificative discourse from a semiotic point of view we refer to Barley (1983).

Table 1 - Facts arranged according to the term set

| | Terms | | | A | | | | B | | | | C | | | |
a	b	c	d	e	f	g	h	e	f	g	h	e	f	g	h
VVL	1	1	0,11	F9,F15	2	2	0,29		0	0	0,00		0	0	0,00
VL	1	2	0,22	F3	1	3	0,43	F4	1	1	0,14		0	0	0,00
L	1	3	0,33	F13,F14	2	5	0,71		0	1	0,14	F6,F12	2	2	0,33
A-	1	4	0,44		0	5	0,71	F5,F14	2	3	0,43		0	2	0,33
A	1	5	0,56	F7	1	6	0,86		0	3	0,43		0	2	0,33
A+	1	6	0,67		0	6	0,86	F8	1	4	0,57	F7	1	3	0,50
H	1	7	0,78		0	6	0,86	F1,F10	2	6	0,86	F5	1	4	0,67
VH	1	8	0,89	F5	1	7	1,00	F3	1	7	1,00	F2	1	5	0,83
VVH	1	9	1,00		0	7	1,00		0	7	1,00	F11	1	6	1,00

The results of the semiotic analysis are represented in Table 1, where the facts evoked are arranged according to the term set. The first set is the term set, the others are the facts related to the categories A,B,C. For what regards the term set, we assume that the terms used to evaluate the facts have the same characteristics. This hypothesis is expressed in the column *b* by the same number 1. The practical result of this assumption is that the distance on a scale among the terms is the same. The position of the terms on the scale is calculated by the cumulative sum in column *c* (absolute values) and in column *d* (relative values). In the remaining columns data concerning the relationships between facts and categories are represented. In this example we do not consider intermediate categories. Thus we represent the direct relationships between facts and the final explicative categories. The columns *f* contains the number of the facts used to explain the judgment. The cumulative sums are reported in column *g* (absolute values) and columns *h* (relative values). The data of columns *d* and *h* let us build the first three diagrams, representing a particular membership function, that is the membership function of the individual in the categories A , B and C). The area behind the functions gives us a direct measure of the membership value of the individual in the prototypical category. In our example the values are the following: $\mu_A (X) = 0, 365$; $\mu_B(X) = 0, 607$; $\mu_C (X) = 0, 667$.

The membership degree evaluated in this example corresponds to a sort of average value. But we can have more sophisticated measures evaluating other aspects present in the justification discourse. For example we can evaluate the importance that the evaluator attribute to each fact. The most interesting aspect of this approach is that we can aggregate the evaluations of the facts to reach the final judgment, just iterating the method. In the example, the final judgment is represented by the fuzzy membership degree $\mu_P (X) = 0, 546$. It is possible to convert this result in a verbal judgment, using the scale of the Table 1, column d. In our case the individual X has been judged "around the average" for the position P.

4   Conclusion and further developments

The approach here presented is built taking in mind the justificative discourse as the starting point to realize an organizational procedure of evaluation. As the evaluator builds several justificative discourses for several questioners we may improve the judgment considering the coherence among them. This means considering another level of aggregation of the judgments. The advantage of this method is to represent the human judgments without forcing the degree of formalization. In the organizational context this quality is highly appreciated, because everyone can manage and understand it.

Many other aspects of the justificative discourse have not been considered yet. It is our opinion that appropriate semiotic and fuzzy concepts should let us extract from the justificative discourse the elements needed to build effective organizational procedure of evaluations.

References

Barley, S.R. (1983), "Semiotics and the Study of Occupational and Organizational Cultures", *Administrative Science Quarterly*, 28, pp. 393-413.

Capaldo, G. and Zollo, G. (1993), "Modelling Individual Knowledge in the Personnel Evaluation Process", *EIASM '93 Int. Workshop on Managerial and Organizational Cognition*, Brussels, May 13-14.

Dubois, D. and Prade, H. (1980), *Fuzzy Sets and System: Theory and Applications*, New York, Academic Press.

Eco, U. (1976), *A Theory of Semiotics*, Bloomington, University of Indiana Press.

Schank, R.C. (1986), *Explanation Patterns: Understanding Mechanically and Creatively*, Hillsdale (NJ), Lawrence Erlbaum.

Zimmermann, H.J. (1987),.*Fuzzy Sets Decision Making and Expert Systems*, Boston, Kluwer.

FUZZY LOGIC VERSUS NEURAL NETWORK TECHNIQUE IN AN IDENTIFICATION PROBLEM

G.Cammarata [1], S.Cavalieri[2], A.Fichera[1]

(1) Istituto di Macchine - (2)Istituto di Informatica e Telecomunicazioni
Facoltà di Ingegneria - Università di Catania
V.le Andrea Doria, 6 - 95125 - Catania - ITALY
fax +39 95 338887 email ad@iit.unict.it

Abstract.
The aim of this paper is to compare the behaviour of a neural model and a fuzzy one in a problem of identifying urban traffic noise, for the solution of which both models would seem to be suitable. Many researchers have been involved with the problem of noise pollution in order to understand the phenomenon and to fully describe it. Several correlations have been found in literature by using statistical approaches; most of the proposed relations are linear and this characteristic represents a serious limitation to their capability in describing real situations. The work presented focuses on description of the approaches followed in creating the neural and fuzzy models. Analyses of the results obtained in the identification of urban traffic noise highlight the different behaviour of the two approaches in solving the problem.

1.Introduction

In the last decade, scientific activity in several fields of research has been characterized by the use of alternatives to traditional instruments. A case in point is certainly the use of neural techniques and fuzzy logic. Although the origins of both are more remote, it is only recently that they have achieved widespread consensus in various fields of application. The neural approach, with particular reference to mapping networks which are dealt with in this paper, is characterized by a certain comprehension difficulty, caused by the fact that the internal mechanisms which regulate its functioning are not clearly visible to the user. Conversely, the fuzzy logic approach allows the user to obtain a set of rules whose meaning can easily be understood and that can be used in practical applications. On account of their intrinsic characteristics the two instruments offer their best performance in fields which often differ greatly. The aim of this paper is to compare them in an application scenario which would seem suitable for both. The scenario consists of a problem of identifying urban traffic noise. Noise pollution in urban areas compromises the quality of life and can represent even a danger for people's health. This problem has received great interest from the scientific community. Many researchers have been involved with the problem of noise pollution, trying to understand the phenomenon and fully describe it [1]. The capability to model noise is of great importance because it allows action to be taken to limit noise pollution in urban areas. The main source of noise is represented by the flow of motor vehicles, but the influence of this on noise pollution is modified by some other physical parameters. Finding a model of noise pollution means searching for a relationship between the traffic parameters, some road parameters, urban parameters and noise pollution. Many correlations have been suggested in literature and were found by using statistical approaches; most of the proposed relations are linear and this characteristic represents a serious limitation to their capability in describing real situations. In [2][3][4][5] a neural approach to the problem of noise identification was proposed, and it was shown that it offers considerably better performance than that provided by classical approaches. In this paper the authors compare this neural model with a fuzzy one. The nonlinearity of the fuzzy approach allows one to capture complex interactions among the variables that regulate noise pollution. Both models were obtained by using a set of measurements taken in various roads belonging to a typical European town and it can be considered a valid description of the situation of medium-size cities. The work focuses on the description of the approaches followed in the creation of the neural and fuzzy models. Analyses of the results obtained in the identification of urban traffic noise highlight the different behaviour of the two approaches in solving the problem.

2.Definition of the Identification Problem.

Several relationships have been proposed in literature to determine a model which expresses the relationship between noise pollution and the corresponding sources. In particular it has been assumed that it depends on certain urban parameters, among which the number, nature and speed of the vehicles, the characteristics of the urban area, the geometry of the road section (width of the road and height of the buildings), and the kind of road surface could be considered. Such a model, however, has a very complex structure and is hard to handle or even to identify. Therefore a simpler set of parameters must be selected and   noise pollution must modelled as a function of the equivalent number of vehicles n_{eq} (defined as $n_{eq} = n_{cars} + 3 \cdot n_{motorcycles} + 6 \cdot n_{trucks}$), the width of the road w, the average height of the buildings h corresponding to the road section being considered. Noise pollution is generally identified through the *sound equivalent pressure level* L_{eq}. Having defined $L_{eq} = f(n_{eq}, h, w)$, identification of urban traffic noise consists of determining the function f. The neural and fuzzy solutions to the problem of identifying urban traffic noise are based on a collection of noise pollution measurements. These measurements are used for different purposes in the two approaches. The neural solution requires a set of examples so that the neural network can learn the function f, while in the second approach the measurements are needed to prepare the set of rules which allow the function f to be realized. The data measured were obtained in Catania, a medium-size town of in the south of Italy. The outdoor acoustic surveys referred to sixteen locations in residential, commercial and industrial areas. The roads selected were both downtown roads and roads connecting outlying areas with the city. This choice was made in order to consider as various a set of data as possible. A class 1 phonometer was employed to measure the sound pressure level L_i. The measurement interval was 1 sec and each observation interval lasted ten minutes. Consequently, for each observation interval 600 L_i values were measured and subsequently averaged in order to obtain the mean value of L_{eq}.

3.The Fuzzy Model.

In this section the authors show a fuzzy model for the evaluation of the acoustic noise. The fuzzy logic approach was aimed to obtain a set of m rules of the following form:

$$\text{if } x_1 \text{ is } A_1 \text{ and } x_2 \text{ is } A_2 \text{ and } x_3 \text{ is } A_3 \text{ then } y_i = g(x_1, x_2, \dots, x_n)$$

where $y_i(=L_{eq})$ is the variable of the consequence whose value is inferred, $x_1 (=n_{eq})$ x_2 $(=h)$, $x_3 (=w)$ are the variables of the premise that determine also the consequence, A_1, A_2, A_3 are fuzzy sets defined on the x_1, x_2, x_3 variables and with piece wise linear membership functions, and g (assumed as a linear in the case of the proposed model) is a function that implies the value of y_i when x_1, x_2, x_3 satisfies the premise. The final output y, inferred from the m implications is given as the average on all the outputs y_i of the individual rules. To establish which variables must be considered in the premise, to fix which is the shape for the membership function of each fuzzy set, and to define the polynomial associated to the consequence of each fuzzy rule a procedure proposed in [6] has been developed. The model obtained contains two fuzzy sets, labelled *small* and *large* for the x_1 and x_2 variables and no fuzzy set for the x_3 one, hence the model contains four fuzzy rules. The obtained model is reported in Fig.1, where the values of the parameters that identify the fuzzy sets of the premises and the coefficient of the consequence polynomials are reported.

4.The Neural Model.

The neural model used was theBackpropagation Network (BPN) [7], which is particularly suitable for solving problems like the one outlined above. The BPN is, in fact, a mapping network which can approximate a function f from a finite subset of n-dimensional Euclidean space $A \subset R^n$ to a finite subset $f[A] \subset R^m$ of m-dimensional Euclidean space, using a set of examples of correct mapping $((x_1,y_1),...,(x_k,y_k),..$ (with $x_k \in A$ and $y_k = f(x_k)))$. As we said above, the function to be identified is linked to three urban parameters - the equivalent number of vehicles, the average height of the buildings and the width of the road. The BPN therefore has three inputs, composed of these parameters, and a single output represented by the L_{eq}. As regards definition of the remaining parameters characterizing the neural model, we tried to make the neural solution as comparable as possible to the fuzzy one. It was considered that this could be achieved by establishing the following:

• the number of weights in the neural network. This number had to be in the same order of magnitude as the number of parameters used in defining the fuzzy model. On the basis of what was said in the previous section, the number of parameters for each of the four rules is 8, thus giving a total of 32 parameters. For this reason the neural network has 8 neurons in the hidden layer, resulting in a number of weights equal to 32.

• choice of the activation function in the output neuron The linearity typical of the consequences in the fuzzy models led us to choose a linear activation function only for the output neuron. For the hidden neurons a sigmoidal activation function was maintained.

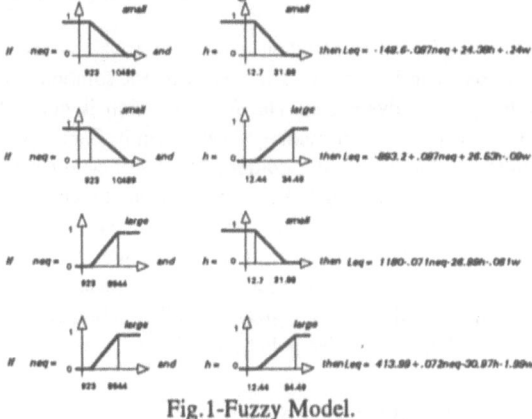

Fig.1-Fuzzy Model.

5.A Comparison between the Fuzzy and Neural Approaches.

In this section the results obtained with the neural and fuzzy models proposed are shown and their behaviour is compared. A first check of the models was performed comparing the *Leq* values, obtained by using the neural approach, the fuzzy model and the measurements, considering different noise pollution measurements from those used previously both for the network learning phase and for tuning of the fuzzy model. The data used for this purpose are still characterized by urban parameters (n_{eq},h,w) belonging to ranges already considered in the model preparation phase. Fig.2 shows the results obtained. A visual inspection of the figure shows a better capability of the fuzzy model to follow the measured data. It seemed to be of great importance to compare the capacity of the two models to identify the sound pressure level on the basis of urban parameters which were qualitatively different from those measured. It was thought that this could be achieved by continuously varying the urban parameters and comparing the capacity for identification the two models offered. Fig.3 shows the identification results obtained, considering the average height of the buildings to be fixed (h=15 meters) and continuously varying both the equivalent number of vehicles (shown on the abscissa) and the width of the road (the curves are parametrized vs. w, where $10 \leq w \leq 50$). As can be seen, variability in the number of vehicles occurs in intervals which depend on the width of the road (e.g. the curves referring to w=10 meters have a n_{eq} variability of less than 4000; this is because in reality a road of this width could not feasibly have a higher number of vehicles in transit) The curves shown in Fig.3 are extremely interesting as they show that the fuzzy model as a certain difficulty in interpolating. This is much more evident when the road width values are 10, 18 and 50 meters. The curves referring to these widths present flexes which are completely unjustified. the neural solution, on the other hand, has an excellent capacity for interpolation, as is shown by the linearity of its trends. It should, however, be noted that it is incapable of identifying high noise pollution values, above all in very narrow roads.

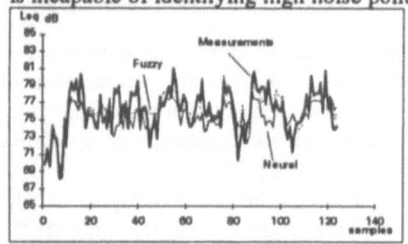

Fig.2 Fuzzy and Neural Identification versus Measured Data.

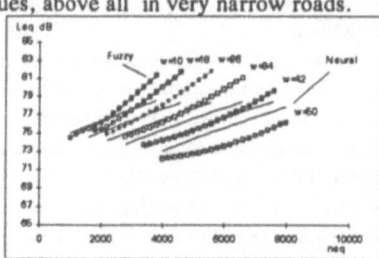

Fig.3-Fuzzy Logic versus Neural Approach

Final Remarks.

In this paper two models for noise pollution evaluation are proposed. They are based on the fuzzy logic and on neural network. The two different solutions were compared, thus highlighting their limits and advantages. The fuzzy solution is certainly characterized by easy comprehension and use, thus providing a user-friendly model for the modelling and identification of urban traffic noise. In addition, its identification capabilities are greater than those typical of the neural model. It has been seen, however, that this capacity is confined to noise pollution scenarios that are similar to those taken into consideration in the design phase. Considering urban parameters which are quite different from those actually measured, the neural model shows a more marked capacity for interpolation. This characteristic gives the neural approach great reliability when the model is used as an instrument for the identification of noise pollution.

References

[1] R. Josse, "Notions d'Acoustique", Ed. Eyrolles, Paris 1972.

[2] S.Cavalieri, A.Fichera, "Exploiting Neural Network Features to Model and Analyze Noise Pollution", 6th Italian Workshop on Parallel Architectures and Neural Networks, May 12-14th 1993, Vietri sul Mare, Salerno, Italy.

[3] Cammarata G., Cavalieri S., A. Fichera, L. Marletta "Noise Prediction in Urban Traffic by a Neural Approach", International Workshop on Artificial Neural Networks, Sitges,Barcelona,Spain, 9-11 June 1993.

[4] G.Cammarata, S.Cavalieri, A.Fichera, L.Marletta, "Neural Networks versus Regression Techniques for Noise Prediction in Urban Areas",World Congress on Neural Networks,Portland,Oregon, July 11-15, 1993.

[5] G.Cammarata, S.Cavalieri, A.Fichera, L.Marletta, "Self-Organizing Map to Fileter Acoustic Mapping Survey in Noise Pollution Analysis", International Joint Conference on Neural Networks, Nagoya, Japan, October 25-29, 1993.

[6] Lofti a. Zadeh, "Outline of a New Approach to the Analysis of Complex Systems and Decision Processes", IEEE Trans. on Syst. Man and Cyb., Vol SMC 3, No.1, 1973.

[7] R. Hecht-Nielsen, "Neurocomputing", Addison-Wesley Publishing Company.

Phoneme Recognition with Hierarchical Self Organised Neural Networks and Fuzzy Systems - A Case Study

N Kasabov+, E Peev++

+ Department of Information Science, University of Otago,
P.O.Box 56, Dunedin, email:nkasabov@otago.ac.nz, New Zealand
++ KZIIT, 25a Acad. Bonchev Str., 1113 Sofia, Bulgaria

1. Introduction

Neural networks (NN) have been intensively used for speech processing (Morgan and Scofield, 1991). This paper describes a series of experiments on using a single Kohonen Self Organizing Map (KSOM), hierarchically organised KSOM, a backpropagation- type neural network with fuzzy inputs and outputs, and a fuzzy system, for continuous speech recognition. Experiments with different non-linear transformations on the signal before using a KSOM has been done. The results obtained by using different techniques on the case study of phonemes in Bulgarian language are compared.

The data base used consists of a small sample of 30 seconds of continuous speech articulated by a male speaker. The speech includes 4 sentences, the ten digits, 10 short words, 20 syllables. The pronounced words contain all 25 phonemes, of them - 6 vowels and 19 consonants. The speech has been digitized with a 20 KHz of frequency.

2. Using KSOM for Phoneme Recognition and the Importance of the Non- Linear Transformations on the Speech Signals

Non-linear transformations on the raw speech signal proved to be advantageous to the final recognition accuracy. Some are experimented here. The first one is log10 transformation of all the 64 Fourier coefficients. Another one is mel-scale filtering. The frequency band is divided into twenty specific bands filtered by corresponding triangular filters, where the first 10 filters are on a linear frequency scale and the other 10 are on a logarithmic frequency scale. The filter outputs are logarithmisised.

Calculating the so called mel-frequency cepstrum coefficients (MFCC) is another non-linear pre-processing transformation. A cosine transformation is calculated on the logirithmisised outputs from the 20 filters. The output vector's dimension could be different, e.g. N=5,10,20 thus having MFCC(5), MFCC(10), MFCC(20). Another non-linear transformation is the calculation of the so called linear frequency cepstrum coefficient (LFCC) which is a cosine transformation over the Fourier spectrum coefficients. A "window" is moving along the time scale and a segment of the signal in the window is taken and transformed by the FFT. The segments overlap, for example on 50% of the window. If the window is 12.8 msec wide, and the discretisation frequency is 20 kHz, then 256 points are taken and weighted through a Hamming window. As the spectrum is taken up to 5kHz, 64 FFT points are used. From the continuous speech sample in the case study, 2050 feature vectors have been extracted and processed.

After having done the pre-processing phase, a 15x15 KSOM has been trained with all the feature vectors; 5000 iterations with a learning coefficient a(0)=0.9 and neighbourhood radius Nr(0)=7 have been done. Different pre-processing methods lead to different phoneme recognition accuracy as illustrated in Table 1. The numbers in the brackets show the dimension of the feature vectors after the non-linear transformation. Obviously the results are far from the best cases reported in the literature on the phoneme recognition task. The reason is that the sample for the study case is a small one. We show here that a non-linear transformation after the FFT increases the accuracy, the MFCC being the best among the tested. The accuracy also depends on the dimension of the feature vectors. A 10-dimensional and 20-dimensional vectors lead to similar results, which are much better than 5-dimensional input vectors. The phonemic KSOM for the case of MFCC(10) is shown in Figure 1.

Table 1.

Method	Accuracy (%)	
	apparent	test
FFT(64)	61	57
FFT +log10(64)	78	74
MEL filters(20)	78	76
MFCC(20)	78	76
MFCC(10)	78	75
MFCC(5)	74	69
LFCC(20)	79	75

Figure 1

3. Hierarchical KSOM

Instead of having a big, and therefore slow to process single KSOM, hierarchical models of KSOM can be used. The first model experimented here uses one 4X4 KSOM at the first level and 16 4x4 KSOM at the second level. Every KSOM at the second level is activated when a corresponding neuron from the first level becomes active. The asymptotic computational complexity of the recognition of the two-level hierarchical model is O(2nm) where n is the number of inputs, m is the size of a single KSOM. This is much less than the computational complexity O(nmm) of a single KSOM with a size of m^2 (m=16 for the experiments). For a general r-level hierarchical model the complexity is O(rnm). The same accuracy as using MFCC and a single KSOM was achieved for the experimented sample data set, but a speed-up of 8 times was achieved here.

Another hierarchical model was developed which uses both time-, and frequency-space representation of the input signals. The first-level KSOM is trained to recognise four classes of phonemes, i.e. pause($), a vocalised phoneme(@), a non-vocalised phoneme(!), a fricative segment(#). The network is trained with three time-features of the speech signal which are: MEAN (the mean value of the energy of the time-scale signal within the segment); ZERO (the number of the crossing of zero for the time-scale signal); NOISE (the mean value of the local extremes of the amplitude of the signal on the time scale). After training a small 5X5 KSOM with instances of the four classes taken from the sample speech data set, the network can recognise the four classes with accuracy of 100%, 99%, 94% and 97% respectively. After the phoneme class is being successfully recognised, the feature frequency-scale vector (MFCC1,MFCC2,..., MFCC10) corresponding to the same time segment, is passed to a KSOM which corresponds to the winning class. The phonemic maps after training are shown in Figure 2.

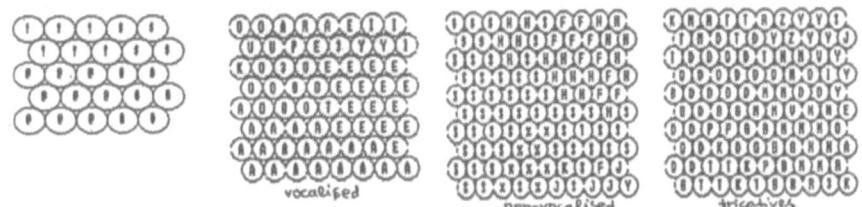

Figure 2

After experimenting with the same training and test sets, the achieved apparent accuracy was 80% and the test accuracy - 78 %. The results show that combining input vectors taken from both time-, and frequency space may give better results.

4. Fuzzy Neuro Systems for Phonemes Recognition

In the previous sections KSOM was used for phoneme recognition. But is the "winner take all" paradigm appropriate here? Would fuzzy inference which produces a fuzzy decision vector instead of a 'winning neuron' be more appropriate?

The frequency input features chosen here are the 20 mel-scale filter coefficients obtained after applying triangular mel-scale filters. Every coefficient is fuzzyfied into three fuzzy values. Their membership functions are defined after calculating the mean of all the mel-filter values for a particular filter band over the training speech segments. A feedforward neural network trained by using the backpropagation algorithm has been used. The network is shown in Figure 3c. Figure 3a gives for the membership functions of the fuzzy terms "low", "medium" and "high" for the fuzzy variable "energy of the speech segment on the mel- scale frequency band 1". Figure 3b gives the membership functions of the output fuzzy terms "phoneme /e/- beginning", "phoneme /e/- middle", " phoneme /e/- end". The outputs in this case provide not only information about the ultimate phoneme the currently processed segment belongs to, but about its fuzzy timing among the whole input phoneme signal.

An output decision block analyses the outputs from the network and 'decides' which phoneme the current speech segment belongs to. The current phoneme is not recognised until the end segment(s) of the phoneme are recognised. This is psychologically plausible as we do not decide upon the heard phoneme or word until we hear the end of it or the beginning of the next one. So, the decision is a 'delayed' one. Having three fuzzy concepts, i.e. "beginning", "middle", "end", which accompany every phoneme, helps to significantly overcome the ambiguity of phonemes recognition. For example, using the KSOM for recognising the word 'sedem' in Bulgarian language we achieve the following sequence of recognised phoneme segments: SSSSSSSSSSS %%%SN3 EEEEEEEEEEEEEE TEE3EH DDDD NI3E33 EEE AEDDG NMMMH MMMMM. For a sake of clarity we have separated the clear phoneme sequences from the begin/end ambiguous ones. When the neural network from Figure 3c is used , the following sequence is recognised: SSSSSSSSSSSSSSS N3 EEEEEEEEEEEEEEEEEEEE H DDDD NI EEEEEEEEE DDG MMMMMMMMMM. In this sequence the correctly recognised segments are more than in the previous experiment. The recognition accuracy on the training set is 90% and on the test set- 86%. This is significantly better than the recognition done by using a KSOM or the hierarchical KSOM model, and slightly better than the one achieved in a hierarchical KSOM - DTW system (Kasabov et al, 1993).

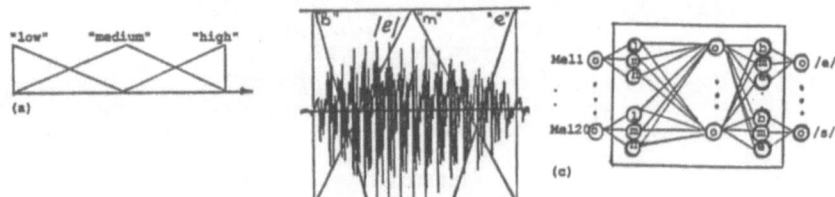

Figure 3

5. Extracting and using fuzzy rules for phoneme recognition

Obtaining (learning) a set fuzzy rules from a trained neural network of the type shown in Figure 3c can be done with the use of the method presented in (Kasabov, 1993). For the case study, 68 fuzzy rules were extracted. Instead of using the neural network, the set of fuzzy rules can be used for the classification phase. When a MAX/MIN composition inference with centroid defuzzification method was used and the same output decision block for solving the ambiguity of the final classification, an accuracy of 88% and 86% was achieved respectively for the same training and test sets of phoneme segments after initial experiments. The fuzzy system provides better generalisation for the case study. The reason may be that the big diversity in the speech signals is better approximated by 'patches' of fuzzy rules rather than by single points in the output space.

6. Conclusions

The experiments on the phoneme recognition task done here with an use of hierarchical KSOM, backpropagation network with fuzzified data and fuzzy inference techniques, suggests that those methods are less computationally heavy and provide better generalisation for continuous speech recognition.

7. References

[1] Morgan, D. and Scofield, C. (1991) Neural networks and Speech processing. Kluwer Academic Publishers

[2]Kasabov, N. (1993) Learning fuzzy production rules for approximate reasoning in connectionist production systems, in: S.Gielen and B.Kappen (Eds) Proceedings of ICANN'93, Springer Verlag, 337-342

[3]Kasabov, N., Nikovski, D. and E.Peev (1993) Speech recognition based on Kohonen Self Organizing Feature Maps and Hybrid Connectionist Systems, in: N.Kasabov (Ed) Artificial Neural Networks and Expert Systems, IEEE Computer Society Press, Los Alamitos, 113-117

NEURONAL NETWORK MODELS OF THE MIND

John G Taylor

Department of Mathematics

Kings College, Strand, London WC2R 2LS, UK.

Summary

An outline is presented of a neural network approach to modelling consciousness and the mind. A global programme is first presented, which is then expanded along three different avenues. One relates to the electrical stimulation experiments of B Libet and to the cortical control structures which may be involved in consciousness. The next discusses 'backward referral in time' discovered by Libet. Finally the relational structures involved in meaning are briefly explored, associated with interactions between various cortical areas, and the possible seat of conscious awareness is discussed in the conclusion.

Introduction

Consciousness is presently at the centre of attention. With the demise of strict behaviourism has come a flood of activity attempting to understand this highest level of human activity. With the development of better understanding of the possible modes of action of neural networks, both at the artificial and living levels, it should be possible to explore models of the mind with a realistic relation to the actual physical structures involved. This would allow better cognizance to be taken of the increasing wealth of data obtained from recordings of dynamic brain activity over the vast range of aggregates, from single cells or cell groups obtained from intercellular electrodes up to areas or even whole brains by PET, EEG, and MEG techniques. These will be mentioned briefly in section 5, after a description of a global model of mind [1,2], and the relevance of the work of Libet [3].

The Relational Theory of Mind

The basic idea behind the approach [4] is that consciousness arises due to the active comparison of ongoing brain activity, stemming from external inputs in the various modalities, with somewhat similar past activity stored in semantic and episodic memory. The mental content of an experience therefore contains, as a component, the set of relations of that experience to stored memories of relevant past experiences. Thus the consciousness of the blue of the sky, as seen now, is determined by the stored memories of one's past experience of blue skies, say on hillsides, at the seaside stretched flat on the sands, or in one's garden sitting in a chair. Not only episodic memory need be involved; there may also be semantic memory of objects of a visual or other scene. The semantic priming thereby involved may itself also lead to further episodic memory activation, in which further relations to past experience are accessed.

The mind, in this approach, may thus be summarised as being created by activation of a set of related memory states by the present input. This relational structure has been explored theoretically elsewhere [1] and more neurophysiologically in [2]. It leads to the relational theory of meaning or intentionality, in which the activation of relevant past memories allows an object to be recognised in terms of its physical structure, actions that can be taken with respect to it, and so on. This approach to meaning is accepted by some neurophychologists [5] and by some cognitive psychologists [6]. The inner, private nature of consciousness is thereby explicable in terms of the internal storage of past experiences, unique to the individual.

There are many questions that arise from this global approach. One of these is as to the nature of the control system which leads to a seemingly unique stream of consciousness. This has been considered in [2,7] and in more detail in [8,9]. We will explore this from a new experimental angle in the next section.

Control Structures

One would expect there to be control structures in the brain which (a) only allow certain memories to be activated, and related to the corresponding input and (b) are involved with assessing the level of discrepancy with new incoming input to that predicted from later parts of activated stored pattern sequences. Both of these aspects were considered in Taylor [2,7] in terms of possible networks which could perform pattern matching. It was suggested that the thalamus-nucleus reticularis thalami(NRT)-cortex complex may support such matching activity, with the NRT functioning as an inhibitory sheet similar in construction and function to the outer plexiform layer. This latter can be modelled as a positive laplacian net [10]. However, since the NRT effectively has negative diffusion (from the GABA-ergic neurons) then it can sustain `bunched' spatially inhomogeneous activity, in which competition between neighbouring thalamic or cortical inputs is occurring. In this manner the NRT may function as a global controller of cortical activity.

Interesting results, now nearly thirty years old [11], appear relevant to the nature of global control. The experiments of Libet were to determine the threshold current for conscious experience when a 1mm diameter stimulating electrode was placed on the post central gyrus and the just conscious experience of what seemed like a localised skin stimulus reported by the patient. The stimulus was delivered as a series of short (of around 0.1 or so msec duration) pulses. There are two features of the data which stand out, which can be summarised as two quantitative laws:

(a) for threshold current to be consciously experienced over a short (<0.5 sec) duration, the applied electrical energy (frequency times duration times square of current) must be greater than a critical value;

(b) for a duration longer than about 0.5 sec the applied electrical power must be large enough to allow the conscious experience to continue. The requirement of enough applied electrical energy to capture, or turn on, conscious awareness in the short term would seem to fit well with the NRT control structure model above. For that is functioning essentially as a resistive circuit, with some non-linearity to provide stability, and such a circuit would be expected to function in terms of electrical energy requirements for capture of the dominant mode. The second result (b) above leads to a need for enough injected power to keep the control system going; there will be a certain amount dissipated, and so power above that critical level will have to be injected to hold the control of consciousness achieved by the earlier injected electrical energy.

It is possible that other approaches may be derived to explain the above results. Thus, they appear similar to many psychophysical results on threshold discrimination levels, such as Bloch's law in vision [12]. This law states that for visual input patterns with persistence less than some 100 msec the discrimination level satisfies a law similar to law 1, where electrical in that law is now replaced by light, primary cortical region by retina and electrode by light surface. For times beyond 100 msec, law 2 applies (again with the same replacements). The simplest explanation of this (and similar) laws is in terms of peripheral summation. Thus rods and cones have a temporal integration period of 200 and 30msec respectively [12]. Such an explanation can not be easily used in the case of direct cortical excitation, since the temporal summation time has been found to be about 50 - 100 msec [11]; this value is very different from the 500 msec or so value of the minimum duration for a pulse at liminal current intensity needed to cause the occurrence of conscious awareness.

<u>Backward Referral in Time</u>.

The Phenomenon of "backward referral in time" has caused a great deal of discussion, as evidenced by the discussion and references in [13]. There are clearly controversial questions of an experimental nature that may still need to be resolved about the phenomenon, but let us accept the results of [12]. We can summarise them as to the nature of the stimulus needed to achieve conscious awareness (a continued train of pulses verses a single pulse) and the presence or not of backdating of the first experience of the stimulus. A peripheral stimulus to the skin only requires a single pulse, which, however gets backdated, which also happens to a pulse train of "neuronal adequacy" duration applied thalamically to VPL or to LM only. The cortical stimulus neither gets backdated nor can be achieved by a single pulse but only by a pulse stream of temporal duration equal to the neuronal adequacy time.

We wish to propose an explanation of these features which is consistent with, and supports, the competitive model of consciousness presented in the previous section. Let us suppose that the negative after-potentials noted in response to a single pulse skin stimulus [14] corresponds to that input having rapidly gained access to the appropriate working memory, and continuing to be active there. That continuation of activity is, indeed, our definition of a *WM*, and agrees with increasing numbers of observations of such continued activity in various brain sites under trial conditions in monkeys [15]. The evidence indicates that more artificial sources of input from *VPL*, *LM* or *C*, are not able to gain direct access to the *WM* that the single skin pulse did. Instead the former inputs have to be injected by creating their own working memory, in other words as a train of pulses, not a single pulse. The mechanism by which the skin pulse can activate the *WM* rapidly (in, say, 50 msec), whilst the other inputs cannot, may be in terms of the mesencephalic reticular activating system *(MRAS)*. This is thought to be excitatory to *NRT*, and as such may convey input directly to the appropriate part of *NRT* so that the skin input, processed through somato-sensory cortex, easily activates the appropriate *WM*. This activity then proceeds to attempt to win the competition on NRT, as discussed in the previous section.

We note, that in the competitive model, however strong an artificial input is to cortex, it still takes a minimal but non-zero time to win the competition. This agrees with the results reported in [14]. Moreover a strong single pulse to *VPL* was reported there as not causing awareness, due to neuronal adequacy of the input being needed to cause such awareness. The same happens in the simulations, when large new input is rapidly switched off almost immediately after being turned on - there is little change, and little chance to win the competition.

The backdating that occurs has a different pattern of distribution from that of the nature of the requisite stimulus (single pulse versus pulse train). A direct sub-cortical input, most likely at thalamic level, appears necessary to achieve backward referral in time. One conjecture as to how that might occur, consistent with the presence of *WM* structures, is that a thalamic-level timing mechanism is turned on by the peripheral or thalamic input on its first appearance. This is then updated constantly, essentially as a counter, until the neuronal adequacy is reached; when the counter is reset to zero. The contents of the counter are used to tag the input time and so allows for the backward temporal referral. If this occurs for two competing inputs, then the first counter has a larger or stronger output, so could give the conscious experience of having occurred before the second, later input. Such a timing mechanism is expected to be hard-wired, since there would appear to be little value in having it adaptive. However the size of the timing interval may be modified by neuromodulatory effects, so leading to the well-known variations of subjective time in, say, emergency situations.

A timer of the above sort has to be able to count up to about 500 msec, and do so with an accuracy of, say, 25 msec. (although other units of time, such as 100 msec, would also do, related to 10Hz oscillations, as compared to 40Hz giving 25 msec). Thus it must be able to count at least 20 different states, each leading to the next as succeeding intervals of 25 msecs pass. That could be achieved by a recurrent network with recurrence time t_o, equal to some fraction of 25 msec. Each recurrence would lead to recruitment of an identical number m of neurons, so that after time nt_o there would be mn neurons active. Different inputs would recruit from different areas of the timing network, and would then compete against each other in their contributions to in puts winning the consciousness competition. Undoubtedly other neuronal models of timing circuits exist, although the one proposed above appears to be one of the simplest. We note that this timed competitive approach to consciousness by *WMs* can also give an explanation of the color-phi and the cutaneous rabbit phenomena described in [13], although there is not space to enlarge on that here.

One of the clear predictions of the above discussions is the existence of timing circuitry, possibly in thalamic regions, although there could be other sites, such as in cerebellum or basal ganglia. This timing mechanism would be activated by a peripheral, *LM* or *VPL* stimulus, and would be predicted to give an increasing input until neuronal adequacy occurred. Such a linear ramp function would have a clear signature, although it might only be seen, for example, by non-invasive techniques, such as by multi-channel *MEG* measurements, at the short time scales involved. A further prediction is that a direct cortical input would not be able to activate the timing circuitry, so not allowing backward referral in time. At the same time there are further predictions as to the activation of suitable *WM* circuitry by a peripheral stimulus, whilst the absence of such activity would be clear if *VPL*, *LM* or *C* stimulation were used. Measurement of *MRAS* activity during such tasks would be important here to probe the activity further.

Relational Structures

We note briefly the nature of the memory structures appropriate for use in scratch-pad and related activity as the source of consciousness. It was suggested in [2] that an associative matrix memory is most appropriate for such feedback, and not a relaxational one. That is due to the lack of ability of a Hopfield-type of net to produce a set of feedback memories, with suitable accuracy, fast enough. The nature of the encoding needs further exploration, however, to settle that point.

The manner in which the feedback memories, aroused by input, are to be used must now be reassessed. One way would be for direct feedback of these memories to the input area or areas where they are still active, as was suggested in section 4 and is observed experimentally in V4 [15]. This could be of value in any verification [6] or decision mode of operation [2], where direct subtraction could be performed. This could occur, for example, in layers 2 and 3 of cortex from feedback activity from higher areas, possibly with support from the thalamic-NRT-cortical system of section 3.If feedback is used to `flesh out,', the input in some manner, there is danger, however, of hallucinations or distortions of the input being experienced if this amplification becomes too strong. This may be avoided by temporal lobe memory excitations (both episodic and semantic) also being used in parallel to the input for later processing. Indeed, such a possibility leads to a clear distinction between the checking mode of operation mentioned above and what may be termed the constrained parallel mode. In the latter, parallel memorial activations lead to extra constraints on future actions and internal limbic or other responses, but not necessarily to modified input per se. That would therefore be still available for verification. Such a difference in use of feedback is necessary in order to be able to handle the information being supplied by the memory activations. Even if predictions or verifications are going according to expectations, there is still a need to make further predictions and to provide additional information, along with the initial input, so that it can be efficiently manipulated to achieve whatever goals of the system are presently most important.

Discussion.

We have presented two "laws" of sensory awareness, in section 3, as deduced from the experimental data of Libet and his colleagues [11,14]. These laws were then shown to be deducible from the NRT-complex competitive net model of conscious control of [2,16]. More details of this latter were then filled out, especially so as to enable more detailed temporal features, associated with "backward referral in time" of the experience, to be explored.

A particularly important question is as to the possible sites of injection of the cortical current which will gain control of consciousness. One proposal is that this occurs either at primary cortices (by injected current) or at sites of working memory, as mentioned in the previous sections. These latter would thereby allow activity to persist longer than TD, and also are known to be crucial in long term memory. Thus an experimental prediction from such a hypothesis would be that suitable injected current in working memory sites might either control awareness or could disrupt it in the appropriate modality (as in the unattended speech effect [17]). Another experimental prediction is that there should be a gradual increase of neural activity in the region of *NRT* relevant to the appropriate winning site of working memory. As applied current is fed into cortex by the surface electrode this activity may be distributed in various cortical areas, but there must be a discernable increase of such activity at some sites in order for the hypotheses presented above to be valid. One place especially suggested in this paper, as already noted, is the *NRT,* but it may occur elsewhere. There are many further prediction which may be made from this model [2].

Finally it is appropriate to turn to the question raised by [18] and mentioned at the beginning of the paper: how can we ever know what it is like to be a bat? As he argues so persuasively [18] ".......every subjective phenomenon is essentially connected with a single point of view, and it seems inevitable that an objective, physical theory will abandon that point of view". As has been discussed in a number of the contributions in [19], there are a number of points being made here, which it is clearly impossible to consider at length in this paper. However the main result we have arrived at, adding to the earlier work of the author summarised in [20], is that a "point of view" of any conscious being is determined both by its semantic and its episodic memory stores in a relational manner, as part of the Relational Theory of the Mind [1].

In detail there are two sets of relations or constraints which the memory structure contributes to any input. One is that arising from the semantic memory related to the *WM* of a given modality. Such memory gives an automatic, pre-aware form of relational structure to input coding, which results in a general species or culture-specific point of view. Thus the input of a stream of sound leads very likely to activation of phonemic nodes which then activate nodes sensitive to certain words. Each word or phrase will itself have a semantic structure imposed on it by means of its relations to other word or phrase centres activated as part of the meaning of the word or phrase. Thus the word "dog" would activate other word centres, such as those for "walk", "lead", or the type and name of one's own dog. These latter centres of activity would not necessarily be so active as to become conscious (although that could happen), but in general would give constraints to further brain activity in terms of the further words made more sensitive by this relational activity (in a predictive manner).

A second, more personal form of coding is that arising from the episodic memory of each individual. That would correspond more closely to the insider point of view, and would give more of the phenomenal content of conscious experience than the semantic memory. It would be the episodic memory of personal experiences with objects or other people which would give the more complete "colour" to the inner life that would be so much more difficult to probe than the more objective semantic coding.

Could one build a virtual reality machine to give one the feeling as to what it would be like to be an X? One could provide an X's retinal transform on inputs, then an X's cortical transform, etc. Continuing in that vein one can see that the virtual reality machine would have in the end to perform a trepanning operation and replaced the wearer's brain by an X's brain. In conclusion the best that one can do is build a machine, based on the relational principles above, with a structure as close as possible to that of an X's brain, and see if the machine responds as one would expect an X to. It is not possible to be an X, to experience "from the inside" what it is like to be an X. But it is possible to understand an X's point of view, from the outside, by recreating that "inside" relationally. If we know all of the relations operating in an X's brain then we know (scientifically) all there is to know about that inside, and certainly more than the X will every know consciously by internal verbal report [21].

References.
1.   Taylor J G (1991), Can neural networks ever be made to think? Neural Network World 1, 4-12.
2.   Taylor J G (1992a), Towards a neural network model of the mind, Neural Network World 2, 797-812.
3.   Libet B, Alberts WW, Wright Jr E W, Delattre D L, Levin G & Feinstein B (1964), Production of Threshold Levels of Conscious Sensation by Electrical Stimulation of Human Somato-Sensory Cortex, J Neurophysiol. 27, 546-578.
4.   Taylor J G (1973), A model of thinking neural networks, Seminar, Institute for Cybonetics, Univ of Tübinjen.
5.   Carlson N R (1991) Physiology of Behaviour. Allyn and Bacon, Boston
6.   Horne P V (1993) The Nature of Imagery, Consciousness and Cognition 2, 58-82.
7.   Taylor J G (1993a), A Global Gating Model of Attention and Consciousness in Neurodynamics and Psychology, ed. Oaksford M and Brown G, Academic Press, New York.
8.   Alavi F and Taylor J G (1992), A Simulation of the Gated Thalamo-Cortical Model, pp 929-932 in Artificial Neural Network 2, ed I Aleksander and J G Taylor, North Holland, Amsterdam.
9.   Alavi F and Taylor J G (1993), in preparation
10.   Taylor J G (1990), A Silicon Model of Vertibrate Retinal Processing. Neural Networks, 3, 171-178.

11.   Libet B, Alberts WW, Wright Jr E W, Delattre D L, Levin G & Feinstein B (1964), Production of Threshold Levels of Conscious Sensation by Electrical Stimulation of Human Somato-Sensory Cortex, J Neurophysiol. 27, 546-578.

12.   Barlow H. B and Mollon J. D. The Senses, Cam.Univ. Press (1982)

13.   Dennett D., and Kinsbourne M. "Time and the Observer", Behavioural and Brain Sciences 15, 183-247 (1992)

14.   Libet B, "Brain Stimulation in the Study of Neuronal Functions for Conscious Sensory Experience", Human Neurobiology. 1,235-242 (1982).

15.   Anderson R., "The Parietal Cortex and Spatial Representations" INNS Neural Networks Symposium Invited Washington DC, Talk, Nov. 1993.

16.   Taylor J.G., "From Single Neuron to Cognition" in Artificial Neural Networks 2, Aleksander I and Taylor J.G.,North-Holland, Amsterdam (1992b)..

17.   Baddeley A.,"Is Working Memory Working ?", Quart. J. Exp. Psych. 44,1-31 (1992).

18.   Nagel, T. "What is it like to be a bat?" Philosophical Reviews 83, 435-450 (1974).

19   Davies M. and Humphreys G. W., "Introduction" pp1-40 in Consciousness, ed Davies M and Humphreys G.W., Blackwells (1993).

20.   Taylor J.G."Modelling the Mind" pp1-20 in Neural Computing Research and Applications, Part One, ed.Orchard G. M. IOP Pub. Co., Bristol (1993).

21.   Taylor J. G. "The Relational Mind", King's College preprint (1994).

The Consciousness of a Neural State Machine

Igor Aleksander

Department of Electrical and Electronic Engineering, Imperial College
London, UK

1   Introduction

There has been a recent renewal of interest in the debate as to whether an explanation of consciousness can (Dennett, 1991) or cannot (Penrose, 1989) be captured by some formal theory. The aim of this paper is to side with the former and to present a theory testable through the concept of a neural state machine. This is a development of a theme first announced in Aleksander (1992). The theory is developed from the point of view of the conditions necessary to synthesize "consciousness" in a manufactured artefact. This is given the name "artificial consciousness" so as to create grounds for a discussion of the difference between the synthetic product and that which is normally thought to be possessed by human beings. During the last 40 years, under the heading of "artificial intelligence", there has been an effort to program computers so as to make them perform functions which, if done by humans, would be said to require intelligence. This endeavour consists of programs that endow the machine with the ability to follow logical rules developed by a programmer. It is not uncommon to point to the poverty of this approach as logical rules do not capture a sense of intentional subjectivity (Searle, 1992) or a sense of "being" (Winograd & Flores, 1986). In this paper it is argued that as a result of novel neural approaches to computation, such properties are not outside the scope of formal theory. A system of postulates is shown to contain the concept of "self awareness" and to support corollaries regarding that special part of human consciousness: natural language.

2   Theory

The theory of artificial consciousness is developed through a series of postulates stated, in turn, first in natural language, followed by a justification, followed by a formal re-statement.

Postulate 1: Framework

The brain of a conscious organism is a state machine whose state variables are the outputs of simple processors called neurons. This implies that a definition of consciousness is developed in terms of the elements of state machine theory.

Justification 1.   Postulate 1 is a statement of intent based on the generality of state machine theory. State machines can model any system with inputs outputs, internal states and input-dependent links between such states. The states and their links form a state structure. Such machines can be probabilistic where links between states are stated as probabilities, they can have a finite or an infinite number of states. The fact that any conscious organism must have something called a brain with an attendant state structure is evidently true and not controversial. The key question is whether enough can be said about the nature of the state structure of organisms that are said to be conscious which explains consciousness itself. This therefore is the task the postulates which follow - to define the characteristics of state structure that are necessary for and specific to organisms that are said to be conscious.

Formalization 1.   In any state machine, five items need to be defined:
i)   The total *input* to the state machine is a vector **i** of input variables i_1, i_2 ...

   $$\mathbf{i}=[i_1,i_2...]$$

   The i_1, i_2 ..variables are the outputs of sensory neurons.

   In living brains the number of such variables, being the number of neurons involved in the first layer of all sensory activity, is very large but finite. There is also some debate about whether it is important for these variables to be considered as binary (firing or not) or real (firing intensity per unit time). While it will be seen that this decision does not alter the course of the theory, the case of these variables being binary is assumed here. This is done without loss of generality but with the gain that, using the methods of automata theory, it becomes possible to develop non-linear models.

   Also, **I** is defined to be the set of all possible input vectors.

ii) The total *output* of the state machine is a vector **z** of output variables z_1, z_2 ...

   $$\mathbf{z} = [z_1, z_2 ...]$$

The z_1, z_2 .. variables are the outputs of 'actuator' neurons.

   Again the variables z_1, z_2.... are considered to be binary, and in living brains would be seen as the outputs of the brain which are responsible for muscular action.

   Also, **Z** is said to be the set of all possible output vectors.

iii) The *inner state* of the state machine is also defined as a vector **q** of variables q_1, q_2 ...

   $$\mathbf{q} = [q_1, q_2 ...]$$

The q_1, q_2 .. variables are the outputs of 'inner' neurons.

Again, variables q_1, q_2 ... are binary, and, in brains, would be the states of neurons neither involved in input sensing nor output generation.

Also, **Q** is said to be the set of all possible input vectors.

iv) The *state dynamics* of the state machine are determined by the equation

   $$\mathbf{q'} = ß(\mathbf{q},\mathbf{i})$$

where **q'** is the "next" state, **q** is the current state and ß a function, which in the case of a finite number of binary variables may be expressed as the mapping,

$$\beta : \quad Q \times I \rightarrow Q.$$

where x is the Cartesian product.

(In the general case this mapping is considered to be probabilistic in the sense that every pair (q,i) of $Q \times I$ maps into every element of Q with some probability.)

v) The *output function* of the state machine is determined by the equation

$$z = \omega (q)$$

where ω is a many-to-many mapping which in the general case is probabilistic.

$$\omega : \quad Q \rightarrow Z.$$

The following three postulates are stated together because their justifications are interleaved.

Postulate 2: Inner Neuron Partitioning

The inner neurons of a conscious organism are partitioned into three sets:

Perceptual Inner Neurons : responsible for perception and perceptual memory.

Auxiliary Inner Neurons : responsible for inner 'labelling' perceptual events.

Functional Inner Neurons : responsible for 'life-support' functions - not involved in consciousness.

Postulate 3: Conscious and Unconscious States

Consciousness in a conscious organism resides directly in the perceptual inner neurons in two fundamental modes:

Perceptual : which is active during perception - when sensory neurons are active;

Mental : which is active even when sensory neurons are inactive.

The activity of the inner perceptual neurons ranges over the same states in for both these modes. The same perceptual neurons can enter semi-conscious or unconscious states that are not related to perception.

Postulate 4: Perceptual Learning and Memory

Perception is a process of the input neurons causing selected perceptual inner neurons to fire and others not. This firing pattern on inner neurons is the inner representation of the percept - that which is felt by the conscious organism. Learning is a process of adapting not only to the firing of the input neurons, but also to the firing patterns of the other perceptual inner neurons. Generalisation in the neurons (i.e. responding to patterns similar to the learnt ones) leads to representations of world states being self-sustained in the inner neurons and capable of being triggered by inputs similar to the learned ones.

Justification 2,3,4.

i. All three postulates stem from the requirement that a conscious organism is conscious through *owning* the sensation-causing firing patterns of its inner neurons.

ii. All three postulates meet the requirement that an organism could not be said to be conscious unless sensations due to sensory input may be sustained in the absence of such sensory input, albeit in reduced detail. (The organism is conscious even with its eyes closed and other senses shut off).

iii. To deal with unconscious function in a brain-like organism, postulate 2 states that perceptual states occur in a subset of inner neurons. That is, not all inner neurons store perceptual memories - some may be encode concepts such as duration, ordinality or even 'mood', while others just keep the organism "alive".

iv. Postulate 3 specifically leaves open the possibility that not all the states of the perceptual inner neurons have direct sensory correlates. This allows the model to account for sleep or anaesthesia.

v. Postulate 4 requires that the formalization should account for the creation of perception-related states by reference to the learning properties of the neuron. Also it calls for a formalization of the process of retrieval of inner perceptual states.

Formalization 2   The inner state variables q are partitioned into three subvectors q^p (perceptual), q^a (auxiliary) and q^f (functional). That is:
$$q = q^p \vee q^a \vee q^f, \quad q^p \& q^a \& q^f = \emptyset \quad \text{(v is conjunction, \& is disjunction)}$$

Formalization 3   A perceptual input i_w has a state correlate in q^p, q_w so that
$$q_w = \beta(i_w, q_w), \text{ the perceptual mode.}$$
Also if i_w is replaced by a neutral input ø
$$q_w = \beta(\emptyset, q_w), \text{ the mental mode.}$$
The set of all input-related states q_w is q^w which is a subset of q^p, the remainder of q^p contains states not related to perception, unconscious and semi-conscious states.

Formalization 4   Learning is the process of first associating an input i_w and an arbitrary state $q_\emptyset$ to form the element of the forward network function ß,
$$q_w = \beta(i_w, q_\emptyset).$$
The key factor is that there is µ, a fixed sampling mapping which transfers some of i_w into q_w, causing q_w to be completely defined by i_w:
$$q_w = \mu(i_w).$$
To create an attractor for i_w, ß is further defined by:
$$q_w = \beta(i_w, q_w).$$
(This is the "iconic" training methodology fully described in Aleksander and Morton, 1993).
The generalization of the system ensures that the requirement of postulate 3:
$$q_w = \beta(\emptyset, q_w), \text{ the mental mode,}$$
is satisfied.

Postulate 5: Prediction

Relationships between world states are mirrored in the state structure of the conscious organism enabling the organism to predict events.

Justification 5   Prediction is one of the key functions of consciousness. An organism that cannot predict would have a seriously hampered consciousness. What needs to be shown formally is that the learning mechanism of postulate 4 extends to postulate 5.

Formalization 5   Say that i_x follows i_w as a result of world state changes. Say that the organism is in state q_w in response to i_w . If the input changes to i_x and iconic learning is taking place, the following element of ß will be added:

$$q_x = ß(i_x , q_w), \quad \text{followed by}$$
$$q_x = ß(i_x , q_x), \text{ where } q_x = \mu(i_x).$$

In the mental mode, the following two transitions become equally probable:

$$q_x = ß(\emptyset , q_w) \ , \ q_w = ß(\emptyset , q_w) \ .$$

This means that, in time, state q_w will lead to q_x in the mental mode, completing the prediction.

Postulate 6: The Self

As a property emerging from iconic learning and feedback between output and the senses, the internal state structure of a conscious organism carries a full representation of the output of the organism and the changes that such outputs effect on world states. This includes a representation of what can and cannot be achieved by the organism itself.

Justification 6   The salient characteristic of self-awareness is first the distinction between changes in world states that are caused by the organism's own actions. A corollary of this and the prediction ability of postulate 5 is that the organism will be able to plan its own actions against some target.

Formalization 6   The objective of this formalization is to include the output of the organism by realising that such output also forms input. Let $z_\emptyset$ be a special "no output" condition and all other z_j output actions that are perceivable at the input. Any input i_w therefore contains two components

$$i_w = \{ j_w , z_j \}$$

(note that if the output is $z_\emptyset$ then $i_w = j_w$.)

It follows from the iconic learning function $q_w = \mu(i_w)$ that

$$q_w = \{ k_w , s_j \}, \qquad \text{where } \{ k_w , s_j \} = \mu \{ j_w , z_j \}.$$

Hence iconic learning leads to parts of states such as s_j which are internal representations of the organisms own actions.

Now suppose that the world is in some state i_a and that this has been learned with z_ϕ, producing therefore a direct iconic representation $q_a = \mu(i_a)$, suppose that action z_1 changes the world state to i_1 where it remains even if the action ceases. The system will learn the following linked internal representations.

$$q_a \rightarrow \{k_a, s_1\} \rightarrow \{k_1, s_1\} \rightarrow \{k_1, s_1\} \ldots \rightarrow q_1$$

Hence iconic learning leads to representations of the way in which the organism's own actions achieve changes in the world state.

3   Corollaries

The postulates seen so far provide the basis of the theory of the consciousness of neural state machines. However, this is by no means complete. Other important properties of consciousness are stated here as a progressive series of corollaries of the postulates stated so far. A complete treatment of these is part of a full paper (Aleksander, 1994).

The Meaning Corollary: When sensory events occur simultaneously in different sensory modalities, iconic learning (Postulate 4) ensures that one can be recalled from the other - verbal meaning may be assigned to objects.
The Naming Corollary: The feedback from output (voiced names) to input as in postulate 6 creates state representations for the voiced output naming of perceived objects.
The Language Corollary: Language is in the repository of a society from which a conscious organism can learn. Following from postulate 5, a conscious organism treats and learns language as a set of sequential relationships of input events in the auditory and visual domains.

4   Conclusion

It has been demonstrated that automata theory applied to a state machine in which the state variables are neurons that learn and generalize, provides the appropriate framework for a formal theory of consciousness that includes self awareness, and an approach to natural language understanding.

References

Aleksander, I. and Morton, H.B. (1993)*Neurons and symbols: the stuff that mind is made of.* London: Chapman and Hall.
Aleksander, I. The case for artificial consciousness (in press) 1994.
Aleksander, I. Capturing Consciousness in a neural system. Proc ICANN 92.
Dennett, D. C. (1991) *Consciousness explained.* London: Allan Lane/Penguin.
Penrose, R. (1989) *The emperor's new mind.* London: Vintage.
Searle, J.R. (1992) *The rediscovery of the mind.* Boston: MIT Press.
Winograd, T. and Flores, F. (1986) *Understanding computers and cognition.* New Jersey: Abelex.

Forward Reasoning and Caianiello's Nets

Ernesto Burattini, Guglielmo Tamburrini

Istituto di Cibernetica C.N.R.
I-80072 Arco Felice - Italy
e-mail: ernb@arco.na.cnr.it

1   Introduction

A production system consists of a *knowledge base* containing *rules*, a *data base* containing *facts*, and a *rule interpreter* to control the inference process. The *rules* we shall consider here are conditional expressions in the Horn clause form

$$p_1 \wedge \ldots \wedge p_k \rightarrow h \qquad (1)$$

with the additional constraint that $p_1, \ldots, p_k$ and h are propositional literals. These restrictions are immaterial in a wide variety of AI applications, in view of the considerable expressive power of (propositional) Horn clause logic (Kowalski, 1979) and the fact that the knowledge base of most expert systems is supposed to capture relations over a finite domain of elements (Torasso et al., 1989), (Gallant, 1993). In this work we illustrate a neural network capable of applying forward search on a neurally implemented knowledge base of facts and rules.

2   The rule model

In a previous paper (Burattini et al., 1992) we used a simple artificial neural model (Caianiello, 1961) for representing production rules. The state equation for a neuron h in this model is given by

$$u_h(t+1) = \mathbf{1} \left[\sum_{j=1}^{n} \sum_{i=0}^{t} a_{j,h} \cdot u_j(i) \cdot \delta_h(t-i) - s_h \right] \qquad (2)$$

where $u_h(i)$ is the state (1 or 0) of the neuron h at time i; $a_{j,h}$ is the coupling coefficient between neurons j and h; $\delta_h(i)$ is a monotone non-increasing function of the discrete time i for neuron h; it represents a time variable memory of the excitation received by h from its neighbours; s_h is the threshold of h; and

$$\mathbf{1}[x] \quad = \quad \begin{cases} 1 \text{ if } x > 0 \\ 0 \text{ if } x \leq 0 \end{cases}$$

Rule (1) can be represented as a net having k neurons $p_1, \ldots, p_k$ connected to a neuron h (see fig.1) with the following settings:

The behaviour of the net formed by the neurons $p_1, \ldots, p_k$ and h

$$\forall j \quad a_{j,h} = 1 \quad (1 \le j \le k)$$

$$s_h = k - \varepsilon \quad (0 < \varepsilon < 1) \tag{3}$$

$$\delta_h(i) = \delta^0(i) \text{ where } \delta^0 \text{ is } \begin{cases} 1 \text{ if } i = 0 \\ 0 \text{ if } i \ne 0 \end{cases} \quad \text{(i.e. there is no memory).}$$

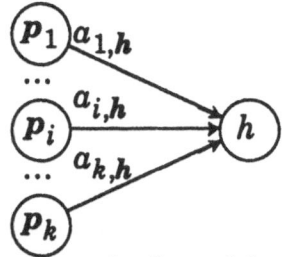

fig.1: *a neural rule model*

reflects faithfully the behaviour of a rule interpreter applied to rule (1). Indeed, by (2) and the settings in (3),

$$u_h(t+1) = 1 \quad \text{iff} \quad \forall j \, u_{p_j}(t) = 1. \tag{4}$$

3 Forward chaining

Using this representation of rules as basic building block, one can design a neural production system, organized into five different layers of neurons, capable of carrying out forward search on a knowledge base of facts and production rules. A specific example of such system, for a set of four rules, is presented in fig.2. In this net all neurons have a decay law as in (2) above.

The first layer (*IN*) accepts external *inputs* to the net. It is formed by as many neurons j_{IN} as different propositional literals appear in the rules (in the example of fig. 2, $1 \le j \le 6$). Each neuron j_{IN} is connected to the neuron j_{DB} representing the same literal in the second layer *DB*. The neurons j_{IN} are activated by an external source and their threshold is ε with $0 < \varepsilon < 1$. The symbol 'ε' will assume throughout an arbitrary value in this range.

The second layer (*DB*) is a partial *data base* formed by as many neurons as in the layer *IN*. These elements store the premises introduced in the *IN* layer: the neuron j_{DB} becomes active at time $t+1$ whenever the neuron j_{IN} is active at time t and preserves this information by self-excitation.

The literals proved by the system during its functioning cycles are stored in the third layer (*KB*) which supports the entire *Knowledge Base*. Here, each rule is represented as in section 2, with the additional condition that if a literal p occurs as the conclusion of z rules, then z distinct neurons — each one representing an occurrence of p in the conclusion part of those rules — have to be introduced in this layer. This additional condition is needed to avoid an incorrect activation of a neuron representing p, which

220

derives from a combination of premises belonging to different rules having
p as conclusion. These z neurons are connected to a neuron p^* which
represents all occurrences of p as premise of production rules and fires on
the consequents of those rules. Since the elements represented by neurons
are propositional *literals*, one may obtain an inconsistency in this layer if
both an atomic proposition and its negation become simultaneously active.
The system is capable of signalling such inconsistencies since each pair of
neurons representing contradictory literals is connected to a neuron
belonging to the control layer which becomes active when both elements of
the pair are active. (see for instance the pair formed by the neural
representatives of d and $\neg d$ in fig.2).

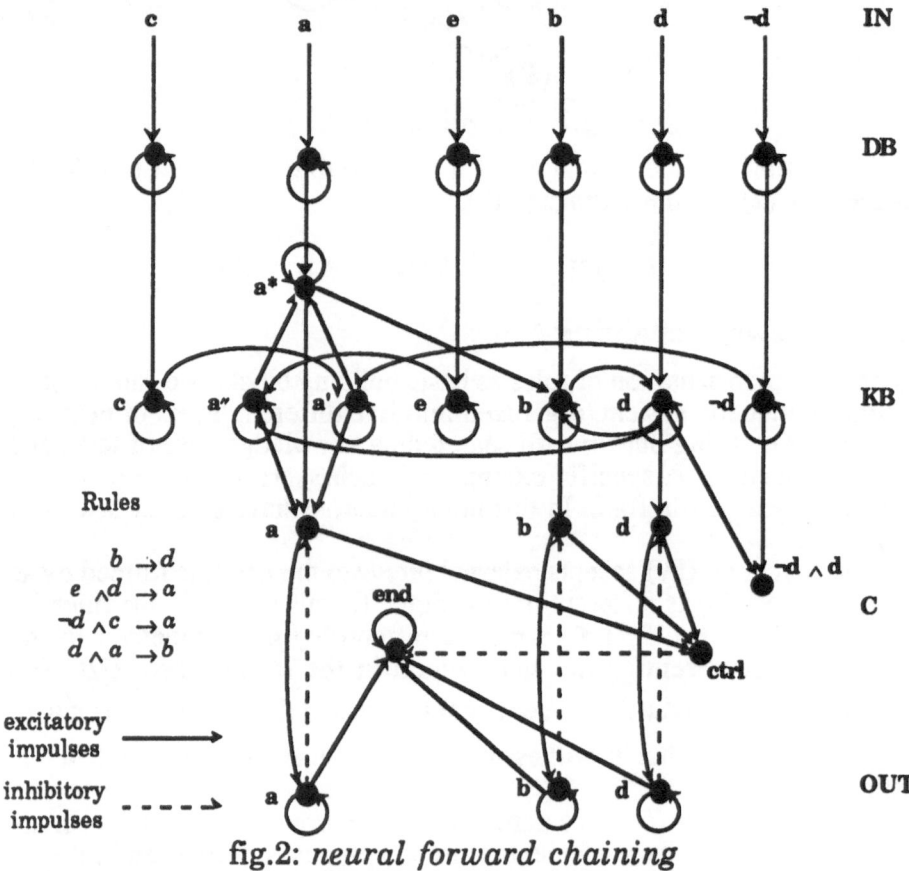

fig.2: *neural forward chaining*

Every neuron in this layer is self-excited. The neuron i_{KB} representing a
literal appearing only as premise of production rules is connected to the
representative i_{DB} of the same literal in *DB* and has a threshold equal to ε.
A neuron i_{KB} representing a literal appearing as conclusion in only one
rule r with q premises is connected to the q neurons in *KB* representing the
premises of r and has a threshold equal to $q-\varepsilon$. When a literal p occurs as
conclusion in z different rules ($z>1$), each neuron i'_{KB} representing one of
these occurrences p' is connected to the q neural representatives of the

premises of p' in KB and has a threshold equal to q-ε. These z i'_{KB} neurons fire on a neuron $i*_{KB}$ which represents all occurrences of p as premise in the production rules. The threshold for such $i*_{KB}$ is equal to ε.

The inferential process is triggered by exciting neurons in IN and its results are codified by active neurons in KB. No scheduling is necessary for carrying out this process, since all rules whose condition part is satisfied in the KB layer are simultaneously applied. Thus, only the following issues have to be settled: (a) how to verify that no more literal can be established on the basis of the available information; (b) how the output is to be read off from the network.

The *control problem* (a) can be solved by introducing a distinct layer C of m neurons, where m is the number of possible different conclusions in the system of production rules. Each neuron i_C in C represents a literal appearing as conclusion in z rules, with $z \geq 1$; it receives impulses from a neuron in KB representing the same conclusion, and activates in the layer OUT the corresponding neuron i_{OUT}. The latter, once excited, sends back an inhibition, equal to $z+1$, to the neuron i_C in the C layer. Moreover, each neuron i_C can fire on the special neuron $CTRL$ which becomes active every time a new conclusion is reached. Another special control neuron END is inhibited by $CTRL$, with strength $m+1$, and is excited by each neuron i_{OUT}. Thus, END is inactive until $CTRL$ is active, i.e., until new conclusions are reached. When nothing else can be proved, $CTRL$ becomes inactive. As a result, END is no longer inhibited and becomes active, thus signalling that the forward process on the input data is terminated.

The *output problem* (b) is solved through a layer OUT of m neurons, where m is the number of possible different conclusions in the system of production rules. Each neuron in OUT is excited by the corresponding neuron in C and is self-excited in order to store this information.

When the forward process terminates, the END neuron becomes active and signals that the process has been completed; the active neurons in the layer DB store the initial input; other active neurons in KB indicate both asserted and proved facts; the active neurons in the layer OUT represent the conclusions of production rules which have been reached by forward chaining under the initial assumptions stored in DB.

REFERENCES

Burattini E. and Tamburrini G. (1992), "A pseudo-neural system for hypothesis selection", *International Journal of Intelligent Systems* 7, 521-545.

Caianiello E. R. (1961), "Outline of a Theory of Thought Processes and Thinking Machines", *Journal of Theoretical Biology* 2, 204-235.

Gallant, S. I. (1993), *Neural Network Learning and Expert Systems*, MIT Press, Cambridge, Massachusetts.

Kowalski, R. (1979), *Logic for Problem Solving*, North-Holland, Amsterdam.

Torasso P. and Console L. (1989), *Diagnostic Problem Solving*, North Oxford Academic, London.

An ANN model of anaphora: implications for nativism.

S. H. Parfitt

Neural Systems, Dept of Elec. Eng.,
Imperial College, Exhibition Rd, London, SW7 2BT, UK.
email: s.parfitt@ic.ac.uk

The argument from the poverty of the stimulus

It is widely believed that acquisition of language involves the formation of successive hypotheses about the adult grammar, and the testing and modification of them with reference to the data. If a wrong hypothesis is formed, negative evidence seems required for it to be discarded. Whilst natural languages are at least context-free, Gold (1967) showed that, given positive evidence alone, only regular languages are learnable. But negative evidence does not appear to exist (Brown & Hanlon 1970, Braine 1992, Hirsh–Pasek et al 1984), and whilst Bohannon and Stanowicz (1988) found evidence of noisy feedback, Gordon (1990) argued that its incomplete nature meant it had no bearing on formal proofs of language learnability, and Marcus (1993) concluded that noisy feedback does not constitute sufficient negative evidence to account for acquisition. Thus, it is argued, the data alone cannot be responsible for enabling the child to decide against the very large population of incorrect hypotheses. This has frequently been cited as proving the need for an innate linguistic knowledge base, and although it has often been suggested that other factors may instead account for the correct choice of hypotheses, no experimental model has yet provided proof. It is this which the present work seeks to address.

An alternative to innate knowledge?

The two main arguments in favour of innate linguistic knowledge are as follows: *(a)* the total number of hypothesis choices needs to be limited so that time is not wasted in examining hypotheses which hold for no language, and *(b)* the child must be prevented from choosing an over-general hypothesis, since the lack of negative evidence would seem to prevent the correction of over-generalisation errors. In standard Government and Binding Theory (GB), this is dealt with by ordering parameter values by markedness as part of the innate specification. Wexler and Manzini (1987) suggest instead that they might be calculated by a learning module using the criterion of generated language size as a yardstick. They formulate this in terms of a "Subset Principle".

At issue here is whether innate knowledge is necessary to deal with the subset problem successfully. Morgan (1986) proved learnability for bracketed input, but needed to assume a d-structure base from which transformations would be learnt. (Subsequent work by Morgan and others (e.g. Altmann 1993) has confirmed the presence of bracketing and other acquisitional cues in the input.) One of the first ANN models to throw light on the matter was Rumelhart and McClelland's(1986) model of the English past tense, which showed a learning device can generate, test and modify appropriate hypotheses with reference solely to positive, *environmental* information. Over-general hypotheses are both generated and discarded whilst the problem space is being explored, and their temporary selection may be seen as a natural part of the learning process. However, knowledge about the past tense is not considered innate. Elman (1991) published the important result that a simple recurrent network (SRN) which initially failed to learn about a linguistic domain containing relative clauses, succeeded when trained incrementally. This involves gradually increasing *either* the complexity of the training data *or* the memory of the network, and performs the function of helping constrain the search space in the early stages to within the area of the adult language. Whilst this work added weight to the possibility of a non-rationalist approach to language acquisition, it did not look directly at any 'innate' linguistic phenomena. Moreover, experiments in language learning, including Elman (1991), have not incorporated tests of whether the final, trained hypotheses are *over*-general; that is to say, models are not tested on data which clearly belongs to a language larger than that which generated the training data: it is therefore not possible to know whether they have formed hypotheses which generate languages "smallest among the languages compatible with the input data", to quote Wexler and Manzini. This is of fundamental importance; showing transfer between a training set and a test set generated from the same language tells us little about the model's ability to deal with the problem central to linguistic theory: that of introducing constraints such that only the set of strings in the language are generated, and none outside it.

The experimental model

We consider here a phenomenon, anaphora, which is typical of those traditionally believed to have an innate basis, and present a model with no in-built knowledge which is predicted to generate appropriate hypotheses, and not to exhibit significant over- or under-generalisation after training. Where anaphora is concerned, over-generalisation will lead to the production and acceptance of utterances in which the anaphor is resolved outside of the correct governing category. In other words, the governing category is larger for the speaker possessing the over-general model than it is for the rest of the population.

All previous related work described as modelling anaphora actually treat *pronouns* and not anaphors (in GB's terminology) (Reilly 1984, Allen 1987, Allen 1989, Coelho 1992). They only use one-clause utterances, whereas the proposed model is trained on utterances of more than one clause, the nature of the task requiring it. Coelho's model employs certain assumptions about the nature of pronoun resolution (Sidner's (1983) focusing structure), whereas non-essential assumptions are avoided in the present work. None of the authors tested their models on ungrammatical utterances, which is a key part of the present experiments.

The present model differs from previous work in several respects.

(a) It is the first, within the ANN framework, to model acquisition of an 'innate' phenomenon by applying to the problem constraints other than innate knowledge. In this case, the constraints on generalisation are those afforded (i) by the back-propagation training algorithm, and (ii) by incremental training.

(b) It is the first model of an 'innate' phenomenon to be explicitly tested for over-generalisation by the use of a test set belonging to a language known to be larger than that to which the training set belongs.

(c) It is the first to be trained on data generated explicitly with reference to parameter values ordered for markedness.

(d) It is the first ANN model of the resolution of anaphors as opposed to pronouns, and the first such model to be trained on utterances more than one clause in length.

(e) It is the first to consider Wexler and Manzini's learning module from the viewpoint of ANNs.

Wexler and Manzini's formulation of the *governing category* parameter gives five values, of which each one in the list generates a language which (for anaphors) is a proper subset of that generated by the value following it. We wish to show that, when an appropriate model with no inbuilt ('innate') knowledge is trained on data, all of which is generated by value q_i, it will not over-generalise to value q_{i+1}. Since the Subset Principle holds for each value, proving that the above is true for any two values q_i and q_{i+1}, proves it also for any other pair. We take the first three values in the hierarchy and use them to generate three data sets. Since English is associated with value A, data generated from B and C will be like English in every respect except that anaphors will be resolvable within larger governing categories. To find whether the model has generated a hypothesis which accounts for the training language, it is tested on reserved data from the same language; to find whether it has produced an over-general hypothesis, it is tested on data generated by the next value down the markedness hierarchy, that is to say, the one generating the next largest language.

The model is an SRN, trained incrementally. It is predicted to succeed at anaphora resolution since the training languages are not more complex than the centre-embedded language used by Elman (1991). Initially, to simplify the task, only the singular anaphors *himself* and *herself* are used, however subsequently experiments use all of *myself, yourself, himself, herself* and *itself*. The pattern for a word is subdivided into portions. One of these contains a unique, random identifier pattern. Others encode features relevant to anaphora resolution, such as gender (in the case of nouns and anaphors) and tense and inflexion (in the case of verbs). Three large corpora of anaphoric utterances are used for training, generated with reference to a vocabulary and three sets of sentence frames. The frames include sentences as follows: transitive, transitive with adjective, prepositional transitive, postpositional transitive, ditransitive, and two clause utterances, some formed using a complementiser and others taking an infinitive.

Training is performed incrementally, starting with simple transitive one-clause utterances, and finishing with two-clause utterances with complementisers. The task is next-word prediction, with the exception of anaphors, where the task is to predict the anaphor's *referent*. The network is first trained on the training portion of set A. It is then tested on reserved portions of sets A and B. The network should correctly predict the referent for the vast majority of anaphors in set A, but fail to do so for a statistically significant portion of anaphors in set B. 'Failure' means either predicting a referent other than the target one, or else predicting a word which is not a legal referent. Similar experiments are performed by training on sets B and C. Since sentences in A are also sentences in B, a network trained on B should correctly predict the referent for the vast majority of anaphors in both of sets A and B, but should fail to do so for set C. Since sentences in A and B are also sentences in C, a network trained on C should correctly find the referent for the vast majority of anaphors in all of sets A, B and C.

Back-propagation and negative evidence

Any acquisition theory put forward as an alternative to innate knowledge must take the no-negative-evidence finding into account. It might be felt that error back propagation does not fulfil this criterion. However, this would be a mistaken view. In the present case, the training data consists solely of information available to the network in its environment: positive evidence about the strings in the language. The errors generated by the learning algorithm are *model-internal*: they are not provided in the data, so they are not subject to the no-negative-evidence problem.

Conclusions and extensions

In this paper, I have presented an ANN methodology for investigating claims of a role for innate knowledge in language acquisition. It seem likely that innate knowledge is just one of several or even many ways of constraining the search for appropriate hypotheses and of preventing over-generalisation. I have argued that these constraints may equally well be provided by a judicious choice of network, training algorithm and training regime. If confirmed, this hypothesis will serve to uphold Wexler and Manzini's notion of a learning module making judgements about hypothesis markedness on the basis of generated language size (however, we would wish to replace the "learning module" with an ANN). If it is found that, given an SRN and positive evidence only, training does result in over-general hypotheses, this will indicate either that there is something extra in the architecture or learning mechanism of the brain, or that there is more information available in the environment than we have allowed for. Bracketed input (Morgan 1986) provides an extra means of constraining the range of possible hypotheses whilst using only information/structure which has been shown to exist in the environment, and a further series of experiments are planned using training data augmented to incorporate it.

References

R.B.Allen (1987). Several studies on natural language and back-propagation. *IEEE 1st Intl. Conf. on Neur. Networks*, 2:II–335–341.

R.B. Allen & M.E. Riecken (1989). Reference in Connectionist Language Users. In R. Pfeifer et al, eds., *Connectionism in Perspective*, pp.301–308, Elsevier.

G. Altmann (1993). Language learnability and the linguists' bootstrap. *Ann. Meeting of the LAGB*.

R. Brown & C. Hanlon (1970). Derivational complexity and order of acquisition in child speech. In J. Hayes, ed., *Cognition and the development of language*, pp.11–53, Wiley, NY.

J.N. Bohannon & L. Stanowicz (1988). The issue of negative evidence: Adult responses to children's language errors. *Dev. Psych.*, 24:684–689.

M. Braine (1992). What sort of innate structure is needed to bootstrap into syntax? *Cognition*, 45.

O.B. Coelho (1992). A connectionist approach to anaphora resolution in task-oriented discourses. In I. Aleksander & J. Taylor, eds., *Artificial Neural Networks*, vol. 2. Elsevier.

J.L. Elman (1991). Incremental learning, or the importance of starting small. *13th Cog. Sci. Soc. Conf.*, pp. 443–448, Chicago.

E.M. Gold (1967). Language identification in the limit. *Inf. & Cntrl*, 16:447–474.

P. Gordon (1990). Learnability and feedback. *Dev. Psych.*, pp.26:217–220.

K. Hirsh–Pasek, R. Treiman & M. Schneiderman (1984). Brown and Hanlon revisited: Mothers' sensitivity to ungrammatical forms. *Jnl. Child Lang.*, 11:81–88.

G.F. Marcus (1993). Negative evidence in language acquisition. *Cognition*, 46:53–85.

J. Morgan (1986). *From simple input to complex grammar*. MIT, Cambs, MA.

S. Pinker (1984). *Language learnability and language development*, Harvard U.P., London.

S. Pinker (1989). *Learnability and cognition*, MIT, Cambs, MA.

R.G. Reilly (1984). A connectionist model of some aspects of anaphora resolution. *10th Intl Conf. on Computational Linguistics*, pp 144–149.

D. Rumelhart & J. McClelland (1986). On learning the past tense of English verbs. In D. Rumelhart & J. McClelland, eds., *Parallel Distributed Processing*, vol. 2, MIT, Cambs, MA.

K. Wexler & M.R. Manzini (1987). Parameters and learnability in Binding Theory. In T. Roeper & E. Williams, eds., *Parameter setting*, pp.41–76. D. Reidel, Dordrecht.

The Spatter Code for Encoding Concepts at Many Levels

Pentti Kanerva
Swedish Institute of Computer Science
Box 1263, S–164 28 Kista, Sweden
e-mail: kanerva@sics.se

Abstract. The *Spatter Code* is a high-dimensional (e.g., $N = 10,000$), random code that encodes "high-level concepts" in terms of their "low-level attributes" so that concepts at different levels can be mixed freely. The binary spatter code is the simplest. It has two N-bit codewords for each concept or item, a "high-level," or *dense,* word with many randomly placed 1s and a "low-level," or *sparse,* word with a few (that are contained in the many). The dense codewords can be used as inputs to an associative memory. The sparse codewords are used in encoding new concepts. When several items (attributes, concepts, chunks) are combined to form a new item, the two codewords for the new item are made from the sparse codewords of its constituents as follows: the new dense word is the logical OR of the constituents (i.e., their sum thresholded at 0.5), and the new sparse word has 1s where the constituent words overlap (i.e., their sum thresholded at 1.5). When the parameters for the code are chosen properly, the number of 1s in the codewords is maintained as new items are encoded from combinations of old ones.

1   Introduction

In what follows, *mind, concept, attribute,* and *chunk* refer to philosophical or psychological entities, *brain* and *pattern* refer to physical entities, and *term* refers to something in between, but rather physical, for example, to a pattern of neural activity that has become established or organized through repetition and that can then be recreated by the brain.

The human mind is born without a rich repertoire of concepts, and the world is not organized into a neat system of objects, properties, and relations when the brain first "sees" it. The brain must find whatever organization there is in the world and it must construct the terms in which it sees the world. The human brain has a remarkable ability to do this; it is built to do it—it cannot help but to do it—which is why the organized world seem utterly natural to us by the time we are able to talk about it. But how the brain gets organized is a big mystery.

We do know some general principles. For example, the brain remembers recurring activity patterns and detects regularities in them. That is why things become familiar to us when we see them over and over, and why we rehearse when we want to learn a physical or a mental skill. Mathematical models for studying such things include associative memory and self-organizing maps (Hassoun, 1993; Kohonen,

1988; Willshaw & von der Malsburg, 1975 & 1979).

The neural activity patterns most readily available in the new-born brain are produced by the sense organs and by emotions. They are the initial raw material, and whatever regularities are found in them, become the brain's initial terms for describing the world.

By age three the human mind has developed a rich store of concepts, built out of other concepts. This is expressed most strikingly in the use of language. The "high-level" terms that the brain now uses are built out of and on top of the brain's initial terms for describing the world, so that they form a kind of a superstructure.

This paper is about building the superstructure. Could it be built from the low-level terms in the same way as the initial, low-level terms are built from patterns of activity from the senses and emotions? For example, if a certain combination of low-level terms occurs repeatedly in some part of the brain, would a new higher level term be produced? It could be, if new activity patterns are generated and regenerated from old ones. Whatever regularities and organization there is in these new patterns could then become the new terms for describing the world.

2 The Binary Spatter Code

The spatter code is a particularly simple way to construct new patterns from combinations of old ones. It does so in a way that would be natural for neurons, and it could generate endless raw material for the building of new, high-level terms. The binary spatter code, which is described here, is the simplest. The idea generalizes readily to nonbinary, and some things that are problematic for the binary code can be avoided.

It has been shown in psychological experiments that our short-term memory, or span of immediate attention, or mental focus has a limited capacity, and that the capacity is best expressed in chunks. The number of chunks is around seven (Miller, 1956), whereas what constitutes a chunk is very flexible. When chunks are made of chunks, we can deal with things in the world both generally and in great detail. The low-level chunks then represent concrete things and detail, and the high-level chunks represent abstract things and the general (Albus, 1991).

The great freedom in defining and combining chunks suggests that chunks of all kinds and at all levels are represented internally in the same way. The representation is neutral as to what it represents. The binary spatter code takes this idea into the extreme: chunks are represented by large, random bit patterns. The main conditions for these patterns—that is, for the codewords for the chunks—are that

1. as several chunks combine to form a new chunk, the codewords for the components should be visible in the codeword for the new chunk (the new codeword should be *analyzable*); and

2. the new chunk should be represented by a codeword that can be used in the construction of codewords for further chunks (the code should be *recursive*).

To satisfy both conditions, the spatter code represents each chunk by two related codewords: a *dense* codeword, to satisfy condition 1, and a *sparse* codeword, to satisfy condition 2. If the code space is large enough (e.g., 10,000-bit code), the sparse codewords by themselves are both analyzable and recursive.

```
    bit 1                              bit 10,000
      /                                   /
w1  0 0 0 0 0 1 0 0 0 0 0 0 0 0 ... 1 ... 0 0 0
w2  0 0 0 0 0 0 0 0 0 0 0 0 0 0 ... 0 ... 0 0 1
w3  0 0 0 0 0 1 0 0 0 0 0 0 1 0 ... 1 ... 0 0 0
w4  0 0 0 0 0 0 0 0 0 0 0 1 0 0 ... 0 ... 0 0 0    0.058  1s
w5  0 0 0 0 0 0 0 0 0 0 0 0 0 0 ... 0 ... 0 0 0
w6  0 0 0 0 0 0 0 0 0 0 0 0 0 0 ... 1 ... 0 0 0
w7  0 1 0 0 0 0 0 0 0 0 0 0 0 0 ... 0 ... 0 0 0
```

```
Sum 0 1 0 0 0 2 0 0 0 0 0 1 1 0 ... 3 ... 0 0 1

W8  0 1 0 0 0 1 0 0 0 0 0 1 1 0 ... 1 ... 0 0 1    0.34  1s
w8  0 0 0 0 0 1 0 0 0 0 0 0 0 0 ... 1 ... 0 0 0    0.058  1s
```

Figure 1. Generating dense and sparse codewords W8 and w8 from the sparse codewords of seven constituents, w1, w2, ..., w7. If the probability of a 1 in the sparse words is 0.058, the code will be recursive and the probability of a 1 in the dense words will be 0.34.

Table 1

The Probability of a 1 in the Sparse (p) and Dense (P) Codewords of a Recursive Spatter Code that Combines n Chunks

n	p	P
3	0.5	0.875
5	0.131	0.504
7	0.058	0.342
9	0.032	0.256
11	0.021	0.204

Figure 1 shows the construction of a (10,000-bit) binary spatter code when seven old chunks are combined into a new chunk. The dense codeword for the new chunk is simply the logical OR of the sparse codewords of the seven constituent chunks. The sparse codeword for the new chunk is built from overlaps: it has 1s where two or more constituent codewords have a 1.

In a recursive code, the probability p of a 1 in the sparse words depends on the number n of chunks that are combined into a new chunk. The recursive p is gotten by solving

$$q^n + npq^{n-1} = q$$

for p, where $q = 1 - p$ (assuming that the constituents are random and independent). The left side of the equation expresses the probability that a bit of the new sparse word will be a 0 (i.e., when all n constituent words have a 0, or when any one of them has a 1 and the rest have a 0). This probability should be the same (q) as it is in the constituent words. The dense codewords will then have 1s with probability $P = 1 - q^n$. Table 1 gives the recursive probabilities for several chunking factors.

3 Discussion

The neural character of the spatter code is evident when the two new codewords are construed as linear threshold functions of their constituents: the (vector) sum of the constituent codewords is thresholded anywhere between 0 and 1 (e.g., at 0.5) to get the dense word, and it is thresholded anywhere between 1 and 2 (e.g., at 1.5) to get the sparse word (see Fig. 1). Furthermore, the sparseness of the sparse words agrees with the low level of activity of many neural circuits, and large patterns and randomness agree with the size and connectivity of many neural circuits.

The code is strongly directional. Given the component codewords, the two new codewords fall out most naturally. However, a codeword's constituents cannot be determined from the codeword, although it is easy to decide (with logical AND) whether one codeword is a constituent of another, and in that sense the code is analyzable. The synthetic and holistic nature of the human mind could be related to such issues of directionality and analyzability.

3.1 Relation to other work

This work was influenced most strongly by the scatter code of Smith and Stanford (1990; Stanford & Smith, 1993 ms.) that uses randomness and logical XOR to map real variables into Hamming space. The other major inspiration was Manevitz and Zemach's (1991) paper on multilevel information processing in an associative memory, which encodes *sequence* data at multiple levels; here, the elements or chunks that go into making higher level chunks are an *unordered* set. Jaeckel's (1989) encoding of simple visual images considers unordered sets of image components and constructs the encoding of the image with the logical OR of very long bit strings that encode image components.

3.2 Acknowledgment

I am most grateful to Derek Smith for his visit with me at the Santa Fe Institute in February of 1993, to tell me about the Scatter Code. I chose the term Spatter Code partly to acknowledge its close kinship to the Scatter Code, and partly to capture the ideas of sparseness, randomness, and hitting some spots more than once when you spatter several (7) times over the same area.

References

Albus, J.S. (1991) Outline for a theory of intelligence. *IEEE Trans. Systems, Man, and Cybernetics* 31(3):473–509.

Hassoun, M.H., ed. (1993) *Associative Neural Memories: Theory and Implementation*. New York: Oxford University Press.

Jaeckel, L.A. (1989) Some Methods of Encoding Simple Visual Images for Use with a Sparse Distributed Memory, with Application to Character Recognition. Report RIACS TR 89.29, Research Institute for Advanced Computer Science, NASA–Ames Research Center.

Kohonen, T. (1988) *Self-Organization and Associative Memory*, 2nd ed. Berlin: Springer–Verlag.

Manevitz, L.M., and Zemach, Y. (1991) Assigning meaning to data: Multi-level information processing in Kanerva's SDM. *Proc. 1991 Israel Conference on AI and Computer Vision (IAICV 8);* 114–130.

Miller, G.A. (1956) The magical number seven, plus or minus two: Some limits on our capacity for processing information. *Psychological Review* 63:71–97.

Smith, D.J., and Stanford, P.H. (1990) A random walk in Hamming space. *Proc. 1990 Int'l Joint Conference on Neural Networks (IJCNN 90)*, vol. 2; 465–470.

Stanford, P.H., and Smith, D.J. (1993 ms.) The Multidimensional Scatter Code: A Data Fusion Technique with Exponential Capacity. Submitted to ICANN 94.

Willshaw, D.J., and von der Malsburg, Ch. (1975) How patterned neural connexions can be set up by self-organisation. *Proc. Royal Society B.*, 194:431–445.

Willshaw, D.J., and von der Malsburg, Ch. (1979) A marker induction mechanism for the establishment of ordered neural mappings: its application to the retinotectal problem. *Phil. Trans. Royal Society B.*, 287:203–243.

Learning in Hybrid Neural Models

A.M. Colla †, N. Longo ‡, G. Morgavi ¶, S. Ridella ‡

† Elsag Bailey – Un'Azienda Finmeccanica spa
Via G. Puccini 2, 16154 Genoa (ITALY)
‡ DIBE University of Genoa
Via all'Opera Pia 11/a, 16145 Genoa (ITALY)
¶ I.C.E. C.N.R.
Via Marini 6, 16149 Genoa (ITALY)

1   Introduction

In this work two different learning strategies for a hybrid neural model are compared. This model has been developed for Pattern Recognition applications, in particular for handprinted character classification.

The most natural solution to classification problems is obtained by supervised algorithms. This solution generally is sub-optimal and can better and better approximate the optimal (Bayesian) solution (Ruck et al. 1990) as the input data representation is ameliorated. Neural models oriented to data representation such as the Self-Organizing Map (SOM) can supply a useful pre-processing to a classification system.

The model we analyze consists of a **Kohonen Self-Organizing Map (SOM)** for pre-processing, followed by a **Multi-Layer Perceptron(MLP)** for classification. The two parts can be trained either *separately* (on the same training set) or *as a whole* by extending the *Error Back Propagation* learning algorithm also to the SOM layer.

The two learning algorithms are compared; experimental results are presented, in terms of both learning trend and generalization results.

2   Learning Strategies

2.1   Separate training

The usual strategy for connecting a SOM to a MLP consists in using the output of the ordered SOM neurons (1 for the winner node, 0 for all the others) as input to the MLP. As a result, the MLP receives just *symbolic* information, that is an indication about which cluster the SOM assigns the current pattern to. In our approach (Colla et al. 1993) the SOM is *not* used for clustering, *but it is used to code* the input patterns into an *analog* representation of the data distribution, in the form of distances from "templates" identified in the learning phase. No hypotheses about input space distribution are required. The hybrid **SOM_MLP** network learning, as proposed in (Colla et al. 1993), is based on the following 3 steps :

- The SOM is trained following traditional techniques (Kohonen 1990). The chosen structure is 2-dimensional, with a *toroidal* topology. A standard

Euclidean distance is adopted, with *circular* neighborhoods. The learning rate is decreased in a piecewise linear way.

- The *analog* outputs of the SOM, i.e. the distances of the current input pattern to *all* the SOM neurons, constitute the input to the MLP net.
- The MLP is trained according to traditional learning techniques (Hertz et al. 1991). The topology is a standard, fully connected one. The neurons' transfer function is the hyperbolic tangent. Learning is *by epoch*, with an accelerating method based on the Vogl algorithm and described in (Drago and Ridella 1991).

SOM learning is completely unsupervised : no labelling phase is required, but it is sufficient to declare the target classes at the time of training the MLP portion of the net. This results in some appropriate filtering of the noisy data. At the same time, a rich representation of the input space is supplied to the MLP classifier if the number of SOM neurons is sufficient.

2.2 Unified training

The hybrid net (**SBP**) is composed of one SOM layer (Kohonen neurons) and of *perceptron* layers with sigmoidal neurons (one or more hidden layers and one output layer). Training of the *whole* SBP network is performed by *extending the Error Back Propagation learning algorithm also to the SOM neurons*, which is made possible by the *analog interface* between the Kohonen layer and the perceptron layers (differentiable propagation and activation rules).

In the learning phase, the weights in the MLP subnet are modified according to the above mentioned procedure.

SOM neurons differ from MLP ones in the activation function (Euclidean distance instead of weighted sum) and in the non-linear transfer function (square root instead of hyperbolic tangent). A straigtforward calculation yields the following weight updating rule for SOM neurons:

$$\Delta \vec{n}(i) = \eta \frac{1}{N_{inp}} \frac{1}{S_i(p)} \sum_{h=1}^{N_{hid}} (\delta_h(p) w_{h,i})(\vec{n}(i) - \vec{x}(p)) \tag{1}$$

where $\vec{n}(i)$ is the weight vector of the i-th SOM node, $\vec{x}$ is the input vector representing pattern p, $S_i(p)$ is the Euclidean distance of input pattern p to the i-th SOM node, $w_{h,i}$ is the weight of the connection from the h-th node in the lowest hidden perceptron layer to the i-th SOM node, $\delta_h(p)$ represents the error back-propagated from the MLP layers, and N_{inp} is the dimension of the pattern space.

In comparison with the standard Kohonen weight updating rule, it can be noticed that in (1) the "learning rate" consists of a term depending on the *single* node distance to the current pattern. The closest nodes undergo larger modifications than the farthest.

3 Experimental Results and Conclusions

Some character classification tests were run on Elsag Bailey's proprietory database *CARATTE28*. This database, consisting of 19,441 *handwritten digits* produced by several different writers, was split into two parts: a balanced *training*

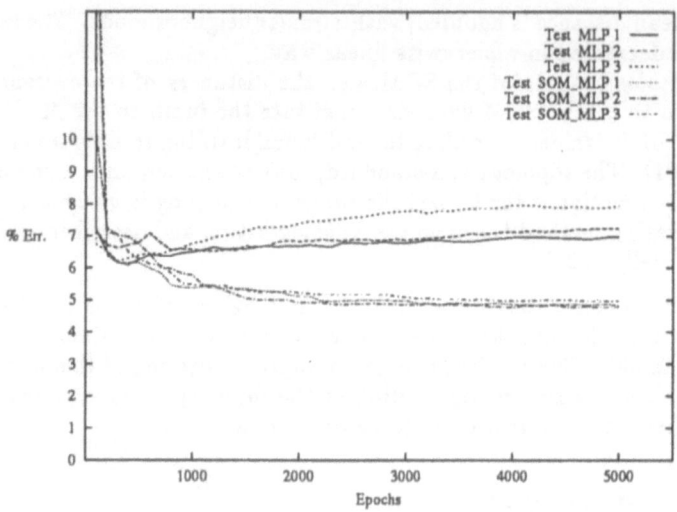

Figure 1: *Generalization tests with* **MLP** *and* **SOM_MLP** *nets.*

set of 6,460 images (646 per class), and a disjointed *test set* containing 12,981 images. Each character in the database was represented by a 28-dimensional array of the normalized percentages of black pixels within small square windows partitioning the image *("gray pixels")*.

Figure 1 shows the generalization improvement with the SOM_MLP net with respect to standard MLP (classification at 0 rejection rate). SOM layer is 16 × 16 nodes in each SOM_MLP net. After 5,000 epochs SOM_MLP nets perform 2–3% better than MLP nets with the same number (20) of hidden nodes, and the error is still decreasing. This improvement is statistically meaningful due to the test set size. Besides, in the best tests shown, the smallest error rate obtained by SOM_MLP is about 1.3% less than by the best MLP.

Figure 1 also shows that the SOM_MLP net is able to improve its generalization with training with practically *no overfitting effect*, which occurs with an equally small MLP net. This is most likely due to the fact that SOM's supply a coarser pattern space tessellation than MLP's, thus providing a comparatively worse description of the training set, but, at the same time, filtering noisy patterns out. The generalization ability of the whole neural system is therefore improved.

Results obtained by unified training (SBP) are, as expected, even more encouraging. In Figure 2 it is possible to notice that SBP performs better, both in learning and generalization, than SOM_MLP with the same structure (10 × 10 nodes in the Kohonen layer, one hidden perceptron layer with 10 nodes). However, the small performance improvement has been obtained by a slower and more computationally expensive learning procedure (each epoch in SBP training requires about 4 times as much computation as in SOM_MLP). Besides, it must be noticed that after a "short" training (less than 4,000 epochs) SOM_MLP performs still better than SBP, while in the long run SBP beats

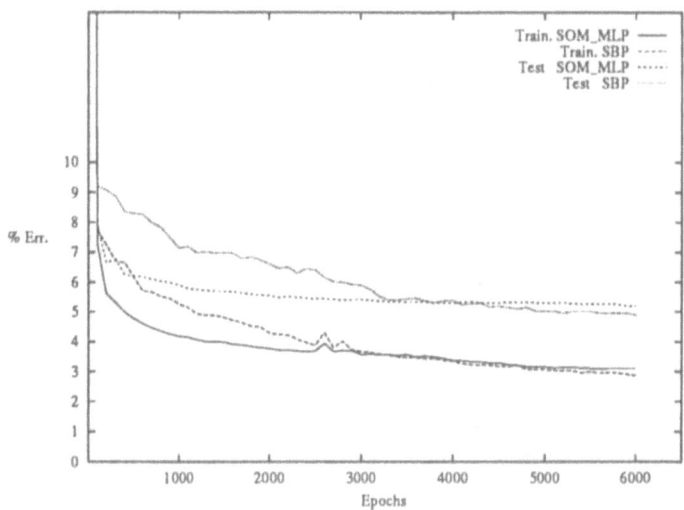

Figure 2: *Learning and generalization with* **SOM_MLP** *and* **SBP**.

SOM_MLP. With more training, SBP has been observed to ameliorate up to 1%. The best result obtained by SBP in this test was 4.3% generalization error after 10,000 epochs.

In conclusion SOM_MLP and SBP nets compare favourably with MLP classifiers of equivalent size in a realistic character recognition task.
A *mixed learning procedure*, starting with separate (SOM_MLP) learning for the initial epochs, and refined by unified (SBP) learning, can be adopted to reduce learning time. Preliminary results show that, after 3,000 SOM_MLP + 3,000 SBP epochs, the same performance can be reached as after 6,000 SBP epochs, with an overall reduction in learning time of about 33%.

References

- Colla, A.M., Longo, N., Morgavi, G., Ridella, S. (1993) *SBP : A Hybrid Neural Model for Pattern Recognition*, Proc. *Sixth Italian Workshop on Neural Networks - WIRN*, Vietri Sul Mare (SA), Italy, May 1993
- Drago, G.P., Ridella, S. (1991) *An optimum weights initialization for improving scaling relationships in BP learning*, in T. Kohonen, K. Mäkisara, O. Simula, and J. Kangas (Ed.s), *ARTIFICIAL NEURAL NETWORKS*, vol. II, pp. 1519–1522, North-Holland
- Hertz, J. Krogh, A. and Palmer, R.G. (1991) *An Introduction to Neural Computation*, Addison-Wesley
- Kohonen, T. (1990) *The self-organizing map.* In *Proceedings of the IEEE* **78**(9), pp. 1464–1480
- Ruck, D.W., Rogers, S.K., Kabrisky, M., Oxley, M.E. and Suter, B.W. (1990) *The Multilayer Perceptron as an Approximation to a Bayes Optimal Discriminant Function*, IEEE Trans. on Neural Networks **1** (4), pp. 296–298

A Connectionist Model for Context Effects in the Picture-Word Interference Task

Peter A. Starreveld and Jan N. H. Heemskerk
Leiden University, Unit of Experimental and Theoretical Psychology
P.O. Box 9555, 2300 RB Leiden, The Netherlands
E-mail: STARREVE@RulFsw.LeidenUniv.nl, Tel: (31) 71/273631

Abstract

In the picture-word interference task, two context effects can be distinguished: the semantic interference effect and the orthographic facilitation effect. A theory is described to explain these effects. This theory was implemented in a connectionist model. The model is able to simulate the time courses of the two context effects and their interaction.

1. Introduction

In the long history of research on Stroop-like effects in naming tasks, the task of naming a picture while ignoring an accompanying word is probably the most widely used. This task can also be used to study the time courses of the observed effects, by varying the time between the presentation of the picture and the presentation of the word (called the Stimulus Onset Asynchrony, SOA). Two effects stand out in the literature. First, naming times are increased when the word is semantically related to the picture, in comparison with naming times with an unrelated word. So, a picture of a CAT is named more slowly when accompanied by the semantically related word *horse*, than when accompanied by the unrelated word *house*. This effect is known as the semantic interference effect and is calculated by subtracting the mean reaction time in the condition with unrelated words from the mean reaction time in the condition with semantically related words. It occurs only in a limited SOA-interval from approximately -100 ms (word first) to +100 ms (picture first, Glaser and & Düngelhoff, 1984). Second, naming times are reduced when the accompanying word is orthographically or phonologically related to the picture's name, again in comparison with an unrelated word. So, a picture of a CAT is named faster when accompanied by the orthographically related word *candy*, in comparison with the unrelated word *house*. This effect has been called the orthographic facilitation effect, and it spans a broad time range from at least -200 ms (word first) to +100 ms (picture first, Rayner & Springer, 1986; see MacLeod, 1991, for a review of the literature on both effects). Rayner and Springer (1986), and Starreveld and La Heij (1993a) have shown that these two effects also show an interaction, in that the semantic interference effect is reduced when there is also an orthographic relation between the words and the name of the picture. So, when naming the picture of a CAT, the semantic interference effect obtained with the orthographically related words *calf* and *candy* is smaller than the one obtained with the orthographically unrelated words *horse* and *house*. This interaction was further investigated (Starreveld & La Heij, 1993b) in an experiment in which the SOA between picture and word was varied from -200 ms to +200 ms, in steps of 100 ms. The results are presented in Figure 1, and can be summarized as follows: a) a semantic interference effect was found in a small SOA-range, from SOA -100 to SOA 0, b) an orthographic facilitation

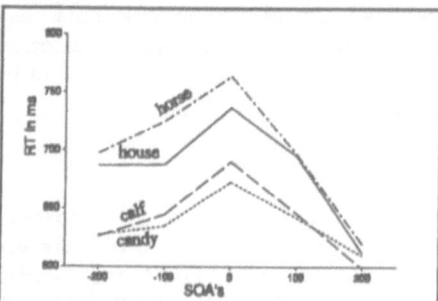

Figure 1. Experimental data. The words are examples of words which were presented with the picture of a CAT in the relevant conditions of the experiment.

effect was found in a much larger SOA-range, from SOA -200 to SOA 100, and c) these effects showed an interaction in that the semantic interference effect was reduced when there was also an orthographic relation between the words and the name of the picture.

2. Theoretical explanation

The effects as described above can be explained when both the effect of a semantic relation and the effect of an orthographic relation between the words and the name of the picture are localized at the same level of word production, the level where the phonological representation of the name of the picture is retrieved (Starreveld & La Heij, 1993a). In naming a picture, two major steps are distinguished (Collins & Loftus, 1975, Glaser & Glaser, 1989). First the picture has to be recognized in a semantic system, in which concepts are represented as nodes and related concepts share connections between their nodes. Second, the phonological representation of the name of the pictured concept has to be retrieved in the lexicon. In this lexicon the phonological representations of words are represented as phonological nodes. There is no semantic processing in the lexicon. The retrieval of the name of the picture can be seen as the selection of the right phonological representation from the lexicon. According to the view proposed by Starreveld and La Heij (1993a), the semantic interference effect is caused by a more difficult selection process, while the orthographic facilitation effect is caused by an easier selection process, as compared to the selection process with an unrelated word.

The semantic interference effect arises because, in comparison with the activation of the phonological node of an unrelated word, the phonological node of a semantically related word receives extra activation. This extra activation is due to the processing of the picture in the semantic system. In this system, the processing of the concept CAT will cause a rise of the activation of all related conceptual nodes (including that of the conceptual node for horse) by means of spreading activation (Collins & Loftus, 1975). These nodes will, in turn, activate their phonological nodes. Now, because the phonological node of a semantically related word receives extra activation from the semantic system, the selection of the right phonological node (of the picture's name) takes more time (see also Glaser & Glaser, 1989, and La Heij, 1988).

The orthographic facilitation effect arises because a presented word does not only activate its own phonological node but also, to a lesser extend, the phonological nodes of orthographically related words. Therefor, when the name of the picture is orthographically related to the presented word, its phonological node will receive extra activation in comparison with the situation with an unrelated word, rendering easier selection (see also Lupker, 1982).

The interaction between these two effects arises because the influence of an orthographic relation is directly reflected in an easier selection process. A semantic relation is only able to influence this process indirectly, via the conceptual level. Therefor, when both relations are present, the influence of a semantic relation is overruled by that of an orthographic relation.

3. A connectionist model

The effects described above can be simulated in a connectionist model, in which a selection process is used similar to the one described above. In the model, the critical variable is the number of iterations it takes to select the phonological node of the picture's name. Selection takes place when the activation of the right phonological node exceeds that of all other phonological nodes by some critical amount c. If the selection process is more difficult, the selection will take more iterations,

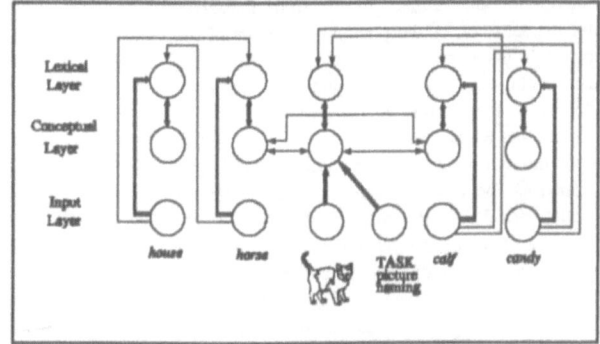

Figure 2. A connectionist model for context effects in the picture-word interference task.

and the simulated reaction time will be longer (Roelofs, 1992).

The model consists of two major layers, a layer of concept nodes and a layer of phonological nodes. Connections between these layers have the same weights (w_{cp}) and are bidirectional. A third layer consists of input nodes, and represents the inputs from the visual system to the concept layer (in case of pictures) and to the phonological layer (in case of words), see Figure 2. The connections to these layers have a weight w_i. An input node representing the task of picture naming is also located at this layer. This task node represents an attention process and connects (with weight w_t) to the concept node of the to be named picture.

Related concepts are represented through connections between concept nodes. In the model, these connections are bidirectional and have the same weight (w_{cc}), but this is not a prerequisite. Orthographically related words are represented through connections (w_{io}) from the input node of the orthographically related word to their phonological nodes. The model in Figure 2 shows the representations of the concepts and phonological forms for *cat* (the name of the picture), *candy* (an orthographically related word), *horse* (a semantically related word), *calf* (a both orthographically and semantically related word), and *house* (an unrelated word).

In each iteration, the activation of each node in the model is updated using the CALM-activation rule (Murre, 1992):

$$A_i(t+1)=(1-k)A_i(t)+\frac{e_i}{1+e_i}[1-(1-k)A_i(t)] \quad \text{with} \quad e_i=\sum_j w_{ij}A_j(t)$$

Using this rule, the minimum activation value of each node is 0, and the maximum value is 1. The first component of the rule represents the autonomous decay of a node. The rate of this decay is determined by k. The second part of the rule squashes the total input each node receives from other nodes to a value between 0 and 1. The third part ensures that the increase in activation of a node diminishes as the node's activation reaches the maximum activation value (Murre, 1992).

The presentation of a picture and an accompanying word can now be simulated by presenting input activation A_i to a) the input node of the picture and b) the input node of the relevant word, and c) the task node. So, to simulate the processing of a picture of a CAT, accompanied by the semantically related word *horse*, input activation should be given to a) the input node for the picture CAT, b) the input node for the word *horse* and c) the task node. In the same way, a picture accompanied by either an orthographically related, or a both orthographically and semantically related, or an unrelated word can be simulated. In each case, the total number of iterations that is necessary to select the phonological node of the picture can be obtained.

Also, the time courses of the effects can be simulated. If the duration of one iteration represents 10 ms (time step $\Delta t = 10$), the effects at for instance SOA 100 can be simulated by presenting the input activation of the picture and the task at iteration $t = 0$, and the input activation of the word at iteration $t = 10$. In this way, the time course of the effects was simulated for SOA -200 (word first) to SOA +200 (picture first) in steps of 100 ms. In all simulations the input duration of the word presentation was fixed at (du_w).

Reaction times can be simulated by multiplying the number of iterations necessary for the selection of the phonological node of the picture's name with the value of the time step and adding a constant amount of time (C_{va}) necessary for a) the visual processing of the picture and b) the processing of the phonological code through the articulatory system into a vocal response. The model is only simulating the selection process of the right phonological representation, which takes place between visual processing and speaking. Simulated reaction times are presented in Figure 3. The values of the parameters of the model are shown in appendix A.

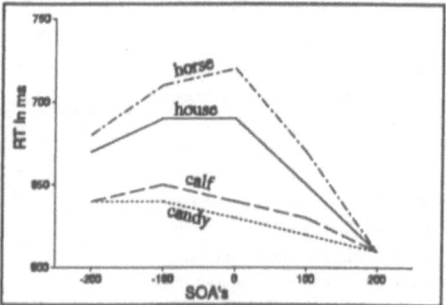

Figure 3. Simulated data. The words were represented by the word input nodes, which were activated together with the input node for the picture of a CAT.

Although not all aspects of the real data were captured, the most important aspects could be simulated: a) a semantic interference effect that was restricted to a small SOA-range (-100 ms to + 100 ms), b) orthographic facilitation in a larger SOA range (-200 to +100 ms) and, c) an interaction of these effects in such a way that the semantic interference effect was reduced when there was an orthographic relation between the words and the name of the picture. Also, the sizes of the effects could be simulated: the largest simulated semantic interference effect was 30 ms (the maximum effect in the real data amounted to 36 ms) and the largest simulated orthographic facilitation effect was 60 ms (the maximum effect in the real data was 62 ms). Finally the sizes of the semantic interference effects necessary to produce the observed interaction could be simulated: the simulated semantic interference effect with orthographically related words was about one third of the simulated semantic interference effect of the orthographically unrelated words; the same pattern can be observed in the real data.

4. Discussion

The presented model is able to simulate the time courses of the semantic interference effect, the orthographic facilitation effect, and their interaction in the picture-word interference task in a qualitative way. To obtain a quantitative fit, reaction time *distributions* should be simulated. To this end, the presented model can be varied in two ways, in order to simulate different subjects and different items. Different subjects could be simulated by using (slightly) different decay values or squashing functions. In addition, for each subject different items could be simulated by slightly changing the connection weights.

Furthermore, it should be stressed that the presented parameter set is not claimed to be the best solution. Other (perhaps better) parameter sets are possible. In fact we found, using other time step values and other activation rules, several parameter sets that produced similar results. This indicates that the model's behavior does not depend on some exceptional property of the presented parameter set. Instead, the model's behavior can be explained in structural terms (see the theoretical explanation).

All computations in the model can be locally performed. It is our conviction that this forms a prerequisite for neurophysiologically and psychologically plausible modelling.

Finally, it has been argued that cognition is state-space evolution in dynamical systems and not just computation (Smolensky, 1988). The presented model simulates cognitive phenomena by stressing the *flow* of activation through the network rather than by focussing on obtained static input/output associations.

Acknowledgements

In carrying out this research, Peter A. Starreveld was supported by the Dutch Organization for Scientific Research (Grant 560-256-067). Jan N. H. Heemskerk was supported by the Foundation for Computer Science in the Netherlands (SION, Grant 612-322-110).

References

Collins, A. M., & Loftus, E. (1975). A spreading-activation theory of semantic processing. *Psychological Review, 85*, 249–277.

Glaser, W. R., & Düngelhoff, F.-J. (1984). The time course of picture-word interference. *Journal of Experimental Psychology: Human Perception and Performance, 10*, 640–654.

Glaser, W. R., & Glaser, M. O. (1989). Context effects on Stroop-like word and picture processing. *Journal of Experimental Psychology: General, 118*, 13–42.

La Heij, W. (1988). Components of Stroop-like interference in picture naming. *Memory & Cognition, 16*, 400–410.

Lupker, S. J. (1982). The role of phonetic and orthographic similarity in picture-word interference. *Canadian Journal of Psychology, 36*, 349–367.

Starreveld, P. A., & La Heij, W. (1993a). Semantic interference, orthographic facilitation and their interaction in naming tasks. *Submitted*.

Starreveld, P. A., & La Heij, W. (1993b). *Unpublished data*.

MacLeod, C. M. (1991). Half a century of research on the Stroop effect: An integrative review. *Psychological Bulletin, 109*, 163–203.

Murre, J. M. J. (1992). *Categorizing and learning in neural networks*. Harvester Wheatsheaf, U.K.

Rayner, K., & Springer, C. J. (1986). Graphemic and semantic similarity effects in the picture-word interference task. *British Journal of Psychology, 77*, 207–222.

Roelofs, A. (1992). A spreading-activation theory of lemma retrieval in speaking. *Cognition, 42*, 107–142.

Smolensky, P. 1988. On the proper treatment of connectionism. *Behavioral and Brain Sciences, 11*, 1–74.

Appendix A

Parameter values used in the simulations

$c = 0.071$	$w_{cp} = 0.025$	$w_i = 0.125$	$w_t = 1.600$
$w_{cc} = 0.008$	$w_{io} = 0.088$	$k = 0.040$	$A_i = 0.025$
$du_w = 31$ iterations	$\Delta t = 10$ ms	$C_{va} = 490$ ms	

Inductive Inference with Recurrent Radial Basis Function Networks

M.Gori, M. Maggini, and G. Soda

Dipartimento di Sistemi e Informatica, Università di Firenze

Via di Santa Marta 3 - 50139 Firenze - Italy

1 Introduction

The difficulties of learning automata exactly have already been shown experimentally by numerous researchers. Our research in this field is illustrated in (Frasconi *et al.*, 1994). For most experiments of inductive inference of regular grammars, the main problem is not that of learning the examples, but that of generalizing the automata behavior to other sequences of the grammar, particularly if they are very long. In fact, the apparent automata behavior arising from the learning of sequences of small length may change to more complex dynamics for longer sequences.

To some extent, the techniques for extracting automata after learning (Giles *et al.*, 1992) are interesting attempts to overcome this problem. However, an implicit assumption for a successful extraction of automata with clustering techniques is that the network state space is fairly well-separated in clusters. Unfortunately, the network dynamics deriving from learning by example can be very complex and hardly approximable with automata.

In order to deal with automata we introduce particular networks, referred to as recurrent radial basis function (R^2BF) networks and a technique for forcing clustered representations of the network states. Using these networks, we report very successful results for inductive inference of regular grammars.

2 The R^2BF architecture

A R^2BF network is composed of four layers: an input layer ($l = 0$), a radial basis function (RBF) layer ($l = 1$), a state layer ($l = 2$) and an output layer ($l = 3$). For sequence parsing tasks, the output layer consists of one sigmoidal neuron only.

The number of neurons per layer is denoted by $n(l)$. Each neuron of layer l is referred to by its index $i(l)$, $i(l) = 1, \ldots, n(l)$. Depending on the kind of neuron, the following processing is performed:

$$a_{i(1)}(t) = \frac{1}{\sigma^2_{i(1)}} \left[\sum_{j(0)=1}^{n(0)} (x_{j(0)}(t) - c_{i(1),j(0)})^2 + \sum_{j(2)=1}^{n(2)} (x_{j(2)}(t-1) - c_{i(1),j(2)})^2 \right]$$

$$a_{i(2)}(t) = w_{i(2)} + \sum_{j(1)=1}^{n(1)} w_{i(2),j(1)} x_{j(1)}(t)$$

where $a_{i(l)}(t)$ is the neuron's activation at step t, $x_{i(l)}$ is the corresponding output, $w_{i(l),j(l-1)}$ denotes the weight of the link between the neurons $i(l), j(l-$

1), $c_{i(1),j(l)}$ are coordinates of the centers of the RBF units, and $\sigma_{i(1)}$ are the widths of the Gaussians. For the RBF units $x_{i(1)} = \exp(-a_{i(1)})$ while for sigmoidal units $x_{i(2)} = 1/(1 + \exp(-a_{i(2)}))$.

The R^2BF architecture is very well-suited for implementing automata. Let us assume that an automaton is given in terms of its next-state function in the following canonical form (*sum of products*) [1]:

$$s_i|_{t+1} = \sum_{j,k} m(S_j, I_k)|_t \quad \forall \ i = 1, \ldots, n \ ,$$

where s_i is the i-th bit of the state code, S_j is the binary code of the j-th state, I_k is the code for the k-th input symbol, and $m(S_j, I_k)$ is the minterm associated with the pair (S_j, I_k) (it produces a *high* value iff its inputs match the codes of S_j and I_k). We can associate a minterm $m(S_j, I_k)$ with each radial basis function neuron locating its center on the associated hypercube vertex and assuming small values for σ_i. Finally, the sigmoidal neurons can perform the "or" of the minterms using the $w_{i(2)j(1)}$ weights.

The assumption of exclusive coding of the states turns out to be very useful, since only one feedback connection from the state units to each RBF unit is needed. This choice reduces the problems of the learning due to neuron saturation which are likely to appear when the unit fan-in increases (Gori *et al.*, 1993). These architectures are referred to as R^2BF1.

In addition to the hint associated with the automata minterms that lead to fix up the centers of the radial basis function units on the verteces of the boolean hypercube, we suggest using a constraint that forces the outputs of the state neurons to be "high" and "low" (e.g.: 0.9,0.1).

In order to force these outputs during the input processing, we introduce the following *penalty function* to be added to the error function:

$$P = \sum_{q=1}^{Q} \sum_{t=1}^{T(q)} \sum_{i(2)=1}^{n(2)} max(0, (x_{i(2)}(t,q) - \rho^-)(\rho^+ - x_{i(2)}(t,q)))$$

where Q is the number of strings used for learning and $T(q)$ is the length of string q. Basically, this function is null when the outputs of the state neurons fall outside the interval $[\rho^-, \rho^+]$, during the input processing. If E is the error function, then we must optimize $V = E + \lambda P$.

3 Experimental results

In this section, we report the experimental results obtained using R^2BF1 networks for inductive inference of regular grammars. We tested extensively the R^2BF1 architecture using the Tomita's languages (Tomita, 1982) as benchmark.

For each Tomita's language, we trained a R^2BF1 network with 5 state neurons using a learning set composed of the original Tomita's training sets incremented with 5% of all the strings with lengths from 1 up to 10. These

---

[1] We deal with automata acting as recognizers and, therefore, their complete specification can be given by the next-state function and the set of accepting states.

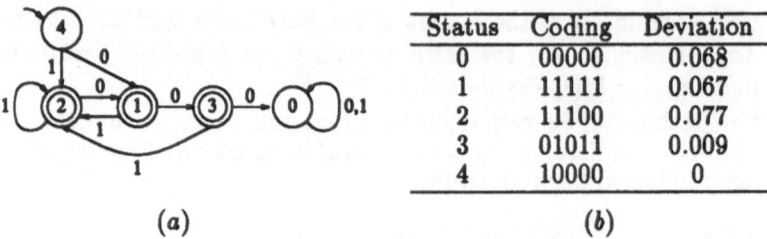

Status	Coding	Deviation
0	00000	0.068
1	11111	0.067
2	11100	0.077
3	01011	0.009
4	10000	0

(a) (b)

Figure 1: Learning Tomita-4 language (a) Extracted automaton. (b) State coding with the corresponding standard deviation of the output distribution of the state layer neurons for the associated cluster.

strings were randomly chosen. We adopted an incremental learning strategy, so that new strings were added to the learning set only when the previous ones had been learned exactly.

The R^2BF1 networks with 5 state neurons ($n(2) = 5$) had 10 radial basis function units ($n(1) = 10$), and one first order sigmoidal output neuron ($n(3) = 1$). The centers were set up in such a way that the RBF units extracted minterms. As a consequence, these units were partitioned in two sets of 5 units with centers in $[0.0, 0.9]$ and $[1.0, 0.9]$, respectively. Each unit in a set received the feedback connection from a different state neuron. The widths of the radial basis functions were initialized to 0.3. Both the centers and the width of these units were kept fixed during training. The state layer and the radial basis function layer were fully-connected. All the state units were connected to the output neuron with learnable weights.

Before feeding the network with a sequence, all the state neurons were initialized to 0.0, apart from neuron 0 that was set up to 1.0.

Each network was trained by forcing automata representations. The gradient of the error E was computed by using the *Backpropagation Through Time* algorithm (BPTT). The target values were 0.9 for grammatical strings and 0.1 for ungrammatical ones. The penalty function P was chosen in order to penalize output values in $[0.1, 0.9]$. These constraints were managed as additional supervisions on the state neurons and, therefore, the gradient of P was still computed by $BPTT$. In all the experiments we found no problems in reaching $E \simeq 0$, whereas sometimes it was hard to satisfy all the constraints. In these cases, after a fixed maximum number of epochs, the training was stopped without reaching the optimal solution also for the constraints.

If we find a global minimum of the function V then the outputs of the state layer can easily be quantized and a finite state automaton can be extracted directly. The extraction of the automaton becomes more difficult when the learning algorithm gets stuck in a local minimum of the function V. In these cases the points of the state trajectory are not necessarily clustered round the hypercube vertices, and consequently, the automata extraction is more involved. In all the cases we extracted an automaton from each trained network using a clustering algorithm based on K-mean (Gori *et al.*, 1993).

We were able to learn exactly all the languages and to attain perfect generalization on all Tomita's languages. Fig. 1a shows the automaton extracted from the network that recognized successfully Tomita-4 language. Fig. 1b re-

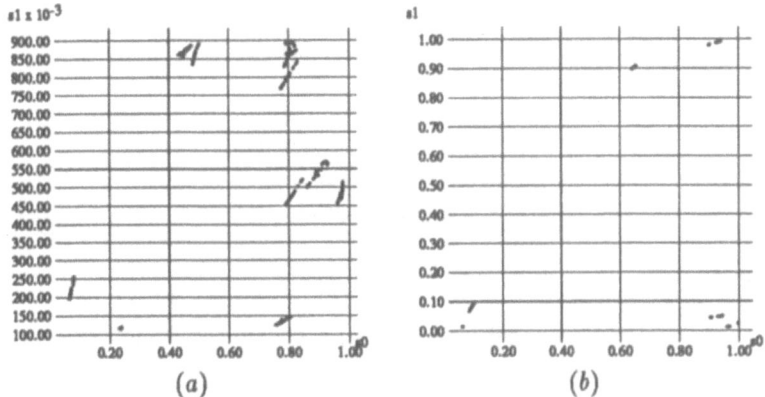

Figure 2: State space trajectories for a R^2BF1 network of size 2 trained on Tomita 4 language. (a) Unconstrained network. (b) Constrained network.

ports the state coding of the extracted automaton and the standard deviation of the associated clusters. The extracted automaton is minimal apart from state 4, that acts as the initial state and that can easily be removed.

Fig. 2a shows the state space trajectories for a R^2BF1 network with 2 state neurons trained without constraints on Tomita-4 language. The diagram reveals the presence of 6 quite large clusters, while Fig. 2b shows the effect of introducing the constraints. Only 4 small clusters appear which correspond with the 4 states of the extracted automaton. Finally, the effect of the constraints was that of improving the performance (100% for the constrained net v.s. 96.0% for the unconstrained one).

References

[1] Frasconi, P., Gori, M., Maggini, M., and Soda, G. (1994) "Unified Integration of Explicit Rules and Learning by Example in Recurrent Networks," *IEEE Trans. on Knowledge and Data Engineering*, (to appear).

[2] Giles, C.L., Miller, C.B., Chen, D., Chen, H.H., Sun, G.Z., and Lee, Y.C. (1992), "Learning and extracting finite state automata with second-order recurrent neural networks," *Neural Computation*, Vol. 4, No. 3, pp.393-405.

[3] Gori, M., Maggini, M., and Soda, G. (1993) "Insertion of Finite State Automata in Recurrent Radial Basis Function Networks," Tech. Rep. DSI 17/93.

[4] Tomita, M. (1982) "Dynamic construction of finite-state automata from examples using hill-climbing," *Proc. of the Fourth Annual Cognitive Science Conference*, Ann Arbor, MI, pp. 105-108.

Neural Networks as a Paradigm for Knowledge Elicitation.

G.P.Fletcher & C.J.Hinde
Dept. Computer Studies
A.A.West & D.J.Williams
Dept. Manufacturing Engineering
University of Technology
Loughborough.

1 Introduction

It has been consistently stated that the hardest part of constructing any knowledge based system is extracting the information from the "expert" (Hart 1986). Rule based systems are one genre of expert systems, in this paper we demonstrate a method for converting a neural network into a set of rules. This maintains the basic advantages of neural networks but redresses some of the disadvantages.

2 The Gluing Machine

The first real application (Chandraker et al. 1990) is the dispensing of adhesive in the manufacture of "mixed technology" P.C.B.s in which through hole and surface mount components are present on the same board. The surface mount components are secured to the board, prior to a wave soldering operation, by a small amount of adhesive. The amount of adhesive dispensed is critically dependent upon several process environment variables.

Messom et al. 1992 produced a sparsely connected 7 input, 5 output neural network system for controlling the adhesive dispensing machine. The trained neural network should therefore be equivalent to a set of rules that could have been learnt by a rule induction package.

Step one in the analysis is to examine the first layer of the network and to express this part as a set of inequalities. This step does not simplify the information but does display it much more naturally than a set of weights on a diagram.

```
IF  rise_time < 1.25          IF  area > 1.0258
THEN rise_timeflag            THEN midnode1
ELSE −rise_timeflag           ELSE −midnode1
       .                             .
       .                             .
```

The remainder of the network can then be converted to a set of Boolean functions (Fletcher et al. 1993a). Substituting the above inequalities gives rules of the form:

```
IF  rise_time < 1.25          IF (pulse_height > 0.975 ∧ pulse_height < 1.025 )
THEN rise_timeflag            THEN pulse_heightflag
ELSE −rise_timeflag           ELSE −pulse_heightflag
       .                             .
       .                             .
```

These rules serve to illustrate the usefulness of the interpretation system for a medium sized control network. This type of result is most useful when it is necessary to check that what the network has learnt is reasonable and when it is necessary for a system to explain its actions, something classic neural networks cannot normally do.

3 Noise Immunity of Neural Network Derived Rules

A good connectionist based solution should exhibit fewer overall problems than the currently available symbolic systems. A useful quantitative measure of noise immunity is the average percentage of the learnt hypothesis that matches the required hypothesis. A correct function is represented by the truth table shown in table 1a, and a noisy version with one output incorrect is represented in table 1b. As these truth tables match in 87% of the outputs the noisy hypothesis is defined to have an integrity of 87%.

INPUTS	00000000 01010101 00110011 <u>00001111</u>	00000000 01010101 00110011 <u>00001111</u>
OUTPUTS	01010101 Table 1a	01010111 Table 1b

Table 1a is a truth table representing a Boolean function, whereas table 1b represents a noisy hypothesis, note the change of the seventh bit in the OUTPUTS line.

Figure 1 shows the responses of a neural network after training for conjunction and parity with differing amounts of noise. This demonstrates the immunity to noise using a neural network induction system, although live applications will be more complex than the illustrations above they will typically be trained on noisy data.

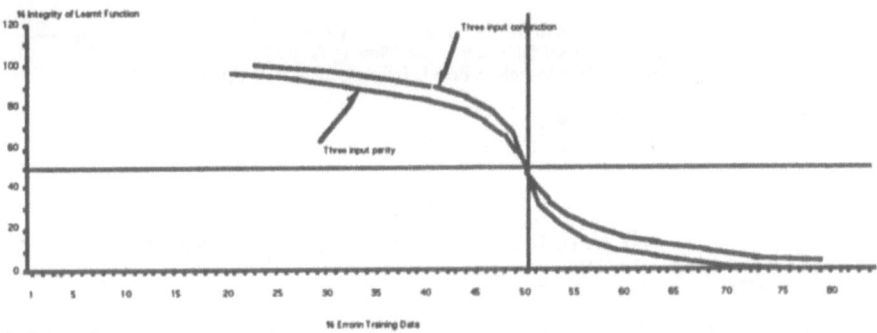

Figure 1 The response of a neural network derived from a training set for conjunction with differing amounts of noise superimposed on the response of the neural network derived from a 3 dimensional parity training set.

4 The Pole Balancing Problem

The pole balancing problem (Michie et al. 1968) is an example of applied adaptive control and has become a standard tutorial problem. The control system must balance a pole on a motorised cart by moving the cart back and forward in a confined space. The implementation of most interest here is the neural network (Zhang et al. 1989) shown in figure 2 which was successful in a balancing the pole under a variety of circumstances.

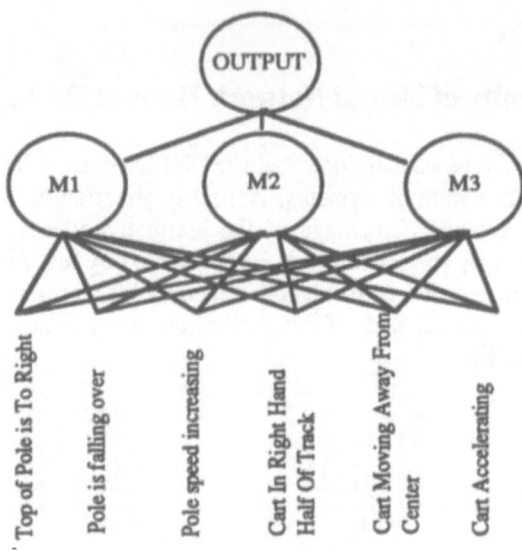

Figure 2. A neural network developed by Zhang and Grant to solve the pole balancing problem.

Fletcher et al. (1993b) describe a method for applying simple image enhancement techniques to Boolean functions in order to expose the underlying emphasis. The analysis of the pole balancing net resulted in the following versions of the output function under different levels of simplification.

IF	(Top of pole is to right ∧ ¬Pole is falling over ∧ ¬Pole Speed Increasing ∧ Cart Accelerating) ∨
	(Top of pole is to right ∧ ¬Pole is falling over ∧ ¬Pole Speed Increasing ∧ Cart In Right Hand Half Of Track ∧ Cart Moving Away From Centre) ∨
	(Top of pole is to right ∧ Pole is falling over ∧ Pole Speed Increasing)
THEN	Apply right force
ELSE	Apply left force

Set of rules generated with no simplification

IF	(Top of pole is to right ∧ ¬Pole is falling over ∧ Pole Speed Increasing) ∨
	(¬Top of pole is to right∧Pole is falling over ∧ ¬Pole Speed Increasing)∨
	(¬Pole Speed Increasing∧¬Cart In Right Hand Half Of Track ∧ ¬Cart Accelerating)
THEN	Apply right force
ELSE	Apply left force

This shows us that the function attaches importance to input variables that should have no bearing on the outcome for the cases analysed. This means that the output is not correctly computed.

IF	(Top of pole is to right ∧ Pole Speed Increasing) ∨
	(Top of pole is to right ∧ Pole is falling over)
THEN	Apply right force
ELSE	Apply left force

This shows the net has implemented an asymmetric function in response to a symmetric problem. This means that there are some areas that will need further training for completeness.

IF	Top of pole is to right
THEN	Apply right force
ELSE	Apply left force

This shows that the network is *basically* correct.

5 Limits of Boolean Rule Conversion

The nature of neurons means that as the number of inputs grows so does the length of the Boolean description. In the worst case the numbers of conjunctions in a minimal disjunction of conjunctions representation grows at a rate of 2^{n-1} where there are n inputs. The time taken to calculate an answer is exactly proportional to its size. The answer becomes intractable to compute for neurons with more than 20 inputs, and meaningless for a human far earlier. Current work is aimed at implementing systems which will enable the user to interactively train and study the hypotheses of large networks (Fletcher et al. 1993c).

6 Conclusion

Results have been demonstrated that illustrate the extraction of Boolean based rules from a neural network. These rules were also used to check that the hypopthesis of the neural network was consistent with the problem. The rule induction system has also been shown to have a high integrity in the presence of noise, making neural networks an ideal method for rule induction in noisy problem domains.

7 References

Chandraker, R., West, A.A., & Williams, D.J., 1990, Intelligent control of Adhesive Dispensing, IJCIM, special issue in Intelligent Control, 3, No 1, pp. 24-34.

Fletcher, G.P. & Hinde, C.J., 1993a, Interpretation of neural networks as Boolean transfer functions, Dept. Computer Studies research report 779, to be published in Knowledge-Based Systems.

Fletcher, G.P. & Hinde C.J., 1993b, Using neural networks as a tool for constructing rule based systems, Dept. Computer Studies research report 781.

Fletcher, G.P. & Hinde C.J., 1993c, Providing evidence for the hypothesis of large neural networks, Dept. Computer Studies research report 793, submitted in revised form to Neurocomputing.

Hart, A., 1986, Knowledge Acquisition for Expert Systems, Kogan Page.

Messom, C.H., Hinde, C.J., West, A.A. & Williams, D.J., 1992, Designing neural Networks for Manufacturing Process Control Systems, Proc. 1992 International Symposium on Intelligent Control, August 1992, I.E.E.E. Publishers.

Michie, D. & Chambers, R. A., 1968, BOXES: An Experiment in Adaptive Control. Machine Intelligence 2, ed. Dale, E. and Michie, D., Oliver and Boyd, Edinburgh University Press.

Zhang, B. & Grant, E., 1989, A Neural Net Approach to Autonomous Machine Learning of Pole Balancing. Proc. IEEE conference on Intelligent Control. pp 123.

Unsupervised Detection of Driving States with Hierarchical Self Organizing Maps

Peter Weierich, Michael von Rosenberg

Bavarian Research Center For Knowledge-Based Systems (FORWISS)
Knowledge Processing Research Group (Head: Prof. Dr.-Ing. H. Niemann)
Am Weichselgarten 7, D-91058 Erlangen, Germany
email weierich@forwiss.uni-erlangen.de Phone: +49-9131-691-134, FAX -185

Abstract

In this paper the unsupervised detection of time dependent states during measurement drives is presented. The two-layered model consists of a dimension reduction of the measured signal by a first SOM (Self Organizing Map). The order of firing neurons is evaluated by a second SOM, which thus learns to recognize longer periods.

1   Problem

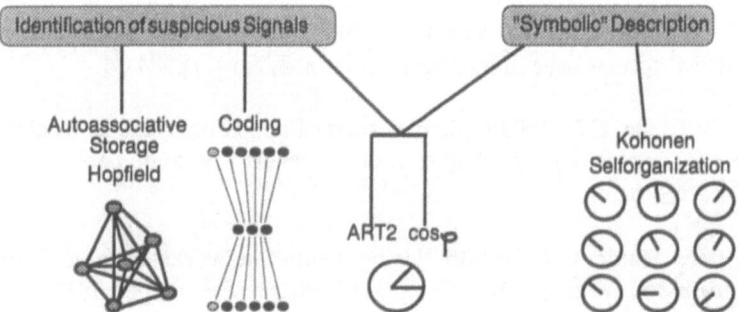

Figure 1: Overview Neural Networks are investigated for two tasks within the framework of the project Messpert for the automated analysis of multivariate measured data.

New assembly groups in car development afford costly methods to prove the fulfillment of requirements of the market (Petersen und Tunker, 1991). Measured data acquired during test runs is characterized by a huge storage requirement. For instance, during one hour of measurement the 64 sensors provide about 100 MB of data. The objective of the Project Messpert[1] is the intelligent support of the evaluation of measured data (see Figure 1. In (Weierich, 1993) an approach was presented to detect "faulty" signals during the measurement.

---

[1] Messpert is funded by AUDI AG (Ingolstadt, Germany) and BMW AG (Munich)

One task of Messpert is to automatically generate a "symbolic" description of the test run. The lack of labeled data demands an unsupervised technique to classify the measured data. Additionally it is necessary to represent the dynamic of the signal over time, too.

2 Self Organizing Maps

2.1 Vector Quantization

To reduce the dimension of the input space (20 measured channels) a vector quantiziation technique is needed. Neural network models as ART2 (Carpenter und Grossberg, 1987) or the Cosine-Classifier (Kratzer, 1991) showed the same problem as classical methods applied in speech recognition in comparision with Self Organizing Maps (SOM) as described in (Kohonen, 1991): They lack a topology preserving structure of the input–output–mapping.

In first experiments SOMs were trained directly with the input data and produced an output which did not take into account the temporal structure of the measured data: Within homogeneous parts of the course, for example curves, frequent changes of the answers of SOMs were observed (Figure 3)

2.2 Temporal Context Processing

There are already attempts to learn a temporal structure with SOMs. For example, (Chappell und Taylor, 1992) proposed a biologically plausible modification to neurons in a Kohonen network which is able to represent temporal pattern in ONE layer without external time delay mechanisms.

In our approach, the temporal context is processed by learning the sequence of answers of the first SOM by a second one. The idea is shown in Figure 2. The second layer is fed with the linear interpolated virtual coordinates of the best matching units of the first layer neurons. This approach was considered to be a good approximation of the use of Jacobian matrices as described in (Ritter et al., 1992).

2.3 Visualization of the Results

The display of results of the unsupervised classification with conventional techniques, for example by entering the coordinates of firing neurons into X-Y-Diagrams was not sufficient for evaluation.

The solution was found in the colour coded display of results in maps of the driving course: The coordinate positions of the best matching units of the Kohonen maps are coded by intensity values in the three colour channels red, green and blue. Therefore, maximally 3-dimensional Kohonen maps can be visualized.

3 Results and Discussion

Firstly, a 3D–SOM was trained with 14.000 frames of measured data (40 frames per second). The colour coded map is shown in Figure 3.

248

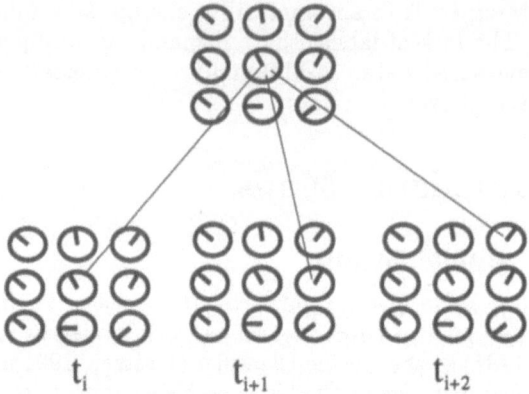

t_i t_{i+1} t_{i+2}

Figure 2: The structure of the temporal processing: Beyond are shown three identical SOMs. Their inputs are unpreprocessed frames of the measured data. The coordinates of the virtual best matching units of, in this case three, subsequent frames form the input for the second layer (above).

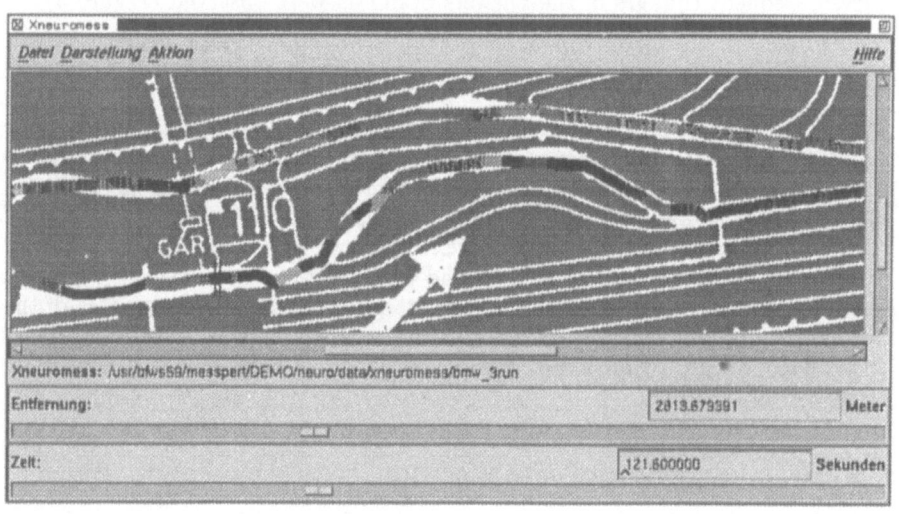

Figure 3: The results of a 3 dimensional SOM with 10 x 10 x 10 Neurons without use of temporal information.

The subsequent processing was carried out with different parameters (number of frames, different topologies of the SOM). In general, the temporal layer produced a better smoothed output, and similar driving states during the drive were mapped to neighbouring neurons, as shown in Figure 4. The number of homogenious regions was reduced from 1990 to 1090 for two minutes of measurement.

The coding of neuron activities with colours has proved to produce a reasonable visualization of the responses: Similar colours code neighbouring neurons.

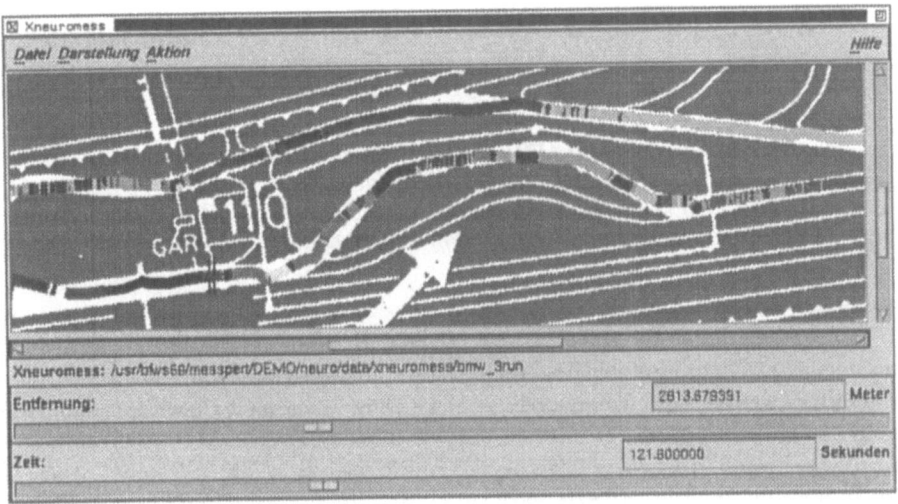

Figure 4: The mapping of the input signal to the context sensitive SOM: Similar driving stages are coded by similar colours. The unsteadiness of the signal on the upper left part of the image is an expected result due to the special properties of the corresponding part of the lane.

References

Carpenter, G. A. und Grossberg, S. (1987). ART2: Self-Organization of Stable Category Recognition Codes for Analog Input Patterns. *Applied Optics*, 26(23):4919–4930.

Chappell, G. J. und Taylor, J. G. (1992). The Temporal Kohonen Map. *Neural Networks*, 6:441–445.

Kohonen, T. (1991). The Hypermap Architecture. In *International Conference on Artificial Neural Networks*, pages 1357–1360, Espoo, Finland.

Kratzer, K. P. (1991). A Neural Approach to Data Compression and Classification. In *EPIA 91, 5th Portuguese Conference on Artificial Intelligence*, Lecture Notes in Artificial Intelligence, pages 250–263. Springer.

Petersen, J. und Tunker, H. (1991). Vom Messaufnehmer zur Datenbank: Systemanalyse, Abläufe, Anforderungen. *Informatik-Spektrum*, 14:69–73.

Ritter, H., Martinetz, T., und Schulten, K. (1992). *Neural Computation and Self Organizing Maps* Addison Wesley Publishing Company, Bonn.

Weierich, P. (1993). Fault Detection in Multivariate Time Series with a Coding Approach. In Gielen, S. und Kappen, B., editors, *ICANN '93, Proc. of the International Conf. on Artificial Neural Networks*, pages 926 – 929, Springer.

Using Simulated Annealing to Train Relaxation Labeling Processes

Marcello Pelillo and Angelo Maffione

Dipartimento di Informatica
Università di Bari
Via G. Amendola, 173 - 70126 Bari (Italy)

1. Introduction

Relaxation labeling processes are a class of parallel distributed processing models quite popular within the pattern recognition and machine vision domains (Kittler and Illingworth, 1985). They attempt to combine local and contextual information in order to remove labeling ambiguities in classification problems, and are currently being employed in such different areas as optimization (Pelillo, 1993) and associative memories (Pelillo and Fanelli, submitted), owing to their interesting dynamical properties (Hummel and Zucker, 1983; Pelillo, submitted). Moreover, it is generally agreed that intriguing similarities exist between relaxation labeling and certain mechanisms in biological visual systems (Ballard *et al.*, 1983; Zucker *et al.*, 1989).

In a recent work (Pelillo and Refice, in press), a learning algorithm for relaxation labeling processes has been developed which involves minimizing a certain cost function with classical gradient techniques. Although the experimental results obtained so far are very encouraging, the gradient algorithm suffers from some inherent drawbacks that could prevent it from being applied to real-world high-dimensional problems of practical interest. These include essentially its inability to escape from local minima as well as its computational complexity.

In this paper we propose the use of simulated annealing (SA) (Kirkpatrick *et al.*, 1983; van Laarhoven and Aarts, 1987) to overcome the difficulties with the gradient method. Some experiments are presented which confirm the effectiveness of the approach, both in terms of quality and speed.

2. Relaxation Labeling and the Learning Problem

Relaxation labeling involves a set of objects $B = \{b_1, \cdots, b_n\}$ and a set of labels $\Lambda = \{1, \cdots, m\}$. The purpose is to label each object of B with exactly one label of Λ. By means of some local measurements it is possible to construct, for each object b_i, a vector $p^{(0)} = (p_{i1}^{(0)}, \cdots, p_{im}^{(0)})^{\mathrm{T}}$, such that $p_{i\lambda}^{(0)} \geq 0$, and $\sum_{\lambda} p_{i\lambda}^{(0)} = 1$ for all $i = 1 \ldots n$. Each $p_i^{(0)}$ is interpreted as the *a-priori* (non contextual) probability distribution of labels for the object b_i. By simply concatenating $p_1^{(0)}, \cdots, p_n^{(0)}$ we obtain an initial weighted labeling assignment for the objects of B, that will be denoted by $p^{(0)} \in \mathbf{R}^{nm}$. Contextual information is represented by a four-dimensional real matrix R of *compatibility coefficients*. The component $r_{ij}(\lambda, \mu)$ measures the strength of compatibility between the hypotheses "λ is on object b_i" and "μ is on object b_j;" high values mean compatibility and low values mean incompatibility. We will find it convenient to "linearize" the compatibility matrix R, and consider it as a column vector which, hereafter, will be denoted by r.

The relaxation algorithm accepts as input the initial labeling $p^{(0)}$ and updates it iteratively taking into account the compatibility model, in an attempt to achieve global consistency. At the t-th step the labeling is updated according to the following formula (Rosenfeld et $al.$, 1976):

$$p_{i\lambda}^{(t+1)} = p_{i\lambda}^{(t)} q_{i\lambda}^{(t)} \left/ \sum_{\mu=1}^{m} p_{i\mu}^{(t)} q_{i\mu}^{(t)} \right. \tag{1}$$

where the denominator is simply a normalization factor, and

$$q_{i\lambda}^{(t)} = \sum_{j=1}^{n} \sum_{\mu=1}^{m} r_{ij}(\lambda,\mu) p_{j\mu}^{(t)} \tag{2}$$

measures the strength of support that context gives to λ for being the correct label for b_i. The process is iterated until convergence.

Now, let $L_1, \cdots, L_N$ be a set of available instances of the problem at hand, where each sample L_γ $(\gamma = 1...N)$ is a set of labeled objects of the form

$$L_\gamma = \left\{ (b_i^\gamma, \lambda_i^\gamma) : 1 \le i \le n_\gamma, \, b_i^\gamma \in B, \, \lambda_i^\gamma \in \Lambda \right\}.$$

For each $\gamma = 1...N$ let $p^{(L_\gamma)} \in \mathbf{R}^{n_\gamma m}$ denote the unambiguous labeling assignment for the objects of L_γ, that is:

$$p_{i\alpha}^{(L_\gamma)} = \left\{ \begin{array}{l} 0, \text{ if } \alpha \neq \lambda_i^\gamma; \\ 1, \text{ if } \alpha = \lambda_i^\gamma. \end{array} \right.$$

Furthermore, suppose that we have some mechanism for constructing an initial labeling $p^{(I_\gamma)}$ on the basis of the objects in L_γ, and let $p^{(F_\gamma)}$ denote the labeling produced by the relaxation algorithm, when $p^{(I_\gamma)}$ is given as input.

Broadly speaking, the learning problem for relaxation labeling is to determine a vector of compatibilities r so that the final labeling $p^{(F_\gamma)}$ be as close as possible to the desired labeling $p^{(L_\gamma)}$, for each $\gamma = 1...N$. To do this, we use a "cost" function E defined as:

$$E_\gamma(r) = n_\gamma - \sum_{i=1}^{n_\gamma} \log_2(1 + p_{i\lambda_i^\gamma}^{(F_\gamma)}(r)) \tag{3}$$

which comes from a novel information divergence measure recently proposed by Lin (1991). Note that $E_\gamma(r) = 0$ if and only if $p^{(F_\gamma)} = p^{(L_\gamma)}$, and attains its maximum value, n_γ, when the relaxation algorithm assigns null probabilities to all the correct labels. The total error achieved over the entire learning set can now be defined as

$$E(r) = \sum_{\gamma=1}^{N} E_\gamma(r) . \tag{4}$$

In conclusion, the learning problem for relaxation labeling can be stated as the problem of minimizing the function E with respect to r.

3. Learning with SA

Simulated annealing is a stochastic optimization algorithm largely inspired from the analogy between optimization problems and the physical annealing of solids. It begins with an initial point r_0, determined either randomly or using heuristic information, and iteratively produces a sequence of points $\{r_k\}$ according to the following algorithm. Given the k-th point r_k, a new point r_{k+1} is tentatively generated by perturbing the preceding one (this is what is called a "transition"); later, the change in the objective function ΔE is recorded: if $\Delta E < 0$ then the new point is retained, otherwise r_{k+1} is accepted with probability $P = \exp(-\Delta E/T)$, where T is a (dynamically decreased) control parameter playing the role of the temperature in physical annealing.

In our implementation, the temperature is initially set to a value T_0 determined so that virtually all transitions are accepted, and is gradually lowered according to the scheme $T_{\text{new}} = \alpha T_{\text{old}}$, $\alpha < 1$. At each temperature, the system is allowed to perform a fixed number of transitions. Annealing terminates when the temperature reaches some predetermined final value. The generation mechanism adopted here consisted of adding to each coefficient $r_{ij}(\lambda, \mu)$, with fixed probability p_m, a fixed-variance Gaussian noise. Doing so only a fraction of the whole compatibility vector is actually perturbed.

It is straightforward to verify that SA performs a number of operations roughly proportional to m^2 on each transition, and this should be compared with the $O(m^4)$ calculations made by the gradient procedure on each step to compute derivatives (Pelillo and Refice, in press).

4. Experimental Results

To test the proposed approach, experiments were carried out over a practical application which involves labeling words with their parts-of-speech. The problem is tackled by first associating each word with a list of potential labels; due to the presence of homographs, a *disambiguation* is then carried out on the basis of context. The task was to train a relaxation labeling process to correctly disambiguate words in context (Pelillo and Refice, in press).

The initial labelings for relaxation were derived using a dictionary look-up which provided, for each word, the list of its possible labels: these labels were then given uniform probability. The following label set was used: verb, noun, adjective, determiner, conjunction, adverb, preposition, pronoun, and a special miscellaneous label. Also, the "context" consisted of the right word only, and relaxation was stopped at the first step. A labeled 1,000-word training set containing 26 Italian sentences (with 148 ambiguities) was employed.

In the implemented SA the value $\alpha = 0.95$ was chosen and exactly 15 transitions were made for each temperature (we observed that larger values did not significantly improve the performance of the algorithm). Moreover, the perturbation probability was $p_m = 0.1$, and the Gaussian variance was set to 1. Ten initial points were chosen randomly and as many runs of both the gradient method and SA were performed. The results are shown in Fig. 1, where the average behavior of the objective function E is plotted for both the algorithms. As can be seen, SA outperformed the gradient procedure in a significant way, and we mention that it was about nine time faster than the gradient algorithm. Also, the best objective function value found by SA was $E = 1.96$, while the best of the gradient procedure was $E = 4.61$, both corresponding to about 99% disambiguation accuracy.

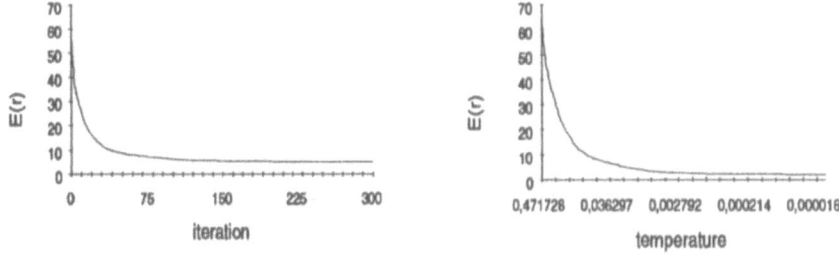

Fig. 1. Behavior of E during training: (left) gradient method, (right) SA.

In addition, generalization performance was assessed over a separate 1,000-word test set (with 37 sentences and 136 overall ambiguities). Relaxation was run using the best points found by the two algorithms. The average objective function value for SA was $E = 13.09$ (91.8% disambiguation accuracy), while for the gradient method an average of $E = 17.17$ was found (91.0% accuracy).

References

Ballard, D. H., Hinton, G. E., Sejnowski, T. J. (1983). Parallel visual computation. *Nature* **306**, 21-26.

Hummel, R. A., and Zucker, S. W. (1983). On the foundations of relaxation labeling processes. *IEEE Trans. Pattern Anal. Machine Intell.* 5, 267-287.

Kirkpatrick, S., Gelatt, C. D., and Vecchi, M. P. (1983). Optimization by simulated annealing. *Science* **220**, 671-680.

Kittler, J., and Illingworth, J. (1985). Relaxation labeling algorithms - A review. *Image Vision Comput.* **3**, 206-216.

Laarhoven, P. J. M. van, and Aarts, E. H. L. (1987). *Simulated Annealing: Theory and Applications.* Kluwer, Dordrecht.

Lin, J. (1991). Divergence measures based on the Shannon entropy. *IEEE Trans. Inform. Theory* **37**, 145-151.

Pelillo, M. (1993). Relaxation labeling processes for the traveling salesman problem. in *Proc. IJCNN-93*, Nagoya, Japan, pp. 2429-2432.

Pelillo, M. (submitted). On nonlinear relaxation labeling. *IEEE Trans. Pattern Anal. Machine Intell.*

Pelillo, M., and Fanelli, A. M. (submitted). An associative memory model based on relaxation labeling processes. *12th ICPR*, Jerusalem, Israel, 1994.

Pelillo, M., and Refice, M. (in press). Learning compatibility coefficients for relaxation labeling processes. *IEEE Trans. Pattern Anal. Machine Intell.*

Rosenfeld, A., Hummel, R. A., and Zucker, S. W. (1976). Scene labeling by relaxation operations. *IEEE Trans. Syst. Man Cybern.* **6**, 420-433.

Zucker, S. W., Dobbins, A., and Iverson, L. (1989). Two stages of curve detection suggest two styles of visual computation. *Neural Comp.* **1**, 68-81.

A neural model for the execution of symbolic motor programs *

Claudio M. Privitera, Pietro Morasso

Department of Informatics, Systems, and Telecommunications
University of Genova, via Opera Pia 11A, 16145 Genova, Italy

1 Introduction

The generation of a Motor Plan is a complex process which transforms a syntactical and synthetical input representation in the corresponding time sequence of patterns defined in the motor space $\Re^n$. Moreover, the elaboration of sensorial information is essential for the correct adjustment of the plan execution to the casual changing of the application environment. The Speech Production is a clear instance of this process: in this case the input representation corresponds to the lexical storing of the word (Wernicke's area) and the problem consists in generating the complex sequence of articulatory-acoustic events corresponding to the voiced realization of the word.

The problem to define neural models for storing and controlling temporal sequences generation appears to be quite complex, not only as regard the complexity of the learning phase (think for instance of back-propagation through time) and the limits about storing capability of neural networks, but especially because it turns out very difficult to contemporaneously obtain the diversified characteristics at the base of biological Serial Order in behavior by a single neural architecture.

We already mentioned to the *compensation* phenomenon that is in other words the capability to perform a specific motor plan in diversified environment conditions (i.e. the production of bite-block phonemes (Gay, 1981)). Another fundamental point is the so called *ambiguousness* problem that is when the appearance of a specific temporal pattern during the plan generation can be followed every time by different patterns (Keele et al., 1990). Finally we underline the need to control the velocity of the sequence generation, both during the transition between two states of the plan, and during the stabilization of the plan in one of its own steps.

2 Storing temporal sequences

In (Privitera & Morasso, 1993) and (Privitera et al., 1993) the problem of storing and generating temporal sequences has been solved defining a neural network structured in two layers of computational maps: the higher level map performs an interpretation of the input vector μ (which represents the syntactical expression of the plan) and consequently produce a set of vectorial parameters able to *trigger* and *guide* on the other map a *chain process* which

---

*This work was supported by MURST-PNRTB-Theme2

causes the sequential activation of all *patterns* composing the sequence (that is the operational semantic of the input expression).

In other words, the approach can be considered quite complementary as regard the model existing in the literature: the *patterns* belonging to the sequences are already topographically represented in a two-dimensional low level map, and the generation of the sequence is equivalent to a sequential activation of *neural islands*. The dynamics is generated by a sort of *constrained* gradient descent defined on a iterative potential field [1] whose conformation is determined by vectorial parameters produced by the higher map.

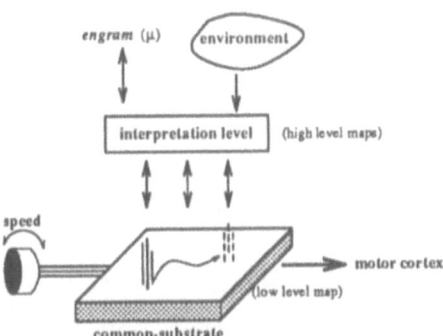

Figure 1: The general structure of the model composed by two order of maps: the higher level performs an interpretation of the symbolic representation of the sequence (the *engram* of the motor plan) whereas the low level represents the substrate exploited both in the generation and observation phase. The generation of a sequence is explicable as a sequence of activation peaks on the surface of the substrate.

A general representation of the model is shown in figure 1: it is worth noting that the interpretation performed by the higher map take into consideration also the set of real time sensorial information in order to resolve the problem of compensation. The knob connected to the lower map represents the possibility to modify the input sensitivity of the neuron [2] composing the map and consequently the speed of the temporal sequence.

---

[1] The potential field on the lower map is given by the following equation:

$$\epsilon(t) = \sum_i \| \tau_{Target}(t) - \bar{\tau}_i \|^2 \, U_i(\tau(t)) \tag{1}$$

where $\tau_{Target}(t)$ is derived from the vectorial parameters determined by the higher map, $\bar{\tau}_i$ represents the prototypes of the neuron i and $\tau(t)$ the coordinates of the neural resonance on the substrate (Privitera & Morasso, 1993). The gradient of ϵ is determinated by the following derivation: $\dot{\tau} \propto - \nabla \epsilon(\tau) = \frac{\partial \epsilon}{\partial \tau}$

[2] In particular the processing elements of the lower map are an extension of the Leaky Integrator neuron model (Reiss & Taylor, 1991).

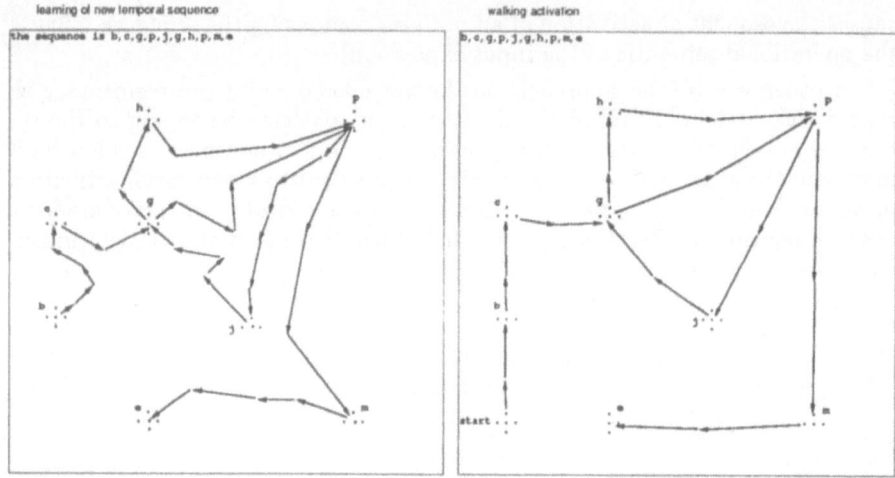

Figure 2:   Learning and generation of a new temporal sequence

2.1   High level and the problem of the compensation

The high level map is structured in a set of prelabeled classes of neurons $M = M_1, M_2, \ldots$ each of them corresponding to a specific symbolic code of a pattern (i.e. the patterns composing the plans are symbolically represented by means of an alphabet of symbols $\mu_1, \mu_2, \ldots$); the training set is composed by a set of triples $(\mu_j, \mathbf{X}, \tau)$ where μ_j represents the symbolic code, $\mathbf{X}$ the sensorial informations and finally τ the corresponding target in the output space. Let $(\tilde{\mathbf{w}}_i, \tilde{\tau}_i)_h$ be the pair representing the prototypes of the generic neuron i belonging to the class M_h, then the learning rule is the following Hebbian rule:

$$
\begin{aligned}
&if \qquad h = j \qquad then \\
&\Delta \tilde{\mathbf{w}}_i = \eta(\mathbf{X} - \tilde{\mathbf{w}}_i)U_i(\mathbf{X}) \qquad and \qquad \Delta \tilde{\tau}_i = \eta(\tau - \tilde{\tau}_i)U_i(\mathbf{X}) \quad (2) \\
&else \qquad \Delta \tilde{\mathbf{w}}_i = \Delta \tilde{\tau}_i = 0
\end{aligned}
$$

where $U_i(\mathbf{X})$ is the competitive activation level of the neuron as a function of the input vector $\mathbf{X}$.

The mapping between the output and the sensorial space is represented by the following population coding:

$$
f(\mu_i, \mathbf{X}) = \sum_{j \in M_i} \tilde{\tau}_j U_j(\mathbf{X}) \tag{3}
$$

In other words, the map associates a symbolic element belonging to a generic alphabet with the corresponding motor target, taking into consideration also the sensorial information $\mathbf{X}$.

2.2   Learning a new plan by the inversion of the model

During the direct execution of the model, the vector μ is sequentially read by the high level map which consequently determines a set of vectorial parameters (the vectors $\tau_{Target}(t)$) able to correctly control the gradient descent (and the sequential activation of the patterns) on the low level map. Learning a new plan means simply to invert the above mentioned process: the low level map is exploited in order to follow an external phenomenon which (as in direct phase) is able to active a sequence of neural regions. By inverting the population coding exploited by the high level map it is consequently possible to determine the symbolic vector μ able to generate again the same temporal sequence. In other words, during the learning phase only the structural contents of the plan is drown out in order to define the corresponding symbolic code; during the generation phase the code is interpretated and arranged with the sensorial conditions by the higher map.

3   Experiments and conclusions

Figure 2 shows the dynamics in the lower map during an execution of the model; the map is composed by a grid of 44×44 neurons and the alphabet contains 16 symbols (the first 16 letters). The plot in the figure is the superimposition of the sequential frames set representing all states of the map during the execution: the rows denote the direction of the gradient descent and the little black asterisks the lighting of the neurons during the gradient descent. In the left plot, the model is observing a new external phenomenon composed by some closed loop (the ambiguous problem: for example the region labeled with g is covered two times during the observation). At the end of the *observation* phase, it is possible to obtain the symbolic code of the sequence just executed (the vector μ). The application of this vector to the model during the direct phase, realizes again the same activation sequence (the right plot).

The possibility to store long temporal sequences defined in a complex output space by means a compact symbolic vectors, the simplicity characterizing the learning of a new plan, the compensation property obtained by the high level (during the generation, a plan is fitted with the new and dynamic sensorial informations) and finally the varied combinations of patterns in the learnable sequences are features which make the approach interesting in many applications, starting from the Speech Production one.

References

Gay, T. (1981). Production of bite-block vowels: Acoustic equivalence by selective compensation. *Acoustical Society of America, 69*, 802–810.

Keele, S., Cohen, A., & Ivry, R. (1990). Motor programs: concepts and issues. In Jeannerod, M. (Ed.), *Attention and performance XIII*. Lawrence Erlbaum Ass., Hillsdale, NJ,.

Privitera, C., & Morasso, P. (1993). A new approach to storing temporal sequences. In *Proceedings of IJCNN'93*, Vol. 3, pp. 2745–2748 Nagoja, Japan.

Privitera, C., Sanguineti, V., & Morasso, P. (1993). Temporal sequences generation and computational maps. In *Proceedings of NeuroNimes'93: Neural Networks and their Industrial and Cognitive Applications*, pp. 55–64 Nimes, France.

Reiss, M., & Taylor, J. (1991). Storing temporal sequences. *Neural Networks, 4*, 773–787.

Evolution of Typed Expressions describing Artificial Nervous Systems

C. Jacob

Lehrstuhl für Programmiersprachen, Universität Erlangen-Nürnberg
Postfach 3429, D-91022 Erlangen, Germany
Email: *jacob@informatik.uni-erlangen.de*

1 Evolution of typed expressions

Evolutionary algorithms implement adaptation or learning in analogy to natural selection over a population of individuals competing in a certain environment. Similar to the genetic programming paradigm introduced by J. Koza [6] who uses LISP-S-expressions our structures undergoing adaptation are hierarchical, typed expressions (terms).[1] The set of generable structures is the set of all possible compositions of typed terms that can be composed recursively from a problem specific set of function symbols $\mathcal{F} = \{f_1, f_2, \ldots, f_N\}$ with arities $\mathcal{A} = \{a_1, \ldots, a_N\}$, $a_i = (minarg_{f_i}, maxarg_{f_i})$. For each function f_i the arity can vary between a minimum and maximum number of arguments. According type descriptions $\mathcal{T} = \{t_1, \ldots, t_N\}$ define the expressions' type structures, where the t_i are regular expressions over $\mathcal{F} \cup \mathcal{S}$ and $\mathcal{S}$ represents a set of standard types.[2]

These terms serve as problem specific codings of individuals which can be computer (LISP-) programs, descriptions of connection structures, partial descriptions of neural networks or definitions of rewrite systems. Decoding routines tell about how the expressions have to be interpreted. Figure 1 shows an example of expressions describing the connection structure of an artificial nervous system by a list of (possibly recurrent) signal paths from input to output neurons via cortex cells.

1.1 Structure generation

Evolution starts with random generation of an initial population of expressions. Each expression is constructed in a recursive manner starting with the outermost term (the axiom).[3] Types have to be respected for all subexpressions. For a given non-standard type t we define the set $\Sigma(t) = \{f_i \in \mathcal{F} | t \in \mathcal{L}(f_i)\}$.[4] Each function symbol $f_i \in \mathcal{F}$ is given a selection probability p_i from $\mathcal{P} = \{p_{f_1}, \ldots, p_{f_N}\}$, with $p_{f_i} \in [0, 1]$ introducing a kind of ranking among suitable functions. Each time a subexpression of type t has to be generated the selection probability of function symbol $f_i \in \Sigma(t)$ is $\dfrac{p_{f_i}}{\sum_{f \in \Sigma(t)} p_f}$. None of the expressions

---

[1] For an alternaitve grammar-based approach see [1].

[2] Note that each function symbol serves as a type symbol. Possible standard types are: integer number, real number, boolean value.

[3] The outermost term constitutes the axiom if the term generation process is described by an equivalent parallel rewriting system.

[4] $\mathcal{L}(t)$ is the language defined by the regular expression t.

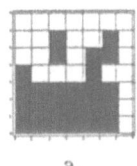 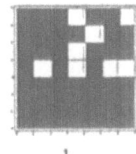 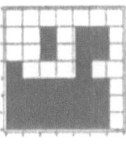

a.   b.   c.

NNTopology[1] =
{PATH[{NEURON[in[2]], {NEURON[cortex[1]], {NEURON[out[1]], {NEURON[cortex[1]],
{NEURON[cortex[1]], {NEURON[cortex[2]], NEURON[cortex[3]], NEURON[cortex[1]],
{NEURON[out[1]], NEURON[out[1]]}}}}}}}, NEURON[out[1]]}]},
NNTopology[2] =
{PATHLIST[{PATH[{NEURON[in[2]], NEURON[out[1]], NEURON[out[1]]}],
PATHLIST[{PATH[{NEURON[in[1]], NEURON[cortex[2]], NEURON[out[1]]}],
...],
PATHLIST[{NEURON[in[2]], {NEURON[cortex[2]], NEURON[out[1]]}, NEURON[out[1]]} ...]}

Figure 1: Generated genotypical redundant representation of the connection structures. Connection matrix (c.) is an overlay of (a.) and (b.)

– considered as trees – should exceed a maximum predefined depth. This is achieved by either only accepting expressions which are within the specified limits or by limiting the number of iterations in the term extension process which can be interpreted as application of parallel rewrite rules (as e.g. in L-systems [7]).

1.2 Genetic operators

Size and shape of the expressions change dynamically during the evolution process through the following genetic operators which we explain by some example expressions describing neural net signal paths (fig. refConnectionMatrices):

1.2.1 Mutation

To perform **mutation** on an individual expression

pathList[path[in[2], cortex[3], out[1]], path[in[1], out[2], cortex[1], out[1]]]]

a function symbol *path* $\in \mathcal{F}$ is selected according to the probabilities from $\mathcal{P}$. With mutation probability μ_m each subexpression with head *path* is then replaced by a newly generated expression with an equivalent head *path* or *pathList* resulting in a modified individual genotype:

pathList[path[in[2], cortex[3], cortex[2], out[1]], pathList[path[in[1], out[1], out[1]]]].

1.2.2 Deletion

The **deletion** operation (deletion probability μ_d) is analogous to mutation. The only difference is that selected subexpressions are completely erased; however, this is only possible as long as the number of arguments of the surrounding expression does not drop below its minimum argument number. Performing deletion on the second expression given above with $a_{pathList} = (1, 3)$, might

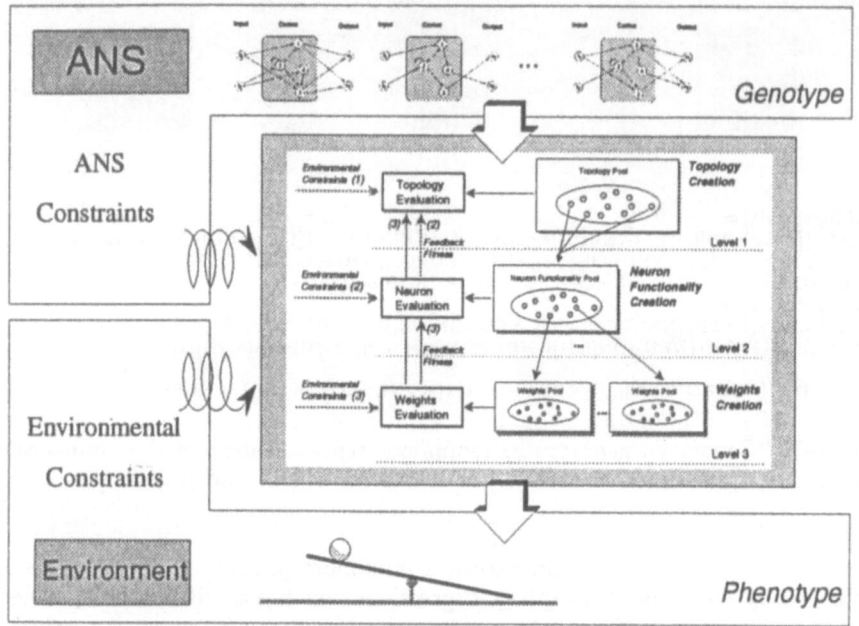

Figure 2: Design hierarchy for artificial nervous systems

result in an expression pathList[path[in[2], cortex[3], cortex[2], out[1]]] where the second *pathList* term is erased.

1.2.3 Crossover

Crossover is used as a recombination operator with probability μ_c which enables exchange of structures of the same type between two individuals. Given the two expressions from above a function symbol *path* $\in \mathcal{F}$ is chosen. Subexpressions with head *path* are selected randomly within each expression and exchanged between the two individuals resulting (e.g.) in the following modified expressions:

pathList[path[in[2], cortex[3], out[1]], path[in[2], cortex[3], cortex[2], out[1]]]]

pathList[path[in[1], out[2], cortex[1], out[1]], pathList[path[in[1], out[1], out[1]]]].

1.2.4 Non-standard operators

Beside the standard operators we experiment with - still problem-independent - high-level operators which should improve the effects of the basic operators: Mainly we use a **template extraction** and **collapse** operation for collecting suitable patterns and for replacing subexpressions by variable symbols.

1.3 Reproduction: from generation to generation

For each optimization problem to be solved within this paradigm specific interpretations and fitness evaluations of the generated structures have to be defined
in order to perform reproduction operations on the populational level. Each
population consists of a collection of expressions serving as partial descriptions
of "organisms" in a problem dependent environment. The organisms have to
solve predefined tasks and receive merit (fitness) according to their ability to
find good solutions. The organisms' fitness values are used for defining probabilities for each individual genotype to be reproduced into the next generation.

2 Hierarchical evolution of connectionist systems

We used the described evolution system of typed expressions for designing
artificial nervous systems.[5] Figure 2 shows an example where neural nets are
used to control a seesaw balancing a ball.[6] Herein we consider neural net
evolution as a threefold development process of net topology, functionality of
cortex neurons and weight settings.[7] Separating net descriptions into several
levels has the advantage that on each level appropriate genotype codings and
interpretations can be applied. Furthermore, the set of basic genetic operators
can be enhanced by level-specific operators.

Figure 3 shows an example of typed function sets used to generate terms
describing connection structures and neuron functionalities for recurrent neural
networks as depicted in figure 2. Figure 4 shows connection matrices of the best
(seesaw controlling) networks evolved over 30 generations. Similar results are
obtained for the evolved neuron functionalities.

Weight settings for the connections are evolved by mapping weight values
into fixed-length bitstrings and performing standard genetic algorithm operations [3] on these bitvectors. For more difficult problems more advanced techniques of weight adaptation for recurrent nets can be used.[8]

Currently we use the expression evolution system to evolve context sensitive parallel rewrite systems (L-systems [7]) describing stepwise generation of
connection structures influenced by environmental signals.

References

[1] Antonisse, H.J. (1991), *A Grammer-Based Genetic Algorithm*, in: Rawlins G
(ed), Foundations of Genetic Algorithms, San Mateo, 1991.

[2] Bornholdt, S., and Graudenz, D. (1992), *General Asymmetric Neural Networks
and Structure Desgin by Genetic Algorithms*, Neural Networks, vol. 5, 1992.

---

[5] A more detailed description can be found in [4].

[6] Another neural net controls a pair of walking legs similar to [3].

[7] Most of the work about evolution systems for neural networks has solely focused on net
connectivity [2], [5], [6].

[8] The modules of each level can be replaced independently. The weight evolution level
might, e.g., be substituted by an evolution module for learning rules.

NNTopology[$n\_Integer, m\_Integer, \_pathList$],
pathList[$\_path, \_pathList$],
pathList[$\_path$],
path[$\_inputNeuron, \_neuronList, \_outputNeuron$],
neuronList[$\_cortexModule \mid \_outputNeuron, \_neuronList$],
neuronList[$\_cortexModule \mid \_outputNeuron$],
inputNeuron[$i\_Integer$] with $i \in \{1,\ldots,n\}$,
cortexModule[$\_cortexNeuron \mid \_NNTopology$],
cortexNeuron[$j\_Integer$] with $j \in \{1,2,\ldots\}$,
outputNeuron[$k\_Integer$] with $k \in \{1,\ldots,m\}$,

NeuronFunction[$\_input, \_activation, \_output$],
input[$\_min \mid \_max \mid \_sum \mid \_prod$],
cortex[$\_sigmoid \mid \_sin \mid \_tanHyp \mid \_lin$],
output[$\_id \mid \_lin \mid \_thre$],
min[], sigmoid[a_Real, b_Real, c_Real], lin[a_Real, b_Real],
...

Figure 3: Typed function set for ANS connectivities and transfer functions

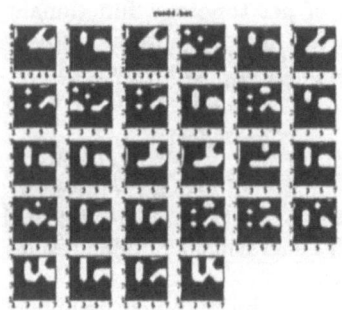 

Figure 4: Best connection (left) and root matrices - representing the connectivity structure common to each generation's individuals - over 30 generations

[3] de Garis, H. (1991), *Genetic Programming: Building Artificial Nervous Systems with Genetically Programmed Neural Network Modules*, in: Soucek, B. (ed.), *Neural and Intelligent Systems Integration*, New York.

[4] Jacob, C., and Rehder, J. (1993), *Evolution of neural net architectures by a hierarchical grammar-based genetic system*. ICNNGA, International Joint Conference on Neural Networks and Genetic Algorithms, Innsbruck, 1993, 72-79.

[5] Kitano, H. (1990), *Designing neural networks using genetic algorithm with graph generation system*, in: Complex Systems, 4:461-476, 1990.

[6] Koza, J.R. (1993) *Genetic Programming, On the Programming of Computers by Means of Natural Selection*, MIT Press, London.

[7] Lindenmayer, A. (1968) *Mathematical models for cellular interaction in development*, in: Journal of Theoretical Biology, 18, 280-315, 1968.

BAR: A Connectionist Model of Bilingual Access Representations

O. Soler and R. van Hoe

Institute for Perception Research

Eindhoven, The Netherlands

1 Introduction

A parallel distributed model of bilingual access representations (BAR) is presented. The model is part of an extended version of the theoretical framework for language processing developed by Schreuder and Baayen (in press) (cf., figure 1). According to this framework, language processing occurs in two steps. In the first step (i.e., segmentation, phonology, and orthography), input from both sensory modalities, that is, phonological and orthographical input, is encoded and presented to the system which builds up intermediate access representations. These intermediate access representations activate in their turn the (final) internal access representations. In the second step (i.e., licensing and composition) the access representations activate concept nodes. The concept nodes contain syntactical and semantical information which is fed into the higher-level systems responsible for discourse processing.

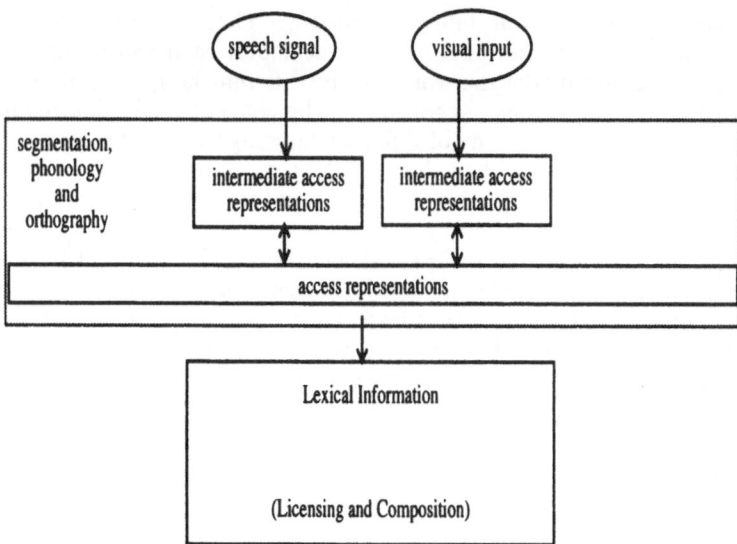

Figure 1: A theoretical framework for bilingual language processing

The main goal of the BAR model is to analyze how access representations are structured when two languages are represented simultaneously, and to study the acquisition of the words of a new language at this level. Our model is concerned with the development of the internal access representations for words of the two languages. In order to implement this level of processing Seidenberg and McClelland's (1989) distributed model of word recognition and naming was adapted. Hence, within the extended framework of Schreuder and Baayen, the BAR model can be considered as a connectionist account of the first step of bilingual language processing. Further research will be needed in order to verify that the same framework can deal with the subsequent, licensing and composition, level.

In the next section, the relevant literature in bilingual lexical research is briefly reviewed in order to introduce the empirical benchmarks of the model. The second section describes the distributed model for word recognition and naming of Seidenberg and McClelland (1989) which is used as the basic architecture of the BAR model. Last but not least, in the third section the bilingual access representations model (BAR) model itself is described more in detail.

2 Bilingual Language Processing

Bilingualism and multilingualism have always attracted much interest in cognition research. The organization of the bilingual lexicon is one of the most frequent studied topics in this domain, and was and still is a controversial research issue.

Two main models for bilingual lexical organization have been proposed: the word-association model (Kirsner, Smith, Lockart, King & Jain, 1984) and the concept-mediation model (Potter, So, von Eckart & Feldman, 1984). Both models assume the existence of an independent language-specific lexicon for each of the languages of the bilingual subject. These lexicons are composed of lexical entries which are local representations of each word. According to the word-association model the words of the different languages are connected directly to their translations. According to the concept-mediation model there are no links between the words of different languages, and the connection between translations is established through their meaning.

Both models are mainly concerned with the semantical relation between two languages. However, studies exploring other aspects of bilingualism, such as form priming or lexical access, provided evidence which cannot be explained by the word-association model nor by the concept-association model. Grainger and Beauvillain (1987) showed that lexical access is a pre-lexical and language-independent process, i.e. the access to the lexical entries takes place before any lexical (i.e., semantic and syntactical) information is available, and also before the language of the input is identified. Studies using the form priming paradigm show a strong relationship between cognates (words of a similar form) with respect to non-cognates (e.g., De Groot & Nas, 1991). These studies suggest that the organization of the access representations is strongly dependent of the form of these representations and the similarities between forms in the two languages, as Cristoffanini, Kirsner & Milech (1986) and Beauvillain (1992) proposed. The models developed by these authors describe two language-specific lexicons that share some of the lexical entries. Cognate words are supposed to share the same lexical entry and hence constitute the shared part of the lexicons.

These findings show the need for a detailed study of the pre-lexical processes. The processes occurring before lexical information is retrieved, suggest a particular organization of the access representations. Access representations seem to be organized in a language-independent structure according to the similarity of their form. Since current models define lexical entries as local nodes, it is very difficult to identify the words shared by the two languages. Similarity of words is not a discrete feature. Furthermore, the current models only consider the visual modality of language perception, but there are phonological similarities as well which can influence the processing of two different languages. The research question which automatically arises then is: how can a 'word form' be defined, and subsequently, how can the similarity of two forms be defined? In order to clarify these aspects, a model for bilingual access representations based on Seidenberg and McClelland's (1989) monolingual word recognition model was developed.

3   Monolingual Word Recognition

Seidenberg and McClelland (1989) developed a model for word recognition and naming that accounts for the effects of interaction between phonological and orthographical coding during auditory and visual word recognition. Several studies provided evidence that both orthographic and phonological representations are activated during the first steps of language processing.

The model described by Seidenberg and McClelland (1989) provides an internal representation for both codings in a multilayer network. Although they proposed a completely framework for language processing, only the part concerning access representations is implemented (cf., figure 2). The two sets of input/output units in Seidenberg and McClelland's model correspond with the intermediate access representations of figure 1. The pool of hidden units connecting both of them constitutes the access representations. An important feature of the model is that, in contradistinction with current local models of word recognition, the words are represented by means of a pattern of activation over a set of hidden units, and not by a single node. Local representation models of word recognition also cannot explain how a new word is acquired (Monsell, 1991). In Seidenberg and McClelland's model the back-propagation learning algorithm is used to model the development of the monolingual lexicon.

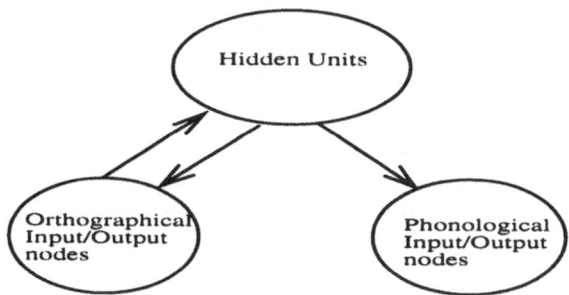

Figure 2: Seidenberg and McClelland's model of word recognition

The model of Seidenberg and McClelland is the most suitable connectionist framework to account for the bilingual lexical processing research issues described in the previous section: the internal distributed representation unifying both coding modalities (speech and visual input) can account for different kinds of cognate words[1] and for the learning of new words in a different language. The description of the bilingual adaptation of Seidenberg and McClelland's model is discussed in the next section.

4   The BAR Model

The architecture of the BAR model is basically the same as Seidenberg and McClelland's (1989) model. The feedback connection between the phonological input/output set of units, which was not implemented in their distributed model, is. incorporated in our model in order to facilitate the learning of the words from the two coding modalities. The phonological and orthographical input is presented to the network using the same coding scheme of Seidenberg and McClelland (1989) in order to replicate their simulation results for the monolingual (Dutch) lexicon. However, as these codings proved not to be the most suitable ones, other coding systems, especially for the phonological input, are currently being tested.

The training procedure of BAR encompasses two main steps: (1) first language acquisition: when Dutch input and output is provided to the network, and (2) learning of the second language: when Dutch input and English output is provided. An extra input unit is needed for the network to discriminate which task is required but this unit is not a language discriminator or switch (see figure 3).

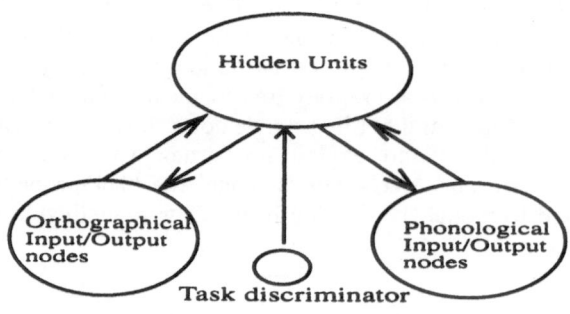

Figure 3: The BAR model

Our main hypothesis is that the network will first build up internal access representations for the Dutch language, and that in the course of the second training phase the hidden units will develop new patterns of activations. It

---

[1] The distinction between different types of cognate words is not explicit in the current literature. However, is not difficult to realize that the words of different languages can be similar either orthographically or phonologically, or both. As an example we can consider three different cases of Dutch and English. 'Fruit' and 'Fruit' are orthographically similar but are phonologically different. 'Voet' and 'foot' are phonologically similar but are orthographically different. Finally, 'hel' and 'hell' are both orthographically and phonologically similar.

is predicted that these new patterns of activations will be similar for cognates in the two languages, and different when the words are dissimilar. In order to study the performance of the model the analysis is focussed on the hidden units, and not on the output units as it was the case in Seidenberg and McClelland's paper. Clustering and freezing techniques are used to explore the internal organization in the hidden units when the access representations are coping with two different languages.

The distributed access representations are possibly going to account for the empirical results on the cognate versus non-cognate distinction, and hence providing a new model for bilingual lexical access organization.

5 References

De Groot, A.M.B., and Nas, G.L.J. (1991). Lexical representation of cognates and non-cognates in compound bilinguals. Journal of Memory and Language, 30, 90-123.

Beauvillain, C. (1992). Orthographic and lexical constraints in bilingual word recognition. In R.J. Harris (Ed.), Cognitive Processing in Bilinguals. Amsterdam: North Holland.

Cristoffanini, P., Kirsner, K., and Milech, D. (1986). Bilingual lexical representation: The status of Spanish-English cognates. The Quarterly Journal of Experimental Psychology, 38A, 367-393.

Grainger, J., and Beauvillain, C. (1987). Language blocking and lexical access in bilinguals. The Quarterly Journal of Experimental Psychology, 39A, 295-319.

Kirsner, K., Smith, M.C., Lockart, R.S., King, M.L. and Jain, M. (1984). The bilingual lexicon: Language-specific units in an integrated network. Journal of Verbal Learning and Verbal Behavior, 23, 519-539.

Monsell, S. (1991). The nature and locus of word frequency effects on reading. In D. Besner and G.W. Humphreys (Eds.), Basic processes in reading. Hillsdale: Lawrence Erlbaum.

Potter, M.C., So, K.F., von Eckart, B., and Feldman, L.B. (1984). Lexical and conceptual representation in beginning and proficient bilinguals. Journal of Verbal Learning and Verbal Behavior, 23, 23-38.

Schreuder, R. and Baayen, R. (in press). Modeling morphological processing.

Seidenberg, M.S., and McClelland, J.L. (1989). A distributed, developmental model of word recognition and naming. Psychological Review, 96(4), 523-568.

An Architecture for Image Understanding by Symbol and Pattern Integration

M. Nishi[1], K. Ohzeki[2], N. Sakurai[3], T. Omori[1]

1 Department of Electronic and Information Science, Faculty of Engineering,
  Tokyo University of Agriculture & Technology, Koganei, Tokyo 184, Japan

2 Fuji XEROX Co., Ltd., Tokyo, Japan

3 ASAHI KASEI JOHO SYSTEM Co., Ltd., Tokyo, Japan

1 Introduction

Symbolic processing is thought to represent human language dependent thinking process as a first order approximation. But it is also known that the representation is only a part of the wide variety of human thinking.

On the other hand, human has the other intelligence of sensory pattern learning and recognition. Distributed information like pattern seems to represent nonverbal and sensory oriented thinking ways and shares large part of brain area. We suppose the pattern processing as a second term of the human intelligence approximation.

In this paper, we propose a symbol-pattern integrated architecture. In the architecture, there are "knowledge representation by pattern" and "operation on the representation". They just correspond to the "knowledge representation by symbol" and "operation on symbol" that also exist in the model. Between them, a mutual transformation system works to realize a dynamic interaction.

Main problem here was the mutual transformation between the image pattern and the symbolic predicates. In this research, we adopted "Selective attention model(SA model)"[1] as the transform system. It realizes symbol and pattern dynamic interaction. We show a computer simulation result of the image understanding process by the architecture.

2 Pattern – Symbol Integrated Architecture

2.1 Dual coding of Recognition & Knowledge

We suppose a sensory event is dually represented in pattern and symbolic form. In the symbolic processing, knowledge is described in symbol, inferred by symbolic operations, and new knowledge description is obtained and interpreted. The operations have mathematical basis of a symbolic logic. In the neural network, a set of connections corresponds to the knowledge, and a pattern operation corresponds to the logical inference.

In spite of logic symbol and pattern discrepancy, human pattern knowledge can be verbalized. Also, we can imagine visual image from verbal representation, and can operate on it[2]. Then, sensory information and verbal information are mutually transformed and applied. They are selected and used depending on a situation.

2.2 Dynamic Interaction of Symbol and Pattern

We can expect a mutual complementary function from the dual representation. If one representation failed to solve a problem, the problem can be mapped to another representation and might be solved easily. The system is composed of three parts of the pattern system, the symbolic system, and a mutual transformation system (Fig.1).

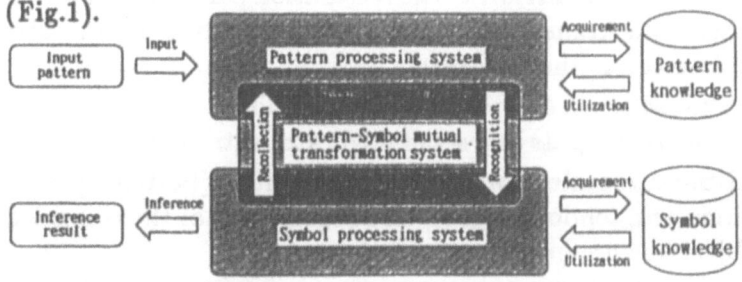

Fig.1 The construction of the system.

A. Symbolic System : Receives a recognition result as a symbol, operates with symbolic knowledge. It operates on the transformation and the pattern system to reflect the symbolic result.

B. Pattern System : Processes pattern information like image. We adopt simple image area masking out from the recognition process as the pattern operation.

C. Symbol–Pattern Transformation system : To reflect operational results, symbol and pattern representations of an event are transformed to each other. The pattern-to-symbol transformation is a "recognition". As a symbol may correspond to many variations of a pattern, it is a many-to-one transformation. The inverse symbol-to-pattern transformation is one to many in general.

3 Application of Symbol–Pattern Integration to image understanding

3.1 Sequential recognition by SA model

SA model is an extention of Neocognitron, an image recognition model[1]. It recognizes an object independent of position shift and shape deformation. It can recollect image features that have contributed to current recognition. The topdown image recollection process extracts all of the image features like edges and their combinations at the same time with symbolic recognition result.

Most of the SA model process can be described as a set of local image computations, such as convolution. We have extended a logic programming language Prolog to include image processing library and image representing predicates. The extended Prolog can sequentially execute the SA model as a logic program. The extended Prolog recognize a given image as following sequence R–1 to R–5.

R–1. **Image recognition** : The recognition part of SA model extracts and learns local features hierarchically until an object is detected at the last layer.

R–2. **Symbolic Recognition** : The symbolic system interprets the neural outputs as a set of predicates with confidence value.

R–3. **Image recollection** : By an interaction of bottom up recognition and topdown signal, an image area that the recognized object has occupied is extracted.

R–4. **Symbolic inference** : From the symbolic recognition result, some information that suggest recognition strategy are inferred through symbolic knowledge. They are used to control the SA model recognition process.

R–5. **Sequential recognition** : The object image area is masked out from subsequent recognition. Other objects are recognized sequentially until all of the objects are masked out.

3.2 Application to circuit diagram recognition

Fig.2 shows a set of learned typical patterns of circuit elements and a target image for the recognition.

Fig.3 shows the recognition process of the Fig.2(c) image. Given an input image, all of circuit parts are recognized and recollected in each place. The recollected patterns are masked out from the input image. Then, only connections between the elements remain in the image. To recognize the connection pattern, the symbolic system designates a recognition location, cuts out an image, and recognizes it.

4 Conclusion

We have presented an architecture of symbol and pattern integration, and shown an application on an image understanding system. The SA model was very important for the symbol-pattern mutual transformation. It was the key concept and key technique of the architecture.

References

[1] K.Fukushima: A Neural Network Model for Selective Attention in Visual Pattern Recognition, Biolog. Cybern., Vol.55, No.1, pp.5–15, Oct.1986

[2] Sakurai N.,Omori T.: Image Recognition with a Cooperation of Symbolic Inference and Neural Networks, Proc. of International Conference on Artificial Neural Networks '92, 1992

[3] Omori T.: Image Transformation by Spatial Inhibiton and Local Association, Proc. of IJCNN '91-Singapore, 640–645, 1991

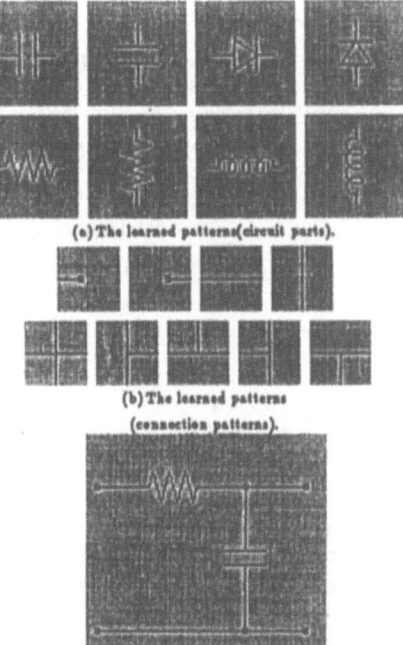

(a) The learned patterns(circuit parts).

(b) The learned patterns (connection patterns).

(c) A target pattern for the recognition.

Fig.2 A set of learned typical patterns of circuit parts and a target image for the recognition.

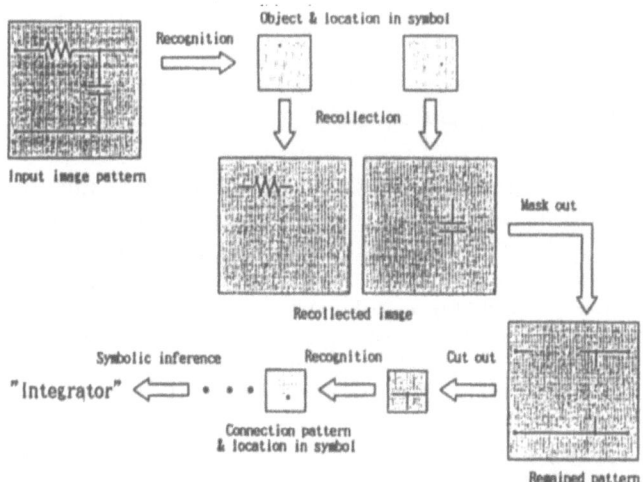

Fig.3 The recognition process of the Fig.2(c)image.

Encoding Conceptual Graphs by Labeling RAAM

M. de Gerlache†, A. Sperduti‡, A. Starita‡

† V.U.B., Prog Lab
Pleinlaan 2, 1050 Brussels, Belgium
‡ Department of Computer Science, University of Pisa
Corso Italia 40, 56125 Pisa, Italy

1 Introduction

The meaning of medical texts is not automatically recognized by computers. A representation of this information is strongly recommanded to allow medical texts databases queries. The conceptual graph formalism developed by Sowa [Sow84] is a knowledge representation language initially designed to capture the meaning of natural language. Conceptual graphs have been used in many natural language understanding works [BRS92, VZB⁺93, Ber91]. In this paper we discuss the possibility to memorize and retrieve natural language sentences and especially medical language sentences given in this kind of formalism with the use of the LRAAM model [Spe93b, Spe93a]. In Section 2 we explain the idea underlying conceptual graphs. In Section 3 we briefly expose the access by content capabilities of the LRAAM and suggest a generalization of the access by content procedures introducing the concept of Generalized Hopfield Network. A discussion on the impact of this generalization on knowledge extraction from a database of conceptual graphs is given in the conclusion.

2 Conceptual Graphs

By definition a conceptual graph CG is a finite, connected and bipartite graph. It consists of two kinds of nodes : concepts and conceptual relations. Concepts refer to discrete units of perception and are connected by conceptual relations. Each conceptual relation has n arcs (≥ 1) each of which must be linked to some concept. The meaning of a subgraph with a concept c1 that is linked by a conceptual relation r to a concept c2 is "the r of c1 is c2". Concepts c and relations r are typed, i.e. there is a function type that maps concepts and conceptual relations to type labels. Thus, for example, the concepts x and y are of the same type if type(x) = type(y). For any concept c and any conceptual relation r, type(c) is different of type(r). A type label may be specified or unspecified. A specific type label refers to a certain individual, an unspecified type label to a variable individual. The conceptual graph of the sentence 'A male patient x of 71 years old has been hospitalized urgently' is represented, in linear form, as:

[GENERAL-TREATMENT : hospitalization]
    → (EXPER)   → [PATIENT: x]
                       → (CHRC) → [SEX : male]
                       → (CHRC) → [AGE : 71]
    → (CHRC)   → [CHARACTERISTIC : urgently]

There is a partial ordering (x<y) defined over the set of concepts type labels which forms the concept type hierarchy. If x<y, then x is called subtype of y; and y is called a supertype of x.

3 Associative Data Access by Labeling RAAM

The Labeling RAAM is an extension of the Recursive Auto-Associative Memory (RAAM) by Pollack which allows one to encode labeled graphs with cycles by representing pointers explicitly. The result of the encoding is that each graph represented in the training set is represented by a fixed pattern, independently of the size of the graph. In this way it is possible to apply neural networks to structured domains, since a structure can be represented by a fixed size pattern.

Information on the components of each graph can be retrieved by decoding the pointers belonging to it, however, data encoded in an LRAAM can be accessed by content as well. Direct access by content can be achieved by transforming the encoder network of the LRAAM into a particular Bidirectional Associative Memory (BAM). In particular, a component of a structure in the training set can be accessed by label, by outgoing pointers[1], or by a combination of both. Statistics performed on different instances of LRAAM show a strict connection between the associated BAM and a standard BAM.

It seems thus appealing to encode conceptual graphs in an LRAAM, since both standard inference techniques and associative access can, in principle, be performed. Moreover, the kind of distributed representations obtained using an LRAAM are suited to be processed by other type of networks, such as multilayer perceptrons. Multilayer perceptrons in the context of conceptual graphs has been proposed by Lendaris (see, for example [Len88]).

3.1 Generalized Access Procedures

In this section we briefly discuss a generalization of the associative access procedures defined on the LRAAM by introducing the concept of Generalized Hopfield Network (GHN). This concept allows the access to data also by using a partially defined connected substructure (*query*) as key. This capability is particularly important in view of its application to a knowledge database of conceptual graphs. Specifically, we give here an example of list query. The extension to a tree query is not difficult, while the case of a graph query needs special treatment.

---

[1]In this case the access is by content, since the pointers are used as keys and not decoded in order to retrieve information.

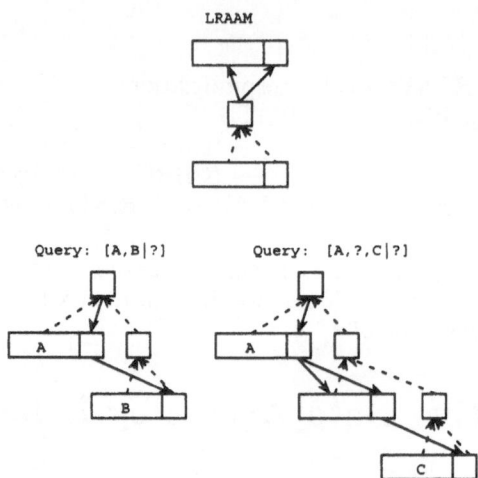

Figure 1: The GHN for the query $[A, B|?]$ and the query $[A, ?, B|?]$.

A GHN is a Hopfield network whose topology is defined according to the topology of the query. Each node of the query corresponds to a set of neurons. Specified information in the query is represented by input units in the GHN. The connectivity of the network is given, depending on the connectivity of the query and on the specified information. The weights on the connections are given by the weights of the LRAAM. An example is given in Figure 1. Given an LRAAM encoding lists, the GHN for the query $[A, B|?]$ is shown. In this case both A and B are specified labels, thus they correspond to input units in the GHN. The pointer to the list, as well as the pointer to the *tail* of the list and the pointer to $[B|tail]$ are not known. They can be considered non instantiated variables and correspond to hidden units in GHN.

The pointer to the list is represented by the set of units at the top of the GHN. Initially the activity of this set of neurons is random. For the first query, the pointer to the list is decoded twice and the result (candidate pointer to *tail*) is then used in the encoding phase. The encoding phase starts by encoding B and the candidate pointer to *tail*. The result is the candidate pointer to $[B|tail]$ which is encoded with A in order to get the candidate pointer to the list. The process is then repeated till the network reachs a stable state.

The network is said to be in a *consistent* stable state if the representations for the pointer to $[B|tail]$ obtained by the decoding and encoding phase match. A consistent stable state is said to be *valid* if the labels obtained by decoding the pointers match the labels used in input. Otherwise, it is called a *wrong* stable state (see [Spe93b]). If a valid stable state is reached, the access procedure is successful and the pointer to the list points to the retrieved list.

In Figure 1, we have given also the GHN for the query $[A, ?, B|?]$. It must be observed that in this case, since the second label is not specified, the corresponding set of units is a set of hidden units. It must be noted that specified information can involve pointers as well.

4   Conclusion

The GHN results to be a very elegant and flexible tool in the context of knowledge extraction from a database of conceptual graphs. In fact, given a database of instantiated conceptual graphs encoded in an LRAAM, the technique discussed above allows one to build in real time a GHN for each type of query the database can support by just composing opportunely the weights of the LRAAM. The main problem with our implementation of the LRAAM is that it uses backpropagation and consequently learning is very slow. One solution to this problem is the use of modular LRAAM [Spe93c], however the impact of this solution on the associative access capability of the model has not yet been assessed.

Once these technical problems are eventually solved, the possibility to exploit both standard inference tools and the GHN to extract information by content, will improve qualitatively the ability of an artificial system to manage knowledge information represented in conceptual graphs. Moreover, the speed of processing can be potentially improved since the GHN will allows also the exploitation of analog hardware when it will be available.

References

[Ber91]   J. Bernauer. Conceptual graphs as an operational model for descriptive findings. In *Proc. of Symposium on Computer Applications in Medical Care*. McGraw-Hill, 1991.

[BRS92]   R. Baud, A. Rassinoux, and J. Scherrer. Natural language processing and semantical representation of medical texts. *Methods of Information in Medicine*, 31:117–125, 1992.

[Len88]   G. G. Lendaris. Conceptual graph knowledge systems as problem context for neural networks. In *IEEE Second International Conference on Neural Networks*, pages 133–140, 1988.

[Sow84]   J.F. Sowa. *Conceptual Structures: Information Processing in Mind and Machine*. Addison-Wesley, 1984.

[Spe93a]   A. Sperduti. Encoding of labeled graphs by Labeling RAAM. In *Neural Information Processing Systems*, 1993. To appear.

[Spe93b]   A. Sperduti. Labeling RAAM. Technical Report 93-029, International Computer Science Institute, 1993.

[Spe93c]   A. Sperduti. *Optimization and Functional Reduced Descriptors in Neural Networks*. PhD thesis, Computer Science Department, University of Pisa, Italy, 1993. TD-22/93.

[VZB+93]   F. Volot, P. Zweigenbaum, B. Bachimont, M. Ben Said, J. Bouaud, M. Fieschi, and J-F. Boisvieux. Structuration and acquisition of medical knowledge : Using ulms in the conceptual graphs formalism. In *Proc. of Symposium on Computer Applications in Medical Care*. McGraw-Hill, 1993.

Hybrid System for Ship Detection in Radar Images[*]

G. Fiorentini† G. Pasquariello‡ G. Satalino† F. Spilotros†

†ALENIA - Sistemi Civili presso I.E.S.I. - C.N.R.

‡ Istituto per l'Elaborazione dei Segnali ed Immagini (I.E.S.I. CNR)

1   Introduction

This paper describes a research activity devoted to verify the feasibility of a neural network approach for automatic detection of naval targets in radar imagery. The activity is part of a more ambitious industrial project concerning the use of advanced image processing techniques for improving the safety in maritime surveillance of harbour traffic. Having in mind this final application, the task of automatic target detection has to fulfil two main requirements: 1) quasi real time response, in the sense that each input radar image of about 1000 by 1000 pixels has to be analyzed in a time comparable with the sweep period of the acquisition system, i.e. three seconds; 2)very high accuracy, i.e. each ship in the scene must be surely detected: this means that the efficiency of the detection system should be not minus than 100%. These two applicative needs have to meet with the high degree of noise (clutter) characterizing the input radar data. In order to face up to the complexity of the exposed goal, the task has been decomposed in a set of sequential subtasks: *noise elimination*, preliminary *object recognition* and classified *image reconstruction*. Moreover, both neural and traditional tools have been used, designing a multi-modular hybrid system (a detailed discussion about the advantages of hybrid architecture can be found, for example, in (F. Folgelman Suolie, 1993)). The core of the neural approach is based on two different architectures: the first one, trained with an unsupervised strategy, performs an image segmentation using a Self Organizing Map; the second one, supervised, classifies objects of interest using a Multi-Layer-Perceptron. The procedure follows an essential line of development: first pre-processes the original radar image with a prefiltering process using one of the SOM's output classes and then classifies objects with the MLP. In section 2 will be shown in more detail the acquisition system; in section 3 will be described the proposed processing system; finally in 4 some preliminary results are discussed.

---

[*] This work has been developed within a grant of the Programma Nazionale di Ricerca per la Bioelettronica assigned by the Italian Ministry of Scientifica and Technological Research to Elsag Bailey

2   Radar images acquisition

The data set of real radar images, has been recorded in Naples port area, on July '93, using the Poseidon radar station of Alenia - Sistemi Civili. The Radar works in X band (10.5 GHz) at the peak power of 45 Kw and leads available, in transmission, three values of PRF (Pulse Repetition Frequency), respectively of 3200, 1600, 800 Hz, with radar divisions of 1/4 NM, 1/2 NM, 1 NM.

The acquisition system supplies for sampling of receiver output signal and for real time A/D conversion. The sampled data are quantized in 4 bits.

The pixel resolution is related to the set range on the receiver. The maximal resolution is obtained with the range of 1 NM (one pixel for every signal sample). The display of radar images is of PPI type upon a 1024 x 1280 monitor and a window based system software supplies a large number of options (such as rescaling, setting of fixed and mobile markers, off-centre).

Using the described acquisition system, many situations of naval traffic, into the port area, have been recorded for an off-line elaboration of radar images.

To obtain a data-set which keeps accounts of detail level in detection using both different transmission modes (short-medium-long pulse) and video thresholds, different images have been recorded varying these parameters.

Furthermore, the registrations are consecutive in time in order to track targets leaving or entering into the port.

3   System analysis

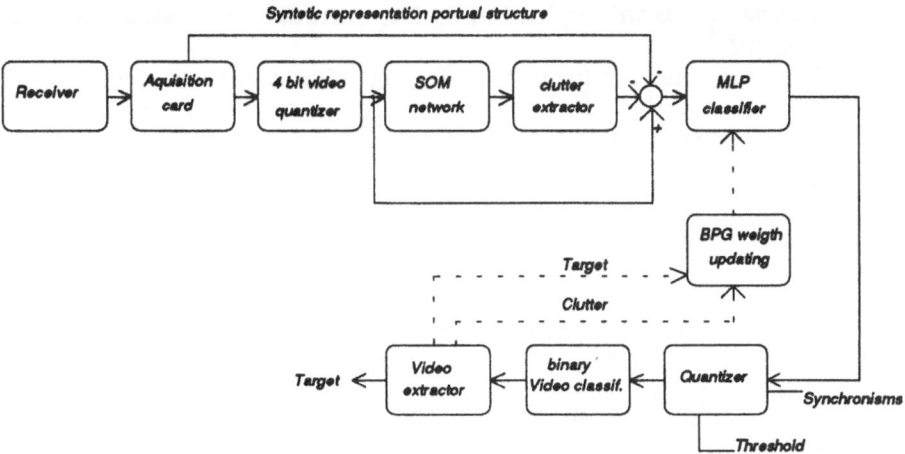

Figure 1 gives an overall description of the multi- modular hybrid system we used: in the following we will enter into details of the principal sub-systems. The Som network module (Kohonen T, 1987) realizes a four classes segmentation of the radar image after video quantization. One of these classes has been recognised as the one associated to clutter (in particular we refer to reflections due to port structures, atmospheric phenomena, etc...); so it is used for filtering the image. The masking operation is accomplished by the clutter extractor module: in the same module are also cleared away the fixed docks structures applying a synthetic representation of the port area. The segmentation network is trained considering the 3-by-3 windows associated with the central grid points of a single typical arbour image and then applied to the new ones. The MLP (Rumelhart et al.,1986) module, is applied to the filtered image in order to classify the radar representation. The architecture we use consists of three fully connected neural layers, with 625 input units, 25 hidden units and one output unit (1 for a pixel belonging to a target, O otherwise). In the learning phase various target and clutter patterns are presented to the network; during the forward phase, for each pixel of the filtered image a 25-by-25 window is presented to the network and the quantizer module performs a thresholding on the output values. The video extractor module operates on the binary image with the aim of eliminating isolated spikes. In this module we adopt a conventional technique commonly employed in radar processing, based on a video correlation which takes care of the distribution of the activated pixels after binarization. Dot lines show the feedback process used during the training of the MLP network: we use both correct and erroneous classifications to give the network a new training set which, ensures a better generalization capability.

4   Experimental results

Figure 2 gives an example of the performance of the system described in the previous section. The input image of fig 2.a is segmented by the Som network already trained, giving as output the four classes image of fig 2.b. The effect of noise removal filter is well depicted by fig 2.c, where the clutter contamination is greatly reduced.
Finally, looking at the last figure, the effect of MLP classification, binarization and reconstruction is shown: in this picture the two crafts entering and leaving the harbour area are clearly singled out.

5   Conclusions

Naval target recognition in cluttered high traffic areas reveals a fundamental importance when high precision of detection is required. A multi-modular system has been proposed as an automatic system for detection and tracking of mobile objects in environments such as port areas. Neural network seems to be powerful tolls for clutter suppression and target recognition
Far from being close to the definitive resolution to the problem, the processing model proposed would be the initial approach to give, to the human operator, an efficient tool of decision and control in all those ambiguous situations that, till now, take the responsibility of accidents and chaotic regulation of maritime traffic.

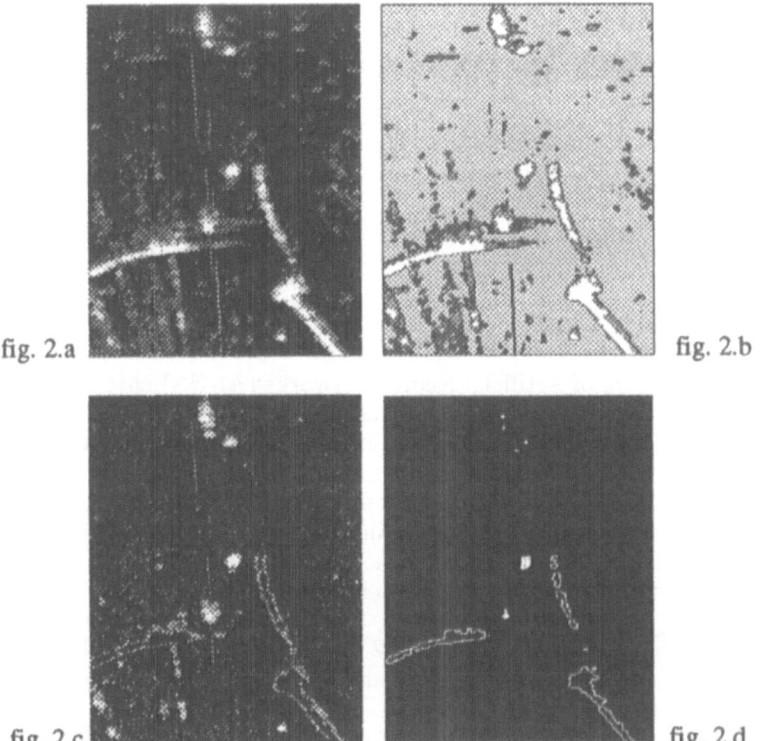

fig. 2.a   fig. 2.b

fig. 2.c   fig. 2.d

References

Folgelman Soulie F., (1993), *Multi-Modular Neural Network-Hybrid architectuires: a Review*, Proc. of 1993 Int. Conf. on Neural Network, Nagoya, 2231-2236.

Hertz J.,Krogh A., Palmer R. G., (1991) *Introduction to the Theory of Neural Computation*. Redwood City, CA: Addison-Wesley.

Rumelhart D. E., Hinton G. E., Williams R. J., (1986) *Learning internal representation by error propagation*. in Parallel Distributed Processing. vol. 1, Rumelhart D. E., Mc-Clelland J. L. Cambridge: MIT Press.

Kohonen T., *Self Organization and Associative Memory*. (1987) 2nd edition Springer Verlag, Berlino.

Using ART2 and BP co-operatively to classify musical sequences

N.J.L. Griffith†,

† Department of Computer Science, University of Exeter

Exeter, UK.

1 Introduction

The modular nature of Artificial Neural Networks (ANN's) encourages the use of composite networks. This research[1] has used ART2 (Carpenter & Grossberg 1987) and BP (Rumelhart, Hinton & Williams 1986) networks, to model the induction of keys and scale degrees in music. General classifications of pitch use - into keys - are used to define the representational granularity of a process classifying the context of pitch classes into degree identities. These identities are used to learn associative mappings between pitch, key and degree, and between pitch context and degree. Two stages of the model will be described separately. In both, statistics of pitch frequency are summed in a memory function and then classified in an ART2 net. A perm of 60 simulations - four memory types, three ζ rates; 0.75-0.25 and five η rates; 0.005-0.5. was run.

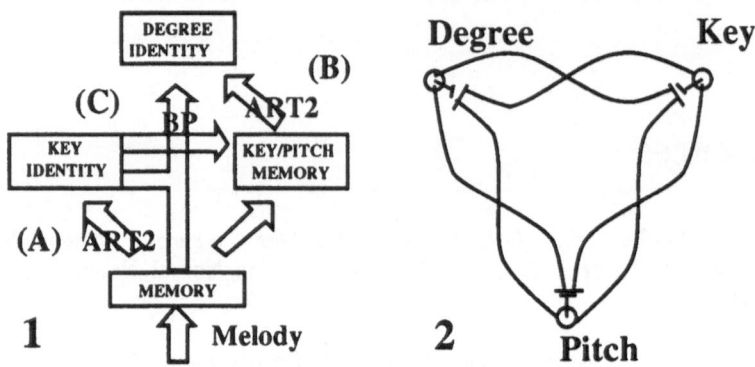

Figure 1: **1.** A model showing the induction of keys and degrees in music. **2.** Mapping between pitch, key and degree.

The model is shown in Figure 1-1. One ART2 (Carpenter & Grossberg 1987) net (A), classifies pitch use into keys. A second ART2 net (B) classifies patterns of intervals used within keys into seven nodes equivalent to the degrees of the scale. These degree identities are used to learn, in a BP net (C), associative mappings between pitch, key and degree; and also between the pitch-by-pitch

---

[1]The research presented in this paper is part of the author's doctoral research, supervised by Noel Sharkey and Henry Shaffer, supported by an SERC studentship.

and end-of-song memories and degree. The functions within the model are bipartite. Firstly, bottom-up processes using ART2 classify patterns of pitch and interval use. Secondly, the identities emerging from this process are used to learn associations of pitch, key and degree. These associations allow the encoding, transposition and recovery of tunes in any key.

2   Inducing Key from Pitch Frequency

In the first network the pattern of pitch[2] use in a sequence is extracted by a tracking memory implemented as a process of slow learning. The memory vector is equivalent to the input vector. Where x_i is the input value, w_i is the memory value, initialised to zero, and the rate of the memory is η, learning is as follows:

$$\text{if } x_i(t) > ZERO : w_i(t+1) \quad = \quad w_i(t) + \eta(x_i(t))(1 - w_i(t)) \qquad (1)$$
$$\text{else } w_i(t+1) \quad = \quad w_i(t)$$

The memory elements move towards a value of 1 at a rate determined by the η value. When the patterns that develop in the memory by the end of a song are classified, the network develops twelve nodes, each equivalent to a key. Table 1 shows that 91.5-98.3% of mappings are to a *home* key, a few are mapped to adjacent keys. The accuracy of the network when tested with the pitch-by-pitch memories drops by 1-13%, see Table 1. The pattern of correct attributions is correlated to the typicality of the pitch use associated with a song. The classifications in this part of the model are at a general level, and, as a result, the pattern within a memory emerges quickly and is relatively stable.

		TRAINING		TESTING			
		Training Set		Training Set		Test Set	
ρ	η	Home	Other	Home	Other	Home	Other
0.925	0.005	98.3	1.7	89.5	7.5	87.8	12.2
0.931	0.01	98.3	1.7	90.3	6.8	88.5	11.6
0.950	0.05	96.9	3.1	91.6	5.5	90.6	9.4
0.940	0.10	96.0	4.0	91.7	5.6	91.5	8.5
0.965	0.50	92.4	7.6	84.0	13.8	89.7	10.3

Table 1: The mappings created by an ART2 network classifying pitch frequency. ρ is the vigilance of the net, η is the memory rate.

3   Inducing Degree from Intervallic Patterns

The second stage of the model encodes degree identities[3]. It extracts the pattern of intervals associated with pitch classes. This allows the direct comparison

---

[2]Each pitch in a sequence is represented as a simple identity vector. The identity is indicated by one element being 1, while the rest are 0.

[3]The abstraction of pitch is widely accepted to be fundamental in the memorisation of melodies.

of patterns of pitch in different scales, and the emergence of functional identities.

The memories comprise an initial trace memory (Grossberg 1978), and a subsequent tracking memory. The initial memory traces the occurrence of intervals over all pitch classes. The tracking memories - twelve of the same form as in equation 1 - track the contents of the first memory for each pitch. Four trace memory types: *AST*, *SAT*, *STR* and *ATR*, use different permutations of adding and shunting elements. The resources for each are identical, and are initialised for each song. The trace memory is $\mathbf{V}$, $\mathbf{x}_j$ is the input value and $\mathbf{v}_j$ is the trace memory value, ζ is the decay rate for the trace memory $\mathbf{V}$.

$$AST : \text{if } \mathbf{x}_j(t) > 0 : \mathbf{v}_j(t) = \zeta(\mathbf{v}_j(t-1) + \mathbf{x}_j(t)) \tag{2}$$
$$\text{else } \mathbf{v}_j(t) = \zeta\mathbf{v}_j(t-1)$$
$$SAT : \text{if } \mathbf{x}_j(t) > 0 : \mathbf{v}_j(t) = (\mathbf{v}_j(t-1) + (\zeta\mathbf{x}_j(t))) \tag{3}$$
$$\text{else } \mathbf{v}_j(t) = \zeta\mathbf{v}_j(t-1)$$
$$STR : \text{if } \mathbf{x}_j(t) > 0 : \mathbf{v}_j(t) = \mathbf{x}_j(t) \tag{4}$$
$$\text{else } \mathbf{v}_j(t) = \zeta\mathbf{v}_j(t-1)$$
$$ATR : \text{if } \mathbf{x}_j(t) > 0 : \mathbf{v}_j(t) = \mathbf{v}_j(t-1) + \mathbf{x}_j(t) \tag{5}$$
$$\text{else } \mathbf{v}_j(t) = \zeta\mathbf{v}_j(t-1)$$

The tracking memories are focused on the use of a pitch in a key - 144 pitch-key pair identities. These identities are mapped in an ART2 net in which the vigilance is set so that only identical vectors are clustered together. The ART2 net is adapted so that each identity is coupled with a set of memory weights. The patterns developed in these interval-use-memories were classified into seven nodes, each node identified with a scale degree. The network was tested with the end-of-song memories. The number of correct attributions is low, 40-50%, and 2-8% were not mapped to any exemplar. The commonest mis-attribution, 16-23%, is to the degree a fifth above, and the next commonest, 9-13% is to the degree a fifth below. The variation in the data set reflects relationships that are tonally significant.

3.1 Associating Key, Pitch and Degree

Having identified degrees, the associations between pitch, key and degree, see Figure 1-2, are learned by a BP net. This facilitates the encoding, transposition and recovery of a tune in any key. The pitch-by-pitch and end-of-song memory patterns were also associated directly with degree identity. The networks were run for 1000 cycles - 90%+ of the patterns had been learned by this time, and subsequently very few more were learned[4]. The results of the simulations are shown in Tables 2 and 3. The networks learn consistently 85-98% of the end-of-song patterns, which is a 40%+ improvement over the attribution of the end-of-song patterns in the ART2 net.

---

[4] The criterion for learning was taken to be when the output required became the maximum value over all the outputs.

END-Of-SONG MEMORIES ⇒ DEGREE									
		AST		SAT		STR		ATR	
η	ζ	Train	Test	Train	Test	Train	Test	Train	Test
0.005	0.25	96.85	73.25	97.32	73.88	96.85	73.88	96.85	73.25
0.05	0.5	97.90	68.15	98.42	75.79	97.90	74.52	98.16	75.16
0.5	0.75	86.64	56.68	86.91	52.86	85.86	57.96	82.98	52.86

Table 2: Mappings learned between end-of-song memories and degree.

PITCH-By-PITCH MEMORIES ⇒ DEGREE									
		P-by-P		E-o-S		P-by-P		E-o-S	
		Train	Test	Train	Test	Train	Test	Train	Test
η	ζ	AST				SAT			
0.005	0.25	89.83	72.90	90.31	71.97	89.95	76.50	89.52	76.43
0.05	0.5	93.46	73.55	93.19	70.06	92.21	79.13	90.83	74.52
0.5	0.75	83.7	58.64	70.42	51.59	57.66	15.96	18.84	12.10
η	ζ	ATR				STR			
0.005	0.25	90.03	73.56	89.79	71.33	53.22	42.77	89.23	74.52
0.05	0.5	92.33	74.13	91.93	70.06	92.29	76.77	92.14	75.80
0.5	0.75	82.00	59.21	75.91	52.23	76.55	51.75	63.87	49.68

Table 3: Mappings learned between pitch-by-pitch patterns and degree, and generalisation over end-of-song memories.

4 Conclusion

The simulations described in this paper have explored the combination of Competitive Learning and Multilayered Perceptron paradigms in a model of key and degree induction in music. The integration was guided by the wish to use functionally appropriate modules. The model is an effective procedure that is both self-organising and incremental. The combination of models is important. In isolation neither Competitive Learning nor Multilayered Perceptrons are capable of completely realising the model's function.

References

Carpenter, G. & Grossberg, S. (1987), 'Art2: Self-organization of stable category recognition codes for analog input patterns', *Applied Optics* **26**(23), 4919–4930.

Grossberg, S. (1978), 'Behavioral contrast in short term memory: Serial binary memory models or parallel continuous memory models', *Journal of Mathematical Psychology* **17**, 199–219.

Rumelhart, D., Hinton, G. & Williams, R. (1986), Learning internal representations by error propagation., *in* D. Rumelhart & J. McClelland, eds, 'Parallel Distributed Processing: Explorations in the Microstructure of Cognition', Vol. 1:Foundations, MIT Press, Cambridge, MA.

Forecasting Using Constrained Neural Networks

Raqui Kane, Maurice Milgram

Laboratoire de Robotique de Paris - Université Pierre et Marie Curie

4, place Jussieu - 75252 Paris cedex 05 - France

1 Introduction

Forecasting techniques have traditionally been studied using probabilistic methods. However, some empirical anomalies remain unexplainable. More recently some investigations have been done in signal analysis [1], [2] and forecasting [3] using the dynamical and inductive property of the neural networks systems to try to solve the problems encountered with the traditional methods. This paper describes two complementary methods. The first one is a forecasting approach applied to bonds market using constrained neural networks (CN). In the CN are two kinds of units: logical-numerical (resp numerical) units which hold the logical (resp numerical) information of the network. The second main point of this paper is rules extraction from trained networks. This property is very intersting as we are able to understand what happens inside the network. Simulation results are reported.

2 The forecasting methodology

The problem here is to predict the seasonal tendency of bonds market from weekly financial data using the CN.

2.1 The logical-numerical and numerical units

The goal is to specify the task of the units during the training, namely to build two kinds of units: units holding some logical or numerical information. Consider a unit i with n inputs $u_1^i, \cdots, u_n^i$, its output o_i is:

$$o_i \; = \; f(\sum_{j=1}^{n}(w_{ij} \star u_j^i - \theta_i)) = f(a_i) \tag{1}$$

Where f is a sigmoid function ($f(u) = tanh(u)$), w_{ij} is the weight of the connection between units j and i, θ_i is the bias of unit i and a_i its activation. A condition for i to be a logical operator is: $\forall \ u_1^i, \cdots, u_n^i \in S = \{-1(false), 1(true)\}^n$, $o_i \in \{-1, 1\}$. A method to build logical units is developed in [4]. We will extend it to get logical-numerical operators. We define a logical-numerical operator i as a unit such that: $\forall \ u_1^i, \cdots, u_n^i \in [-1, 1]^n$, $o_i \in \{-1, 1\}$. To get a logical-numerical unit, we constrain its output to be in $\{-1, 1\}$ during the training by adding a penalty term $P(W)$ to the standard error function $E(W)$ of the back-propagation (BP). The function to minimize becomes $E'(W) = E(W) + P(W)$.

Note 1: Analyzing equation (1), we notice that $sign(o_i) = sign(f(a_i)) = sign(a_i)$, where $sign$ is the function sign. Then for a logical-numerical unit, we obtain the following property: $o_i = +1$ *if* $a_i \geq 0.0$ and $o_i = -1$ *if* $a_i < 0.0$.

2.2 The financial modelisation

The data used are some weekly data of the french bonds market from December 1988 to September 1992. We choose the following model:

$$u(i) \quad = \quad F(W_i, U(i-1)) \tag{2}$$

Where $u(i)$ is the financial data at week i, $U(i-1) = [u(i-1), u(i-2), \cdots, u(i-k)]$, k is the index for data length, W_i is the weight vector and $F(.,.)$ is a vector function representing the network.

2.3 The Architecture

We choose multi-layer feed-forward networks, some units are constrained to be logical-numerical operators. Figure 1 shows an example of a trained network. The architecture of the network is: 4 - 2 - 2 - 1 namely 4 layers with respectively 4, 2, 2 and 1 unit. The units of the second layer are constrained to be logical-numerical operators. Each unit i has an activation a_i and an output u_i.

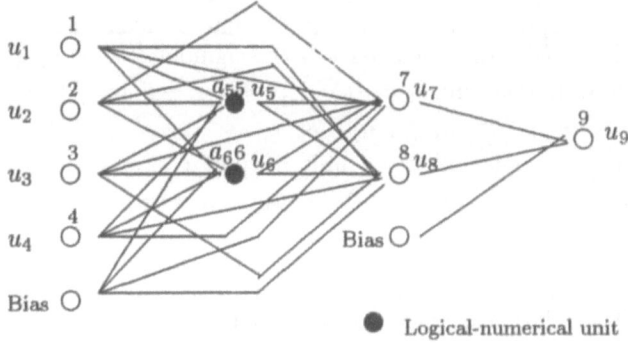

Figure 1: Architecture of a constrained neural network

2.4 Rule extraction

The main features of the rules extraction method follow:

• Find the property of each unit: it is mainly to seek logical-numerical units of the network. This step is very simple as we decide before the training which units should be logical-numerical operators and constrain them to get this property.

• Local extraction: it consists of extracting from each hidden or output unit the realized rule.

• Global extraction: we have to substitute rules extracted from units by their value inside the network to get the global rule extracted from the network.

Example: Let us extract rules from a 3 - 2 - 1 network trained for the bonds market forcasting problem. We use the CN and the financial modelisation described in section 2. The notations are defined in section 2.1 and 2.3.

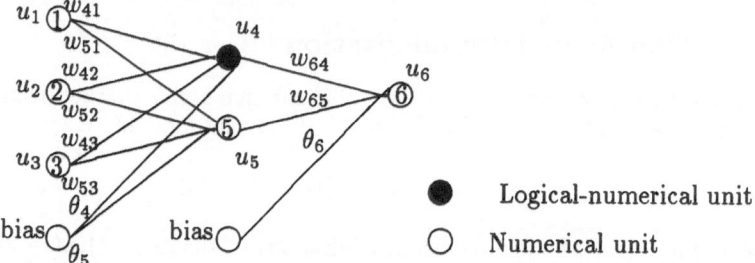

Figure 2: Extracting rules from a sample constrained neural network

● Local extraction

- Extracting dependencies from unit 4: unit 4 is constrained to be a logical-numerical unit. One consequence of constraining a unit to be a logical-numerical operator is to lead the weights connected to this unit to get great absolute values. Using equation (1), u_4 is given by: $u_4 = f(\sum_{j=1}^{3} w_{4j} \star u_j - \theta_4) = f(a_4)$. Using Note (1), we obtain the following relations:

$IF \sum_{j=1}^{3} w_{4j} \star u_j - \theta_4 \geq 0.0 \; THEN \; u_4 = +1.0$

$IF \sum_{j=1}^{3} w_{4j} \star u_j - \theta_4 < 0.0 \; THEN \; u_4 = -1.0.$

- Extracting dependencies from unit 5: unit 5 is an unconstrained and numerical unit, u_5 is in the linear part of the sigmoid. Using equation (1), we obtain the following relation: $u_5 = f(\sum_{j=1}^{3} w_{5j} \star u_j - \theta_5) = f(a_5)$. Using Note (1), u_5 becomes: $u_5 = \sum_{j=1}^{3} w_{5j} \star u_j - \theta_5$.

- Extracting dependencies from unit 6: unit 6 is an unconstrained and numerical unit u_6 is given by: $u_6 = \sum_{j=4}^{5} w_{6j} \star u_j - \theta_6$.

● Global extraction

- Extracting dependencies from the network: the rule extracted from the network is recursively obtained by substituing rules of hidden units in the outputs units. The relation extracted from unit 6 is given by: $u_6 = w_{64} \star u_4 + w_{65} \star (\sum_{j=1}^{3} w_{5j} \star u_j - \theta_5) - \theta_6$. Therefore, substituing u_4 and u_5 by their value, we obtain for the network the relations:

$IF \sum_{j=1}^{3} w_{4j} \star u_j - \theta_4 \geq 0.0 \, THEN \; u_6 = w_{64} + w_{65} \star (\sum_{j=1}^{3} w_{5j} \star u_j - \theta_5) - \theta_6$

$IF \sum_{j=1}^{3} w_{4j} \star u_j - \theta_4 < 0.0 \, THEN \; u_6 = -w_{64} + w_{65} \star (\sum_{j=1}^{3} w_{5j} \star u_j - \theta_5) - \theta_6$

3 Simulation

Several networks of the proposed architecture (see section 2.2) were tested. We use succesive sequences of $k + 1$ inputs data for each trained network. The k first data are the inputs data and the last data is the output of the network. We obtain performant experiments results.

Example: We present here some results for $k = 4$. The chosed architecture is 4 - 2 - 2 - 1, the units of the second layer are constrained to be logical-numerical operators (see section 2.3). Figure 3 shows some results about the forecasting procedure. Figure 3 is divided in two main parts. The first part represents the training area, namely the network is trained with the actual financial data of this part. In the second part, the network is not any more trained, just a test is done to measure the forecasting performance. We have compared the constrained BP (CBP) with the standard BP (SBP). Experiments show that the forecasting is better when we use the CBP. The convergence time is also shorter with the CBP. Simulation results show the accuracy of the rules extracted from trained networks.

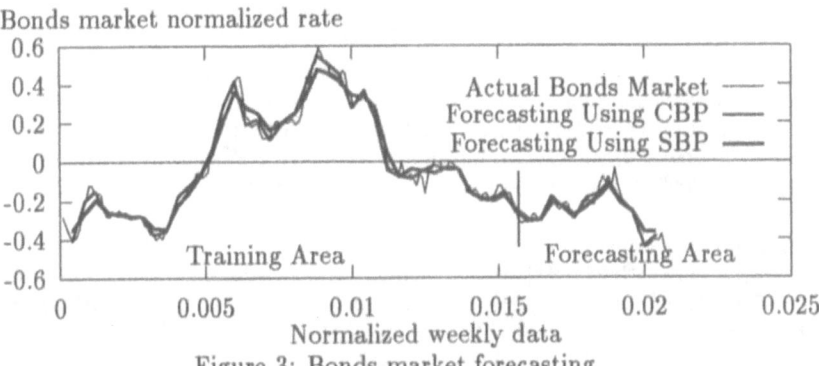

Figure 3: Bonds market forecasting

4 Conclusion

In this paper, we have proposed a new forecasting method using CN. Some units of the network are constrained to hold the logical information of the training data set; the unconstrained units hold the numerical information. This approach shows that to specify the task of the units in networks can help neural networks to be performant pedictive tools. The quality of learning is highly increased and convergence fastly reached. Another important point of this method resides on its capability to extract rules from trained networks. It overcomes the limitation of the conventional black box artificial neural network approach, in which networks don't explain their reasoning.

References

[1] J. Moody and C. J. Darken, "Fast learning in networks of locally tuned processing units," in *Neural computation, 1* 1989.

[2] A. Lapedes and R. Farber, "Nonlinear signal processing using neural networks: prediction and system modelling," Los Alamos Laboratory LA-UR-87-2662 1987.

[3] C. G. Winsor, B. A. Huberman and D. E. Rumelhart, "Predicting the futur: A connectionist approach," in *International Journal of Neural systems, Standford* CA 1990

[4] R. Kane and M. Milgram, "Extraction of semantic rules from trained neural networks," in *Proceedings of ICNN, San Francisco* CA 1993.

The EVALUATIONS of ENVIRONMENTAL IMPACT: COOPERATIVE SYSTEMS.

Alejandro Pazos, Antonino Santos del Riego and Julián Dorado.

Laboratory for Biomedical Applications of Artificial Intelligence.
Department of Computer Science. Faculty of Informatic. University of La Coruña.
15071 La Coruña. SPAIN.

1. INTRODUCTION

In trying to discern between the uncertain and the certain in our environment we must emphasize that if something is known about nature it is that it acts according to a pattern of interdependence between the different components that make up an ecosystem. Influencing in these types of standards is always a complex task. The new technologies of Artificial Intelligence (AI) provide the tools that have been used to develop the present system.

In regard to the subject under discussion a hybrid system has been implemented composed of the Expert System (ES) EEIE [1] which contains three knowledge bases working in parallel with an Artificial Neural Network (ANN) [2] and feeding on a Relational Data Base (RDB) where part of the knowledge that they must handle is represented.

Initially it was necessary to have access to a great quantity of information involved in the realization of the Evaluation of Environmental Impacts (EEI). In the aforementioned RDB we include all the information compiled about actions that produce impacts, Environmental Factors (EF), corrective actions, legislation, etc; as well as the relationships between these. The RDB is improved through the conclusions obtained in the EEI's carried out by the system.

The EEIE Expert System is charged with characterizing and evaluating the hypotheses associated with the impacts suggested by the selfsame ES starting out from the RDB, those suggested from the previously established impacts and those suggested by the ANN.

As has already been pointed out, the ANN takes charge of identifying the impacts produced on the different EF's by each one of the actions to be evaluated. This ANN has a double function, on the one hand, it provides new hypotheses to be considered by the ES, and on the other, increases the confidence level of these hypotheses that have also been suggested by the ES. Therefore this system allows the automatization of part of the most tedious labour involved in the EEI's.

2. THE EEIE EXPERT SYSTEM

This ES presents the following four abstraction levels:
  1.- Projects Level.
  2.- Projects Actions Level.
  3.- Impact Levels (Links between actions and EF's).
  4.- Corrective Actions Level.

The input into the system is composed of the group of projects object of the EEI, these projects suggest a posible group of impact causing actions and these actions can in turn suggest other actions. Starting from the final group of actions, links are established between these actions and the EF's affected (cause-effect). These links evaluate the impacts on the EF's starting from real measures of the environment, to finally suggest corrective actions that can modify the previous reasoning (see Figure 1).

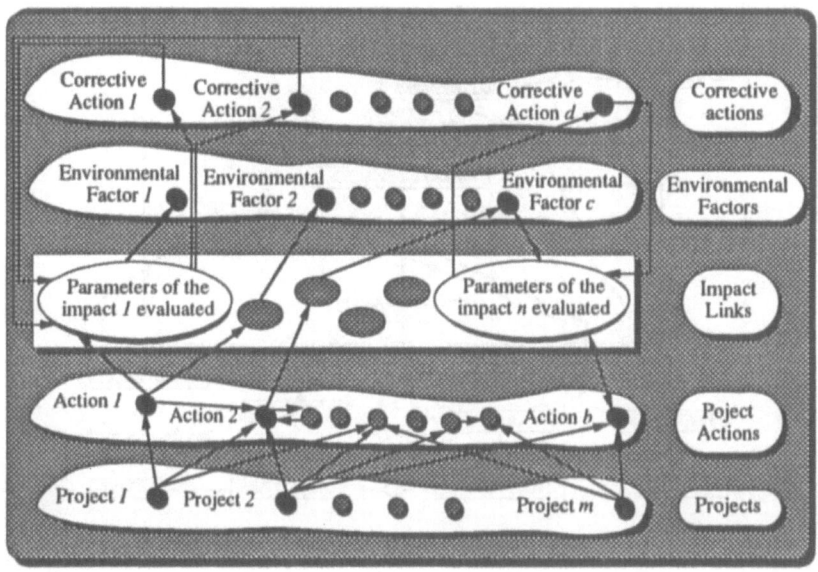

Fig1.- EEIE four abstraction levels.

3. THE ARTIFICIAL NEURAL NETWORK (ANN)

The object of the ANN lies in the identification of impacts. The aforementioned ANN takes as an input the separation of the projects that must carry out EEI according to EEC legislation, the action to be evaluated in each moment and the general characteristics of the environment in which the project is to be situated. With the aim of improving its convergence we have chosen the option of carrying out a codification of these actions, in such a manner that, actions with similar impacts are codified in a similar manner (with a distance of Hamming 1). With this in mind, a broad study of hierarchical clustering has been carried out on the initial training file.

The output of the ANN provides us with a vector with the 96 EF's that can be affected by each input action. The pairs formed by the input action and each one of the positively identified EF's establishes cause-effect type hypotheses, which in turn are used by the ES to complete the final group of hypotheses to be evaluated.

The aforementioned ANN is of the feed-forward type with a backpropagation learning algorithm. The ANN has 3 layers, with 63 Process Elements (PE) in the input layer, 43 PE's in the hidden layer and 96 PE's in the output layer; there exists a complete interconnection between the three layers. For the learning process a Delta Generalized Rule has been used, taking a learning coefficient from the clusterization study [3,4] carried out on the initial training file, till a 0.09 convergence has been obtained. The transference function was the Sigmoidal one.

4. THE RELATIONAL DATA BASE (RDB)

The related information needed to carry out the EEI's is collected in a RDB whose entity-relation model is represented in figure 2 (the attributes of the entities have been omitted due to the lack of space). [5]

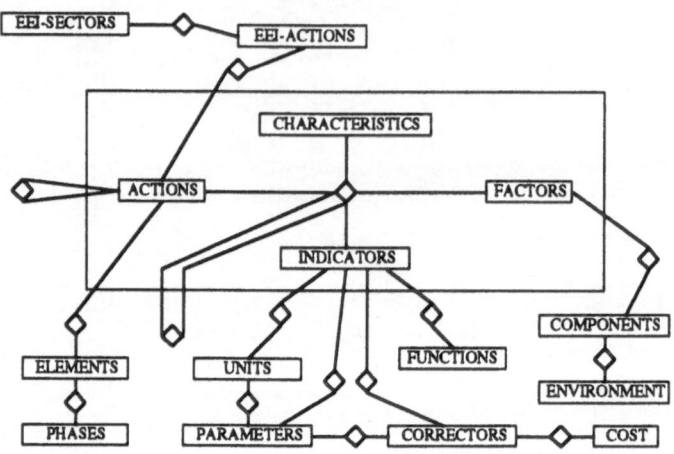

Fig. 2 RDB Entity-Relation Diagram

These entities store information referring to: projects to be evaluated, actions that cause impacts and their relationships, EF's, impact intersections (action-EF), indicators and functions to evaluate the aforementioned impacts, environmental parameters involved, corrective actions, etc.

Also, the results obtained by the system when it carries out new EEI's are stored in this RDB with the objective of using this information as an experience for future EEI's. During the majority of these information recovering operations from/or to the ES, the bridge chooses a simple object to store all the field registers, and the fields are read or written from the objects slot. In this manner the contents of a register are represented by an object and the fields of that register are represented through slots of that object; in this way the register-field relationship is transformed into an object-property relationship in the working memory of the ES. [6]

5. RESULTS

To validate the system some 13 EEI studies have been used, obtaining the percentage of impacts from the ES and the ANN over those identified in the EEI by the experts (Table 1).

	PERCENTAGE OF IMPACTS
only by the EEIE Expert System	9%
only by the ANN	13%
jointly by the EEIE & the ANN	68%
TOTAL	90%

Table 1.- The percentages of impacts obtained from the ES and the ANN.

Therefore, the ES identified some 77% of the impacts and the ANN some 81%, being the result of the collaboration between the ES and the ANN an identification of 90% of the impacts and an improval of confidence in the same of some 68%.

6. DISCUSSION AND FUTURE WORKS

The present system carries out the identification and evaluation of environmental impacts that allow decision making in the EEI (plan the alternatives to the project under evaluation, establish corrective actions, as well as determining the location of the project).

In considering the use a hybrid system linking a connectionist approach with a symbolic one, it allows us to confront the traditionnal limitation of the ES when faced by very wide domains and whose limits are not totally defined. The inclusion of an ANN allows the identification of standards that allow the focalization of the objectives with which the ES is to work with.

The future development of this work has as its aim, on the one hand, completing the knowledge of our system increasing the number of posible projects to be evaluated and, on the other, include new modules in this system which will allow us to treat the problem associated with the dispersion of contaminating agents (in both air and water), passing from the planification themes to the themes of simulation and control in real time.

ACKNOWLEDGEMENTS

Thank you to Mr. Juán Insua Naya for translate and Mss. Bertha Guijarro for her smile.

REFERENCES

1.- A. Pazos, A. Santos, A. Rivas, V. Maojo& J. Segovia. (1993). EEIE: An Expert System for Environmental Impact Evaluation. Proc. IEEE Eng. Med. Biolo. Soc, 2, 632-633.
2.- A. Pazos, A. Santos, J. Dorado. (1994). Linking of an Artificial Neural Network with the EEIE Expert System to identify environmental impacts. Proc. World Congress on ANN (in press).
5.- A. Pazos, Justo Alvarez. (1993). Modelización de una Base de Datos utilizando Oracle para almacenamiento de información referente a EIA. Proyecto Fín de Carrera, Dep. de Computación, Fac. Informática, Univ. de La Coruña. (in Spanish).
4.- Domenico Ferrari, Giuseppe Lerazzi, Alessandro Zeigner. (1972). Measurements and Tuning of Computer Systems, Ed. Prentice-Hall.
3.- Harry A.C. Eaton & Tracy L. Olivier. (1992). Learning Cofficient Depende on Training Set Size. Neural Networks-Pergamon Press, 5-2, 283-288.
6.- Karman Parsaye, Mark Chignell, Lftrag Kroshafian, Harry Wong. (1989). Intelligent Databases. Wiley.

What Generalizations of the Self-Organizing Map Make Sense?

T. Kohonen

Helsinki University of Technology
Laboratory of Computer and Information Science
Rakentajanaukio 2 C, FIN-02150 Espoo, Finland

Abstract. The number of researchers working on the Self-Organizing Map (SOM) for the present is at least on the order of 1500, and many variants of the basic model have already been suggested. This presentation tends to clarify the essence of the SOM, to set up its theory in the most fundamental form, and to point out what the computing functions are or should be (e.g., various structures of the network, variants of the cell function, acceleration of learning etc.). After that it will be easier to see along what lines the modifications should be developed. In particular, generalization of the static SOM into a self-organizing array of dynamic operators for sequential data is discussed.

1 Introduction

The Self-Organizing Map (SOM) (Kohonen, 1982, 1989, 1990, 1993a) is an unsupervised classification or clustering algorithm that maps high-dimensional vectorial input data onto a much lower-dimensional (usually 2-D) array of (neural) cells in an orderly fashion. The SOM is one of the few existing neural-network algorithms that takes into account the spatial order of the processing functions in the network. The brain is known to have this property.

With each cell (node, unit) of the array, a codebook (parameter, reference, weight) vector is associated. The codebook vectors of the array depend on the values of their neighbors, and in an adaptation process where they are made to approximate to the distribution of input vectors, they behave like in a nonparametric *regression*. Thus they may be thought to form the nodes of an "elastic net", and the degree of "stiffness" of this "net" can be controlled by the degree of interdependence of the codebook vectors.. Referring to this illustrative view one may understand that the smoothness of the regression, and the distances of the input samples from the "regression surface" (often wrongly interpreted as errors) are contradictory properties for which no unique optimum can be found. At least there is no sense in stipulating that the "net" should pass through all samples or even all clusters of samples. It is not so important for the SOM to approximate the detailed form of the input density function, but to find its main dimensions!

Since the SOM belongs to the category of the so-called vector quantization (VQ) methods (Makhoul et al., 1985), the starting point in this regression anyway must be the *quantization error in the vector space*. Assume that $x \in \Re^n$ is the input vector and the $m_i \in \Re^n, i \in \{$Lattice of nodes$\}$ are the codebook vectors; let $d(x, m_i)$ define a generalized distance function of x and m_i. The quantization error is then defined as

$$d(x, m_c) = \min_i \{d(x, m_i)\} \, , \tag{1}$$

where c is the index of the "closest" codebook vector to x in the space of input signals.

An even more central function in the SOM, however, is the *neighborhood function* $h_{ci} = h_{ci}(t)$, which describes the interaction of codebook vectors m_i and m_c during adaptation (regression) and is often a function of time t. To this end it will be useful to define the entity called *distortion measure*. Denote the set of indices of the lattice units by L; the distortion measure e is then

$$e = \sum_{i \in L} h_{ci} d(x, m_i) \, , \tag{2}$$

that is, a sum of distance functions weighted by h_{ci}, whereby c is the index of the closest codebook vector to x. If we now form the average expected distortion measure

$$E = \int e p(x) dx = \int \sum_{i \in L} h_{ci} d(x, m_i) p(x) dx \, , \tag{3}$$

one tentative way of defining the SOM is to solve for the set of m_i that globally minimizes E.

Exact optimization of (3), however, is yet an unsolved theoretical problem, and extremely heavy numerically. The best approximative solution that has been obtained so far is based on the so-called *Robbins-Monro stochastic approximation* (Robbins & Monro, 1951): if $\{x(t), t = 1, 2, \ldots\}$ is a sequence of input samples and $\{m_i(t), t = 1, 2, \ldots\}$ the recursively defined sequence of codebook vector m_i, then

$$e(t) = \sum_{i \in L} h_{ci}(t) d[x(t), m_i(t)] \tag{4}$$

is a stochastic variable, and the sequence defined by

$$m_i(t+1) = m_i(t) - \lambda \cdot \nabla_{m_i(t)} e(t) \tag{5}$$

is used to find an approximation to the optimum, as asymptotic values of the m_i. This would then define the SOM algorithm for a generalized distance function $d(x, m_i)$. It must be emphasized, however, that although the convergence properties of stochastic approximation have been thoroughly known since 1951, the asymptotic values of the m_i obtained from (5) only approximately minimize E in (3). *Then, on the other hand, (5) may be taken as another definition of a class of SOM algorithms.*

2 Possible generalizations of the SOM

The above basic SOM defines a regression solution to a class of vector quantization problems and in that capacity does not need any generalization. However, there exist other related problems where the SOM philosophy can be applied in various modified ways. Below are examples of such lines of thought.

2.1 Nonidentical inputs

The input vector x may be composed of subsets of signals of very different nature. They may then connect to different areas of the SOM, whereby the dimensionalities of x and the m_i may also be different. Particularly interesting abstractions of data by the SOM may be obtained in cells that lie at intersections of input connection sets.

2.2 The Hypermap architecture

The "winner" m_c may be defined in a sequential search, as described in Kohonen, 1991. This principle in connection with the SOM should also be called "Hypermap": the central idea is that by means of one subset of input signals only a candidate set of best-matching nodes is defined, after which the "winner" is selected from this subset by means of other inputs. This architecture speeds up searching in very large maps and may have stabilizing effects, especially if different inputs have very different dynamic ranges and time constants.

2.3 Dynamically defined neighborhoods and growing maps

Several authors (cf., e.g. Fritzke, 1992; Szepesvári et al., in press) have suggested that the structure of the network, or definition of h_{ci} should be made dependent on intermediate results (e.g., quantization error during the process). This author too experimented with such ideas around 1981, but abandoned them, because only a better approximation of $p(x)$ is thereby obtainable; and the best approximation, as stated earlier, was never the primary goal. One should also notice that in almost all practical problems the dimensionality n of x and the m_i is very high, and, for instance, to describe a "hyperrectangle" one needs 2^n vertices. It is simply impossible to define any more complicated structures in high-dimensional spaces by a small number of codebook vectors!

If, on the other hand, the map is understood as a regression, it would be a better idea, especially in very-high-dimensional spaces, to let the original map find the optimal orientation along $p(x)$ and thus the main extensions of $p(x)$.

If the speed of convergence is the primary reason for suggesting, e.g., maps to which new nodes are added upon demand (growing maps), it may be emphasized that the same result can be obtained by other, more straightforward means (cf. Sec. 2.4).

On the other hand, in *nonordered vector quantization* (Kangas et al., 1990) one can achieve significantly accelerated learning by defining the neighborhoods in the signal space, not over the array.

2.4 Acceleration of learning in the SOM

By far the most effective means to guarantee fast convergence is to define the initial codebook vector values properly. They may be selected, for instance, as a two-dimensional regular (rectangular) array along the two-dimensional regression hyperplane that is fitted to $p(x)$, and the dimensions of the array may then be made to correspond to the two largest radii of inertia of $p(x)$.

If learning is started with values of λ that are of the order of, say, .25, the codebook vectors quickly start to approximate to $p(x)$.

One problem posed long ago concerns the optimal sequence of the learning rate parameter $\lambda = \lambda(t)$ in (5). In the original Robbins-Monro stochastic approximation the two necessary and sufficient conditions are

$$\sum_{t=1}^{\infty} \lambda^2(t) < \infty, \quad \sum_{t=1}^{\infty} \lambda(t) = \infty . \tag{6}$$

The former of these conditions is easily found necessary for convergence, while the latter guarantees that the convergence limit is unique in the neighborhood of a local optimum. Obviously $\lambda(t) = \text{const.}/t$ satisfies (6).

Next we introduce a fast computing scheme in which the problem about the learning rate is eliminated completely.

If we take $d(x, m_i) = ||x - m_i||^2$ and $2\lambda = \alpha(t)$, we obtain the original SOM algorithm:

$$m_i(t + 1) = m_i(t) + \alpha(t) \cdot h_{ci}(t) \cdot [x(t) - m_i(t)] . \tag{7}$$

In the convergence limit every $m_i = m_i^*$ must satisfy the equilibrium condition

$$E\{h_{ci}(x - m_i^*)\} = 0 , \tag{8}$$

whereby $E(\cdot)$ means the expectation value, and in the averaging over the x space, the subscript c of the "winner" is a function $c = c(x; m_1^*, m_2^*, \ldots)$.

It has been shown that the following simple definition of h_{ci} is effective enough in practice and saves much computing time: $h_{ci} = 1$ if i belongs to some *topological neighborhood set* $N_c = N_c(t)$ of cell c in the cell array, whereas otherwise $h_{ci} = 0$. With this h_{ci} we can write (8) as

$$m_i^* = \frac{\int_{V_i} x p(x) dx}{\int_{V_i} p(x) dx} , \tag{9}$$

where V_i means the following domain of values of x: Let some cell c be selected by values of x that belong to the domain (Voronoi set) V_c around c, and let N_c be the topological neighborhood set of c (as referred to cell indices). If cell i is a common member of several neighborhood sets N_c, the union of the corresponding V_c is then called V_i.

It has to be noted that m_i^* has thereby not been solved explicitly; the set V_i on the right still depends on all the m_i^*. Nonetheless (9) is already in the form in which the so-called *iterative contraction mapping* is applicable. If z is an unknown vector that has to satisfy the (generally nonlinear) equation $f(z) = 0$, then, since it is always possible to write the equation as $z = g(z)$, the successive approximations of the root may be computed as a series $\{z_n\}$ where

$$z_{n+1} = g(z_n) . \tag{10}$$

The iterative process in which a number of samples of x is first classified into the respective V_i regions, and the updating of the m_i^* is made iteratively as defined by (9), can be expressed as the following steps. This algorithm, dubbed *"Batch Map"* (Kohonen, 1992, 1993a), resembles the familiar K-means algorithm (Makhoul et al., 1985), where all the training samples are assumed to be available when learning begins. The learning steps are defined as follows:

1. Define the initial codebook vectors, for instance, by the method suggested in the beginning of this section.

2. For each map unit i, collect a list of copies of all those training samples x, whose nearest reference vector belongs to the topological neighborhood set N_i of unit i.

3. Take for each new reference vector the mean over the respective list.

4. Repeat from 2 a few times.

Definition of the size of the neighborhood set N_i can be similar as in the basic SOM algorithms. "Shrinking" of N_i in this algorithm means that the neighborhood size is decreased while the steps 2 and 3 are repeated . At the last iterations, N_i may contain the element i only, and the last steps of the algorithm are then equivalent with the K-means clustering.

2.5   Operator maps

There are no restrictions, e.g., to definition of the matching of x and m_i in terms of dynamic operators (Kohonen, 1993b): for instance, a finite sequence $X_t = \{x(t-n+1), x(t-n+2), \ldots, x(t)\}$ of input samples might be regarded as one input entity, whereby each cell would correspond to an operator having a set of adaptive parameters. Particularly interesting cases are obtained if the cell function is an estimator G_i that defines the *prediction* $\hat{x}_i(t) = G_i(X_{t-1})$ of the signal $x(t)$ by unit i at time t on the basis of X_{t-1} (cf. Lampinen & Oja, 1989). Each cell shall make its own prediction; let the prediction error $x(t) - \hat{x}_i(t)$ define the degree of matching at unit i:

$$||x(t) - \hat{x}_c(t)|| = \min_i\{||x(t) - \hat{x}_i(t)||\} . \qquad (11)$$

The sample function of the average expected squared prediction error that is locally weighted with respect to the winner may be defined as

$$e(t) = \sum_i h_{ci}(t)||x(t) - \hat{x}_i(t)||^2 , \qquad (12)$$

where the unknown parameter vectors w_i of the estimators occur in both the estimates $\hat{x}_i$ and the subscript c. According to the principle of stochastic approximation, the recursive formula for the parameter vector w_i reads

$$
\begin{aligned}
w_i(t+1) &= w_i(t) - \lambda(t)\nabla_{w_i(t)}e(t) \\
&= w_i(t) + \alpha(t)h_{ci}(t)[x(t) - G_i(X_{t-1})] \cdot \nabla_{w_i(t)}G_i(X_{t-1}) \quad (13)
\end{aligned}
$$

whereby we assume that c is not changed in a gradient step (because the correction of the "winner" is towards input) and $\alpha(t) = 2\lambda(t)$ is a scalar that defines the size of the step.

2.6 Systems of SOMs

A far-reaching goal in self-organization is to create autonomous systems, the parts of which control each other and learn from each other. Such control structures may be implemented by special SOMs: the main problem thereby is the interface, especially automatic scaling of interconnecting signals between the modules for best mapping, and picking up relevant signals for interconnects.

Summary

This article tried to survey and criticize some trends in the past development of the SOM architectures, and to open new avenues, especially for the analysis of dynamic signals.

References

Fritzke, B. (1992). Wachsende Zellstrukturen – ein selbstorganisierendes neuronales Netzwerkmodell. PhD Thesis, Technische Fakultät, Universität Erlangen-Nürnberg, Erlangen, Germany.

Kangas, J., Kohonen, T., and Laaksonen, J. (1990). Variants of self-organizing maps. *IEEE Trans. on Neural Networks*, 1, 93-99.

Kohonen, T. (1982). Self-organized formation of topologically correct feature maps. *Biological Cybernetics*, 43, 59-69.

Kohonen, T. (1989). *Self-Organization and Associative Memory*, 3rd ed. Heidelberg: Springer.

Kohonen, T. (1990). The self-organizing map. *Proceedings of the IEEE*, 78, 1464-1480.

Kohonen, T. (1991). The hypermap architecture. In T. Kohonen et al. (Eds.), *Artificial Neural Networks*, Proceedings of 1991 International Conference on Artificial Neural Networks, Espoo, Finland, vol. 2, pp. 1357-1360.

Kohonen, T. (1992). New developments of learning vector quantization and the self-organizing map. In *SYNAPSE'92*, Symposium on Neural Networks; Alliances and Perspective in Senri 1992, Osaka, Japan

Kohonen, T. (1993a). Things you haven't heard about the self-organizing map. *Proceedings of 1993 IEEE International Conference on Neural Networks*, San Francisco, California, pp. 1147-1156.

Kohonen, T. (1993b). Generalizations of the self-organizing map. *Proceedings of 1993 International Joint Conference on Neural Networks*, Nagoya, Japan, pp. I-457–462.

Lampinen, J. and Oja, E. (1989). Self-organizing maps for spatial and temporal AR models. *Proceedings of the 6th Scandinavian Conference on Image Analysis*, Oulu, Finland, pp. 120-127.

Makhoul, J., Roucos, S., and Gish, H. (1985). Vector quantization in speech coding. *Proceedings of the IEEE*, 73, 1551-1588.

Robbins, H. and Monro, S. (1951). A stochastic approximation method. *Ann. Math. Statist.*, 22, 400-407.

Szebesvári, C., Balázs, L., and Lörincz, A. Topology learning solved by extended objects: a neural network model. *Neural Computation* (in press).

A Novel Approach to Measure the Topology Preservation of Feature Maps

Th. Villmann[†]   R. Der[†]   Th. Martinetz[‡]

[†]University Leipzig, Inst. of Informatics,
D–04109 Leipzig, Augustusplatz 10/11, Germany
villmann,der@informatik.uni-leipzig.d400.de

[‡]Siemens AG, Corporate Research and Development
81730 München, Germany

1   Introduction

Kohonen's self-organizing feature map (SOFM) (Kohonen 1984) creates a topology preserving map from a data manifold $M \subseteq V$ onto a lattice A of neural units i. The topology preserving property can be employed in a variety of information processing tasks, ranging from classification over robotics to data reduction and knowledge processing. To each neural unit i of A a reference or synaptic weight vector w_i is assigned, defining the receptive field or Voronoi polyhedron V_i of each unit i by the set of all data points $v \in M$ which are matched best by this reference vector. This mapping from the data manifold M onto the lattice A is called topology preserving, if neighbouring units i have receptive fields V_i which are adjacent on M. Under certain conditions, i.e., if a topological mismatch between M and A exists, the lattice folds itself into V and the topology preservation may be lost (Ritter et al. 1992).

Various qualitative and quantitative methods for characterizing the degree of topology preservation (Bauer and Pawelzik 1992), (Zrehen 1993), (Der et al. 1993) have been proposed. All these approaches, however, can provide correct results only for linear submanifolds $M \subseteq V$. If the manifold is nonlinear, like it is the case in many practical applications of SOFMs, all these approaches can not distinguish a correct folding due to the folded data manifold from a folding due to a topological mismatch between M and A. Particularly when using the SOFM for non-linear principle component analysis one has to have a means to distinguish between these two cases. In this paper we introduce a method for quantifying topology preservation which can be applied to linear *and* non-linear data manifolds M. Further, this method allows to *quantify* the range of folds. Our approach employes what we call the topographic function, which is defined based on the so-called masked Voronoi polyhedra $\tilde{V}_i = V_i \cap M$ which were introduced in (Martinetz 1993) for defining neighbourhood and topology preservation of feature maps.

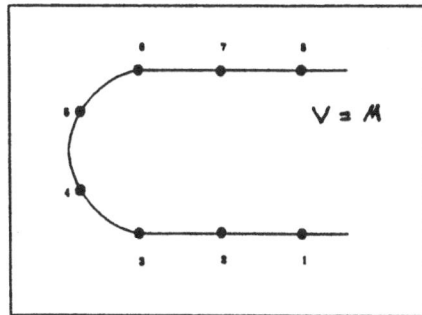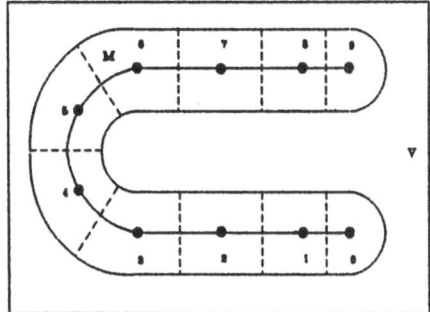

Figure 1: Example of a linear (left, $M = V$) and nonlinear (right, $M \subset V$) data manifolds with the hypothetical positions of the images of the neural units

2 The Topographic Product of a SOFM

The SOFM algorithm defines a map $F : V \supseteq M \longmapsto A$, where the dimension of V is n_V and A is a n_A-dimensional lattice of neural units. With each time step a stimulus vector $v \in M$ is presented. The winner (best matching) unit i^* is defined by

$$\|w_{i^*} - v\|_V \leq \|w_i - v\|_V \quad \text{for all} \quad i \in A, \tag{1}$$

with $\|\cdot\|_V$ denoting the Euclidean distance in V. The reference vectors w_i are adapted in a learning step according to

$$\triangle w_i = \epsilon h_{i^*,i} (v - w_i) \quad \text{for all} \quad i \in A, \tag{2}$$

with the neighbourhood function

$$h_{i^*,i} = \exp\left(-\frac{\|i^* - i\|_A}{2\sigma^2}\right) \tag{3}$$

determining the neighbourhood range in A. $\|\cdot\|_A$ denotes the Euclidean distance in A. ϵ and σ are learning parameters. An interesting quantity for measuring the topology preservation, the topographic product P, has been indroduced by Bauer and Pawelzik (Bauer and Pawelzik 1992). It measures the preservation of the neighbourhood between the neural units i in A and their reference vectors w_i lying on M. However, the topographic product does not consider the neighbourhood relations of the reference vectors lying in M, but only the neighbourhood relations of the reference vectors within the embedding space V. Therefore, an approach based on the topographic product is not able to differentiate between correct foldings arising from a nonlinear data manifold M and incorrect foldings which may result from a dimensional conflict between M and A or an incorrect formation of the map (topological defects, twists, kinks). An example is shown in Fig.1. In both the linear and nonlinear case of M the topographic product has the same value indicating a loss of topology preservation. However in the nonlinear case the map has been formed correctly.

3   The Topographic Function Φ_A^M

In this chapter we introduce the topographic function Φ_A^M for measuring the topology preservation of a SOFM, which considers explicitly the structure of the data manifold M. Following (Martinetz 1993), we define the receptive field of a neural unit i by

$$R_i = V_i \cap M, \tag{4}$$

which corresponds to the masked Voronoi polyhedron $\tilde{V}_i$ in (Martinetz 1993). The basic idea of our approach is that we do not use the reference vectors w_i of the neural units i but their receptive fields R_i to measure neighbourhood relations. In a perfectly ordered SOFM only nearest lattice neighbours i' of a unit i have receptive fields $R_{i'}$ which are adjacent to R_i. If there are other units which have adjacent receptive fields, perfect topology preservation is lost. Let A be a $N_1 \times N_2 \times \ldots \times N_{n_A}$ neuron lattice of dimension n_A. Then neural unit i is indicated by $i = (i_1, \ldots, i_{n_A})$. For each unit i we define

$$f_i(k) = \# \{j \mid \|i - j\|_{\max} > k \; ; \; R_i \cap R_j \neq \emptyset\} \tag{5}$$

with $k = 1, \ldots, N_{\max}$, $N_{\max} = \max_{i=1}^{n_A} |N_i|$. $\# \{\cdot\}$ denotes the cardinality of a set and $\|\cdot\|_{\max}$ denotes the maximum norm. Looking at a neural unit i, $f_i(k)$ determines the number of units j which have receptive fields R_j adjacent to R_i and, at the same time, have a lattice distance to i larger than k. The topographic function is then defined by

$$\Phi_A^M(k) = \sum_{j \in A} f_j(k). \tag{6}$$

Φ_A^M is a monotonically decreasing function, and we obtain $\Phi_A^M \equiv 0$ if and only if the SOFM is perfectly topology preserving. The largest k for which $\Phi_A^M(k) \neq 0$ holds yields the range of the largest fold. As depicted in Fig.1 in the linear case we get $\Phi_A^M(k) \neq 0$ for all k-values, which indicates a mismatch over the range of the whole net. In the nonlinear case we obtain the correct result $\Phi_A^M(k) \equiv 0$.

Choosing a normalized k, i.e., $k^* = k/N_{\max}$, and choosing a normalized Φ_A^M, i.e., $\Phi^{*M}_A = \Phi_A^M / N(N - 3^{n_A})$ with $N = \prod_{k=1}^{n_A} N_k$, allows to compare maps of different size.

4   Computing the Topographic Function Φ_A^M

Computing Φ_A^M requires to determine whether two receptive fields R_i, R_j are adjacent on the given manifold M. A way to determine the adjacency of two receptive fields $R_i = V_i \cap M$, $R_j = V_j \cap M$ has been proposed in (Martinetz 1993). Let $\mathbf{C}$ be a connectivity matrix determining connections between units $i, j \in A$ (in addition to the connectivity matrix defined by the fixed lattice structure). Initially, the elements $\mathbf{C}_{ij}$ of $\mathbf{C}$ are set to zero. Simply by sequentially presenting input vectors $v \in M$ and each time connecting (setting $\mathbf{C}_{ij} = 1$) those two units i^*, j^*, the reference vectors w_{i^*} and w_{j^*} of which are closest and second closest to v, leads to a connectivity matrix $\mathbf{C}_{ij}$ for which

$$\lim_{t \to \infty} \mathbf{C}_{ij} = 1 \quad \Leftrightarrow \quad R_i \cap R_j \neq \emptyset \tag{7}$$

is valid. It can be shown (Martinetz 1993) that the resulting connectivity structure connects units and only units the receptive fields of which are adjacent. This allows to rewrite eq.(5) to

$$f_j(k) = \#\{i \mid \|i - j\|_{\max} > k \; ; \; \mathbf{C}_{ij} = 1\} \qquad k = 1, \ldots, N_{\max}. \qquad (8)$$

After a SOFM has been formed, we then can determine Φ_A^M by the following algorithm:

1. present an input vector $v \in M$ and determine the two nearest reference vectors w_{i^*}, w_{j^*}.

2. connect the units i^*, j^*, i.e., set $\mathbf{C}_{i^* j^*} := 1$ and go to step 1 .

After a sufficient number of input vectors v the algorithm yields a connectivity matrix $\mathbf{C}$ for which eq.(7) is valid. $\mathbf{C}$ can then be used to calculate the topographic function Φ_A^M according to eq.(8) and eq.(6).

5 Conclusion

We presented a novel approach to the problem of measuring the topology preservation of a SOFM. The approach is based on the neigbourhood relations between receptive fields. The introduced topographic function is an improvement over the topographic product suggested in (Bauer and Pawelzik 1992) since it determines the degree of topology preservation by considering explicitly the given input manifold M.

THE REPORTED RESULTS ARE BASED ON WORK DONE IN THE PROJECT 'LADY' SPONSORED BY THE GERMAN FEDERAL MINISTRY OF RESEARCH AND TECHNOLOGY (BMFT) UNDER GRANT 01 IN 106B/3.

References

[1] H.-U. Bauer, K. Pawelzik: IEEE Transactions on Neural Networks 3(4), 570-579, (1992);

[2] R. Der, M. Herrmann, Th. Villmann : Time Behavior of Topological Ordering in Self-Organized Feature Mapping, submitted to Biolog. Cyb., (1993);

[3] T. Kohonen: Self-Organization and Associative Memory, Springer Series in Information Science 8 (Springer, Berlin, Heidelberg 1984);

[4] Th. Martinetz: Competitive Hebbian Learning Rule Forms Perfectly Topology Preserving Maps, Proceedings of the International Conference on Artificial Neural Networks 1993, Eds. St. Gielen and B. Kappen, Springer-Verlag London Berlin Heidelberg, (1993);

[5] H. Ritter, T. Martinetz, K. Schulten: Neural Computation and Self-Organizing Maps. Addison Wesley: Reading, Mass., (1992);

[6] St. Zrehen: Analyzing Kohonen Maps With Geometry, Proceedings of the International Conference on Artificial Neural Networks 1993, Eds. St. Gielen and B. Kappen, Springer-Verlag London Berlin Heidelberg, (1993);

Self-Organized Learning of 3 Dimensions

Cs. Szepesvári†‡ and A. Lőrincz†

† Department of Photophysics, Institute of Isotopes

The Hungarian Academy of Sciences

Budapest, Hungary

‡ Department of Mathematics, Attila József University of Szeged,

Szeged, Hungary

1 Introduction

The system to be described is an application of the artificial neural network architecture we developed for 2 dimensional images [Szepesvári et al., 1993]. Here two of 2-dimensional projections of 3 dimensional objects guide the network to wire in the 3 dimensional geometry of the external world. The basis of the system is a self-organizing competitive artificial neural network [Grossberg, 1987] that receives, as inputs, images of the external world. The primary building block of the algorithm is a 'winner-take-all' network. In such a model every neuron receives every input through its connections or input filter system; this filter system may differ and change during learning. Elements of this filter system shall be called as feedforward connections, denoted by w_{ij}, where indices i and j correspond to the input and the neuron they connect, respectively. The other set of connections that may change during the course of learning are the connections between neurons. This set shall be called as lateral or feedback connections, denoted by q_{kl} ($k, l = 1, 2, \ldots, m$) and indexed by the two neurons they connect: the first index being the index of the neuron that is sending its output and the other index being the index of the neuron that is receiving the output as an input. The change that these systems go through was called learning and it is given by connection update rules. The update rules could be in the form of a numerical algorithm suitable for software implementation. In other cases the update rules are formulated in the form of differential equations that either model realistic neurons [Carpenter and Grossberg, 1987, Földiák, 1991, Grossberg, 1988] or could be models of analog hardware [Fomin and Lőrincz, 1993]. In the following the update rules shall be given in the form that is suitable for software implementation.

2 Forming spatial filters

The numerical procedure of the 'winner-take-all' network is implemented in the following way: first an input is presented to the network and then neurons process their inputs and develop activities in accordance with the equation

$$a_j = F(\mathbf{x}, \mathbf{w}_i)$$

where $\mathbf{x}$ denotes the n dimensional input vector, a_j the input activity of the jth neuron and F is a similarity function Usually $F(\mathbf{x}, \mathbf{w}) = \mathbf{x}\mathbf{w}$, where $\mathbf{x}\mathbf{w}$ denotes the dot inner products of vectors $\mathbf{x}$ and $\mathbf{w}$.

Competition starts. The winner of the competition is the neuron of largest activity. The stored vector of the winning neuron l is then modified with the help of the update rule:

$$\Delta \mathbf{w}_l = \alpha(\mathbf{x} - \mathbf{w}_l)$$

where α is the so called feedforward learning parameter; $0 < \alpha < 1$.

In earlier simulations, we presented 2 dimensional objects of a 2 dimensional space to the network [Szepesvári et al., 1993]. Input vectors were derived by computing the overlap of the local, extended, randomly positioned objects and the pixels of digitization. The inputted local, extended objects as well as the 'winner-take-all' mechanism result in local spatial filters. This is the consequence of the correlations of the objects are forwarding to the neural network about the external world. The competitive process that forces the neurons to find the distinct correlations.

3   Geometry by Hebbian learning

It is an easy task to learn the geometry of the external world with the help of our local filters since in many cases when we present an object to the network, the position of the object is such that more than one neuron shall have non-zero input activities. In this case neurons that assume large input activities have filters that are close to each other, since they are excited by the same object.

In the context of artificial neural networks learning takes place as the development of connections. For this reason we may try to develop connections q_{ij} between neurons i and j according to their input activities:

$$\Delta q_{ij} = \beta(a_i a_j - q_{ij}),$$

where β is the lateral connection learning parameter. The best training results were achieved when only the winning neuron could update its connections:

$$\Delta q_{ij} = \beta(y_i + y_j)(a_i a_j - q_{ij}),$$

where y_k is the output of the kth neuron after competition: the output is 1 for the winning neuron and 0 for the others. In this way $y_i + y_j$ is not zero if and only if either the ith or the jth neuron was winning.

There is one point worth mentioning here, and that is how one can ensure to have a wiring that corresponds to the external world. Consider, for example, a field with an elongated lake in the middle. Assume, that objects can move on the field and not on the lake. Two points on the opposite sides of the lake may be very far from each other if one is restricted to the field and cannot pass the the lake. In our view a representation of the geometry is appropriate if there will be no connection between the spatial filters containing those 'distant' points. We want our q_{ij} connections to represent the geometry in the above sense. This may be formulated as follows: We call two receptive fields of two neurons separated, or distant, if there is no input that could overlap with both. The neurons themselves shall also be called separated, or distant neurons. The condition of not developing connections between distant neurons is that neural activity should be zero if the input does not overlap with the neuron's receptive field and vice versa. The formulation of that condition and the proof of its necessity may be put on a firm mathematical basis [Szepesvári, 1993]. The problem if this condition is met is hidden in the special form of neural activity function F. It has been shown [Szepesvári, 1993], that the inner product function fulfills the said condition for the 2D problem. In our earlier paper we have shown for the 2 dimensional case that local filters close to each other develop connections whereas filters far from each other are not capable of developing connections [Szepesvári et al., 1993].

In another example we present two 2-dimensional projections of 3-dimensional objects to two 2-dimensional (left and right) 'retinas'. It may be shown, that in this special case the separability condition is fulfilled if the neurons sum up their respective inputs from the two retinas as those are transmitted through the neural filters and then multiply the two sums to determine the neural activities:

$$a_j = (\mathbf{w}_j^{(1)}\mathbf{x}^{(1)})(\mathbf{w}_j^{(2)}\mathbf{x}^{(2)})$$

where $\mathbf{w}_j^{(k)}$ $k = 1, 2$ denote the left ($k = 1$) or right ($k = 2$) part of the input filter system and $\mathbf{x}^{(k)}$ $k = 1, 2$ denote the appropriate input vector components. It may be shown that this form of neuron activity fulfills the separability condition, while the previously used inner product function does not [Szepesvári, 1993]. The neural activities are then the subject of competition. The input vector was composed of two

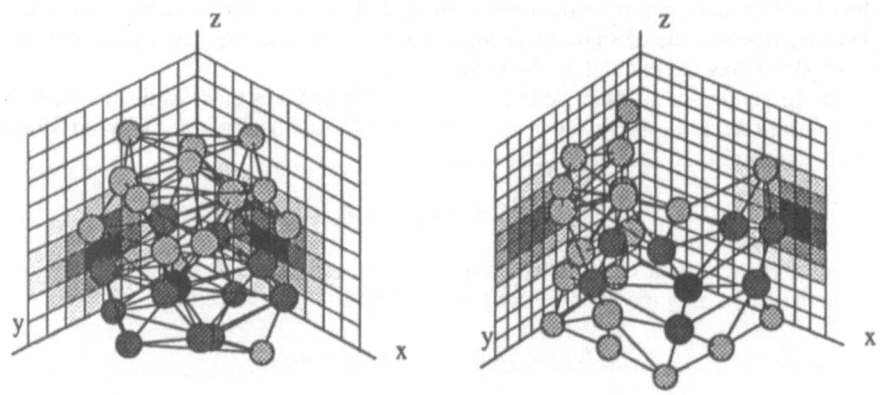

Figure 1: Learnt neigboring relations in 3-dimensions

Two of 2-dimensional projections of randomly positioned 3-dimensional object were inputted to the neural network. Neurons that could develop connections are linked by lines. One neuron (colored by black) with its receptive fields is shown on both sides of the figure. Neighboring neurons – neurons that could develop connections to the black neuron – are colored dark grey. Other neurons are white. Positions and sizes of circles correspond to the centers and sizes of the neurons' receptive fields, respectively.

144-dimensional vectors and we allowed 27 neurons to develop local filters and to learn the geometry of the external world. If the external world had been a 2-dimensional world of 288 pixels then we would now have 18 surviving neurons and 2-dimensional wiring. As the figure shows, all 27 neurons have survived and developed 3-dimensional wiring. It means that our network is capable of finding the geometry of the external world just by 'looking at it'.

To show that the geometry of the space is discovered by the network we presented objects in a U-shaped part of the 3-dimensional world that could be important for robotic applications where the shortest distance may not be feasible for the robot. The result is shown on the righ-hand side of Fig. 1. Connection strengths were thresholded to cut off the weak connections. As can be seen the connections correspond solely to the possible routes of the world.

If the neural activity is not chosen properly - i.e. the separability condition is not satisfied - then the topology of the external world will not be reflected by the connections between the neurons.

4 Conclusions

Competitive networks that are capable of creating spatial filters can develop connections that fit the geometry of the external world. These connections are then capable of producing cooperative neighbor training through neural methods [Szepesvári et al., 1993]. The whole network is self-organizing, and the different self-developing structures can develop simultaneously, in other words: the self-organizing processes work together. It is an easy task for the network to discover the 3-dimensional world from two 2-dimensional orthogonal projections of three dimensional objects.

Acknowledgements

This work was partially supported by the grant of the National Science Research Foundation, Grant No.: 1890/1991.

References

[Carpenter and Grossberg, 1987] Carpenter, G. and Grossberg, S. (1987). *Computer Vision, Graphics, and Image Processing*, 37:54–115.

[Földiák, 1991] Földiák, P. (1991). *Neural Computation*, 3(2):194–200.

[Fomin and Lörincz, 1993] Fomin, T. and Lörincz, A. (1993). *Neural Networks*. submitted.

[Grossberg, 1987] Grossberg, S. (1987). *Cognitive Science*, 11:23–63. and references therein.

[Grossberg, 1988] Grossberg, S. (1988). *Neural Networks*, 1:12–61.

[Szepesvári, 1993] Szepesvári, C. (1993). Master's thesis, Attila József University of Szeged. in Hungarian.

[Szepesvári et al., 1993] Szepesvári, C., Balázs, L., and Lörincz, A. (1993). *Neural Computation*. in press.

A Model of Fast and Reversible Representational Plasticity using Kohonen Mapping

Marianne Andres, Oliver Schlüter, Friederike Spengler, Hubert R. Dinse

Inst. f. Neuroinformatik, Ruhr-Universität Bochum, 44780 Bochum, Germany
e-mail: marianne@neuroinformatik.ruhr-uni-bochum.de

1 Introduction

The adult mammalian cortex maintains a substantial potential for post- ontogenetic plasticity even after the end of the critical developmental period. Long-term cortical reorganization in adult sensory systems are well documented after lesions [6]. Behavioral training [10], classical conditioning [14] and prolonged natural sensory stimulation [1] were also shown to remodel cortical maps and receptive fields, indicating their modificability by use thus extending the impacts of post-ontogenetic plasticity into the fields of higher cognitive functions. Models of representational plasticity generally make use of Hebbian synapses to account for changes of cortical topography [7, 8, 9, 4, 13]. Here, we use Kohonen's approach to model experimental data of fast and reversible post-ontogenetic plasticity observed in somatosensory cortex of adult rats. In search of an operational basis for a description of plastic processes, we choose receptive fields (RFs), because they map physical events onto representations based on neural activity. As plastic changes are reflected in variations of RFs, we incorporated activity graded RF organization in the model.

2 Biological Findings

Standard electrophysical methods were used to study reorganizational plasticity in adult anaesthetised rats [12]. Motivated by the Hebbian postulate to induce plastic changes by temporal coincidence of external events, we used a protocol of paired, simultaneously applied tactile stimulation to two different skin locations (PPTS). Reorganization was measured by comparing RF topography, RF size, the degree of RF overlap, the area of cortical skin field representations and the cortical magnification before and after PPTS. Quantitative analysis of RFs was accomplished by computing response planes based on multiple stimulation points [2]. The fine grained topography mapped under control is lost after PPTS because of fusion of those RFs that were simultaneously stimulated. The analysis of response planes showed focal zones of high activity surrounded by larger zones of less activity displaying a complex spatial activity distribution (Fig.1). After PPTS, enlarged zones of maximal activity emerged that are shifted selectively towards the stimulated RFs. All PPTS induced changes were fully reversible after several hours [1]. Because of the short time scale of the induction and the reversibility, the underlying mechanisms are supposed to be mainly functional modulations of synaptic coupling.

3 Theoretical Model

As a first step we utilize selforganizing feature maps as a tool to produce the paw representation in the somatosensory cortex. The sensory input is derived from a set R which contains 400 receptors distributed with variable density.

RFs are of variable size depending on the location of the paw. The geometric properties of the tactile stimuli are modelled to match the size and shape of the RFs. We define the stimulus $\vec{v} = (v_0, \ldots, v_{n-1})$ as a vector of spatial points (x_i, y_i),

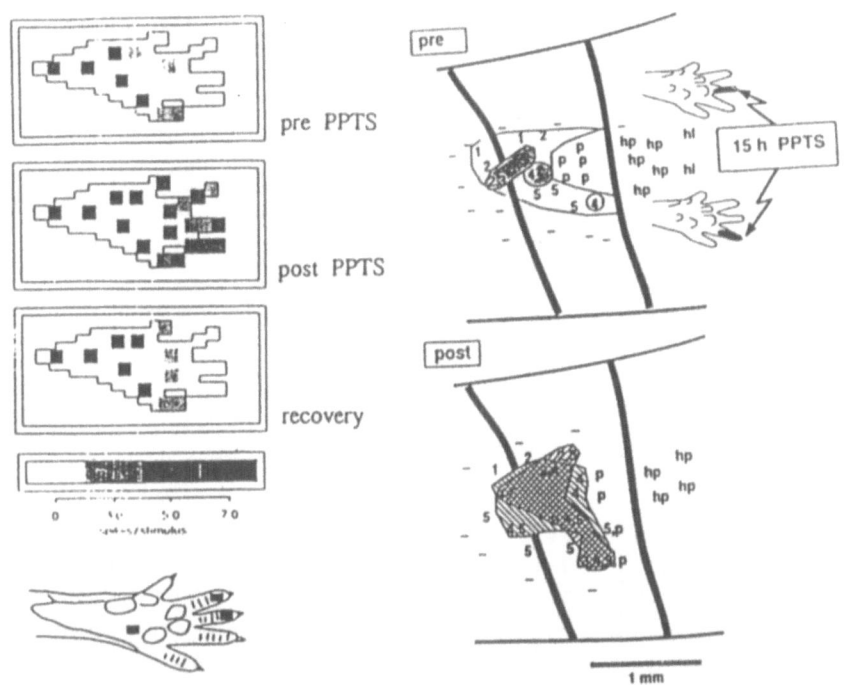

Figure 1: Somatosensory RFs and cortical representation pre- and post PPTS

$i = 0, \ldots, n-1$, of the rat paw, excited with 1 within an ellipsoid input range with center (x_s, y_s) and axes a, b and 0 outside.

$$v_i = \begin{cases} 1 & : \quad v_i \in ellipse(x_s, y_s, a, b) \\ 0 & : \quad v_i \notin ellipse(x_s, y_s, a, b) \end{cases}$$

$$ellipse(x_s, y_s, a, b) := \{(x, y) \in R : \frac{(x - x_s)^2}{a^2} + \frac{(y - y_s)^2}{b^2} \leq 1\}$$

To make realistic assumptions we use experimental data to create new inputs by an interpolation algorithm. The center of the input $c_r = (x_r, y_r)$ is choosen randomly, the size of the ellipse – characterized by (a_r, b_r) – is determined by the distance weighted average of the nearest RFs. The size of these measured RFs is taken from the four quadrants of a hypothetical coordinate system through the input center. The axis a_i and b_i are weighted conversely to their Euklidian distances between the stimulus center c_i and c_r. The size of the $ellipse(c_r, a_r, b_r)$ becomes

$$a_r = \frac{1}{k-1} \sum_{i=1}^{k} (1 - \frac{d_i(c_r, c_i)}{\sum_{i=1}^{k} d_i(c_r, c_i)}) a_i \qquad (1)$$

and analog the axis b_r, with $d_i(c_r, c_i) = \sqrt{(x_r - x_i)^2 + (y_r - y_i)^2}$ and $k = 1, \ldots, 4$.

From this data set a somatosensory map with $N \times N$ neurons is selforganized by the "shortcut" version of Kohonen's algorithm [5, 11] with a Gaussian function for the lateral connections of the neurons. To account for stimulus depending reorganization of this map we assume the existence of a residual plasticity expressed by the setting of the learning rate $\alpha \geq 0$ and the lateral connection $\sigma \geq 6$.

4 Results

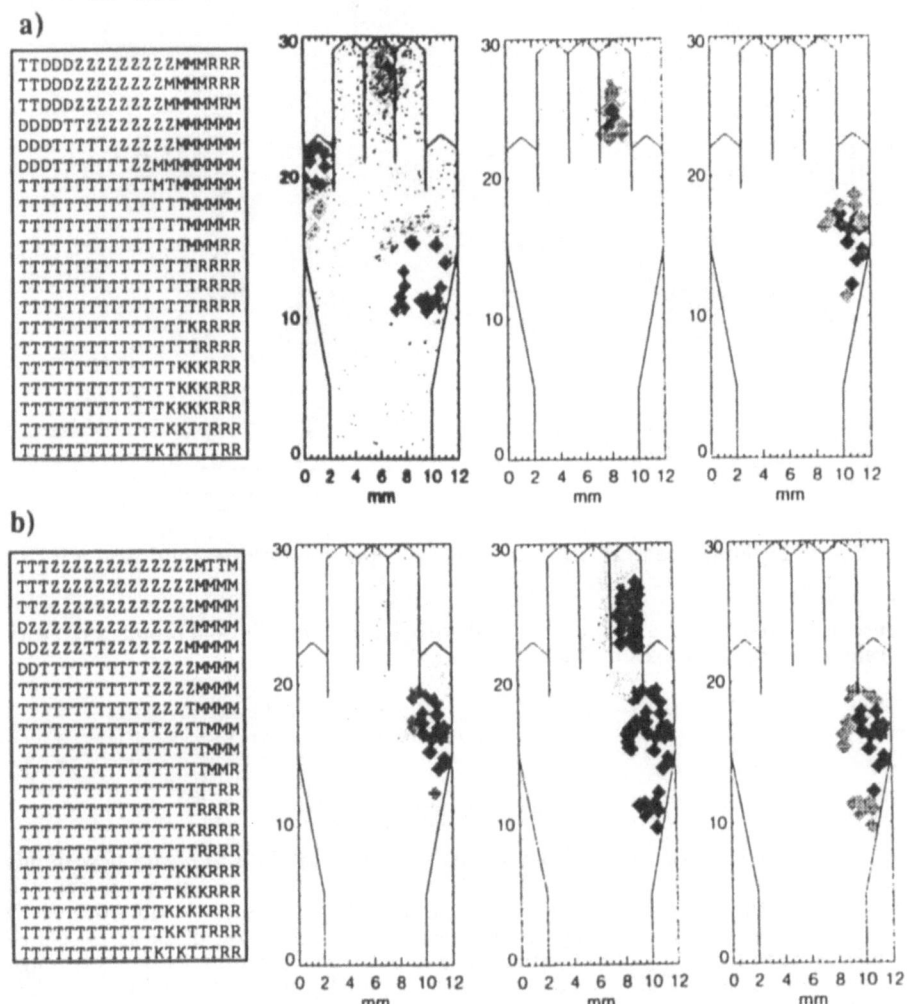

Figure 2: Simulated hindpaw representation and RFs a. pre-PPTS b. post-PPTS

To generate a "normal" topography that can be used to study post-ontogenetic plasticity, we generated according to ontogenetic development a topographic map based on 20×20 neurons using the described interpolation algorithm. The maps obtained in this way were highly ordered and the corresponding RFs had properties resembling those measured experimentally. Their size and shape differed systematically along the digit-pad axis of the hindpaw (Fig.2a). It is important to note that our approach included graded activity profiles of the RFs. Induction of plasticity was simulated by selecting two skin field locations whose RFs were clearly non-overlapping (Fig.2a). Already a small number of learning steps were sufficient to generate plastic changes. RFs of the stimulation sites melted into large RFs comprising both sites. RFs around various distances of the stimulation sites showed also fusion of the stimulated skin field representations, resulting in a general increase of RF size and overlap (Fig.2b). Depending on the distance from the stimulations sites we also observed shifts of the centers of activity within RFs leading to highly asymmetric RF structures and changes of the overall RF shape. RFs located far distant from the stimulation sites were not effected. At the global level of the maps, significant overrepresentations of the stimulation sites emerged, resulting in a distortion of the normally regular lattice of the map which is highly compatible with the experimentally observed map changes. The aspect of reversibility was modelled by switching from the PPTS mode to a random stimulation pattern.

5 Discussion

Using Kohonen's feature map with 400 units and the same number of receptors is sufficient to simulate qualitative changes in somatosensory RF representation close to experimental data. The aim of further studies is the implementation of physiological differences of magnification of RF representation by increasing the number of units and reducing the dimensionality of the input signals by a parametric formulation of the stimulation areas. The presented simulations show that the Kohonen model is not only a tool for simulation dramatic changes in cortical representation following amputation, but for short-term and reversible neural plasticity as well. The behavioral relevance of post-ontogenetic plasticity was documented in human psychophysical experiments [3]. A few hours of PPTS decreases the threshold of the spatial two point discrimination. We believe that a combination of electrophysiology, human psychophysics and modelling turns out benificial to unravel the relationships of cortical representation, information processing and behavioral performance.

References

[1] B. Godde, F. Spengler, and H.R. Dinse. In N. Elsner and M. Heisenberg, editors, *Gen - Gehirn - Verhalten*, page 140. Thieme, 1993.

[2] B. Godde, F. Spengler, and H.R. Dinse. In N. Elsner and M. Heisenberg, editors, *Gen - Gehirn - Verhalten*, page 148. Thieme, 1993.

[3] B. Godde, F. Spengler, and H.R. Dinse. In N. Elsner and M. Heisenberg, editors, *Gen - Gehirn - Verhalten*, page 149. Thieme, 1993.

[4] K.A. Grajski and M.M. Merzenich. *Neural Computation*, 2:71 – 84, 1990.

[5] T. Kohonen. *Biological Cybernetics*, 44:135–140, 1982.

[6] M.M. Merzenich, G. Recanzone, W.M. Jenkins, T. Allard, and R.J. Nudo. In *Neurobiological of Neocortex, Dahlem Konferenzen 1988*. Wiley, 1988.

[7] K. Obermayer, H. Ritter, and K. Schulten. In R. Eckmiller, G. Hartmann, and G. Hauske, editors, *Parallel Processing in Neural Systems and Computers*, pages 71 – 74, Amsterdam, 1990. North-Holland.

[8] K. Obermayer, H. Ritter, and K. Schulten. *Parallel Computing*, 14:381 – 404, 1990.

[9] J.C. Pearson, L.H. Finkel, and G.M. Edelman. *Jour. of Neuroscience*, 7:4209 – 4333, 1987.

[10] G.H. Recanzone, M.M. Merzenich, W.M. Jenkins, K. Grajski, and H.R. Dinse. *Journal of Neurophysiology*, 67:1031 – 1056, 1992.

[11] H. Ritter, T. Martinetz, and K. Schulten. *Neuronale Netze*. Addison-Wesly, Germany, 1990.

[12] F. Spengler and H.R. Dinse. Reversible relocation of representational boundaries of rat somatosensory cortex into previous non-somatic zones by intracortical microstimulation (icms). *submitted*.

[13] S. Wacquant, F. Joublin, F. Spengler, B. Godde, and H.R. Dinse. In S. Gielen and B. Kappen, editors, *ICANN93*, pages 31 – 36. Springer, 1993.

[14] N.M. Weinberger, J.H.Ashe, R. Metherate, T.M. McKenna, D.M. Diamond, and J. Bakin. *Concepts in Neuroscience*, 1:91 – 132, 1990.

Multiple Self-Organizing Neural Networks with the Reduced Input Dimension

Jongwan Kim*, Jesung Ahn*, Chong Sang Kim*, Heeyeung Hwang**, Seongwon Cho***

* Dept. of Computer Engineering, Seoul National University, KOREA
** Dept. of Electronics Engineering, Hoseo University, KOREA
*** Dept. of Electrical and Control Engineering, Hong Ik University, KOREA

Abstract - In this paper, we propose a new multiple self-organizing neural network architecture. Whereas conventional learning methods utilize the full dimension of the original input patterns, the proposed system consists of multiple neural networks with the reduced input dimension. We have developed three consensus schemes so as to judge the classification using multiple neural networks. Each network of ensemble has dynamic properties. The number of output neurons is increased as learning proceeds. Every output neuron has its own class threshold, which represents a class boundary. The class threshold value is tuned according to input pattern distribution. The performance of the multiple self-organizing neural network is compared with those of conventional competitive learning algorithms.

1. INTRODUCTION

Recently, competitive learning has been widely used for solving the pattern recognition problem due to its rapid learning speed. Simple Competitive Learning(SCL) refers to the algorithm that learns only the winner neuron, whose weight vector is closest to the input pattern. Several different methods were previously proposed in order to improve the performance of neural networks using competitive learning. For example, Kohonen proposed the Self-Organizing Feature Map(SOFM) algorithm that changes the weight vectors of the neighbor neurons of the winner as well [1]. Ahalt et al. proposed conscious mechanism where the winning rate of the frequent winning neuron was reduced [2]. In addition, Carpenter and Grossberg proposed ART(Adaptive Resonance Theory) model [3]. ART learns weight vectors dynamically expanding the dimension of its output layer according to the input pattern distribution. But it is difficult to determine the value of the vigilance factor by which an input vector is tested for similarity to the weight vectors.

In this paper, we propose an ensemble of neural networks with consensus schemes. The basic idea of neural network ensemble is to use the reduced input dimension for learning multiple neural networks. Each neural network is learned using a new competitive learning algorithm that minimizes intraclass variance and dynamically creates new output neurons.

2. COMPETITIVE LEARNING WITH DYNAMIC OUTPUT NEURON GENERATION

In the previous paper [4], we have proposed a three-layered neural network using 2-phase learning. In phase 1, learning proceeds by dynamically changing the class region boundary to minimize intraclass variance and creating internal neurons as needed. In phase 2, well learned internal neurons are merged or promoted to output neurons. In this paper, we limit the maximum size of dynamically increasing output layer. For hardware implementation, the fixed size output layer is desirable. Thus, the internal layer is eliminated and one phase learning becomes possible. Learning is stabilized by iteration as in conventional competitive learning algorithms.

The proposed neural network consists of two layers, input layer and output layer. Whereas the size of the input layer is fixed, the number of neurons in the output layer is increased dynamically. The architecture of the proposed network is shown in Figure 1. Each neuron in the output layer has a counter(F_j), a variance(S_j), and a class threshold(σ_j) associated with it. The proposed neural network is different from conventional ones in several aspects. Firstly, the performances of the conventional competitive learning algorithms may be affected by initial weight vectors. In the proposed neural network, the output layer is dynamically increasing and the class threshold of generated output neuron is tuned according to the difference in intraclass variance, which means the proper adaptation to the input pattern distribution. Secondly, the proposed system performs only feedforward weight vector updates whereas ART model performs feedback as well as feedforward weight vector updates. Thirdly, the proposed neural network uses a single class threshold value as the learning parameter, while ART2 has more parameters. Finally, the proposed learning method can be combined with any competitive learning algorithms. For example, we call SCL the Dynamic Simple Competitive Learning(DSCL) when applying our learning method.

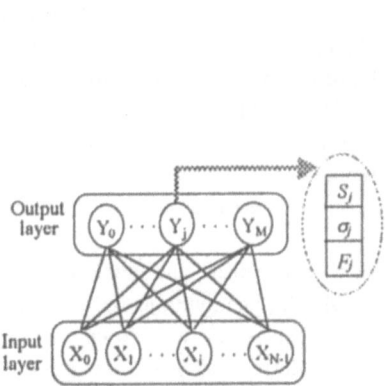

Figure 1. The architecture of the proposed self-organizing neural network.

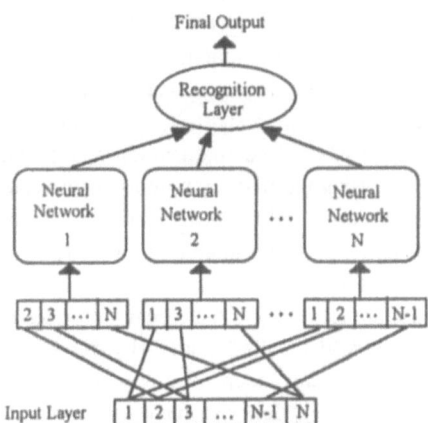

Figure 2. The architecture of multiple neural networks with the reduced input dimension.

3. MULTIPLE NEURAL NETWORKS WITH THE REDUCED INPUT DIMENSION

Conventional learning methods utilize the full dimension of the original input patterns. However, a particular attribute or dimension of the input patterns does not necessarily contribute to classification and may even cause misclassification in certain cases. In addition, each attribute has the different degree of contribution to classification. In order to get around the disadvantage of the conventional method, N neural networks with $N-1$ input size are trained and used for classification, where N is the number of full dimension of input patterns. Figure 2 shows the architecture of multiple neural networks with the reduced input dimension.

The overall classification accuracies of N neural networks are expected to be different. Moreover, the classes of a neural network have different classification accuracies. The collective decision made by N neural networks is less likely to be in error than the decision produced by any of the individual networks. In order to combine N neural networks, we use three consensus schemes. In Scheme 1, the class of an input pattern is determined by the weighted sum of the overall classification rate of each network. Scheme 1 is a sort of plurality approach in which the collective decision is the classification reached by more networks than any other [5,6]. Scheme 2 utilizes the classification information of only the network with the best overall classification accuracy. Scheme 3 is a hybrid method, which uses only the networks having the highest recognition rate for each class.

Differently from the previous multiple neural network methods [5,6] that use the same input vectors with different network parameters, our learning method uses different input vectors in order to take maximum advantage of the information in each dimension of the input patterns.

4. EXPERIMENTS

Experiments were conducted in order to show the synergism by combining the proposed self-organizing neural network and multiple neural network learning method. The results of the synergism are compared with those of SCL algorithm and Frequency-Sensitive Competitive Learning(FSCL) algorithm [2].

A 8-band 8-class remote sensing data set [7] was used as inputs. Training was done with 200 input patterns per class and testing with another 375 input patterns per class. So the total number of training and testing data are 1600 and 3000, respectively. The standard SCL and FSCL used eighty output neurons and the initialization of the weight vectors was carried out by choosing the first ten training input vectors belonging to each class. In the proposed self-organizing neural network, the maximum number of output neurons was limited to eighty and the initial class threshold of 0.5 was used. A linearly decreasing learning rate with the initial value of 0.9 was chosen.

Figure 3-(a) shows the recognition rates of the standard SCL, DSCL, and multiple DSCL for testing data, and Figure 3-(b) shows those of FSCL, DFSCL, and multiple DFSCL. Each experiment was independently conducted. That is, first, the experiment with 10 epochs was conducted, and the experiment with 50 epochs was newly conducted, etc. The performance of DSCL is lower than SCL. However, the inferiority of DSCL was overcome by multiple neural network learning. Figure 3-(b) indicates that learning with dynamic output neuron generation improved the performance of FSCL and the further increased performance was obtained by multiple neural network learning.

5. DISCUSSION

We presented a new self-organizing neural network. We simplified our 2-phase self-organizing neural network [4], which results in faster learning. Because the class threshold of output neuron dynamically is changed according to the difference in class variance, the proposed neural network adapts properly to the input pattern distribution.

The multiple self-organizing neural networks were trained with the reduced input dimension. Each network uses different input vectors and the Recognition Layer combines the outputs of the networks for final classification. The learning time can be reduced because the reduced input dimension is used and each network can be trained in parallel.

From the experimental results, we found that our multiple neural network learning method could improve the recognition accuracy although it is difficult to know which consensus scheme is most suitable for specific neural network algorithm beforehand. The proposed multiple neural network learning method is suitable for recognition of the patterns whose particular attributes lead to recognition ambiguity.

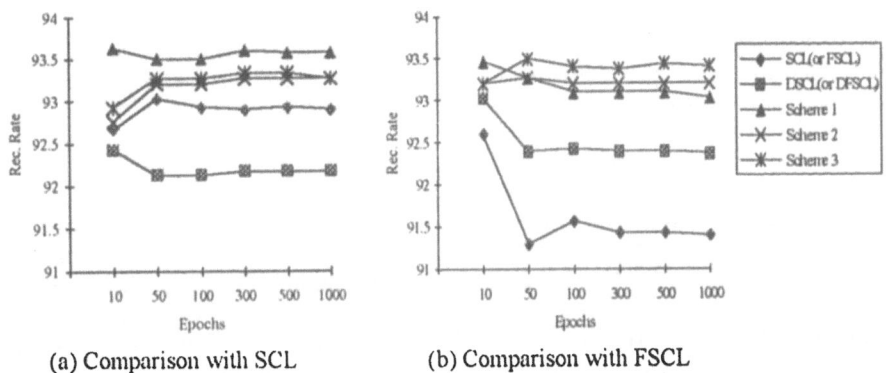

(a) Comparison with SCL   (b) Comparison with FSCL

Figure 3. Recognition rates of the SCL, FSCL, DSCL, DFSCL, multiple DSCL, and multiple DFSCL.

REFERENCES

1. Kohonen, T., Self-Organization and Associative Memory, Springer-Verlag, New York, 1984.
2. Ahalt, S. C., Krishnamurthy, A. K., Chen, P., and Melton, D. E., "Competitive Learning Algorithms for Vector Quantization," Neural Networks, Vol.3, pp.277-290, 1990.
3. Carpenter, G. and Grossberg, S., "A Massively Parallel Architecture for a Self-Organizing Neural Pattern Recognition Machine," Computer Vision, Graphics, and Image Processing, Vol.37, pp.54-115, 1987.
4. Kim, J. W., Ahn, J. S., Hwang, H. Y., "2-Phase Self-Organized Classifying Network with Minimal Intracluster Variance," Proc. of 1993 World Congress on Neural Networks, Portland, Oregon, Vol.II, pp.492-496, 1993.
5. Hansen, L. K. and Salamon, P., "Neural Network Ensembles," IEEE Trans. on PAMI, Vol.12, No.10, pp.993-1001, 1990.
6. Hoffman, J., Skrzypek, J. and Vidal, J. J., "Cluster Network for Recognition of Handwritten, Cursive Script Characters," Neural Networks, Vol.6, pp.69-78, 1993.
7. Cho, S., Ersoy, O. K., and Lehto, M. R., "Parallel, Self-Organizing Hierarchical Neural Networks with Competitive Learning and Safe Rejection Schemes," IEEE Trans. on Circuit and Systems, to appear.

Adaptive modulation of Receptive Fields in self-organizing networks[*]

F. Firenze, P. Morasso

Department of Informatics, Systems and Communications

University of Genoa, Italy

1 Introduction

Most self-organizing neural networks involve units which, similarly to biological neurons, have *Receptive Fields* (RFs) of finite sizes, defined as the regions in the *features-space* where they achieve non-vanishing activation. In some models, the widths of the RFs are slowly restricted during learning in order to achieve global convergence through *weights "freezing"* (Martinetz & Schulten, 1991); in other cases, they are adjusted, either in supervised mode (Reilly et al., 1982) or heuristically (Moody & Darken, 1989), with the aim to favour development of *"locally tuned"* units. We propose a new mechanism of adaptive modulation of the RFs, driven by the local density of the input data distribution, which can be coupled with many self-organizing models, making them more robust and flexible. In particular, we shall focus our attention on two interesting aspects: *"self-stabilization"* of learning parameters during *"on-line" learning* and function approximation with *"adaptive resolution"*.

2 The model

We consider the class of self-organizing networks consisting of a pool of *competitive* units (Reggia et al., 1992), excited by the same n-dimensional input vector $\vec{x} \in \Re^n$. Unit N_i is characterized by two elements: an activation level a_i, determined by the same and common activation function, and a set of synaptic weights which can be conveniently grouped in a weight vector, $\vec{w}_i = (w_{1i}, ..., w_{ni}) \in \Re^n$, interpreted as a *prototype* in the input feature-space. As regards the activation function, it is usually chosen as a bounded non-linear decreasing function γ of the euclidean distance d_i between the current input-vector $\vec{x}$ (randomly extracted from an underlying data distribution) and the prototype $\vec{w}_i$,

$$a_i = \gamma(d_i) \ , \ d_i = \| \vec{x} - \vec{w}_i \| \ . \tag{1}$$

Typical shapes for γ are: *sigmoidal* (Grossberg, 1976) (Firenze & Morasso, 1993), *Gaussian* (Bridle, 1989) (Poggio & Girosi, 1990), and *exponential* (Martinetz & Schulten, 1991). For all of them, we can define a spherical RF for each unit, as the sphere, centered around the prototype $\vec{w}_i$, with radius r_i equal to the distance d_i for which the activation function (1) becomes neglegible. We also assume that some kind of Hebbian learning is used in order to allow the

---

[*]This work has been supported by the Italian Ministry of University and Scientific and Technical Research (MURST) within the "Programma Nazionale di Ricerca sulle Tecnologie per la Bioelettronica" - Theme 2

vector prototypes to cluster around interesting areas of the input distribution. The proposed mechanism of adaptive modulation of the RFs of each neuron is defined as an updating rule of the radius r_i:

$$\Delta r_i = \epsilon(d_i - r_i)e^{-d_i/p} , \qquad (2)$$

where $\epsilon \in (0,1)$ and $p > 0$. It is a simple local rule that can be explained in intuitive terms: the tendency of r_i to decrease, due to the exponential decreasing factor, tends to be balanced by a reactive tendency to increase due to an increase of the number of samples which fall outside the RF; this leads to an equilibrium very close to the radial dispersion of data locally around the center $\vec{w}_i$ of the RF. The parameters ϵ and p weakly affect the equilibrium if the input distribution is stationary, and mainly influence the reaction of the network to statistical transients of the distribution.

3 "Self-stabilization" of learning parameters during "On-line" learning

The adaptive modulation mechanism described above can be coupled with standard self-organizing models with the purpose of automatically tune the learnin-rates. For example, if we consider the gain factor in the *Neural Gas* model (Martinetz & Schulten, 1991), which is decreased during learning according to a heuristic profile externally imposed, then we can automatically perform this function by linking the gain parameter of each neuron to the absolute value of the time derivative of its RF radius. Fig.1 shows a simulation result (normal neural gas: 1a; modified neural gas: 1b). In a stationary environment the two models behave in the same way. However, if there are local statistical transients the standard model remains stuck in the frozen configuration whereas the modulation mechanism (Fig.1c and 1d) is capable of (i) detecting the transient, (ii) tracking it and (iii) settling in the new configuration. In other words, this adaptive mechanism allows self-organized learning to operate on-line.

4 Function approximation with "adaptive resolution"

(Poggio & Girosi, 1990) showed that learning an input-output mapping, $f : \vec{x} \in \Re^n \to \vec{y} \in \Re^m$, from a training set $\{(\vec{x}^{(k)}, \vec{y}^{(k)}), k = 1, ..K\}$, can be accomplished, in the general framework of the *regularization theory*, by a three layers network implementing a weighted sum of Radial Basis Functions (RBF). They also suggested that using different RBFs, such as Gaussians with different widths, leads to *"multiple-scales"* function approximation. In this context, (Moody & Darken, 1989) proposed a heuristic approach to locally tune the widths of the gaussian based on the statistical distribution of input data. The result, roughly speaking, is that many Gaussian units with small width cover high density data regions and few Gaussian units with large width cover low density data regions. The adaptive modulation mechanism of the RF radius r_i allows us to obtain the same result but in a completely automatic way, by simply setting the variance of each Gaussian unit equal to r_i. An example is

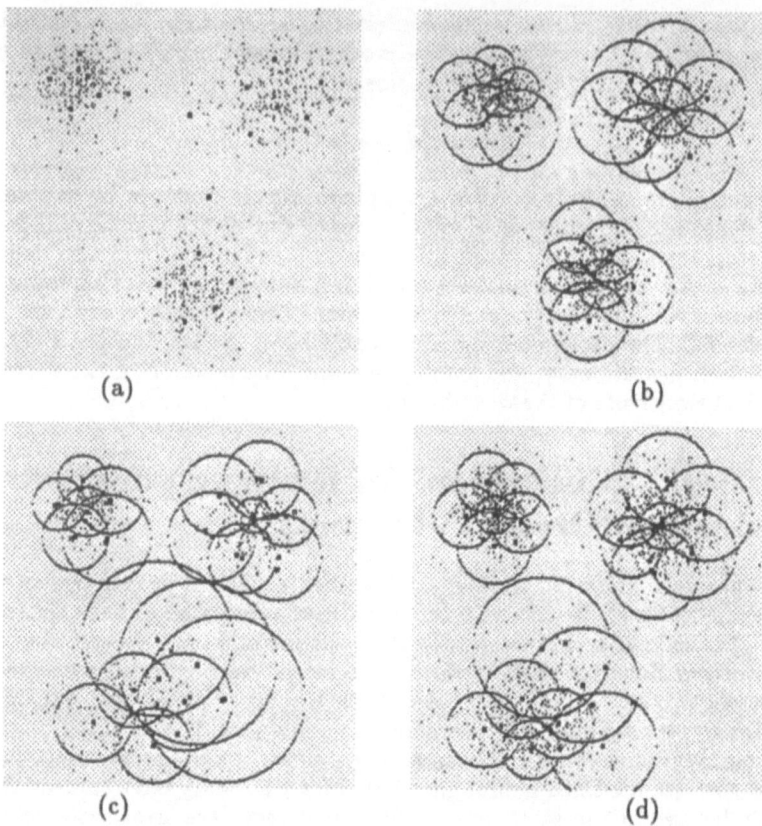

Figure 1: On-line learning with the modified "Neural Gas" network (2D features-space: black dots are data, little squares are the final positions of the neural prototypes; circles are the RFs; patterns of connections are not represented)

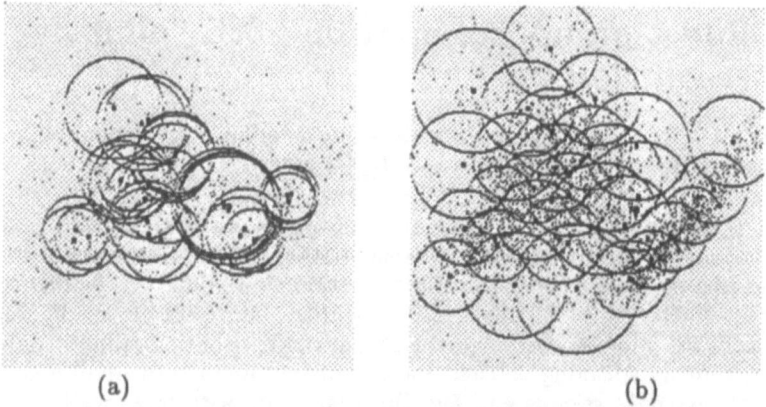

Figure 2: Non-uniform distribution of examples from a function domain adaptively approximated by a distribution of gaussian locally tuned units: (a) intermediate learning situation; (b) equilibrium

shown in Fig.2. This result implies that a function can be locally approximated with a level of resolution adapted to the local density of examples in the function domain.

5   Conclusions

A mechanism of local adaptive modulation of neural Receptive Field has been showed to allow self-organizing networks to self-stabilize learning in response to local time and space density invariants of an input data distribution. In particular, *on-line learning* and function approximation with *adaptive resolution* can be succesfully faced: the first one driven by time density invariants and the second one by space density invariants.

Reference

Bridle, J. (1989). Probabilistic interpretation of feedforward classification network outputs, with relationships to statistical pattern recognition. In Soulie, F. F., & Herault, J. (Eds.), *Neurocomputing*, Vol. NATO ASI F-68, pp. 227–236. Springer Verlag, Berlin.

Firenze, F., & Morasso, P. (1993). The capture effect model: a new approach to self-organized clustering. In *Neuro-Nimes93, Sixth Int. Conf. on Neural Networks and their industrial & cognitive applications*, pp. 65–74 Nimes.

Grossberg, S. (1976). Adaptive Pattern Classification and Universal Recoding,(I). *Biological Cybernetics*, *23*.

Martinetz, T., & Schulten, K. (1991). A 'Neural-Gas' network learns topologies. In Kohonen, T., Makisara, K., Simula, O., & Kangas, J. (Eds.), *Artificial Neural Networks*, pp. 397–402 Amsterdam. North-Holland.

Moody, J., & Darken, C. (1989). Fast learning in networks of locally-tuned processing units. *Neural Computation*, *1*, 281–294.

Poggio, T., & Girosi, F. (1990). Networks for approximation and learning. *Proceedings of the IEEE*, *78*, 1481–1495.

Reggia, J., D'Autrechy, C., Sutton, G., & Weinrich, M. (1992). A competitive distribution theory of neocortical dynamics. *Neural Computation*, *4*, 287–317.

Reilly, D., Cooper, L., & Elbaum, C. (1982). A neural model for category learning. *Biological Cybernetic*, *45*, 35–41.

About the convergence of the generalized Kohonen algorithm

Jean-Claude FORT *   Gilles PAGÈS †

Introduction

The Kohonen algorithm was originally devised and studied by Kohonen in 1982 (see [4], [5]). Unfortunately, as far as mathematical treatment is concerned, rigourous results are not so easy to establish. For instance, any simulation in a one dimensional setting (i.e. with scalar inputs) shows the existence of a self-organization property. Nevertheless, the proof of such a property requires some non trivial Markov material. It was first carried out in [2] for uniformly distributed stimuli and then was extended to general distributions in [1]. Many open questions related to the one dimensional self-organization still remain, and no similar result was proved in higher dimension so far.

The paper is devoted to the study of the usual generalization (see [7]) of the Kohonen algorithm once self-organization occured. More precisely we deal with the converging properties of the algorithm under the "decreasing step assumption". Our aim is to fulfill the assumptions of the Kushner & Clark conditional convergence theorem (see [6]). It requires to study the stability of the equilibrium points of the related Ordinary Differential Equation $\dot{x} = -h(x)$ where h denotes the average function of the algorithm.

1   Definitions and former results

Let consider on one hand a set of units (or *neurons*) represented by a finite subset I of $\mathbf{Z}^d$ equipped with a neighbourhood function $\sigma : I - I := \{i - j, \; i, j \in I\} \mapsto [0, 1]$ satisfying $\sigma(k) = \sigma(-k)$, $\sigma(0) = 1$ and, on the other hand, a stimuli set $[0, 1]^d$ endowed with a Borel Probability measure μ, *strongly* diffuse (*i.e.* no hyperplan is weighed by μ) with $[0, 1]^d$ support. Each unit $i \in I$ has a $[0, 1]^d$-valued weight vector x_i. The Kohonen algorithm has the property of building some weight vectors that both quantify μ and preserve the neighbourhood structure provided by σ. $\| . \|$ will denote the Euclidean norm on $\mathbb{R}^d$ and $(.|.)$ its related inner product.

*Univ. Nancy I, Fac. Sciences, B.P. 239, F-54506 Vandœuvre-Lès-Nancy Cedex & Samos, Univ. Paris I, 90 rue de Tolbiac, 75634 Paris Cedex 13, mail: fortjc@iecn.u-nancy.fr.

†Lab. de Proba., URA 224, Univ. Paris 6, 4 pl. Jussieu, F-75252 Paris Cedex 05 & Univ. Paris 12, 61, av. du Gal de Gaulle, F-94010 Créteil Cedex, mail: gpa@ccr.jussieu.fr.

Let $X^0 = (x_i^0)_{i \in I}$ be the initial weight vector state and X^t be the state of the weight vector at time $t \in \mathbb{N}$. At time $t+1$, X^t is updated as follows :

(i) Computation of the winning unit $i^{t+1} := i(\omega^{t+1}, X^t) = \mathrm{argmin}_k \|\omega^{t+1} - X_k^t\|$.

$$(ii) \ \forall j \in I, \quad X_j^{t+1} = X_j^t - \varepsilon_{t+1} \sigma(i^{t+1} - j)(X_j^t - \omega^{t+1}), \tag{1}$$

where $(\varepsilon_t)_{t \geq 1}$ is a sequence of $(0,1)$-valued real numbers. More synthetically, the algorithm reads

$$\forall t \in \mathbb{N} \qquad X^{t+1} = X^t - \varepsilon_{t+1} H^\sigma(X^t, \omega^{t+1}). \tag{2}$$

Let $D_I := \{x \in ([0,1]^d)^I \ / \ x_i \neq x_j \ \text{iff} \ i \neq j\}$ be the set of $([0,1]^d)^I$-valued vectors with parted components. If $x \in D_I$, then $\mathbb{P}_x$-a.s. $X^t \in D_I$ for every $t \in \mathbb{N}$. To write down the $O.D.E.$, we need to define the Voronoï tesselation:

Definition 1 *The Voronoï tesselation* $\{C_i(x)\}_{i \in I}$ *of* $x \in D_I$ *is defined by* $C_i(x) := \{\omega \in [0,1]^d \ / \ \|x_i - \omega\| < \|x_k - \omega\| \ \text{if} \ k \neq i\}, \ i \in I.$

Then, the $O.D.E.$ displays as

$$\dot{x} = -h^\sigma(x), \ x_0 \in D_I, \quad h_i^\sigma(x) = \sum_{k \in I} \sigma(k - i) \int_{C_k(x)} (x_i - \omega)\mu(d\omega), \ i \in I$$

where $h^\sigma(x) := \mathbb{E}(H^\sigma(x, \omega^1))$ is continuous on D_I. Let us finally recall the well-known result by Kushner & Clark

Definition(-Theorem) 2 *(Kushner & Clark, [6])* *Let* x^* *be a zero of* h^σ *and* K *be a compact subset of its attracting area* Γ_{x^*}. *Then any bounded path of* $(X^t)_{t \geq 1}$, *that visits* K *infinitely often converges to* x^*. X^t *is said to converge "conditionally" a.s. to* x^*.

2 General results in the 1-dimensional setting

We study the 1-dimensional Kohonen algorithm. As $d = 1$ the neighbourhood function is defined by a function $\sigma : \mathbb{N} \mapsto [0,1]$ with $\sigma(0) = 1$. The proposition below shows the existence of absorbing classes and of equilibrium points.

Proposition 1 *(a) If* σ *is non increasing, the two subsets* $F_n^+ := \{x \in [0,1]^n / 0 < x_1 < x_2 < \cdots < x_n < 1\}$ *and* $F_n^- := \{x \in [0,1]^n / 0 < x_n < x_{n-1} < \cdots < x_1 < 1\}$ *are left stable by the algorithm.*
(b) If μ *is diffuse, then there exists at least one equilibrium point* x^* *in the closure* $\overline{F}_n^+$ *of* F_n^+, *i.e. satisfying* $h(x^*) = 0$.
(c) Assume that $\mathrm{supp}(\mu) = [0,1]$, σ *is non increasing and satisfies assumption* $(\mathcal{S}) \equiv (n \geq 2 \ \text{and} \ \sigma(1) < 1) \ \text{or} \ (n \geq 3 \ \text{and} \ \sigma(2) < 1) \ \text{or} \ (n \geq 5 \ \text{and} \ \sigma(3) < 1)$ *then* x^* *lies in* F_n^+.

So, from now on we will always assume that the initial value x lies in F_n^+.

2.1 Stability of an equilibrium point

We suppose now that μ has a density f. The main result of this section is

Theorem 1 *Assume that f is continuous on $[0,1]$ and $f > 0$ on $(0,1)$. Let $x^* \in \overline{F_n^+}$ be an equilibrium point. If σ is non increasing, satisfies (S) and if one of the following assumptions holds:*

1. *(i) $\log f$ is concave on $[0,1]$ and $f(0) + f(1) > 0$,*
2. *(ii) $\log f$ is strictly concave on $[0,1]$.*

   then $x^* \in F_n^+$ and X^t a.s. conditionally converges to x^* (see Definition 2).

The proof is rather lengty and technical (see [3]). It consists in showing that the real parts of the eigenvalues of $\nabla h(x^*)$ are positive. If (S) holds then, $f = 1_{[0,1]}$ being log-concave with $f(0) = f(1) = 1$, theorem 1 implies that every equilibrium point x^* is conditionally stable. Actually this result can be substantially improved.

Theorem 2 *If σ satisfies assumption (S) and $\mu := U([0,1])$ (uniform distribution on $[0,1]$) then the equilibrium point x^* is unique and unconditionally stable, i.e. $\forall x \in F_n^+ \quad X^t \overset{t \to +\infty}{\longrightarrow} x^* \quad \mathbb{P}_x\text{-a.s.}.$*

Once established the uniqueness, the result follows from the classical Robbins-Monro theorem and uses the positivity of the symmetrized of ∇h^σ.

3 Some multi-dimensional results

We consider a $[0,1]^d$-valued stimuli space, $n = n_1 \cdots n_d$ units. For every $l \in \{1, \cdots, d\}$, let $I_l := \{1, 2, \cdots, n_l\}$ so $I := I_1 \times \cdots I_d$. The product neighbourhood function σ is defined by $\forall \, \mathbf{i} = (i_1, \cdots, i_d), \, \mathbf{j} = (j_1, \cdots, j_d) \in I^2 \quad \sigma(\mathbf{i} - \mathbf{j}) := \prod_{1 \leq l \leq d} \sigma_l(|i_l - j_l|)$ where the σ_l's are neighbourhood functions on I_l. When the stimuli distribution μ on $[0,1]^d$ also has a product form $\mu := \mu_1 \otimes \cdots \mu_d$ μ_l probability measures on $[0,1]$, the related $K_d(I, \mu, \sigma)$-Kohonen algorithm is said to have a product structure and is denoted $\otimes_{l=1}^d K_1(I_l, \mu_l, \sigma_l)$.

Proposition 2 *If the x^{l*} are some equilibrium points of $K_1(I_l, \sigma_l, \mu_l)$-Kohonen, $1 \leq l \leq d$, then the grid point $x^{1*} \otimes \cdots \otimes x^{d*} := ((x_{i_1}^1, \cdots, x_{i_d}^d))_{\mathbf{i} = (i_1, \cdots, i_d) \in I}$ is an equilibrium point of the $K_d(I, \sigma, \mu)$-Kohonen. These are the only grid equilibrium points provided that $\text{supp}(\mu_l) = [0,1]$, $1 \leq l \leq d$.*

We need some further notations to calculate the gradient ∇h^σ. For every $\mathbf{i}$, $\mathbf{j} \in I$, we define $\vec{n}_x^{\mathbf{ij}} := \frac{x_{\mathbf{j}} - x_{\mathbf{i}}}{\|x_{\mathbf{j}} - x_{\mathbf{i}}\|}$ and $\lambda_x^{\mathbf{ij}}(d\omega)$ the Lebesgue measure on the median hyperplan of $(x_{\mathbf{i}}, x_{\mathbf{j}})$.

Theorem 3 h^σ *is* C^1 *and (here* μ *needs not to be a product measure):*

$$\frac{\partial h_i^{\sigma,l}}{\partial x_j^l}(x) = \sum_{k \in I} \sigma(i-k)\mu(C_k(x))\delta_{ij}\vec{e}^l + \sum_{k \neq j} (\sigma(i-k)-\sigma(i-j)) \times$$

$$\int_{\overline{C}_k(x) \cap \overline{C}_j(x)} (x_i^l - \omega^l)\left(\frac{1}{2}\vec{n}_x^{kj} + \frac{1}{||x_k - x_j||}(\frac{x_k + x_j}{2} - \omega)\right) f(\omega)\lambda_x^{kj}(d\omega)$$

where $h_i^{\sigma,l}$, $l=1,\cdots,d$ *denotes the* l^{th} *component of* h_i^σ, $(\vec{e}^1,\cdots,\vec{e}^d)$ *is the canonical basis of* $\mathbb{R}^d$ *and* δ_{ij} *is the Kronecker symbol.*

We can now derive some results for the 2-dim grid equilibrium points:

Theorem 4 *(a) A general case of unstability: if* $\sigma_1(1) < 1$ *or* $\sigma_2(1) < 1$, *if* $\sum_{j=1}^{[n_1/2]}\sigma_1(j)=0(n_1^{\alpha/2})$ *with* $0 < \alpha < 1$, *if* $(\frac{n_2}{n_1})=O(n_1^{-\alpha/2})$ *and if* $f_l \geq a_l > 0$, $l = 1, 2$, *then for large enough* n_1, *every* (n_1, n_2)-*grid equilibrium is unstable.*
(b) If $\sigma(i)=0$ *for* $i > 0$ *(zero neighbour algorithm) and* $\mu_1 = \mu_2$ *is the uniform distribution over* $[0,1]$, $x^* = ((\frac{2i_1 - 1}{2n_1}, \frac{2i_2 - 1}{2n_2}))_{(i_1, i_2) \in I}$ *is the unique grid equilibrium. For any* $c > \sqrt{3}$ *there exists* n_0 *such that:* $n_1 \geq n_0$ *and* $n_2 \geq cn_1$ *imply that* x^* *is unstable. For instance, when* $c = 2$, x^* *is unstable as soon as* $n_1 = 6$ *and* $n_2 = 12$.
(c) A stability case: if $n_1 = 1$, $K_2(I, \mu, \sigma)$ *is a Kohonen string. We assume that* x^{2*} *is a stable equilibrium of* $K_1(I_2, \mu_2, \sigma_2)$. *Let* $x^{1*} = \int_0^1 \omega\mu_1(d\omega)$. *Then there exists an* $\eta > 0$, *such that* $Var(\mu_1) < \eta$ *implies* $x^* = x^{1*} \otimes x^{2*}$ *is a stable equilibrium of* $K_2(I, \mu, \sigma)$ *(see [3] for a proof).*

All the simulations agree with these theoritical results.

References

[1] C. Bouton, G. Pagès, Self-organization and convergence of the one-dimensional Kohonen algorithm with non uniformly distributed stimuli, *Stoch. Proc. and their Appl*, **47**, 1993, p.249-274.

[2] M. Cottrell, J.C. Fort, Etude d'un algorithme d'auto-organisation, *Annales de l'Institut Henri Poincaré*, **23**, 1986, n°1, p.1-20.

[3] J.C. Fort, G. Pagès, Sur la convergence *p.s.* de l'algorithme de Kohonen généralisé, note aux CRAS, série I, 1993, **317**, p.389-394 (a detailed paper is scheduled)

[4] T. Kohonen, Analysis of a simple self-organizing process, *Biol. Cybern.*, **44**, 1982, p. 135-140.

[5] T. Kohonen, Self-organizing maps: optimization approaches, *Artificial Neural Networks*, T. Kohonen et al. ed., Elsevier Science Publishers B.V., 1991, p.981-990.

[6] H.J. Kushner, D.S. Clark, Stochastic Approximation for constrained and unconstrained systems, *Applied Math. Science Series*, **26**, Springer, 1978.

[7] H. Ritter, T. Martinetz, K. Schulten, *Neural Computation and Self-organizing Maps*, Addison-Wesley, 1992, New-York, 306p.

Reordering Transitions in Self–Organized Feature Maps with Short–Range Neighbourhood

R. Der and M. Herrmann
Universität. Leipzig, Institut für Informatik
PSF 920, D–04109 Leipzig, F. R. Germany

Abstract

We report a reordering phase transition in Kohonen's self–organized feature maps with short neighbourhood range. The transition is established for the specific case of mapping a two–dimensional square shaped input distribution onto a square lattice of neurons. Below a critical value of the neighbourhood width $\sigma^* \approx 0.49$ the symmetry of the globally ordered state is broken spontaneously. This is predicted theoretically by a stability analysis in terms of a Langevin approach to the dynamics of the system and supported by numerical simulations which clearly show the spontaneous emergence of hexagon–shaped receptive fields below the critical value of σ. Our results show that for a hexagonal lattice of neurons a more reliable mapping is achieved since the phase transition which is accompagnied by faults in the map is avoided.

1 Introduction

Self–organized feature maps generated by Kohonen's algorithm play an important role in many applications of neural nets. Their capability of generating a topology preserving vector quantization of real–world input spaces is of interest in data reduction, feature extraction, and in developing low–dimensional internal representations of high–dimensional input spaces. Quite generally, Kohonen's feature maps may be considered as a non–linear principal component analyzer.

In practical applications of Kohonen's algorithm it is often observed that the properties of the map generated by the algorithm may depend sensitively on the choice of parameters and the topology of the neural net. In the case of the principal component analyzer a phase transition connected with the partial loss of the topografic properties is known to occurr with increasing influence of the minor components, cf. [6] as a consequence of a topological conflict between the data manifold in input space and the net topology.

The present paper reports a new kind of phase transition resulting from a conflict between the shapes of the elementary cell of the neuron lattice and that of the receptive fields (domains of the neurons) which may be viewed as the images of the elementary cells of the neuron lattice as mapped into the input space. The shape of the elementary cells is prefixed by the user in defining the lattice of the neurons whereas the equilibrium shape of the receptive fields largely is the result of a payoff between the elastic forces arising from the neighbourhood cooperation between the neurons and the tendency to decrease the reconstruction error. For a large neighbourhood range $\sigma \gg 1$ the elasticity effects are dominant favouring under suitable conditions a conformal mapping of the elementary cells, whereas for smaller σ those are counterbalanced by the reconstruction error effect favouring hexagonal receptive fields. For $\sigma \leq \sigma^* \approx 0.49$ the latter effect dominates. Hence putting the neurons onto a **square lattice** and shrinking (cooling) σ leads to a series of phase transitions that eventually results in an equilibrium configuration corresponding to **hexagon–shaped domains** of the neurons. This *reordering* transition is not to be confounded with the order–disorder *"melting"* transition occurring in the case of both high values of the learning parameter $\epsilon > \approx 0.5$ and small values of σ, where *any* order is disrupted.

Our results show that by choosing a hexagonal lattice of neurons (hexagonal elementary cell) the phase transition which is accompagnied by faults in the map is avoided what leads to a more reliable mapping. Hence of the two possible lattice types proposed already by Kohonen, cf. [5], our results clearly favour the hexagonal lattice.

2 Kohonen's algorithm

For the purpose of demonstrating the above mentioned effect we consider a Kohonen map [5] of a two–dimensional input space to a two–dimensional square lattice of the neurons.

More specifically there are N^2 neurons situated at sites $\vec{r} = (r_x, r_y)$; $r_x, r_y \in \{1, \ldots, N\}$ that receive randomly chosen inputs $\vec{v}$ from the square shaped data distribution $\vec{v} \in [0,1] \times [0,1]$ via connections $\vec{w}_{\vec{r}} \in \mathcal{R}^2$. The latter evolve according to Kohonen's learning rule that is taken as usual

$$\Delta \vec{w}_{\vec{r}}(t) = -\epsilon h_{\vec{r}, \vec{r}^*}(\vec{w}_{\vec{r}} - \vec{v}) \tag{1}$$

where $\vec{r}^*$ denotes the best matching (*winner*) neuron defined by $\|\vec{w}_{\vec{r}} - \vec{v}\| \geq \|\vec{w}_{\vec{r}^*} - \vec{v}\|$ $\forall \vec{r} \neq \vec{r}^*$. The components of $\vec{w}_{\vec{r}}$ may be viewed as the coordinates of the image of the neuron at lattice site $\vec{r}$. This defines the map of the lattice of neurons into the input space. The algorithm subdivides the input space into a disjunct set of domains of the neurons (*Voronoi tessellation*) the topological order of the input space being conserved as far as possible.

We study short ranged neighborhood functions being defined by

$$h_{\vec{r}\vec{r}^*} = \begin{cases} \exp\left(-\frac{\|\vec{r} - \vec{r}^*\|^2}{2\sigma^2}\right) & \text{if } \|\vec{r} - \vec{r}^*\| \leq 1 \\ 0 & \text{otherwise} \end{cases} \tag{2}$$

3 Analysis of short–range SOFM's

We apply a Langevin approach for the analysis of of the dynamics of the map of a two–dimensional square–shaped input set onto a square array of neurons in the vicinity of the presupposed stationary state $\vec{w}_{\vec{r}}^{(0)} = \vec{r}$. An analysis of the dynamics of the map is given in terms of the Fourier amplitudes of the deviations $u_{\vec{r}} = \vec{w}_{\vec{r}} - \vec{w}_{\vec{r}}^{(0)}$ of the synaptic vectors from their respective equilibrium positions $\vec{w}_{\vec{r}}^{(0)}$, namely $\vec{u}_{\vec{k}} = \frac{1}{N} \sum_{\vec{r}} e^{i\vec{k}\vec{r}} u_{\vec{r}}$

Kohonen's algorithm can be rewritten as a generalized Langevin equation [2] which in the linear region takes the simple form

$$\frac{\partial}{\partial t} \vec{u}_{\vec{k}} = -\mathbf{B}(\vec{k})\vec{u}_{\vec{k}} + \vec{f}_{\vec{k}}(t) \tag{3}$$

where $\vec{f}_{\vec{k}}(t)$ is the noise and $\mathbf{B}$ is a matrix of generalized friction coefficients. The stability of the modes is governed by the signs of the eigenvalues of the matrix $\mathbf{B}$. If at least one of these eigenvalues, which have been calculated analytically [6], is less than zero the corresponding mode is instable. The analysis discovers regions of instability which increase with decreasing values of σ. If $\sigma > \sigma^* \approx 0.49$ the stationary state with square shaped domains is stable with respect to any wavelength. If $\sigma^* > \sigma > \sigma^{\times} \approx 0.45$ patterns of relatively long wavelength arise, unidirectional modes, i.e. with either k_x or k_y small being preferred.

4 Numerical results

Simulations of (1) have been performed at various network sizes $(N^2 = 4^2..64^2)$ and at $\sigma = 0.2$. Since a whole bunch of modes is instable for this value of σ symmetry breaking is to be expected due to the selection of a small number of different $\vec{k}$.

If, in particular, $\sigma < 0.45$ any mode with either $k_x = 0$ or $k_y = 0$ is destabilized which leads to a symmetry breaking scenario, i. e. we may get at least domainwise a shifting of rows (or columns) against each other. The modes with large values of $|k_x|$ (or $|k_y|$) will have the best chances for winning the competition since they correspond to short range periodicitiy in the row shifting pattern. For the extreme case $|k_y| = 2\pi/N$ the mode $\vec{u}_{k_y}$ corresponds to a periodicity of 1, i.e. neighbouring rows interlace, cf. Figure 1b. In our simulations, however, deviations from the square lattice occur domainwise either in vertical or horizontal direction, cf. Figure 1a. It is this effect which causes the faults in the map. The arousal of a global order according to the new symmetry is expected to take place only at very large times. On the other hand, if the network is initialized close to a state with unidirectional symmetry breaking it remains in this state.

5 Conclusion

The analytical as well as the numerical results presented so far are of importance for both the theoretical understanding and practical use of Kohonen maps. A first point to be addressed are implications to the property of topological ordering which has been proved for one–dimensional Kohonen nets in Ref. [4]. We have shown that in the two–dimensional case the existence of instabilities for small values of σ excludes the possibility of globally regular ordering in the parameter region $\sigma < \sigma^*$ in the case that the neurons are arranged on a square lattice. Therefore, $\sigma > \sigma^*$ is suggested as a necessary condition for convergence to a globally ordered state in two and (for an accordingly changed value of σ^* also for) higher dimensions.

Of more practical relevance is the predicted behaviour at small values of σ. Although the reconstruction error further decreases with σ decreasing below σ^*, the topography of the map is deteriorated, where, practically, faults between domains of unidirectional symmetry breaking are more obstaculous than slight deviations in the domain of a neuron. Our results imply that hexagonal neural connectivities which where mentioned already in [5] are more stable at small values of σ than usually applied square connectivities.

Figure 1: *a) Domainwise symmetry breaking. b) Global symmetry breaking.*

Acknowledgement: THE REPORTED RESULTS ARE BASED ON WORK DONE IN THE PROJECT LADY SPONSORED BY THE BFMT (F. R. GERMANY) UNDER GRANT 01 IN 106B/3.

References

[1] I. Csabai, T. Geszti, G. Vattay (1992): Criticality in the one–dimensional Kohonen neural map. *Physical Review A* **46**, R6181.

[2] R. Der, M. Herrmann, Th. Villmann (1993) Spontaneous symmetry-breaking effects in self–organized feature maps: A Ginzburg–Landau approach, To appear in *Phys. Rev. E*.

[3] R. Der, Th. Villmann (1993) Dynamics of self–organized feature mapping. In: J. Mira, J. Cabestany, A. Prieto: *New trends in neural computation.* LNCS 686. Springer, Berlin, Heidelberg, New York. p. 312–315.

[4] E. Erwin, K. Obermayer, K. Schulten (1992) Self–organizing maps: Stationary states, metastability and convergence rate. *Biol. Cybern.* **67** 35–45. & Self–organizing maps: Ordering, convergence properties and energy functions. *Biol. Cybern.* **67** 47–55.

[5] T. Kohonen (1984) *Self–organization and associative memory.* Springer Series in Information Science **8**, Springer, Berlin, Heidelberg, New York.

[6] H. Ritter, T. Martinetz, K. Schulten (1992) *Neural computation and self–organizing maps.* Addison Wesley, Reading, Massachusetts.

Speeding-up Self-organizing Maps: The Quick Reaction

Jürgen Monnerjahn

Zentrum für Kognitionswissenschaften
Universität Bremen, D–28334 Bremen

1 Introduction

One of the most serious problems in the usage of neural networks is their computational cost. In this paper, an algorithm is introduced for speeding-up learning and recalculating of self-organizing feature maps (SOFM) and extensions of this network type (Kohonen 1982, Ritter et al. 1991).

2 The SOFM Algorithm and its Computational Cost

The SOFM learning algorithm contains the following steps:

1. Initialize all weight vectors with random values.
2. Choose an input signal.
3. Select the winner by finding the weight vector with minimal distance to the input signal.
4. Apply learning rule to all weight vectors inside a defined neighbourhood around the winner neuron.
5. Decrease learning rate and neighbourhood radius, continue at step (2.)

For an estimation of the overall cost of the algorithm the steps (1.), (2.) and (5.) can be neglected. (1.) is only applied once, (2.) and (5.) only consist of few simple operations. (3.) and (4.) remain for further consideration. In a typical implementation the selection of the winner (3.) requires a distance calculation and comparison for each weight vector of the network (complete search). The adaptation of the weight vectors (4.) is the most expensive step during the first phase of learning, since the neighbourhood radius is big and most or all neurons have to be adapted. But after this first phase, the neighbourhood radius decreases very quickly and there is no need to apply the learning rule to more than a small area of the network. For most learning steps the computational cost of the adaptation is far smaller than for finding the winner neuron. Thus a quicker algorithm for finding the winner can significantly increase network performance.

3 The Quick Reaction Algorithm

The disordered state of a randomly initialized SOFM-network (figure 1) does not show any possibility for finding the winner without calculating the distance to the input signal for each weight vector. But after a few learning steps the weight vectors have developed a spatial order (figure 2): neighbouring neurons include similar weight vectors. The quick reaction algorithm (QR) uses this order to find the winner without complete search.

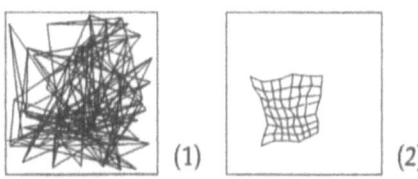

(1)   (2)

The easiest way to demonstrate the principle of the algorithm is to apply it to a small example. Figure 3 shows the weight vector positions of a 10×10-network in its two-dimensional input space. Moreover, the position of an input vector $\mathbf{v}$ is marked. For this input the winner neuron is determined as follows:

First of all, the distance $|\mathbf{v}-\mathbf{w}|$ is computed for a 3×3-grid of neurons (figure 4). The centre of this grid is the centre of the M×N-network: (M div 2, N div 2).[1] The vertical and horizontal distances between grid neurons are determined by a search parameter sp. In an M×N-network sp is assigned the start value sp := max(M,N) div 3, ie in a network of 10×15 neurons sp := 15 div 3. In our example sp := 10 div 3 = 3. The neuron with minimal distance $|\mathbf{v}-\mathbf{w}|$ among these nine is the winner of the first search step. In the next search step this neuron is the new centre for the 3×3-grid. If the winner is the same neuron as in the last step, the search parameter sp (ie the range of the grid) is reduced by sp := sp div 2, and the winner neuron of the new grid is determined. If sp has already reached the value 1, it is not further reduced.

If the new grid centre is near the border of the network, such that not all grid coordinates are inside the network (figure 5), the outside grid points are ignored and one of the existing neurons is selected as the winner.

The shrinking of the grid and the determination of a new grid centre is repeated until: (a) sp = 1 and (b) the winner is the centre of the 3×3-grid (figures 6, 7).

For the n-dimensional case the algorithm can be summarized as follows:

1. centre of a 3×3×...×3-grid in an $N_1 \times N_2 \times ... \times N_n$-network:
   m := (N_1 div 2, N_2 div 2, ... , N_n div 2)
2. m^{old} := m
3. sp := max(N_1, N_2, ... , N_n) div 3
4. determine new grid centre m by the condition $|\mathbf{v}-\mathbf{w}_m| \le |\mathbf{v}-\mathbf{w}_r|$.
   r are the neurons of the 3×3×...×3-grid.
5. if (sp > 1) and (m^{old} = m) then sp := sp div 2
   else if m^{old} = m then m is the winner neuron for the input $\mathbf{v}$.
6. m^{old} := m, continue at step 4.

---

[1]div means integer-division.

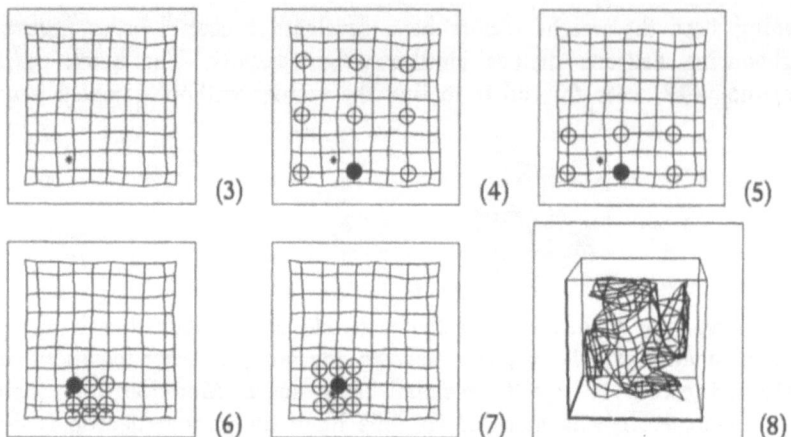

(3)   (4)   (5)   (6)   (7)   (8)

4 Computational Cost of the Quick Reaction

For one size of the search grid typically no move or one move of the grid is necessary. Therefore it takes between 2^n and $2*3^n-2^n$ distance calculations until the grid is shrunk. But how many shrinking operations are necessary for a given start value sp_i? The algorithm runs until $sp = 1$. Thus x can be estimated as:

$$sp_i * (1/2)^x = 1$$
$$x = \ln (1/sp_i) / \ln (1/2)$$

Each learning step takes between $(\ln(1/sp_i) / \ln(1/2) +1) * 2^n$ and $(\ln(1/sp_i) / \ln(1/2) +1) * (2*3^n-2^n)$ distance calculations[2].

In a 10×10-network it takes the quick reaction between 11 and 38 distance caculations (11% to 38%), in a 20×20-network 15 to 52 (3.75% to 13%), and in a 100×100-network 24 to 85 (0.24% to 0.85%). The bigger the network the smaller the percentage of distance calculations.

5 Qualities of the Quick Reaction in Application

As mentioned above, the quick reaction can only find the minimal distance if the weight vectors are ordered. But the algorithm can be used even for an initial network without disadvantage. In this case the algorithm will probably not find the absolute minimum, but this is not really necessary: The network unfolds during the first few learning steps in spite of this error and after that, the weight vectors show the order the quick reaction needs for »correct« results.

Even if the network is completely ordered, there is no guarantee that the absolute minimum is found. Especially in big and folded networks errors are possible. But even in these cases, the quick reaction selects a neuron with minimal distance value in its neighbourhood (local minimum).

For the overall training result it is obviously not crucial that the algo-

---

[2]This estimation does not apply integer division. Thus the computational cost of the worst case is »overestimated«.

rithm sometimes selects a local minimum. Single errors can hardly influence the final result.

Of course, the reliability of the quick reaction had to be examined in several simulations. Randomly initialized networks of different size and different number of dimensions of input space and network structure were trained by 10 000 learning steps with random input vectors. At each learning step, the old and the new algorithm for winner selection were applied, their results were compared and hits and misses were counted.

(a) In various simulations with the same number of dimensions of input space and network structure, the quick reaction produced several errors at the beginning of learning. After a few learning steps, only single errors occurred, and after the first 2 000 steps, all results of the quick reaction were correct. (b) In simulations with more input dimensions than network dimensions (figure 8), the error rate grew with the network size. But in most test applications, no difference in quality between the input space representation of the standard learning algorithm and the modified algorithm could be found. Only in extremely folded networks like one-dimensional networks for approximating a »Peano curve« the QR produced worse results.

The QR has been used extensively for map building and for learning visuo-motor control of a simulated robot (Monnerjahn 1994). In these applications the QR worked with remarkable reliability.

An earlier approach to speeding-up the search task is the Probing algorithm (Lampinen and Oja, 1989) which is based on a kind of gradient descent. This algorithm corresponds to the last phase of the QR, where $sp = 1$ and the centre of the search grid is moved from one neuron to a neighbouring one. A detailed comparison shows that QR and Probing require roughly the same number of distance calculations. A significant difference is that the QR has a far better robustness against local minima while searching in folded networks. Thus the QR could be the method of choice for a wide range of learning problems.

6 References

Kohonen, T. (1982) Self-organized formation of topologically correct feature maps. Biological Cybernetics, 43, 59–69.

Lampinen, J., Oja, E. (1989) Fast Self-Organization by the Probing Algorithm. Proc. Int. Joint Conf. on Neural Networks, Vol. 2, 503–507.

Monnerjahn, J. (1994) Visuomotorische Robotersteuerung mit selbstorganisierenden Karten - Entwicklung und Simulation effizienter Lernalgorithmen. Bericht des Zentrums für Kognitionswissenschaften. Universität Bremen. (forthcoming).

Ritter, H., Martinetz, T., Schulten, T. (1991) Neuronale Netze. 2nd edition. Addison Wesley (Deutschland).

Dynamic Extentions of Self-Organizing Maps

Josef GÖPPERT and Wolfgang ROSENSTIEL
Lehrstuhl für Technische Informatik, University of Tübingen
Sand 13, 72076 Tübingen, Germany

1 Introduction

The self-organizing map (SOM) is a commonly used clustering algorithm, which represents training data, by codebook vectors and creates an ordered map. A typical application is the classification of feature vectors. Though topology preservation is representing a totally new principle in the domain of vector quantisation and function approximation the training is reasonably fast [Göp93b].

The training algorithm was inspired by cortical columns and mutual activation of columns and regions in the cortex. Neurons and cortical columns don't change their activity instantly. They are increasing or decreasing the activity in a smooth way. A closer look to the human cortex reveals also dynamic properties which may lead to completely new properties of artificial neural nets. This comportment is supposed to produce attentional effects.

1.1 Dynamical Behaviour

The SOM takes into account cooperative aspects of neighbouring neurons by using individual activation for all neurons. The activation reflects the distance of one unit to the winner in the grid of neurons and is used for the calculation of the adaptation strength. So, similar adaptation of neighbouring units leads to similar features and in a global view of the map, to a topology preserving mapping of the training vectors. From a biological point of view the activation of a neuron raises its attention to the next stimulus. Why not using this activation in the following step for a modified training?

The distance to the input vector defines the probability of winning. Increasing of attention can be modelled by a reduction of the distance to the input vector according to its activity. This can be done by decreasing the euclidean distance or simply by reducing the set of eventual winners to neurons which had been sufficiently activated in the previous step. The winner position from one step to the next will not change abruptly which leads to a more continuous variation of the excitation (attention) center.

Continuous variation of the attention center also necessitates an ordered presentation of the input vectors. In fact: Random choice of input stimuli is biologically not plausible. Most types of training data have any type of one- or

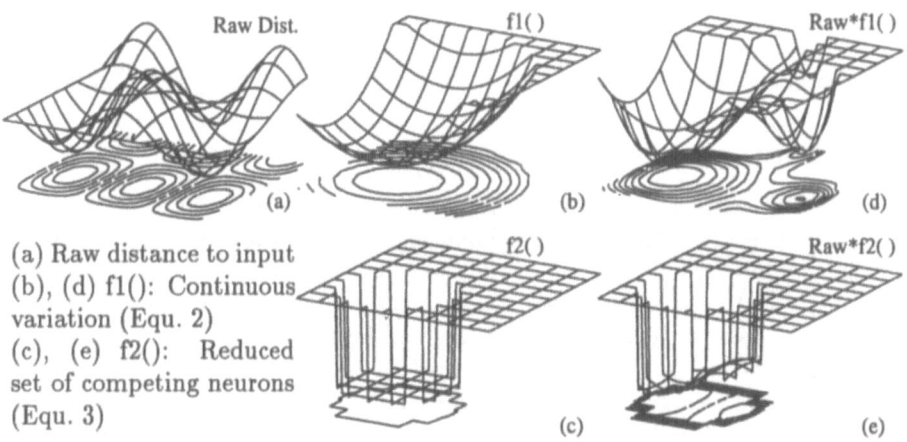

Raw Dist.   f1( )   Raw*f1( )

(a)   (b)   (d)

(a) Raw distance to input
(b), (d) f1(): Continuous
variation (Equ. 2)
(c), (e) f2(): Reduced
set of competing neurons
(Equ. 3)

f2( )   Raw*f2( )

(c)   (e)

Figure 1: Influence of the attention onto the Distance.

multidimensional ordering (temporal order, spatial position, continuous varying parameters, etc..). Normal SOM do not take advantage of this fact, even if it would help in exploring the data space. Adapted methods of presentation order would be to go back and forth (right and left) in the ordered list (matrix) of training vectors, to follow predefined paths or to perform some kind of random walk.

2   Modified Training for a dynamic map

Given the activation center $\mathbf{C}_A$ at time t, which is considered as center of attention. A neuron i is situated in the grid of neurons at position $\mathbf{C}_{Ni}$. The grid distance D_C of one neuron to the attention center acts on the Distance D_i (Equ. 2) that is to say to change its probability to win. Another method with simular effect ist to exclude distant neurons from the competition (Equ. 3):

$$D_i \quad = \quad f(\|\mathbf{C}_A - \mathbf{C}_{Ni}\|) \cdot \|\mathbf{X} - \mathbf{W}_i\| \quad \|\ldots\| \; Euclidean \; Distance \quad (1)$$

$$f(D_C) \quad = \quad 1 + \frac{D_C^2}{D_\sigma^2} \qquad\qquad\qquad Continuous \quad (2)$$

$$or$$

$$f(D_C) \quad = \quad \begin{cases} 1 & if \; D_C < D_\sigma \\ \infty & else \end{cases} \qquad Subset \; of \; Winners \quad (3)$$

$$\mathbf{C}_A \quad = \quad \mathbf{C}_{Nw} \quad w \ni D_{Cw} \leq D_{Ci} \; \forall i \qquad New \; activation \; center \quad (4)$$

This new distance expresses at once the similarity of the input vector to the codebook neuron i and attention of the neurons. The way how this is done is shown in figure 1. The euclidean distances (a) are amplified with the attentional function (b) and (c) and produces the *attentional distances* (e) and (f).

The training principle of the self organizing map stays unchanged [Koh84]. Adaptation of neurons around the winner is defined by the adaptation function.

One supplementary parameter is the width D_σ of the attentional function. Good results have been obtained with big starting values in order to allow global organization of the map and a smooth reduction during the training.

2.1 Context sensitive recognition

The training patterns have to be presented in roughly ordered sequence. So they get trained together with neighbouring vectors in a given context. Due to the modified training, these vectors are also represented by neighbouring neurons, and the contextual order will lead to a *context preserving organization* of the map. Like in SOMs similar vectors will be represented by neighbouring units, but only, if they have been trained in a common context. Dynamic training can be seen as an extension from topological feature space into spatio-temporal space of continuous varying stimuli. A combination of this map with interpolation principles [Göp93a] may lead to a completely new representation of continuous spatio-temporal data in a continuous ordered map.

Context leads also to new principles for the recognition of ambiguous data. If the same stimulus occurs in several ways and in different contexts, the use of some neighbouring inputs may allow to build up a context sensitive recognition procedure, by finding an activation center which best represents a set of input vectors.

3 Simulation Results

To show the performance of this new training method, some topological difficult Lissajous' figures have been chosen. A circle can be represented by two sine curves with a phase shift of $\frac{\pi}{2}$. Presented to a one dimensional SOM, it will be approximated in a good way. If the phase shift is reduced to, say $\frac{\pi}{8}$, the circle is distorted into an ellipse. Trying to approximate this figure by a SOM results either in a stretched map (big neighbourhood) or in a meandered map inside the ellipse (figure 2). Dynamically trained maps lead to a context preserving approximation.

In the second example, two sine waves with different frequencies produces an "∞ like" shape. Here standard SOM produces a quite good approximation of the shape, which changes the direction near the crossing point. This might be correct from a topological point of view, but is a contextual error. A dynamically trained map is able to follow the shape in a contextual correct manner. The crossing point is an ambiguous point, which might also be correctly recognised if presented in its context.

Both examples lead to simular results, using either the most active subset of neurons for the competition (Equ. 3) or the continuous varying attentional function (Equ. 2). In fact, for the left example of figure 2 (ellipse) the subset method was used and for the right one, the continuous method. In both examples the neighbourhood for training was not reduced to zero, to come to a more discriminative graphical representation. A reduction to zero would push

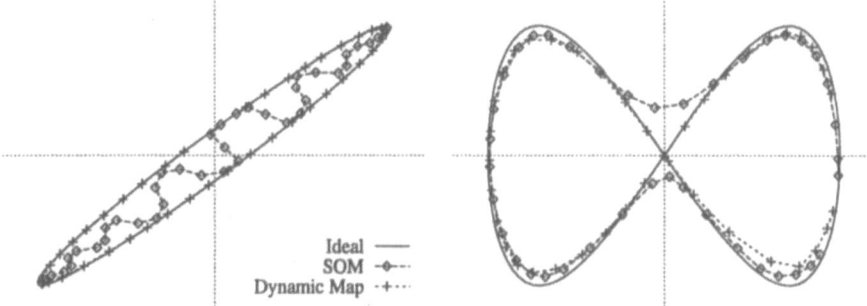

(a) Clustering properties in an Ellice.   (b) Topological vs. contextual order.

Figure 2: Approximation properties of the dynamic training methode.

the prototypes closer to their optimal position, but would not change at all the topological aspects.

4   Conclusions

A new training algorithm for self organizing maps was presented. Inspired by natural neurons and cortical columns an aspect of attention was introduced which produced "Dynamic behaviour". Properties were introduced which modify the topology preservation towards a context preserving training of the map. Some simple example show the basic difference to standard SOM training. It is expected that these properties allow a better representation of context by neural maps and show higher performance in the evaluation of ambiguous or noisy data. The algorithm is adapted to problems of ordered data (e.g. time series: an ellipse is a sampled sine curve with a window size of 2) and may provide a new contextual storage and recognition principle.

References

[Göp93a] J. Göppert and W. Rosenstiel. Topology-preserving interpolation in self-organizing maps. In *Proc. of NeuroNimes 93*, Nanterre, France, 1993. EC2.

[Göp93b] J. Göppert and Rosenstiel W. Self-organizing maps vs. backpropagation: An experimental study. In *Proc. of Workshop on DMMSP* Gliwice, Poland. 1993.

[Koh84]   T. Kohonen. *Self-Organization and Associative Memory*. Springer Verlag Heidelberg New York Tokyo, 1984.

Feature Selection with Self-Organizing Feature Map

Jukka Iivarinen, Kimmo Valkealahti, Ari Visa, Olli Simula

Helsinki University of Technology

Laboratory of Information and Computer Science

Rakentajanaukio 2 C, FIN-02150 Espoo

Finland

1 Introduction

Feature selection is an important phase in pattern recognition. It has also an important role in neural network methods even though it is often neglected. The fundamental function of feature selection is to find a set of features that will represent the pattern vector in a most optimal way. Only the information that is either redundant or irrelevant to the classification task is removed from the pattern vectors. The dimension of the feature vector is usually smaller than the dimension of the pattern vector so the computation time and the memory requirements are greatly reduced.

Feature selection provides also an indication about the discriminatory potential of the features. Thus it could be used as a tool to determine whether it is necessary to look after additional features that would allow the improvement of the performance of the classifier [Devijler et al. 1982].

It is not a trivial problem to find the best set of features or the best transformation. The features must be selected according to the given problem. For feature selection the principal component analysis (PCA), also known as the Karhunen-Loéve expansion, is often applied. However, it is not very illustrative. A covering collection of transformations that are used in image processing applications is found in reference [Jain 1989].

In this paper a new tool to feature selection is described. This approach is based on the visualization capability of the Self-Organizing Feature Map.

2 Feature Selection with Self-Organizing Feature Map

Some neural network methods can also be used in feature selection. The suggested method is based on the Self-Organizing Feature Map [Kohonen 1990].

SOM is used in a feature selection as follows. Take n features. Train the map with a large set of these features. Label the map with a set of preclassified samples. If the classes in the labeled map overlap add a new feature or replace a feature with a new one. If the feature does not improve the separability of the classes, the feature will be replaced. Train the map again with a set of new features and label the map. Repeat this procedure until you have a set of features that can reach the desired classification rate. This is a practical method to find a set of features. It is straightforward and easy to implement,

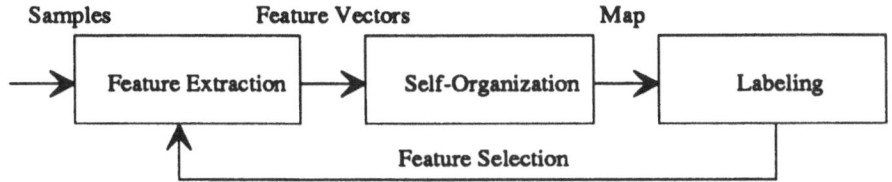

Figure 1: Block diagram of the feature selection with SOM.

and the need of preclassified samples is minimized. The method is illustrated in Figure 1. The features that are chosen to be tested, are found among the features that are suggested in literature.

SOM is a topology-preserving feature map that represents the distribution of features in the feature space. In SOM similar feature vectors are mapped to adjacent map elements. The map is trained in an unsupervised way. During the training phase the map is fed with a large number of unclassified feature vectors.

For a feature vector $\mathbf{x}$ the best matching map element $\mathbf{m}_c$ is presented as an output. The best matching element is determined by

$$\|\mathbf{x} - \mathbf{m}_c\| = \min_i\{\|\mathbf{x} - \mathbf{m}_i\|\} \tag{1}$$

During the training phase the best matching element and its neighborhood are updated. The neighborhood $N_c(t)$ is defined to include all the elements inside a certain distance from the central element. The training process can then be expressed as

$$\mathbf{m}_i(t+1) = \begin{cases} \mathbf{m}_i(t) + \alpha(t)[\mathbf{x}(t) - \mathbf{m}_i(t)] & \text{if } i \in N_c(t) \\ \mathbf{m}_i(t) & \text{if } i \notin N_c(t) \end{cases} \tag{2}$$

where $\alpha(t)$ is a monotonically decreasing scalar function, $0 < \alpha(t) < 1$, that determines the learning rate. The neighborhood $N_c(t)$ is dependent on time t. In the beginning of the learning $N_c(t)$ is quite wide but decays into $N_c(t) = \{\mathbf{m}_c\}$.

After the training phase the map is usually labeled. A set of preclassified feature vectors is fed into the map and the best matching element to each feature vector is calculated. Each map element can be identified to represent a class. The elements that have responded only to one class can be labeled directly to the respective class. If an element gives response to samples of several classes, the labeling can be done according to the majority decision. In this case reselection of features is necessary.

3 Case : Feature Selection for Cloud Classification from NOAA Satellite Images

The suggested method is suitable for problems where exist a lot of samples and a few preclassified samples. The cloud classification is a good example of that.

The cloud images are collected through the Advanced Very High Resolution Radiometer (AVHRR) on board of the NOAA-11 polar orbiting satellite

Cloud Genus	Number of samples	Results with 9 features [%]	Results with 14 features [%]
Cirrus over sea/land	13	84.6	100.0
Cirrus over low clouds	9	88.9	100.0
Thick Cirrostratus	15	66.7	73.3
Altostratus/Altocumulus	14	57.1	92.9
Stratus/Stratocumulus	12	100.0	91.7
Cumulus	9	100.0	100.0
Cumulonimbus	20	95.0	95.0
Nimbostratus	15	60.0	100.0
Total	107	80.4	93.5

Table 1: The test results based on preclassified cloud samples as a function of the number of features.

[Visa et al. 1991]. Experienced meteorologists from the Finnish Meteorological Institute have collected 218 preclassified samples. The samples were collected from 43 summer day images and they contained small well-defined regions of one cloud type. The samples were split into two separate sets, a learning set and a test set. The learning set was used in labeling the map. The same set was also used in testing the map in order to determine the discriminatory potential of the selected features. The goal for the classification rate of the learning samples was 90 per cent. If the goal were set to 100 per cent the map would become too tuned to the learning set.

The suggested features contain both texture and spectral features. Texture features are chosen among the Haralick features [Haralick et al. 1973] and spectral features among the features suggested in reference [Gu et al. 1991].

The size of the Self-Organizing Feature Map is crucial. If the map is too small, there does not exist enough elements to represent each class, and the classes will overlap. The map size was 11-by-14 elements which was obtained by experience.

The training of the map was done in two phases. First the map was topologically ordered with 60000 unclassified samples. Learning rate $\alpha(t)$ was 0.6 and neighborhood $N_c(t)$ was 10. During the second phase the map was fine-tuned with 200000 unclassified samples. Learning rate was 0.2 and neighborhood was 4.

Learning rate $\alpha(t)$ has a great impact during the training phase. If the initial value of $\alpha(t)$ is too small, the map will not be able to organize properly. The learning rate is smaller in the second phase.

The initial size of the neighborhood $N_c(t)$ is also important. If $N_c(t)$ is too small in the beginning, the map is not organized properly. In the first phase, $N_c(t)$ is usually taken almost equal to the diameter of the map. In the second phase, however, it is smaller.

The feature selection is made as in Figure 1. After the evaluation of the suggested texture features, the nine texture features were selected. The classification rate was 80 per cent for the learning samples. The suggested spectral features were evaluated next. Finally the nine texture features and the five spectral features were selected. The classification rate was 94 per cent for the learning samples. This can be considered as a sufficient result. The test results are given in Table 1.

The map was also tested with the test samples. The test samples were classified 79 per cent right with all the features. This shows that the map is not too tuned to the learning samples and it can be used in a real classification task.

4   Conclusions

The suggested method is a practical tool for feature selection. It does not select features but merely tests the discriminatory potential of features. The features that are chosen to be tested should be found by experience. SOM is an illustrative tool in the sense that it shows the clustering effect of features.

Only a few preclassified samples are needed. However, thousands of unclassified samples are needed during the training phase. This is a minor drawback because unclassified samples are usually easy to obtain.

The size of the map, the learning rate $\alpha(t)$ and the neighborhood $N_c(t)$ must be chosen carefully according to empirical experiments.

The suggested method is general in the sense that it is not dependent on the features. It also offers an efficient way to visualize high-dimensional feature vectors.

References

[Devijler et al. 1982]   Devijler P. and Kittler J. (1982), Pattern Recognition: A Statistical Approach. Prentice-Hall International, Inc.,London.

[Gu et al. 1991]   Gu Z. Q., Duncan C. N., Grant P. M., Cowan C. F. N., Renshaw E., Mugglestone M. A. (1991), Textural and spectral features as an aid to cloud classification. Int. J. Remote Sensing, Vol. 12, No. 5, pp. 953-968.

[Haralick et al. 1973]   Haralick R. M., Shanmugan K., Dinstein I. (1973), Textural Features for Image Classification. IEEE Transactions on Systems, Man and Cybernetics, Vol. SMC-3, No. 6, November.

[Jain 1989]   Jain A. K. (1989), Fundamentals of Digital Image Processing. Prentice Hall, Englewood Cliffs.

[Kohonen 1990]   Kohonen T. (1990), The Self-Organizing Map. Proceedings of the IEEE, Vol. 78, No.9, September.

[Visa et al. 1991]   Visa A., Valkealahti K., Simula O. (1991), Cloud Detection Based on Texture Segmentation by Neural Network Methods. Proceedings of IEEE International Joint Conference on Neural Networks, Singapore, November 18-21, Vol. 2, pp. 1001-1006.

Unification of Complementary Feature Map Models

O. Scherf, K. Pawelzik, F. Wolf and T. Geisel

Institut für Theoretische Physik, Universität Frankfurt

60054 Frankfurt/M., Germany

email:klaus@chaos.uni-frankfurt.d400.de

1 Introduction

Selforganizing feature maps serve as models for the organization of primary sensory areas and have many applications in technical pattern recognition. In most models the evolution of receptive field centers $w : X \to \Omega$ is driven by the excitation $e_w(v)$ in the neuronal tissue: $\dot{w} = < (v - w)e_w(v) >$, where $< \ldots >$ denotes the average over the stimulus distribution $P(v)$. The models differ, however, in the way $e_w(v)$ is determined. In the Kohonen model [1] a hard competition for the stimulus takes place and the excitation then spreads into the neighborhood of the winning neuron. This very effcient algorithm, however, generates only simple network excitations, which is biologically unrealistic. Complementary to that, the elastic net algorithm [?] uses a neighborhood in input space. Here, however, the "elastic" neuronal interaction has no obvious biological interpretation. In this contribution we present a model unifying both approaches. The elastic net and the Kohonen model turn out to be asymptotic cases of this convolution model. For the elastic net we obtain the elasticity parameter β directly from a small, but finite neighborhood range in the neuronal layer.

2 Soft Competition and Spead of Activation

We derive our model from two basic principles: soft competition about an input and spread of activation in the neuronal space. On the input side a soft competition governs the response to a stimulus given at v:

$$g_{\sigma_r}(v - w(x)) = \frac{h_{\sigma_r}(v - w(x))}{\int_X h_{\sigma_r}(v - w(x))dx} \tag{1}$$

where h_σ denotes a Gaussian and σ_r the width of the receptive field. In the neural tissue the activation then spreads into an area of width σ_n due to local neuronal interactions, which we model by a Gaussian kernel h_{σ_n} for convenience. These two principles then enter the evolution equation for the receptive fields simply by a convolution:

$$\frac{d}{dt}w(x) = \epsilon \int_\Omega P(v) \int_X g(v - w(x'))h_{\sigma_n}(x - x')dx'(v - w(x))dv \tag{2}$$

For $\sigma_r \to 0$ only the receptive field center $w(x')$ nearest to the stimulus survives and we immediately have

$$\frac{d}{dt}w(X) = \epsilon \int_X P(v)h_{\sigma_n}(x - x')(v - w(x))dv \qquad (3)$$

which is rigorously equivalent to the Kohonen model. For $\sigma_n \to 0$ Eq. 2 yields the first term of the elastic net model [2], corresponding to $\beta = 0$. In the following we discuss the structure formation emerging in our model via linear instability. Thereby we show how the elasticity parameter β relates to the cortical interaction range σ_n.

3 Stability Analysis

Structures as e.g. ocular dominance bands [3] or maps of preferred orientation [4] can be understood by a linear instability of a stationary solution of self organizing feature map models [5, 6]. The resulting patterns emerge by an amplification of unstable modes which are given by the spectrum of positive eigenvalues of the linearized dynamics [7]. While the lengthy but straightforward calculations for our convolution model and for the elastic net will be published elsewhere ([8]), we here only discuss the resulting spectra

$$\lambda(k) = -1 + \frac{<y^2>}{\sigma_r^2}exp(-\frac{k^2\sigma_n^2}{2})(1 - exp(-k^2\sigma_r^2)), \qquad (4)$$

where k denotes the wave number and $<y^2>$ is the variance of the stimulus distribution (Fig.1). For $\sigma_r \to 0$ this reduces to

$$\lambda(k) = -1 + <y^2> k^2 exp(-\frac{k^2\sigma_n^2}{2}) \qquad (5)$$

as has already been shown for the Kohonen model [9]. Structure formation is dominated by the eigenvalue λ^* which first becomes positive. From $max_k(\lambda(k)) = 0$ we obtain from Eq.4

$$<y^2> = \frac{\sigma_n^2 + 2\sigma_r^2}{2}exp(\frac{\sigma_n^2}{2\sigma_r^2}log(1 + \frac{2\sigma_r^2}{\sigma_n^2})), \qquad (6)$$

the condition of marginal stability (Fig.2). In order to compare this to the elastic net we expand Eq.4 around $\sigma_n = 0$ and obtain

$$\lambda(k) \sim -1 + \frac{<y^2>}{\sigma_r^2}(1 - exp(-k^2\sigma_r^2)) - \frac{<y^2> k^2\sigma_n^2}{2\sigma_r^2}. \qquad (7)$$

If the marginal stable k obeys $\sigma_n k << 1$ we have

$$\sigma_r^2 = <y^2> -\frac{\sigma_n^2 <y^2>}{2\sigma_r^2}(1 + log(\frac{2\sigma_r^2}{\sigma_n^2})). \qquad (8)$$

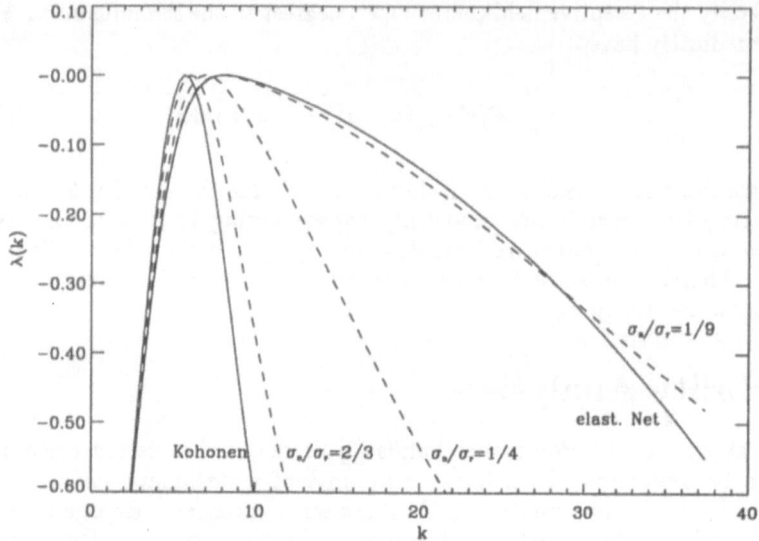

Figure 1: Spectra of modes (Eq.4) of marginal stability ($\max_k(\lambda(k)) = 0$) of the convolution model (Eq.2, dashed lines) compared to the Kohonen model and the elastic net model (lines, the elastic net as in Ref.[2] with $\beta/\alpha = 1/200$ Clearly our model interpolates these complementary approaches at the maximum which is the part of the spectrum relevant for structure formation.

Comparing this to the corresponding condition for the elastic net ([8])

$$\sigma_r^2 = <y^2> - \frac{\beta}{\alpha}(1 + \log(\frac{\alpha}{\beta} <y^2>)) \tag{9}$$

leads to an estimation of the elasticity parameter β from the cortical interaction width:

$$\frac{\beta}{\alpha} = \frac{\sigma_n^2}{2\sigma_r^2} <y^2> . \tag{10}$$

4 Conclusion

We presented a general feature map model which incorporates both, a finite receptive field size σ_r and an arbitrary and discretization independent size σ_n of activation in the neuronal layer. Our convolution model reduces rigorously to the Kohonen case for $\sigma_r \rightarrow 0$ and for small σ_n we derived the otherwise biologically unmotivated elasticity parameter of the elastic net algorithm. In this way our model represents a unifying approach to biological feature map development which because of its simple mathematical structure is accessible to rigorous treatment and might be useful in thechnical applications.

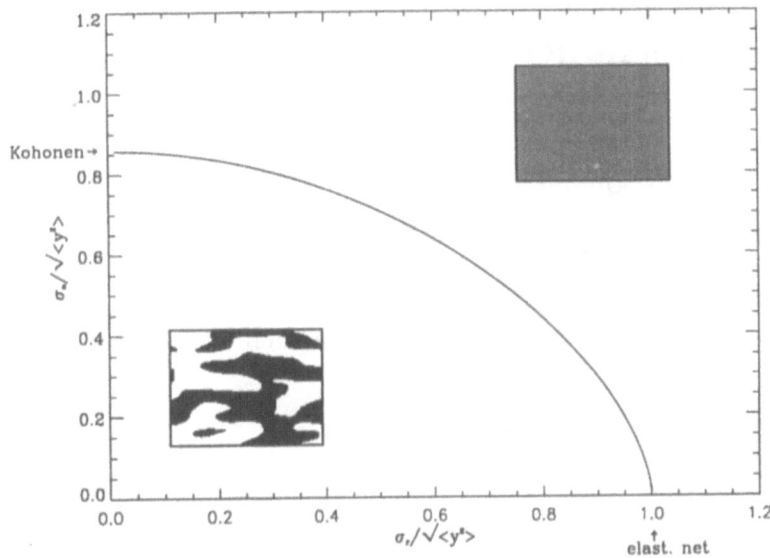

Figure 2: Phase diagram for the interaction ranges σ_r and σ_n normalized to the standard deviation of the stimulus distribution $\sqrt{< y^2 >}$. The line of marginal stability defines the border at which structures emerge.

References

[1] T.Kohonen, *Selforganization and Associative Memory*, Springer, 1984.

[2] Durbin, R., Mitchison, G., Nature **343**, pp 644-647, (1990).

[3] Miller, K.D., Keller, J.B. & Stryker, M.P., Science **245**, pp 605-615, (1989).

[4] Bonhoeffer, T., Grinvald, A., Nature 343,pp 429-431 (1991).

[5] Jones, D.G., Van Sluyters, R.C., Murphy, K.M., J. Neurosci. 11(12), pp 3794-3808 (1991).

[6] Obermayer, K., Blasdel, G.G., & Schulten, K., Phys. Rev. A **45**, pp 7568-7589 (1992).

[7] Ritter, H., Martinez, T., Schulten, K., *Neuronale Netze*, Addison-Wesley, New York, 1990.

[8] O.Scherf, F.Wolf, K.Pawelzik,in preparation.

[9] Ritter, H. & Schulten, K., Biol. Cyb. **60**, pp 59-71 (1989).

Considerations of geometrical and fractal dimension of SOM to get better learning results

H. Speckmann, G. Raddatz, W. Rosenstiel
Lehrstuhl für technische Informatik
Universität Tübingen
72076 Tübingen, Sand 13, Germany
speckman@peanuts.informatik.uni-tuebingen.de

Abstract

In this paper we present a possibility to improve the learning results for different data sets which earlier were difficult or impossible to learn. We choose the calculated fractal dimension of the data set as the geometrical dimension of the selforganizing map for guaranteeing the maps ability of topology preserving. Furthermore we examine the dynamics of the learning process and the final map by exploring the fractal dimension of the space represented by the stored weight vectors and present interesting results.

1   Introduction

Kohonen's selforganizing map (SOM) which has been introduced by T. Kohonen [1] has many applications. So the SOM's algorithm is an efficient alternative to traditional signal processing. But there is still one big problem. There exists no common valid prove of convergence for this efficient algorithm. A couple of theoretic works deal with this problem (e.g. [4]), but they are only valid in the following special cases: input data have to be equally distributed, map's state must be near an equilibrium state, means the adaptation factor has to be already small, and the map's geometrical dimension is one.

In our opinion the input data set's properties have to be considered. So there exist data sets which are very easy to learn using a two dimensional map, e.g. gas spectra or VLSI process data. Others are difficult or impossible to learn, e.g. EEG data or PCB layouts what we are currently trying. Probably these data sets require a larger geometrical dimension of the map.

The SOM's main property is the nonlinear projection to the principal manifolds by finding a n-dimensional layer to approximate the input data. If the geometrical dimensionality of the map is too low, the map tries to approximate the higher dimension by folding itself into the input space. So we have to determine the necessary dimensionality of the input space to avoid this violation of topology preserving. Second a correct dimension of the map leads to better learning results. So for the most evaluation tools of the SOM like spanning trees and component cards, a correct topology preserving is urgent.

2 Fractal dimension and topology measurement

We calculate the information dimension, a method of nonlinear dynamics, of the input space and the space represented by the weight vectors of the map in different learning states. Topology preservation is measured with the waber product introduced in [2].

The first mentioned method bases on the theory of nonlinear dynamic systems, the calculation of the dimension of attractors, characterizing the geometrical structure of an attractor in phase space [3]. Because this number has often a non integer value it is called fractal dimension. We use this method to determine the scaling behaviour of a given data set during decreasing partitioning of the phase space. Transmitting this technique of nonlinear dynamics to a real data set M with finite resolution is done by some simplifications [2].

Measuring the information dimension, one special kind of fractal dimension, the number of points, determined by the vectors, $N_i(l)$ in each volume element $V_i(l)$ are counted and its scaling behaviour in respect to the size of the volume elements is determined. E.g. if the points of a given data set are arranged on a plain $N_i(l)$ is proportional to l^2, leading to an information dimension of 2. Bisection of the length of the edge of the embedding cubes leads to a reduction of factor 4 of points. The scaling behaviour of the information dimension d_i is given as follows. Following the definition of the information by Shannon, the information dimension d_I is given as follows:

$$d_I = \lim_{l \to 0} \frac{(-\sum_{i=1}^{A(l)} p_i(l) \log p_i(l))}{\log(1/l)} \tag{1}$$

where $A(l)$ is the number of volume elements necessary to cover the whole data set and $p_i(l)$ is the probability to find a point of the data in volume element i.

The SOM's topology preservation is measured with the waber product P_{waber}. The main idea of this algorithm is to compare the neighbourhood relation between the neurons with respect to their position on the map on the one hand $(Q_2(j,l))$ and according to their stored weight vectors on the other $(Q_1(j,l))$.

$$P_{waber} = \frac{1}{N(N-1))} \sum_{j=1}^{N} \sum_{k=1}^{N-1} \log((\prod_{l=1}^{k} Q_1(j,l) Q_2(j,l))^{\frac{1}{2k}}) \tag{2}$$

A value of 0 for P_{waber} characterizes perfect topology preservation, a negative value folding into the input space. N is the number of all neurons.

3 Results

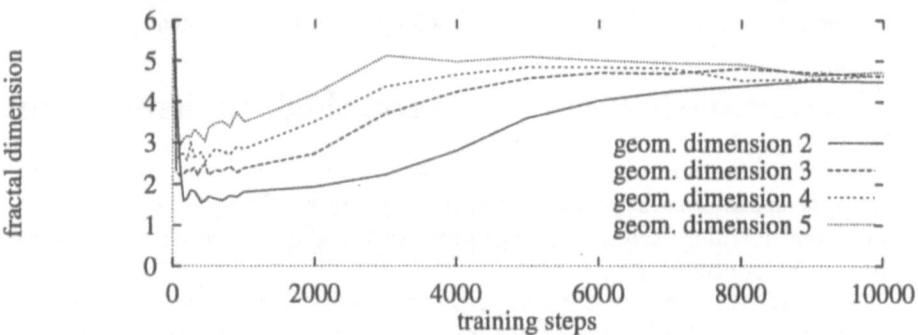

Figure 1: The information dimension with different geometrical dimensions

We explore different data sets of real applications concerning their information dimension. The gas spectra and VLSI process data are easy to learn resulting in good structured spanning trees and component cards. The corresponding information dimensions of these data sets were 1.5 and 1.7, respectively. In contrast to that we tried many training cycles with a two dimensional map for the EEG data and the PCB layouts. Variing the different training parameters like number of processing units, number of training steps, heights and width of adaptation at the beginning, we were not able to get good learning results. The reason is a higher information dimension than 2. The information dimemsion of the PCB layout pictures was 5.4 and of the EEG data set 5.7. The lower dimensional map tries to approximate the higher dimensional input space by folding resulting in unstructered component cards.

Next we learn the EEG data set with SOMs having different geometrical dimensions and calculate the waber product and the information dimensions of the space represented by the stored weight vectors in different learning states.

Considering figure 1 though the information dimension of the EEG data set is 5.7 the 5 dimensional map's weight vectors are only able to achieve an information dimension of 5.1 learning this data. Also a lower dimensional map is able to achieve an information dimension of about 4.75 but at the expense of the preservation of topology (figure 2). Considering the dynamics of learning, while the geometrical dimension of the map corresponds to the fractal dimension of the weight vectors the waber product is about 0. If the fractal dimension increases the waber product becomes negative, indicating, that the folding process begins. The behaviour of the curves at the beginning characterizes the ordering of the weight vectors from the random initialisation. Second a map with a lower geometrical dimension takes more time to achieve this higher information dimension. So it is necessary to consider both, the information dimension and the preservation of topology, to discuss the learning results.

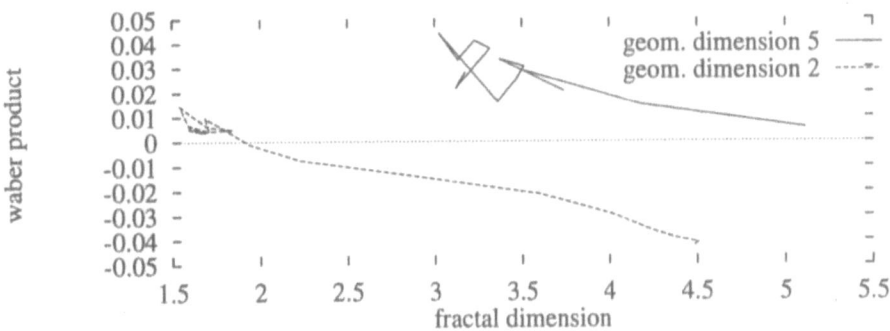

Figure 2: The waber product for maps in different dimension during learning

4   Conclusion and further work

We looked at different data sets and learning states of the SOM calculating the information dimension and determine the ability of topology preservation by using the waber product. Both is necessary to discuss a learning result. Lower dimensional maps approximate a higher dimensional input space at the expense of topology preservation.

In further work we will try to control the training phase. Choosing the corresponding map with a geometrical dimension equal to the information dimension of the data set we will stop the decreasing of width of the adaptation function on the point when the map's stored weight vectors has achieved the necessary fractal dimension and only decrease the heights of the learning function. Simulations have shown, that this strategy leads to the best learning results means the lowest rate of topological defects.

References

[1] T. Kohonen. *Selforganization and associative memory.* Springer Verlag Heidelberg New York Tokyo, 1984.

[2] K. Pawelzik. *Nichtlineare Dynamik und Hirnaktivität.* Verlag Harri Deutsch, 1991.

[3] J. Peinke, J. Parisi, O.E. Rössler, und R. Stoop. *Encounter with chaos.* Springer-Verlag, 1992.

[4] H. Ritter, T. Martinetz, und K. Schulten. *Neuronale Netze.* Addison Wesley, 1990.

On the Ordering Conditions for Self-Organising Maps

Marco Budinich[†] and John G. Taylor

Centre for Neural Networks - King's College London

([†] Permanent address: Dip. di Fisica & INFN, Via Valerio 2, 34127 Trieste, Italy)

Self-organising neural networks have been proposed to model the feature maps of the brain [1,2] but the underlying theory is still not completely understood (see e.g. [3], [4] and references therein). In what follows we focus on Kohonen nets [2].

These nets map a continuous vectorial input space, the space of the patterns $\{ \bar{\mu} \}$, onto a lattice of neurons. Here we consider a one dimensional lattice, i.e. a string of neurons. The k-th neuron has weights $\bar{w}_k$ and its response to a pattern $\bar{\mu}$ is $\bar{\mu} \cdot \bar{w}_k$.

In this view both patterns and neurons can be thought as points in space. Figure 1 contains two different representations of a net with two dimensional input and 5 neurons.

We start by giving a reminder of some facts about Kohonen nets. The learning algorithm is:

1) set the weights to initial random values;
2) select a pattern at random, say $\bar{\zeta}$, and feed it to the neurons;
3) find the output neuron with maximal output, say m[1];
4) train m and its neighbours up to a distance d by the Hebb rule; the training affects $2d + 1$ neurons:

$$\bar{w}_i' = \bar{w}_i + \alpha (\bar{\zeta} - \bar{w}_i) \qquad \forall i : |i - m| \le d \qquad (1)$$

5) update the parameters d and α according to a pre-defined schedule and, if the learning loops are not yet finished, go to 2.

In [2] the net is one dimensional and the author introduced an order parameter D

$$D = \sum_{i=2}^{n} |w_i - w_{i-1}| - |w_n - w_1| \ge 0$$

equality holding if and only if the neurons are sorted; obviously D evolves during learning. The ordering theorem gives necessary and sufficient conditions for $\bar{\zeta}$ to lower D in a learning step. A corollary of the theorem states that if the patterns are ordered $(D = 0)$ repeated applications of the learning algorithm leave D unchanged.

We now arrive at the main result of this paper: a geometrical interpretation that gives a simpler necessary and sufficient condition for the decrease of D. Since this approach applies to the more general case of d dimensional spaces we generalise the definition of D to the case where the weights are vectors:

---

[1]This definition can be tricky unless pattern and weight vectors are somehow normalised. Since both patterns and weights define points in space, the problem can be circumvented by defining the most active neuron for pattern $\bar{\zeta}$ as the neuron whose weights define the nearest point to $\bar{\zeta}$. Simple algebra shows that the two definitions are equivalent.

$$D = \sum_{i=2}^{n} \| \bar{w}_i - \bar{w}_{i-1} \| - \| \bar{w}_n - \bar{w}_1 \| \geq 0 \tag{2}$$

and, as in [2], we calculate the change $\Delta D = D' - D$ produced in a learning step. For brevity we take just 5 neurons ($n = 5$) and suppose that the training affects the third neuron and its immediate neighbours ($m = 3$, $d = 1$; the other cases can be treated similarly and will be not dealt with). From (2) and applying (1) to neurons 2, 3 and 4 we obtain, after simple algebra,

$$\Delta D = \quad \| \bar{w}_2 + \alpha (\bar{\zeta} - \bar{w}_2) - \bar{w}_1 \| - \| \bar{w}_2 - \bar{w}_1 \| + $$
$$(|1 - \alpha| - 1) (\| \bar{w}_3 - \bar{w}_2 \| + \| \bar{w}_4 - \bar{w}_3 \|) + $$
$$\| \bar{w}_5 - \bar{w}_4 - \alpha (\bar{\zeta} - \bar{w}_4) \| - \| \bar{w}_5 - \bar{w}_4 \| .$$

If $0 \leq \alpha \leq 1$ then $|1 - \alpha| - 1 = -\alpha$ and with some other rearrangements:

$$\Delta D = \alpha (\| \bar{\zeta} - \bar{F}_1 \| + \| \bar{\zeta} - \bar{F}_2 \| - 2a) \quad \text{where} \tag{3}$$

$$\bar{F}_1 = \bar{w}_2 + \frac{\bar{w}_1 - \bar{w}_2}{\alpha}, \qquad \bar{F}_2 = \bar{w}_4 + \frac{\bar{w}_5 - \bar{w}_4}{\alpha},$$

$$2a = \frac{\| \bar{w}_2 - \bar{w}_1 \|}{\alpha} + \| \bar{w}_3 - \bar{w}_2 \| + \| \bar{w}_4 - \bar{w}_3 \| + \frac{\| \bar{w}_5 - \bar{w}_4 \|}{\alpha} .$$

It is easy to see that D decreases if and only if $\bar{\zeta}$ lies within the ellipsoid of foci $\bar{F}_1$ and $\bar{F}_2$ and principal axis a. This observation simplifies the ordering theorem of [2] and extends its validity to input spaces with more than one dimension[1]. Figure 2 shows a two-dimensional example.

Simple manipulations of (3) yield the following observations:

- Identical relations hold for any value of d (see (1)) as long as α is the same for all updated neurons (a step neighbourhood function).
- The focus $\bar{F}_1$ ($\bar{F}_2$) rests on the line from $\bar{w}_2$ to $\bar{w}_1$ ($\bar{w}_4$ to $\bar{w}_5$) tending to $\bar{w}_1$ ($\bar{w}_5$) for $\alpha \to 1$ and tending to infinity for $\alpha \to 0$.
- D = 0 if and only if the 5 neurons are aligned and sorted and in that case the ellipsoid shrinks to a segment.
- All the weights $\bar{w}_i$ lay within the ellipsoid, being on the border if and only if the ellipsoid shrinks to a segment.
- The eccentricity ε of the ellipsoid is given by:

$$\varepsilon = \frac{\| \frac{\bar{w}_2 - \bar{w}_1}{\alpha} + \bar{w}_3 - \bar{w}_2 + \bar{w}_4 - \bar{w}_3 + \frac{\bar{w}_5 - \bar{w}_4}{\alpha} \|}{\frac{\| \bar{w}_2 - \bar{w}_1 \|}{\alpha} + \| \bar{w}_3 - \bar{w}_2 \| + \| \bar{w}_4 - \bar{w}_3 \| + \frac{\| \bar{w}_5 - \bar{w}_4 \|}{\alpha}} \leq 1$$

---

[1] The proof reported in [2], that enumerates all the possible sign combination of the terms of D, is correct only in the limit $\alpha \to 0$ when there are no changes of relative position among the w_i due to learning.

being equal to 1 when the weights are aligned and sorted. On the other hand the more "disordered" are the weights the lower is ε.

D and ε account for the known properties of learning in self-organising neural nets (we refer in particular to experiments done with a net devised for the Travelling Salesman Problem [5]). It is well known that learning in these nets goes through two distinct phases [2]. In the first phase (ordering) $\alpha \approx 1$ and the weights evolve from the completely random situation of the start to a condition of maximal order. From that moment the second phase of learning begins (refinement); now $\alpha \approx 0$ and the weights change very slowly to reproduce the distribution of the input patterns as closely as possible. In this phase some of the preceeding order is lost.

Let us see how the proposed geometrical view can interpret these facts (see illustrative simulation results in figure 3). At the start the weights are random, α is large and for most of the learning steps the ellipsoid has a small eccentricity; in this case it is almost spherical and has a comparatively large volume. It follows that with high probability $\bar{\zeta}$ lies within the ellipsoid and D decreases. This accounts for the ordering phase during which the neurons become less and less random until when D reaches a minimum and the neurons have maximal order.

In the refinement phase α is small and consequently ε tends to be larger even if the patterns are disordered (a part from pathological cases). Thin ellipsoids give lower probability of having $\bar{\zeta}$ at their interior and in the second phase D can increase to converge, at the end, to the optimal solution.

We can use the nature of D and ΔD to prove convergence to the ordered state in one dimension, simplifying the argument in Cottrell and Fort [6]. The crucial step in their proof is to show that the elements of the Markov matrix of the learning algorithm are strictly positive; a general theorem then shows that the Markov process will converge with probability one to the absorbing states. Here the non-zero value of the matrix elements follows immediately from the finite size of the ellipse for each point of the chain.

We can also understand why convergence to ordered states does not extend to two or more dimensions. We have seen that when there is precise order the ellipsoid shrinks to a segment. In dimension greater than one, this will have zero volume, so zero probability of $\Delta D = 0$. In other words the ordered state will not be absorbing. In one dimension, however, even when shrunk to a segment, the ellipse covers a finite fraction of the input space giving a finite probability of $\Delta D = 0$, so the ordered state is absorbing.

References:

[1]   Von der Malsburg Ch., Kybernetik **14** 1973 pp. 85-100;

[2]   Kohonen T., *Self-Organisation and Associative Memory*, 1984 (3rd Ed. 1989) Springer-Verlag Berlin Heidelberg;

[3]   Erwin E., Obermayer K. and Schulten K., Biological Cybernetics **67** (1992) pp. 47-55;

[4]   Ritter H., Martinetz T. and Schulten K., *Neural Computation and Self Organizing Maps: An Introduction*, Addison-Wesley, New York, 1992, pp. 303;

[5]   Budinich M., in press;

[6]   Cottrell M. and Fort J.-C., Ann. Inst. Henri Poincaré (Probabilités et Statistiques), **23** (1) 1987 pp. 1-20.

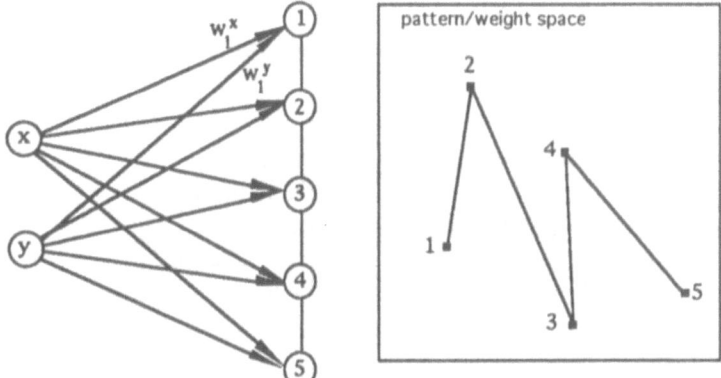

Figure 1 A simple example: a two-dimensional input space mapped to a one dimensional lattice of 5 neurons. The drawings represent the net and its representation in pattern/weight space.

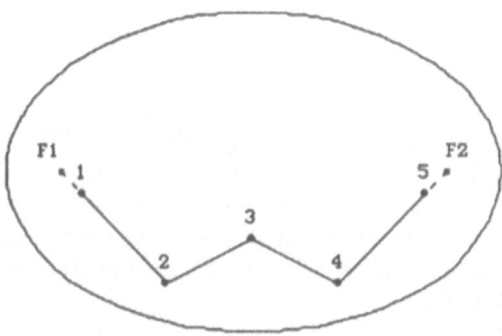

Figure 2 5 neurons in a configuration that forms a "w". If ζ lies within the ellipse $\Delta D < 0$; $\Delta D = 0$ when on the ellipse and $\Delta D > 0$ outside.

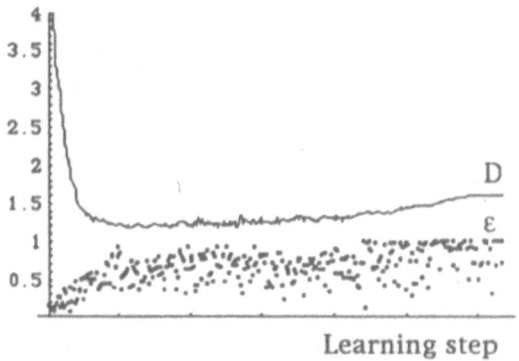

Figure 3 Typical behaviours of D and ε during 11,000 learning steps of a net with 50 neurons; the input patterns are random points in the plane. Here α drops linearly from 0.5 to 0.005 at iteration 10,000 to maintain this last value up to the end. The first part of the horizontal scale is enlarged by a factor 50 to show the details of the ordering phase.

Representation and Identification of Fault Conditions of an Anaesthesia System by Means of the Self-Organizing Map

Mauri Vapola [*], Olli Simula [**], Teuvo Kohonen [**], and Pekka Meriläinen [*]

[*] Datex Division Instrumentarium Corp.
P.O.Box 446, SF-00101 Helsinki, Finland

[**] Helsinki University of Technology
Laboratory of Computer and Information Science
TKK-F, SF-02150 Espoo, Finland

1. Introduction

During large surgical operations the patient is usually anaesthesized, whereby his muscles are relaxed. This blocks the patient's spontaneous breathing and during the operation an anaesthesia machine usually takes care of the breathing function. It transfers the chosen amount of anaesthetic gas mixture to the patient's lungs during the inspiratory phase and transfers approximately the same amount of gas away from there during the expiratory phase.

In the worst case, the cost of a failure in the anaesthesia machine or in the practice of anaesthesia is the death of the patient. This paper describes how the risks of anaesthesia could possibly be reduced by identifying the fault conditions in the breathing circuit of the anaesthesia machine before the fault causes injury to the patient. The monitoring system is based on the Self-Organizing Map [Koh89]. Westenskow has previously used an error-backpropagation network to identify the breathing circuit faults [Wes91].

The basic architecture of our system is similar to Westenskow's system: At first, the existence of any fault condition is detected using a set of so-called absolute features. After that, only if the existence of the fault has been detected by means of absolute features, the fault condition is identified using the differential features. In this work, we have replaced the error-backpropagation network by a self-organizing map and proposed some extensions to Westenskow's model.

Intentional generation of fault conditions in a real operating room environment is not possible, and it is also difficult to collect an extensive set of measurement data in a simulated environment. The greatest advantages offered by our system compared to the system using the error-backpropagation network therefore are that it can be trained to detect the existence of the fault condition by using samples collected during perfectly normal anaesthesia cases, and all the conditions became clearly organized in the visual display.

2. Monitoring system based on the Self-Organizing Map

The anaesthesia system comprises the anaesthesia machine, the patient and the anaesthesia personnel. Each part of the system can affect the values measured by the monitoring system. Therefore, even monitoring the operation of one part of the system is a difficult task. Although there is a lot of data available for monitoring purposes, the descriptive features must be extracted from the measured parameters. In addition, these features should be as invariant as possible. Another problem is collection of data for training the neural network. The data should cover faulty situations as completely as possible. Fault conditions can be simulated, but the simulator does not take into account human behaviour and the amount of measurements from real situations is limited due to low accident rate.

The developed fault detection and identification system is based on a kind of simplified model of anaesthesia. In this model, the operating point is assumed to be stable unless the settings of the anaesthesia machine are changed. The fault conditions are identified based on the deviations of measurements from the operating point. The operating point thus corresponds to the normal situation.

The monitoring system consists of Self-Organizing Maps on two levels. The first level map identifies the operating point from the measured parameters and calculated features. The map, or set of maps, on the second level is used to locate faults based on the deviations in the parameters compared to the operating point. The purpose of the first-level map is to increase accuracy by dividing the operating point space into different parts. The amount of deviation caused by a faulty situation may change remarkably at different operation points. Therefore, each map on the second-level corresponds to a relatively small part of the state space of the first-level map.

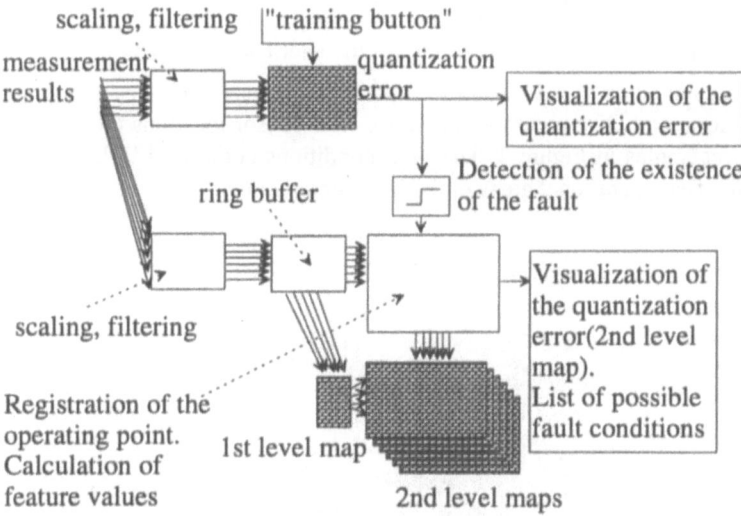

Figure 1. Fault detection (upper part) and identification (lower part) system:

The second-level map is used to detect faults by comparing the measured values and features with the operating point. The operating point is registered, e.g, when the settings of the anaesthesia machine are changed.

Real anaesthesia situations are not as ideal as assumed above. The parameter values corresponding to a normal situation are changing continuously due to changes in the patient and unnecessary alarms would then become a problem. To avoid unnecessary alarms, the fault detection problem is further divided into the following parts:

(1) First the existence of the fault is detected and (2) only after that its reason is identified. The first Self-Organizing Map, called *fault detection map*, is used to detect the existence of the fault. The fault detection map cannot identify the reason of the fault, but it learns the normal settings used in the operating room. By calculating the quantization error on this map, it is possible to detect the existence of a fault: an increase in the quantization error of the best-matching unit indicates a faulty situation. A similar approach has been used in process control [Try91] and in monitoring the operation of devices [Kas92].

Localization of the fault is possible by examining the measured values during a short period preceeding the fault detection. This can be done by storing a number of samples in a buffer. The oldest samples preceeding the fault detection are used to determine the operating point and the latest samples are used to locate the fault. The block diagram of the complete monitoring system is depicted in Figure 1.

3. Implementation of the system

The breathing circuit of the anaesthesia machine is illustrated in Figure 2. In the inspiratory phase the expiratory valve is closed and the ventilator bellow forces the previously expired gas mixture to flow through the CO_2-absorber, which absorbs most of the carbon dioxide produced by the patient. Thereafter, some fresh gas with additional oxygen and possibly anaesthetics is injected into the gas mixture and the mixture flows through the inspiratory valve, inspiratory hose and the intubation tube to the patient's lungs. In the expiratory phase the inspiratory valve is closed and the gas is allowed to flow from the patient's lungs to the ventilator bellow.

In the experimental setup two anaesthesia monitors measured the flow, pressure, CO_2 concentration and O_2 concentration values. The sensor locations are depicted by two black rectangles in Figure 2. The fault conditions comprised leaks and obstructions in the five operational areas of the breathing circuit.

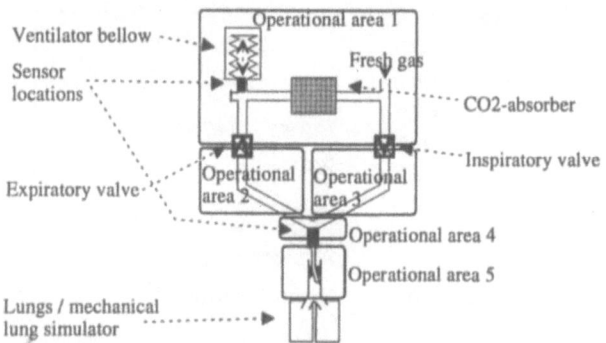

Figure 2. Different operational areas of the anaesthesia machine

4. Experiments

The fault detection map was taught using data collected in a real operating room environment. The training data for the second level map was collected using a mechanical lung simulator. Fourteen different types of fault conditions were generated when collecting this material and the total size of the training set was about 10000 samples. The accuracy of the second level map was further increased using the optimized LVQ1 - and LVQ3-algorithms [Koh90].

The accuracy of recognition for the second-level map was 87% on the average, if the operating point of the test set was relatively close to the operating points of the training set and the test set was collected using the lung simulator. When the location of the operating point deviated considerably from the operating points of the training set, the recognition accuracy decreased to 70 % on the average.

The performance of the map was tested using samples from true situations as well. For example, when the position of the patient was changed, the intubation tube was accidentally obstructed for a short period of time. The responses of the fault detection map and the second level map are presented in Figure 3. Figure 3 (a) illustrates the quantization error of the fault detection map and Figure 3 (b) shows the trajectory of the best matching unit during the situation.

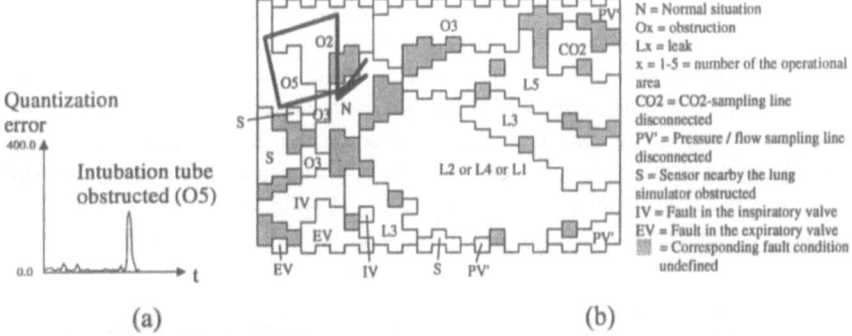

(a)   (b)

Figure 3. Quantization error of the fault detection map (a) and the trajectory of the operation point during faulty situation (b)

References

[Kas92] Kasslin, M., Kangas, J., and Simula, O., "Process State Monitoring Using Self-Organizing Maps", Artificial Neural Networks 2, Proc. ICANN-92, Brighton, UK, September 1992, Elsevier Science Publishers B.V., Amsterdam 1992, pp. 1531-1534

[Koh89] Kohonen, T., *Self-Organization and Associative Memory*, 3rd Edition, Springer-Verlag, Berlin, 1989

[Koh90] Kohonen, T., The *Self-Organizing Mapy*, Proceedings of the IEEE, Vol. 78, N0. 9, 1464-1480, Sept. 1990

[Try91] Tryba, V. and Goser, K., "Self-Organizing Feature Maps for Process Control in Chemistry", Artificial Neural Networks, Proc. ICANN-91, Espoo, Finland, June 1991, Elsevier Science Publishers B.V., Amsterdam, 1991, pp. 847-852

[Wes91] Westenskow, D., "Device and Method for Neural Network Breathing Alarm", Patent Application No. PCT/US90/05250, Sept. 14, 1990, 44 p.

Sensor arrays and Self-Organizing Maps for Odour Analysis in Artificial Olfactory Systems

Fabrizio Davide[O] , Corrado Di Natale*, Arnaldo D'Amico*

- [O] Dipartimento di Ingegneria Elettrica, Università di L'Aquila, Monteluco di Roio, 67100 L'Aquila, Italy
- * Dipartimento di Ingegneria Elettronica, Università di Roma "Tor Vergata", Via della Ricerca Scientifica, 00133 Roma, Italy

1 Introduction

Our studies have regarded bio-inspired adaptive artificial olfactory systems composed of a sensor array for gas sensing and an artificial neural network, Self-Organizing Topology Preserving Map (SOM), introduced by T. Kohonen (Kohonen, 1989). In order to state the main working principles, Fig. 1 shows an overview of the digital version of a system for odour classification. The information flows from the left-hand side to the right-hand side: the gas mixtures in the environment determine the m sensor outputs which are sampled and converted into the digital stream z at each clock time. The module implementing the SOM network accepts a sequence of samples by a delay line, classifies the pattern according to its internal class models, and provides a class label as output (Davide et al.,1992 I,II,III).

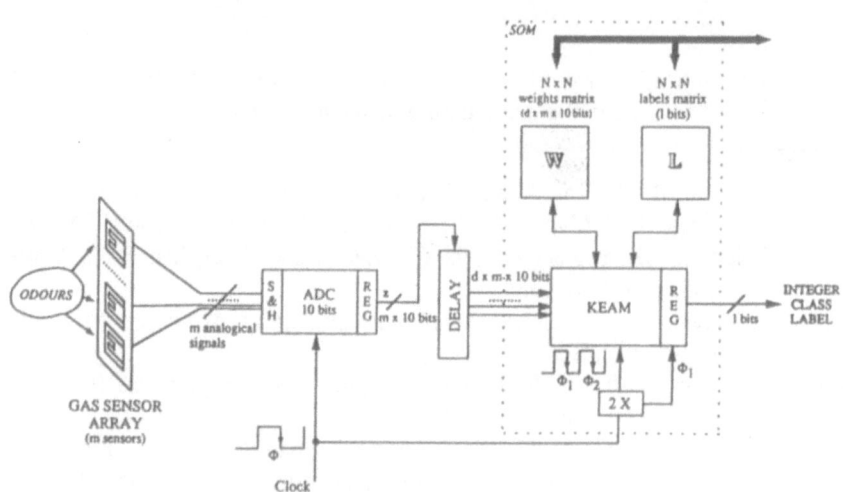

Fig. 1. Stand-alone artificial olfactory system composed of a Kohonen SOM network feeded by a gas sensor array.

Instead of an adaptive system, like a SOM, sensor arrays have been usually connected to a variety of numerical modules for multivariate statistical analysis of gas mixtures (Vahinger et al., 1991). All these modules need a calibration data set and are unmodifiable after the calibration. Unfortunately available sensors for gas sensing are chemically low selective, slow-responding devices, which suffer of unsatisfactory long term stability and reproducibility: thence large calibration data set (size of 10^2 samples at least) and frequent repetitions (every 10^2 days) of the calibration procedure are required [1].

As far as odour classification tasks are concerned, unsupervised artificial neural structures which are able of continuous adaptation to changes, can reduce the costs and make feasible certain applications, e.g. non invasive biomedical analyses, which require a great amount of accuracy and stability. In the following an algorithm of a SOM-based family of clustering techniques devoted to odour classification is introduced and applied to an odour recognition experiment.

2 Odour recognition by extended SOM algorithms

A novel SOM algorithm has been properly developed in order to approach odour recognition problems. It belongs to a family of clustering algorithms which can be regarded as unsupervised versions of the potential function class modelling method, intensively utilized in Chemometrics (Forina et al., 1991).

The computation employs the "updating the difference only" principle, widely adopted in other application fields of neuro-computing (Ritter, 1992) (Kasslin, 1992), and ensures a good class modelling every time a correct mapping is achieved by the original Kohonen algorithm. The comparison with other clustering strategies is favourable, mainly because convexity and connectivity constraints on the clusters are not required and a parallel implementation is easily feasible. Moreover the algorithm has a simple interpretation under the analogy with critical phenomena occurring in thermal systems (Davide et al., 1993 II).

The generalized SOM lattice, composed of up 10^4 neurons with their weights and auxiliary internal variables, is implemented via a dedicated Single Instruction Multiple Data (SIMD) machine, programmed in OCCAM. A single instruction stream issued by a unique controller is broadcast to all the processing units (PEs), placed on a 2-D grid of regular interconnections. A "coarse grain assignment" prescribes each PE to be responsible for the dynamics of a 10X10 neural sub-lattice.

Moreover a general purpose sensor array is adopted which is composed of six sintered SnO_2 Taguchi gas sensors: 822, 813, 815 and three 812 catalytically modified with 1 mg of Pd, Au, Rh respectively. By using the two points method, sensors conductivities were measured at both constant gas flow (150 ml/min) and constant temperature monitored by an infrared pyrometer (IP1, IMPAC). Promising performances of the global olfactory system have been ratified by a long series of experiments, dealing with odour classification tasks, such as recognition of chlorinated, aliphatic, aromatic and OH-containing organic compounds, such as perchloroethylene, trichloroethylene, n-hexane, n-octane, toluene and ethanols. Here an experiment of recognition of two different organic flavours of commercial use, say an "aromatic" one and a "spicy" one, have been tried in a noisy environment. The experimental procedure prescribes the two compounds to be alternatively injected into the test chamber, with unperfect time separation, in order to allow the formation of transient unpredictable mixtures. Each point in Fig. 2 represents a sample of the sensor outputs, projected onto a

---

1 A single run of a calibration procedure may require 50 hours of laboratory time

suitable plane [2]. The two main crowdings correspond to the two different odours; the link between them is determined by transient conditions in which the two odours are simultaneously present. Sparse points are determined by noise in the environment or in the sensor array. This experiment can be considered as a benchmark, because it has both a strong noise and a disturbance effect, due to the transitions between the odours, which limit both the selectivity and accuracy of any classification mechanism.

Fig. 3 reports a single realization of the stochastic process of SOM development for the given environmental statistics: the data topology is clearly revealed, and a good assigment of receptive fields is achieved. The network internal class representation after 3000 clocks is shown in Fig. 4: the two odours, labelled as A and B, have been recognized by the network. This assignment is neighborhood preserving because adjacent neurons have become specialized to the same class.

3 Discussion

Fig. 5 shows the short-range autocorrelation lenght of the neuron density d(p), defined as the inverse of the mean volume of the receptive fields, averaged over the neurons with potential greater than p. It clearly appears that d(p) is locally perturbed by incipient phenomena of organization. The labels mark the phase transition points. PT1 points out the situation of undistinguished classes, because the corresponding value of potential coincides with the level of environmental disturbance, i.e. unpredictable mixtures of the two odours; PT2 marks a local crowding of environmental noise near a class; finally PT3 reveals an accidental crowding of neurons and PT4 signals the birth/death of a second class.

In this situation the neural algorithm under test tries to maximize the number of active neurons, under the constraint of a signal to noise ratio high enough, and it chooses a value of p corresponding to the valley between PT1 and PT2, resulting in a 10% error rate and 80% of labelled neurons (which represents an intrinsic boundary to the efficiency of network exploitation, determined by the requirements of disturbance rejection).

It turns out a remarkable performance for the whole system in comparison with supervised non-adaptive methods appeared in (Gardner, 1992), (Moore et al., 1992).

Further work is required for the long term stability problem which remains one of the main problems for any sensoric system. Some interesting abilities of our system in counteracting drift effects, due to the "residual plasticity" of a SOM, are demonstrated by realistic simulations. Invariance of classification performances with drift and transient noise effects can be seemingly the most evaluable features of this artificial olfactory architecture.

References

Davide, F., Di Natale, C., and D'Amico, A., Pattern Recognition Techniques in Gas Sensing, invited at 1st Int. Workshop on New Develop. in Gas Sensors, Castro Marina, Italy, Sept. 13-14, 1993.

Davide, F., Di Natale, C., and D'Amico, A., Sensor Arrays and Neural Networks in Multicomponent Gas Analysis, to be published on Sensors and Actuators B.

Davide, F., Di Natale, C., and D'Amico, A., Self-Organizing Sensory Maps for Odour Classification Mimicking, invited at 2nd CEC Worksh. on Bioelect., Frankfurt, Germany, Nov. 23-25, 1993.

Forina, M., Armanino, C., Leardi, R., and Drava, G., Journal of Chemometrics, 5 (1991) 435-453.

Gardner, J.W., Sensors and Actuators B, 4 (1992) 109-116.

---

2 In the following all the quantities involved are visualized on the same Fisher plane, nevertheless the classification procedure and the SOM network work in the whole six-dimensional space.

Kasslin, M., Kangas, J., and Simula, O., Proc. ICANN 92: Artificial Neural Networks 2, Amsterdam, North Holland (1992) 1531-1534.

Kohonen, T., Self-organization and Associative Memory, 3rd eds., Springer-Verlag, Berlin, 1989.

Moore, S.W., et al., Sensors and Actuators B, 3 (1992) 37-38.

Ritter, H., Martinetz, T., and Schulten, K., Neural Computation and Self-Organizing Maps, Reading, Addison Wesley Publishing Company Inc., 1992.

Vahinger, S. and Göpel, W., in Chemical and Biochemical sensors part I, eds. W. Göpel, T.A. Jones, M. Kleitz, J. Lundström and T. Seiyama, VCH (1991) 191-237.

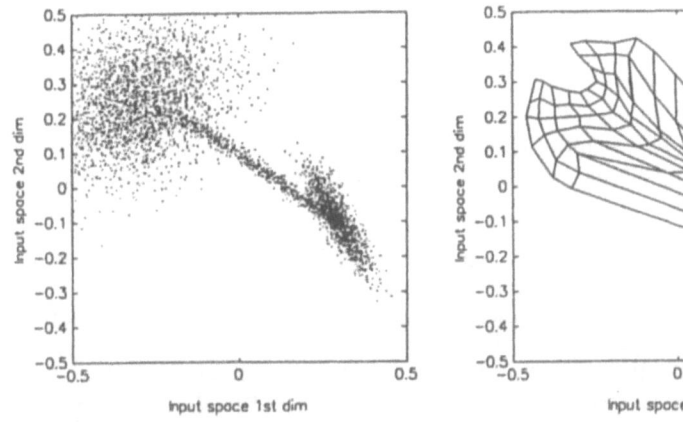

Fig. 2. Samples of six sensor outputs in a two odours recognition experiment, projected onto a representation plane (normalized units).

Fig. 3. The weights of a 10X10 neurons SOM, after 3000 learning steps.

Fig. 4. The internal class representation for the net in Fig. 3.

Fig. 5. Short-range autocorrelation lenght for the density of neurons over a potential threshold. Labels are referred to phase transitions explained in the text.

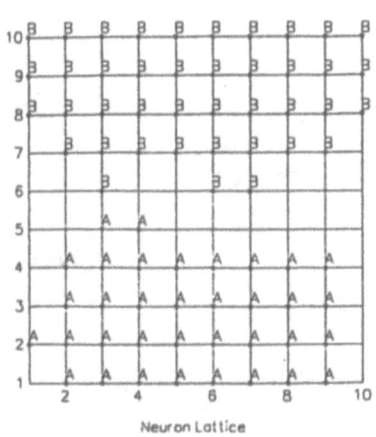

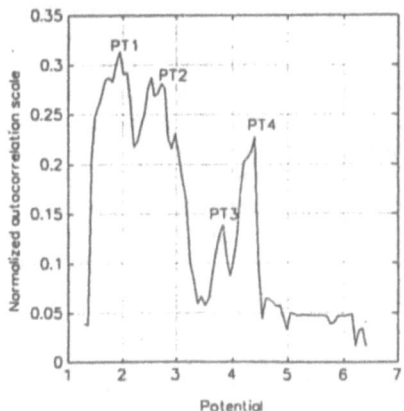

A Self-Organising Neural Network for the Travelling Salesman Problem that is Competitive with Simulated Annealing

Marco Budinich

Centre for Neural Networks - King's College London

(Permanent address: Dipartimento di Fisica & INFN, Via Valerio 2, 34127 Trieste, Italy)

The Travelling Salesman Problem (TSP) has been studied for long times (see e.g. [2]) and, as for all NP-complete problems, all exact solutions take a time that can grow exponentially with the number of cities. Many heuristic algorithms exist for practical instances [3].

In 1987 Durbin and Willshaw [1] proposed a new idea for the approximate solution of the TSP with cities in the plane. They observed that in one dimension the exact solution is trivial: always step to the nearest unvisited city. Consequently, with a clever mapping of the cities to points on a circle, one can easily find the shortest tour for these "image cities" and a visit order for them. This visit order gives also a tour for the original cities. How near this tour is to the shortest one depends on how "good" the mapping is. Durbin and Willshaw positively showed that good maps are those that best preserve local neighbourhood relations between the cities. In this way the original TSP is transformed to the search of a good neighbourhood preserving map.

There are several different ways to search such a map: Durbin and Willshaw did it relaxing a fictitious physical system, an "clastic net". Also a self-organising neural net can find neighbourhood preserving maps. These nets, proposed to model the self-organising feature maps in the brain [4,5], can find good neighbourhood preserving maps through unsupervised learning. In this way the TSP is ultimately solved by unsupervised learning.

In conclusion there are two steps to get a solution for the TSP: the first is to teach the problem to a self-organising neural net that, while learning, builds up the neighbourhood preserving map. In the second step, from the solution for the image cities, one obtains a tour for the original TSP.

In what follows the TSP is given by n cities randomly distributed in the (0,1) square. The network has two input neurons that pass the $(x, y) = \overline{\zeta}$ coordinates of the cities to n output neurons. The output neurons form a ring; the distance $D(i, j)$ between neurons i and j is one plus the minimum number of neurons between i and j. Each output neuron has just two weights: $(w_x, w_y) = \overline{w}$ and, in response to the input (x, y) from city $\overline{\zeta}$, gives an output $o = \overline{\zeta} \cdot \overline{w}$. Figure 1 gives a schematic view of the net.

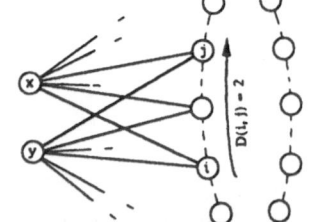

Figure 1 Schematic net: not all connections from input neurons are drawn. In this example $D(i, j) = 2$.

For the neural network the square represents the input space and the cities are the patterns to be learned. After learning, the network maps the two dimensional input space onto the one dimensional space given by the ring of neurons.

The standard learning algorithm for this kind of neural networks is [5]:

1   set the weights to initial random values in [0,1];

2   select a city at random, say $\overline{\zeta}$, and feed its coordinates to the input neurons;

3   find the output neuron with maximal output, say m[1];

4   train m and its neighbours up to a distance d with Hebb rule; the training affects 2 d + 1 neurons:

$$\overline{w_i}' = \overline{w_i} + \alpha (\overline{\zeta} - \overline{w_i}) \qquad \forall\, i : D(i, m) \leq d$$

5   update the parameters d and α according to a pre-defined schedule and, if the learning loops are not yet finished, go to 2.

Some systematic, but by no way complete, study of the learning parameters gave this recipe to obtain the best performances for any number of cities.

- The number of output neurons is fixed and equal to the number of cities n;
- 50 learning loops for each city of the problem;
- the learning constant α starts from 0.8 and its value is decreased linearly at each learning loop reaching zero at the last iteration;
- the most active neuron m is trained with learning constant α, its neighbours are trained with a value of α that decreases linearly with $D(i, m)$ and vanishes for neurons at distance greater than d from m;
- the distance of update d is   6.2 + 0.037 n   at the start and is decreased linearly to 1 in the first 65% of the iterations. For the successive 35% of the learning loops d remains constant at 1.

After learning, the net maps neighbouring cities to neighbouring neurons: for each city its image is given by the neuron with maximal activity. One easily finds the shortest tour for the images of the cities that, in turn, gives a tour for the TSP.

This straightforward version of the learning algorithm has a major flaw. The map it produces is not injective: many cities can be mapped to the same neuron (this happens for a fraction between 45% and 50% of the total number of cities n for $10 \leq n \leq 1000$). When two or more cities are mapped to the same neuron one cannot say which of them has to come first in the tour and this problem substantially reduces the performances of this algorithm [6,7].

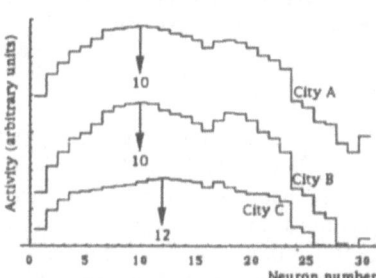

Figure 2 3 activity profiles for a net with 30 neurons obtained presenting three different cities (of the 30 of the original TSP). Cities A and B both give maximal activity in neuron number 10.

A decisive insight comes from the "activity profile" of the net: a plot of the neuron outputs in response to a given city. Figure 2 contains the activity profiles for 3 cities on a net with 30 neurons. Cities A and B both produce the maximal activity on the 10th neuron.

When mapping cities to neurons we associate to each city a coordinate along the ring. The standard choice of the maximal acting neuron produces an integer

---

[1]This definition is ambiguous unless city and weight vectors are somehow normalised. Since both cities and weights define points in the plane, the problem can be circumvented by defining the most active neuron for city $\overline{\zeta}$ as the neuron which weights define the nearest point to $\overline{\zeta}$. Simple algebra shows that the two definitions are equivalent.

coordinate that is the modal value of the activity profile. A better choice turns out to be an averaging process on the neurons to obtain a real coordinate: for example a weighted coordinate calculated with the maximal active neuron and its nearest neighbours using their activities as weights. In this way each city is mapped to a point with a real coordinate along the ring, the map becomes injective and the ambiguities in the tour disappear.

This new mapping produces substantially shorter tours assuring that there is valuable information also in the neurons near to the most active one. The average not only uses this additional information but is also a better and more plausible estimator of the city image than the modal value since it is a linear function of the neuron activities and it makes easier to imagine a further layer of neurons doing this job.

Since the best tour was not known the I compared the results of this algorithm with those produced by simulated annealing [8]. The implementation of simulated annealing was that of [9] that is tailored to the TSP and uses exchange terms inspired by Lin and Kernigham heuristics [10]. An annealing factor of 0.9 gives average performances equal to those quoted by Durbin and Willshaw [1] (that can be improved with annealing factor = 0.95). The performances of this implementation of simulated annealing remain uniformly within 10% of the theoretical bound of

$0.749 \sqrt{n}$ when the number of cities n varies between 10 and 1000.

I took 10 runs of each algorithm on the same problem and then calculated the average tour length. With this choice simulated annealing gives its best results [3] and can beat also the most reputed heuristics [10] (even if in longer times).

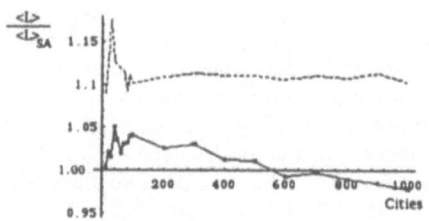

Figure 3 contains the ratio of the average tour lengths of this algorithm over those obtained from simulated annealing. The neural net produces shorter tours than those of simulated annealing for problems with more than n = 500 cities.

Figure 3 Ratio of the average tour length obtained in 10 runs from this algorithm and from simulated annealing with number of cities varying between 10 and 1000. The upper, dashed, curve shows the results obtained from [7].

The memory space used grows as O(n) while the time employed grows at most like $O(n^2)$ since both the number of learning loops and the search of the most active neuron are linear in n. Nevertheless this is probably an upper bound since there are indications that the number of iterations does not need to increase as n to have constant performances. Typically, for 100 cities, this algorithm is an order of magnitude faster than simulated annealing and it is 3 times faster for 1000 cities and in this case gives better average performances. It is remarkable that this net can perform better than simulated annealing in its best achievements (average length) compensating at the same time its main weakness: slowness.

With the parameters tuned for 50 cities, the average performances over 5 runs are substantially equal to those obtained by Durbin and Willshaw [1] whose algorithm uses the same principle of the topological preserving map. Table 1 contains a comparison of the performances done on identical problems. Whereas Durbin and Willshaw report a time similar to that employed by simulated annealing this net is an order of magnitude faster.

City set	Durbin and Willshaw [1]	This algorithm
1	5.98	5.975
2	6.03	6.110
3	5.74 [5.70]	5.737
4	5.90 [5.86]	5.830
5	6.49	6.583

Table 1 Comparison of average tour length in 5 runs for the same problems.

Even if this is the first time a neural network algorithm proves to be competitive with simulated annealing, deeper analysis is needed to establish if it has any relevance for the TSP itself. Just to mention one limitation it is well known that the TSP's with random cities in the plane are relatively easy.

Nevertheless the results show that this net works satisfactorily on a provably difficult problem and especially in large sizes: a not so frequent quality in neural networks. Even more important is that the results derive from a well established learning procedure applied to a previously unstructured net. This poses this approach in a favourable position when compared to those that solve the TSP relaxing a finely pre-tuned neural network [11].

It is also intriguing that the theory behind this algorithm, intimately connected to the self organising processes [12], is today not really understood (even if the ordering theorem of Kohonen [5] can be extended to the d to 1 dimensional case [13]).

References:

[1]   Durbin R. and Willshaw D., Nature **336** 1987 pp. 689-691;
[2]   Lawler E.L., Lenstra J.K., Rinooy Khan A.G. and Shmoys D.B. (editors), *The Traveling Salesman Problem*, 1985 J. Wiley New York;
[3]   Johnson D., in: *Proceedings of the 17th Colloquium on Automata, Languages and Programming*, 1990 Springer-Verlag New York, pp. 446-461;
[4]   Von der Malsburg Ch., Kybernetik **14** 1973 pp. 85-100;
[5]   Kohonen T., *Self-Organisation and Associative Memory*, 1984 (3rd Ed. 1989) Springer-Verlag Berlin Heidelberg;
[6]   Angéniol B., de La Croix Vaubois G. and Le Texier J.-Y., Neural Networks **1** 1988 pp. 289-293;
[7]   Favata F. and Walker R., Biological Cybernetics **64** 1991 pp. 463-468;
[8]   Kirkpatrick S., Gelatt C.D. Jr and Vecchi M.P., Science **220** 13 May 1983 pp. 671-680;
[9]   Müller B. and Reinhardt J., *Neural Networks*, 1990 Springer-Verlag Berlin Heidelberg;
[10]   Lin S. and Kernigham B.W., Oper. Res. **21** 1973 pp. 498-516;
[11]   The first algorithm of this kind was introduced by: Hopfield J.J. and Tank D.W., *"Neural" Computation of Decisions in Optimization Problems*, Biological Cybernetics **52** (1985) pp. 141-152;
[12]   Erwin E., Obermayer K. and Schulten K., Biological Cybernetics **67** (1992) pp. 47-55;
[13]   M. Budinich and J. G. Taylor in preparation.

ACKNOWLEDGEMENTS: I want to acknowledge the many, extremely fruitful, discussions with John G. Taylor at King's College. I also thank warmly King's College, the British Council and the "Consiglio Nazionale delle Ricerche" for supporting my visit at King's and D. Willshaw and M. Simmen for kind hospitality and for providing the TSP's used in [1].

Learning Attractors as a Stochastic Process

Daniel J. Amit

INFN, Section of Rome, Institute of Physics, University of Rome
and Racah Institute of Physics, Hebrew University, Jerusalem

1 Introduction

Plastic attractors for adaptive behavior

Learning attractors: One may go as far as to suggest that some of what is most impressive about 'brainy' performance is its ability to organize to perform in an unfamiliar situation. Or, at least, to adjust an approximate behavior, using ambient information to arrive at qualitatively novel behavior. The attractor structure of an ANN may be considered as an internal representation of the classes of stimuli that are classified by the network. In other words, the variety of external stimuli which drive the network to the same attractor is represented by that attractor, provided these attractors are stable on relatively long time scales. This seems to be consistent with biological observations. See e.g. (Amit, 1992). But as the nature of the data arriving to the network changes significantly in time, it may be desirable for the proper functioning of the network, as a device or as a brain component, to modify the attractor structure, either by deforming the attractors or by creating new ones. To have the ANN capture the large scale statistics of the inflowing data, synapses must be plastic. It seems also reasonable to assume that learning of attractors is an unsupervised process. See e.g. ref. (Amit & Fusi, 1994).

The role of attractors: The main role of attractors is to serve as a mechanism for shifting a passive memory from the synaptic structure into a working state, or working memory, by a class of stimuli that should be treated in an equivalent way. See e.g. refs. (Miyashita & Chang, 1988) for discussion of similar ideas in the biological context. In other words, the memory imprinted in synaptic efficacies can be actualized and communicated only via neural activities. Attractors, therefore, make it possible to operate on a memory elicited by a stimulus long after the stimulus has disappeared. The actual form of the attractor may be immaterial. What matters is that functionally equivalent stimuli lead, following the removal of the stimulus, to the same attractor, i.e. to the same neural activity distribution.

Short vs long time scale synaptic behavior: There may be an essential role for the analog behavior of a synapse on the short time scale. It is one way in which internal time scales interact with external ones. In other words, if there is a time scale on which the learning behavior of a synapse becomes markedly non-linear, that time scale defines a dividing rate for the arrival of stimuli. Stimuli arriving rapidly on this time scale can be superposed, while stimuli arriving more slowly are separated by the non-linear effect.

It is reasonable to suppose that on short time scales a synapse can indeed be analog, i.e. it can maintain a finely grained set of values and can move between them. What makes it so plausible is that it can be implemented on a capacitor. This simple device can maintain its acquired analog values for times shorter than its RC constant. Most importantly, its acquired value can modulate a conductivity. On longer time scales this analog behavior is much less likely. We are not familiar with a device that can vary continuously and maintain the analog values for long times.

The simplest assumption about learning beyond the short time scale is that a synapse can maintain for long times a discrete set of values, and moreover, that the discrete synaptic efficacies are universal: all synapses have approximately (up to fluctuations) the same set of discrete values. This way, given a flow of stimuli, the synaptic values are continuously modified, with the intervention of the 'refreshing' mechanism that drives the synaptic value toward one of the stable values. It depends though on the actual synaptic value. When the stream of stimuli stops for a long period, each synapse maintains, one of the set of stable synaptic values. As soon as stimuli begin arriving again, those values begin to drift, either up or down, depending on whether the synapse 'sees' a correlated or anti-correlated pair of neuronal activities. If in the course of this drift a synaptic threshold is crossed, the synapse is attracted toward a different stable value. The new value may be depends on the covariance of the activities of the neurons connected by it. This is the way the network learns and forgets.

Learning parameters: In the present description information is coded in stimuli, or in the resulting attractors, as spike rates of neurons. When the stimulus is imposed on the network, it induces a certain distribution of activities among the neurons of the nework. Yet even among neurons whose activities are well above the spontaneous level, the rates may be distributed in a wide range. If learning is assumed to be driven by the covariances of neural activities, then the source driving the learning in a given synapse will depend on the magnitudes of the rates of the two neurons. This is one variable affecting the speed of learning of a given stimulus on each synapse.

Another variable affecting the speed of learning is the temporal length of each presentation. If it is short the synaptic value may not reach the threshold for the corresponding transitions between stable states. For a short presentation only synapses connecting neurons with the highest rates may change, if any. The shorter the presentation, relative to the integration time constant of the synapse, the fewer the synapses on which the stimulus will leave a long term effect.

It is rather likely that the refresh mechanism is not deterministic, either because the threshold for transitions between stable states is noisy. Or, because the neural rate variables are noisy. Consequently, the transitions from one synaptic state to another will be a stochastic process.

Speed of learning vs memory capacity: We shall see that if synapses have a discrete, finite set of stable states, in which too many synapses change in a single presentation of a stimulus, the number of memories that can be recalled is very limited(Amit & Fusi, 1994). The levers of control are principally the coding levels of the stimuli, i.e. the fraction of the neurons of the network driven by the stimulus and the transition probabilities of the refresh (discretizing) mechanism.

At a given coding level, and with sizable synaptic transition probabilities, per typical presentation time interval, a stimulus presented even once will be learned. That means that after a single presentation the changes in the synaptic structure will create an attractor similar in activity distribution to the stimulus. The higher the probabilities the higher the signal to noise ratio in the attractor and the easier it is to detect. But in such a case, the number of patterns that can be learned and recalled with such intensity, grows at most logarithmically with the number of neural elements. On the other hand, if the synaptic transition probabilities induced in a single presentation of a stimulus are very low, the signal to noise ratio for recall of the learned memory may be too low. The stimulus is not learned. There exists(Amit & Fusi, 1994) an intermediate regime in which stimuli are learned with synaptic transition probabilities high enough to allow the creation of attractors upon a single presentation, and yet low enough to allow for high storage. Low synaptic transition probabilities can be interpreted as slow learning. To compensate for slow learning longer presentations or multiple presentations can be used. Speeding up learning becomes counterpoised to high storage capacity. It should be pointed out that the low capacity that goes along with fast learning is not due to synaptic instability or decay, but rather to the elimination of older memories by newer ones. It acts as a palimpsest and the rate of erasure of old memories by new ones is intimately connected to the speed with which the memories are learned.

Speed of learning vs prototype extraction: The speed of learning is also related to the processing capability of the network, in other words, to the structure of the resulting set of attractors. If in the stream of stimuli there are groups of similar, i.e. highly overlapping patterns, with larger distances (in the Haming sense) between patterns of different groups, one would like the resulting attractors to create a relatively stable representative for each group. The structure of such an attractor can depend only on the patterns whose trace is still in the synaptic structure.

If learning is fast, erasure of the traces of previous patterns is fast and only the traces of a small number of patterns are available. If many patterns of a given group appear together, i.e. are within the same logarithmic memory span, the corresponding attractor for the group will have traces of them all. Though the last presented pattern will have a very pronounced weight. It will still represent the group, in the sense that stimuli similar to it will flow into it. But, this representative will be very unstable.

On the contrary if learning is slow, the memory span is long and the change in the synaptic structure induced by each stimulus is small. The attractor will be some mean of the stimulus group, and it can be considered a good prototype for all the patterns of the group. After each presentation of a new pattern of the same group, the attractor will move very slowly toward this new pattern preserving the information about all the other stimuli belonging to the same group which had been learnt previously.

2 Learning as a stochastic process and palimpsest memory

We denote the information arriving on a given neuron by a binary variable ξ, indicating whether the corresponding neuron does or does not carry information. Upon the arrival of a pair of information coding discrete variables a synapse will undergo a transition with probability which may be lower than unity. As a consequence the presentation of a sequence of uncorrelated stimuli induces a Markovian process on the set of values of the $N(N-1)$ synapses. More formally, the probability that a synapse make a transition $J \rightarrow J'$ is a product of $p_1(\xi, \tilde{\xi})$, the probability of the arrival of the pair $\xi, \tilde{\xi}$ on the two neurons connected by the synapse and the probability that given that pair the transition takes place, $p_2(J \rightarrow J'|\xi, \tilde{\xi})$. We shall further assume that a given pair $\xi, \tilde{\xi}$ can produce a transition between a single pair of neighboring synaptic states, or no transition at all.

The resulting markovian process is described by the probability distribution function of the synaptic values. The conditional distribution function $\rho_J^p(\xi, \tilde{\xi})$ of obtaining the value J following the presentation of p patterns the first of which imposed $\xi, \tilde{\xi}$ on the synapse, satisfies:

$$\rho_J^p(\xi, \tilde{\xi}) = \sum_K \rho_K^1(\xi, \tilde{\xi})(M^{p-1})_{KJ}, \tag{1}$$

in which M_{KJ} is the transition matrix whose elements are determined by the probabilities discussed above and the index J runs over all the stable synaptic values. Generically this dynamics is ergodic:

$$\rho_J^p(\xi, \tilde{\xi}) \rightarrow \rho_J^\infty,$$

which is independent of $\xi, \tilde{\xi}$. This makes a memory of this type a *palimpsest*(Nada *et al.*, 1986). In other words, patterns learned far in the past are erased by new patterns learned subsequently in sharp contrast to memories of the Hopfield or the Willshaw. For a network which is to be available for learning for indefinitely long periods, the generic initial distribution on top of which learning takes place, is the asymptotic distribution ρ_J^∞. Having an asymptotic distribution is a necessary condition for palimpsest behavior. It is not sufficient. The asymptotic distribution must allow the learning process to imprint new stimuli upon it.

To have a functioning learning network with a finite number of synaptic states, the presentation of a *given* new stimulus must change the conditional distribution $\rho_J(\xi, \tilde{\xi})$. Following the presentation of a given pattern consecutive presentations drive the conditional distribution back toward the asymptotic form, making the effect of the initial pattern progressively weaker. The question of the number of patterns that can be retrieved reduces therefore to the question about the age (distance into the past) of the oldest pattern that can still be retrieved, despite the effect of the subsequent patterns. Given the palimpsestic nature of the process, younger patterns can be retrieved a fortiori.

The findings: If all parameters, such as number of states per synapse; coding level in stimuli and transition probabilities of a synapse for a given pair of neural activity variables, are independent of the number of neurons in the network, then at most $\log N$ patterns can be retrieved.

If the number of synaptic states increases with N, as fast as $\sqrt{N}$, then one can reach a storage of order N. This was also observed in ref. (Nada *et al.*, 1986). Going beyond $\sqrt{N}$ in the number of states destroys the palimpsest behavior.

If the coding level in the arriving stimuli is $\log N/N$, one can reach storage capacities as high as $N/\log N$. For this type of patterns it was found(Willshaw *et al.*, 1969) that a network with two state synapses could have the optimal storage of $N^2/\log^2 N$ patterns. This is the price paid here for uninterrupted learning. Yet, when the intrinsic synaptic transition probabilities compensate for the coding level to make the mean number of up transitions (potentiation) of the same order as the number of down transitions (depression), one recovers optimal storage and enjoys continual learning.

3 Criteria for retrieval

In the simple case of auto-associative memory the possibility of retrieving a memory is determined by the distribution of depolarizations among the neurons in the network, upon the presentation of one of the previously memorized patterns. If that distribution is such that there exists a threshold which separates the depolarization of neurons which had been active in the learned pattern from those which had been quiescent, retrieval is in principle possible.

If the sequence of afferent stimuli to be learned is $\xi_i^1, \xi_i^2 \ldots \xi_i^p$, then the synaptic input to neuron i when ξ^1 is presented, following the learning of the entire sequence is:

$$h_i^p = \frac{1}{N} \sum_{j=1}^{N} J_{ij}(p) \xi_j^1,$$

where $J_{ij}(p)$ is the synaptic efficacy following the learning of the p patterns. If the neural activity is coded by a binary variable ($\xi_i = \zeta_1, \zeta_2$), there will be two distributions of synaptic inputs: one for neurons which saw the value ζ_1 when ξ^1 was presented and another for those that saw ζ_2. The values of the input in each class have a conditional mean:

$$\langle h_i^p \rangle_\xi = \sum_{\tilde{\xi}} \rho_{\tilde{\xi}} \sum_J J\tilde{\xi} \rho_J^p(\xi, \tilde{\xi}) \tag{2}$$

where the conditional expectation $\langle \ldots \rangle_\xi$ is defined as $\langle f \rangle_\xi = \mathrm{E}(f | \xi_i^1 = \xi)$ and ρ_ξ is the probability that a neuron had activity ξ when the network was stimulated by ξ^1. The expectation is over all the ξ_j^μ with $\mu > 1$ and $j = 1, \ldots, N$, and on ξ_j^1 with j different from i. The signal, for the binary case can be defined as:

$$S = \langle h_i^p \rangle_{\zeta_1} - \langle h_i^p \rangle_{\zeta_2} = \sum_{\tilde{\xi}} \rho_{\tilde{\xi}} \sum_J J\tilde{\xi}[\rho_J^p(\zeta_1, \tilde{\xi}) - \rho_J^p(\zeta_2, \tilde{\xi})], \tag{3}$$

where the sum on $\tilde{\xi}$ extends over the two possible values of the activity ξ_j^1 and J runs over all n values of the stable synaptic states.

The fluctuations of the two h_i^p are estimated by: $R^2(\xi) = \langle (h_i^p - \langle h_i^p \rangle_\xi)^2 \rangle_\xi$ If the random variables h_i^p are gaussian, then total noise is: $R^2 = \frac{1}{2}[R^2(\zeta_1) + R^2(\zeta_2)]$. The variances are minimal when the two sets of J's are assumed independent(Amit & Fusi, 1994). Then

$$R^2(\xi) = \frac{1}{N} \left(\langle J_{ij}^2 \xi_j^2 \rangle_\xi - \langle h_i^p \rangle_\xi^2 \right) = \frac{1}{N} \left(\sum_{\tilde{\xi}} \rho_{\tilde{\xi}} \sum_J J^2 \tilde{\xi}^2 \rho_J^p(\xi, \tilde{\xi}) - \langle h_i^p \rangle_\xi^2 \right) \tag{4}$$

Retrieval is possible if the ratio S/R is large enough. If one requires that the probability of an error on any neuron tend to zero with increasing N, then the square of the signal-to-noise ratio must grow at least as $\log N$ (Weisbuch & Fogelman-Souliè, 1985).

4 The logarithmic constraint

The matrix M has a single eigen-value 1, corresponding to the ergodic part. Writing equation (1) in terms of the eigen-values, λ_α, of M, one has:

$$\rho_J^p(\xi, \tilde{\xi}) = \sum_K \rho_K^1(\xi, \tilde{\xi}) M_{KJ}^{p-1} = \rho_J^\infty + \sum_{\alpha > 1} \lambda_\alpha^{p-1} \sum_K \rho_K^1(\xi, \tilde{\xi}) u_K^\alpha v_J^\alpha \qquad (5)$$

where u^α and v^α are respectively the right and the left eigenvectors associated to eigenvalues λ_α. $\lambda_1 = 1 > \lambda_2 = \lambda_M \geq \lambda_3 \geq ...\lambda_n$. Note that the terms multiplying λ_α^{p-1} for $\alpha > 1$ depend on the initial conditional distribution and on the eigenvectors of M, corresponding to λ_α. They are independent of p and N. Substituting ρ in Eq. (2) one finds

$$\langle h_i^p \rangle_\xi = h_\infty + \sum_{\alpha > 1} \lambda_\alpha^{p-1} F_\alpha(\xi) \qquad (\xi = \zeta_1, \zeta_2) \qquad (6)$$

where h_∞ is the term due to the asymptotic part of the distribution ρ_J^∞: When $\lambda_2(= \lambda_M)$ dominates S is obtained by substituting Eq. (6) in Eq. (3) (the asymptotic part h_∞ cancels) For fixed $\mathcal{P}$, λ_M dominates and

$$S = [\lambda_M(\mathcal{P})]^{p-1} \cdot C_2(\mathcal{P})$$

in which C_2 and λ_M depend on N only via the parameters which affect the learning dynamics (e.g. coding level of patterns, transition probabilities, presentation rate, number of stable synaptic states).

In a similar way one computes the uncorrelated part of the variances, which is an upper bound. See e.g. ref. (Amit & Fusi, 1994). For large p only the asymptotic terms in the variances survive and the signal to noise ratio behaves as:

$$\frac{S^2}{R^2} = \lambda_M^{2(p-1)} N \cdot C(\mathcal{P}) \quad \text{in which} \quad C(\mathcal{P}) = \frac{[C_2(\mathcal{P})]^2}{\langle J^2 \xi^2 \rangle_\infty - h_\infty^2} \qquad (7)$$

If we impose that in the limit $N \to \infty$ the ratio S^2/R^2 grow at least as $\log N$, then we obtain a bound on p:

$$p_c < \frac{1}{-2 \log \lambda_M(\mathcal{P})} \log \left(\frac{N \cdot C(\mathcal{P})}{\log N} \right). \qquad (8)$$

This result makes, of course, sense only if the argument of the logarithm is greater than unity. Or that

$$NC(\mathcal{P}) > \log N, \qquad (9)$$

Setting $p=1$ in Eq. (7), this condition is seen to be equivalent to the condition that the ratio of signal to noise will allow the recall of the most recently learned pattern ($\lambda_M < 1$).

5 Possible escapes

The logarithmic constraint on the number of retrievable patterns concerns a very wide class of networks with dynamic synapses. If one allows the parameters, $\mathcal{P}$, contained in λ_M to vary with the size of the network, so that $\lambda_M \to 1$, then it is possible to go beyond the logarithmic constraint. Specifically, for small x

$$\text{if} \quad \lambda_M = 1 - x(\mathcal{P}) \to p_c \sim x^{-1}. \tag{10}$$

Here we discuss two types of parameters $\mathcal{P}$ which affect the learning dynamics and that may depend on N:

Stochastic refresh mechanism. The transition probabilities of a synapse for given input can be made to decrease with N.

Coding level of the patterns. The fraction of information carrying bits per stimulus can be made to decrease with increasing N.

6 Stochastic learning of sparsely coded patterns

The most interesting results appear already in the case of 0–1 neurons, with a low mean fraction f of 1's and synapses with 2 stable states $(0,1)$. This case is exactly soluble. An imposed stimulus can produce the following transitions at a synapse:

If $J_{ij} = 0$ and $\xi_i^\mu = \xi_j^\mu = 1$, $0 \to 1$ occurs with probability q_+. Hence the probability of potentiation is $f^2 q_+$.

If $J_{ij} = 1$ and a mismatched pair of activities arrives, then depression $1 \to 0$ occurs with probability $q_-(10)$ for $\xi_i^\mu = 1, \xi_j^\mu = 0$ and $q_-(01)$ when $\xi_i^\mu = 0, \xi_j^\mu = 1$. Denoting $q_- = q_-(10) + q_-(01)$: the total probability for depression is $f(1-f)q_-$.

A pair of inactive neurons leaves the corresponding synapse unchanged.

The resulting transition matrix is:

$$M = \begin{pmatrix} 1 - f(1-f)q_- & f(1-f)q_- \\ f^2 q_+ & 1 - f^2 q_+ \end{pmatrix}. \tag{11}$$

The two eigen-values are 1 and:

$$\lambda_M = 1 - f^2 q_+ - f(1-f)q_-. \tag{12}$$

The asymptotic distribution, the left eigen-vector belonging to the eigen-value 1, is:

$$\rho^\infty = \left(\frac{f^2 q_+}{f^2 q_+ + f(1-f)q_-}, \frac{f(1-f)q_-}{f^2 q_+ + f(1-f)q_-} \right) \equiv (p_+, p_-) \tag{13}$$

where p_+ is the fraction of synapses=1. Note that in the Willshaw model (Willshaw et al., 1969) $q_-=0$. Hence, $\rho^\infty = (1,0)$: all synapses become 1.

For the present case, since $\rho_0^p = 1 - \rho_1^p$, Eqs. (3) become

$$\langle h \rangle_{+1} = f\rho_1^p(1,1) \quad \langle h \rangle_0 = f\rho_1^p(0,1)$$

The conditional probabilities are given by Eq. (5) as:

$$\rho_1^p(\xi, 1) = \lambda_M^{p-1} \sum_K \rho_K^1(\xi, 1) u_K v_1 + \rho_1^\infty$$

where the eigen-vectors corresponding to λ_M are given by: $u_K = (p_-, -p_+)$ and $v_K = (1, -1)$.

Starting from asymptotic distribution, we have that the conditional probabilities following the presentation of the oldest pattern ξ^1 are:

$$\rho_1^1(1,1) = p_+ + p_- q_+ , \qquad \rho_1^1(0,1) = p_+(1 - q_-) \tag{14}$$

When calculating the signal, the asymptotic parts cancel and the leading term, is proportional to λ_M^{p-1}.

$$S = \lambda_M^{p-1}(q_+ p_- + q_- p_+) f \tag{15}$$

Note again that for the Willshaw model $\rho^1 = \rho^\infty$ and $S=0$, i.e. no learning is possible on top of the asymptotic distribution.

For the calculation of the uncorrelated part of the noise, Eq. (4), we need $\langle h^2 \rangle$ which is the same as $\langle h \rangle$ with J_+ replaced by J_+^2 and J_- by J_-^2. One finds that:

$$R^2 = \frac{f}{N} \left[p_+ - (p_+)^2 f + \mathcal{O}(\lambda_M^{p-1}) \right] \tag{16}$$

For small f we keep only terms of leading order in f and, for large p, the signal to noise ratio is:

$$\frac{S^2}{R^2} = \lambda_M^{2(p-1)} N f \left(\frac{(q_+ p_- + q_- p_+)^2}{p_+} \right)^2 . \tag{17}$$

Extremal cases and the return of optimal storage

Lowest coding level: First we take the coding level $f \sim \log N/N$ (as in ref. (Willshaw et al., 1969)) keeping the transition probabilities q_+ and q_- fixed and both different from zero. Eq. 12 gives $\lambda_M \sim 1 - x$ with

$$x = f^2 q_+ + f(1 - f) q_- = \mathcal{O}(f) = \mathcal{O}(\log N/N), \quad \rightarrow \quad p_c = \mathcal{O}\left(\frac{N}{\log N} \right)$$

In fact, $f \sim \log N/N$ is as low as f can be without violating the bound (9). In other words, this network performs much worse than Willshaw's (Willshaw et al., 1969), which for the same coding level gives $p_c \sim N^2/\ln^2 N$. This is a price for continual learning.

Optimal storage recovered: If we take $f \sim \ln N/N$ and $q_- = f q_+$, from Eq. 12 we have $x \sim (\frac{\log N}{N})^2$, and provided the bound 9 is not violated one has the optimal storage

$$p_c \sim \left(\frac{N}{\log N} \right)^2 .$$

In ref. (Amit & Fusi, 1994) it is shown that the retrieval bound, (9) is respected.

Intermediate cases: A coding rate of $\frac{\log N}{N}$ is rather extreme. One can trade off some of the N dependence of f for an N dependence of q_+, which has been assumed finite in the limit of large N. The constraint on $C(\mathcal{P})$ implies that if

$$f = \left(\frac{\log N}{N}\right)^{\beta}, \quad \text{then} \quad q_+^2 = \left(\frac{\log N}{N}\right)^{1-\alpha}$$

with $\beta \in [0,1]$ ($f < 1$ implies that $\beta > 0$ and $q_+ < 1$ gives the upper bound $\beta < 1$). For x of Eq. (10) we have

$$x \sim f^2 q_+ \sim \left(\frac{\log N}{N}\right)^{\frac{1}{2}+\frac{3}{2}\beta} \quad \text{and hence} \quad p_c < \frac{1}{x} = \left(\frac{N}{\log N}\right)^{\frac{1}{2}+\frac{3}{2}\beta}.$$

The case $\beta=0$, i.e. fixed finite coding level f, reproduces the result of Tsodyks(Tsodyks, 1990), with a capacity of $\mathcal{O}(\sqrt{N})$.

7 Simulations

We have carried out extensive simulations to test the predictions of the theoretical estimates in the most extreme case, that of optimal storage in 2-state synapses and 0–1 neurons. We have taken

$$J_+ = 1, \quad J_- = 0, \quad f(N) = A\frac{\log N}{N}, \quad q_+ = 1, \quad q_- = f \quad (18)$$

with fixed $A = 4$. The signal and the noise are estimated for each choice of parameters N and p in the following way: A sequence of $N_p = 500 + p$ random N-bit words is generated, the stimuli to be learned. p is the maximal age of a pattern to be tested. Each word is generated by assigning 1's, at random, with probability f. All $500 + p$ patterns are presented consecutively to the network. Upon the arrival of each pattern ξ^μ, learning takes place, modifying the synaptic matrix according to the learning rule described at the beginning of Section 6. Following the learning of ξ^μ ($p < \mu < p + 500$) the state of the network is set to the pattern of age p, $s_i = \xi^{\mu-p}$, i.e. the stimulus to be retrieved. Then, with the new synaptic matrix J_{ij}^μ, we calculate the average of the post-synaptic input over the foreground and the background neurons in order to estimate the conditional mean of Eq. (3), i.e.

$$\langle h^p \rangle_\zeta(\mu) = \frac{1}{N_\zeta} \sum_{i:(s_i=\zeta)} \sum_{j\neq i} J_{ij}^\mu \xi_j^{\mu-p},$$

where the index i runs over all neurons for which $\xi_i^{\mu-p} = \zeta(= 0, 1)$; N_ζ is the number of neurons with $s_i = \zeta(=0,1)$ in the pattern presented. From this data we compute the square of the signal as the average over all presentations, i.e.

$$S^2 = \frac{1}{N_p} \sum_{\mu=1}^{N_p} [\langle h^p \rangle_1(\mu) - \langle h^p \rangle_0(\mu)]^2$$

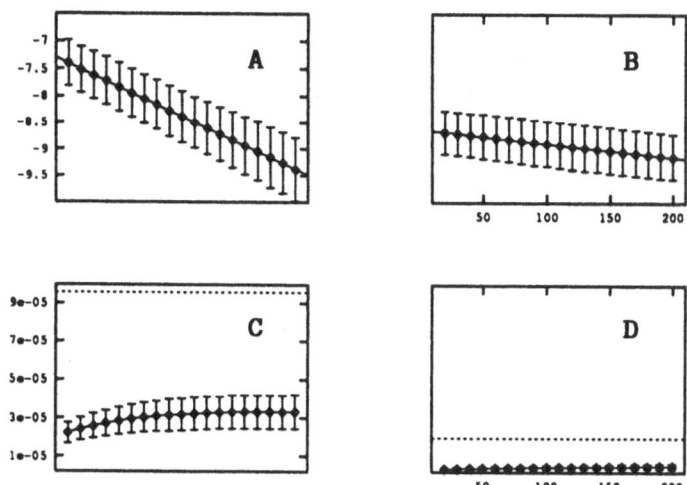

Figure 1: $\log S^2$ (A, B) and R^2 (C, D) vs number of memories p for fixed N: (A,C) N=600, (B,D) N=1400. Error bars are RMS in measured signals.

And the noise R^2 is calculated as half the sum of the standard deviations of h around $\langle h^p \rangle_1$ and $\langle h^p \rangle_0$. These results are then compared to the theoretical estimates. In particular we have tested the dependence of S^2 on N for fixed p and its dependence on p for fixed N. The theoretical expectation for the square of the signal, Eq. (15), are:

$$S^2(p, N) \simeq [\lambda_M(N)]^{2(p-1)} f^2(N) \left(\frac{2 - f(N)}{2 - 3f(N)} \right)^2 , \qquad (19)$$

where $f(N)$ is defined in (18). With the present choice of parameters $\lambda_M = 1 - 3f^2(N) + 2f^3(N)$, and from equation (16):

$$R^2(p, N) = \frac{1}{N} \left[f(N)p_+(N) + \lambda_M^{p-1} \frac{f(N)}{2}(1 - p_+(N)) - \mathcal{O}(f^2) \right] < \frac{f(N)}{2N}[1 + p_+(N)] \qquad (20)$$

If (19) and (20) are verified in the regime of the asymptotic behavior in N, then the number of storable and retrievable patterns can grow as $N^2/\log^2 N$. Equation (19) is written in the form:

$$y_1 = \log S^2 = a_1(N)p + b_1(N) , \quad \text{with} \quad a_1(N) = 2\log[1 - 3f^2(N) + 2f^3(N)]. \qquad (21)$$

Fitting $\log S^2$ vs p to simulation data for N=600,800,1000,1400, we find values for $a_1(N)$ which agree well with the theoretical theoretical values given by Eq. (21) for several values of N.

The behavior of S^2 vs N, for the same value of A, is presented in Figure 1(A,B), where S^2 is plotted as function of p for (N=600,1400). The continuous line represents the theoretical estimate while the points are simulation results.

The agreement improves with increasing N although, even for small N, the theoretical lines pass through the error bars. The upper bound on R^2 vs p, for the corresponding values of N, is tested in C and D. It tends to its asymptotic value, and is always below the straight dotted line which represents the upper bound (Eq.(20)).

References

Amit DJ (1992) In defence of single electrode recording, *NETWORK* **3** 385.

Amit DJ and Fusi S (1994) Learning in neural networks with material synapses, Neural Computation, in press.

Miyashita Y and Chang HS (1988) Neuronal correlate of pictorial short-term memory in the primate temporal cortex, *Nature*, **331** 68.

Weisbuch and Fogelman-Souliè F (1985) Scaling laws for the attractors of Hopfield networks, *J. Physique Lett.*, **2**, 337.

Nadal J-P, Toulouse G, Changeux JP and Dehaene S (1986) Networks of formal neurons and memory palimpsests, *Europhys. Lett.*, **1**, 535; Parisi G 1986 A memory which forgets, *J. Phys.*, **A19** L617.

Willshaw D Buneman OP Longuet-Higgins HC (1969) Non-holographic associative memory, Nature, London, **222**, 960.

Tsodyks M (1990) Associative memory in neural networks with binary synapses, *Modern Physics Letters*, **B4**, 713.

Nominal Color Coding of Classified Images by Hopfield Networks

P. Campadelli*, P. Mora*, R. Schettini°
* Dipartimento di Scienze dell'Informazione
Università degli Studi di Milano
via Comelico 39/4, I-20135 Milano (Italy)
° ITIM, CNR
Via Ampere 56, I-20131 Milano (Italy)

1 Introduction

The experimental evidence that color greatly increases the observer's under-standing of the information contained in a picture and his capacity for remem-bering it, has led to assigning color a fundamental role in conveying qualitative information in graphical environment.

Associating a set of colors with a set of items to express the significance of each is called "nominal coding" (Bertin, 1983). Nominal color coding is widely used by the image processing community to represent the output of a classification-segmentation process. In this context a serious problem arises because the chosen colors must be displayed together and assigned to classes composed of regions of different size and morphology. The user must take into account the characteristics of the image (the number of classes, the links between them, and the geometric and topological features of the regions that belong to the different classes) so that the association of classes with colors produces a readable, pleasant coded image. Although some systems have been proposed to support the user in color coding, these are mainly based on a trial-and-error approach and demand a concerted effort on the part of the user (Della Ventura et al., 1993). In order to automate color-class association we have proposed a suitable description scheme for both the color set to be used and the image to be coded. Exploiting this description we have then defined a suitable energy function for Hopfield's neural networks (Hopfield, 1984). The aim is to assign more "evident" colors to less "visible" classes, attributing highly contrasting colors to classes with an high "adjacency".

2 Image and Color Description

We assume that the image to be coded contains n classes, that each classe is composed of a variable number of regions having different characteristics, and that the color set $C = \{c_1, \ldots, c_n\}$ to be used in coding has been already selected by the user according to some criteria. An automatic evaluation of the image to be coded obviously requires a description of the image itself. The data structure used to represent regions and their relationships is the Region Adjacency Graph (Rosenfeld et al., 1982), in which each node represents a region, and two nodes are joined by an arc if the corresponding regions are adjacent. We associate with each node a description of the region (relative area, perimeter, compactness and thickness) and a class label, while with each arc the degree of adjacency of that pair of regions (relative perimeter that the two regions share).

A Class Adjacency Graph (here each node represents a class) is then derived from the Region Adjacency Graph. Each node i of the graph is labeled by its class visibility v_i, which represents the weighted sum, normalized to 1, of the properties of the regions belonging to the class i. To each arc $\{i, j\}$, the Adjacency A_{ij}, a value in the range of $[0, 1]$, is defined as the maximun of the relative adjacency between the regions belonging to the classes i and j.

The elements of the color set to be used in coding may be represented in several ways, all of which require three mutually independent features. The RGB color space, which is embodied in the electronic medium, specifies colors as three-component vectors along the red, green, and blue color axes. This representation is not consistent with the way the user is accustomed to thinking of color, i.e. in terms of the psychological dimensions of hue, saturation (chroma), and brightness (lightness). Moreover the RGB space is not perceptually uniform, i.e. the same physical distances do not correspond to equal perceptual differences. To overcome these drawbacks the colors of the set are mapped in the CIELUV color space (Della Ventura, 1990a), where the color difference between the two colors c_1 and c_2, indicated with $D(c_1, c_2)$, is equal to the Euclidean distance between the two points representing those colors in the space, and correlates of chroma ($C^\star$) and lightness ($L^\star$) are defined. Since the lightest and most highly saturated area of a color display immediately draws the user attention, we define the evidence of a color c_i as:

$$\epsilon(c_i) = \frac{C^\star(c_i) + L^\star(c_i)}{\max_{k=1,n}(C^\star(c_k) + L^\star(c_k))}$$

Evidence may be interpreted as the probability that a color will remain discriminable in a cluttered display, while the visibility of a class may be interpreted as the probability that the class will be recognizable in the coded image, no matter what color is associated with it.

3 The energy function

Color-class association has been transformed into a problem of minimizing a suitable energy (Lyapunov) function for symmetric neural networks of analog neurons (Hopfield, 1982,1984). The network, in its evolution from an initial, arbitrary condition, finds a minimum of the energy function.

To design the energy function, and thus the network structure, we must choose a representation scheme which allows the output state of the neurons to be decoded into the chosen set of couples ⟨color, class⟩. A feasible solution (one which assigns a different class to each color) can be represented by a $n \times n$ permutation matrix. Each row of the matrix corresponds to a particular color, while each column corresponds to a particular class. To obtain feasible "good" solutions the following requirements must be satisfied:

1. stable states corresponding to a permutation matrix must be favored;

2. permutation matrices representing "good" associations must be favored. A color-class association is "good" if the most evident colors are attributed to the least visible classes (and viceversa), and if highly contrasting colors are attributed to classes with a high adjacency.

A simple way to take into account both requirements is to express the energy function as the sum of two terms $E(V) = P(V) + f_{ob}(V)$, where $P(V)$ is a term which penalizes unfeasible solutions (i.e. solutions which do not satisfy point 1.), and $f_{ob}(V)$ is the formalization of the concept of "good" association. Following Hopfield and Tank (Hopfield et al. 1985), a penality term which satisfies 1. is:

$$P(V) = A \sum_x \sum_i \sum_{i \neq j} V_{xi} \cdot V_{xj} + B \sum_i \sum_x \sum_{y \neq x} V_{xi} \cdot V_{yi} + C (\sum_x \sum_i V_{xi} - n)^2$$

where the indices x and y represent colors, the indices i and j represent classes, V_{xi} is the output of the neuron xi, and A, B, and C are suitable positive constants.

The function $f_{ob}(V)$ is described by the following expression:

$$f_{ob}(V) = F \sum_x \sum_i \Phi_1(x, i) \cdot V_{xi} + G \sum_x \sum_{y>x} \sum_i \sum_{j>i} \Phi_2(x, y, i, j) \cdot (V_{xi} \cdot V_{yj} + V_{xj} \cdot V_{yi})$$

where F, and G are suitable positive constants, $\Phi_1(x, i)$ is defined in terms of the evidence $\epsilon(x)$ of a color and the visibility v_i of a class, $\Phi_2(x, y, i, j)$ is a function of the Adjacency A_{ij} between couples of classes and of the difference $D(c_x, c_y)$ between couples of colors. Letting D_{min} express $\min_{\{x,y\}} D(c_x, c_y)$ we have:

$$\Phi_1(x, i) = k(\epsilon(x) + v_i - 1)^2 \quad k > 1$$

$$\Phi_2(x, y, i, j) = h(A_{ij} + D_{min} \cdot \frac{1}{D(c_x, c_y)} - 1)^2 \quad h > 1$$

We note that Φ_1 and Φ_2 assume values close to 0 for "good" color-class associations as defined above, while they assume a value near k or h respectively for those which are "bad" (i.e. the most highly visible classes assigned the most evident colors, and couple of classes with a high adjacency displaying similar colors).

4 Experimental results and conclusions

Once the network connection matrix has been derived from the expression of the energy function, the connection matrix can be used to generate the equations of motion. The choice of parameters A, B, and C, is critical; on an empirical basis we have set A at 200, B at 200, C at 100. The parameters F, and G, used to control the weight of the functions Φ_1 and Φ_2, have both been set at 10, while h and k have been set at 1.5. Equal values, plus or minus a certain amount of noise, have been assigned to the states of the neurons initially; the mean value was determined by having $\sum_x \sum_i V_{xi} = n$. A "freezing" criterion was used to determine when to stop a particular simulation: after each update the new values were compared, and when no value changed more then a threshold the network was assumed to have reached a final stable state.

The most significant test of the network's performance was performed on a satellite image having 2157 regions and 12 classes. The network found feasible solutions in seven out of ten trials; each network simulation required about 10 seconds on a Sony NWS-3860 workstation (20Mhz).

The effectiveness of the network solutions can not be quantified, but only evaluated heuristically by the users themselves on the basis of their experience in color coding. Judging from the experimental results, the authors believe that the neural network approach proposed here can deal successfully with the complexities of nominal color coding. A more sistematic evaluation of the method is in progress.

5 References

Bertin, J. (1983) Semiology of graphics. Un. of Wisconsin Press, Madison.

Della Ventura, A., and Schettini, R. (1990a) Perceptual color scales for data display. Automatika, 31, A.61-A.71.

Della Ventura, A., Padula, P., and Schettini, R. (1990b) Communicating with the help of color: syntactic, semantic and pragmatic aspects. Automatika, 31, 5-12.

Della Ventura, A., and Schettini, R. (1993) Computer aided color coding. Communication with Virtual Word (N.M. Thalmann, D. Thalmann eds), Spriger-Verlag, Tokyo, 62-75.

Hopfield, J.J. (1982) Neural networks and physical systems with emergent collective computational abilities. Proc. Natl. Acad. Sci. USA 79, 2554-2558.

Hopfield, J.J. (1984) Neurons with graded response have collective computational properties like those of two-states neurons. Proc. Natl. Acad. Sci. USA 81, 3088- 3092.

Hopfield, J.J., and Tank, D.W. (1985) "Neural" computations of decisions in optimization problems. Biol. Cybernetics 52, 141-152.

Rosenfeld, A., and Kak, A.C. (1982) Digital picture processing. Academic Press Inc, Orlando, Florida, 2, 148.

Does Terminal Attractor Backpropagation Guarantee Global Optimization?

M. Bianchini, M. Gori, and M. Maggini

Dipartimento di Sistemi e Informatica, Università di Firenze

Via di Santa Marta 3 - 50139 Firenze - Italy

1 Introduction

Recently, Wang *et al.* (Wang *et al.*, 1991) have introduced two new learning algorithms, called *TABP* and *HTABP* [1], that are based on the properties of terminal attractors. These algorithms were claimed to perform global optimization of the cost in finite time, provided that a null solution exists. In this paper, we prove that, unfortunately, there are no theoretical guarantees that a global solution will be reached, unless the learning process begins in the domain of attraction of the global minimum. When a local minimum basin is entered, quite random jumps in the weight space take place that may led to cycles. Moreover, when approaching local minima, overflow errors may also occur that force the learning to stop. Finally, particular care must be taken in order to avoid numerical problems that may occur even when approaching global minimum.

2 The TABP and HTABP algorithms

When the Lipschitz condition is violated for a differential equation at a given equilibrium point, such point becomes a *terminal attractor* (Zak, 1989). Let us illustrate the idea of terminal attractors with an example:

$$\dot{x} = -x^k, \quad 0 < k < 1. \tag{1}$$

This equation has an equilibrium point at $x = 0$ violating the Lipschitz condition ($\frac{d\dot{x}}{dx} = -kx^{k-1} \to -\infty$ as $x \to 0$). Being $x_0 \geq 0$ the initial condition , the closed form solution to eq. (1) is

$$x(t) = \begin{cases} (x_0^{1-k} - (1-k)t)^{\frac{1}{1-k}} & t \leq t_e \\ 0 & t > t_e, \end{cases} \tag{2}$$

where $t_e = \frac{x_0^{1-k}}{1-k}$. In the finite time t_e the transient starting at $x(0) = x_0$ reaches the equilibrium point $x = 0$, which is a terminal attractor.

In (Wang *et al.*, 1991) the following application of the terminal attractor concept to the neural network learning was proposed through *TABP* and *HTABP*

---

[1] Terminal Attractor Backpropagation and Heuristic Terminal Attractor Backpropagation.

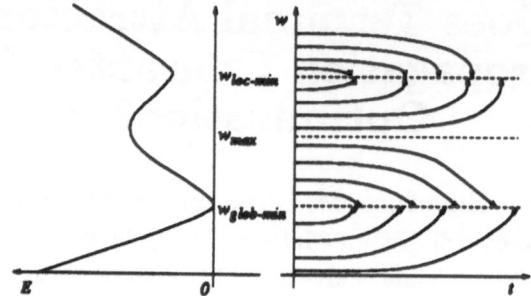

Figure 1: Trajectories of $E(w)$ having local minima.

algorithms. Let $E(W)$ be the cost function. The weight updating rule is

$$\frac{dW}{dt} = -\gamma \nabla_W E, \tag{3}$$

where $W \in \mathcal{R}^n$ is the weight vector and γ is a time-varying gain. Let us choose γ as:

$$\gamma \doteq \frac{E^k}{\| \nabla_W E \|^2} \quad 0 < k < 1. \tag{4}$$

Then, the dynamics of the error function becomes

$$\frac{dE}{dt} = (\nabla_W E)^T \frac{dW}{dt} = (\nabla_W E)^T \left(-\frac{E^k}{\| \nabla_W E \|^2} \nabla_W E \right) = -E^k, \tag{5}$$

which is characterized by the presence of a terminal attractor solution for $E = 0$. Thus, it seems that the learning process takes place by following eq. (5) independently of the kind of problem we are dealing with. For this reason, Wang et al. (Wang et al., 1991) concluded that their algorithm leads to a global solution.

3 TABP: the paradox of global convergence

Concerning eq. (5), we notice that the substitution of $\frac{dW}{dt}$ is not allowed when γ is undefined. Such a situation occurs whenever a local minimum is reached. We show that local minima are attractors for eq. (3) also when using γ as defined by eq. (4), that is the evolution $E(t)$ follows eq. (5) only in the basin of attraction of each minimum.

For the sake of simplicity, let us consider the unidimensional case (Fig. 1). Let $\tilde{w} = w - w_{min}$ be. In the neighbourhood of the local minimum w_{min} the weight update law can be written as

$$\frac{d\tilde{w}}{dt} \simeq -\frac{E^k(w_{min})}{\frac{\tilde{w}^{2m-1}}{(2m-1)!}\frac{d^{2m}E}{dw^{2m}}\big|_{w=w_{min}} + \mathcal{O}(\tilde{w}^{2m})} \simeq -\frac{h_0}{\tilde{w}^{2m-1}}, \tag{6}$$

where we approximate the values of E at w by $E(w_{min})$ and $2m - 1$ is the multiplicity of the first derivative zero at the same point. Thus the closed form solution to eq. (6) may be written as

$$\tilde{w}(t) \simeq \pm(\tilde{w}_0^{2m} - 2mh_0 t)^{\frac{1}{2m}}, \qquad (7)$$

which is defined only for $t \leq \frac{\tilde{w}_0^{2m}}{2mh_0}$, when the local minimum is reached [2].

The time-discretized version of the weight update rule (3) is

$$W_{n+1} = W_n + \eta\gamma\nabla_W E_n, \qquad (8)$$

where η is the integration step. When a local minimum is approached, γ increases and, if no overflows occur, the system will jump from the local minimum point causing a "restart" of the E dynamics. We notice that the direction and the extension of such jumps are not guaranteed to lead closer to the global minimum; the trajectory may get stuck into the local minimum basin, giving an oscillatory solution for E, or may escape from the basin and jump to another minimum [3].

Moreover, for some choices of the exponent k, we may have jumps even in the neighbourhood of the global minimum. In fact, depending on the multiplicity of the zero of function E, the trajectories may approach $E = 0$ with an infinite slope which, combined with the use of a constant integration step, can cause large increments for the weights. For the unidimensional case, using Taylor expansions for the numerator and the denominator of eq. (3) and discarding the higher order terms, we derive

$$\frac{d\tilde{w}}{dt} = \mathcal{O}(\tilde{w}^{(2mk - 2m+1)}), \qquad (9)$$

where $\tilde{w} = w - w_g$, and $2m$ is the multiplicity of the global minimum w_g. From the previous relationship we see that, if $k < 1 - \frac{1}{2m}$, $\frac{dw}{dt} \to \infty$ when $w \to w_g$.

The following example shows experimentally the problems of convergence described from a theoretical point a view. Let us consider the unidimensional error function $E(w) = w^6 - 2w^4 + \frac{28}{27}w^2$, which has a global minimum in $w = 0$ and two symmetrical local minima in $w = \pm\frac{\sqrt{6+2\sqrt{2}}}{3}$. Fig. 2 depicts undesirable behaviour of the *TABP*. In Fig. 2a the w dynamics first oscillates in the neighbourhood of the positive local minimum and then jumps into the basin of the negative one. In Fig. 2b, the trajectory jumps out the global minimum basin, showing the worst pathology the algorithm may be affected by.

4 Conclusions

In this paper, we have shown that, unfortunately, the "jumps" of *TABP* and *HTABP* are essentially random and that they do not lead necessarily to the

---

[2]The choice of sign in eq. (7) depends on the initial condition $\tilde{w}_0$.

[3]The transients are strongly dependent on the parameter values, the starting point and the machine precision. In particular, overflow errors may occur quite unpredictably as the minimum is approached.

380

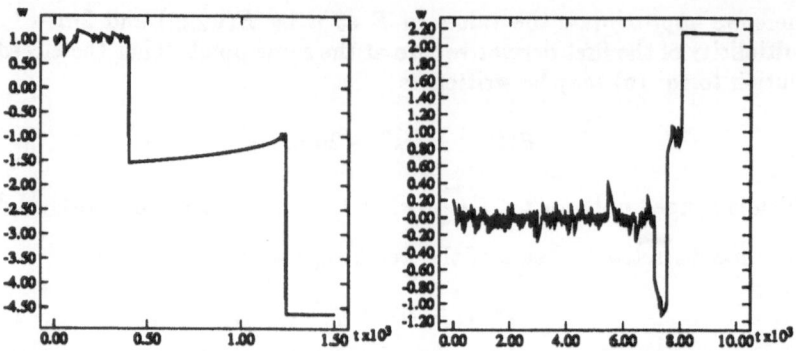

Figure 2: Diagrams of $w(t)$. (a) Example of a jump from a local minimum to the other one ($w_0 = 1.0$, $\eta = 0.005$, $k = 0.3333$, $nepochs = 1500$). (b) Jumping out the global minimum basin ($w_0 = 0.2$, $\eta = 0.001$, $k = 0.2$, $nepochs = 10000$).

global minimum. This clarifies the paradox arising from the comparison of Wang *et al.*'s claims on the time required for performing global optimization with Judd's results on the computational complexity of the loading problem (Judd, 1990). However, the criticisms contained in this paper concern the theoretical claims on global convergence only and not the actual effectiveness of *TABP* and *HTABP* for learning. On the opposite, the idea behind these algorithms is interesting and deserves further investigations (see e.g. (Jones *et al.*, 1993))

Acknowledgments
We thank S. Fanelli for his very useful comments and suggestions.

References

[1] Jones, C. R., and Tsang, C. P. (1993), "On the Convergence of Feed Forward Neural Networks Incorporating Terminal Attractors," *Proc. of the IEEE-ICNN93*, San Francisco, CA, vol. 2, pp. 929-935.

[2] Judd, J. S. (1990), "Neural network design and the complexity of learning," *The MIT Press*, Cambridge, London.

[3] Wang, S. D., and Hsu, C. H. (1991), "Terminal Attractor Learning Algorithms for Back Propagation Neural Networks," *Proc. of International Joint Conference on Neural Networks* (Singapore), IEEE Press, pp. 183-189.

[4] Zak, M. (1989), "Terminal Attractors in Neural Networks," *Neural Networks*, Vol. 2, pp. 259-274.

LEARNING and RETRIEVAL in ATTRACTOR Neural Networks with Noise above Saturation

R. Erichsen Jr.† and W. K. Theumann‡

†Instituut voor Theoretische Fysica, K. U. Leuven, B-3001 Leuven
Belgium
‡Instituto de Física, UFRGS, Caixa Postal 15051, 91501-970 Porto
Alegre, RS, Brazil

1. Introduction

The attractor neural network model consists of a system of N binary neurons interacting via synaptic couplings. The state of neuron i is represent by the variable S_i that assumes values $+1$ or -1, depending if the neuron is active or idle, respectively. The element of matrix J_{ij} represents the interaction between neurons i and j.

The model has the property of associative memory. Starting from a given configuration, it evolves dynamicaly until to be attracted by one of the patterns $\{\xi_i^\mu, 1 \le i \le N, 1 \le \mu \le p\}$. In this context, *learning* signifies to change properly the interactions J_{ij} in order to make the states $\{\xi_i^\mu\}$ attractors in the phase space of the system.

The pattern μ is correctly stored in the site i if the local stability λ_i^μ satisfies the condition

$$\lambda_i^\mu = \frac{1}{\sqrt{N}} \sum_{i=1}^{N} J_{ij} \xi_i^\mu \xi_j^\mu > \kappa. \tag{1}$$

κ is a positive parameter, introduced to assure a finite basin of attraction around the pattern. To each learning algorithm is defined a cost function $V(\lambda)$ that penalizes violations to (1). Learning is thought as a dynamical evolution in the space of the J_{ij} governed by the cost function $V(\lambda)$.

In recent years, optimal properties of networks generated deterministicaly by several algorithms have been studied (Amit et al., 1990; Griniasty and Gutfreund, 1991). In the present work, we propose to clarify what kind of properties result if learning is supposed non-deterministic, i. e., it occours in the presence of a *learning noise* quantified by the learning temperature T.

In what follows, we will briefly review the formal analysis. Detailed results for cost functions of Gardner-Derrida and perceptron will be shown. At the end, dynamical properties of sparcely connected networks will be discussed, as well as preliminary results for the adatron cost function.

2. Formal analysis

Following Gardner and Derrida (Gardner and Derrida, 1988), we explore the physical properties given by the free energy

$$-\beta G(\beta) = \langle \ln Z \rangle_\xi, \tag{2}$$

where β is the inverse temperature and the partition function is

$$Z = \int \prod_{j=1}^{N} dJ_{ij} \delta(N - \sum_{j=1}^{N} J_{ij}^2) \prod_{\mu=1}^{p} \exp[-\beta V(\lambda)]. \tag{3}$$

Angular brackets indicate average over the patterns $\{\xi_i^\mu\}$, assumed to be independent random variables assuming values $+1$ or -1 with the same probability. The δ-function imposes the spherical constraint $\sum_j J_{ij}^2 = N$.

When using the replica method to perform the pattern average it is introduced the parameter $q_{\rho\sigma} = \sum_j J_{ij}^\rho J_{ij}^\sigma / N$, representing the overlaps between replicas. The calculation is already well known, and details will be omitted. In the replica-symmetric framework, where $q_{\rho\sigma} = q$, $\forall$ $\rho < \sigma$, free energy (2) becomes

$$-\beta G(\beta) = \alpha \int Dt \ln \psi + \frac{1}{2}\ln(1-q) + \frac{1}{2(1-q)} + \frac{1}{2}\ln 2\pi, \tag{4}$$

with $\alpha = p/N$ and ψ being defined as

$$\psi = \int_{-\infty}^{+\infty} \frac{d\lambda}{\sqrt{2\pi(1-q)}} \exp\left[-\beta V(\lambda) - \frac{(\lambda - \sqrt{q}t)^2}{2(1-q)}\right]. \tag{5}$$

$Dt = dt \exp(-t^2/2)/\sqrt{2\pi}$ is a gaussian measure.

We access network properties by calculating the distribution of local stabilities according to Kepler and Abbott (Kepler and Abbott, 1988). For the replica symmetric solution we have

$$\rho(\lambda) = \int_{-\infty}^{+\infty} \frac{Dt}{\sqrt{2\pi(1-q)}\,\psi} \exp\left[-\beta V(\lambda) - \frac{(\lambda - \sqrt{q}t)^2}{2(1-q)}\right]. \tag{6}$$

3. Results for specific cost Functions

The basic quantity to calculate is the function ψ. It has the form

$$\psi = \frac{1}{2}\left(1 - \mathrm{erf}\frac{K}{\sqrt{2}}\right) + \frac{1}{2}\left(1 + \mathrm{erf}\frac{\overline{K}}{\sqrt{2}}\right)\exp(-\varphi), \tag{7}$$

where $K = (\kappa - \sqrt{q}t)/\sqrt{1-q}$. For Gardner-Derrida, $\overline{K} = K$ and $\varphi = \beta$, while for the perceptron $\overline{K} = K - \beta\sqrt{1-q}$ and $\varphi = (K^2 - \overline{K}^2)/2$. The saddle-point value of q is given by eq. (5). Figures 1a and 1b show the probability distribution of local stabilities for $\alpha = 0.55$ and $\kappa = 1.0$ at several temperatures, for both algorithm. At zero temperature, reusults are those of Griniasty and Gutfreund (Griniasty and Gutfreund, 1991).

4. Dynamics of sparsely connected networks

The time evolution of the network is characterized by its instantaneous overlap with the stored patterns, $m_\mu(t) = \sum_i S_i \xi_i^\mu / N$. For sparsely connected networks, where connectivity of each neuron is C, with $C \approx \ln N$ and $\alpha = p/C$, synchronous evolution of the overlap is described, in the absense of synaptic noise, by the

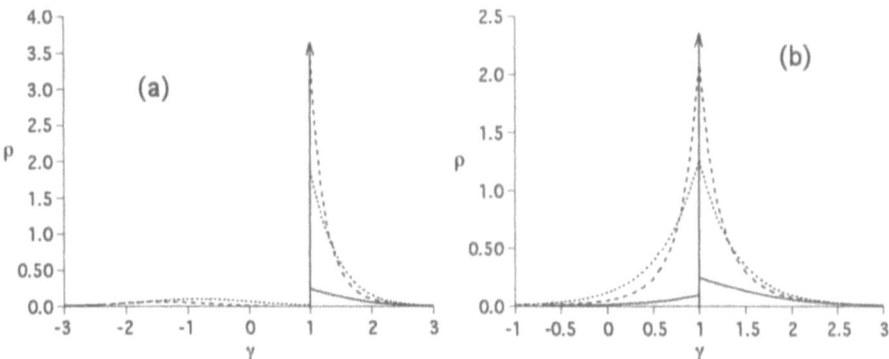

Figure 1: Probability distribution of local stabilities for $\alpha=0.55$ and $\kappa=1.0$; full curve: T=0; broken curve: T=0.1; dotted curve: T=0.2; (a) Gardner-Derrida; (b) perceptron.

expression

$$m(t+1) = \int_{-\infty}^{+\infty} d\lambda\, \rho(\lambda)\, \mathrm{erf}\, \frac{m(t)\lambda}{\sqrt{2(1-m^2(t))}}. \tag{8}$$

Calculating the fixed point $m*$ of the equation above at a given learning temperature, we obtain some information about the effect of learning noise over retrieval properties of the network. Figures 2a and 2b show phase-diagrams over the plane (α,κ) for both cost functions at $T = 0.1$. Results for zero temperature (Griniasty and Gutfreund, 1991) are also shown. Three regions respective to retrieval properties are observed: non retrieval ($m* = 0$), partial retrieval, ($m* \neq 0$, $m_0 > 0$) and full retrieval ($m* \neq 0$, $m_0 = 0$). m_0 is the size of the basin of attraction. κ plays an important role in stabilising the network at non-zero temperature. In fact, at $T = 0$ the maximum retrieval capacity is 2 for both algorithms, at $\kappa = 0$. At $T = 0.1$, the retrieval capacity is zero at $\kappa = 0$. For Gardner-Derrida, at this temperature, the maximum ($\alpha \approx 0.69$) occours at $\kappa \approx 0.18$. At the same temperature, the perceptron has a zero retrieval capacity for

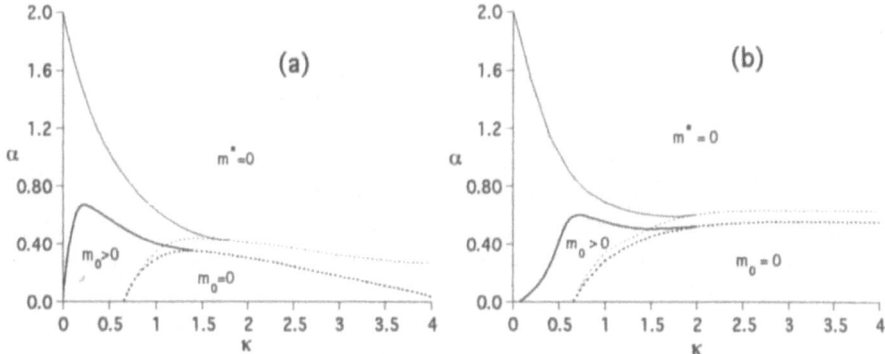

Figure 2: Phase diagrams for sparcely connected networks at $T=0$ (thin lines) and $T=0.1$ (thick lines); $m_0=0$: no retrieval; $m*>0$, $m_0=0$: full retrieval; $m*>0$, $m_0>0$: partial retrieval; full(dotted) lines: discontinuous(continuous) phase transitions; (a) Gardner-Derrida; (b) perceptron.

384

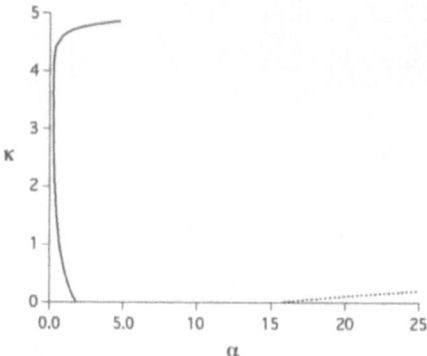

Figure 3: AT lines for Gardner-Derrida (full line) and perceptron (dotted line) at T=0.1. Replica symmetric solution for each cost-function is stable at the left side of the corresponding AT line.

$\kappa < 0.08$ and the maximum ($\alpha \cong 0.61$) occours at $\kappa \cong 0.70$. Results for the adatron reveal that the shift is still greater: retrieval capacity is zero for $\kappa < 0.63$, and the maximum ($\alpha \cong 0.71$) occurs at $\kappa \cong 1.10$.

In order to verify the stability of replica symmetry solution, de Almeida-Thouless analisys was performed (de Almeida and Thouless, 1978). Figure 3 shows the AT lines for Gardner-Derrida and perceptron, for $T = 0.1$. For the adatron, replica symmetry solution is stable at this temperature.

5. Conclusion

In this work we have shown that the model of attractor neural network is robust in the presence of noise during the learning process. It was also shown that maximum retrieval at $\kappa = 0$ is marginal at zero temperature, and non-zero κ is important to assure finite retrieval capacity.

This work was supported by CNPq (Conselho Nacional de Desenvolvimento Científico e Tecnológico, Brazil).

References

de Almeida, J., and Thouless, D. J. (1978) "Stability of the Sherrington-Kirkpatrick solution of a spin glass model". *J Phys A: Math. Gen.* **11** 983-990.

Amit, D. J., Evans, M. R., Horner, H., and Wong, K. Y. M. (1990) "Retrieval phase diagram for attractor neural networks with optimal interactions". *J. Phys A: Math. Gen.* **23** 3361-3381.

Gardner, E., and Derrida, B. (1988) "Optimal storage properties of neural networks models". *J. Phys. A: Math. Gen.* **21** 271-284.

Griniasty, M., and Gutfreund, H. (1991) "Learning and retrieval in attactor neural networks above saturation". *J. Phys. A: Math. Gen.* **24** 715-734.

Kepler, T. B., and Abbott, L. F. (1988) "Domains of attraction in neural networks". *J. Phys . France* **49** 1657-1662.

A Method of Teaching a Neural Network to Generate Vector Fields for a Given Attractor

N. H.-R. Goerke,   R. Eckmiller

Department of Computer Science VI - Neuroinformatics, University Bonn

Römerstr. 164, D-53117 Bonn, F. R. Germany

Tel.:++49–228–550–247   FAX: ++49–228–550–425

e–mail: goerke@nero.uni-bonn.de

Introduction

Consider a finite n-dimensional space, with an n-dimensional vector S completely describes the "state", in which a nonlinear system resides; to know the vector of state-changes is a complete representation of the system dynamics provided it is deterministic and of first-order. The totality of all state-changes form a vector field that contains the entire system dynamics. A trajectory in state space is formed by pieceing together infinitesimal steps in the direction indicated by this vector field. The result is a curve whose tangent at each point is always aligned with the vector field (Thompson and Stewart, 1986).

If a neural network can be trained to learn this vector field, – to associate the actual state vector S at the input with the vector of state-change – the network establishes a representation of the dynamic system, equivalent to a set of nonlinear first order differential equations.

Various neural networks with generalization capabilities have been proposed in the literature recent (Werntges, 1991) for the task of vector field approximation.

In this paper we present a method of teaching a multi layer perceptron (MLP) to generate a vector field in state space that describes the dynamic of a nonlinear system, when only knowledge of the *attractor* to which the system evolves is available. Starting with an initial state, the dynamic of the network governs the sequence of state vectors $S(t)$, and generates a state space trajectory. The application of such a method of mapping desired trajectories onto embeddings in vector fields, can bes considered for path planing modules with online obstacle avoidance features.

The Network

We decided to use an MLP because of its generalization capabilities and because a varietey of established learning methods including a large number of simulation tools are available.

The MLP receives the n-dimensional state vector S as input, and maps it to an n-dimensional output vector ΔS interpreted as direction of state-change. Adding the presumed state-change to the actual state vector should yield the new state of the system $_{new}S =_{old} S + \Delta S$. The network is now used within an outer feedback loop forming a recurrent network architecture. To demonstrate the capabilities of our approach we concentrated on a 2-dimensional state space $S(t) = (x(t), y(t))$. A standard 3 layer feed forward network with 2 inputs and 2 outputs, sigmoidal transfer-function for the hidden layer neurons and either sigmoidal or linear function for the output alternatively, was sufficient. The network input is (x_t, y_t) and the output is $(\Delta x_t, \Delta y_t)$. During recall the systems evolution to the new state is calulated by the feedback loop to: $S_t = (x_{t+1}, y_{t+1}) = (x_t, y_t) + (\Delta x_t, \Delta y_t)$

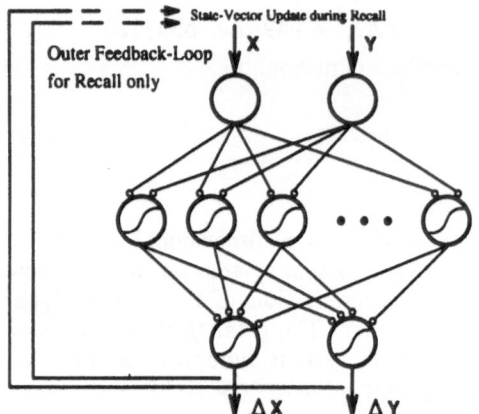

Fig. 1: The MLP network,

2 input units: X, Y

a hidden layer with sigmoid transfer-function,

2 output units to generate ΔX, ΔY, with sigmoid or linear transfer-function alternatively.

Expanding the Training Data

As training data we used different sequences of state-space vectors (trajectories), that were sampled with a fixed temporal stepsize. The desired network output (teacher) is calculated using the differences between two succecive steps. We want the network to establish a global vectorfield with the given trajectory as the *attractor* the system should converge to. Assuming that this trajectory allready represents the *attractor*, we can request that a small pertubation from this trajectory has to vanish under the systems dynamic. Unfolding a local vectorfield in the direct neighbourhood of the given trajectory will force the neural network to learn the major property of *attractors*.

Starting at the position S_t the original vector of state-change ΔS_t is calulated to: $\Delta S_t := S_{t+1} - S_t$. To obtain a new set of training data a displacement vector D is added to the original state, yielding the new virtual position S_t^D. The attracting property of the system requires that the displaced state-change vector ΔS_t^D points to the next original state vector S_{t+1} (Fig. 2). The displacement D is a random vector, equaly distributed drawn from the neigbourhood of the original state S_t. The size of the neighbourhood used in our simulations was choosen in the same range as the average state-change. This calculation is performed for every state in the original training data set multiple times (at least 10), to generate a cluster of new vectors (Fig.2 b) forming the larger set of training data. The output data was then normalized to a range from 0.1 to 0.9 to fit the requirements for sigmoid output units. A typical trajectory consists of about 40 original training points, expanded to 400 new vectors and shuffled to a random sequence.

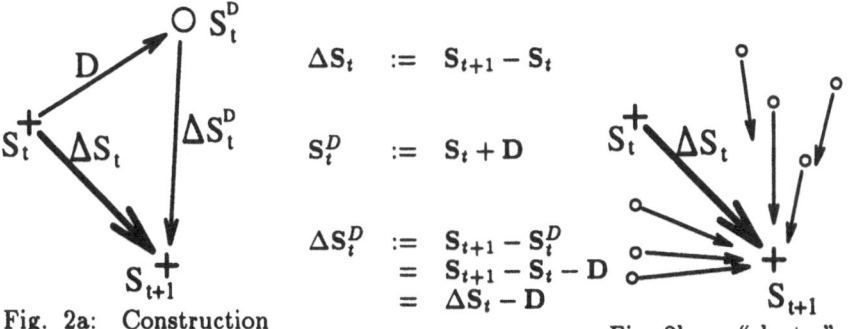

$$\Delta S_t := S_{t+1} - S_t$$

$$S_t^D := S_t + D$$

$$\Delta S_t^D := S_{t+1} - S_t^D$$
$$= S_{t+1} - S_t - D$$
$$= \Delta S_t - D$$

Fig. 2a: Construction of a displaced state-change vector ΔS_t^D

Fig. 2b: a "cluster" of novel vectors.

Learning

To train the network with the new larger data set, several learning paradigms have been tested and found to be adequate for the desired task. Pure *back-propagation of error* (BP) - with and without momentum term, cumulative *backprop*, *quickprop* and other BP-acceleration techniques, with sigmoid and linear output function were used to train the network sucessfully.

During learning a error of zero could not be reached because some training vectors are inconsistent to others, caused by the naive generation of the training data: it is possible, that two different teacher vectors are assigned to the same input state. This did not effect the qualitative learning result due to the generalization capability of the network.

Simulation results

Testing the network several times during the learning procedure with different learning paradigms, parameters, or number of hidden neurons, we found some rules of thumb: longer learning yields better results (overfitting was not observed), a smaller learning rate (≤ 0.001) gives "smoother" vectorfields, more hidden neurons generate "sharper edges".

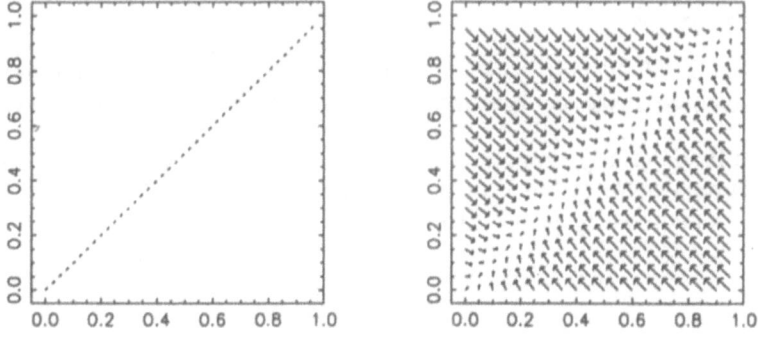

Fig. 3: A line as *attractor* and the network generated vectorfield.

To visualize the generated vectorfields, the network was tested in the unit square on a uniform 2-dimensional grid, the output was depicted as an arrow pointing in the direction the system tends to go (Fig. 3 and Fig.4). An additional plot shows the *attractor* the systems should converge to.

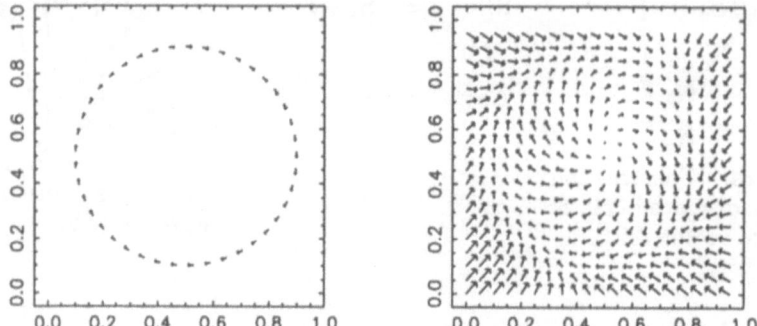

Fig. 4: A circle as *attractor* and the network generated vectorfield.

Discussion and Conclusion

Only based on a given *attractor*, the presented preprocessing of the training data into local clusters, allows MLPs to generate a corresponding vector field that represents the dynamics of a nonlinear differential equation. It is the essential feature of this vector field that its *attractor* is identical with the training data. First results proved to be independent in a wide range of the learning rule.

We are currently investigating the question what kind of nonlinear vector fields can be generated and learned by the presented method as well as the developement of adequate evaluation and learning criteria. Of special interest for control tasks are the capabilities and limitations for learning and generating vectorfields in higher dimensions.

Such trained networks can serve as modules for various tasks including obstacle avoidance, prediction and control.

This work was supported by Federal Ministry for Research and Technology (BMFT) under grant 01 IN 105 G/6 (SENROB).

References

Thompson, J. M. T. and Stewart H. B., (1986), Nonlinear Dynamics and Chaos: Geometrical Methods for Enineers and Scientists, John Wiley and Sons.

Werntges H.W. (1991), Approximation von Vektorfeldern durch Selbstorganisation neuronaler Netze mit Fehlerrückführung am Beispiel der inversen Kinematik eines planaren 4-Gelenk-Arms, Ph.D. Thesis, University Düsseldorf.

RECURRENT NEURAL NETWORKS WITH DELAYS

J. Guignot*, P. Gallinari **

* LIPN, Univ. Paris 13, France, guignot@laforia.ibp.fr
** Laforia-IBP, Univ. Paris 6, France, gallinari@laforia.ibp.fr

1   Introduction

Recurrent neural networks (RNN) have been introduced for the modelization of non linear dynamic systems. In spite of their potential ability to implement systems of arbitrary complexity, they are often avoided because of large training times. For many problems, local recurrent networks or time delay feed forward architectures perform well enough. They are thus often preferred to more general systems, even though they have limited capabilities for handling sequences and can only build short term memory of past events [Frasconi et al. 92] .

We will deal here with non local RNN and show that the addition of tapped delay lines to feed forward and recurrent connections may improve in several ways the behaviour of these systems. These Time Delay recurrent neural networks (TD-RNN) do have the same computational capabilities as other RNN, but their convergence time and performances may show important improvements over RNN with no delay (RNN for short). Time delays have been used for some years in feed-forward or locally dynamic NN, they have proved to be a valuable tool for e.g. large speech recognition tasks [Waibel et al. 87, Bengio et al. 92]. This is because they allow to capture short term dynamics of sequences. When used in non local recurrent systems, delays may also memorise sequential events by encoding simple relations between activations of different cells. Many relations may be encoded easily in such a way with no need for learning complex recurrent structures. To make things simple, TD-RNN can be considered to perform learning at two different levels : simple relations are directly learned into delay weights, this eases the task for the dynamic architecture which is free to configure itself so as to encode more complex relations. For a fixed cell number, learning complexity is increased compared to RNN. On the other hand, the ability of the network being considerably increased, the architecture may be much simpler. Additional cells necessary to encode information through their dynamic interaction are thus exchanged with additional connections which are much easier to learn. This brings a reduction of complexity for a wide class of problems of practical interest.

We present the algorithm in 2 and in 3 tests which have been performed on benchmark problems allowing to illustrate the behaviour of the algorithm.

2   Algorithm

In a TD-RNN, there may be several delayed connections between input lines and any cell or between any pair of cells in the network. Let max_delay_input and max_delay_internal denote respectively the maximum number of such connections. The maximal connectivity of the network is (max_delay_internal*n_cells^2 + max_delay_input*n_cells*n_inputs) where n_cells is the number of cells of the network and n_inputs the number of external inputs. Every cell memorises its max_delay_internal last outputs. Let w_{ij}^k denotes the connection from cell j to i with delay k, $x_i(t)$ and $s_i(t)$ be the state and output of cell i :

$$x_i(t) = \ f(s_i(t)) = f\left(\sum_k \sum_j w_{ij}^k \ x_j^k(t) \right)$$

where f is the sigmoïd function.

TD-RNN may be trained through a variety of learning rules. We have used here the RTRL rule [Robinson & Fallside 87, William & Zipser 89] because of its generality. It may be used for any RNN and allows to handles sequences of undetermined length, its drawback being that its computational and memory requirements forbid its use for training large networks. For convenience we recall below some basic notations. As for a general RNN, any unit may receive external inputs or desired outputs. Teaching is defined on $T*U$, where T denotes a set of discrete time instants $\{1,..,T\}$ and $U = \{U_1, ..., U_T\}$, U_t being the set of units with a desired output at time t. Let the error at time t be :

$$E(t) = \sum_{k \in U_t} (d_k(t) - x_k(t))^2$$

Where $d_k(t)$ is the desired output at t for unit k. The global error over the sequence will be

$$E = \sum_{t \in T} E(t)$$

RTRL attempts to minimise E through on line minimisation of E(t). E could be alternatively minimised by accumulating errors through the whole sequence. We have used here an enhanced gradient method which modifies the weights in proportion to the derivative of the total error with respect to these weights.

Choosing the right architecture is a difficult problem for RNN, we have implemented a weight pruning strategy so as to eliminate useless parameters. The ratio of information conveyed by weight $w_{ip}{}^q$ at time t is measured as $w_{ip}{}^q x_p{}^q(t)/ s_i(t)$. A connection whose ratio over several cycles is below a given threshold is eliminated. Pruning is not performed on line, but at certain time steps corresponding to local minima of the error.

3   Tests

TD-RNN include RNN as a special case, their theoretical capabilities to handle sequences are those of RNN, although they may behave differently. Tests have been performed on some classical small problems. They are intended to show that RTRL works well with TD-RNN, to analyse the difference of behaviour between RNN and TD-RNN and evaluate the potential of the latter.

3.1 Prediction problem

The NN having been taught with pairs (f (t) , f (t+1)), f being any function, the task is then to guess the value of a f(t) over several cycles. During the test, after the initialisation, there is no more external input and the output is fed back to the input of the network.

In the following experiment we have used a network with two cells and one input line to learn the function $f(t) = sin(t / 30)$. After 4000 learning cycles, learning can be considered as perfect. Fig. 1 shows the test for some cycles, slight differences between desired and computed outputs occur after 20.000 test cycles. The number of delays depends upon the frequency of the sine function : the higher the frequency, the more delays are needed to obtain perfect prediction. Although it is easy to produce sine waves with linear units, this is not the case with logistic cells and the result below cannot be obtained with a classical RNN trained with RTRL [Williams & Zipser 89].

Learning two dimensional functions can be performed in the same way, except that the convergence takes much more time. We have performed tests on simple cycloid or eight shaped figures. Networks with only two cells do not provide perfect identification in this case. A higher number of cells and TD connections is required with a corresponding increase in training time.

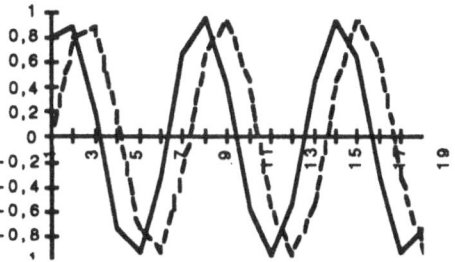

Figure 1 : prediction of the TD-RNN over some test cycles. Plain curve is the sine function and dotted line is the computed output. The two curves have been slightly moved one from the other to show the perfect prediction, otherwise they overlap exactly. The rough shape is only due to a raw discretization.

3.2   Sequence recognition

A classical recurrent net can detect a particular sequence of inputs, but the size of the network must be greater than or equal to the length of the sequence to be detected. This problem may be simplified with time-delays : any sequential event in the input can be learnt and detected using a single connection if the length of the sequence to be detected is smaller than the temporal delay of the cell. In the experiment reported here, we have taught networks to recognise the event "baac" in infinite sequences built from these three letters. Table 1 shows the performances of a classical recurrent NN with 4 fully connected cells and 4 external inputs (Ext0=a,Ext1=b,Ext2=c,Ext3=bias) and a TD-RNN with one cell, 4 external inputs, max_delay_input = 4, max_delay_internal =2.

RNN	TD-RNN
(98.1,)	(99.1,)

Table 1 : performances of a RNN and a TD-RNN, result (,) is the averaged performance and standard deviation for 15 tests.

Results are 1% better for TD-RNN, but the difference is not very significative, the important point is that in this case, TD-RNN learns the simplest network configuration for this problem : it eliminates recurrent connections and learn to be a TD feed-forward net with input delayed connection corresponding to the identification of the sequence. Thus for learning sequences of short events, TD-RNN may configurate its TD connections to learn the basic events (e.g. "baac" ...) and use the recurrent part of its architecture to learn the succession of events.

3.3   Recognizing a regular grammar

The problem of recognizing wether a string belongs to a given regular grammar can be solved with NNs. NN can simulate finite-state automata (FSA), and it is easy to recognize wether a FSA generates a given grammar. We have dealt here with the following problem :

* the grammar is known only by strings it has generated.

* learning must be on-line and the NN is not taught when a word begins or ends.
This kind of recognition can be very useful for speech recognition problems, where the boundaries of phonemes are unknown.
We have performed tests on the simple Reber grammar [Williams & Zipser 89], the FSA has no initial or final state and generates an infinite string of characters.

The network has to recognize the character(s) which can follow the present character. We have tested this problem with a TD-RNN composed of one input (max_delay_input = 2) three hidden cells, three output cells (max_delay_internal = 3) which allows to encode in a straightforward way the seven states of the grammar. For the test, network output is fed back to the input. Learning is considered perfect is the network does not fail to predict the next character over a period of 1000 characters.

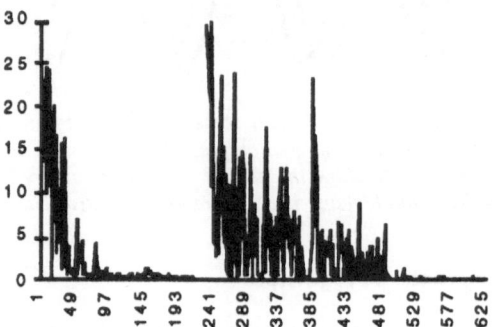

Figure 2 : Prediction error over several training epochs. Curve to the right corresponds to RNN with 2 hidden cells and 7 output cells (learning begins at 240 on the x axis) .and left curve to the TD-RNN with 3 hidden units and three output cells (learning begins at 0).

Figure 2 compares the performance of a 9 units RNN and the above TD-RNN and shows that convergence is strongly improved with a TD-RNN. This is because many simple temporal relations do exist between the successive characters of the string (e.g. 'e' is always followed by 't' or 'p'). These relations are coded in some delayed connections (a character can inhibit a state, or activate a set of states). As for the other relations, the TD-RNN behaves like a classical recurrent NN.

4   Conclusion

We have investigated the use of time delays in fully recurrent neural networks. They allow to encode several relations between events in a much simpler way than non delayed recurrent network. We have also presented some experiments to illustrate how this system learned to use its delay lines in order to simplify the global computation. The resulting architecture allows to deal with problems for which RNN will always fail to converge. Pruning strategies have been implemented to simplify the architecture during training.

5   References

Bengio Y, De Mori R., Gori M. (1992) : learning the dynamic nature of speech with back propagation with sequences. Patterns recognition letters 13, 375-385.
Frasconi P., Gori M., Soda G.(1992) : Local Feedback Multilayered Networks", Neural computation, Vol 4, N° 1, 120-130.
Pearlmutter B. A. (1990) : dynamic recurrent neural networks, TR CMU-CS-90-196.
Robinson A.J., Fallside F. (1987) : static and dynamic error propagation with application to speech coding, NIPS 632-641.
Waibel A., Hanazawa T., Hinton G., Shikano K., Lang K. (1987) : phoneme recognition using time delay neural networks, TR-1-006, ATR, Japan.
Williams R.J., Zipser D. (1989) : experimental analysis of the real time recurrent learning algorithm, connection science, vol 1, N°1, 87 - 111.

EMAN : EQUIVALENT MASS ATTRACTION NETWORK

Mahmut H. Erdem   Gamze Baskomurcu   Yusuf Ozturk
Ege University, Department of Computer Engineering, İzmir Turkey

ABSTRACT

This paper introduces a new neural network model for binary pattern classification. In a recent work we have proposed a network model, namely, MAN (Mass Attraction Network) [1,2] which can be used as an autoassociator. In MAN, memory items have been considered as masses at the corners of a hypercube. Exploiting Newton's mass attraction theory, a recall scheme utilizing "attraction forces" between memory items and input patterns has been developed. EMAN is the consequence of efforts to extent the concept to do classification. The main idea in EMAN is to create an equivalent mass instead of two close masses. After introducing MAN and EMAN concepts, some improvements will be presented. This paper concludes with simulation results.

1. MASS ATTRACTION CONCEPT and MAN

In a recent paper a neural network model capable of associative recall has been proposed [1]. Inspired from Newton's mass attraction theory, MAN has been shown to be superior to Hamming net, Hopfield Network and Harmony Theory[1]. In the model, memory items have been considered as masses at the corners of a hypercube. During recall, all stored masses (patterns) attract the input pattern (which assumes a unit mass) and the one that attracts it most (the one with biggest force) becomes the recalled pattern.

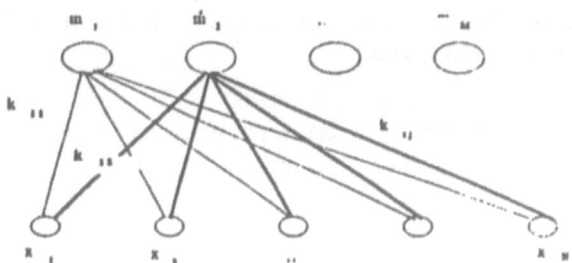

Figure 1: MAN Architecture

Fig.1 depicts the fully connected network architecture of MAN. The lower layer is the input/output layer and the upper layer holds the stored masses. The patterns are stored at the connections, which can assume a value of either 1 or 0. During learning phase, bits are assigned to the connections and the masses, which represent the probability of the patterns, are accumulated in the upper layer nodes. During recall, whenever an input pattern is presented to the network through the lower layer nodes, all upper layer nodes compute the "attraction force" by assuming the input pattern has a unit mass,

$$F_i = \frac{m'_i}{D_i} \qquad (1)$$

where

$$m'_i = b \ (1 - e^{-\frac{m_i}{T}}) \qquad (2)$$

Here m_i is the number of times a pattern is observed, b and T are constants , HD_i is the Hamming distance between the input pattern and connections of the ith upper layer node. Then every upper layer node updates its state according to:

$$x_j = 1 \text{ if } E_j >= 0 \qquad (3)$$

$$= 0 \text{ if } E_j < 0$$

where $\qquad E_j = \Sigma (2k_{ij} - 1) F_i^n \qquad$ here n : magnification factor $\qquad (4)$

This network is extremely fast. The recall phase consists of two steps (forward and backward pass). Detailed analysis regarding these issues can be found in [1]. EMAN is based on the same concept but it has been reformulated to do classification. The main idea here is to create an equivalent mass instead of two close masses. Ultimately, the system produces an equivalent mass in the (gravitational) center of each cluster. This takes place in the learning phase. The recall phase of EMAN is identical to MAN. Also the network architecture is the same.

2. EQUIVALENT MASS

Suppose we have two masses m_1 and m_2 with a distance r between them. We would like to create an equivalent mass M instead of two close masses. M will be the sum of m_1 and m_2. The place of the new mass must be between m_1 and m_2, closer to the one with greater mass. See fig.2.

$$M = m_1 + m_2 \quad \text{and} \quad r = r'_1 + r'_2 \qquad (5)$$
$$m_1 r'_1{}^2 = m_2 r'_2{}^2 \qquad (6)$$

Considering binary patterns, m_1 and m_2 are located at two different corners of a hypercube. Therefore the squared distance between m_1 and m_2 will be equal to the Hamming distance ($r^2 = HD$), Thus;

$$(r'_1 + r'_2)^2 = HD \qquad (7)$$

However , the equivalent mass, M, can not be placed at the geometrically correct location, because this point will be in the interior of the hypercube. Since binaryvalues

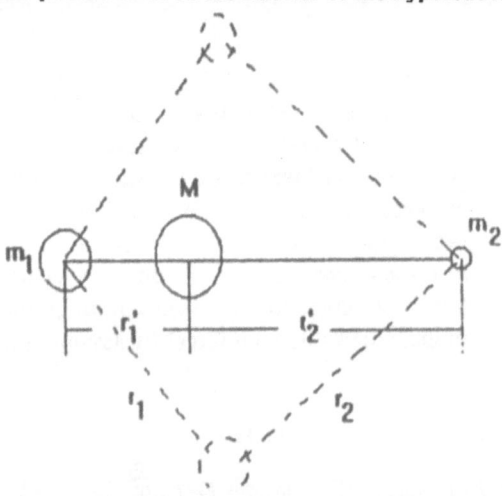

Figure 2 . Creating an equivalent mass instead of two masses

are employed, only the corners can be used. What can be done now is, to place M at the closest corner to the correct position. From equations 6 and 7,

$$r_1^2 = \frac{HD}{1+\frac{m_1}{m_2}} \tag{8}$$

Here r_1^2 and r_2^2 are the Hamming distances between patterns (m_1,M) and (m_2,M) respectively. In an N-D hypercube, there are N corners which have equal Hamming distances to a particular corner. There are two such corners in the plane, see fig.2. Consequently, in order to create an equivalent mass, from m_1, r_1^2 bits of m_1 must be inverted. These bits must be randomly selected among different bits between m_1 and m_2. Also, the mass must be incremented $(M= m_1 + m_2)$. For example, suppose two patterns with masses m_1 and m_2 are given, and the Hamming distance between them is 10. Assume $r_1^2 =3$ by eq.(8). Three bits among those different 10 bits of m_1 are randomly selected and inverted. Therefore a pattern is reached whose HD to m_1 is 3, and to m_2 is 7.

3. UNSUPERVISED LEARNING

Assume there is a population of stimulus patterns and each stimulus pattern is presented with some probability. The system is supposed to classify them properly. Before we start the classification procedure, there must be P memory items each with mass one stored in upper layer nodes. The patterns can be chosen randomly, by selecting P patterns which are assumed to be best exemplars of each class, or by storing the first P patterns presented to the network. Whenever an input pattern is presented to the network, forces (F) acting on the input pattern are computed and the mass (class) producing the maximum F is found. Then the process is the same as in supervised learning. The network changes relevant connections of winning class (upper layer node) using the methodology given in section 2. Thus, the class center moves towards the input pattern.

After learning is completed, equivalent masses will be created at the center of classes in which their masses denote the probability of the items merged into this cluster center. If there is a structure in the stimulus patterns, the equivalent masses will break up the patterns along structurally relevant lines. This means that, the system will find clusters if they are there. The more the stimuli highly structured, the more the stabilization of classification will be. If the stimuli are not well structured then classifications are more variable, thus a given stimulus pattern will be responded to by first one and then another equivalent mass.

4. RESERVED MASS

An input pattern may not be able to move the cluster center because of two reasons. First of all if the input pattern is very close to the cluster center then it is quite obvious that the cluster center can not be modified. Moreover the total occurrence probability of patterns assigned to a particular class might be very big, forming a stationary cluster center hence the second reason. In both of the cases above eq.8 will yield $r_1^2 < 0.5$. Therefore, an input pattern (unit mass) can never move a big equivalent mass. To overcome this problem, here we propose the usage of temporary cluster centers which will be called as reserved masses throughout this

396

study. To implement this improvement, 2P (wheras P : number of classes) upper layer nodes are required. Every extra node is the reserved mass of an equivalent mass. Input patterns that can not move an equivalent mass, are accumulated in its reserved mass by the same equivalent mass creating method. Then a reserved mass and its equivalent mass are merged. There is also a danger that, an input mass can not move an equivalent mass and its reserved mass. In this case the input pattern is ignored. Simulations indicate that, these situations are very rare when a suitable value of T in eq.2 is selected.

5. SIMULATION RESULTS AND DISCUSSION

Experiments have been performed on a transputer system. Here we report an example in which, the learning algorithm given above is employed. In the experiment, unsupervised classification property of the system is tested by presenting to it the patterns which are not clearly separable into two clusters. The patterns presented can be thought as points in the interior of two spheres which are overlapping, see fig.4.

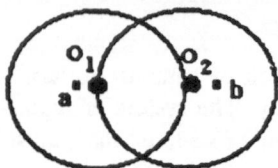

Figure 4. Illustration of input patterns

Since patterns with 30 bits are used, the network is composed of 30 lower layer nodes. The upper layer consists of two nodes corresponding to two classes. A random pattern is selected as the center of a sphere O_1, then 10 randomly selected bits of this pattern are inverted for creating the second center O_2. Later by randomly inverting i (i=1,2...,10) bits of the two patterns the input patterns are produced. The cluster centers must be near O_1 and O_2 but outwards (a and b in fig.4). In this simulation T was 2 and no pattern is discarded. About half of the patterns went to one of the equivalent mass and the remaining went to the other one. The cluster centers produced by the system have a Hamming distance of 2 to O_1 and O_2 and Hamming distance of 12 to each other. Thus, satisfying the expectations.

EMAN is a classifier network for binary patterns. It has similarities to competitive learning [3] and Kohonen's self organizing feature map [4].EMAN essentially does not learn associations, but it uses a statistical learning scheme. We believe that it can be used for various applications by supervised or unsupervised way.

REFERENCES
[1]Erdem M. H. and Ozturk Y.,"MAN : Mass Attraction Network", will be published in ISCAS-94

[2]Buskomurcu G, Erdem M.H. and Ozturk Y.,"Parallel Implementation of Mass Attraction Network",submitted to ICANN 94 for possible publication.

[3]Rumelhart D.E. and McCleland J.L.,Parallel Distributed Processing, Chap.5 Competitive Learning,MIT Press, 1986.

[4]Kohonen T.,"Adaptive, Associative and Self Organizing Features in Neural Computing",Applied Optics, Dec. 1987.

Analysis of an Unsupervised Indirect Feedback Network

M. D. Plumbley†

† Department of Computer Science, King's College London
Strand, London WC2R 2LS, UK

1 Introduction

Earlier work by the present author suggested the use of feedback structures
to optimize information under conditions of noise at the output, or noise at
both input and output (Plumbley, 1993a). Although these models were not
designed as models of biological signal processing, the use of a feedback is
highly suggestive for the role of the feedback pathways known to exist in the
cortex.

In this paper we present a modified feedback network which attempts to
address some of the criticisms that can be levelled at these earlier feedback
networks as models of biological perceptual processing. Analysis of this new
network and algorithm set shows that it has a number of very convenient properties and possible generalizations.

2 Earlier feedback networks

Previous work by the author on feedback networks was motivated by an information optimization criterion, related to Linkser's *Infomax* (Linsker, 1988).
Suppose that the output of a network has limited power, and is exposed to
equal-variance independent additive Gaussian noise. Then information is maximized if the output signal from the network is also an equal-variance decorrelated (i.e. orthonormal) Gaussian signal (Plumbley, 1993b). Two network
arrangements were suggested to perform this orthonormalization: one with
direct lateral inhibitory feedback connections (Fig. 1(a)); and one with feedback through a set of inhibitory interneurons (Fig. 1(b)). Both of these used
Hebb-like algorithms with types of weight decay to achieve the desired output
characteristic. Fig. 1(c) shows an alternative arrangement (reduced to one dimension) of the network of Fig. 1(b). Here the interneurons z are considered

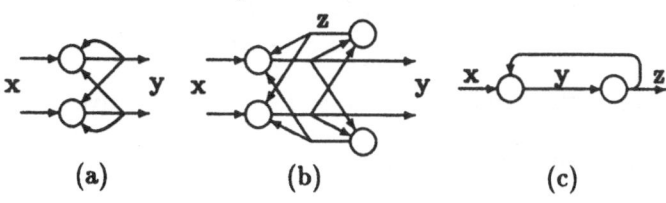

(a)   (b)   (c)

Figure 1: Earlier linear feedback networks.

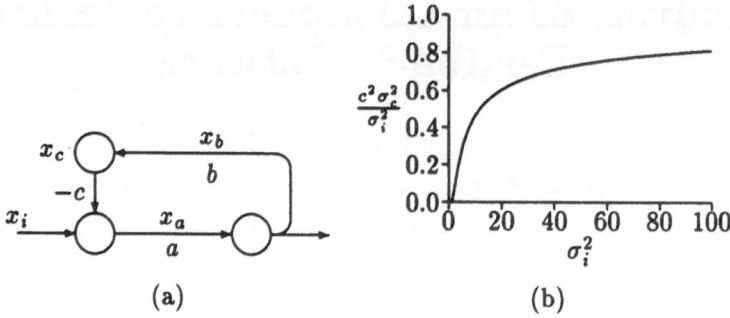

Figure 2: Indirect feedback network.

to be the inputs to a subsequent stage of processing, but still ensure that the feed-forward activations **y** are decorrelated with equal variance.

From the perspective of biological plausibility, a number of objections can be raised against this network. The network elements are linear, and use graded activations instead of spike-based activations. However here we shall concentrate on more architecture related problems: specifically that in biological systems (a) inhibitory connections do not normally adapt; (b) neurons do not normally produce both inhibitory and excitatory connections; and (c) feedback loops do not normally have direct feedback of the form given in Fig. 1(c) (Kuffler et al., 1984). We therefore introduce an additional interneuron stage into our network to overcome these objections.

3 Indirect feedback network

We consider first the one-dimensional case (Fig. 2(a)). This has excitatory forward and backward connections a and b respectively, with an inhibitory relay with weight $-c$. In this network, when activations have settled to their equilibrium conditions, we have

$$x_b = ax_a \qquad x_c = bx_b \qquad x_a = x_i - cx_c \qquad (1)$$

so, for example, we have

$$x_a = (1 + abc)^{-1}x_i. \qquad (2)$$

The algorithms we use for the connection weights a and b are

$$\Delta a = \eta_a(x_a x_b - k_a a) \qquad (3a)$$
$$\Delta b = \eta_b(x_b x_c - k_b b) \qquad (3b)$$

where η_a, η_b are positive update factors, and k_a, k_b are positive constants. Under certain conditions (see for example (Hornik & Kuan, 1992)), these are closely approximated by the o.d.e. pair

$$\dot{a} = \eta_a(\sigma_a^2 - k_a)a \qquad (4a)$$
$$\dot{b} = \eta_b(\sigma_b^2 - k_b)b \qquad (4b)$$

where $\sigma_a^2 = E(x_a^2)$ is the variance of x_a, and similarly for x_b. We assume for the moment that the inhibitory connection $-c$ remains fixed.

By inspection of (4a) and (4b), we can see that the system is stationary when

$$\sigma_a^2 = k_a \quad \text{or} \quad a = 0 \tag{5a}$$
$$\sigma_b^2 = k_b \quad \text{or} \quad b = 0. \tag{5b}$$

Stability analysis gives a the following hessian matrix

$$\mathbf{H} = \left[\begin{array}{cc} -2\eta_a abc(1+abc)^{-1}k_a & -2\eta_a c(1+abc)^{-1}k_b \\ 2\eta_b ab(1+abc)^{-1}k_a & -2\eta_b abc(1+abc)^{-1}k_b \end{array} \right] \tag{6}$$

which has

$$\det(\mathbf{H}) = 4\eta_b\eta_b k_a k_b abc(1+abc)^{-1} \tag{7}$$
$$\text{Tr}(\mathbf{H}) = -2(\eta_a k_a + \eta_b k_b)abc(1+abc)^{-1} \tag{8}$$

so both eigenvalues of $\mathbf{H}$ are negative, and thus stability holds, if $abc > 0$ or $abc < -1$. Only the first of these is compatible with stable settling of the activations in (1), which leaves us with the stability condition $abc > 0$. From (2) at this point we have

$$k_a = \sigma_a^2 = (1+abc)^{-2}\sigma_i^2 \tag{9}$$

where σ_i^2 is the variance of the input signal x_i, so for stability we must have

$$k_a < \sigma_i^2 \tag{10}$$

i.e. that the target forward variance must be less than the input variance.

Therefore provided this stability condition is satisfied, the forward and backward variances σ_a^2 and σ_b^2 after adaptation are determined by the parameters k_a and k_b in the learning algorithms. Further analysis shows that if σ_i^2 drops below k_a, the forward and backward variances will collapse to zero. It is not necessary for the inhibitory connection $-c$ to adapt.

A little manipulation gives us the variance of the inhibitory relay x_c, which varies according to

$$c^2\sigma_c^2 = \left(\sigma_i - \sqrt{k_a}\right)^2 \tag{11}$$

so $\sigma_c^2 \to 0$ as $\sigma_i^2 \to k_a$, and $c^2\sigma_c^2/\sigma_i^2 \to 1$ as $\sigma_i \to \infty$. This behaviour is plotted in Fig. 2(b).

4   Generalizations

If desired, the inhibitory connection $-c$ can be made adaptable, for example with an algorithm similar to the excitatory connections, but with opposite sign:

$$\Delta c = \eta_c(x_c x_a - k_c c). \tag{12}$$

More manipulation shows that the variance of the inhibitory relay now follows

$$\sigma_c^2 = k_c \left(\sigma_x / \sqrt{k_a} - 1 \right) \tag{13}$$

which varies with the root of σ_x^2 for large values, instead of proportionally, as before.

In addition, this network can be generalized in the obvious manner to multiple inputs, outputs, and relay units x_a, x_b, x_c, and connection matrices $A, B, -C$, so that

$$x_b = Ax_a \qquad x_c = Bx_b \qquad x_q = x_i - Cx_c. \tag{14}$$

In the special case of equal numbers of units, the forward and backward covariance matrices will converge to

$$\Sigma_a = k_a I \qquad \Sigma_b = k_b I \tag{15}$$

provided $\Sigma_i - k_a I$ is positive definite. More generally, provided the number of inhibitory relay units x_c is large enough, the output will extract the space spanned by the principal components of the input (the *principal subspace*). As in the one-dimensional arrangement, the inhibitory connections $-C$ need not be adaptable, but must have sufficiently high rank. Space limitations preclude more details being given here: this analysis will be expounded in a later paper.

5 Discussion and conclusions

We have briefly analyzed a modification to an earlier feedback network considered by the author, again using simple local Hebb-like learning algorithms with weight decay. This network has excitatory forward and backward connections, with inhibitory relay units similar to the arrangement found in e.g. the feedback loop between the visual cortex and the LGN. The inhibitory connections are restricted to special relay units, and need not be adapted. This therefore successfully answers our earlier architectural objections (a)-(c), and offers an interesting suggestion for a biological information processing architecture.

References

Hornik, K. & Kuan, C.-M. (1992). Convergence analysis of local feature extraction algorithms. *Neural Networks*, 5, 229–240.

Kuffler, S. W., Nicholls, J. G., & Martin, A. R. (1984). *From Neuron to Brain.* Sunderland, MA: Sinauer Associates, second edition.

Linsker, R. (1988). Self-organization in a perceptual network. *IEEE Computer*, 21(3), 105–117.

Plumbley, M. D. (1993a). Approximating optimal information transmission using local Hebbian algorithms in a double feedback loop. In S. Gielen & B. Kappen (Eds.), *Proceedings of the International Conference on Artificial Neural Networks, ICANN'93, Amsterdam, The Netherlands*, (pp. 435–440). Springer-Verlag.

Plumbley, M. D. (1993b). Efficient information transfer and anti-Hebbian neural networks. *Neural Networks*, 6, 823–833.

Hopfield Energy of Random Nets

Erol Gelenbe
Department of Electrical Engineering,
Duke University, Durham, N.C. 27709-0291

1 Introduction

The seminal papers of the early eighties (Hopfield 1982, Cohen and Grosberg 1983, Hopfield and Tank 1985) have created a large body of work on using neural networks to approximately solve intractable optimization problems (AArts and Korst 1989, Abe 1001, Aiyer et al. 1990, Gelenbe et al. 1993).

The Random Neural Network (RNN) model (Gelenbe 1990) has the nice property of being analytically solvable and computationally fast, so that any application of the model is based on obtaining solutions to a simple system of fixed-point equations. Here we define the Hopfield Energy for the RNN model and prove that it is minimized at each fixed-point iteration used in solving the RNN. We illustrate this via a heuristic solution to the Minimum Node Covering Problem (MCP) for graphs.

In the RNN signals in the form of spikes of unit amplitude circulate among the neurons. Positive signals represent excitation and negative signals represent inhibition. Each neuron's state is a non-negative integer called its potential, which increases when an excitation signal arrives to it, and decreases when an inhibition signal arrives. Information is carried by the *frequency* at which spikes travel. Neuron j, if it is excited, will send spikes excitation (resp. inhibition) spikes to neuron i at a frequency ω_{ji}^{+} (resp. ω_{ij}^{-}). q_i the stationary probability that neuron i is excited is then computed using the fixed-point iteration:

$$q_i(s+1) = \frac{\Lambda(i) + \sum_{j=1}^{n} q_j(s)\omega_{ji}^{+}}{\lambda(i) + r(i) + \sum_{j=1}^{n} q_j(s)\omega_{ji}^{-}} \tag{1}$$

where the $\Lambda(i)$, $\lambda(i)$ are external excitation and inhibition frequencies to neuron i, and $r(i) = \sum_{j=1}^{n}[\omega_{ij}^{+} + \omega_{ij}^{-}]$.

2 Energy and RNN Optimization

The energy minimizing properties of the RNN are summarized in the following definition and result.

Definition The Hopfield Energy $H(s)$ of the RNN at step $s = 0, 1, 2 \ldots$ is defined as:

$$H(s) = -\sum_{i=1}^{n}[q_i(s)\Lambda(i) + \sum_{j=1}^{n} q_i(s)q_j(s)\omega_{ji}^{+} - \sum_{j=1}^{n} q_i^2(s)q_j(s)\omega_{ji}^{-} - q_i^2(s)(r_i + \lambda_i)] \tag{2}$$

This energy function differs somewhat from the classical approach of Hopfield. Note the *additional* terms which are squared in one state variable and linear in the other. For optimization problems in the variables $\{\,0, 1\,\}$ a cost function with such terms is identical to a cost function with only quadratic terms, since $V^2 = V$ if $V = 0$ *or* 1. Therefore, the above energy function can correspond to a a quadratic cost function:

$$C = -\sum_{i=1}^{n}[V_i\Lambda(i) + \sum_{j=1}^{n} V_iV_j\omega_{ji}^{+} - \sum_{j=1}^{n} V_iV_j\omega_{ji}^{-} - V_i(r_i + \lambda_i)] \tag{3}$$

Theorem The RNN minimizes its Hopfield Energy and its state converges to a minimum energy value, *i.e.* $[H(s+1) - H(s)] \leq 0$, for each $s = 0, 1, \ldots$.

To illustrate the use of the Hopfield Energy, consider the well known minimum cover problem (MCP) for graphs. Given a non-directed graph G whose nodes are $N = \{1, \ldots, n\}$, described by its binary adjacency matrix $A = \{a_{ij}\}$. A cover of G is any subset K of N such that for any arc of the graph represented by $a_{ij} = 1$, either node i or node j is in K. Thus K contains at least one of the "end nodes" of any arc of the graph. A cover is said to be minimal if there is no smaller cover (in number of nodes). Let us now construct a heuristic solution for the MCP with a 2n-neuron RNN. Each node i in the graph is "represented" by two neurons, i, and i^*: the first will be "on" if the RNN recommends that i be included in the cover, while the second will have the opposite role. Then:

- The firing rate for neuron i is set to $r(i) = 2n$, and $p^-(i, i^*) = 1$, so that weights are $\omega_{i,i^*}^{-} = 2n$ and all other $\omega_{i,j}^{-} = \omega_{i,j}^{+} = 0$.

- The firing rate for neuron i^* is $r(i^*) = D(i)$, where $D(i)$ is the degree of node i. Furthermore $p^+(i^*, j) = 1/D(i)$ for any neighbour j of i, *i.e.* with $a_{ij} = 1$.

- $\Lambda(i) = D(i)$, so that neuron i is being excited proportionally to node i's degree – this is the aspect of the random network which mimics the greedy algorithm discussed below. $\Lambda(i^*) = 1$, to provide external excitation to neuron i^*. All other weights and rates are set to zero.

The equations for the q_i , q_{i^*} become:

$$q_i = \frac{D(i) + \sum_{j=1}^{n} q_{j^*} a_{ji}}{2n}, \quad q_{i^*} = \frac{1}{D(i) + 2nq_i}.$$

As a consequence, the Hopfield Energy of the network is:

$$H(s) = -\sum_{i=1}^{n}[q_i(s)D(i)+q_{i\bullet}+\sum_{j=1}^{n} q_j(s)q_{i\bullet}(s)a_{ij}-2nq_i^2(s)q_{i\bullet}(s)-2nq_i^2(s)-D(i)q_{i\bullet}^2($$

$$(4)$$

which, using (3) corresponds to a quadratic cost function C to be <u>minimized</u> by binary variables V_i, $V_{i\bullet}$:

$$C = -\sum_{i=1}^{n}[V_i D(i) + V_{i\bullet} + \sum_{j=1}^{n} V_{i\bullet} V_j a_{ij} - 2nV_i V_{i\bullet} - 2nV_i - D(i)V_{i\bullet}] \qquad (5)$$

This is a plausible cost function for the MCP. Indeed it may be written as:

$$C = \sum_{i=1}^{n}\{2nV_i - V_i D(i) + [D(i) - 1]V_{i\bullet} - \sum_{j=1}^{n} V_{i\bullet} V_j a_{ij} + 2nV_i V_{i\bullet}\} \qquad (6)$$

The first term states that there should be as few nodes as possible in the minimum cover, while the second and third state that we should favour nodes which have a large degree. The fourth term states that we would like to have only one of each end node of an edge in the cover. The last term states that we cannot have the same node both in and out of the cover. Though this cost function is more elaborate than the usual quadratic form, it provides a more detailed representation of the problem's constraints.

The procedure we use for deriving an approximate solution to the MCP is then: **Step 1.** Solve (??) by a fixed point iteration, and select the node i whose probability q_i is highest; that node is included in the cover. Then **Step 2.** remove that node from the graph, and remove all edges which originate at that node. If the resulting graph has no edges, stop. If the resulting graph still has edges, then for the <u>new reduced graph</u> go back to Step 1.

This has been tested on a set of randomly generated graphs, and compared when possible with the exact enumerative search for the optimum, and with the standard Greedy Algorithm. The number of arcs in the graphs has been varied: $n = 20, 50, 100$. For each fixed n, we generate 25 graphs at random with probability π that for any pair of nodes i, j there is an edge (i, j). The two heuristic are compared with respect to the following criteria: "Mini" the percentage of cases where the method obtains the minimum cover, "Exc" is the average number of nodes in excess of the minimum for a given (n, π) pair, T is the computation time on a workstation. The RNN outperforms the greedy heuristic in all cases. We have observed that this improvement is substantial when the graph is "sparse" (*i.e.* small values of π), though we omit these results for lack of space.

References

[1] Aarts E.H.L., Korst J.H.M. (1989) "Boltzmann machines for traveling salesman problems", *European Journal of Operational Research*, No. 39, pp 79-95.

[2] Abe S. (1991) "Determining weights of the Hopfield neural networks", *Proc. ICANN'91*, Helsinki, pp 1507-1510.

[3] Aiyer S.V.B., Niranjan M., Fallside F., "A Theoretical Investigation into the Performance of the Hopfield Model", *IEEE Transactions on Neural Networks*, Vol. 1, No. 2, pp 204-215, June 1990.

[4] Cohen M.A., Grossberg S.(1983) "Absolute stability of global pattern formation and parallel memory storage by competitive neural networks", *IEEE Trans. Sys. Man Cyber.*, Vol. 13, No. 5.

[5] Gelenbe E. (1990) "Stability of the random neural network model", *Neural Computation*, Vol. 2, No. 2, pp. 239-247.

[6] Gelenbe, E., Koubi, V., Pekergin, F. (1993) "Dynamic random neural approach to the Traveling Salesman Problem", *Proc. of the IEEE Annual Symposium on Systems, Man and Cybernetics (SMC)*,La Baule.

[7] Hopfield, J.J. (1982) "Neural networks and physical systems with emergent collective computational abilities", *Proc. National Acad. Sci. USA*, Vol. 79, pp. 2554-2558.

[8] Hopfield, J.J., Tank, D. (1985) " "Neural" computation of decisions in optimization problems", *Biological Cybernetics*, Vol. 52, pp. 141-152.

Method	Mini	Exc	T	Mini	Exc	T	Mini	Exc	T
n=20	$\pi = 0.5$			$\pi = 0.25$			$\pi = 0.125$		
GR	72%	0.28	0s	68%	0.40	0s	84%	0	0s
RN	100%	0	3s	96%	0.04	0s	100%	0	0s
n=50	$\pi = 0.5$			$\pi = 0.25$			$\pi = 0.125$		
GR	16%	1.20	0s	24%	1.28	0s	12%	1.52	0s
RN	52%	0.52	32s	60%	0.4	20s	80%	0.2	9s
n=100	$\pi = 0.5$			$\pi = 0.25$			$\pi = 0.125$		
GR	20%	1.28	0s	0%	2.04	0s	4%	2.72	0s
RN	60%	0.52	3'40s	48%	0.56	2'00s	56%	0.52	1'10s

Table 1: Comparison of RNN, Greedy and Exact Solutions

Multiple Cueing of an Associative Net

M. Budinich[†], B. Graham, D. Willshaw

Centre for Cognitive Science, Edinburgh University, United Kingdom

[†] permanent address: Physics Department & INFN, Trieste University, Italy

1   Introduction

The associative net is a matrix model of memory for associative storage and retrieval. Previous work has characterised its basic properties and has shown how the completely connected net can be used efficiently (Willshaw et al., 1969; Willshaw, 1971; Buckingham, 1991; Buckingham and Willshaw, 1992; Buckingham and Willshaw, 1993). In this paper we consider the case of using noisy cues. We show how repeated probing of the system by many noisy examples of the same cue can lead to improved performance.

The associative net consists of a set of N_A *input units* each of which is connected to each of a set of N_B *output units* by a modifiable binary synapse (or *weight*). Pairs of patterns are stored in the net, represented by the set of binary valued vectors $((\mathbf{A}_r, \mathbf{B}_r), r = 1, ..., R)$ where M_A components of each *input* vector of type A and M_B of each *output* vector of type B have value 1, the remaining components being set to 0. In the storage of any particular pattern pair, the value of each previously unmodified synapse between an input unit and an output unit which are both in the active state is changed from 0 to 1. This is a simple expression of the Hebb rule (Hebb, 1949). In recall, an input pattern is used as a *cue* to retrieve its associated output pattern, or *target*. A count is made of the number of modified synapses that are on active input lines. If this quantity (the *dendritic sum*) exceeds a threshold value then the output unit is placed in the active state. Output units can make two types of error: units that should be inactive (*low*) may be active; units that should be active (*high*) may be inactive. Setting the threshold involves discriminating between the distributions of the dendritic sums for low and high units. For fully connected nets and noise-free cues, these distributions have a simple form, and optimal threshold setting is straightforward. For noisy cues in fully or partially connected nets, the situation is more complicated.

We consider the case when each output unit has synapses from $Z N_A$ different input units ($Z \leq 1$), chosen at random. The cues used in recall are noisy versions of previously stored input patterns. In each cue a certain proportion s (*spurious*) of the M_A components of value 1 in the original pattern are changed at random to 0, and an identical number of components of value 0 changed to 1. The components in the cue that remain unchanged at 1 are called *genuine*.

Buckingham and Willshaw (1993) calculated that for an output unit that has been active in the storage of r out of the total R associations and should be in the low state in recall, the probability that the dendritic sum d_l has value x is

$$P(d_l = x) = \binom{M_A}{x} (Z\rho[r])^x (1 - Z\rho[r])^{M_A - x} \qquad (1)$$

for $d_l > 0$, where $\rho[r] = 1 - (1 - \alpha_A)^r$ is the probability of any synapse being modified, $\alpha_A = M_A/N_A$ and $\alpha_B = M_B/N_B$.

For an output unit that should be in the high state, a good approximation to the probability distribution of its dendritic sum d_h is

$$P(d_h = x) = \sum_{x_g=0}^{x} \binom{m_g}{x_g} Z^{x_g}(1 - Z)^{m_g - x_g} \binom{m_s}{x_s} (Z\rho[r])^{x_s}(1 - Z\rho[r])^{m_s - x_s}$$

(2)

where x_g and x_s are the contributions to the dendritic sum made by the genuine and spurious input lines respectively; $x_s + x_g = x$, $m_s = sM_A$, $m_g = (1-s)M_A$.

2 Multiple Cueing

Efficient threshold setting involves separating these two distributions in an optimal fashion. Suitable strategies range from simple winners-take-all to more complex ones which exploit knowledge of the parameter values of the distributions, such as Z, s and unit usage ρ (Willshaw and Buckingham, 1990; Buckingham, 1991; Buckingham and Willshaw, 1993). An alternative approach to producing good recall from noisy cues is to investigate methods for narrowing these distributions before the threshold is applied. Here we explore the consequences of achieving this using multiple cueing.

The idea is to feed n_{cue} different noisy versions of the same input pattern into the memory and average their effects. For each output unit, n_{cue} different dendritic sums d_i ($i = 1, 2...n_{cue}$) are obtained, from which the average $\bar{d}$ of the d_i is computed. This can be thought of as using a neuron whose dendritic sum is averaged over some time during which the input cue is constant and the noise is mutable.

The advantage is manifest when we consider the distribution of $\bar{d}$. This has the same mean as the distribution for a single d_i but with smaller variance. Thinner distributions have smaller overlap that in turn gives better memory performance. In other words, a given threshold value now produces fewer errors. Figure 1 supports this observation with numerical evidence, suggesting also that the improvement in recall tends to an asymptotical value as the number of cues grows large.

To quantify the improvement we note that the standard deviation $\sigma_{\bar{d}}$ of the distribution of $\bar{d}$ for a low unit when averaged over n_{cue} cues all with the same noise s decreases as a function of n_{cue}. If ρ_c is the correlation coefficient of two d_i and σ is the standard deviation of the distribution of a single d_i,

$$\sigma_{\bar{d}} = \frac{\sigma}{\sqrt{n_{cue}}}\sqrt{1 + \rho_c(n_{cue} - 1)}.$$

(3)

The correlation coefficient is calculated as

$$\rho_c = 1 - s\left(2 - \frac{s}{1 - \alpha_A}\right)$$

(4)

For example, for $n_{cue} = 40$, $s = 0.4$ and $\alpha_A = 0.03$ (as in Figure 1) the standard deviation of the low unit is reduced to 62% of its original value.

We have also investigated the effects of varying the amount of noise, s. For zero noise, ρ_c equals 1 and no improvement is achieved since $\sigma_{\bar{d}} = \sigma$. Maximum reduction in $\sigma_{\bar{d}}$ is obtained when ρ_c takes its minimal value at $s = 1 - \alpha_A$. However, at this noise level the distributions of $\bar{d}$ for high and low units coincide, and so no amount of reduction of the variance will improve memory performance. So at both $s = 0$ and $s \geq 1 - \alpha_A$, using multiple cues does not improve memory performance. Thus there should be an optimal noise level, $0 < s_{opt} < 1 - \alpha_A$ at which the gain in memory performance by using n_{cue} cues is maximal. This optimum represents the trade-off between increasing effectiveness in reducing the variance and increasing overlap between the low and high distributions as the noise level increases. An example is shown in Figure 2 to support this.

3   Conclusion

Use of a very simple cueing strategy for an associative net with noisy cues improves the efficiency of recall markedly. Our preliminary investigations of varying the noise demonstrate that a deleterious effect of up to 30% noise in the cues can be almost eliminated (Figure 2). The fact that at constant noise a maximum correlation is reached asymptotically suggests that in the limit of very frequent cueing the expressions (1) and (2) for the dendritic sums for low and high units have a much simpler form. Using this fact, it may be possible to derive simpler and more efficient threshold-setting strategies than those used hitherto. This point and possible ways of quantifying our finding of an optimal level of noise are currently under investigation.

Acknowledgement We thank the UK Medical Research Council for supporting this work under Programme Grant PG9119632.

References

Buckingham, J. T. (1991). *Delicate Nets, Faint Recollections: a Study of Partially Connected Associative Network Memories*. PhD thesis, University of Edinburgh.

Buckingham, J. T. and Willshaw, D. (1992). A note on the storage capacity of the associative net. *Network*, 3:407–414.

Buckingham, J. T. and Willshaw, D. (1993). On setting unit thresholds in an incompletely connected associative net. *Network*, 4:441–459.

Hebb, D. O. (1949). *The Organisation of Behaviour*. Wiley, New York.

Willshaw, D. J. (1971). *Models of Distributed Associative Memory*. PhD thesis, University of Edinburgh, Edinburgh.

Willshaw, D. J. and Buckingham, J. T. (1990). An assessment of Marr's theory of the hippocampus as a temporary memory store. *Philosophical Transactions of the Royal Society of London, Series B*, 329:205–215.

Willshaw, D. J., Buneman, O. P., and Longuet-Higgins, H. C. (1969). Non-holographic associative memory. *Nature*, 222:960–962.

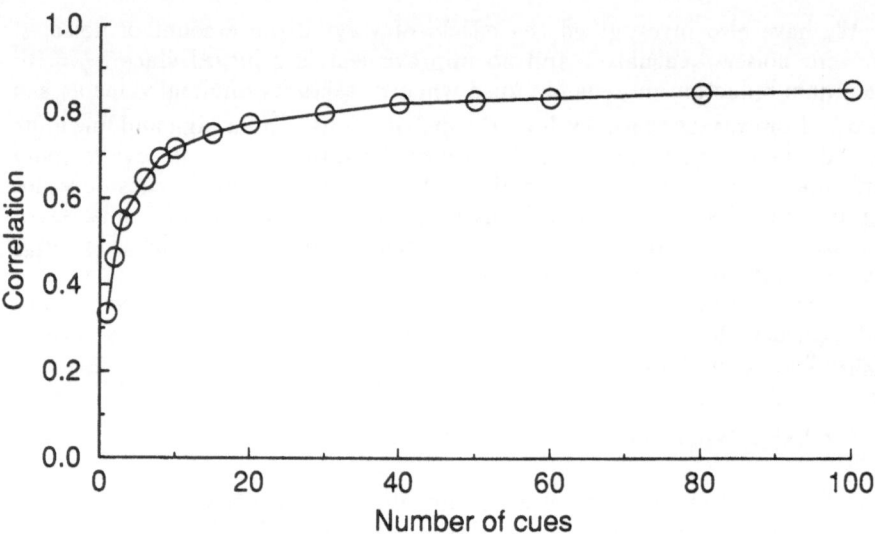

Figure 1: Correlation between stored and recalled output patterns depending on the number of cues n_{cue}. Correlation is measured as the cosine of the angle between recalled vector and the corresponding target vector, with a value of 1 if the vectors are identical and 0 if they are orthogonal. Parameter values are: $N_A = 8000, N_B = 1024, M_A = 240, M_B = 30, Z = \frac{2}{3}$, 3500 stored patterns ($R$), noise ($s$) = 40%. A winners-take-all threshold setting strategy was used in which the dendritic sum for each unit was scaled according to unit usage. Each data point is the average of 100 recalls and has a negligibly small standard deviation

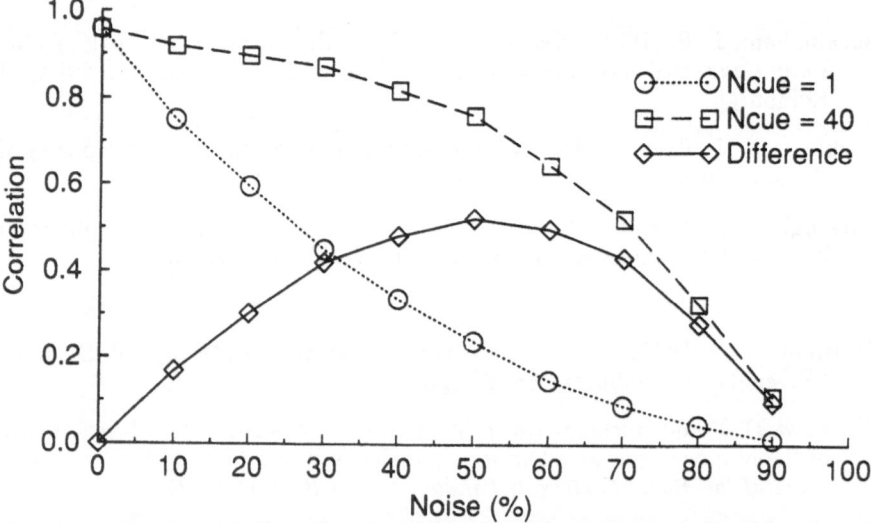

Figure 2: Correlation between stored and recalled output patterns for $n_{cue} = 1$ and $n_{cue} = 40$ at different noise levels. Same parameters as for Figure 1. The difference between the results, shown by the solid line, indicates an optimal noise level of about 50%.

Programmable Mixed Implementation of the Boltzmann Machine

V. Lafargue, E. Belhaire, H. Pujol, I. Berechet, P. Garda.

Institut d'Electronique Fondamentale - Bat 220 - Universite de Paris Sud

91405 Orsay France

1   Introduction

Boltzmann Machines show very high recognition rates (Kohonen et al., 1988), but their simulations are desperately slow. Mixed analog/digital implementations were therefore described, whose learning algorithm is hardwired. But new learning algorithms were recently proposed, specifically for Synchronous Boltzmann machines. In this paper we introduce a new analog/digital architecture for this Boltzmann Machine. Its learning algorithm is programmable, and a prototype implementation is described.

2   Boltzmann Machines

A Boltzmann Machine is operated as follows. Let x_i be the state of the neuron u_i, which may have values of 1 or 0. Let w_{ij} be the synaptic weight from the neuron u_j to the neuron u_i. Let V_i be the network contribution to u_i, computed according to the equation (1). Then the state of the neuron u_i is tossed at random with the probability given by the equation (2).

$$V_i = \sum_{j \neq i} w_{ij}\, x_j \quad (1) \qquad\qquad P(x_i = 0) = \frac{1}{1 + \exp(V_i/T)} \quad (2)$$

In the asynchronous Boltzmann machine (Hinton et al., 1984), when a neuron updates its state, all the other neurons have to keep their state unchanged. On the contrary in the Synchronous Boltzmann machine (Azencott 1989), all the neurons update their state simultaneously at each sweep. This significantly decreases the duration of each sweep.

Now let us describe their learning algorithms. Let us choose input, output and hidden neurons among the network. The weight update process is repeatedly performed for all the patterns. For each of them, it consists in a clamped and a free phase. During each phase, hundred of sweeps are performed. A cooccurrence counter is associated to each weight, according to one of the next equations:

(Hinton et al., 1984)	asynchronous (3)	$p_{ij} = \dfrac{1}{N} \displaystyle\sum_{n=M}^{N+M} x_i{}^n . x_j{}^n$
(Azencott 90)	synchronous (4)	$p_{ij} = \dfrac{1}{2N} \displaystyle\sum_{n=M}^{N+M} x_i{}^n . x_j{}^{n-1} + x_i{}^{n-1} . x_j{}^n$
(Lacaille 92)	asymmetric (5)	$p_{ij} = \dfrac{1}{N} \displaystyle\sum_{n=M}^{N+M} x_i{}^n . x_j{}^{n-1}$

During the clamped phase, a pattern is imposed both on the input and output neurons, while the hidden neurons are left free to update their states, and the cooccurrence p_{ij}^+ is computed. Whereas during the free phase, the input pattern is presented to the input neurons, while the output and hidden neurons are left free to update their states, and the cooccurrence p_{ij}^- is computed. After these clamped and free phases, the weights are updated according to one of the next rules. The gain h, the momemtum coefficient $\alpha < 1$ and the temperature T are positive parameters.

(Hinton et al., 1984)	linear (6)	$\Delta w_{ij} = \dfrac{h}{T} \cdot \left(P_{ij}^+ - P_{ij}^- \right)$
(Hinton et al., 1984)	threshold (7)	$\Delta W_{ij} = h \cdot \text{sgn}\left(P_{ij}^+ - P_{ij}^- \right)$
(Lacaille 92)	momemtum (8)	$\Delta W_{ij}^{\ n} = \alpha . \Delta W_{ij}^{\ n-1} + (1-\alpha).h. \left(P_{ij}^{n+} - P_{ij}^{n-} \right)$

3   Architecture of the synaptic cell

Several mixed analog/digital implementations were described by (Kreuzer et al 1988) (Alspector et al 1990) (Arima et al 1991). They all included a hardwired learning algorithm with the threshold weight update rules of asynchronous network according to the equations (3) and (6). On the contrary, we introduce a new synaptic cell whose learning algorithm is programmable. It includes an analog part, a digital part and an analog-to-digital interface, as shown in Figure 1. (Lafargue 1992)

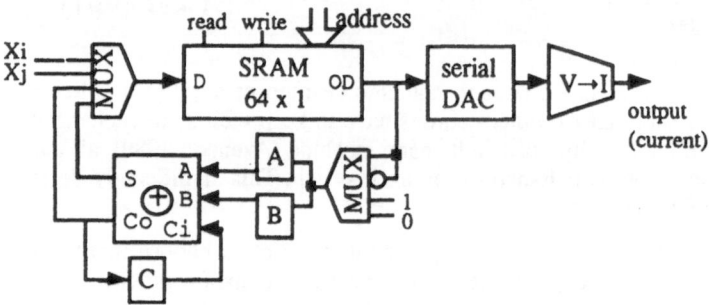

Figure 1: Architecture of the synaptic cell

The analog part is involved in the computation of the network contribution according to the equation (1) by a single rail current summator. For this purpose it includes a linear voltage-to-current converter driven by a weight capacitor, described in (Belhaire et al 1993).

The digital part includes a programmable processor and a digital static memory. The former is a bit-serial processor with three 1-bit registers and a 1-bit wide ALU. It is very efficient for medium precision integer additions, subtractions, and scaling by powers of 2. These arithmetic operations are those needed by equations (3) to (7)

The digital SRAM includes 64 1-bit wide words. It stores the integer representation of the concurrence counter P_{ij}, the weight update DW_{ij}, and the weight W_{ij}. Observe that the neuron states X_i and X_j are also stored in the memory.

The weight W_{ij} is represented as an integer in the digital part and as an analog voltage in the analog part. A digital to analog converter is used to refresh the analog representation from the digital one. It is a serial converter, and this fits very well with the architecture of the bit-serial processor.

4   Implementation of a prototype chip

A prototype circuit, called MBAT11, was designed, including an array of 4x4 synaptic cells. The four cells $(C_{ij})_{0 \leq j \leq 3}$ of the i-th row share the same output wire, which carries a current representing the network contribution $\sum_{0 \leq j \leq 3} w_{ij} x_j^n$. On the other hand the chip has two groups of 4 input pads which carry the states $(x_i)_{0 \leq i \leq 3}$ and $(x_j)_{0 \leq j \leq 3}$ respectively. Then each of the state x_i (resp. x_j) is broadcast to the 4 cells of the i-th row (resp. j-th column).

All the bit-serial processors are operated in SIMD mode under the control of a single common sequencer. The global parameters h, a and T, needed by the weight update rules, are stored in the sequencer which broadcasts them to all the PE. Through this control, they perform the computations of the cooccurence counter P_{ij} according to equations (3) or (4) or (5), and the update of the weight W_{ij} according to equations (6), (7) or (8).

The digital part of the circuit was designed with standard cells, whereas the analog part was designed in full custom. The chip was realised in the MIETEC/ALCATEL 2.4µm DLM DLP CMOS process on a 3.9x4.4 μm^2 surface. It was simulated with a 20 MHz clock and it was successfully operated at 11 MHz. Thus an arithmetic operation on 16 bits integers takes less than 2 µs. This speed is compatible with that of the analog update of the neurons states which takes 3 µs.

This circuit was used in an experimental system. It included 4 input neurons, whose state were clamped by the sequencer. It also included 4 updating neurons, whose state were either updated by the system or clamped by the sequencer, depending on the training phase. The system included a full connection from the input neurons to the updating neurons, and a full connection between the updating neurons. This required a 8x4 synaptic matrix and a column of 4 neuron cells. It was realised on a Printed Circuit Board hardwiring two MBAT11 chips with 1 MBA2 chip, the latter providing 4 neuron cells (Belhaire et al., 1993).

A Digital Analysing System DAS 9200 (Tektronix) was used as a sequencer for the system. It provided also the input and output patterns for the training sessions.

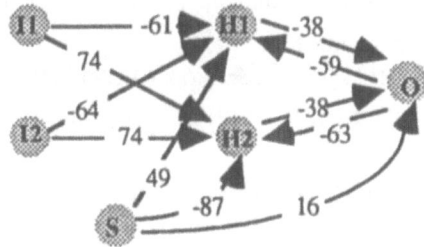

Figure 2: Example of a learned asymmetric XOR 2-2-1 network

This system was used to perform a XOR 2-2-1 network. An asymmetric synchronous linear learning algorithm was used (equations 5 and 6), with P_{ij} estimated over 32 iterations per pattern. The P_{ij} and W_{ij} were computed over 8 bits. The number of weight updates varied between 100 and 300 for a successful training, and an experimental set of asymmetric weights is shown in Figure 6.

5   Conclusion

In this paper, we have described the architecture of a new synaptic cell for the mixed digital/analog implementation of the Boltzmann Machines, and we have introduced several original points. This synaptic cell is programmable, and the learning algorithm may be selected according to the difficulty of the learning. Its architecture combines a bit-serial processor, and a bit-serial DAC. It achieves therefore an excellent trade-off between silicon area and processing speed for low-precision integer arithmetic. Moreover, as the algorithms are performed in a bit-serial way, their duration is proportional to the complexity of the algorithm. Finally the precision of the variables which are used for the learning, such as the cooccurrence counter or the weight, may easily be adapted to the training task. We showed though the use of an experimental system that the coupling between analog and digital parts has no influence on the learning, and that the speed of these two parts are balanced.

6   Acknowledgements

This work has been supported by D.R.E.T. under Contract 87/292 and by C.N.R.S., and M.R.E.S. under P.R.C.-A.M.N., Cognisciences and G.C.I.S. project RA. We thank Robert Azencott and Jérome Lacaille, from D.I.A.M.E.N.S. in Paris, Jacques-Olivier Klein and Zhu Yi Min from I.E.F., for fruitful discussions.

7   References

Alspector J. et al., "Relaxation Networks for Large Supervised Learning Problems", *Procs N.I.P.S.*, 90

Arima et al, "A 336-Neuron, 28K-Synapse, Self-Learning Neural Network Chip with Branch-Neuron-Unit Architecture",*JSSC*, vol. SC-26, pp. 1637-164491

Azencott R., "Synchronous Boltzmann Machines and their learning algorithms", *Neurocomputing*, Les Arcs, 1989, Springer-Verlag , Vol. F-68

Azencott R., "Boltzmann Machines: high-order interactions and Synchronous learning", *Procs Stochastic models,,* Ed. by P.Barone and A.Frigessi, Lecture Notes in Statistics, Springer-Verlag, 1990

Belhaire E. & al., "A scalable analog implementation of Multilayer Boltzmann Machines", *ESSCIRC 93*, Sevilla, September 22-24 1993

Hinton G. & al., "Boltzmann Machines", *C.M.U. Tech Report CMU-CS-84-119*, Carnegie Mellon University, 1984

Kreuzer I. & al., "A modified model of Boltzmann Machines for W.S.I. realization", *Signal Processing IV*, 1988

Lacaille J , "Machines de Boltzmann, Théorie et Applications" , Ph D Thesis, june 25th 1992, University of Paris Sud

Lafargue V., "Machines de Boltzmann, Théorie et Applications" , Ph D Thesis, june 25th 1992, University of Paris Sud

The Influence of Response Functions in Analogue Attractor Neural Networks

N. Brunel†, R. Zecchina‡

† INFN/Dip. di Fisica, P.le Aldo Moro 2,
I–00185 Rome, Italy

‡ Politecnico di Torino/Dip. di Fisica, C.so Duca degli Abruzzi 24,
I–10129 Torino, Italy

1 Introduction

In the context of Attractor Neural Networks (ANNs), we will be concerned with the following issue: assuming the interaction (synaptic) matrix to be the simple Hebb-Hopfield correlation matrix, we discuss how the storage performance of an ANN may depend on the equilibrium analogue neural activities reached by the dynamics during memory retrieval. In both discrete and analogue Hopfield-like attractor neural networks, the phase transition of the system from associative memory to spin-glass, is due to temporal correlations arising from the static noise produced by the interference between the retrieved pattern and the other stored memories. The introduction of a suitable cost-function in the space of neural activities allows us to study how such a static noise may be reduced and to derive a class of simple response functions for which the dynamics stabilizes the 'ground-state' neural activities, i.e. the ones that minimize the cost function, up to a number of stored patterns equal to $\alpha_* N$ (N =number of neurons), where $\alpha_* \in [0, 0.41]$ depends on the average activity in the network.

2 The model

The neural network model is assumed to be fully connected and composed of N neurons whose activities $\{V_i\}_{i=1.N}$ belong to the interval $[-1, 1]$. The global activity of the network is defined by

$$\gamma = \frac{1}{N} \sum_i \epsilon_i \quad , \tag{1}$$

where $\epsilon_i = |V_i|$ and we denote by $\mathcal{E} = [0, 1]^N$ the space of all ϵ_i. A macroscopic set of $P = \alpha N$ random binary patterns $\Xi \equiv \{\{\xi_i^\mu = \pm 1\}_{i=1,N}, \mu = 1, P\}$, characterized by the probability distribution $P(\xi_i^\mu) = \frac{1}{2}\delta(\xi_i^\mu - 1) + \frac{1}{2}\delta(\xi_i^\mu + 1)$, is stored in the network by means of the Hebb-Hopfield learning rule $J_{ij} = \frac{1}{N} \sum_{\mu=1}^P \xi_i^\mu \xi_j^\mu$ for $i \neq j$ and $J_{ii} = 0$.

At site i, the afferent current (or local field) h_i is given by the weighted sum of the activities of the other neurons $h_i = \sum_j J_{ij} V_j$. We consider a continuous dynamics for the depolarization I_i at each site i ($\tau \dot{I}_i(t) = -I_i(t) + h_i(t)$), in which the activity of neuron i at time t is given by $V_i(t) = f(I_i(t))$, where f is the neuronal response function. No assumptions are made on f, except that it is such to align the neural activity with its afferent current ($xf(x) \geq 0$ for all x).

We consider the case in which one of the stored pattern, $\mu = 1$ for example, is presented to the network via an external current which forces the initial configuration of the network, thus we take $V_i = \epsilon_i \xi_i^1$. The current arriving at neuron i becomes, in terms of the overlaps of the network state with the stored patterns $m_\mu \equiv \frac{1}{N} \sum_j \xi_j^\mu \xi_j^1 \epsilon_j$,

$$h_i = m_1 \xi_i^1 + \left(\sum_{\mu > 1} \xi_i^\mu m_\mu - \alpha \epsilon_i \xi_i^1 \right) \tag{2}$$

Notice that the global activity is equal to the overlap of the network configuration with the retrieved pattern ($m_1 = \gamma$).

The first term in the r.h.s. of equations (2) is the signal part, whereas the second term, when the ϵ_i are fixed and for N large, is a Gaussian random variable with zero mean and variance $\sigma^2 = \sum_{\mu > 1} m_\mu^2$ and represents the static noise part, or cross-talk, due to the interference between the stored patterns and the recalled one. In the following we will be interested in minimizing the overlaps of the network state with the stored memories $\mu \neq 1$ which are not recalled. If one succeeds in finding such states ($m_\mu = 0$ for all $\mu \neq 1$), then the second term in the r.h.s. of Eq. (2) reduces to $-\alpha \epsilon_i \xi_i^1$, and the interference effect vanishes. We will see that this is indeed the case in a finite range of the parameter α and that it is also possible to derive a class of effective response functions realizing such a minimization of the interference and thus leading to an improvement of the network storage capacity.

3 Interference Reduction

In order to study how the interference may be reduced, we define a cost function $E(\Xi, \vec{\epsilon})$ depending on the neural activities and the set of stored patterns, proportional to σ^2, i.e. to the sum of the squared overlaps of the network state with all the stored memories except the retrieved one

$$E(\Xi, \vec{\epsilon}) = \frac{1}{N} \sum_{\mu \neq 1} \left(\sum_j \xi_j^\mu \xi_j^1 \epsilon_j \right)^2 \tag{3}$$

and we study its ground states. If no constraints are present on $\vec{\epsilon}$, the minimum is always $E = 0$ for $\vec{\epsilon} = 0$; obviously, in this case no information is obtained when one presents the pattern and therefore we impose the constraint $\gamma = K$ on the average activity, where $K \in [0, 1]$.

In order to determine the typical 'free energy' F_0, we compute $\langle \ln Z \rangle_\Xi$, where $\langle \ldots \rangle_\Xi$ stands for the average over the quenched random variables Ξ and Z is the partition function given by $Z = Tr_{\vec{\epsilon}}[\exp(-\beta E(\Xi, \vec{\epsilon}))] \delta(\gamma(\vec{\epsilon}) - K)]$ ($T = 1/\beta$), using the replica method. Depending on the storage level α one obtains the following results: For $\alpha < \alpha_0(K)$ we have $F_0 = 0$. At $\alpha = \alpha_0$, the space of neural activities such that $F_0 = 0$ shrinks to zero, and F_0 becomes positive. $\alpha_0(K)$ is given in Fig. 1.

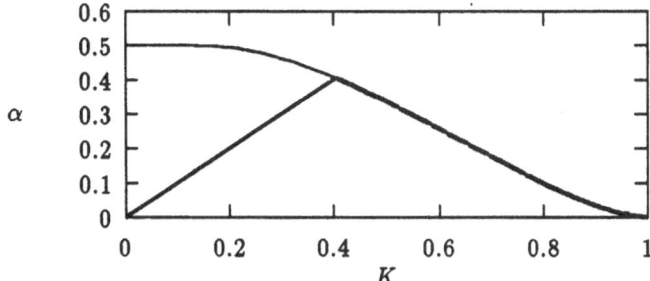

Figure 1: Bold curve: α_* versus K. Light curve: α_0 versus K.

The study of the gradient flow associated to a smooth version of the energy function (3),

$$E_\lambda(\Xi, \vec{\epsilon}) = \frac{1}{N} \sum_{\mu \neq 1} \left(\sum_j \xi_j^\mu \xi_j^1 \epsilon_j \right)^2 + \frac{\lambda}{N} \left(\sum_i \epsilon_i - KN \right)^2 , \qquad (4)$$

allows us on the one hand to find the relation between activities and afferent currents in the ground state, and on the other to check the outcome of the (stable) RS solution (one finds a remarkable agreement between the analytic solution and the numerical simulations).

4 Critical Capacity and Response Functions

If the minimum of $E_\lambda(\Xi, \vec{\epsilon})$ – which is quadratic form positive semi–definite – is zero in $[0, 1]^N$, then its gradient vanishes (which implies $\gamma = K$). This leads to a very simple relation between the activity and the stability $\Delta_i \equiv h_i \xi_i$ at each site

$$K - \Delta_i = \alpha \epsilon_i , \qquad (5)$$

which also coincides with the expression of the afferent currents, eq. (2), in which $m_\mu = 0$ for $\mu > 1$. For $\alpha > \alpha_0(K)$, due to the non-linear constraint on the bounds of the activities, the flow reaches a fixed point which does not coincide with the minimum of E_λ in R^N.

Under the initial assumption on the neuronal transfer function ($x f(x) \geq 0$), the condition for a 'ground-state' configuration correlated with the presented pattern to be a fixed point of the dynamics is to have positive Δ's at all sites. This indeed happens if the storage level α satisfies $\alpha < \alpha_*(K)$ where $\alpha_*(K)$ is the critical line identified by $\alpha_*(K) = \min(K, \alpha_0(K))$ and shown in Fig. 1.

It follows from eq.(5) that, when $\alpha < \alpha_*(K)$, the ground state activities are fixed points of the network dynamics with the (non-monotonic) transfer function f

$$f(h) = \begin{cases} \text{sgn}(h) & \text{if } |h| \in [0, K - \alpha] \\ \text{sgn}(h)(K - |h|)/\alpha & \text{if } |h| \in [K - \alpha, K] \\ 0 & \text{if } |h| > K \end{cases} \qquad (6)$$

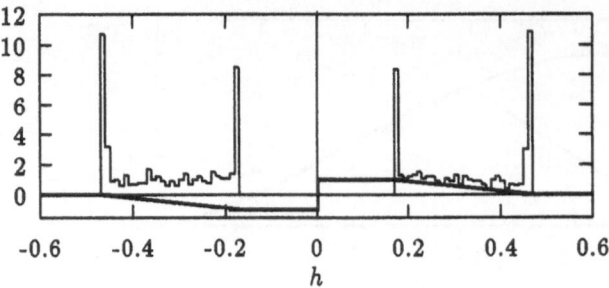

Figure 2: Bold curve: neuronal transfer function for $\alpha = 0.3$ and $K = 0.47$. Light curve: distribution of the currents at equilibrium, for a network of size $N = 1000$ with a neuronal transfer function characterized by the same parameters and after presentation of one of the stored patterns ξ^ν ($V_i(t = 0) = \xi_i^\nu$ for all i).

shown in Fig. 2. In the same figure we also report, for $\alpha < \alpha_*(K)$, the equilibrium distribution of the local currents obtained by numerical simulations performed on a network with such a dynamics. All the Δ's belong to the interval $[K - \alpha, K]$, as expected.

Hence, for $\alpha < \alpha_*(K)$, a network with the discussed dynamics is capable to stabilize a configuration with activity K highly correlated with the retrieved pattern: the optimal activity is $K_{opt} = 0.41$ for which we have $\alpha_*(K_{opt}) = 0.41$. Usually, the critical capacity is defined as an upper bound for the presence of retrieval states correlated with the presented pattern. In the case of a sigmoid transfer function, the critical capacity is obtained with a finite σ^2 and yields ~ 0.14 (Amit et al.,1985). Here α_* is derived as an upper bound for the presence of retrieval states with $\sigma^2 = 0$. This means that α_* is a lower bound for the critical capacity, which is expected to be higher in the region where the Δ's are strictly positive at all sites, i.e. $K > \alpha_*$ or $K > 0.41$. Interestingly enough, our results on the maximal storage capacity $\alpha_*(K_{opt})$ is very similar to an estimate obtained in (Yoshizawa et al., 1993) by a completely different method on a particular non-monotonic transfer function.

All the details concerning the above results can be found in (N. Brunel and R. Zecchina, 1993); further results on continuous 0-1 neurons, as well as on the exact study of the dynamics in the diluted case, will be published elsewhere.

References

[1] .J. Amit , H. Gutfreund and H. Sompolinsky, (1985) *Phys. Rev. Lett.* **55**, 1530.

[2] . Brunel, and R. Zecchina, (1993) preprint POLFIS-TH. 25/93, submitted.

[3] .J. Hopfield, (1982) *Proc. Natl. Acad. Sci. USA* **79**, 2554.

[4] . Yoshizawa , M. Morita and S. Amari, (1993) *Neural Networks*, **6**, 167–176.

Improvement of Learning in Recurrent Networks by Substituting the Sigmoid Activation Function*

J.M. Sopena†, R. Alquezar‡

† Departament de Psicologia Basica, Universitat de Barcelona
Barcelona (Spain)

‡ Institut de Cibernetica, Univ. Politecnica de Catalunya - CSIC
Barcelona (Spain)

1   Introduction

Several recurrent network architectures have been devised in recent years to deal with sequential tasks. One such model is the Simple Recurrent Network (SRN) proposed by Elman (Elman, 1988). The backpropagation rule was employed for learning in the former published works with SRNs, e.g. (Cleeremans et al., 1989). Later on, full gradient learning schemes, such as RTRL and BPTT, have been proposed for learning in fully-connected recurrent networks. These algorithms can also be used to train the weights of the recurrent hidden layer in SRNs.

On the other hand, the use of the sigmoid activation function has been common, not only in SRN works, but also within other architectures (Smith and Zipser, 1989; Fahlman, 1991). However, unbounded activation functions should better record an entered input sequence in the hidden units (Sopena, 1991), since a sequence is stored as a sum of terms recursively filtered by a nonlinear function, which may cut useful data if its input saturates, as it can occur with the sigmoid. In this work, we propose an unbounded nonlinear activation function to be used in the recurrent units of a SRN: the antisymmetric logarithm. Moreover, we suggest to replace the sigmoid function by a sinusoidal activation function in the non-recurrent (non-input) units of the SRN. This substitution is motivated by the derivative profile: while the sigmoid derivative is almost null for the most part of its domain, the sinus derivative takes significant values for all points except for those around the extremes of each period. Hence, it is reasonable to expect that the landscape of the error function with respect to the weights be much smoother if a sinusoidal activation function is used, thus avoiding the plateau regions that are typical in error landscapes of sigmoid-based networks.

In order to test both changes, SRNs have been trained, using a full gradient descent algorithm, to learn the prediction task for the two Reber grammars studied in some previous works (Cleeremans et al., 1989; Smith and Zipser, 1989; Fahlman, 1991). The comparative results have revealed an impressive improvement in learning performance when the sigmoid function is substituted by antisym-log and sinusoidal functions in recurrent and non-recurrent units, respectively.

*This work has been partially supported by a grant of the Government of Catalonia

2   Architectures and learning algorithm

The layered recurrent network architectures shown in Fig.1 were used for the study. Three different combinations of activation functions were tested: (Sigmoid/Sigmoid), (Antisym-log/Sigmoid) and (Antisym-log/Sinusoidal), where, for each pair, the former function corresponds to the fully-connected recurrent layer, and the latter is used in the higher layers. Sigmoid and sinusoidal functions were ranged between 0 and 1. The antisym-log function is defined as

$$g_{al}(x) \;=\; sgn(x)\; log(1 + k|x|) \tag{1}$$

where $sgn(x) = +1(-1)$ for $x \geq 0$ ($x < 0$), respectively; log denotes the natural logarithm; and k is a parameter, the slope at the origin (which was set to 2).

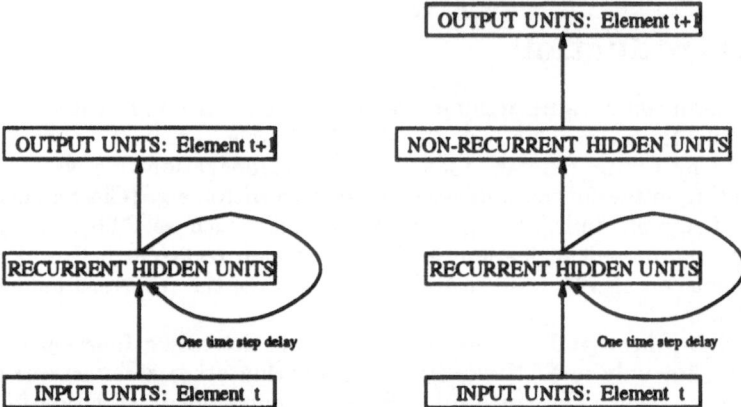

Fig.1: Two- and three-layer networks with a single fully-connected recurrent layer.

The backpropagation rule was used to update weights for higher layers and to supply the error values for the recurrent hidden units. The learning scheme for recurrent units was an efficient version of RTRL due to (Schmidhuber, 1992). In order not to introduce additional variability, the learning rate (α) and momentum (β) parameters were fixed for all tests ($\alpha = 0.025$, $\beta = 0$).

3   Methods and results

Firstly, the simple Reber grammar was selected to generate random sequences of valid strings, by distributing a uniform probability among the possible successor symbols. Two specific models were trained to predict the next symbol in a sequence of legal transitions: network **A** consisted of 6 input units (one for each grammar symbol, including an end-of-string symbol), 6 recurrent hidden neurons, and 6 output units, again one for each symbol; network **B** had an additional layer of 6 feed-forward hidden units between the recurrent layer and the output layer. As in (Cleeremans et al. 1989), a prediction was considered correct when the activation values of output units corresponding to the possible successors were above some threshold whereas the rest were below it; the prediction threshold was set to 0.25. Valid strings were continuously generated for on-line training, but the recurrent unit activations were reset to 0.1 after

Rec. Layer Act. Func. / High. Layers Act. Func.	Sigmoid / Sigmoid	Antisym-Log / Sigmoid	Antisym-Log / Sinusoidal
Network **A**	10 **116.6 ± 18.6** [90 − 146]	10 **9.0 ± 4.0** [6 − 18]	10 **7.2 ± 3.7** [4 − 16]
Network **B**	10 **195.0 ± 71.4** [138 − 388]	10 **49.8 ± 22.6** [18 − 78]	10 **9.4 ± 3.3** [6 − 14]

Table 1: Learning performance figures for simple Reber grammar.

each string end. Learning was regularly stopped after groups of 2,000 training transitions, when 1,000 transitions were tested for the prediction task. Training ended either when all predictions in a test phase were correct, or when a maximum number of 2 million training transitions were reached without success. Ten trials with different random initial weights were run for each model and each pair of activation functions.

The results of learning the prediction task for the simple Reber grammar are shown in Table 1. Three performance figures are placed inside each entry, from top to bottom: a) number of successful runs (up to 10), b) mean and standard deviation (over the successful runs), and c) minimum and maximum, of the number of training transitions (in thousands) required to reach 0% prediction error. All the 60 runs were successful. The (antisym-log/ sinus) configurations learnt 16 and 20 times faster than the all-sigmoid nets, for 2- and 3-layers, respectively. The upgrade due to the sigmoid to sinus replacement in non-recurrent units was moderate in the 2-layer net (20% faster), but it was quite large in the 3-layer net (5.3 times faster). It is helpful to compare our results with those reported elsewhere for the same task and grammar: (Smith and Zipser, 1989) reported to learn the task after 19,000 to 63,000 string presentations with a single-layer recurrent net and 2 hidden units, using RTRL and sigmoid function; (Fahlman, 1991) reported an average of 25,000 string presentations using the constructive recurrent cascade-correlation approach and sigmoid functions. Our two-layer net with 6 hidden units, trained by RTRL in the recurrent layer, needed an average of 17,000 strings with sigmoid functions, but only 1,030 with the (antisym-log/ sin) pair!

Next, the prediction task for the more complex symmetric Reber grammar was tested using two larger models: network **C** consisted of 6 input units, 12 recurrent hidden units, and 6 output units; network **D** had an additional layer of 12 feed-forward hidden units. The training-test procedure was the same as before. Table 2 displays the learning performance measures that were obtained. The (antisym-log/ sin) function pair contributed to a good learning performance: 100% success, with an average of 26,000 and 16,000 string presentations for networks **C** and **D**, respectively. SRNs with only sigmoids performed poorly. Again, the learning improvement due to sigmoid/sinus change in non-recurrent units is notable in the 3-layer net. (Smith and Zipser, 1989) learnt the same task using RTRL, sigmoid function, and a fully-connected recurrent net, in some unspecified fraction of attempts (best result: 25,000 training strings with 12 hidden units). (Fahlman, 1991) needed an average of 182,000 string presentations using recurrent cascade correlation (7 − 9 hidden units), and perfect learning was achieved in just half the trial runs.

Rec.Layer Act.Func. / *High.Layers Act.Func.*	Sigmoid / Sigmoid	Antisym-Log / Sigmoid	Antisym-Log / Sinusoidal
Network C	1 **1970.0** [1970]	9 **238.4 ± 81.4** [120 − 402]	10 **235.0 ± 122.8** [122 − 480]
Network D	6 **1487.3 ± 373.0** [980 − 1920]	10 **265.2 ± 232.3** [102 − 846]	10 **144.0 ± 33.5** [80 − 192]

Table 2:   Learning performance figures for symmetric Reber grammar.

4   Discussion

It is widely accepted that a full gradient training scheme is required for a recurrent network to learn complex tasks, such as inferring the symmetric Reber grammar (Smith and Zipser, 1989). However, this is not the whole story. We have shown that using an unbounded activation function (the antisymmetric logarithm) in recurrent units yields a great improvement in network learning performance with respect to a similar network with sigmoid function. We hypothesize that useful information, which is represented in the sum of terms associated with the sequence past history, is lost when bounded activation functions are used in recurrent units, thus reducing the short-term memory capacity of the network.

It has been shown that a second source of learning upgrade in SRNs is obtained by the replacement of sigmoid by sinusoidal function in the output units. The improvement is much more remarkable when the substitution is also applied to an intermediate hidden layer. So, it may be concluded that sinusoidal activation functions better back-propagate the error values, and this result should be valid for multi-layered feed-forward networks. An informal explanation has been given in terms of derivative function and error-weights landscape. However, further experimental and theoretical work is encouraged to confirm the conjectures presented here.

References.

Cleeremans,A., Servan-Schreiber,D., McClelland,J.L. (1989) "Finite-State Automata and Simple Recurrent Networks". Neural Computation, 1, 372-381.

Elman,J.L. (1988) "Finding Structure in Time". CRL Technical Report 8801, University of California, San Diego, Center for Research in Language.

Fahlman,S.E. (1991) "The Recurrent Cascade-Correlation Architecture". Advances in Neural Information Processing Systems 3, Lippmann,R.P., Moody,J.E., and Touretzky,D.S., eds., Morgan Kaufmann, San Mateo CA, 190-196.

Schmidhuber,J. (1992) "A Fixed Size Storage O(n3) Time Complexity Learning Algorithm for Fully Recurrent Continually Running Networks". Neural Computation, 4, 243-248.

Smith,A.W., Zipser,D. (1989) "Learning Sequential Structure with the Real-Time Recurrent Learning Algorithm". Int.Journal of Neural Systems, 1, 125-131.

Sopena,J.M. (1991) "ERSP: A Distributed Connectionist Parser that Use Embedded Sequences to Represent Structure". Tech. Rep. UB-PB-91-1, Dep. Psicologia Basica, Univ. de Barcelona.

Attractor Properties of Recurrent Networks with Generalising Boolean Nodes

Antônio de Pádua Braga[1]
Imperial College/Dept. of Electrical and Electronic Engineering
London SW7 2BT, England

1 Introduction

Recurrent networks with Generalising Boolean nodes, which are also known as GNUs (Aleksander, 1990a), are studied in this work in terms of their retrievability and attractor properties in the state space. Figure 1 shows a Generalising Boolean node with 4 inputs that has learned to fire 0 when the pattern at the inputs ABCD is more similar to 0000 and to fire 1 when more similar to 1111. The generalization in this case is defined by the similarity (measured here by Hamming Distance) between an unknown input and the two known ones. The Boolean function performed is seen in Figure 1.(a) in a Karnaugh Map form. It can also be seen that some entries are equally similar to both stored patterns (marked with ? in the map) which can put the node under hesitation. Such "don't care" entries are randomized 0 or 1 at each slot of time. Therefore, the node performs a different Boolean function at each slot of time and can fire different outputs for the same undefined input. Such Boolean nodes are known as G-RAMs (*Generalising Random Access Memories*) as one of the ways found to implement them was to use conventional Random Access Memories and by randomizing the undefined entries in execution time (Aleksander, 1990b). Some work on storage capacity and temporal properties of such networks was done before (Wong and Sherrington, (1992), Braga, (1993a) and Braga (1993b)).

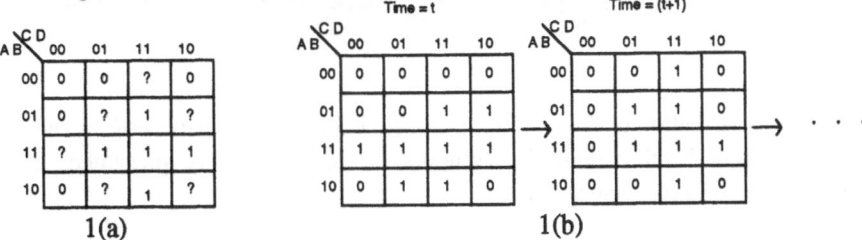

Figure 1. Node's Boolean function 1.(a) Original function with undefined entries. 1.(b) A possible sequence of Boolean functions executed by the node.

Recurrent networks with G-RAM nodes have a stochastic behaviour rather than a deterministic one. In the fully connected auto-associative configuration studied in this work, each node of the network has feed-back connections from the outputs of the other neurons as well as feed-forward ones from the input or external world field. The storage of an association between the

---

[1] On leave from Universidade Federal de Minas Gerais, Brazil. Supported by CNPq grant 202286/91.6

external world and the network's internal representation consists on producing re-entrant states in which the network stabilises in the retrieval phase. The presentation of a known pattern at the external inputs makes the network wander in a sequence of internal states until it reaches the one which corresponds to the internal representation of the presented pattern. Such stochastic behaviour, observed empirically, is demonstrated theoretically in this work.

2 Retrieving associations

The retrieval of one known association depends on how the patterns are distributed among themselves in the Boolean space and on how such distribution affects the number of undefined entries on a node's function. If an external world pattern ξ w is present at the external inputs, the hesitation of the network depends on the distance or radius of the current internal state in relation to the target attractor. If the correspondent internal state ξ i is an effective attractor, for all the possible initial internal states, the network stabilises on it in a finite sequence of steps. As an extension of Kanerva's work (Kanerva, 1988), it is demonstrated here that the distribution of patterns in the n-dimensional Boolean space in relation to two others that are separated by a fixed distance h in such space can be modelled by a Bivariate Normal Distribution (Anderson, 1958). Figure 2 shows such distribution for two patterns separated by distances 10 and 15 in the space $\{0,1\}^{20}$. The random variables r and δ represent, respectively, distances to patterns ξ 1 and ξ 2.

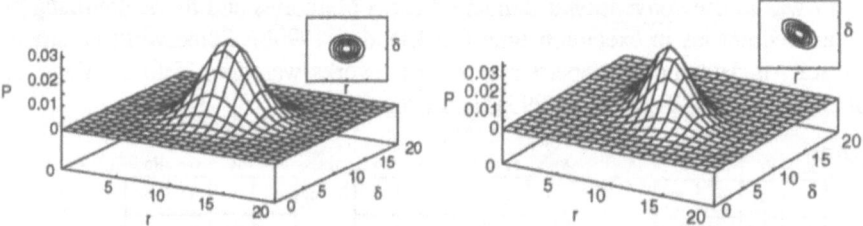

Figure 2. Distribution of the Boolean space in relation to two patterns ξ 1 and ξ 2. In the right hand corner, the contours of the surface are presented. 2.(a) Distance h is 10. 2(b) Distance h is 15.

From the distribution of Figure 2, if a pattern is chosen at random at a fixed distance r_1 to ξ 1, the chance of this pattern to be closer to ξ 1 than to ξ 2 can be calculated by the conditional probability distribution function of δ , given $r=r_1$ (Anderson, 1958):

$$f_\delta(\delta|r_1) = \frac{\int\limits_{\delta > r} e^{\{\frac{-1}{2(1-\rho^2)\sigma_r^2}[\delta - \rho\frac{\sigma\delta}{\sigma_r}(r_1 - m_r) - m_\delta]^2\}} \, d\delta}{\sqrt{2\pi\sigma_r^2(1-\rho^2)}} \qquad \begin{cases} m_r = m_\delta = \frac{n}{2} \\ \sigma_r = \sigma_\delta = \sqrt{\frac{n}{8}} \end{cases}$$

(1)

The constant ρ in expression (1) is the correlation coefficient between the random variables δ and r. The constants m_r, m_δ, σ_r and σ_δ are, respectively, the means and standard deviations of the two random variables. The best match for the correlation coefficient found was $\rho = 0.98 - 1.98\, h/n$, where n is the size of the space and h is the distance between the two patterns. For this specific case, expression (1) estimates the contribution of a pattern at distance h from ξ_i to make the node fire a wrong output when the network has moved to a radius r_1 from ξ_i. Considering a training set chosen at random, the distribution of the Hamming Distances h among the selected patterns can be approximated by a Normal Distribution with mean $n/2$ and standard deviation $\sqrt{n/4}$ for large values of n (Kanerva, 1988). Therefore, expression (1) was applied to all the patterns in the distribution to estimate the probability of a node firing with hesitation for each possible radius r in relation to the target attractor ξ_i. The results of such estimation are shown in next section.

3 Results

Figure 3 shows the main pathways of the probabilistic state machine generated in a fully connected auto-associative network with 8 nodes, 8 external inputs when storing 8 orthogonal patterns (Kanerva, 1988) and one of the external representations is present at the inputs. The attractor properties of the internal state space are visually seen in the figure. When a pattern is randomized at distance greater than 4, there is a great chance that the network will move to a distance between 1 and 4 in the next step. Once at a distance between 1 and 4 it will move to a distance 0 (or that it will move to the attractor) in the next 2 steps.

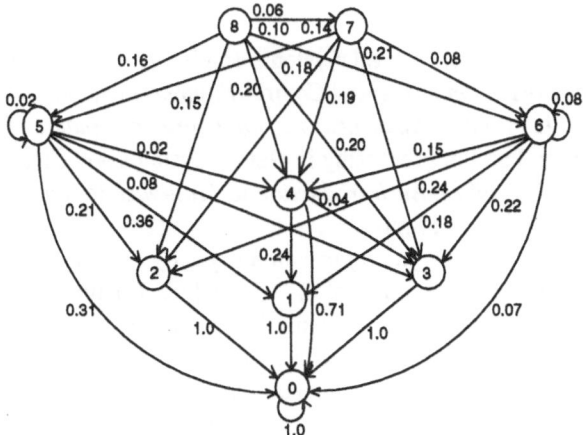

Figure 3. Estimated probabilistic state machine for a network with 8 nodes when storing 8 orthogonal patterns. Each number inside the graph's nodes represent a radius in relation to the target attractor ξ_i.

424

Figure 4 compares the theoretical prediction with simulations of the retrieval time for a network with 16 nodes when one known external pattern is present at the inputs.

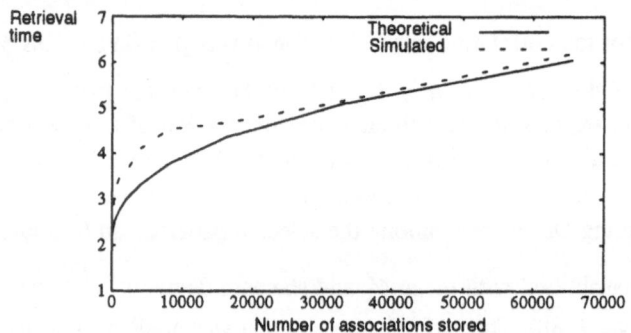

Figure 4. Retrieval time as a function of the number of associations stored.

4 Conclusions

This work was accomplished in two steps: the first one was to model the distribution of the space as a Bivariate Normal Distribution and the second one was to apply the results of such model to estimate the probability of a node firing with hesitation. The first step represent an important contribution to the previous work of Kanerva on what was called "the distribution of the third side of the triangle". It was demonstrated formally that the hesitation of the network increases with the radius of the current internal state in relation to the target attractor and that the retrieval time tend to increase as a logarithmic function of the number of associations stored. The results of both steps fit well with computer simulations.

References

Aleksander, I. (1990a). Neural Systems Engineering : towards a unified design discipline? *IEE Computing and Control Engineering Journal,* 1,259-265.

Aleksander, I. (1990b). *Ideal neurons for neural computers.* Proceedings of the Int. Conf. on Parallel Processing in Neural Systems and Computer.

Anderson, T.W. (1958). *An Introduction to Multivariate Statistical Analysis.* New York: Jonh Wiley and Sons.

Braga, A. P. (1993a). *On the information capacity of auto-associative RAM-based neural networks.* Proceedings of the ICANN-93, London:Springer Verlag.

Braga, A.P. (1993b). Predicting contradictions in the storage process of recurrent Boolean Neural Networks, *IEE Electronis Letters* (Accepted for publication).

Kanerva, P. (1988). *Sparse Distributed Memory.* Cambridge, Mass.: MIT Press.

Wong, K.Y.M. and Sherrington, D. (1992) Storage properties of randomly connected Boolean neural networks for associative memory. *Europhysics Letters,* 7(3):197-202.

ON A CLASS OF HOPFIELD TYPE NEURAL
NETWORKS FOR ASSOCIATIVE MEMORY

Bartlomiej Beliczynski

Warsaw University of Technology, Institute of Control and Industrial Electronics,
ul. Koszykowa 75, 00-662 Warszawa, Poland

Abstract: A class of Hopfield type neural network is investigated and a
new algorithm for transition (weighting) matrix design ensuring almost
zero diagonal is presented.

1. Introduction

In [1] and [2] Hopfield presented models for neural networks that seek
local minima of a certain energy function. Later the Hopfield original idea was
modified in many instances, however its main structural feature being one layer
recurrent network still remains. We will be referring to it as a Hopfield type
network. Its equation in a discrete-time form can be written as follows

$$v(k+1) = f(Tv(k)+s(k)), \qquad k = 0,1,... \qquad (1)$$

where k denotes instances of time, v and s are n-element signal vectors, T is a $n \times n$
transition (named also connection or weighting) matrix and f is a nondecreasing
nonlinear function. In fact, the vector s varying in time appeared much later [3]
and it was a natural generalization of the Hopfield idea.

When (1) is used as an associative memory model, two assumptions are
usually made: the s vector is not a function of k, i.e. $s(k)=s$, the function f is
bounded and reaches its bounds for finite arguments. Points of the minima of
energy (Lyapunov) functions are chosen by the designer usually as binary vectors.
These vectors are then used to determine the matrix T and vector s. There are two
types of function f in use: a sign function and a saturated linear function shown in
Fig.1. The first was extensively studied in [4] and recently [5], the second one
appeared in number of publications for ex. [6], [7].When linear with saturation
function is used one can avoid limit cycles, but the convergence is much slower

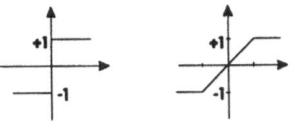

than when the sign function is applied. If f is a sign,
and if additionally the network transition matrix is
symmetric, (1) reaches either an equilibrium point or
a periodic solution (named also limit cycle) with

Fig.1   Commonly   used   period two [4]. If the transition matrix is symmetric
nonlinearities for associative   and positive semidefinite then the network has no
memory   applications.   limit cycles [4]. By introducing negative eigenvalues
to the transition matrix, one can eliminate some spurious (unwanted) equilibria,
but limit cycles will probably appear [8]. They can be avoided by monitoring and
executing serial mode when limit cycle is about to take place [8].

Well-designed associative memory should fulfil the following conditions:
- All desired memory vectors should actually be the network equilibria,
- No assumption about memory vectors should be made ,
- Limit cycles must be eliminated,

- Number of spurious equilibria should be kept as small as it is possible.

In this paper a class of Hopfield type networks with sign nonlinearity will be considered and a new algorithm for transition matrix design will be presented. The paper is organized as follows. In section 1 the network model is derived and discussed. In section 2 the new algorithm is presented and in section 3 conclusions are drawn.

2. The network model

Let us consider a discrete-time binary model as the following

$$v(k+1) = sgn(Tv(k)+s), \quad k = 1,..,m \tag{1}$$

where $v = [v_1,...,v_n]^T$, $v_i \in [-1,+1]$, $i = 1,...,n$, T is nxn symmetric matrix, s is a constant vector, and sgn is a function as follows

$$sgn(x) = \begin{cases} -1 \text{ when } x < 0 \\ +1 \text{ when } x \geq 0 \end{cases} \tag{2}$$

Let $v(0)$ indicates an initial known vector to be corrected. Let also Q be a matrix of desired memory vectors, i.e.

$$Q = [q^1, q^2,...,q^m], \tag{3}$$

where q^i, $i = 1,...,m$, $m<n$ is a n-element vector with -1 or +1 entries. All memory vectors must be equilibria of (1), i.e.

$$sgn(Tq^i + s) = q^i, \quad i = 1,...,m \tag{4}$$

Formula (4) can also be written as follows

$$Tq^i + s = \alpha q^i, \quad i = 1,...,m \tag{5}$$

where α is a nxn matrix such that

$$(\alpha q^i) * q^i \geq 0, \quad i = 1,..,m \tag{6}$$

and "*" denotes a tensor's product (i.e. if $a*b=c$ then $a(i)b(i)=c(i)$). Now again (5) can be written in a more suitable form

$$s = \alpha q^k - Tq^k$$
$$T(q^i - q^k) = \alpha(q^i - q^k), \quad i = 1,..,m, i \neq k \tag{7}$$

The (7) is equivalent to (5) with respect to arbitrarily chosen vector q^k. As pointed out in [7], where a linear with saturation element was used, the network features do not depend on q^k selection. We will be applying k=1.

There are two obvious solutions to (6): α is a constant positive scalar or α is a diagonal matrix with positive entries. Despite many possible solutions to (6), in the remaining part of this paper only one, i.e. $\alpha=1$ will be taken. Finally (7) can be rewritten as follows

$$s = q^1 - Tq^1 \tag{8}$$
$$T(q^i - q^1) = (q^i - q^1), \quad i = 2,..,m \tag{9}$$

Simply by inspection of (9) one can notice that T is a matrix having $m-1$ eigenvalues equal to 1 and the rest of them are not important, as far as equilibria of (1) are concerned.

If one defines
$$Q_r = [q^2 - q^1, q^3 - q^1, ..., q^m - q^1]. \tag{10}$$
and applies SVD factorization to it
$$Q_r = L\Sigma R^T \tag{11}$$
where $L^T L = I^{nxn}$, $R^T R = I^{(m-1)x(m-1)}$,
then T can be expressed as follows
$$T = L \begin{bmatrix} I^{lxl} & 0 \\ 0 & D \end{bmatrix} L^T \tag{12}$$
where $l \leq m-1$ ($l=m-1$ when all memory vectors are linearly independent), I denotes a unity matrix, D is a diagonal matrix with the diagonal entries
$$(d_{l+1}, d_{l+2}, ..., d_n) \tag{13}$$

Remark 1
      When a network is designed based on (8) and (9), one can freely choose the eigenvalues (13) of the transition matrix T.   ∎

      The L matrix can further be decomposed as follows
$$L = [L_1^{nx(m-1)} \ L_2^{nx(n-m+1)}], \tag{14}$$

Two important practical cases
$$D=0 \ \Rightarrow \ T = L_1 L_1^T \tag{15}$$
$$D=dI, \ d=const \ \Rightarrow \ T = L_1 L_1^T + dL_2 L_2^T = L_1 L_1^T + d(I - L_1 L_1^T) \tag{16}$$
Remark 2

      In both (15) and (16) the network can be represented by the matrix L_1 instead of T. Such representation for m<<n, offers huge saving in terms of number of elementary algebraic operations and number of stored coefficients. Moreover L_1 can be calculated incrementally [9], [10].   ∎

3. The new algorithm

      Since original idea of the network [1] and [2], in many publications there is pronounced need that the diagonal of the transition matrix should be equal to zero. This is not obeyed by the design presented in this paper. However by an appropriate selection of the eigenvalues (13), the diagonal of T can be made as close to zero as it is possible in the least squares sense. Additionally at least nonnegative diagonal is necessary to implement serialization when a limit cycle is detected [8].

      Using (12) and (14) and equating the diagonal to zero, one can write
$$diag(T) = L_1^{*2} h + L_2^{*2} diag(\overline{D}) = 0 \tag{17}$$
where L_1^{*2} denote squaring of each element of L_1 and h is an n-element vector consisting of ones.

      Solution of (17) in the least squares sense is the following
$$diag(\overline{D}) = -(L_2^{*2})^+ L_1^{*2} h \tag{18}$$
where "+" denotes pseudoinverse in Moore-Penrose's sense.
With (18) one can however obtain negative values on the diagonal of T, so one more adjustment is needed.

428

If
$$\rho = \min(I - L_2^{*2}(L_2^{*2})^+)L_1^{*2}h < 0 \qquad (19)$$
then
$$D = (\overline{D} - \rho + \varepsilon)/(1 - \rho + \varepsilon), \quad \varepsilon > 0 \qquad (20)$$
substituted into (12) ensures that every diagonal element of T is not less then ε.

At least **the full algorithm determining** T **and** s can be written:
- Given L_1 and L_2
1. Calculate $\overline{D}$ (18) and ρ (19)
2. If $\rho < 0$ then calculate D (20) else $D = \overline{D}$.

Examples run by the author demonstrate that the algorithm gives fewer spurious equilibria and fewer limit cycles than competitive methods.

4. Conclusions

In this paper a class of Hopfield type neural network was investigated and a new algorithm of transition (weighting) matrix design ensuring almost zero diagonal was presented.

References
[1] J.J. Hopfield, "Neural networks and physical systems with emergent collective computational abilities", in Proc.Natl. Acad. Sci. USA, vol.79, pp.2554-2558, 1982.
[2] -- "Neurons with graded response have collective computational properties like those of two-state neurons", in Proc. Natl. Acad. Sci. USA, vol.81, pp.3088-3092, 1984.
[3] F.J. Pineda, "Dynamics and architecture for neural computation", Journal of Complexity, 4, pp.216-245, 1988.
[4] E. Goles, F. Fogelman and D. Pellegrin, "Decreasing energy functions as a tool for studying treshold networks", Discrete Appl. Math., vol.12, pp.261-277, 1985.
[5] Y.Shrivastava, S.Dasgupta and S.M.Reddy, "Guaranteed Convergence in a Class of Hopfield Networks", IEEE Trans. on Neural Networks, vol.3, No.6, November 1992.
[6] .Y. Yen , A.Michel, " A Learning and Forgetting Algorithm in Associative Memories: Results Involving Pseudo-Inverses", IEEE Trans. on Circuits and Systems, vol.38, No.10, pp.1193-1205, 1991.
[7] A.N. Michel, J. Si and G.Yen, "Analysis of a Class of Discrete-TimeNeural Networks Described on Hypercubes", IEEE Trans. Neural Net., vol. 2, No. 1, pp. 32-46, January 1991.
[8] A.Lukianiuk and B.Beliczynski: "An Algorithm for Elimination of Limit Cycles in a Hopfield-type Neural Network", World Congress on Neural Networks, Portland, pp.II 323-326, July, 1993.
[9] B. Beliczynski and A. Lukianiuk, "Fast Algorithm of Learning and Recall in Recursive Neural Network for Associative Memory", Second International Conference on Automation, Robotics and Computer Vision, September, 1992.
[10] B.Beliczynski and A.Lukianiuk: "Fast Algorithms for a Class of Hopfield-type Neural Network with Application to Associative Memory", World Congress on Neural Networks, Portland, pp.II 244-248, July 1993.

Storage Capacity of Associative Random Neural Networks

C. Hubert

UFR Mathématiques-Informatique
Université René Descartes-ParisV
Paris, France

1 Introduction

In order to compare the capabilities of the Random Neural Network (RNN), a new neural network model introduced by Gelenbe (1989), with those of other models in associative memory applications, we examine here its storage capacity. The theoretical capacity is first derived for the single-layer fully-connected RNN and then for the three-layer N-M-N architecture. Experiences have been performed on single-layer networks of different size using RPROP learning algorithm (Riedmiller, 1992) to evaluate the corresponding critical storage capacity.

2 The basic Random Neural Network model

The Random Neural Network (Gelenbe, 1989) consists of a set of N neurons which exchange positive (excitatory) and negative (inhibitory) signals. Referring to queuing theory, each neuron is modelled by a server and a queue which receives signals from outside the network or from other neurons. Exogenous signals arrive at neuron i, following Poisson process of rate $\Lambda(i)$ if they are positive and $\lambda(i)$ if they are negative. In each queue, one negative signal cancels one positive signal. If the number of positive signals in its queue is strictly positive, neuron i is excited and emits signals which depart from the network with probability $d(i)$ or head for neuron j as positive signals with probability $p^+(i,j)$ or as negative signals with probability $p^-(i,j)$. Intervals between signal emissions by neuron i follow an exponential distribution with mean $1/r(i)$.

The state of neuron i is given by its steady-state excitation probability:

$$q_i = \frac{\lambda^+(i)}{r(i) + \lambda^-(i)} \tag{1}$$

$$\lambda^+(i) = \sum_{j=1}^{N} q_j w^+(j,i) + \Lambda(i) \qquad \lambda^-(i) = \sum_{j=1}^{N} q_j w^-(j,i) + \lambda(i)$$

$$r(i) = \sum_{j=1}^{N} [w^+(i,j) + w^-(i,j)]/(1 - d(i))$$

with the synaptic weights $w^+(i,j) = r(i)p^+(i,j)$ and $w^-(i,j) = r(i)p^-(i,j)$.

3 Theoretical storage capacity

We set out to store $S = \{X_1, \ldots, X_P\}$ composed of P vectors $X_p = (x_{1p}, \ldots, x_{Np})$ with nb grey levels, by autoassociative RNN networks.

3.1 Single-layer fully-connected architecture

We first consider a single-layer RNN of N fully interconnected neurons (Hubert, 1993a). Each vector component x_{ip} is associated to neuron i and converted into arrival rates of positive and negative external signals, $\Lambda_p(i)$ and $\lambda_p(i)$. The network is recurrent and the probability of departure is $d(i) = 0$. The excitation probability of neuron i is set to $q_{ip} = x_{ip}/(nb - 1)$.

The autoassociation problem is solved if the weight matrices $W^+ = [w^+(i, j)]$ and $W^- = [w^-(i, j)]$ are adjusted such that, for $(i, p) = (1, 1), \ldots, (N, P)$.

$$q_{ip} = \frac{\sum_{j=1}^{N} q_{jp} w^+(j, i) + \Lambda_p(i)}{\sum_{j=1}^{N} q_{jp} w^-(j, i) + \lambda_p(i) + \sum_{j=1}^{N} [w^+(i, j) + w^-(i, j)]} \quad (2)$$

Equation (2) can be reformulated under the linear form :

$$-q_{1p} w^+(1, i) - \ldots - q_{Np} w^+(N, i) \quad + q_{ip} w^+(i, 1) + \ldots + q_{ip} w^+(i, N)$$
$$+ q_{ip} q_{1p} w^-(1, i) + \ldots + q_{ip} q_{Np} w^-(N, i) \quad + q_{ip} w^-(i, 1) + \ldots + q_{ip} w^-(i, N)$$
$$= \Lambda_p(i) - q_{ip} \lambda_p(i) \quad (3)$$

If weights are not symmetric, we obtain the vectorial form:

$$\boldsymbol{A}_{pi}\, \boldsymbol{w}_i = \boldsymbol{b}_{pi} \qquad \text{for } (p, i) = (1, 1), \ldots, (P, N) \quad (4)$$

where $\boldsymbol{A}_{pi}$ is the i-th row of A_p: $(N \times 4N)$ matrix, $\boldsymbol{w}_i$ is the i-th colomn of W: $(4N \times N)$ matrix and $\boldsymbol{b}_{pi}$ is the i-th element of b_p: N-vector. System (4) is a linear system of PN equations with $2N^2$ unknowns. We thus deduce:

Theorem 1: If the weight matrices W^+ and W^- are not symmetric, the single-layer Random Neural Network of N fully interconnected neurons is guaranteed to store any set of $P = 2N$ vectors.

If weights are symmetric, we obtain another linear system of PN equations with N^2 unknowns. Thus, capacity is reduced:

Theorem 2: If the weight matrices W^+ and W^- are symmetric, the single-layer Random Neural Network of N fully interconnected neurons is guaranteed to store any set of $P = N$ vectors.

3.2 Three-layer N-M-N architecture

We consider now a feedforward three-layer RNN (Hubert, 1993b) with N neurons in the input and output layers, and M neurons in the hidden layer. The input layer is entirely connected to the hidden one with the weights $w_e^+(i, j)$ and $w_e^-(i, j)$. Similarly, the hidden layer is entirely connected to the output one with the weights $w_s^+(j, k)$ and $w_s^-(j, k)$. The probability of departure are respectively for the three layers: $d^{(1)}(i) = 0$, $d^{(2)}(j) = 0$ and $d^{(3)}(k) = 1$.

Each vector component x_{ip} is associated to the input neuron i whose excitation probability is set to $q_{ip}^{(1)} = x_{ip}/(nb-1)$. The excitation probability of the hidden neuron j is

$$q_{jp}^{(2)} = \frac{\sum_{i=1}^{N} q_{ip}^{(1)} w_e^+(i,j) + \Lambda_p^{(2)}(j)}{\sum_{i=1}^{N} q_{ip}^{(1)} w_e^-(i,j) + \lambda_p^{(2)}(j) + \sum_{k=1}^{N}[w_s^+(j,k) + w_s^-(j,k)]} \qquad (5)$$

where $\Lambda_p^{(2)}(j)$ and $\lambda_p^{(2)}(j)$ are the arrival rates of positive and negative exogenous signals. Finally, the excitation probability of output neuron k is:

$$q_{kp}^{(3)} = \frac{\sum_{j=1}^{M} q_{jp}^{(2)} w_s^+(j,k) + \Lambda_p^{(3)}(k)}{\sum_{i=1}^{N} q_{jp}^{(2)} w_s^-(j,k) + \lambda_p^{(3)}(k) + r^{(3)}(k)} \qquad (6)$$

where $\Lambda_p^{(3)}(k)$ and $\lambda_p^{(3)}(k)$ are the arrival rates of positive and negative exogenous signals and $r^{(3)}(k)$, the firing rate with appropriate value.

The association problem is to adjust weights such that $q_{kp}^{(3)} = q_{kp}^{(1)}$ for $(k,p) = (1,1), \ldots, (N,P)$. From (5) and (6), we obtain two linear systems:

$$\text{Equation (5)} \rightarrow C_{pj} w_j' = e_{pj} \qquad \text{for } (p,j) = (1,1), \ldots, (P,M) \qquad (7)$$

where C_{pj} is the j-th row of C_p: $(M \times 4N)$ matrix, w_j' is the j-th colomn of W': $(4N \times M)$ matrix and e_{pj} is the j-th element of e_p: M-vector.

$$\text{Equation (6)} \rightarrow F_{pk} w_k'' = g_{pk} \qquad \text{for } (p,k) = (1,1), \ldots, (P,N) \qquad (8)$$

where F_{pk} is the k-th row of F_p: $(N \times 2M)$ matrix, w_k'' is the k-th colomn of W'': $(2M \times N)$ matrix and g_{pk} is the k-th element of g_p: N-vector.

We have $P(M + N)$ equations with $4MN$ unknown values of weights and PM unknown values of excitation probability of hidden neurons $q_{jp}^{(2)}$.

Theorem 3: The three-layer N-M-N Random Neural Network is guaranteed to store any set of $P = 4M$ vectors.

4 Learning with RPROP algorithm

To solve our autoassociation problems, we used the learning algorithm RPROP (Riedmiller, 1992) easily applicable to RNN model (Hubert, 1993). Here, we present experimental results obtained with single-layer fully-connected networks and binary patterns ($x_{ip} \in \{0, 1\}$).

Let $m = \langle \frac{1}{N} \sum_{i=1}^{N}(2x_{ip} - 1)(2s_{ip} - 1) \rangle$ with $s_{ip} = 1$ if $q_{ip} > 0.5$ and $s_{ip} = 0$ else, be the mean overlap over all patterns and experiments. On Fig.1, m is plotted against the capacity $\frac{P}{N}$, for $N = 16$ and $N = 36$. The critical storage capacity is around 0.75.

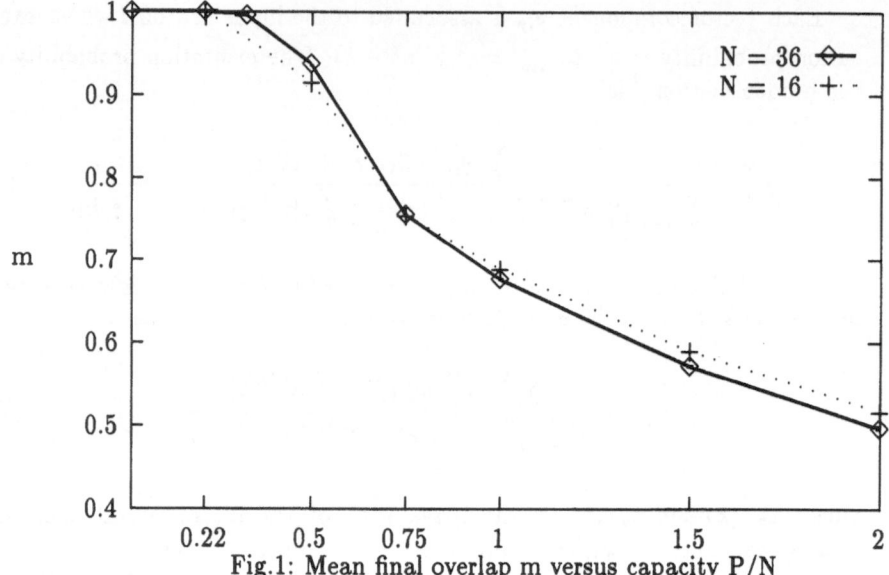

Fig.1: Mean final overlap m versus capacity P/N

5   Conclusion

The theoretical study shows that the single-layer Random Neural Network with N fully connected neurons can store at least $2N$ patterns (N patterns with symmetric weights) and that the N-M-N network can store at least $4M$ patterns. Concerning the single-layer RNN, the experimental critical storage capacity is about 0.75. Further experiences - e.g. on the size of attractors - will help us to compare this model with Hopfield model.

6   References

Atalay, V. and Gelenbe, E. (1993), "Stockage et reconstruction de textures d'images par apprentissage", Comptes Rendus Académie des Sciences, Paris.

Atiya, A. and Abu-Mostafa, Y. S. (1993), "An analog feedback associative memory", IEEE Trans. Neural Networks, 4(1), 117-126.

Forrest, B. M. (1988), "Content-addressability and learning in neural networks", J. Phys. A: Math. Gen., 21, 245-255.

Gelenbe, E. (1989), "Random Neural Networks with negative and positive signals and product form solution", Neural Computation, 1(4), 502-510.

Gelenbe, E. (1993), "Learning in the recurrent Random Neural Network model", Neural Computation, 5(1), 154-164.

Hubert, C. (1993), "Pattern completion with the Random Neural Network using the RPROP learning algorithm", Proc. IEEE Conf. Systems, Man and Cyb., France, 2, 613-617.

Hubert, C. (1993), "Learning internal representations with the N-M-N Random Neural Network", Proc. Int. Conf. Neuro-Nimes93, Nimes, France, 45-54.

Stiefvater, T. and Müller, K. R. (1992), "A finite size scaling investigations for Q-state Hopfield models", J. Phys. A: Math. Gen., 25, 5919-5931.

A generalized bidirectional associative memory with a hidden orthogonal layer

F. Ibarra-Picó, J.M. García-Chamizo
Department of Técnología Informática y computación
University of Alicante
Apdo. 99. Alicante. Spain
email : ibarra@s04.dtic.ua.es ; juanma@s04.dtic.ua.es

1 Introduction

Hopfield (Hopfield,1984a and 1984b) introduced a first model of one-layer autoassociative memory. The Bidirectional Associative Memory (BAM) was proposed by Kosko (Kosko,1988a) and generalizes the model to be bidirectional and heteroassociative. The BAMs have storage capacity limitations (Wang,1990a).

It has been proposed several improvements (Adaptive Bidirectional Associative Memories (Kosko,1988b), multiple training (Wang,1990a) y (Wang,1990b), guaranteed recall (Wang,1991) and a lot more besides). One-step models without iteratión has been developed too (Orthonormalized Associative Memories (MAON), (Garcia,1992) and the Hao´s associative memory (Hao,1992) which uses a hidden layer). In this paper, we propose a new model of associative memory which can be used either bidirectionally or in one-step mode. This model uses a hidden layer, proper filters and orthogonality to increase the storage capacity and reduce the noise effect arising from linear dependences between patterns. Our model, that we call Bidirectional Associative OrthoGonal Memory (BAOGM) , goes beyond the BAM capacity. BAM and MAON models are particular cases of it.

2 Topology and learning process

Let a set of q patterns pairs (a_i, b_i) belonging to the vectorial spaces R^n and R^m. We will build two learning matrices in the following way :

$$A = \left[a_{ij} \right] \text{ and } B = \left[b_{ik} \right] \quad \text{for } i \in \{1,..,q\} \ j \in \{1,..,n\} \ k \in \{1,..,m\}$$

The BAOGM is built as a neural network with two sinaptic matrices (conatining hebbian correlations) W and V, which are computed as $W = AQ^t$ y $V = QB^t$. Where Q is an intermediate orthogonal matrix (Walsh, Householder, and so on) of dimensions qxq. The q_i vectors of Q are an orthogonal base of the vectorial space R^q. This characteristic of the q_i vectors is very important to make accurate associations including noise patterns (Pao,1989). The BAOGM topology and its feedback mode with the transpose W^t and V^t matrices is shown in Figure 1. These transposed matrices be able its bidirectional iteration in the recalling phase.

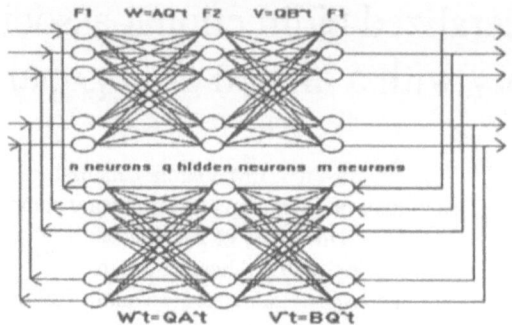

Figura 1. BAOGM network

3 Recalling phase and basic filters

The associations between patterns could be done in only one-step (forward or backward) or in bidirectional mode. One-step recall :

• Let a_i the input pattern, the output b_i is $b_i = f_1\left[f_2\left(a_i^t \cdot W\right) \cdot V\right] = F\left(a_i\right)$

• Let b_i the input pattern, the output a_i is $a_i = f_1\left[f_2\left(b_i^t \cdot V^t\right) \cdot W^t\right] = F^{-1}\left(b_i\right)$

In the bidirectional mode : the patterns are fed forward and backward (feedback) into the BAOGM in a similar BAM style while the energy is falling in a minimun of its energy surface. The process continues until a maximum number of iterations or a convergency desired grade (ε) is reached.

$$a_i^t \rightarrow F\left(a_i^t\right) \rightarrow b_i^{t1}$$

$$b_i^{t1} \rightarrow F^{-1}\left(b_i^{t1}\right) \rightarrow a_i^{t1}$$

Convergency condition $k < K$

$$\cdots\cdots\cdots$$

or $\left|1 - \dfrac{b_i^{tk-1} \cdot b_i^{tk}}{\left|b_i^{tk-1}\right| \cdot \left|b_i^{tk}\right|}\right| < \varepsilon$

$$a_i^{tk-1} \rightarrow F\left(a_i^{tk-1}\right) \rightarrow b_i^{tk} \approx b_i^t$$

In the input and output layer, the network uses the classical bipolar filter f_1 (patterns are coded in bipolar mode). In the hidden layer, the BAOGM computes the filter f_2, (where q^1 and q^2 are the two possible values of the Q components).

$$f_2(x) = \begin{cases} q^1 & x \geq 0 \\ q^2 & x < 0 \end{cases}$$

4 Stability

We propose the following Liapunov function in order to demonstrate the convergence of the system :

$$E\left(a, q, b\right) = -\left(a^t \cdot W \cdot q + q^t \cdot V \cdot b\right)$$

It can be proved that in the bidirectional recalling process the energy function will be falling into a minimum of its energy surface and that this energy surface is enclosed by an absolute minimum (Ibarra-Chamizo,1993). Furthermore, each pair (a_i,b_i) is stored in a minimum ($E(a_i,q_i,b_i) = -(n+m)$) during the learning process.

5 Particular models : BAM and MAON

If, in the BAOGM memory, the intermediate filter function is $f_2(x)=x$ and as we also know that $Q^t=Q^{-1}$ because Q is an orthogonal matrix then the BAOGM model is reduced to the BAM. Moreover, if $Q=I$ (identity matrix) and that the $f_2(x)$ filter is searching for the winner neuron in the hidden layer then we have the MAON model (Ibarra-Chamizo,1993).

6 Extended filters

The BAOGM response could be optimized by extended filters in the intermediate layer. It can be done by introducing threshold values and lateral synapsis. We put two thresholds T_1(fordward) and T_2 (backward) which set up firing conditions (these thresholds filter some linear dependences between patterns). Also, the lateral synapsis S_1 and S_2 between hidden layer neurons, which are shown in Figure 2 where each neuron is inhibited by its two nearest neighbours, increases the filter power.

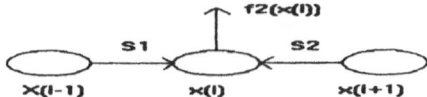

Figura 2. Lateral synapsis

The threshold values can be computed as the winner neuron value when a_i or b_i pattern is present :

$$T_1 = \frac{q-2}{q} - \frac{2}{q} max \left\{ \sum_{\substack{j=1 \\ j \neq i}}^{q} \cos(a_i^t . a_j) \right\}_{i=1}^{q} \qquad T_2 = \frac{q-2}{q} - \frac{2}{q} max \left\{ \sum_{\substack{j=1 \\ j \neq i}}^{q} \cos(b_i^t . b_j) \right\}_{i=1}^{q}$$

The inhibition synapsis values (S_1 and S_2) of a x_i neuron in the hidden layer are calculated as $s_1=k(x_i/x_{i-1})$ and $s_2=k(x_i/x_{i+1})$; where k is a constant that evaluates the tolerance to the noise. So, the proposed extended filter is:

$$f_2(x_i) = \begin{cases} q^1 & (x_i \geq T) \text{ y } (x_i \geq s_1 \cdot x_{i-1 \bmod q}) \text{ y } (x_i \geq s_2 \cdot x_{i+1 \bmod q}) \\ q^2 & \text{other case} \end{cases}$$

This is not the only interesting filter. We are proving and looking for other high-degree filters that could increase the capacity and, in the other hand we are introducing aproximate reasoning in the network by using fuzzy logic.

7 Computer simulations

The BAOGM net was training with 10 pairs of patterns. The imput patterns were the first ten digits (size 4 x 7) and the output was the first ten letters of the alphabet (size 5 x 7). See Figure 3. The synaptic weights are built by using a Householder matrix Q of 10x10, the thresholds and the tolerance constants are $T_1=0.07$ ($a_i \rightarrow b_i$), $T_2=0.06$ ($b_i \rightarrow a_i$) and k=0.5.

Figure 3. Learning pairs

In one-step (forward direction) the network goes to its minimum of energy for all of the learning patterns E=-(28+35)=-63. If a noise pattern is present, the network starts in level of energy that is not a minimum and, in general, after several iterations (typically one or two) the network goes converges to the minimum energy. For instance, the pattern 5 was presented to the network with several noise levels (0, 5, 10, 15, 20, 25 and 30%). In Figure 4, is shown these patterns, the initial and final energy level, and the number of iterations in each of them.

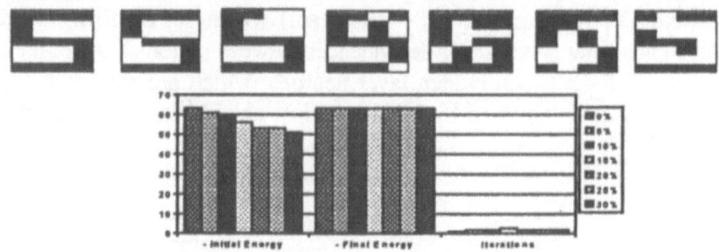

Figure 4. Pattern 5 and energy levels in the (5,E) pair recalling

8 References

Garcia-Chamizo J.M., Crespo-Llorente A. (1992) "Orthonormalized Associative Memories". Proceeding of the IJCNN, Baltimore, vol 1, pg. 476-481.

Hao J., Wanderwalle J. (1992) "A new model of neural associative memoriy" Proceedings of the JJCNN92, vol 2, pg. 166-171.

Hopfield J.J. (1984a) "Neural Networks and physical systems with emergent collective computational abilities". Proceedings of the National Academy of Science, vol 79, pg. 2554-2558.

Hopfield J.J. (1984b) "Neural networks with graded response have collective Computational properties like those of two-state Neurons". Proceedings of the National Academy of Science, vol 81, pg. 3088-3092.

Ibarra-Picó F., Garcia-Chamizo J.M. (1993) "Bidirectional Associative Orthonormalized Memories". Actas AEPIA, vol 1, pg 20-30.

Kosko, B. (1988a) "Bidirectional Associative Memories". IEEE Tans. on Systems, Man & Cybernetics, vol 18.

Kosko, B. (1988b) "Competitive adaptative bidirectional associative memories".Procedings of the IEEE first International Conference on Neural Networks, eds M. Cardill and C. Butter vol 2. pp 759-766.

Pao You-Han. (1989) "Adaptative Pattern Recognition and Neural Networks". Addison-Wesley Publishing Company, Inc. pg 144-148.

Wang , Cruz F.J., Mulligan (1990a) "On Multiple Training for Bidirectional Associative Memory ". IEEE Tans. on Neural Networks, 1(5) pg 275-276.

Wang , Cruz F.J., Mulligan. (1990b) "Two Coding Strategies for Bidirectional Associative Memory ", IEEE Tans. on Neural Networks, pg 81-92.ang , Cruz F.J.,

Finding Correspondences between Smoothly Deformable Contours by Means of an Elastic Neural Network

F. Labonté, P. Cohen

Perception and Robotics Laboratory, Ecole Polytechnique de Montréal

Montréal, Canada

1 Introduction

One important problem related to planar shapes involves the establishment of point correspondences between two successive views of a smoothly deformable contour. In this paper, we propose to use an elastic neural network, which was proposed as a heuristics to find a good solution to the traveling salesman problem (Durbin and Willshaw, 1987), to establish these correspondences. We propose to use the elastic network to deform the first view of an object until it corresponds to the second one. For computational efficiency only, inflection points and curvature extrema are used to establish the correspondences. Our method not only permits to establish the desired correspondences, but also to order the points so that joining them by straight lines creates a good approximation of the original contour. The details of the method are given in section 2 and experimental results are presented in section 3. A short discussion concludes the paper.

2 Proposed Method

In order to establish correspondences between the points on the contour of two successive views of a deformable object, the first view is reduced and then gradually deformed to correspond to the second one. From now on, the contour of the first view will be referred to as the *deformable contour* and the one of the second view as the *target contour*. Let $N = \{\mathbf{x}_1, \ldots, \mathbf{x}_n\}$ denote the position of the n points on the target contour and $M_t = \{\mathbf{y}_{1,t}, \ldots, \mathbf{y}_{m,t}\}$ the position of the m points on the deformable contour $(m > n)$ at iteration t. The displacement, at each iteration t, of each point $\mathbf{y}_{j,t}$, is calculated according to:

$$\Delta \mathbf{y}_{j,t} = \alpha \sum_{i \in N} \omega_{ij,t}(\mathbf{x}_i - \mathbf{y}_{j,t}) + \beta K(\mathbf{y}_{j-1,t} - 2\mathbf{y}_{j,t} + \mathbf{y}_{j+1,t}), \qquad (1)$$

where $\mathbf{y}_{j-1,t}$ and $\mathbf{y}_{j+1,t}$ are previous and next neighbours of point $\mathbf{y}_{j,t}$ on the deformable contour. K is a parameter whose value decreases with the number of iterations, and α and β are parameters controlling the relative strength of the two forces acting to displace the $\mathbf{y}_{j,t}$'s, namely the attracting force of the $\mathbf{x}_i$'s of the target contour, and the attracting force of the neighbouring points $\mathbf{y}_{j-1,t}$ and $\mathbf{y}_{j+1,t}$ of the deformable contour. $\omega_{ij,t}$ represents the normalized influence of $\mathbf{x}_i$ on $\mathbf{y}_{j,t}$ at iteration t and is calculated according to:

$$\omega_{ij,t} = \frac{\phi(|\mathbf{x}_i - \mathbf{y}_{j,t}|, K)}{\sum_{k \in M} \phi(|\mathbf{x}_i - \mathbf{y}_{k,t}|, K)}, \quad \phi(d, K) = \exp(-d^2/2K^2). \quad (2)$$

A decreasing sequence of values for the parameter K is used. For each value of K, a fixed number of iterations is accomplished. At every iteration t, the $\mathbf{y}_{j,t}$'s are displaced according to equation 1. The algorithm terminates when, for each $\mathbf{x}_i$, at least one $\mathbf{y}_{j,t}$ is close enough according to a pre-specified tolerance criterion ϵ. References to articles on the behaviour and on improvements to the elastic network can be found in (Boeres and de Carvalho, 1992). Our contribution in this paper consists first in demonstrating that the elastic network can be used in a different context than the one for which it was proposed, and second in using an arbitrary closed shape as a deformable contour, instead of a circle[1]. With our method, for the first image of the sequence, since no initial estimate of the contour is available, a circle is used as the deformable contour, as in (Durbin and Willshaw, 1987). Starting from the second image, the deformable contour obtained in the previous image when the network converges is used as the deformable contour in the current image.

One requirement for the network to converge is that $m > n$, in order to ensure that every point of the target contour finds a match in the deformable contour. Let I^1 and I^2 be two successive images of the contour of a deformable object which possess respectively n^1 and n^2 inflection points or curvature extrema. n^1 is not necessarily equal to n^2 but is close to it because of the assumption of smooth deformations. The n^2 points of I^2 constitute the target contour. The deformable contour should be a reduced-size version (by a scale factor) of the n^1 points of I^1 plus some other points, in order to make m^2, the number of points on the deformable contour, greater than n^2. In practice, a reduced version of the final state of the deformable contour in I^1 is used.

The proposed method produces a correspondence from m^2 points on the deformable contour to n^2 points on the target contour ($m^2 > n^2$). Some of the points on the deformable contour might correspond to points at intermediate positions between the points on the target contour, and some of them might even correspond to the same point on the deformable contour. One can only be certain that at least n^2 of the m^2 points on the deformable contour correspond to the n^2 points on the target contour. Even if one-to-one correspondences do not hold all the time, the established correspondences are however neighbour-preserving. Experimental results in the next section show that reasonable correspondences are obtained. The proposed method also permits to order inflection points and curvature extrema so that joining them by straight lines provides a good approximation of the original shape.

3 Experimental Results

The proposed method has been tested on a sequence of real images of a bending arm. Figure 1.a shows the first image of the sequence and figure 1.b illustrates the results obtained after applying a Difference Recursive Filter for contour

---

[1]We must acknowledge that Durbin and Willshaw (1987) mention that "The algorithm therefore provides a general method for matching a set of arbitrarily connected points to a second set of points located in a geometrical space of any dimensionality". However, they do not elaborate further than that.

detection to Figure 1.a. Undesired edges inside the contour of the arm shown in Figure 1.b have been manually removed to give one of the contours in Figure 1.c. This figure shows the superposition of the 3 contours used to establish the correspondences. Inflection points and curvature extrema were selected manually, and are represented as dots on the target contours in Figures 1.d, 1.e and 1.f. For each of the 3 contours, 5 points were added manually at positions on the right of the contour to indicate that a junction between the top right point and the bottom right point is desired in order to close the contour. The network parameters that were used for the experimentation were: $\alpha = 0.8$, $\beta = 2.0$, $\epsilon = 0.05$, and $K = 0.10$ with a reduction rate of 95% every 2 iterations.

Figures 1.d, 1.e and 1.f respectively illustrate by dotted lines the correspondences established between a small circle and the first image, a reduced version of the first image and the second one, and a reduced version of the second image and the third one. One should observe that the points on the target contour which were originally unordered have been joined by straight lines, giving a good approximation of the original shape. It is important to observe that points on the deformable contour correspond either to points on the target contour or to intermediate values between them. Also, multiple correspondences are sometimes produced. One can see that the correspondences are reasonable.

In Figure 1.f, at the extreme left, on the bottom of the contour, a cycle is created in the deformable contour. This can be explained by the presence of many close points that create strong forces of attraction that make the points on the deformable contour move toward them. In these situations, the network minimizes the distance on a very local basis since the attraction force of distant points is masked by the stronger force created by the closer ones. From other experiments not reported here because of lack of space, it seems that the elastic network is quite sensitive to the distribution of the inflection points and curvature extrema on the contour.

4   Conclusion

We addressed the problem of establishing point correspondences between two successive views of a smoothly deformable contour. We proposed to use the elastic neural network to deform the first view of an object until it corresponded to the second one. Our contribution consisted first in showing that the elastic network can be applied in a different context than the one for which it was proposed, and second in using an arbitrary closed shape as a deformable contour. For computational efficiency, inflection points and curvature extrema were used to establish the correspondences. The resulting correspondences were not necessarily one-to-one, but were neighbour-preserving. Experimental results have shown that the correspondences are reasonable but not necessarily optimal.

References

[1] Boeres, M.C.S., and de Carvalho, L.A.V. (1992). A faster elastic-net algorithm for the traveling salesman problem. *Proc. IJCNN*, II, 215–220.

[2] Durbin, R. and Willshaw, D. (1987). An analogue approach to the travelling salesman problem using an elastic net method. *Science*, 327:689–691.

440

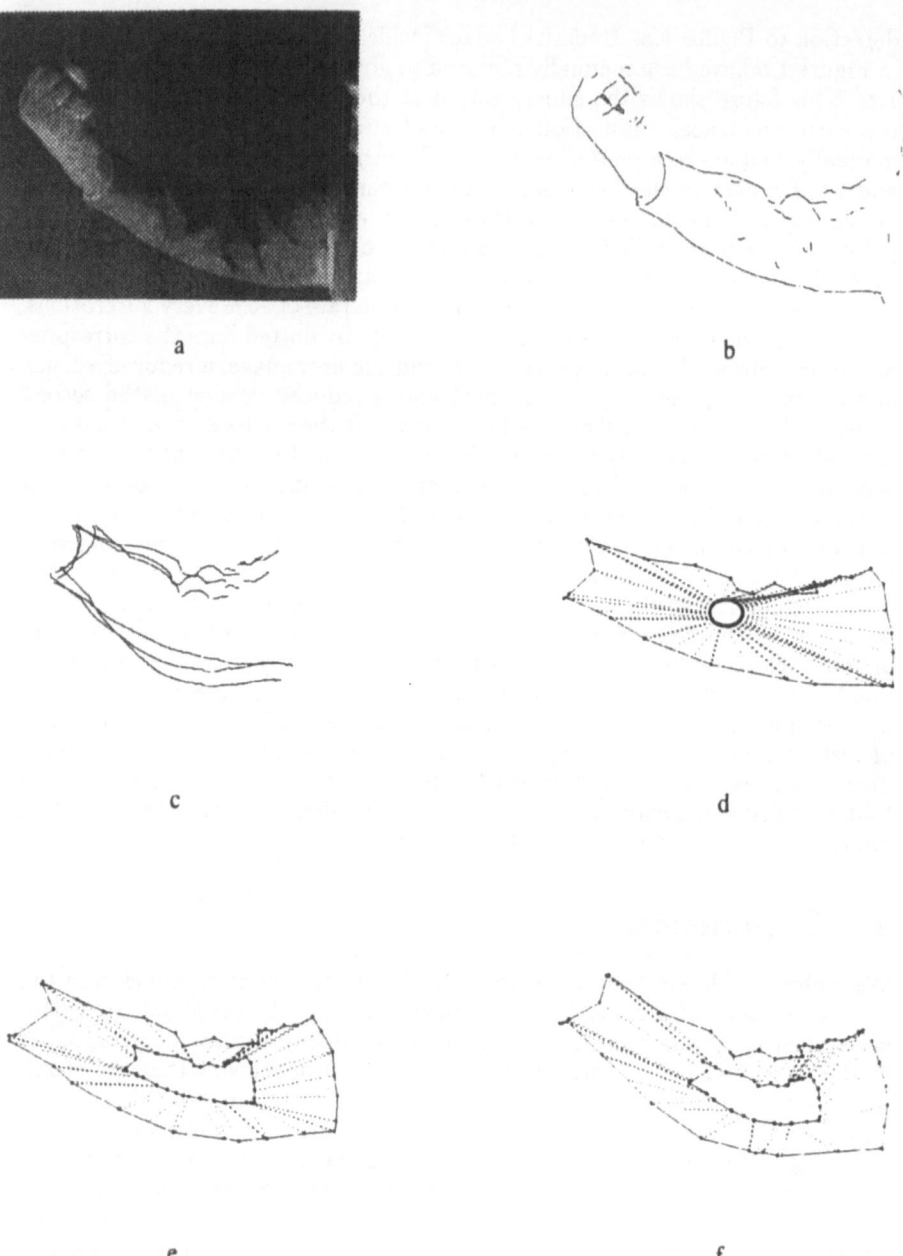

a

b

c

d

e

f

Figure 1: a) first image of the bending arm sequence, b) contour detection on Figure 1.a. c) superposition of the 3 contours of the bending arm sequence, d) correpondences between a circle and the first contour, e) correspondences between the first contour (reduced) and the the second one, f) correspondences between the second contour (reduced) and the the third one.

MINIMIZATION OF NUMBER OF CONNECTIONS
IN FEEDBACK NETWORKS

Valery Tereshko

*Low Temperature Physics Department
and International Laser Center
Moscow State University, Moscow 117234, Russia*

A method allowing to reduce a number of connections in feedback networks from $(n^2 - n)/2$ to n where n is a number of elements is proposed.

Let us consider the following dynamical system:

$$dx_i/dt = f_i(x) = a_i(x_i) - x_i b_i(x),$$

(1)

$$x = x_i, x_i > or = 0, a_i(x_i), b_i(x) > 0(< 0),$$

$$b_i(x) = \sum_{j=1}^{n} f_j(x_j), i = 1, .., n, j not = i.$$

Such system describes the network the state of each element of which is defined by excitment (ingibition) of this element (function $a_i(x_i)$) and negative (positive) feedback acting from side of other elements. We have so-called competitive system if $df_i/dx_j < or = 0$ for any i not = j and cooperative system in opposite case [1].

For definitness we will consider the competitive system which is a special kind of Lotka-Volterra equations:

$$dx_i/dt = x_i(e_i - d_i \sum_{i=1}^{n} x_i), i = 1, .., n$$

(2) System (2) models the dynamics of fully connected network the i-th element (mode, neuron) of which is connected with all others by connections with weight d_i.

In assuming that the ratios e_i/d_i are different for all modes, the stationary state of system (2) is:

$$x_k = e_k/d_k, x_m = 0, m = 1, .., n, m not = k$$

(3) where $e_k/d_k = max e_i/d_i, i = 1, .., n$.

Let us suppose that e_i/d_i is proportional to $< II_i >$, where the right part is a scalar product of the normed functions I and I_i corresponding to the analyzed pattern and prototype, respectively. Then only the mode corresponding to that prototype which is the most similar to the analyzed pattern survives during evolution. Thus the considered system does is able to recognize patterns [2].

To construct really functioning information processing network one should to minimize the number of connections which is $(n^2 - n)/2$ for n each-to-each interacting elements. General theoretical approach to the minimization can be based on thermodynamics of interacting systems [3]. The basic idea of approach consists in introduction of some mean field. All elements of system interact only with this field. That allow to reduce a number of connections up to n when interaction matrix takes the special forms [3].

The example of such minimization is considered below.

Let us suppose that the elements of networks (2) do not interact with each other and do only with some common system called reservoir, the dynamics of which is described by variable r. We will suppose that each x_i growths proportionally to r and decays with constant rate. Then the system dynamics obeys to the following equations:

$$dx_i = x_i(d_i r - 1)$$

(4)

$$r = g(r, x)$$

Let us suppose that system (4) is closed. The consequence from this is the conservation low of the common quantity of matter in the system. So, the integral of motion take place:

$$dr/dt + \sum_{i=1}^{n} dx_i/dt = 0$$

(5) Hence it follows:

$$r + \sum_{i=1}^{n} x_i = r(0) + \sum_{i=1}^{n} x_i(0) = c$$

(6) Taking into consideration equality (6) system (4) reduces to system (2) where $e_i = d_i c - 1$ and equation describing the dynamics of reservoir is:

$$dr/dt = -r \sum_{i=1}^{n} d_i x_i + \sum_{i=1}^{n} x_i$$

(7) Thus systems (2) and (4,7) posses the same dynamics. Essential difference of them is that a number of connections of elements is far less for the second system then for the first one.

1. M. Hirsh / Neural Networks, 1989, 2, 331.

2. A.S. Mikhailov, V.M. Tereshko / Matematicheskoe modelirovanie, 1991, 1, 37.

3. V.M. Tereshko, in preparation.

An Efficient Method of Pattern Storage in the Hopfield Net

S. Coombes and J. G. Taylor
Centre for Neural Networks,
Kings College,
London,
Strand,
WC2R 2LS.

Abstract

We discuss a new a method of endowing the Hopfield net with the properties of an associative memory. A set of N patterns (biased or unbiased) may be stored in a Hopfield network of N spins with a set of connections called inverse-Hebb couplings. Furthermore, an algorithm exists called the quadratic Oja algorithm which can enhance the basin of attraction of a subset of these stored patterns. Simulations show that the combination of the quadratic Oja algorithm with initial conditions given by the inverse-Hebb rule leads to a successful alternative to the traditional Gardner algorithm. Lastly, we introduce the hardware capable of a fast implementation of the inverse-Hebb rule.

Introduction

A Hopfield network [1] may be regarded as a thermodynamic system consisting of N interacting spins s_i connected by symmetric weights $w_{ij} = w_{ji}, i, j = 1, 2, 3 \ldots N$. Memory storage is achieved by making a set of patterns the fixed point attractors of the Hopfield energy function $H[s] = -1/2 \sum_{i \neq j} w_{ij} s_i s_j$.

The original model of Hopfield uses the Hebb rule $w_{ij} = N^{-1} \sum_{\mu=1}^{P} \zeta_i^\mu \zeta_j^\mu \equiv C_{ij}$ to store P patterns in a network of N spins. This model has two main limitations. Firstly, in the thermodynamic limit $P, N \rightarrow \infty$ its storage capacity $\alpha = P/N \approx 0.138$ lies well below the theoretical upper limit of 1.0 for random unbiased patterns. Secondly, it cannot store biased patterns; ones with non-zero mean.

The Gardner algorithm overcomes these difficulties [2], and is capable of ensuring the maximum possible level of stability for patterns. It might therefore be thought that there is no need for alternative algorithms to this one. However, it does suffer from a number of drawbacks. First, prior to learning, a supervisor is required to determine the level of stability that the network should attain. So the algorithm may not converge if too large a value is chosen. Secondly, even if it does converge, learning times are known to be exceptionally large at high capacities. Thirdly the algorithm does not have the ability to operate independently of an external supervisor during the learning process, effectively eliminating its chances of being biologically plausible. We shall now describe an alternative method for achieving associative memory in a Hopfield net which does not suffer from these problems [3][4].

A New Method

Simulations have shown that a set of weights capable of storing N patterns (biased or unbiased) in a Hopfield network of N spins are proportional to the inverse of the pattern correlation matrix ie $w_{ij} = -(C^{-1})_{ij}$ - the inverse-Hebb rule [3]. This has also been established using a *replica symmetric* analysis [4].

Unfortunately this rule is useless for associative memory because each pattern then has a zero basin of attraction. The usefulness of the inverse-Hebb rule is only apparent when one has a mechanism for increasing the stability of a subset of the stored patterns. The quadratic Oja algorithm [4] has just this property. The quadratic Oja learning rule may be written in the form $w_{ij} \rightarrow w_{ij} + \delta w_{ij}$ where

$$\delta w_{ij} = r(1 - \delta_{ij}) \left\langle H[\zeta](2H[\zeta]w_{ij} - \zeta_i \zeta_j) \right\rangle_\zeta$$

$\langle ... \rangle_\zeta = N^{-1} \sum_{\mu=1}^{P} ...$ denotes an expectation with respect to the patterns to be stabilised and r is a learning rate typically $O(N^{-2})$. If one regards the energy of the net as the output of a *quadratic* neuron [3] then the algorithm uses only local information and is unsupervised, apart from the usual decision of when to terminate the learning process. This should occur when one of the patterns in the subset chosen to be stabilised becomes de-stabilised due to competition from other patterns. Unlike the Gardner algorithm this algorithm does not require a choice of stability parameter prior to learning, since the stability of a set of already stored patterns is increased to some maximum

A Hopfield net of 100 spins was loaded with 100 random unbiased patterns using the inverse-Hebb rule and the weights of the network were adapted using the quadratic Oja algorithm with a given subset of the original 100 patterns. Our results for basin size R as a function of α show are shown in the figure below. A transition to $R = 1$ occurs at $\alpha \approx 0.21$ and for higher capacities the basin size rapidly falls away to zero, much like the theoretical prediction of Kohring [5] for the case of a net with *independent* weights.

The quadratic Oja algorithm can stabilise a pattern set very quickly - typically less than *10* iterations in the high capacity regime and *20-30* iterations in the low capacity regime. In the low capacity regime where the Gardner algorithm may be expected to do better, that algorithm required *50-60* iterations to achieve the maximum basin size. Apart from the time required to store the patterns using the inverse-Hebb rule this scheme is an effective method of pattern stabilisation even at low capacities.

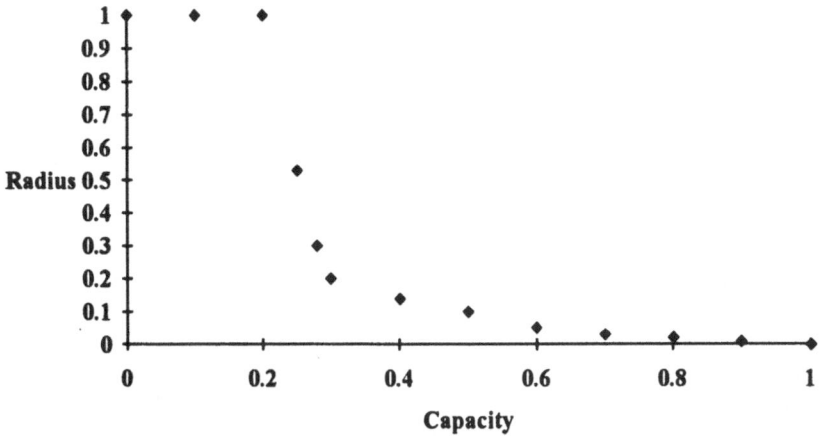

Implementation of the Inverse-Hebb Rule

Matrix inversion may be cast as an optimisation problem [6]. Remarkably, this has led to the construction of a hardware version of a graded response Hopfield network capable of matrix inversion We shall now review this process for the construction of the inverse-Hebb rule.

Consider a set of N energy functions $G_i(V)$, each a function of a set of N parameters $V_{ij}, i, j = 1, 2, 3...N$, defined by

$$G_i(V) = 1/2 \sum_{j=1}^{N} \left(\sum_{k=1}^{N} C_{jk} V_{ki} + \delta_{ji} \right)^2$$

A set of graded response networks that converge to the minima of the cost functions given above ensure a rapid means of obtaining the inverse-Hebb rule. A set of Hopfield learning rules of the form $\dot{u}_{ij} \propto -\partial G_j(V)/\partial V_{ij}$; $V_{ij} = g_{ij}(u_{ij})$ will minimise the above energy functions (denoting differentiation with respect to time with a dot). We identify u_{ij} as the input voltage of the ith neuron in the jth network, V_{ij} as its output and the function g_{ij} as the input-output relationship. Unfortunately this learning rule for the construction of the inverse-Hebb rule is still non-local in that it requires a matrix multiplication :

$$\dot{u}_{ij} = -\eta(\sum_k T_{ik}V_{kj} + C_{ij}) \quad ; \quad T_{ij} = \sum_k C_{ik}C_{kj}$$

A local implementation of this algorithm is discussed elsewhere [4]. Experiments in hardware have shown that this method of matrix inversion is very fast, typically requiring only a few times the characteristic time constant of the network. Furthermore for non-singular matrices the inversion process is relatively insensitive to initial network conditions and to the precise form of the input-output relationship.

Conclusions

It is interesting that many of the features present in our scheme for achieving associative memory in a Hopfield net have been discussed in the model of learning by selection [7]. This model attempts to describe memory storage during development and in the adult using metaphors borrowed from statistical mechanics. It proposes the stabilisation of pre-existing valleys in an energy landscape rather than the creation of new ones. The suggestion is also made that the learning rule should be local using only small synaptic changes. The existence of a non-trivial valley structure prior to learning is obviously essential for this model to work. It would seem that the inverse-Hebb rule provides us with such an energy landscape. This can then be shaped using the quadratic Oja algorithm which is considered local and works well, with small synaptic changes. Furthermore, this scheme for achieving associative memory in a Hopfield net is realisable in either electrical or optical hardware.

References

[1]   Hopfield, J., J. (1982). Neural Networks and Physical Systems with Emergent Collective Computational Abilities, Proc. Natl. Acad. Sci. USA, Vol. 79, 2554-2558.

[2]   Gardner, E .(1988). The Space of Interactions in Neural Network Models. J. Phys. A, 21, 257-270.

[3]   Coombes, S., & Taylor, J., G. (1993). Using Generalised Principal Component Analysis to Achieve Associative Memory in a Hopfield Net, Network, To Appear.

[4]   Coombes, S., & Taylor, J., G. (1993). The Inverse-Hebb Rule, KCL Preprint.

[5]   Kohring, G. A. (1990). Neural Networks with Many-Neuron Interactions, Le Journal de Physique, 51, 145-155.

[6]   Jang, J., Lee, S., & Shin, S. (1988). An Optimisation Network for Matrix Inversion, Neural Information Processing Systems, Ed. D. Z. Anderson, New York, 397-401.

[7]   Toulouse, G., Dehaene, S., & Changeux, J. (1986). Spin Glass Model of Learning by Selection, Proc. Natl. Acad. Sci. USA. Vol. 83, 1695-1698.

Recursive Learning in Recurrent Neural Networks with Varying Architecture

D. Obradovic

Siemens AG, Corporate Research and Development, ZFE ST SN 41
Otto-Hahn-Ring 6, 81739 Munich, Germany

1.0 Introduction

This paper presents a novel on-line procedure for training dynamic neural networks with input-output recurrences whose topology is continuously adjusted to the complexity of the target system dynamics. The latter is accomplished by changing the number of the elements of the network hidden layer whenever the existing topology cannot capture the dynamics presented by the new data. The training mechanism developed in this work is based on the suitably altered Extended Kalman Filter (EKF) algorithm (Gelb, 1974) which is simultaneously used for the network parameter adjustment and for its state estimation. The network itself consists of the single hidden layer with the Gaussian Radial Basis Functions (GRBF) and of the linear output layer. The choice of the GRBF is induced by the requirements of the on-line learning. The latter implies the network architecture which permits only local influence of the new data point in order not to forget the previously learned dynamics. The continuous topology adaptation is implemented in our algorithm in order to avoid memory and computational problems of using a regular grid of GRBF functions which covers the network input space. Furthermore, we show that the resulting parameter increase can be handled "smoothly" without interfering with the already acquired information. In the case when the target system dynamics are changing over time, we show that a suitable forgetting factor can be used to "unlearn" the not-any-more relevant dynamics. The quality of the presented recurrent network training algorithm is demonstrated on the identification of a nonlinear dynamic system.

2.0 Neural Network Model Structure

The dynamic systems treated in this paper are assumed to be time discrete and of the following form:

$$x(k+1) = f(x(k), u(k))$$
$$y(k) = g(x(k), u(k)) \tag{1}$$

where $x \in \Re^n$, $u \in \Re^m$, $y \in \Re^p$, and f and g are the functions of appropriate dimensions which are assumed to be differentiable everywhere. The system presented in (1) is in the state-space form where x is the state variable vector and y is a vector containing measurable states. In addition, we assume that the states $x(k)$ correspond to the physical variables, not necessarily measurable, whose magnitude bounds are known. According to the assumption on the unknown dynamic system, the neural network model is given as follows:

$$\hat{x}(k+1) = f_{net}(\hat{x}(k), u(k), \phi_f)$$
$$\hat{y}(k) = g(\hat{x}(k)) \tag{2}$$

where ϕ_f are the parameters of the recurrent network used to approximate the function f. The network contains a single hidden layer with GRBF as activation functions. When the centers and widths of the GRBF are kept constant, the adjustable parameters of the model are the heights which, therefore, are the elements of the parameter vector ϕ_f.

3.0 Extended Kalman Filter with Continuous Topology Adaptation

By augmenting the dynamics of the neural network model in the (1) with extra states corresponding to the unknown parameters and assuming the presence of additive white noise, we derive an over-all model representation of the following form:

$$\hat{x}(k+1) = f_{net}(\hat{x}(k), u(k), \phi_f) + \xi(k)$$
$$\phi(k+1) = \phi(k) + \mu(k) \tag{3}$$
$$\hat{y}(k) = g(\hat{x}(k)) + \upsilon(k)$$

where $\phi = \phi_f$ is the parameter vector and ξ, μ, υ and are zero-mean, independent Gaussian noise sequences. These sequences represent process, parameter, and measurement noises whose covariance matrices are given as follows: $E\{\xi(k)\,\xi^T(j)\} = Q_x\,\delta_{k,j}$; $E\{\mu(k)\,\mu^T(j)\} = Q_\phi\,\delta_{k,j}$; $E\{v(k)\,v^T(j)\} = R\,\delta_{k,j}$ where $\delta_{k,j}$ is the Kroenecker function. Furthermore, let P(0) be the covariance matrix that captures uncertainty in the initial condition $\hat{x}(0)$.

The system in (3) can be rearranged by defining the augmented state vector as $z(k) = [\hat{x}(k)\,\phi(k)]$ and the augmented process noise vector as $w(k) = [\xi(k)\,\mu(k)]$. The covariance matrix Q(k) of the latter is block diagonal with Q_x and Q_ϕ as its principal elements. The dynamics of the augmented system can, therefore, be represented as follows:

$$z(k+1) = \begin{bmatrix} f_{net}(\hat{x}(k), u(k), \phi(k)) \\ \phi(k) \end{bmatrix} + \begin{bmatrix} \xi(k) \\ \mu(k) \end{bmatrix} = F(z(k), u(k)) + w(k) \tag{4}$$

$$\hat{y}(k) = g(\hat{x}(k)) + \upsilon(k) = G(z(k), u(k)) + \upsilon(k) \tag{5}$$

The system of (3) is now in the form suitable for the EKF algorithm application (Gelb, 1974).

In the case where the GRBF are placed on a regular grid, the size of the error covariance matrix P remains the same for every "k" and it is not influenced by the actual distribution of augmented system inputs $h(k) = [z(k), u(k)]$. Therefore, if GRBF could be placed to cover only the area of the actual network input space $h1(k) = [\hat{x}(k), u(k)]$ where the input is present during the training, a significantly smaller number of the activation functions would suffice. Since the training is done on-line, the future distribution of actual inputs in the input space is not known and it can be estimated only from the already available data. The only available information is the size of the actual input space which is the consequence of the *a priori* known magnitude bounds on the signals u(k) and the states x(k). The corresponding subspace of the actual network input space is, hence, a hypercube **B**. Due to the space limitation, we present only the modification of the EKF algorithm (Gelb, 1974) that dynamically updates the network structure by on-line estimation of the input distribution.

The philosophy of the dynamic update of the network complexity can be formulated as follows:

* Whenever the part of the current network input vector h1(k) falls in the region of the input space **B** "not covered" by the existing GRBF, a new activation function is added to the existing network architecture such that its center coincides with the input data. On the other hand, if h1(k) falls outside of the hypercube **B**, a new GRBF is placed on its projection on **B**. Let S1(h(k)) be a set containing all the centers of the already assigned GRBF which are within the ball of radius d whose center coincides with the current augmented network input h(k). Furthermore, let S2(h(k)) be a set of centers of already assigned GRBF whose activation due to the input h(k) exceeds some positive scalar α in magnitude. The actual criterion for assigning new activation function is as follows:

if $S1(h(k)) \cap S2(h(k)) = \{ \varnothing \}$ ==> assign one new GRBF with the center at h(k) (6)

* Once when the hidden layer is augmented with a new GRBF, the existing state vector and covariance matrices are updated as follows:

$$z'(k) = [x(k) [\phi(k) \Phi]] = [z(k) \Phi] \tag{7}$$

$$P'(k) = \begin{bmatrix} P(k) & 0 \\ 0 & \pi \end{bmatrix} \quad Q'(k) = \begin{bmatrix} Q(k) & 0 \\ 0 & q \end{bmatrix} \tag{8}$$

where Φ is the initial value of the height of the added GRBF, the scalar π defines uncertainty associated with this value, and the scalar q defines the covariance of the corresponding process noise. The structure of the matrices in the equations (11) and (12) is based on the fact that the newly added parameter can be assumed to be uncorrelated with the previously defined augmented state vector z(k).

The presented change to the EKF training algorithm is based on the assumption that the dynamics of the target system do not vary over time. In the case when the target dynamics are not stationary, it is useful to forget the information older the time constant of the system variation. In order to achieve the latter, we storage the information about the sampling instant when the particular GRBF was assigned. Now, whenever some of the activation function is not activated within the time constant T, it is removed from the hidden layer. Therefore, the size of the corresponding correlation matrices has to be reduced by deleting the row and the column associated with the height of this particular GRBF. To be able to guarantee that the parameter reduction will not affect the information about the "recent" dynamics contained in the error covariance matrix, we associate an exponential weight decay on the available measurements (Jazwinsky, 1970). The resulting changes appear only in update steps of the EKF algorithm (Gelb, 1974) which now take the following form:

$$K(k) = P(k|k-1) C^T(k-1) [C(k-1) P(k|k-1) C^T(k-1) + e^{-(\Delta/\tau)} R]^{-1}$$

$$P(k|k) = e^{(\Delta/\tau)} [I - K(k) C(k-1)] P(k|k-1)$$

where Δ is the sampling period and $\tau = \tau(T)$ is a decay factor corresponding to time constant T. In addition, the matrix C(k-1) results from linearization of the function G, i.e. $\hat{y}(k)) \approx C(k) z(k)$. Hence, the exponential time decay factor guarantees that the omission of the parameters updated "earlier than time T" in the past does not effect relevant information contained in the remaining part of the covariance matrix.

4.0 Example of the EKF Learning with Varying Architecture (EKFVA)

In this section we present an application of the learning algorithm to a first order SISO system. The system is described as follows:

$$x(k+1) = 0.01x(k) + 0.4\sin((3x(k)) + 0.5u(k))$$

$$y(k) = x(k)$$

(9)

The identical example was treated in Livstone et al. (1992) where the region of the network input space **B**, defined as $\hat{x} \in [-1.25, 1.25]$ and $u \in [-1.25, 1.25]$, was uniformly covered with 625 GRBF placed on the nodes of a regular grid. In the same example the local property of the activation function was used to identify the 4 nearest neighboring nodes that were further used to approximate the response of the whole network. Nevertheless, the resulting covariance matrices were of the size 625x625.

In order to compare the application results of the EKFVA with the results presented in Livstone et al. (1992), we set the widths σ of the GRBF to the same value as in the latter. Furthermore, we also choose the same values for covariance matrices, i.e. P(0)=100 I, Q=10 I, and R=0.1. For the EKFVA we assume that the added parameters are characterized with π=100 and q=10, according to the equation (8). Therefore, our recurrent network model has two inputs, single output, and a hidden layer whose size is dynamically expanding during the learning. The parameters of the topology adaptation criterion in the equation (6) are α=0 and d=2σ. The resulting number of the GRBF placed during the 6000 sampling periods was 203, which is a significant reduction comparing with the 625 in Livstone et al. (1992). The accuracy of the trained network is checked by comparing its response with the response of the target system with respect to a sinusoid input signal which is shown in Figure 1.

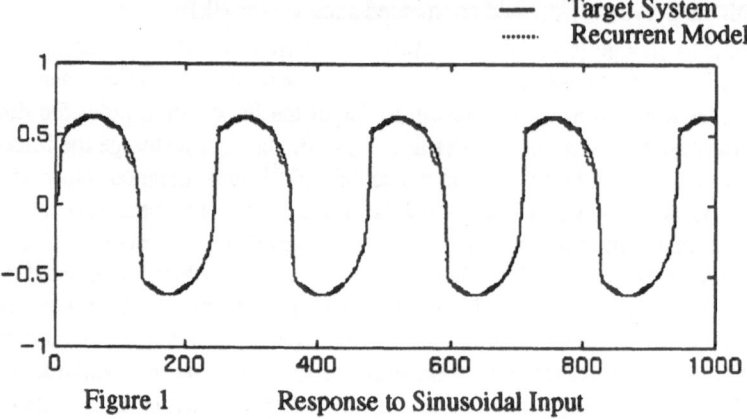

Figure 1   Response to Sinusoidal Input

Reference

- A. Gelb, Ed., "Applied Optimal Estimation," MIT Press, 1974.
- A.H. Jazwinski, "Stochastic Processes and Filter Theory," Academic Press, New York, NY, 1970.
- M.M. Livstone, J.A. Farrell, and W.L. Baker, "A Computationally Efficient Algorithm for Training Recurrent Connectionist Networks," Proc. 1992 ACC, vol. II, pp. 555-561, Chicago.

Pruning in Recurrent Neural Networks

G. Castellano, A. M. Fanelli[(*)], and M. Pelillo

Dipartimento di Informatica
Università di Bari
Via G. Amendola, 173 - 70126 Bari (Italy)

1. Introduction

Recurrent neural networks are attracting considerable interest within the neural network domain especially because of their potential in such problems as pattern completion and temporal sequence processing (Almeida, 1987; Hertz *et al.*, 1991). As for feed-forward networks, in virtually all problems of interest the proper number of hidden units is not known in advance, and usually this turns out to be a trade-off between generalization and learning abilities (Hertz *et al.*, 1991). One popular way of solving this problem involves training an over-dimensioned network and then *pruning* excessive units (Sietsma and Dow, 1988).

In this paper we propose a method of pruning a recurrent neural network, which is a generalization of an algorithm previously developed for feed-forward architectures (Pelillo and Fanelli, 1993; Castellano *et al.*, 1993). The method is based on the idea of removing hidden units and adjusting the remaining weights in such a way that the overall input-output network's behavior is kept approximately unchanged over the entire training set. Experiments demonstrate the effectiveness of our approach.

2. The Pruning Method

Let the network be represented by a directed graph $N=(V,E)$, where $V=\{1,..,n\}$ is the set of units and $E\subseteq V\times V$ is the set of connections. Each connection $(i,j)\in E$ is associated with a weight $w_{ij}\in R$. Furthermore, let V be divided into a subset V_A of input units, a subset V_H of hidden units and a subset V_O of output units. Clearly, we have $V_H\cap(V_A\cup V_O)=\varnothing$. Moreover, the input units have a non-zero bias term I_i defined as the i-th component of the external input. For each unit $i\in V$, let us define its "projective" field as $P_i=\{j\in V\,|\,(i,j)\in E\}$ and its "receptive" field as $R_i=\{j\in V\,|\,(j,i)\in E\}$.

The dynamics of the network is described by the following system of differential equations:

$$dx_i\,/\,dt = -x_i + f_i(u_i) + I_i \tag{1}$$

where

$$u_i = \sum_{j\in R_i} w_{ji}x_j \tag{2}$$

[(*)] Author to whom correspondence should be addressed.

is the net input of unit i, x_i represents the output of unit i, and f_i is a nonlinear differentiable activation function, here assumed to be the logistic function for each unit. The network evolves according to (1) until a fixed point $\hat{x}$ is reached.

Now, suppose that unit $k \in V_H$ has somehow been identified to be removed. Our approach to network pruning involves adjusting the weights incoming into k's projective field in such a way that the net input of every unit $i \in P_k$ remains approximately unchanged. This amounts to requiring that the following relation holds:

$$\sum_{j \in R_i} w_{ji} \hat{x}_j^{(p)} = \sum_{j \in R_i - \{k\}} (w_{ji} + \delta_{ji}) \hat{x}_j^{(p)} \tag{3}$$

for each unit $i \in P_k - \{k\}$ and for each pattern p in the training set, where the δ_{ji}'s are appropriate adjusting factors for the weights w_{ji}'s. Simple algebraic manipulations yield

$$\sum_{j \in R_i - \{k\}} \delta_{ji} \hat{x}_j^{(p)} = w_{ki} \hat{x}_k^{(p)} \tag{4}$$

which is a (sparse) linear system in the unknowns δ_{ji}'s. System (4), which can be conveniently represented as $Ay=b$, is solved in the least-square sense by a very efficient preconditioned conjugate-gradient method proposed by Björck and Elfving (1979). It begins with an initial solution y_0 and iteratively produces a sequence of points $\{y_t\}$ so as to decrease the residuals

$$r_t = \|Ay_t - b\| \ . \tag{5}$$

This naturally suggests a sub-optimal criterion of choosing the to-be-removed units: pick the unit for which the initial residual r_0 is minimum. Since the initial point y_0 is usually chosen to be the null vector, this amounts to selecting the unit for which the norm of b is minimum.

Summarizing, the proposed algorithm proceeds as follows. For each pruning step the "minimum norm" unit is located and the corresponding system is solved (this can be thought of as a relearning phase). The process iteratively continues until some stopping criterion is met. To avoid that the algorithm produces "useless" units, the selected unit $k \in V_H$ should satisfy the following conditions: $P_i - \{k\} \neq \emptyset$ for each $i \in R_k$, and $R_i - \{k\} \neq \emptyset$ for each $i \in P_k$.

3. Experimental Results

To gauge the effectiveness of the proposed method some experiments were carried out over a simulated two-class two-feature pattern recognition task with known probability distributions given by (Niles et al., 1989):

$$p_1(x,y) = N(x,0,\sigma^2)N(y,\mu_y,4\sigma^2)$$

$$p_2(x,y)=[N(x,\mu_x,\sigma^2)+N(y,-\mu_y,\sigma^2)]\,N(y,-\mu_y,\sigma^2)/2$$

where $N(t,\mu,\sigma^2)$ is a Gaussian with mean μ and variance σ^2. In our simulations the values $\mu_x=2.30\sigma$, $\mu_y=2.106\sigma$, and $\sigma=0.2$ were used. A training set of 100 samples and a separate test set of 1,000 samples were generated, both containing equal number of examples from the two classes.

An initial network architecture, consisting of ten fully connected units, was used, with $|V_A|=2$, $|V_H|=7$ and $|V_O|=1$. Ten independent such networks (labeled from 'A' to 'J') were trained, starting with random initial weights, by the recurrent back-propagation algorithm (Pineda, 1988), using a learning rate $\eta=0.01$. The training was stopped after 1,000 steps, and the networks were found to have an average misclassification rate of 8.1%.

The pruning algorithm was later applied over each trained network and was stopped when the misclassification rate became worsen than the original ones of 1.0% or more. The results are summarized in Table I where, for each pruned network, the misclassification rate and the MSE are shown.

Table I

network	no. hidden units	misclass. (%)	MSE
A	1	8.0	0.0327
B	1	8.0	0.0328
C	1	8.0	0.0328
D	1	8.0	0.0328
E	1	9.0	0.0332
F	1	8.0	0.0341
G	1	8.0	0.0333
H	1	8.0	0.0330
I	1	8.0	0.0339
J	1	8.0	0.0337
average	1	8.1	0.0332

As can be seen, the method appears to be extremely robust as, in all of the ten trials, a solution network with just one hidden unit was obtained.

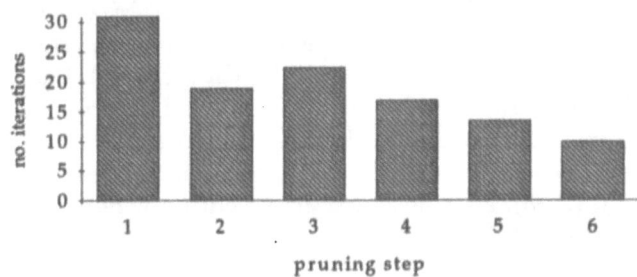

Fig. 1. *Median number of iterations required for the relearning phase.*

Moreover, the relearning phase was able to achieve good solutions very quickly. This can be seen in Fig. 1, where the median number of iterations of the Björck-Elfving procedure is plotted for each pruning step. We emphasize also that the time required to perform a relearning iteration is much less than that needed for a back-propagation epoch.

In addition, ten different 4-unit fully-connected networks were trained to achieve the same performance as the pruned ones (e.g. 8.0% misclassification rate), and their generalization ability was compared over the 1,000-exemplar test set. In Table II the resulting (average) misclassification rate and MSE are shown.

Table II

network	no. units	misclass. (%)	MSE
pruned	4	7.49	0.0299
original	4	9.14	0.0328

4. Conclusions

In this paper a method of reducing the size of recurrent neural networks has been developed. Experimental results have been reported which show the effectiveness of the proposed method, both in quickly finding minimal network solutions, and in improving generalization performance. Further work is in progress to test the feasibility of the approach in other application domains.

References

Almeida, L. B. (1987). A learning rule for asynchronous perceptrons with feedback in a combinatorial environment. in *Proc. ICNN*, San Diego, CA, vol. 2, pp. 609-618.

Björck, A., and Elfving, T. (1979). Accelerated projection methods for computing pseudoinverse solutions of systems of linear equations. *BIT* 19, 145-163.

Castellano, G., Fanelli, A. M., and Pelillo, M. (1993). An empirical comparison of node pruning methods for layered feed-forward neural network. in *Proc. IJCNN-93*, Nagoya, Japan, pp. 321-326.

Hertz, J., Krogh, A., and Palmer, R. G. (1991). *Introduction to the Theory of Neural Computation*. Addison-Wesley, Redwood City, CA.

Niles, L., Silverman, H., Tajchman, G., and Bush, M. (1989). How limited training data can allow a neural network to outperform an optimal statistical classifier. in *Proc. ICASSP-89*, Glasgow, vol. 1, pp. 17-20.

Pelillo, M., and Fanelli, A. M. (1993). A method of pruning layered feed-forward neural networks. in *New Trends in Neural Computation*, J. Mira, J. Cabestany, and A. Prieto, eds., pp. 278-283. Springer-Verlag, Berlin.

Pineda, F. J. (1988). Dynamics and architecture for neural computation. *J. Complexity* 4, 216-245.

Sietsma, J., and Dow, R. J. F. (1988). Neural network pruning - Why and how. in *Proc. ICNN*, San Diego, CA, vol. 1, pp. 325-333.

Making Hard Problems Linearly Separable – Incremental Radial Basis Function Approaches

B. Fritzke

Institut für Neuroinformatik
Universität Bochum, Germany

1 Introduction

Networks of localized (e.g., Gaussian) units are an interesting alternative to networks of global (e.g., sigmoidal) units, both for classification and regression problems. In this contribution we like to give some geometric insight into the way those networks operate. We first develop a method how a network can be constructed without any training to correctly handle a given classification problem. This method is fast and simple but it has some drawbacks concerning the network size and the generalization to new data. Therefore, we will discuss two incremental algorithms. They have the remarkable ability to successively transform a general classification problem into an equivalent one in which the classes are *linearly separable*[1] and can thus be distinguished by a single layer perceptron. An example will be given from a method proposed by us recently.

In the following we are considering radial basis function networks similar to those proposed by Moody & Darken (1989) or Poggio & Girosi (1990). Such networks consist of one layer L of Gaussian units. Each unit $c \in L$ has an associated vector $w_c \in R^n$ indicating the position of the Gaussian in input vector space and a standard deviation σ_c. For a given input datum $\xi \in R^n$ the activation of unit c is described by

$$D_c(\xi) = \exp\left(-\frac{\|\xi - w_c\|^2}{\sigma_c^2}\right). \qquad (1)$$

On top of the layer L of Gaussian units there are one ore more single-layer perceptrons (depending on the output dimensionality of the problem). Without loss of generality we assume in that we have a classification problem involving only two classes. In this case one output unit is sufficient.

2 Projection into a Unit Cube

How can a radial basis function network deal with problems which are not linearly separable? The answer must lie in the pre-processing done by the localized units in layer L since they constitute the only difference to a single-layer perceptron.

---

[1] n-dimensional patterns from two different classes are said to be *linearly separable* if there exists an $(n-1)$-dimensional hyper-plane that separates one class from the other.

In the following we make a simple, but instructive statement about the nature of this pre-processing. Let l be the number of localized units in L. Then every input signal $\xi \in R^n$ is converted into an l-dimensional activation vector

$$\xi' = (D_1(\xi), \cdots, D_l(\xi)) \tag{2}$$

which in turn forms the input of the single-layer perceptron. From eqn. 1 follows that for each Gaussian unit c and for every n-dimensional vector ξ the activation of c is inside the unit interval:

$$0 \leq D_c(\xi) \leq 1. \tag{3}$$

Consequently, the vector ξ' lies always inside the l-dimensional *unit cube*.

3 A Brute Force Approach

It is possible to construct a layer L of localized units such that for a given set of data every partition into two classes is – after being transformed by L as described – linearly separable:

- create one localized unit c_i for each data point $i \in R^n$.

- Set the center of c_i equal to i.

- Set the standard deviation σ_i of unit c_i to a small, positive fraction α of the distance to the nearest other unit.

Then for each data point i exactly one localized unit is maximally activated (the unit with its center at i) whereas for all other units c the following holds:

$$D_c(i) \leq \exp(-1/\alpha). \tag{4}$$

This again means that for small values of α all but one components of the activation vector ξ' are very close to zero. In the limiting case ($\alpha = 0$) the activation vectors lie on those corners of the l-dimensional unit cube which have a distance 1 to the origin. For a particular partition of the data into two classes A and B one can then simply choose the weights from the localized units corresponding to A to have the value 1 and the weights from units corresponding to B to have the value 0.

It is not necessary to set α arbitrarily close to zero: Let l again be the number of localized units (which is equal to the number of data points in this case). Let the network including the weights be constructed as described above. A data point is classified as belonging to A, if the output of the single output unit is larger than 0.5, else as belonging to B. Then no misclassifications for any data point of the training data can occur if for α the condition

$$0.5 < (l - 1) \exp(-1/\alpha) \tag{5}$$

is satisfied or equivalently:

$$\alpha < -\log(0.5/(l - 1)). \tag{6}$$

Stated otherwise: if (6) is fulfilled then for each training pattern the classification derived from its *corresponding* unit can not be changed by the combined activity of all $(l - 1)$ other units.

4 Incremental Methods

The method just described gives some kind of upper limit for the number of units a radial basis function network should have. It has, however, serious drawbacks:

- The number of units scales linearly with the amount of training data.

- The diameters of the localized functions tend to be very small, leading possibly to poor generalization

First, one would prefer a dependency of the network size on the difficulty of the problem. Second, it is not at all necessary that – as in the method described above – a representation is constructed where all possible partitions of the data are linearly separable. Rather, for each particular classification problem it is sufficient to find a representation in which the *given* partition of the data into classes is linearly separable. To achieve this goal it is tolerable if one Gaussian unit covers several data points as long as they are from the same class. This makes it possible to have both smaller networks and better generalization.

A practical problem is the construction of a suitable set of radial basis function units. One could start with a network as described above and successively merge radial basis function units thereby making them larger. This, however, does not work for on-line learning problems with an potentially infinite set of patterns since the starting point is not defined.

Another possibility is to start with a small network and successively add new units. Platt (1991) describes such a model which he mainly uses for regression problems such as time-series prediction. He allocates a new unit whenever one data item is mapped to a poor output value *and* is sufficiently distant from existing radial basis function units. Thereby the width of existing units is not changed. A problem of his approach is that (for noisy data) every outlier generates a new unit which makes the method susceptible of over-fitting.

Recently we proposed a method (Fritzke, 1993) which is similar in spirit to Platt's but differs from it in important points:

- New units are not allocated on the basis of one poorly mapped pattern but rather on error information accumulated over a number of patterns. *This leads to insertion of units only in those areas of the input space where statistical evidence justifies it.*

- If a new unit is inserted its position is interpolated from existing ones. The diameters of neighboring units are slightly reduced. *This automatically leads to a partitioning of the input space with a limited overlap between neighboring units.*

- The centers of the Gaussians are fine-tuned by competitive learning.

This strategy seems to lead to small networks with a very strong generalization ability. An example how such a network grows and successively masters a particular classification problem is given in fig. 1.

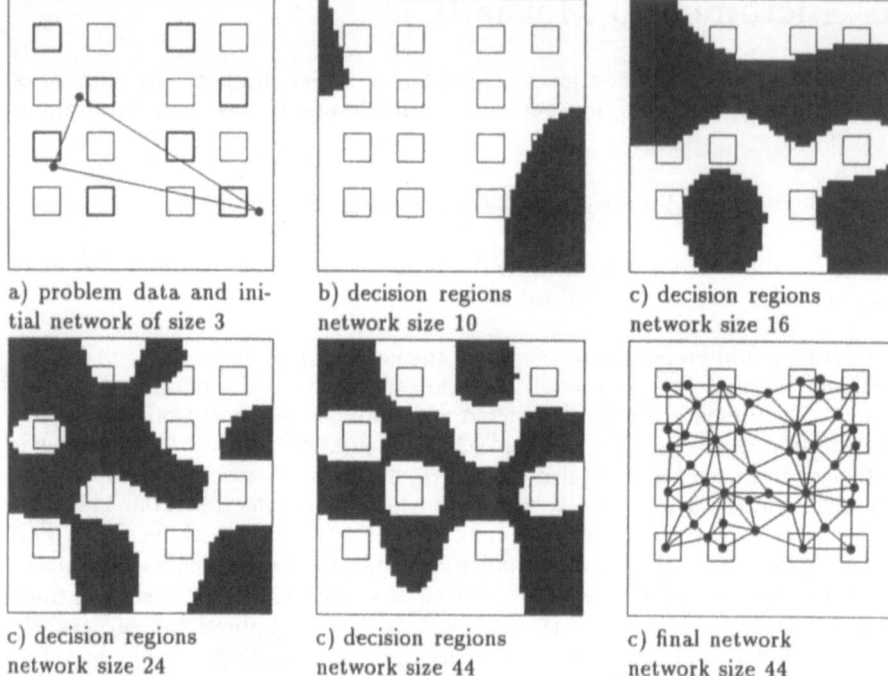

a) problem data and initial network of size 3

b) decision regions network size 10

c) decision regions network size 16

c) decision regions network size 24

c) decision regions network size 44

c) final network network size 44

Figure 1. Development of a network structure for a classification problem. The training data comes uniformly distributed from the thin rectangles (class A) and the thick rectangles (class B). The initial network, the development of the decision regions and the final network are shown. Despite the potentially unlimited number of training patterns a small network is found which maps all training data (inside the squares) correctly and generalizes reasonably well over new data (in between the squares).

5 Discussion

We interpreted networks of localized units in a new geometrical way. This led to a construction for a network which is able to handle general classification problems but has some drawbacks with respect to network size and generalization. From this we derived an *a posteriori* motivation for constructive methods, discussed the particular merits of a recently proposed model of ours, and gave an example.

Fritzke, B. (1993), Growing Cell Structures – a self-organizing network for unsupervised and supervised learning, International Computer Science Institute, TR-93-026, Berkeley.

Moody, J. & C. Darken (1989), Learning with Localized Receptive Fields, in *Proceedings of the 1988 Connectionist Models Summer School*, D. Touretzky, G. Hinton & T. Sejnowski, eds., Morgan Kaufmann, San Mateo, pp. 133–143.

Platt, J. C. (1991), A Resource-Allocating Network for Function Interpolation, *Neural Computation*, 3, pp. 213–225.

Poggio, T. & F. Girosi (1990), Regularization Algorithms for Learning That Are Equivalent to Multilayer Networks, *Science*, 247, pp. 978–982.

'Partition of Unity' RBF Networks are Universal Function Approximators[*]

J. Hakala†, C. Koslowski†, R. Eckmiller†

† Dept. of Comp. Science VI, Neuroinformatics, University of Bonn

Bonn, F. R. Germany

Tel.: ++49–228–550–349   FAX: ++49–228–550–425

e-mail: hakala@nero.uni-bonn.de

1   Introduction

Networks with 'Partition of Unity' Gaussian radial basis functions (RBF) have achieved considerable attention for their capability to incorporate rule–based knowledge into the network and to extract rule–based knowledge from the network (Tresp et. al., 1993). On the other hand, RBF networks are known for fast adaptation as being used for function approximation. Combining these properties, e. g. prestructuring the network with domain knowledge and refining it with rapid training, is a desirable goal. However, the capability of 'Partition of Unity' RBF networks to approximate arbitrary functions arbitrarily close on a bounded domain, the essential requirement for universal function approximators, could not be guaranteed so far.

Multilayer perceptrons with a single hidden layer (Cybenko, 1990) have the universal function approximator property. For RBF networks, there exist proofs for the special case of Gaussian RBF networks (Hartman et. al., 1990). For more general activation functions (Park et. al., 1993) obeying the additional condition $\int K(x)dx > 0$, even for basis functions with different width $K(x/\sigma_i)$ it was shown, that these functions are dense in the function space L^p with regard to the p–norm.

On the other hand normalized basis functions (Specht, 1990; Schløler et. al., 1992) are used to approximate functions, based on the Parzen–window approach. The network approximates the joint probability of the combined input– and output–space using a basis function for each datapoint. If the joint probability is modelled using separabel basis functions, the output is generated as conditional expectation of the output given the input. It was shown (Schløler et. al., 1992) that the probability to encounter an error larger than an arbitrary positive number converges to zero for the special case of Gaussian RBFs.

Since the number of parameters for these models grows linearly with the number of datapoints, (Tresp et. al., 1993) proposed to use mixtures of Gaussians to model the joint probability $P(\vec{x}, y)$ of input– and output–space. Adopting the probabilistic interpretation the output is the conditional expectation:

$$y(\vec{x}) = E(y|\vec{x}) = \frac{\int y P(\vec{x}, y)dy}{\int P(\vec{x}, y)dy} = \frac{\sum_i y_i P(\vec{x}|s_i)P(s_i)}{\sum_i P(\vec{x}|s_i)P(s_i)} \tag{1}$$

[*]Supported by Ministry for Research and Technology (BMFT–Project SENROB), under grant 01 IN 105 G/6

where $y_i = \int y P(\vec{x}, y|s_i) dy$ and $P(\vec{x}|s_i) = \int P(\vec{x}, y|s_i) dy$. The parameters $P(s_i), y_i$, center and width of the Gaussian mixture components can be found using gradient based methods or EM (Tresp et. al., 1993) for a given dataset. But it was not yet clear if these networks are capable to approximate functions arbitrarily close.

2 Gaussian Radial Basis Functions and Domain Knowledge

A question of large interest is the rule–extraction and incorporation within the framework of neural networks. We restrict our focus on networks, which make predictions about a state y given a state $\vec{x}$. Symbolic knowledge is present as if-then rules applicable in a certain state (situation). The description of the region of statespace where the rule is applicable is transferred to a basis function, which represents a description of the area, e.g. 'conclusion is applicable, if state is close to the region'. Predictions are assumed to be representable as real numbers. If an expert can tell us a set of rules applicable in certain situations these rules can be incorporated into the network assuming the probabilities can be modeled as mixture of Gaussians (Tresp et. al., 1993) as $P(s_i)$ the weight of the rule, $P(\vec{x}|s_i) = N_i(\vec{x}, \vec{c}_i, \sigma_i)$ the probability that $\vec{x}$ occurs in state s_i as mixture component and y_i the conditional expectation of y given state s_i to be independent of $\vec{x}$.

This procedure is reversable interpreting the network parameters as constituents of rules, where each Gaussian corresponds to one rule.

There is a close relationship to fuzzy systems (Tagaki et. al., 1985) with center of gravity defuzzification.

3 'Partition of Unity' Networks are Dense in the Space of Continuous Functions

The Stone–Weierstraß theorem e. g. (Schönhage, 1971):
Let domain S be a compact space of D dimensions, and let M be a set of continuous real–valued functions on S, fulfilling

a) for each $\vec{x} \in S$ exists $y \in M$ with $y(\vec{x}) = 1$,

b) for all pairs $\vec{x}_1 \neq \vec{x}_2 \in S$ there exists $y(\vec{x}) \in M$ which satisfies $y(\vec{x}_1) = 1$ and $y(\vec{x}_2) = 0$,

c) M is algebraically closed under $+, \times$,

then M is dense in $C(S)$, the set of continuous real–valued functions on S.
For a domain S let the set of normalized basis functions, which are a partition of unity by definition, be of the form

$$y(x) = \frac{\sum_{i=1}^{N} y_i \alpha_i K \left(\frac{\|\vec{x} - \vec{c}_i\|}{\sigma_i} \right)}{\sum_{i=1}^{N} \alpha_i K \left(\frac{\|\vec{x} - \vec{c}_i\|}{\sigma_i} \right)} \tag{2}$$

with $\alpha_i > 0, \sigma_i > 0$ and let the basis functions K be non–constant and positive on $\mathbb{R}^+$ satisfying a product rule

$$K\left(\frac{\|\vec{x} - \vec{c}_i\|}{\sigma_i}\right) K\left(\frac{\|\vec{x} - \vec{c}_j\|}{\sigma_j}\right) = L(c_i, c_j, \sigma_i, \sigma_j) K\left(\frac{\|\vec{x} - \hat{\vec{c}}_{ij}\|}{\hat{\sigma}_{ij}}\right) \quad (3)$$

with $L > 0$ and $\hat{\vec{c}}_{ij} \in S$. We proof that these basis functions fulfill the conditions of the Stone–Weierstraß theorem.

a) For arbitrary $\vec{x}, \vec{c}_1 \in S, \alpha_1 > 0, \sigma_1 > 0$:

$$y(\vec{x}) = \frac{\alpha_1 K\left(\frac{\|\vec{x} - \vec{c}_1\|}{\sigma_1}\right)}{\alpha_1 K\left(\frac{\|\vec{x} - \vec{c}_1\|}{\sigma_1}\right)} = 1.$$

b) Given $\vec{x}_1 \neq \vec{x}_1 \in S$ there exists $y \in M$ with $N = 2$ and $\vec{c}_1 = \vec{x}_1$ and $\vec{c}_2 = \vec{x}_2$, $\alpha_1 = \alpha_2 = 1$ and since $K(r)$ is not constant on $\mathbb{R}^+$ it is possible to chose $\sigma_1 = \sigma_2$ such that $K(0) \neq K(d/\sigma_1)$ where d is the distance between $\vec{x}_1$ and $\vec{x}_2$. That is sufficent to ensure the existence of weights y_1, y_2 such that $y(\vec{x}_1) = 1$ and $y(\vec{x}_2) = 0$.

c) For simplicity we drop the explicit vector notation.
Algebraic Closure (+):

$$y_1(x) + y_2(x) = \frac{\sum_{i=1}^{N} y_i \alpha_i K(\frac{x - c_i}{\sigma_i})}{\sum_{i=1}^{N} \alpha_i K(\frac{x - c_i}{\sigma_i})} + \frac{\sum_{j=1}^{P} y_j \alpha_j K(\frac{x - c_j}{\sigma_j})}{\sum_{j=1}^{P} \alpha_j K(\frac{x - c_j}{\sigma_j})} \quad (4)$$

$$= \frac{\sum_{i=1}^{N} \sum_{j=1}^{P} (y_i + y_j) L(c_i, c_j, \sigma_i, \sigma_j) \alpha_i \alpha_j K(\frac{x - \hat{c}_i}{\hat{\sigma}_i})}{\sum_{i=1}^{N} \sum_{j=1}^{P} L(c_i, c_j, \sigma_i, \sigma_j) \alpha_i \alpha_j K(\frac{x - \hat{c}_{ij}}{\hat{\sigma}_{ij}})}$$

which is element of S.
Algebraic closure ($\times$):

$$y_1(x) \times y_2(x) = \frac{\sum_{i=1}^{N} y_i \alpha_i K(\frac{x - c_i}{\sigma_i})}{\sum_{i=1}^{N} \alpha_i K(\frac{x - c_i}{\sigma_i})} \times \frac{\sum_{j=1}^{P} y_j \alpha_j K(\frac{x - c_j}{\sigma_j})}{\sum_{j=1}^{P} \alpha_j K(\frac{x - c_j}{\sigma_j})} \quad (5)$$

$$= \frac{\sum_{i=1}^{N} \sum_{j=1}^{P} (y_i y_j) L(c_i, c_j, \sigma_i, \sigma_j) \alpha_i \alpha_j K(\frac{x - \hat{c}_{ij}}{\hat{\sigma}_{ij}})}{\sum_{i=1}^{N} \sum_{j=1}^{P} L(c_i, c_j, \sigma_i, \sigma_j) \alpha_i \alpha_j K(\frac{x - \hat{c}_{ij}}{\hat{\sigma}_{ij}})}$$

which is element of S.
As we have shown the partition of unity networks satisfy the requirements of the Stone–Weierstraß theorem hence they can approximate continuous functions arbitrarily close.

As an example we consider Gaussian radial basis functions with $K(r) = e^{-r^2}$ and the euclidian norm. The resulting parameters for the product rule are:

$$\hat{\vec{c}}_{ij} = \frac{\sigma_j^2}{\sigma_i^2 + \sigma_j^2} \vec{c}_i + \frac{\sigma_i^2}{\sigma_i^2 + \sigma_j^2} \vec{c}_j \quad \text{and} \quad \hat{\sigma}_{ij}^2 = \left(\frac{1}{\sigma_i^2} + \frac{1}{\sigma_j^2}\right)^{-1}$$

and the scaling factor

$$L(\vec{c}_i, \vec{c}_j, \sigma_i, \sigma_j) = K \left(\frac{1}{\sigma_j^2 \sigma_i^2} \frac{\|c_i - c_j\|}{\hat{\sigma}_{ij}^2} \right)$$

Note that it is necessary that $\hat{c}_i = c_i + t(c_j - c_i)$ for some $t \in [0, 1]$ to ensure closure, which on the other hand requires convexity of S. The network with Gaussian radial basis functions is formally equivalent to eq. 1 with $\alpha_i = P(s_i)$ and $P(\vec{x}|s_i) = e^{-\|\vec{x} - \vec{c}_i\|_2^2/\sigma_i^2}$ and weights y_i.

4   Conclusions

We could proof that a class of 'partition of unity' radial basis function networks can approximate continuous functions arbitrarily close. This class contains networks with Gaussian radial basis functions, which have an interpretation in terms of Gaussian mixture models, and can in turn be used to incorporate rules into– and to extract rules from these networks. Combining these properties the network is capable to represent rules and to approximate continuous functions without any theoretical limitation.

Concerning more practical considerations it could be advantageous to use node allocation for efficent use of resources and better generalization (Hakala et. al., 1993).

References

Cybenko G., (1990), Approximation by superposition of a sigmoidal function, *Mathematics of Control, Signals and Systems*, vol. 2, pp. 303–314.

Hakala J. and Eckmiller R., (1993), Node allocation and topographical encoding NATEnet for inverse kinematics of a 6–DOF robot arm, in *Artifical Neural Networks* (S. Gielen and B. Kappen, eds.), pp. 309–313.

Hartman E., Keeler J. D., and Kowalski J., (1990), Layered neural networks with gaussian hidden units as universal approximators, *Neural Computation*, vol. 2, pp. 210 – 215.

Park J. and Sandberg I. W., (1993), Approximation and radial–basis–function networks, *Neural Computation*, no. 5, pp. 305–316.

Schløler H. and Hartmann U., (1992), Mapping neural network derived from the parzen window estimator, *Neural Networks*, vol. 5, pp. 903–909.

Schönhage A., (1971), *Approximationstheorie*. W. de Gruyter, Berlin.

Specht D. F., (1990), Probabilistic neural network, *Neural Networks*, vol. 3, pp. 109 – 118.

Tagaki T. and Sugeno M., (1985), Fuzzy identification of systems and its application to modeling and control, *IEEE Transactions on Systems, Man and Cybernetics*, vol. 15, no. 1, pp. 116–132.

Tresp V., Hollatz J., and Ahmad S., (1993), Network structuring and training using rule–based knowledge, in *Adv. in Neural Inf. Proc. Sys.*, pp. 871–878.

Optimal Local Estimation of RBF Parameters

STEFANIA MARCHINI* and N. ALBERTO BORGHESE**

*Dipartimento di Scienze dell'Informazione, University of Milano, I.
**Institute Neuroscience Bioimages C.N.R., Via Mario Bianco 9, 20131 Milano, I

1 Introduction

The reconstrunction of a continuos curve $g(x)$ starting from a set of points, by means of a parametric function $f(x) = F(x,w)$, with w a set of unknown parameters, is an ill-posed problem that allows an infinite number of solutions. To obtain the solution optimal to the problem, a classical approach is to introduce soft constraints that do not specify exact desired values of the function $F(x,w)$ but only a tendency of the function in the definition domain. One of the most promising classes of reconstructing functions has been recently proposed by Poggio, Girosi, (1989); it allows to write the $F(x,w)$ approximation function as a linear combination of radial (gaussian) functions:

$$F(x,w) = \sum_{i=1}^{N} c_i G\left(x; d_i/\sigma_i\right) \qquad (1)$$

where c_i, d_i e σ_i are the parameters w to be determined, respectively: the amplitude coefficient, the center and the variance of the gaussian G_i, and N is the number of different gaussian functions employed; $N \leq M$, where M is the number of the sampled points $P_i = \left(x_i; y_i\right)$.

In this paper we propose a method for the automatic setting of these parameters as a function of the distances between the points to achieve the optimal reconstructing function. Its behaviour has been tested using a set of randomly generated points. This algorithm can be easily extended to multi-dimensional functions and it is well suited for a hardware neural network implementation.

2 Method

First, the optimal number of gaussian functions is determined, then the coefficeints c_i could be determined analytically imposing that $F(x,w)$ passes through the M sampled point.

$$\sum_{i=1}^{M} c_i G\left(x_i; d_i / \sigma_i\right) = y_i \qquad (2)$$

If the matrix G is close to singular the obtained solution for the coefficients c_i is heavily affected by computational errors. It is therefore mandatory to reduce the number of sampled points, prevents overfitting the data: $F(x,w)$ will not follow the ripples caused by the noise on the points and a better approximation of the original signal is produced. In Figure 1 the reconstructing curve is shown for $N=M$ and different values of the variance σ. Notice the high value of the coefficients and the high oscillations in the curve, above all when σ assumes a low value.

For these reasons it is numerically efficient to reduce the number of gaussian basis functions adopted and the number of coefficeints c_i. The problem can be summarized as: *How many and which gaussian should be eliminated and how?*

464

We propose here an iterative procedure to determine the number and the values of the centers. Initially the abscissas of the centers are set coincident with the abscissas of the sampled points. At every iteration step one gaussian is eliminated: the pair of gaussians whose centers are the nearest ones is determined and a gaussian with the abscissa equal to mean value of the abscissas of the two centers is substituted to them. This procedure ends when all the centers are separated by a distance grater than a predetermined threshold, related to the desired degree of smoothness for the curve.

Alternatively, the optimal number of gaussian is determined using the properties of the singular value decomposition. This analytical technique decomposes the matrix G (NxM) into the product of three matrixes: $G = UWV^T$ where U and V are orthonormal (NxM and MxM) and W is a diagonal matrix (MxM) which contains the singular values (Golub and Van Loan, 1989). The number of effective centers is the range of G that is equal to the number of singular values of W, significatively different from zero. Some ripples that can be easily attributed to noise, are filtered out as can be seen in Figure 2.

Once the number of gaussians N, has been determined, the optimal value of the coefficients center are determinated as (Poggio e Girosi, 1989):

$$C = \left(G^T G + \lambda g\right)^{-1} G^T Y \qquad (3)$$

with C, Y and G are the vectors and matrix defined in equation (2) and g the matrix defined as G but using only the abscissas of the centers. The parameter λ regulates the degree of smoothness of the reconstructing function. The smaller is the value of λ, the nearer to the points is the reconstructing function F(x,w); it will threfore undergo to undesiderable oscillations that will yield a poor generalization capability. The bigger is the value of λ, the smoother is the funtion. In this case the frequency content of the reconstructed siggnal will be reduced (Oppeheim and Shafer, 1975) operating a low-pass filtering that will possibly filter out the ripples.

The parameter λ is threfore global over the entire definition domain of F(x,w). Alternatively, we can play with the value of the variance of the gaussians that is used in the reconstructing function. It is also related to the degree of smoothness in the reconstructing curve. As can be noted from Figure 1, the curve becomes smoother as the variance of the gaussians increases.

The variance can be automatically computed following the heuristic of *global first nearest-neighbor* proposed by Moody and Darken (1989). The variance is set to the mean of the minimal distances between each center and its nearest point P_i.

The sum of the mean square errors on the sampled points can give an idea on the performance of the reconstructing function. In Table 1, this errors is plotted as a function of both λ and of the variance of the gaussians; it increases with the increase of both λ and σ but the curve becomes evidently smoother.

We may get a better approximation of the function considering that the frequency property of the signal to be reconstructed may not be equal over all the definition domain. This can be obtained by choosing gaussians of different variances; each variances σ_i will be a function of the distances between the sampled points and it is set equal to the mean distance of its center from those points that falls into a certain region around its center; this region has the function of a receptive field for each gaussian. The reconstructing curve obtained with gaussian of different variances is plotted in figure 2 and the mean square distance reported in Table 2. It should be remarked that the advantage of this procedure becomes apparent for large set of data.

3 Conclusion

Although λ can, alone, regulate the degree of smoothness of the reconstructing function, it can be used when analytical solutions are feasible. The reconstruction of a function from large sets of data (surface, multi-dimensional temporal sequences), requires the use of numerical solutions. Gaussian basis functions allow to naturally partition the definition domain into regions (*receptive fields*), that ease numerical solutions. The parameters to be tuned are affected only from the behaviour of the data belonging to a small sub-region of the definition domain. Taking advantage of this property, our method can achieve an optimal reconstruction of functions of large data sets.

4 References

G. H. Golub and C. F. Van Loan, (1989) Matrix Computation, 2nd ed. Baltimore: Johns Hopkins University Press.

J. Moody and C. Darken, (1989) Neural Comput. 1, 281-294.

T. Poggio and S. Edelman, (1990) Nature 343, 263-266.

T. Poggio and F. Girosi, (1989) Artif. Intell. Memo 1140, Artificial Intelligence Laboratory, Massachussets Institute of Technology, Cambridge.

T. Poggio and F. Girosi, (1989) Scienze 247, 978-982.

Oppenheim and Shafer, (1975) Digital signal processing, Prentice Hall.

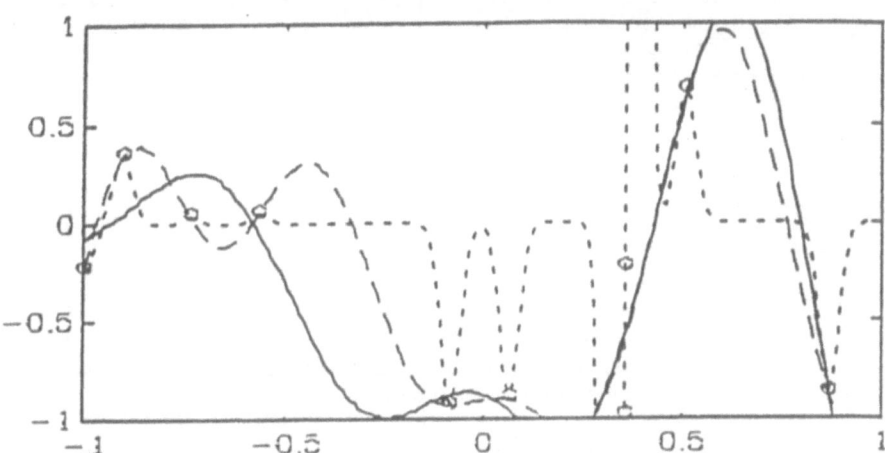

Figure 1: Three different reconstructing functions F(x,w) with different values of the variance σ. With solid line is reported the approximating curve with σ =0.1; relative coefficients are: [3.34; -6.06; 3.19; 0.67; -11.32; 18.12; -12.77; -12.72; 21.36; -5.63]. With dashed line is reported the approximating curve with σ =0.05; relative coefficients are:[-7.01; 10.55; -6.30; 6.62; -1.82; 1.83; -1.54; -1.54; 2.91; -1.07]. With dot line is reported the approximating curve with σ =0.0006; relative coefficients are:[-0.014; 0.021; 0.002; 0.004; -0.057; -0.055; -37.167; 37.130; 0.041; -0.053]. Points coordinates: are: [(-1.00, -0.23); (-0.74, 0.04); (0.51, 0.66); (-0.08, -0.93); (0.06, -0.89); (-0.56, 0.06); (-0.91, 0.34); (0.04, -0.98); (0.36, -0.23); (0.09, -0.87)].

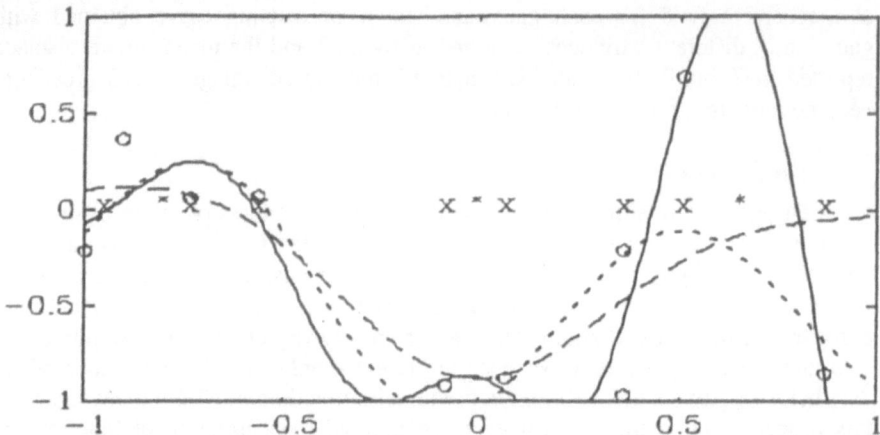

Figure 2: *Approximation of a curve starting from the same set of points as in Figure 1, through a reduced set of gaussians. The sampled points are represented with* ○ . *Solid line represents the approximation function obtained with 8 gaussians centered in* ✕ : *{-0.74; 0.51; -0.08; 0.07; -0.56; 0.87; 0.36; -0.95} with relative coefficients:{-2.06; 22.78; -13.59; 20.84; 3.36; -5.88; -27.62; 0.26} and variance* σ = 0.1 *equall for all gaussians. Dashed line represents the approximation function obtained with 3 gaussian centered in* ✳ : { -0.01; -0.8; 0.65} *with relative coefficients:{-0.70; 0.11; -0.03} and variance* σ = 0.1 *equall for all gaussians. dot line represent the approximation function obtained with the same 8 gaussians of figure 1, centered in* ✕, *with relative coefficients: {6.2530; 0.8655; -2.3784; -0.9216; -2.3840; 4.0225; -3.6141} and variance reported in Table 2.*

λ variance σ	0	0.1	0.5	1	5
0.05	0.4238	0.4136	0.767	1.0083	1.8422
0.1	0.4449	0.7937	1.1747	1.3734	2.1057
0.25	1.2398	1.3326	1.7243	1.9103	2.4464
0.3845	*1.1634*	*1.688*	*2.0079*	*2.14*	*2.5707*
0.5	2.3728	1.9056	2.1401	2.2485	2.635
0.75	2.4181	2.1323	2.2938	2.3801	2.7209
1	2.5269	2.2534	2.3768	2.4548	2.7738

Table 1: *The data of this table are referred to the function F(x,w) constituted of 8 gaussians used to reconstruct the curve through a set of 10 points as shown in figure 2. Mean square error as a function of the smoothness parameter* λ *and of the variance of the gaussians are reported. In italics the variance as computed using global first nearest-neighbor is highlighted.*

Variance σ	0.2023	0.2775	0.3773	0.2948	0.359	0.3582	0.1775	0.1752
Distance	1.3461							

Table 2: *The data of this table are referred to the function F(x,w) constituited of 8 gaussian used to reconstruct the curve through a set of 10 points as shown in figure 2. Mean square error as a function of the variances of the gaussian are reported. The smoothness parameter* λ *is set to 0. The amplitude of the interval to determine the variance is set to* ± 0.5.

Acceleration of Gaussian Radial Basis Function Networks for Function–Approximation*

J. Hakala†, J. Puzicha†, R. Eckmiller†

† Dept. of Comp. Science VI, Neuroinformatics, University of Bonn

Bonn, F. R. Germany

Tel.: ++49–228–550–349   FAX: ++49–228–550–425

e–mail: hakala@nero.uni-bonn.de

1   Introduction

Neural Networks consisting of localized basis functions are used for approximation of continuous functions. They are known for fast adaptation (Moody et. al., 1989) compared to global networks of multilayer perceptrons. Due to the locality of the Gaussian radial basis functions (GRBF) it is possible to take only a small fraction of all neurons into account without a significant change in network performance. This observation is the reason to search for efficent data structures to find the small fraction of necessary neurons quickly. There are data structures to accelerate the search for the k nearest neighbors of a data point like k–d trees, ball–trees and bump–trees e. g. (Omohundro, 1991). But they all lack the guarantee that all of the k nearest neighbors are found using a single leave of the tree. On the other hand they do not take the finite area of influence of the localized basis functions with the possibility of strongly varying width into account. There exist algorithms for non–radial basis functions, which are better suited for tree based acceleration of recall and learning (Sanger, 1991).

2   Accelerating the update of GRBF networks

Localized radial basis function networks compute their output using linear combination of the activities of the neurons. We restrict our focus to partition of unity Gaussian activation functions (Hakala et. al., 1993), but similar activation functions including usual Gaussian activation functions without normalization can be easily adopted. Let

$$\vec{y}(\vec{x}) = \frac{\sum_{i=1}^{N} \vec{y}_i a_i(\|\vec{x} - \vec{c}_i\|)}{\sum_{i=1}^{N} a_i(\|\vec{x} - \vec{c}_i\|)} \quad \text{with} \quad a_i(d) = exp(-d^2/\sigma_i^2) \tag{1}$$

be defined on some input space X. In most applications the basis functions $a_i(d)$ are almost zero for most parts of the input space resulting in the idea to

*Supported by Ministry for Research and Technology
(BMFT–Project SENROB), under grant 01 IN 105 G/6

consider the activity $a_i(d)$ of a neuron to be equal to zero, if $a_i(d)$ is below a threshhold T and to evaluate only neurons with non-zero activity. The following algorithms are heuristic approaches to achieve this goal.

Let $(B_j)_{j \in J}$ be a system of disjunct subspaces (boxes) satisfying $\cup_{j \in J} B_j = X$. For each box B_j and each neuron N_i calculate the distance $d = \min_{\vec{x} \in B_j} ||\vec{x} - \vec{c}_i||$. If $a_i(d) > T$ store the index of the neuron in a list L_j associated with the box. Note that for a given net the system of lists has to be calculated only once. To evaluate the function $\vec{y}$ at a given point $\vec{x}$ first calculate B_j satisfying $\vec{x} \in B_j$, then computation reduces to

$$\vec{y}(\vec{x}) = \frac{\sum_{i \in L_j} \vec{y}_i a_i(||\vec{x} - \vec{c}_i||)}{\sum_{i \in L_j} a_i(||\vec{x} - \vec{c}_i||)}$$

For the description of boxes consider the input space to be an axiparallel hyper–rectangular. Due to practical considerations divide some or all of the axis of the input space into equally sized disjunct partitions inducing a partition of the input space into a system of disjunct axiparallel hyper–rectangulars. This allows fast derivation of the distances between the boxes and a given neuron during construction of the lists L_j due to fast projection methods. During recall, the allocation of a given point to the corresponding box is achieved by a fast, recursive (in dimension of input space) algorithm.

A second approach is induced by the observation, that in case a_i is a Gaussian $a_i(d) < T$ for $T > 0$ implies $d^2 > -\sigma_i^2 \ln T$. Since $\ln T$ is a negative constant it is possible to decide by one multiplication and one compare operation whether or not the exponential function has to be evaluated. This approach can be further elaborated since $\sum_{k=0}^{l}(\vec{x}_k - \vec{c}_{i,k})^2$ is monotonically increasing in l. So $\sum_{k=0}^{l}(\vec{x}_k - \vec{c}_{i,k})^2 > -\sigma_i^2 \ln T$ for some l implies $a_i(x) = 0$, possibly saving multiplications at the cost of extra compare operations. The method with all compare operations will be called 'inner cut–off method', with a single comparison before evaluation of exponentials 'outer cut–off method'.

3 Experimental Results

To show the properties of the two approaches we use a network generated by the Node Allocation and Topographical Encoding (NATEnet) scheme (Hakala et. al., 1993) for robotarm inverse kinematics. The network has a five dimensional input space normalized to the hypercube $[-1, 1]^5$. The network consists of 18549 neurons with 36–dimensional output.

Without any acceleration methods 18549 calculations of exponentials have to be calculated each time an input is present. Using the cut–off strategy alone with differing cutting parameter T, fig. 2 shows that the average number of calculations of exponentials per recall is basically proportional to $(\ln T + c)^{dim/2}$ and small corrections due to edge effects for small T where c is constant.

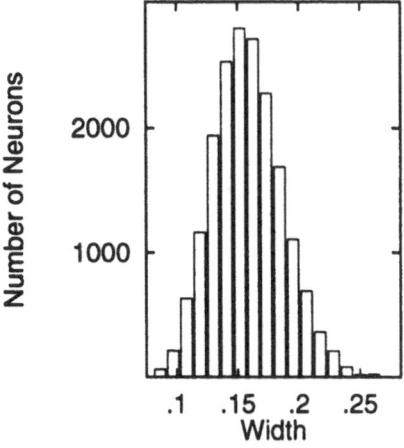

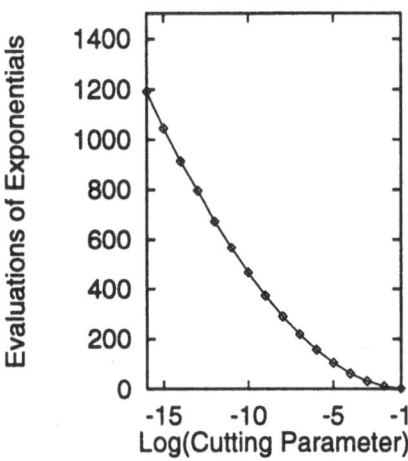

Figure 1 : The distribution of widths σ_i for the specific network reflects the average distance to the nearest neighbors.

Figure 2 : The 'cut–off method' without boxes leads to a reduced number of calculations of exponentials compared to straight–forward evaluation of all neurons, which would require 18549 calculations of exponentials. The exact number depends on the cutting parameter T

The evaluation of the 'boxes method' uses projections onto 2, 3, 4 or 5 dimensions. We use rectangular boxes with equal sidelengths for each dimension. The average number of multiplications for the 'outer cut–off method' and 'inner cut–off method' varies with the number of dimensions of the boxes and the number of boxes used. The calculations needed for box selection are negligable.

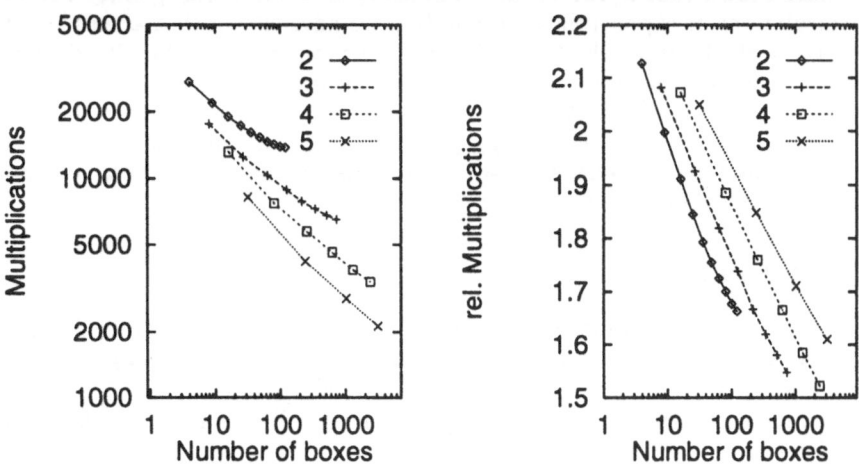

Figure 3 : The number of multiplications required for a complete update of the network depending on the number of boxes for 2, 3, 4, 5 dimensional boxes using the 'inner cut–off method' (left) and the relative increase using the 'outer cut–off method' for $\log(T) = -3$. The update using all neurons requires 92745 multiplications.

470

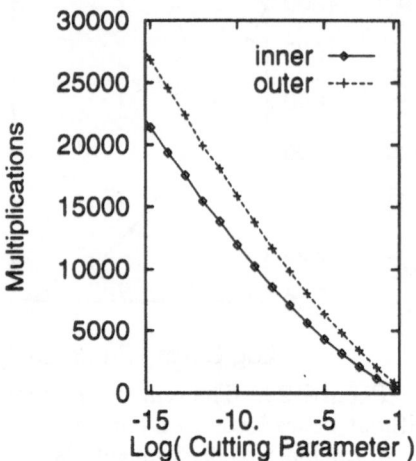

Figure 4 : Number of multiplications for a network update depending on the cut-off parameter using the 'inner' and 'outer cut-off method'.

The parameter $T = 10^{-3}$ was chosen using cross–validation with respect to the residual error due to cut–off. Fig. 3 shows benefical use of higher dimensional boxes. If the required compare operations need less computing time than a multiplication the use of the 'inner cut–off method' is advisable. The overall result is to use the 'cut–off method' with boxes in the full dimensions of space X. The number of multiplications for 5^5 boxes in 5 dimensions depending on the cutting parameter T for both 'cut–off' methods is low even for small T (Fig. 4). This method is apt for accelerated update on MIMD architectures distributing equally sized parts of each list on each processors. Experimental results indicate that the requirement for data of a single neuron being present on a single processor is fulfillable for a moderate number of processors.

4 Conclusions

We have shown that a 'cut–off method' leads to a considerably lower number of calculations of exponentials. In addition we have shown that a simple data structure called 'box method' significantly lowers the number of multiplications needed to perform a complete update of the network. Bringing both methods together enables fast and reliable network updates, where the accuracy in terms of activity is chosen by the cutting parameter T. The approach is suited to accelerate updates on multiprocessors as each list can be split into equally sized parts and distributed onto the array of processors.

5 References

J. Hakala and R. Eckmiller, (1993), Node allocation and topographical encoding NATEnet for inverse kinematics of a 6–dof robot arm, in *Artifical Neural Networks* (S. Gielen and B. Kappen, eds.), pp. 309–313.

J. Moody and C. Darken, (1989), Learning with localized receptive fields, *Neural Computation*, vol. 1, pp. 281–294.

S. M. Omohundro, (1991), Bumptrees for efficient function, constraint, and classification learning, in *Adv. in Neural Inf. Proc. Sys.* (Lippman, Moody, and Touretzky, eds.), pp. 693–699, Morgan Kaufman, San Mateo.

T. Sanger, (1991), Basis–function trees as a generalization of local variable selection methods, in *Adv. in Neural Inf. Proc. Sys.* (Lippman, Moody, and Touretzky, eds.), pp. 701–706, Morgan Kaufman, San Mateo.

Uniqueness of Functional Representations by Gaussian Basis Function Networks

Věra Kůrková, Roman Neruda

Institute of Computer Science

Prague, Czechia

1   Introduction

In most applications dealing with pattern recognition, neural networks are trained to minimize an error function over all parametrizations corresponding to a chosen architecture. The problem of multiple global minima is related to the question of redundancy of a network parametrization. Hecht-Nielsen (1990) pointed out that characterization of functionally equivalent (i.e., determining the same input-output function) weight vectors might speed up some learning algorithms. Several authors studied functionally equivalent weight vectors for perceptron-type networks with hyperbolic tangent as an activation function (Sussmann, 1992, Chen et al., 1993). Recently, also the problem for perceptron-type networks with more general activation functions was studied; Albertini and Sontag (1993), Kůrková and Kainen (1993), Kainen et al. (1993).

In this paper, we show that in the case of standard RBF networks with Gaussian radial function, the input-output behavior uniquely determines network parametrization (weights, widths and centroids) up to a permutation of hidden units. Applying this result, we describe cannonical parametrizations of Gaussian basis function networks and discuss consequences for learning.

2   Cannonical Representations of Functions by Gaussian RBF Networks

We study feedforward networks with one hidden layer containing RBF units with a radial function $\varphi : \mathcal{R}_+ \to \mathcal{R}$ and norm $\| \cdot \|$ on $\mathcal{R}^n$ (n is the number of input units) and with a single linear output unit. Such networks compute functions $\mathcal{R}^n \to \mathcal{R}$ of the form

$$\sum_{i=1}^{k} w_i \varphi \left(\frac{\|\mathbf{x} - \mathbf{c}_i\|}{b_i} \right),$$

where k is the number of hidden units, and $w_i, b_i \in \mathcal{R}, b_i > 0$, and $\mathbf{c}_i \in \mathcal{R}^n (i = 1, \dots, k)$ are parameters (called weights, widths and centroids, respectively). The set of parameters $\{w_i, b_i, \mathbf{c}_i; i = 1, \dots, k\}$ is called an RBF or more precisely a $(\varphi, \| \cdot \|)$ RBF *network parametrization*. If additionally, for every $i = 1, \dots, k$ $w_i \neq 0$ and for every $i \neq j$ either $\mathbf{c}_i \neq \mathbf{c}_j$ or $b_i \neq b_j$, it is called a *reduced parametrization*.

Two $(\varphi, \| \cdot \|)$ RBF network parametrizations $\{w_i, b_i, c_i; i = 1, \ldots, k\}$ and $\{w'_i, b'_i, c'_i; i = 1, \ldots, k'\}$ are called *functionally equivalent* if they determine the same input-output function, i.e.

$$\sum_{i=1}^{k} w_i \varphi \left(\frac{\|\mathbf{x} - \mathbf{c}_i\|}{b_i} \right) = \sum_{i=1}^{k'} w'_i \varphi \left(\frac{\|\mathbf{x} - \mathbf{c}'_i\|}{b'_i} \right).$$

Two network parametrizations are called *interchange equivalent* if $k = k'$ and there exists a permutation π of the set $\{1, \ldots, k\}$, such that $w_i = w'_{\pi(i)}$ and $b_i = b'_{\pi(i)}$ and $\mathbf{c}_i = \mathbf{c}'_{\pi(i)}$, for each $i \in \{1, \ldots, k\}$.

The standard choice of a radial function is Gaussian and the most popular norms are those induced by various inner products. We will show that in this case, an input-ouput function determines the network parametrization uniquely up to a permutation of hidden units. First, we prove two technical lemmas.

Lemma 2.1 *If for every function $\phi : \mathcal{R}_+ \to \mathcal{R}$, for every positive integer n and for every norm $\| \cdot \|$ on $\mathcal{R}^n$ there exists no reduced $(\varphi, \| \cdot \|)$ RBF network parametrization generating the constant zero function, then any two reduced $(\varphi, \| \cdot \|)$ RBF network parametrizations are functionally equivalent if and only if they are interchange equivalent.*

Proof: Suppose that there exist two different reduced $(\varphi, \| \cdot \|)$ RBF network parametrizations determining the same function

$$\sum_{i=1}^{k} w_i \varphi \left(\frac{\|\mathbf{x} - \mathbf{c}_i\|}{b_i} \right) = \sum_{i=1}^{k'} w'_i \varphi \left(\frac{\|\mathbf{x} - \mathbf{c}'_i\|}{b'_i} \right).$$

Subtract the right side of the above equation from the left one and put together the terms having the same centroid and width. This leads to a reduced parametrization which generates the constant zero function. $\square$

Lemma 2.2 *Let $\| \cdot \|$ be a norm induced by an inner product $\cdot$ on $\mathcal{R}^n$, $\{\mathbf{d}_i; i = 1, \ldots, k\}$ be a set of distinct non-zero vectors in $\mathcal{R}^n$. Let $j \in \{1, \ldots, k\}$ be such that $\|\mathbf{d}_i\| \leq \|\mathbf{d}_j\| \ \forall i = 1, \ldots, k$. Then $\mathbf{d}_i \cdot \mathbf{d}_j < \mathbf{d}_j \cdot \mathbf{d}_j$ for every $i \neq j$.*

Proof: To prove it by contradiction suppose that there exists an index i such that $\mathbf{d}_i \cdot \mathbf{d}_j = \mathbf{d}_j \cdot \mathbf{d}_j$. By Schwartz's inequality:

$$\|\mathbf{d}_j\|^2 = \mathbf{d}_i \cdot \mathbf{d}_j \leq |\mathbf{d}_i \cdot \mathbf{d}_j| \leq \|\mathbf{d}_i\|\|\mathbf{d}_j\|.$$

Hence $\|\mathbf{d}_j\| \leq \|\mathbf{d}_i\| \leq \|\mathbf{d}_j\|$ and so $\|\mathbf{d}_i\| = \|\mathbf{d}_j\|$

Let α_{ij} be an angle between $\mathbf{d}_i$ and $\mathbf{d}_j$. Since $\mathbf{d}_i \cdot \mathbf{d}_j = \|\mathbf{d}_i\|\|\mathbf{d}_j\| \cos(\alpha_{ij})$, and $\mathbf{d}_i \cdot \mathbf{d}_j = \mathbf{d}_j \cdot \mathbf{d}_j = \|\mathbf{d}_j\|^2$, we have $\cos(\alpha_{ij}) = 1$. It implies $\mathbf{d}_i = \mathbf{d}_j$, which gives a contradiction. $\square$

Theorem 2.3 *Let n be a positive integer, $\| \cdot \|$ norm on $\mathcal{R}^n$ induced by an inner product, $\gamma(t) = \exp(-t^2)$. Then any two reduced $(\gamma, \| \cdot \|)$ RBF network parametrizations are functionally equivalent if and only if they are interchange equivalent.*

Proof: By Lemma 2.1 it is sufficient to prove that no reduced $(\gamma, \|\cdot\|)$ RBF network parametrization $\{w_i, b_i, c_i; i = 1, \ldots, k\}$ generates the zero function. Suppose to the contrary that for every $x \in \mathcal{R}^n$

$$\sum_{i=1}^{k} w_i \exp\left(-\frac{\|x - c_i\|^2}{b_i^2}\right) = 0. \tag{1}$$

Additionally, we may suppose that $1 = b = \max\{b_i; i = 1, \ldots, k\}$ (otherwise we change scale) and that the b_i are arranged in such a way that $1 = b_1 = \ldots = b_m > b_i$ for every $i = m + 1, \ldots, k$.

Multiplying (1) by $\exp\left(\|x - c_1\|^2\right)$, we get

$$0 = w_1 + \sum_{i=2}^{m} \hat{w}_i \exp\left(2x \cdot (c_i - c_1)\right) +$$

$$+ \sum_{i=m+1}^{k} \hat{w}_i \exp\left(\|x\|^2 \left(1 - \frac{1}{b_i^2}\right) - 2x \cdot \left(c_1 - \frac{c_i}{b_i^2}\right)\right)$$

where $\hat{w}_i = w_i \exp\left(\|c_1\|^2 - \frac{\|c_i\|^2}{b_i^2}\right)$.

Since for every $i = m + 1, \ldots, k$ $1 - \frac{1}{b_i^2} < 0$, for every $u \in \mathcal{R}^n$ the following holds:

$$\lim_{t \to \infty} \sum_{i=m+1}^{k} \hat{w}_i \exp\left(\|tu\|^2 \left(1 - \frac{1}{b_i^2}\right) - 2tu \cdot \left(c_1 - \frac{c_i}{b_i^2}\right)\right) = 0.$$

Then, for every $u \in \mathcal{R}^n$

$$-w_1 = \lim_{t \to \infty} \sum_{i=2}^{m} \hat{w}_i \exp\left(2tu \cdot (c_i - c_1)\right) \tag{2}$$

Put $d_i = c_i - c_1$ and choose among $\{d_2, \ldots, d_m\}$ the vector d_j with the maximal $\|d_j\|$, i.e. $\|d_j\| \geq \|d_i\|; \forall i = 2, \ldots, m$. By Lemma 2.2, for every $i \neq j : d_i \cdot d_j < d_j \cdot d_j$.

Substituting $x = d_j$ into (2) we get

$$-w_1 = \lim_{t \to \infty} \sum_{i=2}^{m} \hat{w}_i \exp\left(2td_j \cdot d_i\right) \tag{3}$$

Multiplying (3) by $\exp\left(-2t\|d_j\|^2\right)$ we obtain

$$-\lim_{t \to \infty} w_1 \exp\left(-2t\|d_j\|^2\right) = \hat{w}_j + \lim_{t \to \infty} \sum_{\substack{i=2 \\ i \neq j}}^{m} \hat{w}_i \exp\left(2t(d_j \cdot d_i - d_j \cdot d_j)\right).$$

As both terms $-\|d_j\|^2$ and $(d_j \cdot d_i - d_j \cdot d_j)$ are negative (the first one is obvious, the second one follows from Lemma 2.2), the limits on both sides equal 0. Thus, we get $\hat{w}_j = w_j e^{-\|c_j\|^2} = 0$. Hence $w_i = 0$, which contradicts the assumption that the parametrization is reduced. □

Theorem 2.3 enables us to describe easily a cannonical representation of a network computing a particular function. Since no parameter transformations other than interchanges leave the input-output function unchanged, we only have to choose a cannonical representation of a set of networks with permuted hidden units. One possible choice is to impose the condition on a parametrization that weight vectors corresponding to hidden units are in increasing lexicographic order. Represent a parametrization $\{w_i, b_i, c_i; i = 1, \ldots, k\}$ as a vector $\mathbf{p} = \{\mathbf{p}_i, \ldots \mathbf{p}_k\} \in \mathcal{R}^{k(n+2)}$, where $\mathbf{p}_i = \{w_i, b_i, c_{i1}, \ldots, c_{in}\} \in \mathcal{R}^{n+2}$ is a weight vector corresponding to the i-th hidden unit. Let $\prec$ denote the lexicographic order on $\mathcal{R}^{n+2}$, i.e. for $\mathbf{p}, \mathbf{q} \in \mathcal{R}^{n+2}$ $\mathbf{p} \prec \mathbf{q}$ if there exists an index $m \in \{1, \ldots, n+2\}$ such that $p_j = q_j$ for $j < m$ and $p_m < q_m$. For a norm $\|\cdot\|$ induced by an inner product, we call a $(\gamma, \|\cdot\|)$ RBF network parametrization $\mathbf{p}$ *cannonical* if $\mathbf{p}_1 \prec \mathbf{p}_2 \prec \ldots \mathbf{p}_k$.

In this terminology, Theorem 2.3. guarantees that for every $(\gamma, \|\cdot\|)$ RBF parametrization, there exists a cannonical RBF network parametrization determining the same input-output function. Chen and Hecht-Nielsen (1991) proposed to study subsets of weight spaces containing exactly one representative of each class of functionally equivalent weight vectors – they called them *minimal search sets*.

Corollary 2.4 *For every positive integer n, for every norm $\|\cdot\|$ on $\mathcal{R}^n$ induced by an inner product, the set of cannonical representations forms a minimal search set.*

Some learning algorithms like genetic learning can operate only with cannonical parametrizations. Since there are $k!$ permutations of k hidden units, genetic algorithms can be restricted to search only within $1/k!$ of the whole weight space, which allows considerable time reduction.

For other radial functions, the essential uniqueness of network parametrization may not hold, and search sets can be more complicated than for the Gaussian case. These questions will be investigated in future work.

References

Albertini, F., Sontag, E.D. (1993). For neural networks, function determines form. *Neural Networks* (in press).

Chen, A. M., Haw-minn, L., Hecht-Nielsen, R. (1993). On the geometry of feedforward neural network error spaces. *Neural Computation* 5, 6.

Hecht-Nielsen, R. (1990). On the algebraic structure of feedforward network weight spaces. In *Advanced Neural Computers*, 129-135. Elsevier.

Kainen, P., Kůrková, V., Kreinovich, V., Sirisengtaksin, O., (1993). A New Criterion for Chooseing an Activation Function: Uniqueness Leads to Faster Learning. *Neural Parallel and Scientific Computations* (in press).

Kůrková, V., Kainen, P. (1993). Functionally equivalent feedforward networks. *Neural Computation* (in press).

Sussmann, H.J. (1992). Uniqueness of the weights for minimal feedforward nets with a given input-output map. *Neural Networks* 5, 4, 589-594.

A Dynamic Mixture of Gaussians Neural Network for Sequence Classification

M. Ceccarelli†, J. T. Hounsou‡

† Istituto per la Ricerca sui Sistemi Informatici Paralleli, IRSIP-CNR
Naples, ITALY

‡ International Institute for Advanced Scientific Studies, IIASS
Vietri sul Mare (SA), ITALY

1 Introduction

In the last years the inclusion of the variable time inside neural network models led to the study of dynamic neural networks for time-series forecasting, control plant modelling, spatio-temporal pattern recognition. Both TDNN (Waibel *et al.* 1991) and recurrent networks (Williams *et al.* 1989) try to incorporate contextual information, the dynamics of the unknown input-output mapping, by considering delayed or recurrent connections such that the output of a given unit at a certain time instant depends on the output of other neurons in the net at a previous time instant. Most of the learning algorithms for this kind of networks make use of the gradient descent method for the weight adjustment, requiring that the activation function of each neuron be differentiable. A typical choice is a sigmoid function of a weighted sum of the inputs to a given neuron. On the contrary, with the use of locally-tuned neurons, *i.e.* neurons with a radial basis activation function (RBF), it is possible to train a static network in a layer-by-layer fashion making use of vector quantization (competitive hebbian learning) followed by least mean square optimization (Moody *et al.* 1989).

Our paper is dicated to the study of a recurrent network of locally-tuned neurons. Specifically, the RBF model is extended in order to include recurrent connections in the hidden layer. This is achieved by localizing the activation of the hidden neurons both on the input space and on the network activation space. Thus the dynamics of each neuron is no more an RBF but a mixture of two RBFs; the specific function considered are Gaussian functions. An exact on-line learning procedure is derived and experimented on a task of sequence recognition. A layer-by-layer learning procedure is also investigated.

2 The MOG Model

The inclusion of delayed connections into the RBF model is a natural way to incorporate contextual information in this model and gave apprecciable results for sequence recognition (Ceccarelli *et al.* 1993). However, the ability of a neural system to recognize sequence can be improved by considering recurrent connections. In this way the network evolves on the basis of the current input and its internal state. Considering that a given input sequence produces a certain trajectory of activations in the hidden layer of an RBF network, our idea is based on the fact that the above goal can be achieved by making the activation of the hidden nodes dependent on the current input and on their state at a previous time slice through the use of a mixture of gaussians (MOG). In this way, the activation of the hidden nodes is localized on both the input and the activation spaces. As we shall see, with this choice the resulting network can also be trained with a sequence of learning steps each consisting into the application of a competitive hebbian learning rule.

Let us consider a simplified situation of a three layer neural network with L output nodes and M hidden nodes. By clamping the input nodes at time t with an input vector $\mathbf{x}(t)$ the output neurons produce the following results:

$$f_i(t) = F_i(\sum_{j=1}^{M} w_{ij}^1 \varphi_j(t)) \quad 1 \leq i \leq L \tag{1}$$

where $F_i(x) = x$ or $F_i(x) = \frac{1}{1+e^{-x}}$ and

$$\varphi_j(t) = \alpha_0 exp(-||\mathbf{x}(t) - \mathbf{w}_j^0||^2) + \alpha_1 exp(-||\mathbf{\Phi}(t-1) - \mathbf{W}_j^0||^2) \tag{2}$$

with $\mathbf{\Phi}(t-1) = (\varphi_1(t-1), \ldots, \varphi_M(t-1))$, $\alpha_0 + \alpha_1 = 1$ and $\alpha_0, \alpha_1 \geq 0$. In this way, the j-th hidden neuron receives signals from the input layer at the current time instant and from the hidden layer at the previous time instant: the ON-region of its activation function $\varphi_j(t)$ is localized in both spaces.

2.1   Exact On-line Learning

In order to exploit the ability of this kind of networks, let us derive a gradient descent learning algorithm for them. Given a sequence of input vector signals at discrete times $\mathbf{x}(t_0), \ldots, \mathbf{x}(t_f)$ and a sequence of target vector signals for the output layer $\mathbf{d}(t_0), \ldots, \mathbf{d}(t_f)$, the overall error on the whole sequence is given by

$$E = \sum_{t=t_0}^{t_f} E(t) = \sum_{t=t_0}^{t_f} \frac{1}{2} \sum_{i=1}^{L} [d_i(t) - f_i(t)]^2.$$

Following the gradient descent method, the adaptation of a given weight must be proportional to the derivative of E with respect to it. For each output unit i, $1 \leq i \leq L$ call $net_i(t)$ its potential at time t, i.e. $net_i(t) = \sum_{m=1}^{M} w_{im}^1 \varphi_m(t)$ $t_0 \leq t \leq t_f$. Then for the weights w_{ij}^1 we have

$$-\frac{\partial E(t)}{\partial w_{ij}^1} = [d_i(t) - f_i(t)] \frac{\partial f_i(t)}{\partial w_{ij}^1} = [d_i(t) - f_i(t)] F_i'(net_i(t)) \varphi_j(t)$$

Thereafter by using the usual chaining rule we obtain:

$$-\frac{\partial E(t)}{\partial w_{jk}^0} = \sum_{i=1}^{L} [d_i(t) - f_i(t)] \frac{\partial f_i(t)}{\partial w_{jk}^0} = \sum_{i=1}^{L} [d_i(t) - f_i(t)] F_i'(net_i(t)) \sum_{r=1}^{M} w_{ir}^1 \frac{\partial \varphi_r(t)}{\partial w_{jk}^0}.$$

Calling $G_r^0(t) = exp(-||\mathbf{x}(t) - \mathbf{w}_r^0||^2)$ and $G_r^1(t) = exp(-||\mathbf{\Phi}(t-1) - \mathbf{W}_r^0||^2)$, it is easy to show that

$$\frac{\partial\varphi_r(t_0)}{\partial w_{jk}^0} = \delta_{rj}\alpha_0 G_r^0(t_0)2(x_k(t_0) - w_{rk}^0),$$

and for $t_0 < t \leq t_f$

$$\frac{\partial\varphi_r(t)}{\partial w_{jk}^0} = \delta_{rj}\alpha_0 G_r^0(t)2(x_k(t) - w_{rk}^0) + \alpha_1 G_r^1(t)2\sum_{s=1}^{M}(\varphi_s(t-1) - W_{rs}^0)\frac{\partial\varphi_s(t-1)}{\partial w_{jk}^0}$$

with δ_{rj} being the Kroneker delta function. In the same way, the derivatives of $\varphi_r(t)$ with respect to the weights W_{jk}^0 are given by

$$\frac{\partial\varphi_r(t_0)}{\partial W_{jk}^0} = 0,$$

$$\frac{\partial\varphi_r(t_0 + 1)}{\partial W_{jk}^0} = \delta_{rj}\alpha_1 G_r^1(t_0 + 1)2(\varphi_k(t_0) - W_{rk}^0),$$

and for $t_0 + 1 < t \leq t_f$

$$\frac{\partial\varphi_r(t)}{\partial W_{jk}^0} = \alpha_1 G_r^1(t)2\sum_{s=1}^{M}(\varphi_s(t-1) - W_{rs}^0)\frac{\partial\varphi_s(t-1)}{\partial W_{jk}^0}.$$

It is worth noting that the rule just derived resembles the classical William and Zipser forward propagation. Its cost, $O(M^4)$, borns from the fact that for each time slice a set of equations must be iterated for each incoming weight to hidden neurons, since such weights affect directly or indirectely all the network dynamics.

2.2 Iterated Vector Quantization

Another less expensive way to train RBF classifiers is the use of vector quantization for the computation of center location. However, in the case of MOG functions this is not immediate since quantization of the weights W_{jk}^0 must be performed of the basis of the patterns of activation of the hidden layer, but these activations depend on these weights. Therefore, there is the need of an iterated vector quantization procedure in which the patterns of activations of the hidden layer, at each iteration, are almost constant. A sort of "annealing" can be used in the following way. Suppose to start with a standard RBF network, $\alpha_0 = 1$ and $\alpha_1 = 0$; then the weights W_{jk}^0 can be set by clustering the pattern of activations of the hidden layer. At this point, a smooth increase of α_1 and a smooth decrease of α_0, i. e. $\alpha_0 \leftarrow \alpha_0 - \epsilon$ and $\alpha_1 \leftarrow \alpha_1 + \epsilon$, produce a smooth change in the pattern of activation of the hidden nodes. These patterns of activation can be used for a new clustering step. This process can be continued unitil α_0 and α_1 reach the desired values. Once the centroids of the gaussians have been determined using the above procedure, the weights w_{ij}^1 can be computed by using the delta rule or the pseudoinverse rule. A large number of simulations showed that this procedure is significatively faster than the gradient descent method and allows to use networks with a larger number of hidden neurons.

3   Simulation Results

The MOG networks described in the previous section seem particularly suited for sequence classification tasks. It is our aim, in this section, to show how the classification ability of an RBF network is improved by the addition of the recurrent weights W_{jk}^0 by using both the learning procedures described above. We considered the problem of classification of patterns of variable lenght, each pattern containing one of three different shapes: a sin wave, a square wave and a triangular wave. These shapes were randomly shifted inside the patterns and randomly time-scaled.

With the use of a network with 16 hidden neurons trained with the on-line exact algorithm we obtain the following recognition performance: 99.7% of correct classification for training patterns and 97.7% of correct classification of the test patterns. A standard time delay RBF network on the same task reaches respectively 99.0% and 96.9% on the training and test sets. Moreover, this improvement is much more evident in the case of layer-by-layer learning. Indeed, an RBF network with 64 hidden units, trained with kohonen's competitive learning for the computation the weight vectors $\mathbf{w}_j^0$ and delta-rule for the computation of the weight vectors $\mathbf{w}_k^1$, reaches 72.0% of recognition on the training and 68.8% of recognition on the test set. On the contrary a MOG network trained with the iterated vector quantization procedure described above, each step using Kohonen's rule, reaches 84.2% of correct classification on the training set and 83.3% of correct classification on the test set. Indeed, the networks trained with the gradient descent, for the example shown, have more discriminatory power, but it must be considered that the computational cost of the learning algorithm is much greater than that of iterated vector quantization. In addition, for applications such as speech, where the two methodologies reach comparable results (Ceccarelli et al. 1993) the improvement produced by MOG functions may be even more evident.

Aknowledgements

This work was supported by MURST 40% unita' INFN of Salerno and by contratto quinquennale IIASS-CNR.

References

Ceccarelli, M. and Hounsou, J. T. (1993) RBF Networks vs. Multilayer Perceptrons for Sequence Recognition. *Proceedings of IEEE Conference on System man and Cybernetics '93*, IEEE Press.

Moody, J. and Darken, C. J. (1989) Fast Learning in Networks of Locally Tuned Processing Units, *Neural Computation*, 1, 281-294.

Waibel A., Hanazawa T., Hinton G., Shikano K. and Lang K. J. (1991) Phoneme Recognition Using Time-Delay Neural Networks, *IEEE Transaction on ASSP*, 24, 1085-1091.

Williams, R. J. and Zipser D. (1989) A Learning Algorithm for Continually Running Fully Recurrent Neural Networks. *Neural Computation*, 1, 270-280.

Hierarchical Mixtures of Experts and the EM Algorithm

M. I. Jordan†, R. A. Jacobs‡

† Massachusetts Institute of Technology
Cambridge, MA, USA
‡ University of Rochester
Rochester, NY, USA

1 Introduction

In the statistical literature and in the machine learning literature, divide-and-conquer algorithms have become increasingly popular. The CART algorithm (Breiman, et al., 1984) and the MARS algorithm (Friedman, 1991) are well-known examples. These algorithms fit surfaces to data by explicitly dividing the input space into a nested sequence of regions, and by fitting simple surfaces (e.g., constant functions) within these regions. The advantages of these algorithms include the interpretability of their solutions and the speed of the training process.

In this paper we present a neural network architecture that is a close cousin to architectures such as CART and MARS. As in our earlier work (Jordan & Jacobs, 1992) we formulate the learning problem for this architecture as a maximum likelihood problem. In the current paper we utilize the Expectation-Maximization (EM) framework to derive the learning algorithm.

2 Hierarchical mixtures of experts

The algorithms that we discuss in this paper are supervised learning algorithms. We explicitly address the case of regression, in which the input vectors are elements of $\Re^m$ and the output vectors are elements of $\Re^n$. Our model also handles classification problems and counting problems in which the outputs are integer-valued. The data are assumed to form a countable set of paired observations $\mathcal{X} = \{(\mathbf{x}^{(t)}, \mathbf{y}^{(t)})\}$. In the case of the *batch* algorithm discussed below, this set is assumed to be finite; in the case of the *on-line* algorithm, the set may be infinite.

We propose to solve nonlinear supervised learning problems by dividing the input space into a set of regions and fitting simple surfaces to the data that fall in these regions. The regions are nested, yielding a multiresolution analysis. The regions have "soft" boundaries, meaning that data points may lie simultaneously in multiple regions. The boundaries between regions are themselves simple parameterized surfaces that are adjusted by the learning algorithm.

The hierarchical mixture-of-experts (HME) architecture is shown in Figure 1.[1] The architecture is a tree in which the *gating networks* sit at the

---

[1] To simplify the presentation, we restrict ourselves to a two-level hierarchy throughout

480

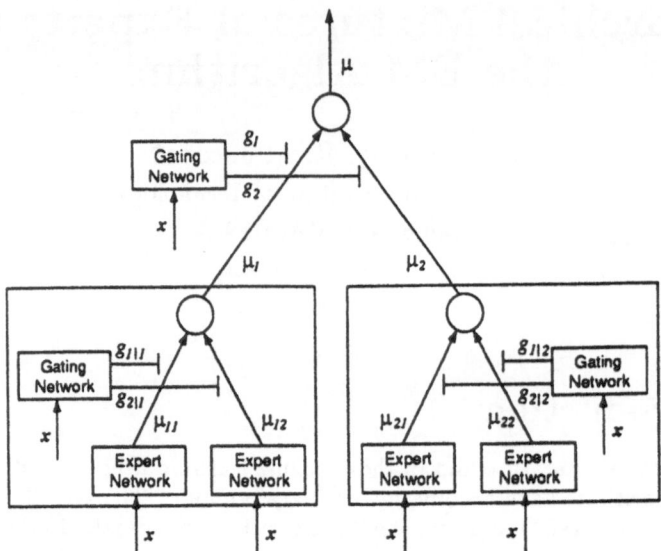

Figure 1: A two-level hierarchical mixture of experts.

nonterminals of the tree. These networks receive the vector **x** as input and produce scalar outputs that are a partition of unity at each point in the input space. The *expert networks* sit at the leaves of the tree. Each expert produces an output vector μ_{ij} for each input vector. These output vectors proceed up the tree, being multiplied by the gating network outputs and summed at the nonterminals.

All of the expert networks in the tree are linear with a single output nonlinearity. We will refer to such a network as "generalized linear," borrowing the terminology from statistics (McCullagh & Nelder, 1983). Expert network (i, j) produces its output μ_{ij} as a generalized linear function of the input **x**:

$$\mu_{ij} = f(U_{ij}\mathbf{x}), \tag{1}$$

where U_{ij} is a weight matrix and f is a fixed continuous nonlinearity. The vector **x** is assumed to include a fixed component of one to allow for an intercept term.

For regression problems, $f(\cdot)$ is the identity function (i.e., the experts are linear). For binary classification problems, $f(\cdot)$ is generally taken to be the logistic function, in which case the expert outputs are interpreted as the log odds of "success" under a Bernoulli probability model. Other models (e.g., multiway classification, counting, rate estimation and survival estimation) are handled readily by making other choices for $f(\cdot)$. These models are smoothed piecewise analogs of the corresponding generalized linear models (GLIM's; cf. McCullagh & Nelder, 1983).

The gating networks are also generalized linear. At the top level, define linear predictors ξ_i as follows:

$$\xi_i = \mathbf{v}_i^T\mathbf{x}, \tag{2}$$

the paper. All of the algorithms that we describe, however, generalize readily to hierarchies of arbitrary depth.

where $\mathbf{v}_i$ is a weight vector. Then the ith output of the top-level gating network is the "softmax" function of the ξ_i.

$$g_i = \frac{e^{\xi_i}}{\sum_k e^{\xi_k}}. \tag{3}$$

Note that the g_i are positive and sum to one for each $\mathbf{x}$. The gating networks at the lower level are defined similarly, yielding outputs $g_{j|i}$ that are obtained by taking the softmax function of linear predictors $\xi_{ij} = \mathbf{v}_{ij}^T \mathbf{x}$.

The output vector at each nonterminal of the tree is the weighted output of the experts below that nonterminal. That is, the output at the ith nonterminal in the second layer of the two-level tree is:

$$\mu_i = \sum_j g_{j|i}\mu_{ij}$$

and the output at the top level of the tree is:

$$\mu = \sum_i g_i \mu_i.$$

Note that both the g's and the μ's depend on the input $\mathbf{x}$, thus the total output is a nonlinear function of the input.

2.1   A probability model

The hierarchy can be given a probabilistic interpretation. We suppose that the mechanism by which data are generated by the environment involves a nested sequence of decisions that terminates in a regressive process that maps $\mathbf{x}$ to $\mathbf{y}$. The decisions are modeled as multinomial random variables. That is, for each $\mathbf{x}$, we interpret the values $g_i(\mathbf{x}, \mathbf{v}_i^0)$ as the multinomial probabilities associated with the first decision and the $g_{j|i}(\mathbf{x}, \mathbf{v}_{ij}^0)$ as the (conditional) multinomial probabilities associated with the second decision. We use a statistical model to model these probabilities; in particular, our choice of parameterization (cf. Eqs. 2 and 3) corresponds to a *log-linear* probability model (see Jordan & Jacobs, 1994). A log-linear model is a special case of a GLIM that is commonly used for "soft" multiway classification (McCullagh & Nelder, 1983). Under the log-linear model, we interpret the gating networks as modeling the input-dependent, multinomial probabilities of making particular nested sequences of decisions.

Once a particular sequence of decisions has been made, output $\mathbf{y}$ is assumed to be generated according to the following generalized linear statistical model. First, a linear predictor η_{ij} is formed:

$$\eta_{ij}^0 = U_{ij}^0 \mathbf{x},$$

where the superscript refers to the "true" values of the parameters. The expected value of $\mathbf{y}$ is obtained by passing the linear predictor through the *link function f*:

$$\mu_{ij}^0 = f(\eta_{ij}^0).$$

The output $\mathbf{y}$ is then chosen from a probability density P, with mean $\boldsymbol{\mu}_{ij}^0$ and "dispersion" parameter ϕ_{ij}^0. We denote the density of $\mathbf{y}$ as:

$$P(\mathbf{y}|\mathbf{x}, \boldsymbol{\theta}_{ij}^0),$$

where the parameter vector $\boldsymbol{\theta}_{ij}^0$ includes the weights U_{ij}^0 and the dispersion parameter ϕ_{ij}^0. We assume the density P to be a member of the exponential family of densities (McCullagh & Nelder, 1983). The interpretation of the dispersion parameter depends on the particular choice of density. For example, in the case of the n-dimensional Gaussian, the dispersion parameter is the covariance matrix Σ_{ij}^0.[2]

Given these assumptions, the total probability of generating $\mathbf{y}$ from $\mathbf{x}$ is the mixture of the probabilities of generating $\mathbf{y}$ from each of the component densities, where the mixture components are multinomial probabilities:

$$P(\mathbf{y}|\mathbf{x}, \boldsymbol{\theta}^0) = \sum_i g_i(\mathbf{x}, \mathbf{v}_i^0) \sum_j g_{j|i}(\mathbf{x}, \mathbf{v}_{ij}^0) P(\mathbf{y}|\mathbf{x}, \boldsymbol{\theta}_{ij}^0). \tag{4}$$

Note that $\boldsymbol{\theta}^0$ includes the expert network parameters $\boldsymbol{\theta}_{ij}{}^0$ as well as the gating network parameters $\mathbf{v}_i^0$ and $\mathbf{v}_{ij}^0$. Note also that we can utilize Eq. 4 without the superscripts to refer to the probability model defined by a particular HME architecture, irrespective of any reference to a "true" model.

2.2 Posterior probabilities

In developing the learning algorithms to be presented in the remainder of the paper, it will prove useful to define posterior probabilities associated with the nodes of the tree. The terms "posterior" and "prior" have meaning in this context during the training of the system. We refer to the probabilities g_i and $g_{j|i}$ as *prior* probabilities, because they are computed based only on the input $\mathbf{x}$, without knowledge of the corresponding target output $\mathbf{y}$. A *posterior* probability is defined once both the input and the target output are known. Using Bayes' rule, we define the posterior probabilities at the nodes of the tree as follows:

$$h_i = \frac{g_i \sum_j g_{j|i} P_{ij}(\mathbf{y})}{\sum_i g_i \sum_j g_{j|i} P_{ij}(\mathbf{y})} \tag{5}$$

and

$$h_{j|i} = \frac{g_{j|i} P_{ij}(\mathbf{y})}{\sum_j g_{j|i} P_{ij}(\mathbf{y})}, \tag{6}$$

where we have dropped the dependence on the input and the parameters to simplify the notation.

We will also find it useful to define the joint posterior probability h_{ij}, the product of h_i and $h_{j|i}$. This quantity is the probability that expert network (i, j) can be considered to have generated the data, based on knowledge of both

---

[2] Not all exponential family densities have a dispersion parameter; in particular, the Bernoulli density has no dispersion parameter.

the input and the output. Once again, we emphasize that all of these quantities are conditional on the input $\mathbf{x}$.

In deeper trees, the posterior probability associated with an expert network is simply the product of the conditional posterior probabilities along the path from the root of the tree to that expert.

2.3 The likelihood

We treat the problem of learning in the HME architecture as a maximum likelihood estimation problem. The log likelihood of a data set $\mathcal{X} = \{(\mathbf{x}^{(t)}, \mathbf{y}^{(t)})\}_1^N$ is obtained by taking the log of the product of N densities of the form of Eq. 4, which yields the following log likelihood:

$$l(\theta; \mathcal{X}) = \sum_t \ln \sum_i g_i^{(t)} \sum_j g_{j|i}^{(t)} P_{ij}(\mathbf{y}^{(t)}). \tag{7}$$

We wish to maximize this function with respect to the parameters θ.

2.4 Applying EM to the HME architecture

In this section we develop a learning algorithm for the HME architecture based on the Expectation-Maximization (EM) framework (Dempster, Laird, & Rubin, 1977). We derive an EM algorithm for the architecture that consists of the iterative solution of a coupled set of iteratively-reweighted least-squares problems.

To develop an EM algorithm for the HME architecture, we must define appropriate "missing data" so as to simplify the likelihood function. We define indicator variables z_{ij} such that one and only one of the z_{ij} is one for any given data point. These indicator variables have the interpretation as labels that specify which expert in the probability model generated the data point. This choice of missing data yields the following complete-data likelihood:

$$l_c(\theta; \mathcal{X}) = \sum_t \sum_i \sum_j z_{ij}^{(t)} \ln\{g_i^{(t)} g_{j|i}^{(t)} P_{ij}(\mathbf{y}^{(t)})\}. \tag{8}$$

Note the relationship of the complete-data likelihood in Eq. 8 to the incomplete-data likelihood in Eq. 7. The use of the indicator variables z_{ij} has allowed the logarithm to be brought inside the summation signs, substantially simplifying the maximization problem. We now define the E step of the EM algorithm by taking the expectation of the complete-data likelihood:

$$Q(\theta, \theta^{(p)}) = \sum_t \sum_i \sum_j h_{ij}^{(t)} \ln\{g_i^{(t)} g_{j|i}^{(t)} P_{ij}(\mathbf{y}^{(t)})\}, \tag{9}$$

where we have used the fact that:

$$
\begin{aligned}
E[z_{ij}^{(t)}] &= P(z_{ij}^{(t)} = 1 | \mathbf{y}^{(t)}, \mathbf{x}^{(t)}, \theta^{(p)}) \\
&= \frac{P(\mathbf{y}^{(t)} | z_{ij}^{(t)} = 1, \mathbf{x}^{(t)}, \theta^{(p)}) P(z_{ij}^{(t)} = 1 | \mathbf{x}^{(t)}, \theta^{(p)})}{P(\mathbf{y}^{(t)} | \mathbf{x}^{(t)}, \theta^{(p)})}
\end{aligned}
$$

$$= \frac{P(\mathbf{y}^{(t)}|\mathbf{x}^{(t)}, \boldsymbol{\theta}_{ij}{}^{(p)}) g_i^{(t)} g_{j|i}^{(t)}}{\sum_i g_i^{(t)} \sum_j g_{j|i}^{(t)} P(\mathbf{y}^{(t)}|\mathbf{x}^{(t)}, \boldsymbol{\theta}_{ij}{}^{(p)})}$$

$$= h_{ij}^{(t)}.$$

The M step requires maximizing $Q(\boldsymbol{\theta}, \boldsymbol{\theta}^{(p)})$ with respect to the expert network parameters and the gating network parameters. Examining Eq. 9, we see that the expert network parameters influence the Q function only through the terms $h_{ij}^{(t)} \ln P_{ij}(\mathbf{y}^{(t)})$, and the gating network parameters influence the Q function only through the terms $h_{ij}^{(t)} \ln g_i^{(t)}$ and $h_{ij}^{(t)} \ln g_{j|i}^{(t)}$. Thus the M step involves the following separate maximizations:

$$\boldsymbol{\theta}_{ij}^{(p+1)} = \arg\max_{\boldsymbol{\theta}_{ij}} \sum_t h_{ij}^{(t)} \ln P_{ij}(\mathbf{y}^{(t)}),$$

$$\mathbf{v}_i^{(p+1)} = \arg\max_{\mathbf{V}_i} \sum_t \sum_k h_k^{(t)} \ln g_k^{(t)},$$

and

$$\mathbf{v}_{ij}^{(p+1)} = \arg\max_{\mathbf{V}_{ij}} \sum_t \sum_k h_k^{(t)} \sum_l h_{l|k}^{(t)} \ln g_{l|k}^{(t)}.$$

Note that each of these maximization problems are themselves maximum likelihood problems, given that P_{ij}, g_i and $g_{j|i}$ are probability densities. Moreover, given our parameterization of these densities, the log likelihoods that we have obtained are weighted log likelihoods for generalized linear models (GLIM's). An efficient algorithm known as iteratively reweighted least-squares (IRLS) is available to solve the maximum likelihood problem for such models (McCullagh & Nelder, 1983). (See Jordan and Jacobs (1994) for a discussion of IRLS.)

In summary, the EM algorithm that we have obtained involves a calculation of posterior probabilities in the outer loop (the E step), and the solution of a set of weighted IRLS problems in the inner loop (the M step). We summarize the algorithm as follows:

HME Algorithm 1

1. For each data pair $(\mathbf{x}^{(t)}, \mathbf{y}^{(t)})$, compute the posterior probabilities $h_i^{(t)}$ and $h_{j|i}^{(t)}$ using the current values of the parameters.

2. For each expert (i, j), solve an IRLS problem with observations $\{(\mathbf{x}^{(t)}, \mathbf{y}^{(t)})\}_1^N$ and observation weights $\{h_{ij}^{(t)}\}_1^N$.

3. For each top-level gating network, solve an IRLS problem with observations $\{(\mathbf{x}^{(t)}, h_k^{(t)})\}_1^N$.

4. For each lower-level gating network, solve an IRLS problem with observations $\{(\mathbf{x}^{(t)}, h_{l|k}^{(t)})\}_1^N$ and observation weights $\{h_k^{(t)}\}_1^N$.

5. Iterate using the updated parameter values.

Architecture	Relative Error	# Epochs
linear	.31	1
backprop	.09	5,500
HME	.10	35
CART	.17	NA
CART (oblique)	.13	NA
MARS	.16	NA

Table 1: Average Values of Relative Error and Number of Epochs Required for Convergence for the Batch Algorithms.

2.4.1 Simulation results

We tested the algorithm on a nonlinear system identification problem. The data were obtained from a simulation of a four-joint robot arm moving in three-dimensional space. The network must learn the *forward dynamics* of the arm; a mapping from twelve coupled input variables to four output variables. This mapping is rather smooth and we expect the error for a global fitting algorithm like backpropagation to be small; our main interest is in the training time.

We generated 15,000 data points for training and 5,000 points for testing. For each epoch (i.e., each pass through the training set), we computed the relative error on the test set. Relative error is computed as a ratio between the mean squared error and the mean squared error that would be obtained if the learner were to output the mean value of the outputs for all data points.

We compared the performance of a binary hierarchy to that of the best linear approximation, a backpropagation network, the CART algorithm and the MARS algorithm. The hierarchy was a four-level hierarchy with 16 expert networks and 15 gating networks. Each expert network had 4 output units and each gating network had 1 output unit. The backpropagation network had 60 hidden units, which yields approximately the same number of parameters in the network as in the hierarchy. The MARS algorithm was run with a maximum of 16 basis functions, based on the fact that each such function corresponds roughly to a single expert in the HME architecture.

Table 1 reports the average values of minimum relative error and the convergence times for all architectures. As can be seen in the Table, the backpropagation algorithm required 5,500 passes through the data to converge to a relative error of 0.09. The HME algorithm converged to a similar relative error in only 35 passes through the data. CART and MARS required similar CPU time as compared to the HME algorithm, but produced less accurate fits. (For further details on the simulation, see Jordan & Jacobs (1994)).

3   Conclusions

We have presented a novel tree-structured architecture for supervised learning. This architecture is based on a statistical model, and makes contact with a number of branches of statistical theory, including mixture model estimation and generalized linear model theory. The learning algorithm for the architecture is an EM algorithm.

The major advantage of the HME approach over related decision tree and multivariate spline algorithms such as CART, MARS and ID3 is the use of a statistical framework. The statistical framework motivates some of the variance-decreasing features of the HME approach, such as the use of "soft" boundaries. The statistical approach also provides a unified framework for handling a variety of data types, including binary variables, ordered and unordered categorical variables, and real variables, both at the input and the output.

Acknowledgements

This project was supported by grants from the McDonnell-Pew Foundation, ATR Human Information Processing Research Laboratories, and Siemens Corporation.

4   References

Breiman, L., Friedman, J. H., Olshen, R. A., & Stone, C. J. (1984). *Classification and Regression Trees*. Belmont, CA: Wadsworth International Group.

Dempster, A. P., Laird, N. M., & Rubin, D. B. (1977). Maximum likelihood from incomplete data via the EM algorithm. *Journal of the Royal Statistical Society. B, 39*, 1-38.

Friedman, J. H. (1991). Multivariate adaptive regression splines. *The Annals of Statistics, 19*, 1-141.

Jordan, M. I. & Jacobs, R. A. (1992). Hierarchies of adaptive experts. In J. Moody, S. Hanson, & R. Lippmann (Eds.), *Advances in Neural Information Processing Systems 4*. San Mateo, CA: Morgan Kaufmann.

Jordan, M. I. & Jacobs, R. A. (1994). Hierarchical mixtures of experts and the EM algorithm. *Neural Computation, 6*, 181-214.

McCullagh, P. & Nelder, J.A. (1983). *Generalized Linear Models*. London: Chapman and Hall.

Outline of a Linear Neural Network and Applications[*]

Eduardo R. Caianiello   Maria Marinaro
Salvatore Rampone   Roberto Tagliaferri
I.I.A.S.S.
Via G.Pellegrino, 19
I-84019 Vietri SM (SA) - Italy
Università di Salerno
Via S. Allende
I-84081 Baronissi (SA) - Italy

1 Introduction

The aim of this paper is to show that additional perspectives are added to research on linear Neural Nets by utilizing a new definition of product, recently introduced (Caianiello, 92) in the context of Neural Nets defined on semirings rather than number fields. This we do by exhibiting its use for the design "on inspection" of a Neural Net which stores as one-step attractor any number of given patterns, with input basins of exactly specified tolerance. The paper is organized in the following way. In the next section we specify the notation and define the Neural Net model. In section three we describe applications in structured and not structured domains.

2 The Network.

The standard rules of vector and matrix calculus are maintained. We consider an NN containing N "neurons" capable of assuming two states "0" and "1" (numbers of a boolean field); a "state" of NN is represented by an N-vector $\mathbf{x}$ with components 0 and 1 (also named "pattern", "word", etc.).

The patterns to be memorized as fixed attractors, in number of P, are denoted as the vectors $\mathbf{w}_1, \mathbf{w}_2, \ldots, \mathbf{w}_P$, columns of the $N \times P$ matrix

$$\mathbf{W} = (\mathbf{w}_1, \mathbf{w}_2, \ldots, \mathbf{w}_P) \tag{1}$$

Complement $\overline{\mathbf{x}}$ of $\mathbf{x}$ is the vector having as components the set-theoretical complements of the components of $\mathbf{x}$

Antidouble $\tilde{\mathbf{X}}$ of a $N \times Q$ matrix $\mathbf{X}$ (including the case $Q = 1$, the vector $\mathbf{x}$), is the $2N \times Q$ matrix:

$$\tilde{\mathbf{X}} = \left(\frac{\mathbf{X}}{\overline{\mathbf{X}}} \right) \tag{2}$$

---
[*]This work was supported in part by C.N.R., Progetto Strategico Reti Neurali, by MURST 40% unità I.N.F.M. Università di Salerno, by Contratto Quinquennale C.N.R. - I.I.A.S.S.

The symbol $\odot$ is a generalized scalar product defined by:

$$\mathbf{x} \odot \mathbf{y} = \overline{\mathbf{x}}^r \mathbf{y} + \mathbf{x}^r \overline{\mathbf{y}} = \sum_{l=1}^{N} (\overline{x_l} y_l + x_l \overline{y_l}) \tag{3}$$

The definition (3) of the $\odot$ product satisfies the property

$$\mathbf{x} \odot \mathbf{y} = d(\mathbf{x}, \mathbf{y}) \tag{4}$$

where $d(\mathbf{x}, \mathbf{y})$ is Hamming distance between $\mathbf{x}$ and $\mathbf{y}$, i.e. the number of discordant components.

The $\odot$ product permits to transfer into 2^N-space, as we are going to see next, some useful properties of N-space orthogonalization.

We assume that our patterns satisfy the equation

$$\mathbf{w}_r \odot \mathbf{w}_k \geq 2K + 1 \qquad \forall r, k \tag{5}$$

and construct the Neural Net defined by the equations

$$\mathbf{x}(t + \tau) = \mathbf{W} 1 \left[\mathbf{K} + \varepsilon - \mathbf{W} \odot \mathbf{x}(t) \right] \tag{6}$$

where $1[f(x)]$ is the Heaviside function, $\mathbf{K}$ and ε are P-vectors with components $K_1 = K_2 = \ldots = K_P = K$ and $\varepsilon_1 = \varepsilon_2 = \ldots = \varepsilon_P = \varepsilon$, $0 < \varepsilon < 1$. The Neural Net (6) acts as one step attractor, any input vector is attracted by one of the memorized vectors $\mathbf{w}_h$ or by null vector $\mathbf{x}$, the last case is verified iff $\mathbf{w}_r \odot \mathbf{x} \geq 2K + 1, \forall r$

(Thus "learning" in this example cannot be Hebbian, but only arises from connection switching).

The introduction of the $\odot$ product permits to disentangle the notion of "tolerance", which appears as a threshold, from the specificity of individual patterns, and rather surprisingly reduces all neuronal couplings to the boolean values 0, 1 while retaining linearity (the passage $N \to 2N$ is a small price for this).

3 Applications

3.1 Structured Data

In information transmission over channels subject to noise disturbances, for example a telephone line, a high frequency radio link, channel noise may corrupt the transmitted signal. The problem is usually faced by encoding the message selected at the source in a redundant systematic way. This allows the decoder, that represents the processing of the channel output, to control the received information. The decoder processing makes use of a priori information about the coding (McEliece, 77, Gallager, 68, Peterson et al., 72, MacWilliams et al., 81).

In a previous paper, a neural net model for decoding noisy patterns received from binary symmetrical channels has been developed by us (Esposito et al., 92). The application of that network, based on the the threshold tuning, was limited by a constraint on the maximum Hamming weight of each codeword.

The introduction of the $\odot$ product, by disentangling the threshold from the codeword weight, overdues the problem.

It is immediate to see that by using the Neural net defined by equation (6) and identifying the columns of the $\mathbf{W}$ matrix with the codewords, it is possible to construct a decoder which allows the exact identification of the transmitted words if the channel noise, η, satisfies the constraint $\eta < \frac{K}{N}$.

Furthermore let us suppose to receive the output of a noisy channel before the decoder, and, for any reason, assume that the code used by the sender is unknown. Then we have to infer the code only by means of the noisy patterns.

As we defined it (Rampone, 93), the problem is to find a set of reproduction vectors such that a given criterion for the total distortion is minimized; therefore it is a clustering optimization, or, equivalently, a vector quantizer design problem (Duda et al., 73, Huseyin, 90). We proved (Rampone, 93) the difference between the original and the inferred code decreases exponentially with high probability as function of the number of the training patterns. The Neural Net under consideration performs the task in a very natural way by taking the Hamming sphere radius equal to K, and by identifying $\odot$ as a distortion measure (Shannon, 59).

Let us denote with $\mathbf{v}_1, \mathbf{v}_2, \ldots, \mathbf{v}_S$ S patterns received as output of a noisy binary symmettric channel, we construct the matrix $\mathbf{V}$ whose columns are the vectors $\mathbf{v}_1, \mathbf{v}_2, \ldots, \mathbf{v}_S$. By using the matrix $\mathbf{V}$ we can divide the vectors $\mathbf{v}_1, \mathbf{v}_2, \ldots, \mathbf{v}_S$ in classes. The class C_m contains all the vectors such that

$$\xi_m = 1\left[\mathbf{K} + \varepsilon - \mathbf{V} \odot \mathbf{v}_{m_i}\right] \tag{7}$$

$$\mathbf{v}_{m_i} \in \mathbf{V} \qquad m = 1, \ldots, P$$

where ξ_m is a vector whose components are all null except the m-th. We define the new vector

$$\mathbf{w}_m = 1\left[\frac{2}{t_m}\left(\sum_{i=1}^{t_m}\mathbf{v}_{m_i}\right) - \mathbf{I}\right] \tag{8}$$

where t_m is the cardinality of the class C_m and $\mathbf{I}$ is the vector whose components are all equal to 1. The $\mathbf{w}_m$'s $(m = 1, \ldots, P)$ are the inferred codewords. The following theorem allow us to determine the minimum number of training patterns to use in order to have inferred codewords correct with probability larger than $1 - \delta$.

Theorem The inferred codeword is correct with probability larger than $1 - \delta$ if the number of examples for each class is

$$2L + 1 \geq 2\frac{lgN - lg\delta}{(1 - 2\eta)^2}$$

To demonstrate the theorem we assume that 1) the distance between the codewords of the original unknown code is $2K + 1$; 2) the channel noisy, η, is such that $N\eta \leq \frac{K}{2}$.

The assumptions 1) and 2) assures that the patterns belonging to the class C_m are corrupted patterns generated by a unique unknown codeword. This allows us to treat the different classes separately. Then from the equation (8) the probability that the r-th component of $\mathbf{w}_m$ be wrong is given by

$$Pr = \sum_{h=\lceil \frac{2L+1}{2} \rceil}^{2L+1} \binom{2L+1}{h} \eta^h (1-\eta)^{2L+1-h} \leq e^{-(1-2\eta)^2(L+\frac{1}{2})} \qquad (9)$$

The previous expression follows from the Hoeffding's inequality

$$\sum_{h\lceil(s+\epsilon)N\rceil} (1-\epsilon)^{N-h} \epsilon^h \binom{N}{h} \leq e^{-2s^2 N}$$

In order to have the codewords correct with probability larger than $1 - \delta$ we have to ask that

$$(1 - Pr)^N \geq \left[1 - e^{-(1-2\eta)^2(L+\frac{1}{2})}\right]^N = 1 - \delta$$

This is verified if

$$2L + 1 = 2\frac{lgN - lg\delta}{(1-2\eta)^2}$$

3.2 Not-structured Data

The previous results hold also when the data have not a natural structure, as in the most of cases afforded by the connectionist community. Then our model can memorize binary patterns as prototypes, and retrieve them when the inputs are corrupted images of them. We store patterns as attractors of basins containing vectors which differ each other at most for K bits, and the stored patterns are the centers of Hamming spheres with radius K. We consider as centers of Hamming spheres the P input patterns $\mathbf{v}_1 \ldots \mathbf{v}_P$ each of N bits. We fix the initial value of K. For example, if we have patterns of size 100 and we think it is useless to correct patterns with more then 30% of errors , then we initially assign $K = 30$ for all the patterns. The following algorithm uses Hamming spheres with different radii so that to obtain three important goals: 1) the spheres are all disjoint; 2) given an initial value to K the spheres hold the maximum input space; 3) a corrupted input pattern will go either into the basin of the nearest attractor or into the basin of the unclassified patterns. We take a single layer net with binary inputs and outputs. The network have $max(P, 2N)$ neurons, which act as input, output and hidden neurons. To the i-th neuron we associate a binary weight vector $\tilde{\mathbf{h}}_i = \left(\frac{\mathbf{h}_i}{\bar{\mathbf{h}}_i}\right)$, a threshold K_i and a temporary threshold K'_i.

To store a pattern $\mathbf{v}_i$ and to make it an actractor of all the patterns with distance $\leq K_i$ we have to put $\mathbf{h}_i = \mathbf{v}_i$, $\bar{\mathbf{h}}_i = \bar{\mathbf{v}}_i$. The two steps of calculating K_i's and constructing the weight matrix can be accomplished in only one stage as follows.

We store the patterns one by one. To memorize the first pattern $\mathbf{v}_1$ we put $\mathbf{h}_1 = \mathbf{v}_1$, $\bar{\mathbf{h}}_1 = \bar{\mathbf{v}}_1$ and $K_1 = K$.

To store the i-th pattern $\mathbf{v}_i$ we put $\mathbf{h}_i = \mathbf{v}_i$, $\bar{\mathbf{h}}_i = \bar{\mathbf{v}}_i$. Then we compare $\mathbf{v}_i$ with all the $\mathbf{h}_j$'s with $j = 1, \cdots, i-1$, and we calculate for each neuron j:

$$K'_j = \begin{cases} K & \text{if } \mathbf{v}_i \odot \mathbf{h}_j \geq 2K+1 \\ K - \left\lceil \frac{2K+1-(\mathbf{v}_i \odot \mathbf{h}_j)}{2} \right\rceil & \text{if } \mathbf{v}_i \odot \mathbf{h}_j \leq 2K \end{cases}$$

Therefore the threshold associated to the neuron i is $K_i = \min_{1 \leq j \leq i-1} K'_j$ while for all the other neurons j, with $j = 1, \cdots, i-1$, $K_j = min(K_j, K'_j)$. It is ease to prove that the obtained net reaches the above goals and it can be used to classify or to recognize patterns. The recognition stage is divided into two steps: In the first one (classification step) we assign to the $2N$ input neurons the values $\mathbf{x}$ and $\overline{\mathbf{x}}$ and compute

$$x_i(t+\tau) = 1\left[K_i + \varepsilon - \mathbf{h}_i \odot \mathbf{x}(t)\right] \qquad i = 1, \cdots, P \qquad (10)$$

Note that this implies that P neurons of the net at time $t+\tau$ act as hidden nodes and at most one of them is equal to 1 (the neuron s for which $\mathbf{h}_s \odot \mathbf{x}(t) < K_s$. In the second step h_{ij} becomes the connection from neuron i to neuron j and we compute for all the N output neurons (for all the others the value is 0):

$$x_i(t+2\tau) = \sum_{j=1}^{P} h_{ji}x_j(t+\tau)$$

Since between the $x_j(t+\tau)$'s there is at most one with value 1 (neuron s of the classification step) then $x_i(t+2\tau)$ is equal to 1 if $h_{si} = 1$ and to 0 if $h_{si} = 0$. In this way the output vector $\mathbf{x}(t+2\tau) = (x_1(t+2\tau) \cdots x_N(t+2\tau))$ reconstructs the vector $\mathbf{v}_s$ which is the center of the sphere to which $\mathbf{x}(t)$ belongs. With this model it is possible to store until P patterns with a number of neurons equal to $max(P, 2N)$ and of connections (with value 1) equal to NP. This model has the great advantage to be easy implemented in hardware since the connections and the outputs are binary while the thresholds are integers. Furthermore the neural net permits to increment the number of stored patterns until $P = 2^N$ simply adding each time a new neuron with its connections with input and output neurons.

References

- Caianiello, E.R. (1992) Do All Neurons Know Algebra? *Art. Neur. Networks II*, I. Alexander and J. Taylor ed.s, **1**, 221-227, Elsevier.

- Duda, R.O., Hart, P.E. (1973) *Pattern Classification and Scene Analysis*, NewYork: J.Wiley and Sons.

- Esposito, A., Rampone, S., Tagliaferri, R. (1992) A Neural Network for Error Correcting Decoding of Binary Linear Codes. *IIASS tech. rep I-9301, in press on Neural Networks*.

- Gallager, R.G. (1968) *Information Theory and Reliable Communication*. New York: J.Wiley and Sons.

- Huseyin, A. (1990) *Vector Quantization*. New York: IEEE Press.

- MacWilliams, F.J., Sloane, N.J.A. (1981) *The Theory of Error Correcting Codes.* New York: North Holland.

- McElice, R.J. (1977) The Theory of Information and Coding. In Gian-Carlo Rota ed., *Encyclopedia of Mathematics and its Applications,* **3.** London: Addison Wesley.

- Peterson, W.W., Weldon, W.J. (1972) *Error Correcting Codes.* Cambridge: The MIT press.

- Rampone, S. (1993) Linear Codes Interpolation from Noisy Patterns by Means of a Vector Quantization Process. IIASS Tech. Rep. I-9303, submitted for pubblication.

- Shannon, C.E. (1959) Coding Theorem for a Discrete Source with a Fidelity Criterion. *Ist. of Radio Eng., Int. Conv. Rec.,* **7,** 142-163.

Numerical Experiments on the Information Criteria for Layered Feedforward Neural Nets

Katsuyuki Hagiwara, Shiro Usui

Department of Information and Computer Sciences,
Toyohashi University of Technology,
Toyohashi 441, JAPAN.

1 Introduction

Our purpose is to provide a guideline for applying information criteria and to analyze the behavior of them for neural nets through some numerical experiments. In this paper, we focus on the criteria, AIC[1] (Akaike's Information Criterion), NIC[5] (Network Information Criterion), BIC[3][4] (Bayesian Information Criterion) and MDL[6] (Minimum Description Length) for nonlinear regression model (NLRM) and nonlinear autoregressive model (NLARM) using three layered feedforward neural nets. Moreover, we introduce PMDL[7] (Predictive MDL) and EIC[8] (Extend Information Criterion).

2 Information Criteria

For a given data $y = \{y_i; 1 \le i \le N\}$, we estimate the model, $f(Y, \theta_k)$, in a specified class. θ_k is a set of parameters and k is the number of parameters. Most of information criteria can be described by

$$IC(k, N) = -\log f(x, \widehat{\theta}_k) + C(k), \tag{1}$$

where $\widehat{\theta}_k = \widehat{\theta}_k(Y)$ denotes the maximum likelihood estimator (MLE). We may select the model which minimizes (1) in a specified class.

If $C(k) = -\text{tr}\widehat{J}^{-1}\widehat{I}$, then $IC(k, N)$ becomes TIC (Takeuchi's modification of AIC), where

$$\begin{aligned} \widehat{J} &= \left. \partial^2 \log f(Y, \theta_k)/\partial\theta_k\partial\theta_k^T \right|_{\theta_k=\widehat{\theta}_k}, \\ \widehat{I} &= \left. \partial \log f(Y, \theta_k)/\partial\theta_k \cdot \partial \log f(Y, \theta_k)/\partial\theta_k^T \right|_{\theta_k=\widehat{\theta}_k}. \end{aligned}$$

The general form of TIC is NIC in which the empirical loss need not to be negative of log-likelihood. If the true distribution is in a family of models, then $-\text{tr}\widehat{J}^{-1}\widehat{I} \simeq k$ and $IC(k, N)$ becomes AIC in which $C(k) = k$. If $C(k) = k/2\log N$, $IC(k, N)$ becomes MDL which is the same as BIC.

To overcome the drawbacks of AIC and TIC , the criterion called EIC in which $C(k)$ is estimated by boot-strap has been proposed [8]. To estimate $C(k)$, we repeat the learning for boot-strap samples which are randomly drawn with replacement from an empirical distribution that is the proxy of the true one. We denote the estimated $C(k)$ in EIC by $C^*(k)$. On the other hand, the criterion PMDL (Predictive MDL) is given by PMDL $= -\sum_{t=0}^{N-1} \log f(x_{t+1}|x^t, \widehat{\theta}_k(x^t))$ for the ordered data $x^N = (x_1, \cdots, x_N)$. To simply obtain PMDL, we divide the data into several segments. For every segment, we calculate the prediction error with the model estimated for all of the past data. PMDL is the sum of the prediction error for all segments. Although PMDL does not have $C(k)$ which penalizes the model complexity, it is added automatically in this procedure. EIC and PMDL may have the advantage for neural nets in the meaning that the analytical solution for $C(k)$ does not need.

In the next, we consider NLRM and NLARM using three-layered neural nets with m hidden units and r inputs, which is denoted by NLRM(r, m) and NLARM(r, m) respectively.

3   Numerical Experiments on NLRM$(1, m)$

3.1   Numerical Experiments

Here, we consider numerical experiments in which the data is 400 samples of Gaussian noise sequences, in which the mean and variance are 0 and 1, of 256 points. In NLRM$(1, m)$, $\theta_k = (\omega_m, \sigma^2)$, where ω_m is a set of connection weights, σ^2 is a variance and $k = 3m + 1$ $(1 \leq m \leq 5)$. In this case, the true number of hidden units is 0. Because we adopted the backpropagation[9] in learning, the estimator of θ_k is the least mean square estimator, which is MLE in this case. Here, we applied AIC, TIC and BIC/MDL. In our case, TIC is the same as NIC.

In this case, we can estimate $C(k)$, because the true distribution is known [10]. Fig.1 shows the estimates of $C(k)$ for NLRM$(1, m)$, in (a), and for polynomial regression model with s coefficients which is denoted by PRM$(1, s)$, in (b). These are denoted by $C_N(k)$ and $C_P(k)$ respectively. In this figure, $s = 3m$ at each m. Also, we show the $C(k)$ in each $IC(k, N)$. Here, $C^*(k)$ was estimated for one sample. Fig.2 shows the frequency of the number of hidden units selected by applying AIC, TIC/NIC and BIC/MDL for 400 samples. In this figure, (a) shows the selected frequency of the NLRM$(1, m)$ and (b) shows that of PRM$(1, 3m)$.

3.2   Discussion

We have shown that the $C(k)$ in AIC can not be derived as k for neural nets[10]. Also, we have numerically shown that the number of hidden units selected by applying AIC tends to be large for NLRM$(1, m)$. Moreover, we pointed out that it is caused by the data fitting capability of neural nets. We, here, discuss the behavior of TIC/NIC, BIC/MDL and EIC except AIC.

In Fig.2 (a) and (b), most of cases of TIC/NIC for NLRM$(1, m)$ fails the correct selection, while that of PRM$(1, 3m)$ is almost the same as the case of AIC. As pointed out in [10], if N goes to ∞ then $\widehat{J}$ becomes singular because of the nonuniqueness of connection weights. This means that there is always no particular point around which the estimator appears with high probability in the weight space. Hence, even if N is finite, this can be true when a network begins to fit to the noise by increasing m. Consequently, the statistical properties of $\widehat{J}$ and $\widehat{I}$ for NLRM$(1, m)$ is quite different from that for PRM$(1, 3m)$ and $\widehat{J}$ for NLRM$(1, m)$ is nearly singular in practical situation. Fig.1 shows that when we calculated $\widehat{J}^{-1}$, $C(k)$ in TIC is not stable any longer and tends to be smaller than k. Although BIC/MDL could select the true m as shown in Fig.2 (a), there is a possibility that the criteria prefers simple structures because the $C(k)$ in BIC/MDL is highly larger than $C_N(k)$. On the other hand, $C^*(k)$ for one sample gives a good approximation of $C_N(k)$. In this case, EIC selected the true m. In EIC, $C(k)$ can be adaptively estimated, we expect that EIC selects the optimal even in other cases.

4 Numerical Experiments on NLARM(r, m)

4.1 Numerical Experiments

Now, we consider the AR process which is generated from NLARM(r^*, m^*), where $(r^*, m^*) = (2, 2)$. In this case, we need to search the optimal structure which minimizes the prediction error by varying both of r and m. In our experiments, as we take $1 \leq r, m \leq 5$, there exists the true structure in the specified class of models. For this case, to examine the selected frequency by AIC, BIC/MDL and PMDL, we generate 10 samples of the NLAR process. When we take 800 data points for training, the prediction error for the new data of 2000 points generated successively after the training data was minimized by the estimated network with (r^*, m^*) for all samples. Therefore, the selected structure according to above criteria should be consistent with the true one in case of using this training data. In Table.1, we show the results by the form; *selected structure (r,m) : selected times.*

Table.1 the selected frequency by each $IC(k, N)$ $(k = r(m + 2) + 1, N = 800)$

AIC	(5,5) : 4	(4,5) : 4	(5,4) : 1
BIC/MDL	(2,2) : 8	(2,1) : 2	–
PMDL	(2,2) : 5	(2,1) : 4	(2,3) : 1

4.2 Discussion

The results in Table.1 show that AIC selected the largest or nearly largest network. This means that $C(k)$ in AIC which corrects the fitting to the noise is useless for NLARM(r, m). In contrast to AIC, however, BIC/MDL and PMDL select the true structure most frequently. However, these tends to select simple structures in which the tendency of PMDL is notable compared with BIC/MDL. In PMDL, because the number of training data is small in early portion of the data, the overfitting arises in the portion. Therefore, even if the model is not so complex, the penalty can be large. Consequently, PMDL prefers the simple model. Also, we have done the numerical experiments for the case the optimal structure depends on the number of data. The result is that, again, simple structures tend to be selected by BIC/MDL and PMDL as increasing the number of data.

5 Conclusion

We discussed the behavior of information criteria. Through some numerical experiments, we found that the estimate of AIC, TIC for the complexity of the network is small while that of BIC/MDL and PMDL is large. In this meaning, EIC is of interest. However, in EIC, there is a problem that it takes much time to repeat the learning for the boot-strap samples. From the theoretical viewpoints, as shown in [10], we can regard that the behavior depends on the data fitting capability. And the capability is benefited by the property that the basis function can be adaptively estimated in neural nets while the ordinal linear function can not be. In this meaning, we can say that neural nets are powerful. However the capability accompanies the difficulties in analyzing the statistical properties. This is the hurdle in thinking of the optimal information criterion for neural nets.

References

[1] Akaike,H. : *In 2nd International Symposium on Information Theory*, B.N.Petrov and F.Csáki eds., Akadémia Kiado, Budapest, pp.267–281, 1973.

[2] Takeuchi,K. : *Mathematical Science*, **153**, pp.12–18, 1976 (in Japanese).

[3] Schwarz, G. : *Ann. Statist.*, **6**, pp.461–464, 1976.

[4] Akaike,H. : *Applications of Statistics*, P.R.Krishnaiah, ed., Noth-Holland, Amsterdam, pp.27–41, 1977.

[5] Murata,N., Yoshizawa,S., Amari,S. : METR 92–05, Technical Reports, University of Tokyo, 1992.

[6] Rissanen,J. : *Ann. Statist.*, **11**, No.2, pp.416–431, 1983.

[7] Rissanen,J. : *Ann. Statist.*, **14**, No.3, pp.1080–1100, 1986.

[8] Kitagawa,G., Ishiguro,M., Sakamoto,Y. : *Technical Report of IEICE*, **IT92-133**, pp.43–62, 1993 (in Japanese).

[9] Rumelhart,D.E. and McClelland,J.L. : "Parallel distributed processing", **Chap.8**, Cambridge,MA:MIT Press., 1986.

[10] Hagiwara,K., Toda,N., Usui,S. : Proc. of IJCNN, Nagoya, **3**, pp.2263–2266, 1993.

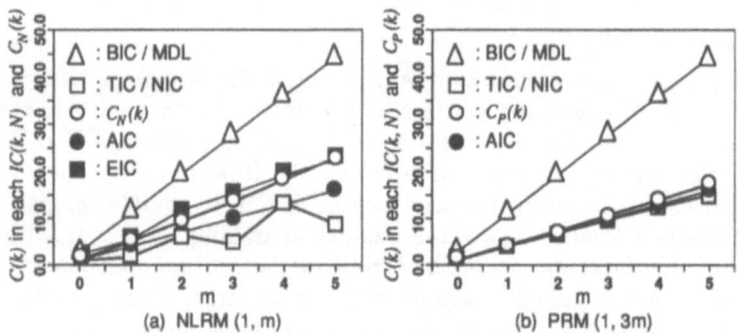

Figure 1: $C(k)$ in each $IC(k, N)$ and $C_N(k)$, $C_P(k)$

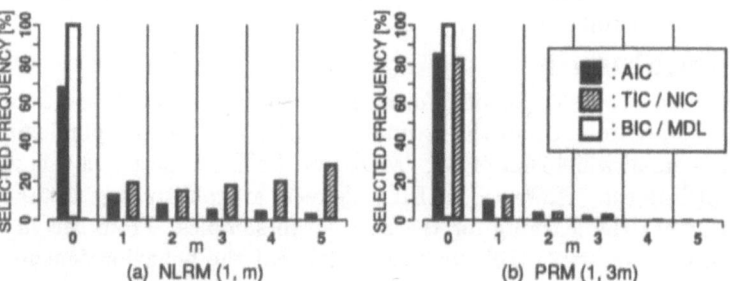

Figure 2: The frequency of the selected number of hidden units

Quantifying a Critical Training Set Size for Generalization and Overfitting using Teacher Neural Networks

R. Lange†, R. Männer†‡

† Computer Science V, University of Mannheim

Mannheim, Germany

‡ Interdisciplinary Center for Scientific Computing, University of Heidelberg

Heidelberg, Germany

1 Introduction

Neural networks are trained to map a given training set of input-output pairs as accurate as possible. In most applications, the training set models a task, which the neural network has to complete — it should not just learn the training data. This *generalization* is estimated by measuring how accurate the neural network maps an independent testing set. However, it is hard to know in advance how well a neural network can generalize. Two major questions arise: Is sufficient data available to model the task? Which architecture fits the given task best? Typically, both questions must be answered by trial and error due to the lack of sound knowledge that would allow to predict the number of training samples required and the optimal architecture.

We have tackled these questions by means of *teacher neural networks*. This is an experimental approach which is sufficiently general to provide meaningful insights. The results reported here show, to start with, that there exists a critical training set size at which both the generalization error and overfitting drop to zero and, secondly, that this critical training set size is proportional to the number of parameters of the neural network.

2 Experiments

Teacher. A teacher is a neural network with fixed weights[1] which is employed to generate the training and testing sets. Hence, the teacher defines the task, instead of a (real or synthetic) data set. There are several advantages of this approach. First, one knows that a perfect solution of the task exists. Second, as many input-output pairs as desired can be produced. Third, the complexity of the task can be scaled by varying the architecture of the teacher.

Our teacher has an n–n–n architecture: an input layer, one hidden layer, an output layer; each n neurons. The weights are uniformly drawn from $[-1, +1]$. The transfer function is $tanh$. Pairs for the training and testing set are produced by presenting random input $\vec{\xi} \in \{-1, +1\}^n$ to the teacher in order to get the corresponding target output $\vec{\zeta} \in [-1, +1]^n$. Let p^t and p^g be the size of the training and testing set.

---

[1] Here and in the following weights include thresholds.

Pupil. The *pupil neural network* is trained on the training set produced by the teacher using online back propagation (Rumelhart et al., 1986). The pupil and teacher architectures are identical. We have scanned a wide range of training lengths. For reasons of lucidity, none of the well-known modifications that increase convergence speed have been applied. Thus, only one parameter is left, the learning rate γ.

Error measure. The training error is $\epsilon^t = \frac{1}{np^t} \sum_{\mu=1}^{p^t} \sum_{i=1}^{n} |z_i^\mu - \zeta_i^\mu|$, where z_i^μ is the actual output of the pupil for input $\vec{\xi}^\mu$. Regardless of n and p^t, this error measure is bounded to $[0, 2]$. Analogously, the generalization error ϵ^g is defined on the testing set.

Training set size. Our interest is to quantify how generalization depends on the number of training samples. The training samples are used to adjust the parameters of the pupil neural network. Therefore, we expected the training set size to scale with the number of pupil parameters to be fixed, i.e. the number $w = 2n(n + 1)$ of weights. This choice is upheld by bounds on the generalization $\epsilon^g(p^t)$ found for linear threshold networks (Baum and Haussler, 1989). Accordingly, we have measured the training set size in units of w, in order to investigate comparable ranges for different neural network sizes.

Parameter settings. We have studied $\epsilon^g(p^t)$ for 13 different n–n–n networks, with $n = 5, 10, \ldots, 65$. The size of the training set varied over the range $[0, w]$, $\frac{p^t}{w} = \frac{1}{16}, \frac{2}{16}, \ldots, \frac{16}{16}$. The testing set must be large enough to measure generalization accurately; we used $p^g = w$. The pupil was trained with a learning rate $\gamma = 0.01$. Weights were initialized with random values uniformly drawn from $[-0.01, +0.01]$.

3 Results

Quantities. We measured the training error ϵ^t and the generalization error ϵ^g during the training process, at $\tau = 0, 1, 2, 4, \ldots, 131072 = 2^{17}$ weight updates. So, instead of the usual choice of *epochs* (presentations of the entire training set), τ gives the number of presentations of input-output *pairs* that are presented to the pupil. The reason for this is that we wanted to compare the results for different training set sizes; this would not be possible if we measured training time in epochs, because in this case a larger training set size would mean longer training.

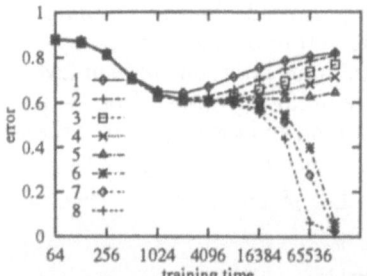

Figure 1: Generalization error as a function of training time for different training set sizes $\frac{16p^t}{w} = 1, 2, \ldots, 8$. The first out of 40 runs for $n = 50$ is shown.

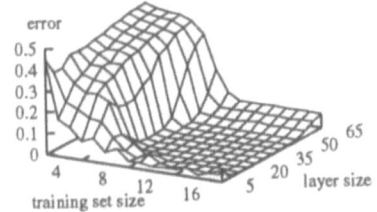

Figure 2: Perfect generalization above critical training set size. The generalization error ϵ^g vanishes.

Generalization and overfitting. Figure 1 shows the generalization error during training for a 50-50-50 network. The different curves correspond to different training set sizes p^t, and illustrate that the behaviour depends dramatically on this parameter; one has either overfitting and bad generalization $(p^t \leq \frac{5}{16}w)$ or no overfitting and perfect generalization $(p^t \geq \frac{6}{16}w)$. This is true for all neural network sizes that we have observed.

Scaling behaviour. For the results presented now, we picked the smallest values ϵ_{opt}^t and ϵ_{opt}^g and recorded these together with the corresponding times τ_{opt}^t and τ_{opt}^g. The difference $2^{17} - \tau_{opt}^g$ between the entire training time and the time for optimal generalization is then a measure for overfitting. This becomes clear in Figure 1: the training time for optimal generalization τ_{opt}^g differs from the training length 2^{17} for curves 1-5; overfitting has occured.

The results for $\epsilon_{opt}^g and \tau_{opt}^g$ presented in the following are the mean of 40 values measured for different weight initializations.

Figure 2 shows the dependence of the generalization error on the training set size for different neural networks. Over a wide range of layer sizes n, the neural networks show the behaviour discussed above for Figure 1. A critical training set size $p_c^t \approx \frac{6}{16}w$ exists with no generalization below and perfect generalization above. The results deviate from this idealized behaviour for both the smallest and the largest neural networks ($n \leq 10$ and $n \geq 55$).

First, we discuss the graph in the region of perfect generalization. The curves are rather ragged for small n, due to bad statistics; for large n, the residual error increases. The latter effect has two reasons. The first reason is trivial; the training length is not sufficient large, as Figure 1 indicates already. The second reason is not that evident; the choice of $\gamma = 0.01$ is not adequate. We know from other experiments we have carried out with teacher neural networks that larger networks require smaller γ. To check on that, we have rerun the simulation for $n = 65$ with $\gamma = 0.004$. This yielded perfect generalization after approximately 2^{19} pairs.

Secondly, considering the region with low generalization, the residual error increases with the layer size n, particularly for small n. This suggests better generalization for small n, which is not really true, because the condition for measuring generalization — independent training and testing set — is no longer fulfilled[2].

Figure 3 displays the time at which generalization is optimal for different neural networks. Indirectly, this shows the dependence of overfitting on the

---

[2]For small n, the training set size $p^t \propto 2n(n + 1)$ becomes comparable to 2^n, the number of possible inputs. Hence, the lower n, the more input-output pairs exist that reside in both the training and testing set.

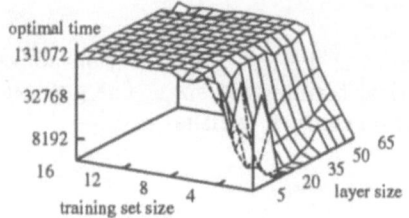

Figure 3: No overfitting above critical training set size. $\tau_{opt}^g = 2^{17}$. The smallest generalization error ϵ^g is at the very end of training.

training set size. Be aware that the training set size axis is reversed compared to Figure 2! Again, over a wide range of layer sizes n, the neural networks show the behaviour described above for Figure 1. A critical training set size $p_c^t \approx \frac{6}{16}w$ exists with no overfitting above and clear overfitting below. As to the smallest training set, for example, the best generalization is achieved at approximately 4096 pairs, i.e. only the 256th part of the entire training time is exploited.

4 Discussion

We have trained neural networks on tasks defined by teacher neural networks. It has been shown how generalization and overfitting depend on the training set size. We expected that the generalization error decreases as the number of training samples is increased. This has been observed for tasks defined by (real or synthetic) data, too. Surprisingly, our studies revealed that there are two distinct regions — one with low generalization and severe overfitting and one with perfect generalization without overfitting — with a sudden transition between them. This transition identifies a critical training set size, which we could show to be proportional to the number of weights in the pupil neural network. Although tasks defined by teacher neural networks do not compare directly with real applications, teacher neural networks are a valuable tool to investigate neural networks beyond the scope of specific applications. Our experience with neural networks used for signal processing proved this conjecture. Regarding a critical learning rate, our results from teacher experiments applied directly to the application. Further results dealing with the question of the optimal architecture are promising, too.

References

[1] E.B. Baum and D. Haussler. What size net gives valid generalization? *Neural Computation*, 1:151–160, 1989.

[2] D.E. Rumelhart, G.E. Hinton, and R.J. Williams. Learning internal representations by error propagation. In D.E. Rumelhart and J.L. McClelland, editors, *Parallel Distributed Processing*, volume 1, chapter 8, pages 318–362. MIT Press, Cambridge, 1986.

Formal Representation of
Neural Networks

R. Freund†, ‡, F. Tafill‡

† Department of Computer Science, University Magdeburg
Magdeburg, Germany

‡ Institute for Computer Languages, Technical University Wien
Wien, Austria

December 9, 1993

1   Introduction

In the fourties of this century, Warren McCulloch and Walter Pitts proposed a neuronal model for perception and nervous activity (McCulloch and Pitts, 1943) that was inspired by biological considerations, but can also be seen as a first theoretical model within the great variety of what nowadays are called neural networks. As Marvin L. Minsky, who in 1951 built the first neurocomputer, pointed out in his famous book on computation (Minsky, 1967), the neuronal networks of McCulloch and Pitts have the same computational power as finite-state machines (Perrin, 1990). These observations established an interesting connection between the theory of neural networks and the theory of formal languages and automata.

Cellular automata also came up from biological considerations. As a parallel model for computation, they also can be seen as a kind of predecessors of modern neural networks. On the other hand, from the very beginning of theoretical investigations of cellular automata, questions related to formal language theory have attracted the interest of important researchers. For example, already in 1958 John von Neumann dealt with the problem of self-reproducibility in cellular automata (von Neumann, 1958). The universal computational power of two- and one-dimensional cellular automata was shown by E.F. Codd (Codd, 1968) and A.R. Smith III (Smith III, 1971). Smith III also investigated the relations between cellular automata or so-called tessellation structures and formal languages (Smith III, 1970). Although some further papers on cellular and tessellation automata can be found in the literature of formal language theory, the bonds between the theory of neural networks and the theory of formal languages and automata seemed to have been cut in the middle of the seventies, when the interest of the scientific community in cellular automata vanished, too.

The model of *perceptrons* was studied by Frank Rosenblatt and his group (Rosenblatt, 1958). As for the simplest class of perceptrons without intermediate layers a convergent learning algorithm could be proved, perceptrons promised to be useful machines for many problems in artificial intelligence. Yet when Marvin L. Minsky and Seymour Papert revealed the severe limitations of these simple neural networks (Minsky and Papert, 1969), most of the computer scientists lost their interest in the neural network paradigm, until it revived in the eighties and nowadays has become a flourishing field for research as well as for many applications (Ritter et al., 1990).

Yet also during this rather quiescent period in the late seventies and the beginning of the eighties, some researchers added new interesting concepts to the field of neural computation (compare the overviews in Hertz et al., 1991, as well as Hecht-Nielsen, 1990). The ideas of Eduardo Caianiello (Caianiello, 1961), who integrated ideas from statistical mechanics and the learning theory of Hebb (Hebb, 1949) were taken up again by Hopfield, who added the physical notions of energy functions (Hopfield,1982). Stephen Grossberg introduced the adaptive resonance theory (Grossberg, 1976).

Going back to Rosenblatt's perceptrons, the learning algorithm for adjusting the weights in successive layers of multi-layer perceptrons which has become known as *(error) backpropagation*, was one of the most influential developments in the field of neural networks (Rumelhart et al., 1986). Another major theme of interest was associative content-addressable memory, in which different input patterns become associated with one another if sufficiently similar. In this area, Kohonen's topological feature maps (Kohonen, 1982; Kohonen, 1984) have become most important in many applications fields like biology and robotics (Ritter et al., 1990).

2 Formal frameworks for the representation of neural networks

The new flourishing period of the neural network paradigm in the past few years has lead to a great variety of different models, which lack a unifying theoretical framework. Using new results from formal language theory it is possible to capture important structural and algorithmic features of various neural network models in an elegant and comprehensive way. The advantages of such an approach being based on the theory of n-dimensional parallel array grammars will be exhibited in this section.

In the late eighties, new links between neural network theory and formal language theory were found. Going back to the ideas of McCulloch and Pitts respectively the idea of cellular automata, Christel Kemke modelled neural networks by means of finite automata (Kemke, 1987), and the theory of automata networks evolved (Choffrut, 1986). Goles and Martinez worked out a general definition for automata networks and neural networks that was based on a graph representation of general cellular automata (Goles and Martinez, 1990).

In searching for a unifying basic definition of neural networks, attributed graphs can be chosen as the underlying structure, where the nodes of the graph represent the cells of the neural network and the edges represent the connections between the cells (Hecht-Nielsen, 1990). By capturing the dynamic features of different neural network models in their rules, suitable attributed graph grammars turn out to be an adequate unifying formal framework for the neural network paradigm (Freund et al., 1992).

Yet when implementation aspects in real-world computer architectures are considered, the representation of neural networks by (attributed) graphs leaves some problems open. Then array structures turn out to be a more realistic underlying structure for neural network architectures. Based on new results in the formal language theory of n-dimensional attributed parallel array grammars (Freund, 1993), we have introduced another new formal framework for representing various neural network models (Freund and Tafill, 1993).

Using this array model, which has its origins in the cellular automata paradigm, we can represent various models of neural networks in a formal but comprehensive way, and moreover, we can establish new algorithms within this formal framework. For example, for Kohonen's self-organizing feature maps we can prove that his global adaption rule for the weights of the cells, which requires total connectivity of the cells, can be simulated in another adequate network with locally bounded neighbourhood in linear time.

This result, which shows how global rules can be implemented on real network architectures with locally bounded neighbourhoods is also important for other neural network models and their implementation. As both the original network and the simulating network can be represented within the same formal framework, formal proofs become possible.

3 Conclusion

In the beginning of the history of neural networks, close relations between the theory of neural networks and the theory of formal languages and automata could be found. After having drifted apart from each other, these two fields recently could be linked together again. As one of these new ties we have presented n-dimensional attributed parallel array grammars, which among other advantages offer

- a theoretical formal framework for representing various neural network models and their algorithmic features, and

- a formal implementation platform for the implementation of specific neural networks on different computer architectures.

4 References

Caianiello, E.R. (1961) Outline of a theory of thoughtprocesses and thinking machines. Journal of Theoretical Biology, 1(2), 204-235.

Codd, E.F. (1968) Cellular Automata. Academic Press.

Choffrut, C. (ed.) (1986) Automata Networks. Proceedings LITP Spring School on Theoretical Computer Science. Lecture Notes in Computer Science, 316, Springer-Verlag, Berlin-Heidelberg-New York.

Freund, R., Haberstroh, B., Bischof, H. (1992) Tools for Dynamic Network Structures: GRAPE Grammars. Proceedings IJCNN'92, Baltimore.

Freund, R. (1993) Aspects of n-dimensional Lindenmayer systems. In: Salomaa, A. (ed.) (1993) Developments in Language Theory. World Scientific, Singapore, to appear.

Freund, R., Tafill, F. (1993) Modelling Kohonen networks by attributed parallel array systems. Proceedings SPIE'93-Symposium on Neural Networks, to appear.

Goles, E., Martinez, S. (1990) Neural and Automata Networks. Volume 58 of Mathematics and Its Applications, Kluwer Academic Publisher, Dordrecht.

Grossberg, S. (1976) Adaptive pattern classification and universal recording: I. Parallel development and coding of neural feature detectors. Biological Cybernetics, 23, 121-134.

Hebb, D.O. (1949) The Organization of Behaviour. Wiley, New York.

Hecht-Nielsen, R. (1990) Algebraic Theory of Processes. MIT Press Series in the Foundations of Computing, MIT Press, Cambrigde, Massachusetts.

Hertz, J.A., Krogh, A.S., Palmer, R.G. (1991) Introduction to the Theory of Neural Computation. Santa Fe Institute, Addison Wesley, Reading, MA.

Hopfield, J.J. (1982) Neural Networks and Physical Systems with Emergent Collective Computational Abilities. Proceedings of the National Academy of Science, USA, 79, 2554-2558.

Kemke, C. (1987) Modelling Neural Networks by Means of Networks of Finite Automata. Proceedings of the IEEE - First International Conference on Neural Networks, San Diego, CA,21-24.

Kohonen, T. (1982) Self-Organized Formation of Topologically Corrected Feature Maps. Biological Cybernetics, 43, 59-69.

Kohonen, T. (1984) Self-Organization and Associative Memory. Springer-Verlag, Berlin-Heidelberg-New York.

McCulloch, W. S. Pitts, W, (1943) A logical calculus of the ideas immanent in nervous activity. Bulletin of Mathematical Biophysis, 5, 115-133.

Minsky, M. L. (1967) Computation: Finite and Infinite Machines. Prentice Hall, Englewood Cliffs, NJ.

Minsky, M. L., Papert, S. (1969) Perceptrons: An Introduction to Computational Geometry. MIT Press, Cambridge, MA.

Neumann, J. von (1958) The Computer and the Brain. Yale University Press, New Haven.

Perrin, D. (1990) Finite automata. In: Handbook of Theoretical Computer Science B, 1-57. Elsevier, MIT Press, Cambridge, MA.

Ritter, H., Martinez, T., Schulten, K. (1990) Neuronale Netze. Addison-Wesley, Reading, MA.

Rosenblatt, F. (1958) The perceptron: a probabilistic model for information storage and organization in the brain. Psychological Review 65, 386-408.

Rumelhart, D.E., Hinton, G.E., Williams, R.J. (1986) Learning Representations by Back-Propagating Errors. Nature, 323, 533-536.

Smith III, A.R. (1970) Cellular Automata and Formal Languages. Proceedings 11th Conference on Switching and Automata Theory.

Smith III, A.R. (1971) Simple computation-universal spaces. Journal of the ACM, 18, 339.

Training-dependent Measurement

Zhengrong Yang

Shanghai Institute of Metallurgy, Academia Sinica, 865 ChangNing Rd.,
Shanghai, 200050, CHINA

1. Introduction

Neural networks have been applied to many fields with more successes [1]~[7]. But one feels awkward to some specious output of a trained net, such as 0.45 or 0.55. How to distinguish genuine from sham is important to the correct reasoning. In fact, the training provides us two important results. One is the trained net, which represents the knowledge hidden in the training patterns and can be used for reasoning. The other is the probabilistic distribution of actual outputs responding to the training input patterns. Because the training patterns prepared by human will not reflect all the characters of the real world or will have some noise with them, so the actual output of a trained net can not reach the target at 100%. Based on this reason, the probabilistic distribution of actual output is an important information. We call this probabilistic distribution the training distribution. By contrast, the test distribution is composed of the test responses. Most applications just utilised the trained net and did not use the training distribution. This paper will argue this issue and present some experimental results.

2. Individual measurement

The individual measurement is a common measure method: when a trained net outputs a response to a test input pattern, one makes his reasoning by check if this response close to 0 or 1. One can define a false threshold and a true threshold, F and T respectively. The measurement is carried out as follows:

$$r = false \qquad if \quad r \le F$$
$$r = true \qquad if \quad r \ge T$$

The problem is that the reasoning based on the individual measurement largely dependent on the experience of human and the values of thresholds. This is not reasonable.

3. Probabilistic measurement

The probabilistic measurement is another kind of measure method: when a trained net outputs a response to a test input patterns, one makes his reasoning based on the probabilistic value. Given the normal distribution for i^{th} output bit and j^{th} code (false/true) of a net is:

$$p_j(r_i, \mu_j, \sigma_j) = \frac{1}{\sqrt{2\pi}\sigma} e^{-\frac{(r_i - \mu)^2}{2\sigma^2}}.$$

where,

506

$$\mu \approx \hat{\mu} = \frac{1}{n}\sum r_i \quad - \quad sample \quad centre$$

$$\sigma^2 \approx \hat{\sigma}^2 = \frac{1}{n}\sum (r_i - \hat{\mu})^2 \quad - \quad sample \quad var iance$$

For each bit, there are two discriminations: true or false. The measurement standard is the value of probability. Let "p_{false}" represents the false probability and "p_{true}" the true probability. For example, in figure 1, if $r_i = 0.4$, because of $p_{false}(0.4) > p_{true}(0.4)$, so r_i should be regarded as false. Probabilistic measurement increases the computational time, but increases the believable degree of the reasoning. This is quite important.

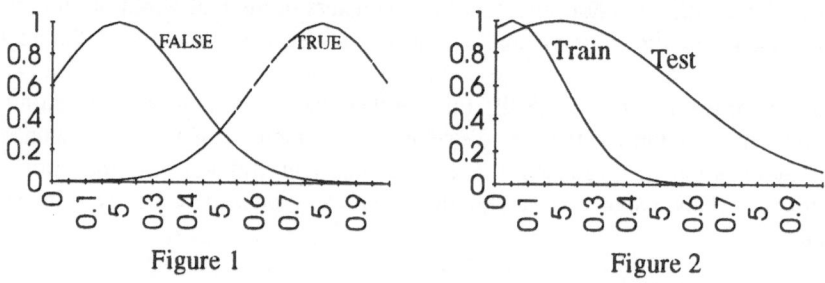

Figure 1   Figure 2

4. Sample-relative distortion

Now the problem is how to choose a suitable sample space. There are two sample-spaces during the measurement using the trained net. One is the training-dependent sample space, which is composed of the actual outputs. The other is the test-dependent or training-independent sample space, which is composed of the test responses. So, there are two probabilistic measurement methods. One is the training-independent measurement, which uses the test distribution. The other is the training-dependent measurement, which uses the training distribution. The sample-spaces of actual outputs and test responses will have different probabilistic distributions. Figure 2 gives an example. Clearly, the test distribution is different from the training distribution and has larger specious field. If the test distribution is used for real reasoning, there will be larger distortion. For example, as in figure 2, some reasoning will deviate from true to false if the test distribution is used. This distortion, which varies with different sample-space, is called the sample-relative distortion.

5. Experiments and conclusion

In this section, two experiments are given. From them, we will agree that the training-dependent measurement should be a better method. The first experiment was a classification problem with a BP 2-4-5-2 structure. Three groups of patterns are listed as table 1.

Table 1

No.	Size	Usage
1	200	Training
2	50	Testing
3	10000	Testing

After training and testing, three related sample-spaces formed and their probabilistic parameters were shown in table 2. In table 2, we can see that the deviation on the first bit of the second sample space reaches 900%.

Table 2

No	$\mu_1(0,1)$	$\sigma_1(0,1)$	deviation 1	$\mu_2(0,1)$	$\sigma_2(0,1)$	deviation 2
1	0.03, 0.93	0.13, 0.15	87%	0.07, 0.97	0.15, 0.13	115%
2	0.07, 0.98	0.18, 0.02	900%	0.02, 0.93	0.02, 0.18	11%
3	0.10, 0.97	0.25, 0.08	312.5%	0.03, 0.90	0.08, 0.25	32%

Table 3

No.	Target	Actual Output	Reasoning 1	Reasoning 2	Reasoning 3
1	0 1	0.64 0.36	Wrong	Correct	Wrong
2	0 1	0.67 0.33	Wrong	Correct	Wrong
3	1 0	0.85 0.15	Correct	Wrong	Correct
4	1 0	0.67 0.33	Correct	Wrong	Correct
5	0 1	0.78 0.22	Wrong	Correct	Wrong
6	0 1	0.62 0.38	Wrong	Correct	Wrong

From this table, it can be concluded that the larger the size of test patterns is, the smaller the deviation will be. For example, when the size increases to 10000 from 50, the deviation on the first bit on the related sample-space decreases to 312.5% from 900%. So, the real reasoning will ask us to prepare very larger size of test patterns. It is impossible and not necessary. Then, the training-dependent measurement should be considered without any doubt. Table 3 gives part of simulated reasoning results of this experiment. In table 3, "reasoning 1" means using the training-dependent measurement, "reasoning 2" means using the training-independent measurement (the second sample space), and "reasoning 3" means using the individual measurement. The experimental results proved that the results of the training-dependent measurement are more realistic than the results of the training-independent measurement. In this experiment, because of less noise in the training patterns, the training was more complete and the distributions of "true" and "false" were separated from each other very clear, that is, the "true" was more close to 1, and "false" was more close to 0, (see table 2, $\mu_1(0) = 0.03$ and $\mu_1(1) = 0.93$). Then the individual measurement could obtain the same results as the training-dependent measurement.

The second experiment was a medical diagnostic example with a BP 18-36-12-2 structure. Table 4 shows part of the experimental results. In this experiment, because of more noise in the training patterns, the training was incomplete and the distributions of "true" and "false" were not separated very clear, that is, the "true" was not close to 1, and "false" was not close to 0. Then the individual measurement could not obtain the same result as the training-dependent measurement sometimes (see the No 4 in table 4).

Table 4

No	Target	Actual Output	Reasoning 1	Reasoning 2	Reasoning 3
1	1 0	0.23 0.77	Wrong	Correct	Wrong
2	1 0	0.20 0.80	Wrong	Correct	Wrong
3	0 1	0.61 0.39	Wrong	Correct	Wrong
4	0 1	0.45 0.55	Correct	Wrong	Wrong

The experimental results have proved that the training-dependent measurement can result in more realistic solution to the real problem. Just because a trained net can not reach the target at 100%, the training distribution is considered closely. With the training-dependent measurement, the reasoning result is more believable.

Acknowledge

The author would like to thank Professor G. Musgrave (Head of dept. of EE & E, brunel university) for his guidance.

Reference

1. Z. Yang, G. Musgrave, "Applying Artificial Neural Network to Fault Diagnosis of Analogue Circuit", the 3rd International Conference on VLSI and CAD -Korea, Nov. 15-19, 1993

2. C. Rodriguez, et al, "A modular approach to the design of neural networks for fault diagnosis in power systems", 1992 International joint conference on neural networks, ppIII-16-23

3. Chong Ho Lee, et al, "Prediction of monthly transition of the composition stock price index using recurrent back-propagation", Artificial neural networks, Vol. 2, 1992, pp1629-1632

4. He Yu, et al, "Knowledge acquisition and resoningbased on neural network - the research of a bridge bidding system",1990 International neural network conference, pp416-423

5. Ah Chung Tsoi, "Application of neural network methodology to the modelling of the yield strength in a steel rolling plate mill", Advances in neural informationa processing systems Vol. 4, 1992, pp698-705

6. Richard Fozzard, et al, "A connectionist expert system that actually works", Advances in neural informationa processing systems Vol. 1, 1989, pp248-255

7. Venkataesh V. Murty, "Sparse neural networks for system identification", 1993 International neural network conference, ppIII-305-309

Genetic Algorithms as Optimisers for Feedforward Neural Networks

L. Vermeersch†, F. Dumortier‡, G. Vansteenkiste†

† Dept. Appl. Math., Biometrics and Process Control, University of Gent
Gent, Belgium

‡ Dept. Process Control and Automatisation, University of Gent
Gent, Belgium

1 Introduction

Currently, multilayer feedforward neural networks with backpropagation learning rule are the most popular artificial neural networks for handling optimisation and learning problems. This success is largely based on the simplicity to implement these networks. However, there are some major drawbacks : there is no guarantee for training in finite time and there is no guarantee that the optimal solution will be reached. In the sequel, the application of genetic algorithms as learning rule (thus, as optimiser) is compared to the error backpropagation learning rule (i.e., gradient descent optimiser).

The sequel of the article is structured as follows. In Section 2 a short overview is given on different families of feedforward neural networks. A brief introduction on the working of genetic algorithms is presented in Section 3. The basic concepts and operators are explained. In the next section the nontrivial linking of genetic algorithms as learning rule for feedforward networks is shown. Finally, some details on the object-oriented implementation of the genetic algorithms are revealed and the results of learning the XOR problem by means of a genetic algorithm based neural network are given.

2 Neural Networks

Artificial neural networks (ANN's) are inspired on models of the human brain, put forward by neurophysiologists, neurobiologists, ...By means of simulation of the human brain dynamics, these artificial neural networks try to solve problems that are hard to tackle by conventional programming paradigms. Image recognition, optical character recognition (OCR) and voice recognition are just mentioned here as commonly known examples of the vast space of tasks that are "easily" performed by humans but are hard to accomplish by computers.

However, ANN's are not a panacea. In most practical cases it is difficult to prove that an ANN will converge to a solution. A second problem is the lack of sufficient compute power provided by contemporary computers. Thirdly, the real intricacies of the human brain are not yet known. Due to these problems,

ANN's are by no means as powerful as the human brain. They are merely a pale shadow of it.

The basic building blocks are neurons just as in the human brain. The organisation of the neurons is similar to the anatomy of the human brain. In this way, one hopes to find neural structures that can perform useful tasks. It is estimated that 10^{11} neurons take part in the neural activities of a human being and that these neurons are interconnected by means of 10^{15} connections, called dendrites. It is clear from these astronomical figures that by no means contemporary computers can make an attempt to simulate such complex structures.

The structure of a neuron is quite simple. It is made up of a number of input channels, a netfunction that combines the inputs to a single value and an activation function F that transforms the netvalue to an output value. By connecting a number of neurons in a structured manner an ANN is put together.

Frequently constraints are imposed on the organisation of the neurons, namely, neurons are collected in layers, the layers are sequentially ordered and the outputs of a layer are the inputs for the next layer. In this way three different types of layers can be considered :

- The input layer, which in numerous works is not considered as a real neural layer, serves as a sort of multiplexer to the rest of the neural system. Its activation function is just a linear ("pass-through") function and its neurons have only one input channel.

- The hidden layers are placed between the input and the output layer. In most of the layered neural networks all of the recognition and classification work is performed by the hidden layers. The number of hidden layers varies from zero to the amount needed for the job.

- The output layer gives the actual output to the environment.

There are many different kinds of layered neural networks. One of the early architectures was the perceptron [8]. These ANN's were especially suited for classifying input patterns. However, a severe limitation is the linear separability constraint on the input patterns [6]. A second type of layered ANN's are the backpropagation based ANN's [9]. These ANN's are the most widely used ANN's. They are applied in image recognition systems, speech manipulation systems, ...The backpropagation NN's tackle the problem of learning multi-layer ANN's. The counter propagation network, developed by R. Hecht-Nielsen [4], can reduce the training time for some classes of problems and this compared to the backpropagation networks. However, the counterpropagation network is not as generally applicable as the backpropagation network. Yet another interesting layered ANN is the cognitron and the neocognitron put forward by K. Fukushima [1, 2]. The recognition accuracy of these ANN's is impressive. However, they require substantial computational resources.

The main problem of ANN's is that they have to be trained, which actually means looking for a global minimum of a cost function in a multidimensional hyperspace. Furthermore, if a global minimum can be found, if it is albeit possible, it is not sure that it will produce a satisfactory solution. In most real-life applications the search procedure has to look for a global minimum in a hyperspace filled with local minima. A widely deployed learning paradigm is error backpropagation, which is actually a steepest descent algorithm. It is well known that this method is not robust with respect to local minima. Therefore, genetic algorithms are proposed as an alternative training paradigm for neural networks.

3 Genetic Algorithms

The fundamental concepts of genetic algorithms (GAs) are based on genetic phenomena found in nature, such as the combination of the survival of the fittest idea with a structured though random information exchange between genetic information strings. It is hoped that, in this way GAs provide a more robust search algorithm than the classical search methods. However, GAs are not always the better solution. Problem oriented solutions will outperform GAs with respect to convergence speed and probably also with respect to the accuracy of the result. Therefore, in general optimisation problems GAs can be applied as a first "smart" guess. This solution can then be used as an initialisation for a classical, problem oriented search algorithm.

The main differences between more traditional optimisation algorithm and GAs are [3]:

- GAs manipulate a coded parameter set and not the parameters themselves. The search space can be determined by means of the coding scheme.

- GAs work with a population of points, in contrast with this classical search algorithms operate on a single point. This point to point calculation is an ideal prescription to converge to a local minimum. By considering a population of points, GAs climb multiple hills at a time.

- GAs make use of objective, optimisation functions. Derivatives are not taken into account, which is an advantage for complex problems where derivatives have to be computed artificially. Furthermore, this approach makes GAs independent of the problem to tackle. However, by neglecting problem dependent information the convergence speed is decreased.

- Probabilistic transition rules are deployed between two populations. This does not mean that GAs are simple random search techniques.

The data structures of GAs are inspired on the natural model. The terminology for the same data components are different. They are summarised in Table 1. The basic operators working on the data structures of simple GAs are reproduction, crossover and mutation. A cycle (i.e. the successive operation of reproduction, crossover and mutation) is repeated until a sufficient optimisation is attained.

GAs commence working on an initial population of strings (i.e. chromosomes). Out of this initial population a new population is selected. The probability, that a string of the initial population is selected to be copied to the new population, is proportional to the measure of compliance of the string to the objective function of the GA. So, strings with a high fitness value are more likely to be selected for the next generation (i.e. new population). Obviously, this

Natural	GA
gene	feature, character, or detector
allele	feature value
chromosome	string
locus	string position
genotype	structure
phenotype	parameter set, alternative solution
	a decoded structure

Table 1: Data Structures of Genetic Algorithms

operator simulates the "Survival of the Fittest" phenomenon. This selection of strings is called reproduction.

The strings in the new population are mated at random after this reproduction phase. For each pair a crossing site is chosen at a random place in the two strings. All genes in front of the crossing site remain at their place, whereas the other genes are mutually exchanged between the two strings. This crossover operator simulates a commutation of short schemata, building blocks, which can be interpreted as representation of basic ideas or notions. In this way, notions with a high fitness value are repeatedly tested and interchanged as to achieve even higher optimisation values.

The last operator in a simple GA, that acts upon a population, is the mutation operator. Mutation is responsible for an allele change at a random place in the new strings. This operator implements a kind of random walk in the "string space" and in doing this it insures against the possible loss of valuable genetic information. Experience has shown [3] that the number of mutations should be less than 1 per thousand cycles.

It has to be noted that a lot of variations exist for these basic operators. They all have their respective advantages and disadvantages. Until now, only organisms with one chromosome were considered, i.e. "haploids". However, in nature most organisms have more than one chromosome, e.g. "diploids" or "polyploids". For these organisms, a more complex operator such as dominance has to be considered.

4   Linkage of Genetic Algorithms and Neural Networks

In this Section attention will be focussed on the coupling between artificial neural networks and genetic algorithms [10]. The global structure of the ANN is determined at the beginning of the learning phase. So, during the optimisation (learning) executed by the GA, the composition of the ANN is not changed only the weight and bias values are adapted. Currently, the weights and biases are stored in a linear vector (cf. Figure 1), since the chromosomes are essentially linear, ordered collection of genes. To store the linear lists in strings, the parameters are translated into a binary representation. The ANN's parameter values could also be saved as a linear list of real-valued genes [7]. However, in this case a lot of the similarities between a GA and "real" genetic life is lost since a natural gene can only have a limited countable number of alleles.

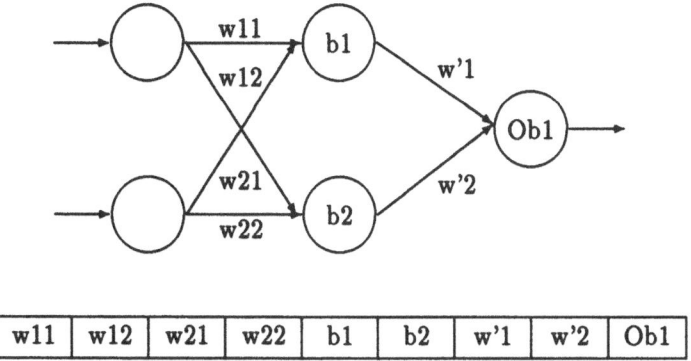

| w11 | w12 | w21 | w22 | b1 | b2 | w'1 | w'2 | Ob1 |

Figure 1: Artificial Neural Network Coding

Moreover, the operators of a GA have to be redefined.

Based on this coding scheme an alternative crossover operator is proposed. Since weights and biases are translated into binary sequences and then concatenated to strings (chromosomes), it is worthwhile to test the effect of a crossover operator with a crossing site per parameter (cf. Figure 2).

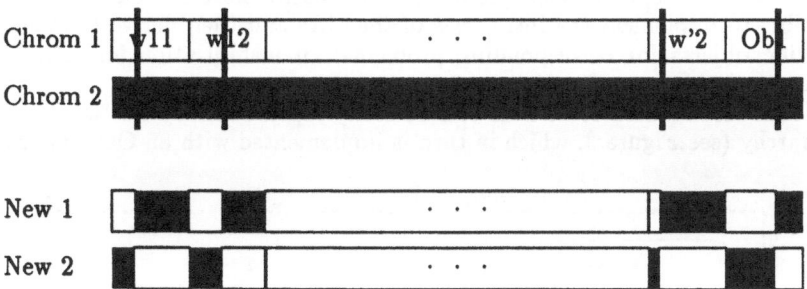

Figure 2: Alternative Crossover for ANN

The choice of the objective function is of major importance to achieve good learning results with an ANN controlled by a GA. The objective function has to reward the "good" strings (good learning) and has to punish the "bad" strings. Also, the fitness determination function has to be positive definite (the function can be zero in limited number of points). Frequently, the "total sum of squared errors" is taken as the basis for an objective function. To ensure the positive definiteness of the objective function the maximum values of the outputs of the output neurons has to be known, which is not always easy to determine[1]. Moreover, the objective function based on the "total sum of squared errors" makes no difference between an ANN with one completely incorrectly learned pattern and an ANN that learned all patterns rather good (but not excellent) if they both give the same total sum of squared errors. A "good" ANN should give a rather good output for all patterns. Therefore, an

---

[1] The objective function can now be defined as the subtraction of the sum of squares of maximum output values and the total sum of squared errors

objective function based on a multiplication of the absolute values of all errors is suggested. The objective function should also be sufficiently selective. So, the following objective function is proposed:

$$objective(x) = \prod_i error(|target_i - output_i|) \qquad (1)$$

with

$$error(x) = 1 - e^{-0.01/x^3} \qquad (2)$$

The first objective function can be seen as "or-operator" on the errors generated by the ANN, whereas the second objective function is more like an "and-operator" on the errors.

5 Genetic Algorithm Simulation Language

An object oriented programming (OOP) library was implemented to test the ideas put forward in the previous sections. The main reasons for choosing an object oriented approach was maintainability, extensibility, reusability and an natural way of programming [5]. The goal of the Genetic Algorithm Simulation Language (GASL) was to build a general purpose library, that could be utilised for all kinds of optimisation problems. Thanks to the OOP paradigm it was feasible to implement the intricacies of the GAs and then derive (instantiate) specific objects for corresponding problems, an instantiation for ANN's, an instantiation for cost functions, ...

The data structures of GAs can be easily represented by an "is-a-part-of" hierarchy (see Figure 3, which in turn is implemented with an OOP approach.

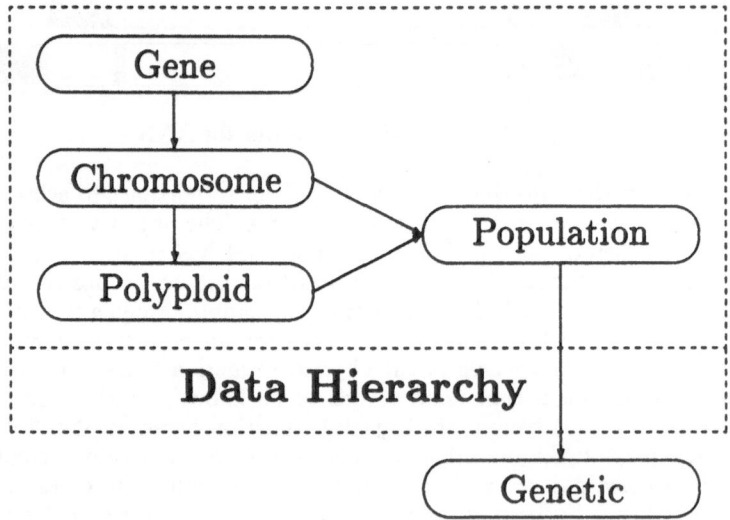

Figure 3: Hierarchical Representation of a Genetic Algorithm

- Gene : forms in a natural manner the base class for GASL, since a gene is the fundamental building block of a GA.

- Chromosome : is an ordered set of genes. In the simplest case an instance of this class is already a usable entity, i.e. a haploid.

- Polyploid : this class models a set of chromosomes. These chromosomes constitute a logical entity, i.e. genotype.

- Population : is a set of chromosomes or polyploids.

- Genetic : is the base class of a GA, either a simple GA or a complex GA.

The selection operator is implemented as a member function of the class Genetic, because it is closely related to the populations of a GA. The other operators, crossover and mutation, are not tightly coupled to the populations. They manipulate chromosomes and genes and hence are defined as separate classes. In this manner it is fairly easy to implement additional variants on these operators by extending their class hierarchy.

6 Results

The ideas presented supra were tested on the well-known XOR problem, i.e. an ANN that has to learn to react as XOR gate. The considered artificial neural network architecture was a 2 layer neural network with 2 linear input neurons, two sigmoid hidden neurons and one sigmoid output neuron. The neurons in the hidden layer and the output layer had a bias factor. As could be expected no satisfactory results were obtained with the conventional objective function and the crossover operator as defined by Goldberg. On the contrary, with the new objective function and the new crossover operator a well trained ANN was the result. On average it took between 200 and 300 generations before a well trained ANN was obtained, whereas with a standard backpropagation network it it took much more epochs.

7 Conclusion

The application of GAs as optimisers for ANN's seems to be promising. However, care has to taken when comparing epochs and generations. Although it took the GA less generations than the number of epochs needed by the backpropagation algorithm, the calculation time was much longer since providing a new generation is much more compute intensive than calculating one epoch. In contrast with this, Goldberg states in [3] that a GA an implicit parallel algorithm is. Hence, by means of parallel computer architectures it could become faster than the backpropagation approach, but it still has to be investigated.

Also, further research efforts are needed to investigate the effects of different coding schemes, new operators, other variants of operators, the magnitude of the population ...

References

[1] K. Fukushima. Cognitron : A self-organizing multilayered neural network. *Biological Cybernetics*, 20:121–136, 1975.

[2] K. Fukushima. Neocognitron : A self-organizing neural network model for a mechanism of pattern recognition unaffected by shift in position. *Biological Cybernetics*, 36:193–202, 1980.

[3] D.E. Goldberg. *Genetic Algorithms in Search, Optimization and Machine Learning*. Addison-Wesley Publishing Company, Inc., Massachusetts, 1989.

[4] R. Hecht-Nielsen. *Neurocomputing*. Addison-Wesley Publishing Company, Inc., Massachusetts, 1993.

[5] S.B. Lippman. *C++ Primer*. Addison-Wesley Publishing Company, Massachusetts, USA, 1990.

[6] M.L. Minsky and S.A. Papert. *Perceptrons : an Introduction to Computational Geometry*. MIT Press, Cambridge, Massachusetts, 1969.

[7] D.J. Montana and L. Davis. Training feedforward neural networks using genetic algorithms. *Machine Learning*.

[8] F. Rosenblatt. The perceptron : A probabilistic model for information storage and organization in the brain. *Psychological Review*, 65(6), 1958.

[9] D.E. Rumelhart, G.E. Hinton, and R.J. Williams. *Learning Internal Representation by Error Propagation*, volume 1 : Foundations, pages 318–362. The MIT Press, Cambridge, Massachusetts, 1988.

[10] L. Vercauteren, B. Sieben, and D. Thierens. Enhanced mapping : an extension of topological mapping to form internal representations and spatial mappings. In T. Kohonen, K. Makisara, D. Simula, and Kangas J., editors, *ICANN 91 Proceedings*, pages 1327–1332, Helsinki, 1991. ENNS, Elsevier Science Publishers B.V. (North Holland).

Selecting a Critical Subset of Given Examples during Learning

Byoung-Tak Zhang

German National Research Center for Computer Science (GMD)

D-53757 Sankt Augustin, Germany

E-mail: zhang@gmd.de

1   Introduction

The quality of training examples is one of the most important factors for effective and efficient training of neural networks, provided the network architecture and the weight modification rule are fixed. Earlier work [4] shows that the perceptron learning can be significantly accelerated by utilizing specific instances. For classification problems, these examples are the border patterns, i.e. the patterns that lie closest to the separating hyperplane. For real applications, however, it is difficult to estimate *in advance* the exact utility of examples, since the ultimate criticality of examples depends not only on the task, but also on the network configuration.

In this paper we describe an example selection method that finds a critical subset of given examples which achieves as good a performance as the whole set. It is efficient because the selection is performed in one trial *while* the learning proceeds. We select examples incrementally, an idea introduced first in [8] and developed further in [5]. The incremental selection is robust against possible errors since the inclusion of some non-critical examples at some stage can be automatically corrected at subsequent stages.

The method differs from [1, 2] in that we are using a computationally inexpensive squared error selection criterion instead of computing second derivatives. Our method is applicable to both classification and function approximation problems of high-dimensional inputs and outputs.

In the remainder of this paper we summarize the algorithm and show some results to demonstrate the effectiveness and generality of the incremental selection method.

2   Algorithm

The selection proceeds incrementally during learning. Given a data set B, the training set D is initialized to contain a small number of examples randomly chosen from B. The rest of B is used as a candidate set C. During learning, D is increased by selecting examples from C. We use a superscript s, as in $D^{(s)}$, to denote the sth training set. The weights of the network are initialized randomly with values from the interval $-\omega \leq w_{ij} \leq +\omega$.

The initial network is trained with the training set $D^{(0)}$. The trained network is used to expand the training set for the next generation $D^{(1)}$ which is again used to find and train the network. Training and selection is repeated until an acceptable generalization is achieved on the candidate data set. Notice that for each s the following conditions are always satisfied

$$D^{(s)} \cup C^{(s)} = B, \quad D^{(s)} \cap C^{(s)} = \emptyset, \quad \text{and} \quad D^{(s)} \subset D^{(s+1)}. \tag{1}$$

In the training phase, the weights are updated using the examples in the training set. If we denote by $\mathbf{w}^{(s,t)}$ the weight vector of the network for the t-th sweep through the training set $D^{(s)}$, the weights are modified by

$$\mathbf{w}^{(s,t+1)} = \mathbf{w}^{(s,t)} - \epsilon \nabla E_s|_{\mathbf{w}=\mathbf{w}^{(s,t)}} + \eta \Delta \mathbf{w}^{(s,t-1)} \tag{2}$$

where E_s is the total sum of the errors for $D^{(s)}$

$$E_s = E(D^{(s)}|\mathbf{w}^{(s,t)}) = \sum_{m=1}^{N_s} \left(\mathbf{y}_m - f(\mathbf{x}_m; \mathbf{w}^{(s,t)}) \right)^2 \tag{3}$$

and the error gradient $\nabla E_s|_{\mathbf{w}=\mathbf{w}^{(s,t)}}$ is approximated by a back-propagation procedure [3] or possibly by any other weight modification rule. ϵ and η are the step size and the momentum factor, respectively.

At every Δt epochs we check the convergence of the error minimization. If the total error for the current training set is reduced to a specified error tolerance level,

$$E(D^{(s)}|\mathbf{w}^{(s,t)}) \leq \frac{1}{\tau} \{(I+1) \cdot H + (H+1) \cdot O\} \tag{4}$$

the training phase terminates and the example selection phase follows. Here I, O and H_g are the number of input, output and hidden units of the network. The constant τ determines the error sensitivity per connection.

In the selection phase, the generalization accuracy of the current network is tested on the original data, $B = D^{(s)} \cup C^{(s)}$:

$$G_s = \frac{1}{N} \sum_{(\mathbf{x}_q, \mathbf{y}_q) \in B} \Theta \left(\mathbf{y}_q, f \left(\mathbf{x}_q; \mathbf{w}^{(s,t)} \right) \right) \tag{5}$$

where the function $\Theta(\cdot, \cdot)$ is some measure of correctness. If G_s satisfies the desired performance level ℓ, say 99%, then the entire algorithm stops. If $C^{(s)}$ is empty, the algorithm also stops. Notice that halting with a nonempty $C^{(s)}$ means the network has generalized correctly to the candidate data set. Otherwise, the criticality with respect to the current model $\mathbf{w}$ is computed.

The criticality $(\mathbf{x}_c, \mathbf{y}_c)$ of an example is defined as

$$e_\mathbf{w}(\mathbf{x}_c) = \frac{1}{\dim(\mathbf{y}_c)} \left(\mathbf{y}_m - f(\mathbf{x}_m; \mathbf{w}^{(s,t)}) \right)^2 \tag{6}$$

which has a value $0 \leq e_\mathbf{w}(\mathbf{x}_c) \leq 1$ if the sigmoid activation function is used at the output layer. Then the training set is increased by selecting λ candidate examples, $(\mathbf{x}_c, \mathbf{y}_c)$, which are most critical:

$$D^{(s+1)} = D^{(s)} \cup \{(\mathbf{x}_c, \mathbf{y}_c)\}, \quad \text{and} \quad C^{(s+1)} = C^{(s)} - \{(\mathbf{x}_c, \mathbf{y}_c)\}. \tag{7}$$

In case of $|C^{(s)}| < \lambda$, all the remaining candidate examples are selected into $D^{(s+1)}$. Using the expanded training set, the next cycle of training and selection is done.

3 Performance

The performance of the algorithm is demonstrated on the following function:

$$y(x_1, x_2) = \begin{cases} 1 & \text{if } (x_1 < 0.5 \text{ and } x_2 > 0.5) \text{ or } (x_1 > 0.5 \text{ and } x_2 < 0.5) \\ 0 & \text{otherwise,} \end{cases}$$

where $0.1 \leq x_1, x_2 \leq 0.9$. This function, called four-quadrant, is a real-input variant of the XOR problem. Due to its graphical characteristics the problem allows an easy analysis of the learning results.

We used a feed-forward network with four hidden units and a total of 400 (20×20 resolution) examples. The training set was initialized with four seed examples at the four corners. In each selection step, additional four examples ($\lambda = 4$) were added to the existing training set. Figure 1 shows the graphes of the approximated function and the corresponding training points used at $s = 0, 5$, and 10.

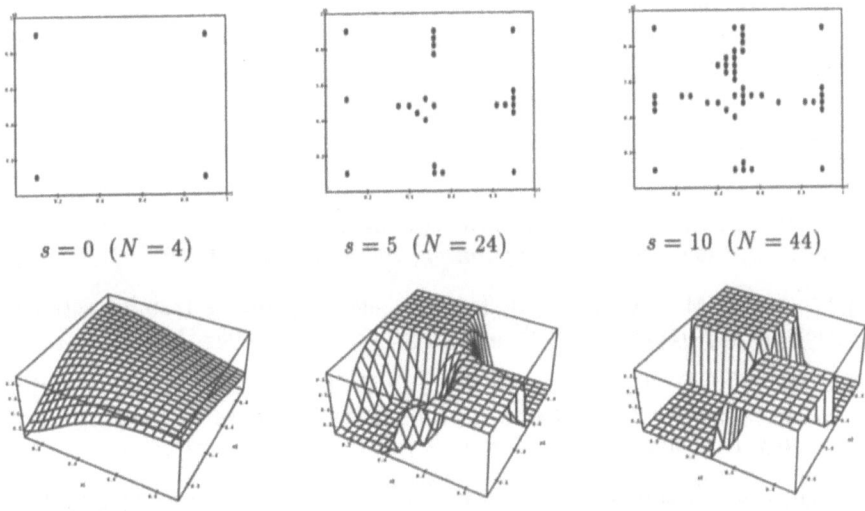

$$s = 0 \ (N = 4) \qquad s = 5 \ (N = 24) \qquad s = 10 \ (N = 44)$$

Figure 1: Selective incremental learning of the four-quadrant function

Notice that preferred examples lie on the separating lines of output 0 and 1. These are the border patterns and thus critical to solving this problem. Notice also that 44 selected examples, corresponding to 11% of the whole data, are sufficient for learning the function.

Next we applied the method to digit recognition in which training patterns consisted of 15×10 bitmap scanned from 20 sheets written by 10 persons. Starting from 10 examples and increasing the training set by 50 examples in each selection step, the method found about 1400 examples on average whose performance achieved the same generalization performance as the whole training set of 3400 examples.

The method has also been tested on several discrete problems, including majority and contiguity, and continuous functions such as the forward and inverse kinematics of a robot arm [5]. In most applications a moderate amount of training examples was sufficient to achieve a reasonable performance.

4 Concluding Remarks

We presented a method for selecting a critical subset of given examples during training the network. The method starts with a small training set which is increased incrementally after training. This incremental approach is based on the observation of close interdependence between the network performance and the training set: the performance of the network can be best improved by being trained on a good training set, whereas the quality of the training set can be maximally enhanced by taking account of the current state of the network.

In effect, the selective incremental learning performs an adaptive global-to-local search not only on the weight space but on the example space as well. This is contrasted with the plain backpropagation training procedure that "pushes" the network to load all the data at the same time, leading frequently to a delayed convergence. Our experimental results suggest the incremental learning generalizes much better and thus converges faster to good solutions than a non-incremental one. The improvement is proportional to the data reduction ratio by example selection.

References

[1] MacKay, D. J. C. (1992) "Information-based objective functions for active data selection," *Neural Computation*, vol. 4, pp. 590–604.

[2] Plutowski, M. and White, H. (1993) "Selecting concise training sets from clean data," *IEEE Trans. Neural Networks*, vol. 4, 305–318.

[3] Rumelhart, D. E., Hinton, G. E., and Williams, R. J. (1986) "Learning internal representations by error propagation," in *Parallel Distributed Processing*, Rumelhart, D. E. *et al.* (eds.), MIT Press, pp. 318–362.

[4] Volper, D. J. and Hampson, S. E. (1987) "Learning and using specific instances," *Biological Cybernetics*, vol. 57, pp. 57–71.

[5] Zhang, B. T. (1992) *Learning by Genetic Neural Evolution*, Ph.D. thesis in German, Universty of Bonn, ISBN 3-929037-16-5, Infix-Verlag.

[6] Zhang, B. T. (1993) "Accelerated learning by active example selection," to appear in *International Journal of Neural Systems*.

[7] Zhang, B. T. (1993) "Self-development learning: Constructing optimal size neural networks via incremental data selection," *GMD-Arbeitspapiere* No. 768, German Nat'l Res. Ctr. for Comp. Sci., Sankt Augustin.

[8] Zhang, B. T. and Veenker, G. (1991) "Focused incremental learning for improved generalization with reduced training sets," in *Artificial Neural Networks*, Kohonen, T. *et al.* (eds.), North-Holland, vol. I, pp. 227–232.

On the Circuit Complexity of Feedforward Neural Networks[1]

Valeriu Beiu[†,‡], Jan A. Peperstraete[†], Joos Vandewalle[†]
and Rudy Lauwereins[†,2]

[†] Katholieke Universiteit Leuven, Department of Electrical Engineering
Division ESAT, Kardinaal Mercierlaan 94, B-3001 Heverlee, Belgium

[‡] on leave of absence from "Politehnica" University of Bucharest
Computer Science, Spl. Independentei 313, 77206 Bucharest, Românîa

1. Introduction

In this paper a *network* will be considered an acyclic graph. It has several input nodes (*inputs*) and some (at least one) output nodes (*outputs*). The nodes of the network are characterized by *fan-in* (the number of incoming edges) and *fan-out* (the number of outgoing edges), while the network has a certain *size* (the number of nodes) and *depth* (the number of edges on the longest input to output path). If with each edge a synaptic *weight* is associated and each node computes the weighted sum of its inputs to which a nonlinear activation function is then applied (*artificial neuron*), the network is a *neural network* (NN). Similarly, a *Boolean circuit* (BC) is a network of gates implementing elementary Boolean functions (AND, OR, NOT).

2. Previous Circuit Complexity Results

The *circuit complexity* of a function is the size complexity of the family of minimal Boolean circuits that compute that function [Smolensky 1987, Wegener 1987]. For example the classes NC^k (AC^k) represent the functions with polynomial circuit complexity that can be computed by constant (unbounded) fan-in BCs of depth $O(\log^k n)$. Another interesting class is TC^k which represents the family of functions realized by polynomial size threshold NNs (the activation function being the threshold function) with unbounded fan-in and depth $O(\log^k n)$.

For placing neural networks in this context of BC complexity the class NN^k has been defined [Shawe-Taylor 1992] to be *"those functions which can be computed by a family of polynomially sized neural networks with weights and activation values determined to b bits of accuracy, fan-in equal to Δ and depth h, satisfying* $\log(\Delta) = O(\sqrt{\log n})$, $b \log\Delta = O(\log n)$ *and* $h \log\Delta = O(\log^k n)$*"*.

While many results concerning threshold NNs have been lately discovered [Beiu 1993a, Bruck 1990, Bruck 1992, Siu 1990, Siu 1991a] and their complexity discussed [Roychowdhury 1991, Williamson 1991], little is known for the case of sigmoid activation functions. One important result [Shawe-Taylor 1992] states that:

$$NC^k \subseteq NN^k \subseteq AC^k. \tag{1}$$

[1] This research work was partly carried out in the framework of a **Concerted Research Action** of the Flemish Community, entitled: *"Applicable Neural Networks"*. The scientific responsibility is assumed by the authors.

[2] Senior Research Associate of the Belgian National Fund for Scientific Research.

If $NC^k \subseteq NN^k$ can be easily proven by directly converting the AND and OR gates to equivalent threshold gates, the second part $NN^k \subseteq AC^k$ is proved in three steps (see [Shawe-Taylor 1992] for more details):

- negative weights are eliminated, increasing the size by at most a factor of 2;
- the NN with positive weights is converted to a threshold NN by taking 2^b copies of each node; it has the same depth, the size and the fan-in being increased by 2^b, while the accuracy has been doubled $2b$;
- using techniques of communication complexity [Karchmer 1988], the conversion of a linear threshold neuron to an equivalent BC, having depth $2\log\Delta$, fan-in $2^{b+\log\Delta}$ and size at most $2\Delta^{b+\log\Delta+1}$, is performed.

The overall result after taking these three steps is that a neural network of size N, depth h, maximum fan-in Δ and accuracy of b bits can be converted to an equivalent BC with depth $h(2\log\Delta + 1)$, fan-in $2^{2b+\log\Delta}$ and size at most:

$$2 \cdot N \cdot 2^b \cdot (2\Delta)^{2b+\log\Delta+1}. \tag{2}$$

3. *ADDITION* and Neural Network Conversion

Some of the adders build out of AND-OR bounded fan-in logic gates have (we use *delay* instead of depth, and *gates* instead of size, as in the original articles):

- $2n-1$ *delay* and $5n-3$ *gates* (school method);
- $4\log n$ *delay* and $35n-6$ *gates* (carry-lookahead adders [Chang 1992]);
- $4\log n$ *delay* and $14n-\log n-10$ *gates* [Brent 1982];
- $3\log n$ *delay* and $3n\log n+12.5n-8$ *gates* (conditional sum adders [Kelliher 1992]);
- $(2+\varepsilon)\log n$ *delay* and $(2+\varepsilon)n\log n+5n$ *gates* [Wei 1990];
- $2\log n+1$ *delay* and $3n\log n+10n-6$ *gates* (conditional sum adder [Wegener 1987]);
- $2\log n$ *delay* and $2.5n\log n+5n-1$ *gates* [Kelliher 1992];
- $2\log n+2k+2$ *delay* and $n(8+6/2^k)$ *gates* (prefix algorithms for $0\leq k\leq \log n$ [Ladner 1980]);
- $\log n+7\sqrt{2\log n}+16$ *delay* and $9n$ *gates* (Krapchenko [Wegener 1987]).

By using adders having size $O(n)$ and depth $O(\log n)$ we can prove that adding m words of β bits each can be done by a fan-in 2 BC having size $O(m\beta)$ and depth $O[\log m(\log\beta + \log\log m)]$. Based on the knowledge that:

- *at least* on the order of K^2 threshold gates of fan-in 2 are needed to replace one fan-in K threshold gate [Abu-Mostafa 1989], and
- the needed accuracy for the weights of a fan-in K threshold gate is on the order of $K\log K$ bits [Hong 1987, Raghavan 1988],

and taking $m = K$ and $\beta = K\log K$, the size becomes $O(K^2\log K)$, which is just a logarithmic factor larger than the theoretical optimum $\Omega(K^2)$ of [Abu-Mostafa 1989].

The constructive proof is based on a binary tree of adders: $m/2$ adders of size $\lambda\beta$ and depth $\mu\log\beta$ in the first layer, followed by $m/4$ adders of size $\lambda(\beta+1)$ and depth $\mu\log(\beta+1)$ in the second layer and so on until the last (the $\log m^{\text{th}}$) layer having one $(m/2^{\log m})$ adder of size $\lambda(\beta+\log m-1)$ and depth $\mu\log(\beta+\log m-1)$. Here λ and μ are constants depending on the adder one has chosen (e.g. $\lambda = 14$ and $\mu = 4$ for the Brent and Kung adder [Brent 1992]). We can now compute the size:

$$size\,(m,\beta) = \lambda \sum_{i=1}^{\log m} \frac{m\,(\beta + i - 1)}{2^i} = \lambda m\beta \sum_{i=1}^{\log m} \frac{1}{2^i} + \lambda m \sum_{i=1}^{\log m} \frac{i-1}{2^i} =$$

$$= \lambda m\beta \frac{m-1}{m} + \lambda m \sum_{i=2}^{\log m} \left(\sum_{j=i}^{\log m} \frac{1}{2^j} \right) \qquad = \lambda\beta(m-1) + \lambda m \sum_{i=2}^{\log m} \frac{m-2^{i-1}}{m \cdot 2^{i-1}} =$$

$$= \lambda\beta(m-1) + \lambda m \frac{m-\log m - 1}{m} \qquad = O(\beta m). \tag{3}$$

The depth is:

$$depth(m,\beta) = \mu \sum_{i=1}^{\log m} \log(\beta + i - 1) \qquad < \mu \left[\log\beta + \sum_{i=1}^{\log m - 1} (\log\beta + \log i) \right] =$$

$$= \mu\log\beta\log m + \mu \sum_{i=1}^{\log m - 1} \log i \qquad = \mu\log\beta\log m + \mu\log[(\log m - 1)!]$$

and using Stirling's formula $k! = \sqrt{2\pi k} \left(k/e \right)^k e^{\theta/12k}$ where $0 < \theta < 1$, we have:

$$depth(m,\beta) < \mu\log\beta\log m + \mu\log \left[\sqrt{2\pi\log m} \left(\frac{\log m}{e} \right)^{\log m} e^{\frac{1}{12\log m}} \right] =$$

$$= \mu\log\beta\log m + \mu \left[\frac{\log 2\pi + \log\log m}{2} + \log m(\log\log m - \log e) + \frac{\log e}{12\log m} \right] =$$

$$= O[\log m(\log\beta + \log\log m)]. \tag{4}$$

As a linear threshold gate does in fact a summation of (some of) the incoming weights, such an adder, having fan-in $m = 2^b\Delta$ and accuracy $\beta = 2b$, can be used in the third step of the conversion from [Shawe-Taylor 1992]. A NN of size N, depth h, maximum fan-in Δ and accuracy of b bits converted in this way will have an equivalent fan-in 2 BC of size at most:

$$2 \cdot N \cdot 2^b \cdot \lambda \left[2b\, 2^b\Delta - 2b + 2^b\Delta - \log(2^b\Delta) - 1 \right] \approx \lambda N\, 2^{2b+1} \Delta (2b+1). \tag{5}$$

Taking logarithms of eq. (5), and requesting that the BC be polynomially sized:

$$\log N + 2b + \log\Delta + \log(2b+1) + \lambda + 1 = O(\log n) \tag{6}$$

shows that $N = O[poly(n)]$, $\Delta = O[poly(n)]$ and $b = O(\log n)$ (as $\log(2b+1) < 2b$) are the conditions to be met by this class NN_Δ^k which, due to the relaxed fan-in condition, strictly contains NN^k. The depth of this BC is:

$$h\,\mu \left[\log(2b)\log(2^b\Delta) + \log(2^b\Delta) \log\log(2^b\Delta) + \dots \right] \tag{7}$$

which due to the parenthesis is $O(\log n \log\log n)$. Remembering that $h = O(\log^k n)$ we have the proof that the depth is $O(\log^{k+2} n)$ which implies:

$$NC^k \subseteq NN^k \subset NN_\Delta^k \subset NC^{k+2}. \tag{8}$$

Restricting the fan-in to $\log\Delta = O(\log^{1-\epsilon} n)$ and the accuracy to $b = O(\log^{1-\epsilon} n)$ a new class $NN_{\Delta,\epsilon}^k$ can be defined. In this case the parenthesis from eq. (7) is now bounded by $O(\log^{1-\epsilon} n \log\log n)$, so:

$$NC^k \subseteq NN^k \subset NN_{\Delta,\epsilon}^k \subseteq NC^{k+1}. \tag{9}$$

It can be observed that *there are interesting fan-in dependent depth-size tradeoffs* when trying to digitally implement sigmoid NNs (see also [Siu 1992]) which should be related to the depth-size tradeoffs of threshold NNs [Beiu 1993b, Siu 1991b]. Putting all these results together the following inclusions are known:

524

$$NC^k \subseteq NN^k \begin{cases} & \subseteq AC^k \quad \text{[Shawe–Taylor 1992]} \quad (10) \\ \subset NN_{\Delta,\varepsilon}^k \begin{cases} \subset NN_{\Delta}^k \subset NC^{k+2} \quad \text{this paper} \\ \subseteq NC^{k+1} \quad \text{this paper} \end{cases} \end{cases}$$

4. Conclusions

The paper has examined the circuit complexity of feedforward neural networks and presented a constructive proof on the lines of [Shawe-Taylor 1992], by *relaxing the logarithmic fan-in condition to a polynomial fan-in one*. The class of sigmoid activation feedforward neural networks which can be implemented in polynomial *size* Boolean circuits is thus substantially enlarged. This has been done on the expense of more layers than [Shawe-Taylor 1992], but with a constant fan-in of 2.

References

Abu-Mostafa, Y. (1989) Complexity in Neural Systems. In C.A. Mead: *Analog VLSI and Neural Systems*, Addison Wesley, Reading, 353-358.

Beiu, V., Peperstraete, J.A., Vandewalle, J., and Lauwereins, R. (1993a) Efficient Decomposition of COMPARISON and Its Applications. *Proc. ESANN'93*, Brussels, Dfacto, 45-50.

Beiu, V., Peperstraete, J.A., Vandewalle, J., and Lauwereins, R. (1993b) Overview of Some Efficient Threshold.Gate Decomposition Algorithms. *Proc. CSCS'93*, Bucharest, 1, 458-469.

Brent, R.P., and Kung, H.T (1982) A Regular Layout for Parallel Adders. *IEEE Trans. Comp.*, C-31(3), 260-264.

Bruck, J. (1990) Harmonic Analysis of Polynomial Threshold Functions. *SIAM J. on Disc. Math.*, 3(2), 168-177.

Bruck, J., and Smolensky, R. (1992) Polynomial Threshold Functions, AC^0 Functions and Spectral Norms. *SIAM J. Comput.*, 21(1), 33-42.

Chang, P.K., Schlag, M.D.F., Thomborson, C.D., and Oklobdzija, V.G. (1992) Delay Optimization of Carry-Skip Adders and Block Carry-Lookahead Adders Using Multidimensional Programing. *IEEE Trans. on Comp.*, C-41(8), 920-930.

Hong, J. (1987) On Connectionist Models. *Tech. Rep.* 87-012, Dept. CS, Univ. of Chicago.

Karchmer, M., and Widgerson, A. (1988) Monotone Circuits for Connectivity Require Super-Logarithmic Depth. *Proc. ACM Symp. on Theory of Computing*, 20, 539-550.

Kelliher, T.P., Owens, R.M., Irwin, M.J., and Hwang, T.-T. (1992) ELM A Fast Addition Algorithm Discovered by a Program. *IEEE Trans. on Comp.*, C-41(9), 1181-1184.

Ladner, R.E., and Fischer, M.J. (1980) Parallel Prefix Computations. *J. ACM*, 27(4), 831-838.

Raghavan P. (1988) Learning in Threshold Networks: A Computational Model and Applications. *Tech. Rep.* RC-13859, IBM Research.

Roychowdhury, V.P., Siu, K.-Y., Orlitsky, A., and Kailath T. (1991) On the Circuit Complexity of Neural Networks. *Proc. NIPS'90*, Denver, Morgan Kaufmann, San Mateo, 953-959.

Shawe-Taylor, J.S., Anthony, M.H.G., and Kern, W. (1992) Classes of Feedforward Neural Nets and Their Circuit Complexity. *Neural Networks*, 5(6), 971-977.

Siu, K.-Y., and Bruck, J. (1990) Neural Computation of Arithmetic Functions. *Proc. IEEE*, 78(10), 1669-1675.

Siu, K.-Y., and Bruck, J. (1991a) On the Power of Threshold Circuits with Small Weights. *SIAM J. on Disc. Math.*, 4(3), 423-435.

Siu, K.-Y., Roychowdhury, V., and Kailath T. (1991b) Depth-Size Tradeoffs for Neural Computations. *IEEE Trans. on Comp.*, C-40(12), 1402-1412.

Siu, K.-Y. (1992) On the Complexity of Neural Networks with Sigmoid Units. *Proc. IEEE-SP Workshop NNSP-92*, Helsingoer, Denmark, 23-28.

Smolensky, R. (1987) Algebraic Methods in the Theory of Lower Bounds for Boolean Circuit Complexity. *Proc. ACM Symp. on Theory of Computing*, 19, 77-82.

Wegener, I. (1987) *The Complexity of Boolean Functions*. Wiley-Teubner, Chichester.

Wei, B.W.Y., and Thompson, C.D. (1990) Area-Time Optimal Adder Design. *IEEE Trans. on Comp.*, C-39(5), 666-675.

Williamson, R.C. (1991) ε-Entropy and the Complexity of Feedforward Neural Networks. *Proc. NIPS'90*, Denver, Morgan Kaufmann, San Mateo, 946-952.

AVOIDING LOCAL MINIMA BY A CLASSICAL RANGE EXPANSION ALGORITHM

D Gorse and A Shepherd
Department of Computer Science, University College London

J G Taylor
Department of Mathematics, King's College London

Conventional classical methods of supervised learning are inevitably faced with the problem of local minima; evidence is presented that fast training algorithms based on conjugate gradient and quasi-Newton techniques are particularly susceptible to being trapped in sub-optimal solutions. A classical technique is presented which by the use of a homotopy on the range of the target outputs allows supervised learning methods to find a global minimum of the error function in almost every case. The method is straightforward to apply and avoids the high computational overheads associated with stochastic techniques such as simulated annealing and genetic algorithms.

Introduction

The problems to which neural computing techniques are most frequently applied involve the supervised learning of an input-output mapping, defined implicitly by a set of P input patterns together with their desired outputs. The error E is a function of all the parameters (weights and thresholds) of the network; this parameter list can be written as a multidimensional vector w. The problem is to change w so as to avoid those solutions of the minimisation condition $\partial E/\partial w = 0$ which do not correspond to the lowest value of E, the local minima of the error-weight surface. The most commonly used supervised training technique, error backpropagation (BP) (equivalent to gradient descent with a fixed step length) is well known to have difficulties with local minima, especially for non linearly separable problems [1]. What is less well known is that the neural implementations of more efficient classical minimisation algorithms, such as conjugate gradients (CG) or the quasi-Newton method (QN), are even more likely to be trapped in sub-optimal solutions. Table 1 shows the percentage success in reaching a global minimum for 100 (2-2-1) networks learning to solve the XOR problem.

XOR	% reaching global minimum	
method	sigmoid in final layer	linear in final layer
on-line BP	85	95
batched BP	75	96
CG	51	80
QN	34	66

Table 1

XOR is a useful benchmark because it is a non linearly separable problem with known local minima [2], but one which can be solved by a small network with only 9 adaptive weights. Linear outputs in the second layer (as opposed to sigmoidal squashing for both computational layers) improve the percentage success, but there is a clear trend toward worsened performance for the more sophisticated algorithms. Simple on-line BP (without momentum) performs best; this may be due to the method's stochastic features, as discussed in [3].

Trapping in local minima can also be observed for continuous function learning problems. McInerney et al [4] have discovered (by exhaustive search of the error-weight surface) local minima in a (1-2-1) network (with a linear output node) learning the sine function. This problem was also investigated, using the same training set as in [4], and the results are summarised in Table 2.

sine	% reaching global minimum
batched BP	100
CG	96
QN	87

Table 2

These results do not show as high a probability of trapping in local minima as in the XOR example, but there is still a significant correlation between the probability of failure and the convergence speed of the method; the quasi-Newton method, with a 13% failure rate, would probably not be a good choice unless multiple restarts were acceptable.

It is commonly believed - though we do not know of any 'no-go theorem' to this effect - that the only techniques guaranteed to converge to a global minimum with a probability approaching 1 are stochastic in character, with methods based on simulated annealing and, currently, genetic algorithms being among the most popular. However these techniques can be very slow and must be applied carefully in order to ensure a good solution. Is there a way to retain the fast convergence of techniques like conjugate gradients and the quasi-Newton method whilst improving the robustness of these algorithms in the face of local minima? We will present here a purely classical method which is guaranteed to succeed in avoiding local minima in almost all cases.

Expanded range approximation (ERA)

The basic idea underpinning this algorithm is that of a homotopy on the range of the target values d_p (for simplicity we consider just one output node). This range is modified by compressing these values down to their mean value $<d> = \frac{1}{P} \sum_{p=1}^{P} d_p$ and then progressively expanding these compressed targets back toward their original values (hence the epithet 'expanded range approximation', or ERA, we have coined for this approach). We define a modified training set for inputs x_p

$$S(\lambda) = \{x_p, d_p(\lambda)\} = \{x_p, <d> + \lambda(d_p - <d>)\}$$

where the $d_p(\lambda)$ are the new, compressed, targets. The problem defined by $S(0)$ is easy for the network to solve (the corresponding error-weight surface can be shown to have only a global minimum); $S(1)$ is the original problem with training set $\{x_p, d_p\}$. The homotopy parameter λ interpolates between these extremes. A λ-parametrised error function can be defined during training on each of the sets $S(\lambda)$ by

$$E(\lambda) = \frac{1}{P} \sum_{p=1}^{P} [<d> + \lambda(d_p - <d>) - z_p(\lambda)]^2$$

where the $z_p(\lambda)$ are the actual network outputs during this procedure. The ERA method involves first solving the problem $S(\lambda_1)$ for small λ_1, then the problem $S(\lambda_2)$ with $\lambda_2 > \lambda_1$, and so on up to the original problem $S(1)$. We have usually chosen to increase λ by uniform steps of η; an 'N-step ERA' method refers to the progressive solution of the N problems $S(\lambda_n = n\eta)$ for n = 1..N=1/η ('1-step ERA' (η=1) is the conventional single step training technique).

As a first example, the ERA method was applied to the same 100 XOR networks (with sigmoidal output in the final layer) as in Table 1, using the CG algorithm. With 10-step ERA (η=0.1), the success rate improves dramatically from 51% when η=1 to 94%. If the step size η is decreased, the percentage success improves still further: 100-step ERA (η=0.01) is 100% successful in solving the XOR problem. As a second example, 2-step ERA (η=0.5) was applied to the sine problem of Table 2, using the QN method. In this case - a continuous as opposed to binary problem, a linear as opposed to sigmoidal output in the final layer, a different training method - there was also a very significant improvement, from 87% success when η=1 to 100% for 2-step ERA.

All the initial simulations suggested a special role for η, the size of the first step. In order to try to get some further insight into the process, we looked at the trajectories in output space followed for the XOR problem by the $z_p(\lambda_1 = \eta)$, the first-step responses to the four patterns p = 00, 01, 10, 11. Since the initial weights are randomly chosen (from the interval [-1,1]) the trajectories in these experiments begin at some arbitrary point inside the hypercube $[0,1]^4$. The target for η=1 is the point (0,1,1,0); the targets for $\eta < 1$ lie on a line joining this point to (½,½,½,½). By taking pairs of these responses we were able to plot trajectories in the six 2-dimensional $(z_{p_1}(\eta), z_{p_2}(\eta))$ subspaces during CG training. Figures 1a-d illustrate trajectories in $(z_{00}(\eta), z_{10}(\eta))$ space for η=1.0 (Figure 1a), η=0.3 (1b), η=0.2 (1c), η=0.1 (1d). A coordinate transformation

$$(x, y) = \frac{1}{\eta}(z_{00} - \tfrac{1}{2}(1-\eta), z_{10} - \tfrac{1}{2}(1-\eta))$$

is used in plotting the diagrams so that the scales are identical, and in each case the target is the top left hand corner. The midpoint on the y-axis represents a local minimum.

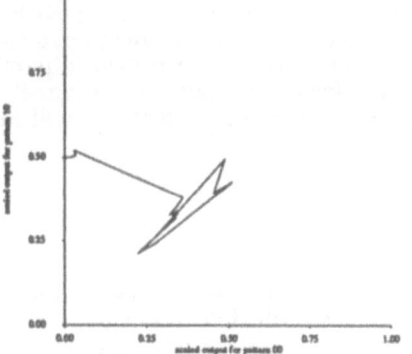

Figure 1a

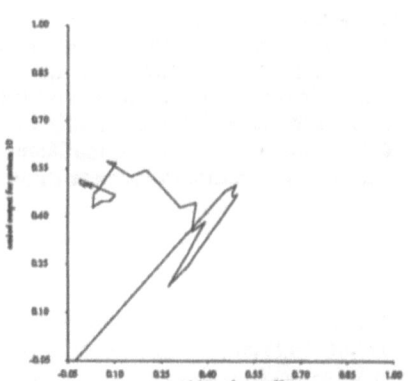

Figure 1b

Figure 1a shows a conventionally trained (η=1) network which fails to solve the XOR problem, becoming trapped in the local minimum corresponding to a final response to the four patterns of z = (0, ½, ½, 0). Figure 1b also shows a failure, for η=0.3, but notice that the trajectory appears to almost escape from the local minimum. Figure 1c shows a success at η=0.2, but the trajectory still spends a lot of time in the vicinity of the local minimum before escaping. Finally, Figure 1d, with η=0.1, shows a trajectory which entirely avoids the vicinity of the local minimum, heading more or less directly for the global minimum of $E(\lambda_1 = \eta)$. There appears to be a progressive change of behaviour as η is decreased; this progression is most marked for small values of η.

528

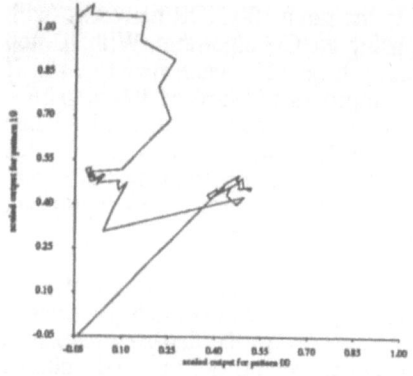

Figure 1c

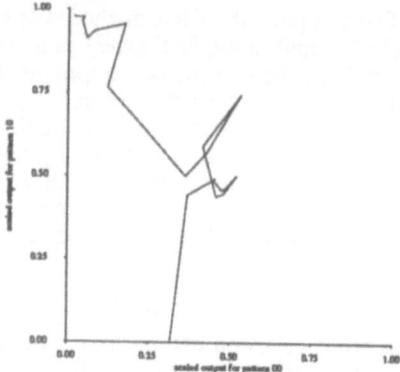

Figure 1d

Discussion

This paper has presented some results which violate the widespread belief that the only way to avoid local minima in supervised learning problems with complex error-weight surfaces is to use computationally expensive stochastic procedures like simulated annealing or genetic algorithms. Further simulations also support the ERA hypothesis [5] but cannot be presented here for reasons of space. If the results presented here can be shown to be securely founded, and the ERA method shown to have wide applicability, there could be a significant changes in the way that supervised learning tasks are approached. We believe that it is possible to construct a rigorous mathematical proof that the ERA method will work in all but pathological (and rare) cases. This material is currently in preparation, and a mathematical analysis of the ERA algorithm will be presented in the literature in the near future.

References

[1]   M Gori and A Tesi, "On the problem of local minima in backpropagation", *IEEE Trans. on Pattern Analysis and Machine Intelligence*, 14, 76-86 (1992).

[2]   E K Blum, "Approximation of Boolean functions by sigmoidal networks: Part I: XOR and other two-variable functions", *Neural Computation*, 1, 532-540 (1989).

[3]   C Darken and J M Moody, "Towards faster stochastic gradient search", in: *Advances in Neural Information Systems 4*, Morgan Kaufmann, San Mateo, CA, 1009-1016 (1991).

[4]   J M McInerney, K G Haines, S Biafore and R Hecht-Nielsen, "Error surfaces of multi-layer networks can have local minima", UCSD Tech. Rep. CS89-157, October 1989.

[5]   D Gorse, A Shepherd and J G Taylor, "A classical algorithm for avoiding local minima", submitted to WCNN '94 San Diego.

LEARNING TIME SERIES BY NEURAL NETWORKS

D. W. Allen and J. G. Taylor

Centre for Neural Networks,
Mathematic Department,
King's College London,
London WC2R 2LS.

0. INTRODUCTION

Neural networks are used extensively to learn time series of a variety of forms. Theoretical analysis of such series tends to only consider any possible noise component as additive. However experimental time series can be influenced by noise which is not of a simple additive form. We present here a method of modelling these influences by the addition of an extra layer to the usual network architecture. The behaviour of the additional layer has a simple interpretation and allows both quantitative and qualitative methods of evaluating hypotheses of additive and multiplicative noise. If sigmoidal output functions are employed the moment generating function of the resulting distributions can be calculated explicitly in terms of the network parameters.

Many authors have investigated methods of noise reduction by choices of coordinates or transformation of Takens-style delay embedding [1,2,3]. The approach here however will be one of identification and quantification by using a combination the usual feed forward networks and kernel estimation of probability densities [4].

Consider for simplicity a univariate time series assumed to arise from observations of an n-dimensional discrete time Markov process for which :

$$X_{t+1} = G_{t+1}(X_t, \dots, X_{t-n}) \qquad (1)$$

Where $\{G_t\}$ is a family iid distributed functions. The problem we wish to solve is that of constructing a good estimate of the conditional probability distribution function

$$P(X_t + 1 < x \,|\, X_t = y_1, \dots, X_{t-n} = y_n) \qquad (2)$$

This problem has been considered by others (eg. [5]) but not by means of the recent powerful developments associated with the universal approximation powers of neural networks [6]. We propose to approach the problem by describing the (in general stochastic) process (1) by learning the distribution function (2). This is expected to be easier in noisy cases than direct inspection of the time series.

2. A NEURAL NETWORK APPROACH

The various so called universal approximation theorems of neural networks [7,8] demonstrate

the ability of simple feedforward networks to approximate mappings $\mathbf{R}^n \to \mathbf{R}^m$ including their derivatives [9,10,11]. We propose here an additional layer of the form

$$F(x|\boldsymbol{y}) = \sum_{i=1}^{n} a_i k\left(\frac{x - c_i}{v_i}\right) \qquad (3)$$

The vector (C_i) is the output from a (minimally single hidden layer) feedforward network

with input $\boldsymbol{y}$ represented here by the expression

$$o_i(\boldsymbol{y}) = \sum_{j=1}^{m} a_{ij} f(\boldsymbol{u}_j \circ \boldsymbol{y} - t_j) \qquad (4)$$

f and $k : \mathbf{R} \rightarrow \mathbf{R}$ being possibly distinct functions. And the extra parameters

(a_i , v_i) are constrained to make (3) a distribution function. In the case of k

being the sigmoid function this can be achieved, without reducing the space of functions

available [8], by assuming $0 \le a_i \le 1$, $\sum_{i=1}^{n} a_i = 1$ and $v_i > 0$. It is necessary

at least to have one non-linear layer in the network preceding this additional layer in order to
make the moments (and in the limit of low noise the outcome) of the distributions nonlinear

functions of the conditionals $\boldsymbol{y}$. This limit is approached as

$$(v_i \rightarrow 0, b_{ij} \rightarrow b_j) \forall i : a_i \neq 0 \qquad (5)$$

because in this case $F(x|\boldsymbol{y}) \rightarrow \theta(x - G(\boldsymbol{y}))$.

In the case of k being a sigmoid function the moment generating function $M(t)$ of
each of the distributions may be calculated and (after some algebra) is

$$M(t) = \Pi t\left(\sum_{i=1}^{n} a_i v_i e^{c_i t} \mathrm{cosec}\,(\Pi v_i t)\right) \qquad (6)$$

The first moment of x_{t+1} is therefore

$$E(x_{t+1}) = \sum_{i=1}^{n} a_i c_i(\boldsymbol{y}) \qquad (7)$$

And by (4) has universal approximator form (each distribution can thus approximate any classical or noisy value). In the noise free case this reduces to the classical mapping

$$x_{t+1} = G(x_t, \dots, x_{t-n}) \tag{8}$$

Also from (6) it can be shown that the higher order moments of x_{t+1} are

$$E\big((x_{t+1})^m\big) = \big(E(x_{t+1})\big)^m \tag{9}$$

as expected.

3. LEARNING TIME SERIES

A natural error criterion for a supervised learning method for these systems is the log likelihood criterion

$$\mathcal{E} = -\sum_{t=n}^{p} \ln F'(x_{t+1} | x_t, \dots, x_{t-n}) \tag{10}$$

with p the length of the series observed and F' denoting the partial derivative of

F with respect to its first argument (which if n=1 and k is the cumulative normal distribution function reduces to the mean squared error). The constrained optimization in the case of sigmoid output neurons can be transformed into an unconstrained problem by means of a substitution of the constrained variables in (3) for squares and dividing it by

$$\sum_{i=1}^{n} (a'_i)^2$$ (where $(a'_i)^2$ has been substituted for a_i and $(v'_i)^2$ for

v_i to preserve their positivity) so that gradient descent or other unconstrained methods can be applied.

4. EXAMPLE

An example currently under investigation is the learning of the stochastic dynamical system in which the parameter of the logistic map is allowed to vary randomly in an interval. Results will be presented.

5. CONCLUSIONS

We have presented a very general method by which the transfer functions of a large family of Markov processes may be modelled. The concentration here has been on univariate distributions. However higher order analogues using multivariate distributions or products of marginal distributions are clearly possible and offer further research. In the case of directly multiplicative noise, the absence of a need to introduce extra nodes in the layer preceding the output layer, if testing such an hypothesis against one of additive gaussian noise, offer tractable possibilities in the case of Bayesian model selection. The accessability of the moment generating function, simple interpretation of the function performed by the output layer, and relative weakness of the assumptions made about the nature of dynamical noise make the approach suitable for the tentative stages in a modelling procedure.

6. REFERENCES

[1] Casdagli et al, State space reconstruction in the presence of noise. (1991) Physica D Vol 51 pp 52-98.

[2] Farmer and Sidorowich, Optimal shadowing and noise reduction noise reduction. (1991) Physica D Vol 47 pp 373-392.

[3] Landa and Rosenblum, Time series analysis for system identification and diagnosis. (1991) Physica D Vol 48 pp 232-254.

[4] Silverman, Density estimation for statistics and data analysis. (1986) Chapman and Hall ISBN 0 412 246620 1

[5] Grabec, Modeling of chaos by a self organizing neural network. (1991) Artificial Neural Networks pp 151.

[6] Hornik, Multilayer feedforward networks are universal approximators. (1989) Neural Networks Vol 2 pp 359-366.

[7] Blum and Kwan Li, Approximation theory and feedforward networks. (1991) Neural Networks Vol 4 pp 511-515.

[8] Hornick et al, Universal approximation of an unknown mapping and its derivatives using multilayer feedforward networks. (1990) Neural Networks Vol 3 pp 551-560.

[9] Albertini and Sontag, Uniqueness of weights for neural networks. (????)

[10] Hornick, Approximation capabilities of multilayer feedforward networks. (1991) Neural Networks Vol 4 251-257.

[11] Gallant and White, On learning the derivatives of an unknown mapping with multilayer feedforward networks. (1992) Neural Networks Vol 5 pp 129-138.

[12] Approximation of a function and its derivatives with a neural network. (1992) Neural Networks Vol 5 pp 207-220.

[13] Ito, Approximation of functions on a compact set by finite sums of a sigmoid function without scaling. (1991) Neural Networks Vol 4 pp 817-826.

The Error Absorption for Fitting an Under-fitting (Skeleton) Net

Zhengrong Yang

Shanghai Institute of Metallurgy, Academia Sinica, 865 ChangNing Rd.,
Shanghai, 200050, P. R. CHINA

1. Introduction

In past few years, a lot of researchers have paid attention to the improvement of the generalisation performance of neural networks [1]~[6]. They skeletonized an over-fitting neural network with different measurements. There are two problems associated with the skeletonization technique. The first is difficult to choose an initial (over-fitting) net. For example, if 100 hidden units is an optimal solution, the initial net with 300 hidden units will have longer process of skeletonization than the initial net with 200 hidden units. The second is that there are some abysses and local minimum's in the training process [7]. If one chooses a larger initial net, there will be larger possibilities that the training sinks into an abyss or sticks at a local minimum such that the final net still has lower generalisation performance.

Sethi [8] has developed a new type of skeleton neural network. This net has no redundant weights for error absorption. If one trains this skeleton net, the generalisation performance will not be higher because of the absence of redundant weights for error absorption. Our experiments have shown that if we train this skeleton net with some added redundant weights, the generalisation performance will be improved to some degree. A skeleton net is called the under-fitting net and the process of adding some redundant weights on the skeleton net is called the fitting process. In this paper, a detail analysis of the error absorption mechanism and the experimental results are presented.

2. The Macro Mechanism of Error Absorption

The mechanism of error absorption is composed of two parts: macro mechanism and micro mechanism. The structure information of a net is described as:

$$a_j = f_j(x_j) \qquad x_j = \sum_h w_{hj} a_h \qquad \& \qquad a_h = f_h(x_h) \qquad x_h = \sum_i w_{ih} a_i \,,$$

where, $A = \{\{a_i\},\{a_h\},\{a_j\}\}$ is the activate-space of the input units, hidden units, and output units, $W = \{\{w_{ih}\},\{w_{hj}\}\}$ is the weight-space of the weights between the input layer and hidden layer and the weights between the hidden layer and output layer, and $F = \{\{f_h\},\{f_j\}\}$ is the function-space of the hidden units and output units. The uniform activate function is the sigmoid function. The weight-updating formulas deduced from Rumelhart [9] are:

$$\Delta w_{hj} = \eta \frac{\partial E}{\partial w_{hj}} = \eta \frac{\partial E}{\partial x_j} a_h = \eta \frac{\partial E}{\partial a_j} f_j' a_h$$

$$\Delta w_{ih} = \eta \frac{\partial E}{\partial w_{ih}} = \eta (\sum_j \frac{\partial E}{\partial a_j} f_j' w_{hj}) f_h' a_i .$$

From these weight-updating formulas, one can find out that the error occurring at an output unit will be absorbed through the weight-space by the units at the lower layers. In fact, the error occurring at any output unit is synthesised from the units at the lower layers through the weight-space during the forward-propagation. The weight-updating is a back-error propagation process, which propagates the errors occurring at the output units back to the units at the lower layers and modifies the weight-space by the gradient descent method.

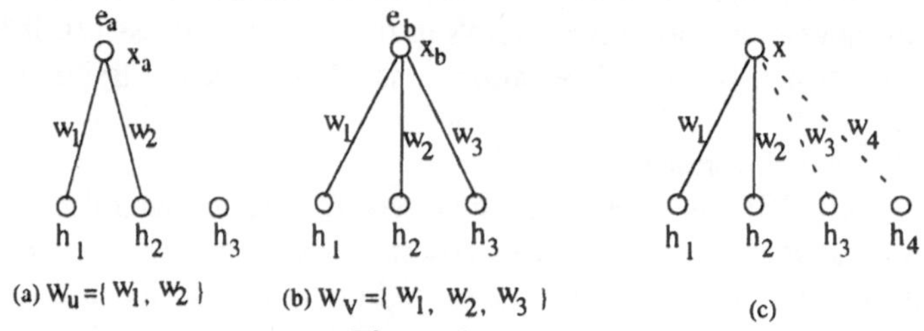

Figure 1

Consider a net, which just has one output and two layers as shown in figure 1. In figure 1 (a) the output error e_a is absorbed by the hidden units h_1 and h_2 through the weight-space W_u. While in figure 1 (b), the output error e_b is absorbed by the hidden units h_1, h_2 and h_3 through the weight-space W_v. The activate values of x_a and x_b at the next time step can be obtained as follows:

$$a_{x_a}^{t+1} = \frac{1}{1 + e^{-[\eta \frac{\partial E}{\partial x_a^t}(a_{h1}^2 + a_{h2}^2)]} * e^{-\Sigma}} = \frac{1}{1 + k_a * e^{-\Sigma}}$$

$$a_{x_b}^{t+1} = \frac{1}{1 + e^{-[\eta \frac{\partial E}{\partial x_b^t}(a_{h1}^2 + a_{h2}^2 + a_{h3}^2)]} * e^{-\Sigma}} = \frac{1}{1 + k_b * e^{-\Sigma}}$$

here, k_a and k_b are weight-independent coefficients. Because of

$$k_b > k_a > 1 \qquad if (\partial E / \partial x < 0)$$
$$k_b < k_a < 1 \qquad if (\partial E / \partial x > 0),$$

then $e_b^{t+1} < e_a^{t+1}$. That is, the error at x_b will decay faster than at x_a. It can be concluded that larger weight-space can improve the error absorption more than smaller weight-space.

3. The Micro Mechanism of Error Absorption

If there are more than one output units in a net, the output unit with the largest error will have the largest contribution to the net error and, this kind of output unit should be selected for adding redundant weight at first. This is because through the redundant weight, the error at this output unit can be absorbed partially by a unit at the lower layer. After the output unit is chosen, which unit at the lower layer has to be selected to connect to this output unit should be considered. Consider the situation shown in figure 2: Given the wanted activate

values of h_3 and h_4 are $a_{h_3}^{()}$ and $a_{h_4}^{()}$ respectively. If $da_{h_4} > da_{h_3}$, a_{h_4} will have

larger deviation from $a_{h_4}^{()}$, and a_{h_3} will have smaller deviation from $a_{h_3}^{()}$ (The

derivatives on the units at the lower layer can be obtained by BP method). We can deduce a derivation on x:

$$da_x^{t+1}(h_i) = \frac{k_{h_i} * e^{-\Sigma}}{(1 + k_{h_i} * e^{-\Sigma})^2} * (w_i - 2a_{h_i}) * da_{h_i}.$$

Given the approximate condition that there is no big difference among w_i s and

among a_{h_i} s. Because of $da_{h_4} > da_{h_3}$, then,

$$da_x^{t+1}(h_3) < da_x^{t+1}(h_4).$$

That is, connecting to the unit with the smallest derivation at the lower layer will result in better error absorption.

4. The Experiments and conclusion

Below, we present two experiments and their simulation results. Figure 2 (a) and (b) give the experimental results of partitioning the plane points into two fields. Figure 2 (a) presents the simulation result with the error absorption mechanism. Figure 2 (b) gives the simulation result without this mechanism. The fitting epochs are five for both them. Clearly, with the error absorption mechanism, the generalisation performance (GP) has increased up to 90%, as shown in figure 2 (a). But without the error absorption mechanism, the generalisation has decreased

536

down to 76%. Figure 2 (c) and (d) shows the results of finding a minimum value among some values. Figure 2 (c) and (d) illustrate the simulation results with and without the error absorption mechanism respectively. The fitting epochs are six both for them. In figure 2 (c), the total tendency of generalisation performance has increased up to 98.5%, while in figure 2 (d), the generalisation performance oscillations greatly.

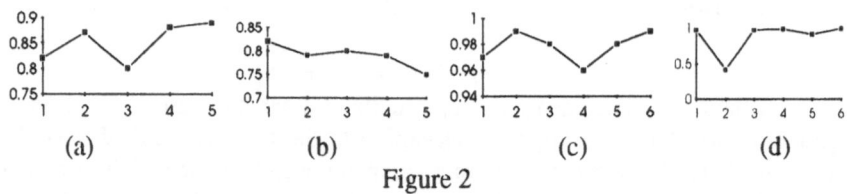

(a)   (b)   (c)   (d)

Figure 2

   This paper has presented a new method for improving the generalisation performance of neural networks. The mechanism of error absorption was developed for fitting an under-fitting (skeleton) net. The experimental results have shown that it is very useful for one type of problems, where the initial skeleton net can be constructed by the expert knowledge, and the training technique is used to overcome the drawback of incomplete knowledge of expert. With this new method, sinking into the abyss can be avoided and the generalisation performance improvement time is obviously reduced.

Acknowledgement
  The author would like to thank Professor G. Musgrave (head of dept. of EE & E, brunel university) for his guidance and thank Dr. Richard Neville and Mr Thomas A. Tawiah (dept. of EE & E, brunel university) for their kindly discussion and support about this research.

Reference
1. Michael C. Mozer, Paul Smolensky, "Skeletonization: a technique for trimming the fat from a network via relevance assessment", Advances in Neural Information Processing Systems, Vol. 1, 1989, pp107-115
2. Yann Le Cun, John S. Denker, Sara A. Solla, "Optimal Brain Damage", Advances in Neural Information Processing Systems, Vol. 2, 1990, pp598-605
3. Andreas S. Weigend, David E. Rumelhart, Bernardo A. Huberman, "Generalisation by weight-elimination with application to forecasting", Advances in Neural Information Processing Systems, Vol. 3, 1991, pp875-882
4. Anders Krogh, John A. Hertz, "A simple weight decay can improve generalisation", Advances in Neural Information Processing Systems, Vol. 4, 1992, pp950-957
5. Sowmya Ramachandran, Lorien Y. Pratt, "Information measure based skeletonization", Advances in Neural Information Processing Systems, Vol. 4, 1992, pp1080-1087
6. F. Hergert, W. Finnoff, H. G. Zimmermann, "A comparison of weight elimination methods for reducing complexity in neural networks", 1992 International Joint Conference on Neural Networks, ppIII-980-987
7. Sankar K. Pal, et al, "Multilayer perceptron, fuzzy sets, and classification",, IEEE trans. on neural networks, vol. 3, no. 5, September 1992, pp683-697
8. Ishwar K. Sethi, "Entroy Nets: From Decision Trees to Neural Networks", Proc. IEEE 1990, pp1605-1613
9. D. E. Rumelhart, "Parallel distributed processing, vol. 1, foundations", Cambridge, MA:MIT press, 1986

Fast Backpropagation using Modified Sigmoidal Functions.

F.C.Morabito

Dipartimento di Ingegneria Elettronica e Matematica Applicata
Università degli Studi di Reggio Calabria, Italy

1. Introduction.

Backpropagation is the most commonly known algorithm for the adjustment of the weights of an Artificial Neural Network (ANN). It is well known that, in this method the partial derivatives of a criterion function with respect to the weights of a multilayer ANN are determined and the weights of the connections are adjusted pursuing a gradient descent in the weights space. At the same time, the standard function used to introduce some nonlinearity in the model is a *sigmoid* (or *squashing* function). The derivative of this function plays a relevant role in the correction process along with the actual output error and the relaxation coefficient of the procedure. Actually, the backpropagation method is a computationally efficient method to train ANNs with differentiable nonlinearities, such as sigmoidal functions. However, often the method converges very slowly, if not at all. In this paper we shall analyse a possible modification of the standard sigmoids which seem to be speeding up the training in a number of experiments we have carried out. The modification is based on analytical rather than on heuristic considerations. This correction explains why a trick used in (Fahlmann, 1988) to accelerate convergence gives good results. For the reason described we need to roughly present the backpropagation algorithm. The technique here presented allows some considerations about the choice of the learning rates as well as about the initialization of the weights in training ANNs. Although the observations made are of general interest, we will use them in problems of functions approximation: some results concerning electromagnetic identification problems are presented to substantiate the approach.

2. Backpropagation algorithm for approximation in static problems.

Multilayer feedforward ANNs consists in sets of *neurons* arranged so that each neuron in a layer can be connected to neurons in adjacent layers. Connections within the same layer are generally prohibited. Each connection is assigned a weight measuring the importance of the correlation between the neurons connected. Before training is started , each weight is randomly initialized. In what follows we shall make use of self-explaining notations.
We can describe the ANN by the following equations:

$$y_i = f_0\left(\sum_{j=1}^{N_h} \omega_{ji} h_j - t_i^{'}\right) = f_0(y_i^{'}) \qquad , i = 1, N_0$$

$$h_j = f_h\left(\sum_{\ell=1}^{N_i} \upsilon_{\ell j} x_\ell - t_j^{'}\right) = f_h(h_j^{'}) \qquad , j = 1, N_h$$

(1)

where fh, fo are the activation functions which can be selected different from layer to layer.
In particular, f_h is generally a sigmoid, i.e.:

$$f_h(h^{'}) = \frac{1}{\left(1 + e^{-kh^{'}}\right)}$$

(2)

while f_0 can be selected as a linear function: $f_0(y') = \alpha y'$

This choice derives from the existence of theorems which guarantee the so called "universal approximation" property for ANNs using sigmoidal functions for hidden layers and linear outputs (Hornik et al., 1989). The sigmoids are continuous, non decreasing, bounded functions which show a sufficient degree of smoothness. These functions are also a suitable extension of the soft limiting nonlinearities used previously in neural computing. One property of sigmoids is that their derivatives have a simple expression in terms of the function itself:

$$f_h^l(h') = f_h(h')\left[1 - f_h(h')\right]$$

Since these functions are differentiable we can refer to backpropagation to train the corresponding ANN. We shall now report the backpropagation algorithm for the approximation of static mapping (Rumelhart, 1986). We consider multilayer feedforward models with a single hidden layer. The same procedure could be recursively applied to any number of layers. During the development we shall emphasize only the aspects of interest for our purposes. If one introduces a criterion function of the type:

$$E\left(\underline{\omega}, \underline{\upsilon}\right) = \frac{1}{N_p} \sum_{p=1}^{N_p} E_p\left(\underline{\omega}, \underline{\upsilon}\right) \tag{3}$$

where $E_p(\underline{\omega}, \underline{\upsilon})$ is the total squared error for the p-th pattern:

$$E_p\left(\underline{\omega}, \underline{\upsilon}\right) = \frac{1}{2N_0} \sum_{i=1}^{N_0} \left[y_i\left(\underline{\omega}, \underline{\upsilon}\right) - d_i\right]_p^2 \tag{4}.$$

A standard way to determine the optimal values of $\underline{\omega}$, $\underline{\upsilon}$ is to minimise E w.r.t. $\underline{\omega}$, $\underline{\upsilon}$ This strategy of minimisation is usually referred to Least Mean Square (LMS) criterion in linear systems. The weight matrices $\underline{\omega}$, $\underline{\upsilon}$ can be recursively adjusted by minimising the error function along the gradient-descent direction. We achieve convergence toward optimal values for weights and biases by taking corrections $\Delta\underline{\omega}$ $\Delta\underline{\upsilon}$ proportional to (-∂E / $\partial\underline{\omega}$)(-∂E / $\partial\underline{\upsilon}$):

$$\Delta\omega_{ji} = -\eta_0 \frac{\partial E}{\partial \omega_{ji}} \quad , \quad \Delta\upsilon_{\ell j} = -\eta_h \frac{\partial E}{\partial \upsilon_{\ell j}} \tag{5}$$

where η_0, η_h are properly selected relaxation coefficients, namely learning rates. Then, we have:

$$\omega_{ji}(k+1) = \omega_{ji}(k) + \Delta\omega_{ji} = \omega_{ji}(k) - \eta_0 \frac{\partial E}{\partial \omega_{ji}} =$$

$$= \omega_{ji}(k) - \eta_0 \frac{1}{N_p} \sum_{P=1}^{N_p} \frac{\partial E_p}{\partial \omega_{ji}} = \omega_{ji}(k) - \frac{\eta_0}{N_0 N_p} \sum_{P=1}^{N_p} (y_i - d_i)_p \frac{\partial y_{ip}}{\partial y_{ip}} h_j \tag{6}$$

For the matrix $\underline{\upsilon}$, as known, the corrections depend on the *deltas* of the preceding layers. Some aspects of this approach deserves special attention:

1) A standard approximation is typically introduced by replacing the summation Σ_p with an estimate of the gradient, based on a reduced number of training examples: $N_p \Rightarrow k$ with k<N_p. In this case a block of k cases is referred to as *epoch*. When k=1 we speak of pattern learning. The size of the epoch strongly affects the processing. On the other hand, some strategies of learning require k=N_p.

2) Sometimes other terms are added to the corrections $\Delta\omega_{ji}$, $\Delta\upsilon_{\ell j}$, attempting to adapt the learning rate as a function of the local curvature of the error surface (Jacobs,1988). By far the most common heuristic is named *momentum*, and in this

case the term introduced takes the form: $\alpha[\omega_{ji}(k) - \omega_{ji}(k-1)]$ where $0<\alpha<1$. It is rarely precised that the choice of α is related to the values of η;

3) It is clear from (6) that both N_p and N_o can be incorporated in η_o, but it is very interesting to note that in this case the learning rate for the output layer depends on the epoch size as well as the number of outputs. Similar considerations can be made for η_h;

4) As known, the corrections $\Delta\omega_{ji}$ depend on η_o, h_i, f'_o, and the error on the corresponding output. $\Delta\upsilon_{ij}$ depends on η_h, x_j, f'_h, the weights connecting the j-th hidden node to the outputs, the errors on all of the outputs, and f'_o;

5) Some methods which introduce additional terms to the corrections (5) can be easily interpreted in terms of different criterion functions. In particular, the *weight decay method* is obtained by introducing to (3) a term of the type:

$$\left(\frac{\ell_\omega}{2} \sum_{i=1}^{N_\omega} \omega_i^2 + \frac{\ell_\upsilon}{2} \sum_{j=1}^{N_\upsilon} \upsilon_j^2 \right)$$

In this work we focus our attention on the functions f_o, f_h. If f_o is a linear one, we have: $f'_o(y'_i)=\alpha$ while for a sigmoid, we have: $f'_o(y'_j)=y_j(1-y_j)$.

It is worth noting that the use ofsigmoids is believed to have a stabilizing effect on the convergence properties of the method because the corrections are larger when the output is about 0.5. This is understandable if we consider a binary classification problem. Indeed, the condition y=0.5 reveals an high uncertainty in the decision. Weights which are connected to units in their midrange are the most changed. In a sense, these units are still uncommitted. But the question is what happens when $y_j=1$ and $d_j=0$, or vice versa. We have $y_j(1-y_j)=0$, and $\Delta\omega_{ji}=0$. That is to say: under these conditions the weights do not change at all, while the output error is maximum! By using a partial batch learning this effect is a bit masked. In addition, we rarely achieve an output of either 0 or 1. Anyway, the inconsistency remains. The same conclusion we hold for a nonlinearity of the type:

$$f_o(y) = \tanh(y) = [2/(1+e^{-ky})-1].$$

In our view, this is one reason for the slow convergence in some backpropagation experiences. Also, this gives some hints about another commonly observed effect in training ANNs, namely the dependence from initial conditions. Indeed, by choosing very small initial weights we reduce the possibility that the described effect can have an impact. Of course, this kind of choise is also useful for other reasons.

By using a linear output the corrections just depend on the output errors and then the effect disappears. On the other hand, the best behaviour of linear output layers is generally masked by the fact that in this case we have to use a smaller η_o w.r.t. sigmoidal output layer to stabilize the convergence of backpropagation. This concept is not pointed out in the literature. Of course, the corrections $\Delta\upsilon_{ij}$ are strongly affected by this effect because they depend on the *deltas* of the output layer. In addition they suffer the effect due to the presence of f'_h. Fahlmann has described a number of methods to speed up backpropagation. These include modifying the derivatives of the sigmoid by adding a small positive offset to them, prior to scaling the local error. He explains the good results obtained by considering that when the incoming weights of a neuron become large, the net input is large too and consequently the activation values become saturated. In these cases, the derivatives become zero and the scaled error is zero too. Adding a positive offset to the derivatives alleviates the problem. The reason of this is partially different and of structural nature, as above described.

540

We propose to reduce the problem by using a function of the type:
$$y_j = f_o(y_j') = [1/(1+e^{-ky_j'}) + \alpha y_j'].$$
i.e. by considering the function shown in fig.1. In this case $f'_o = ky_j (1-y_j) + \alpha$ (see

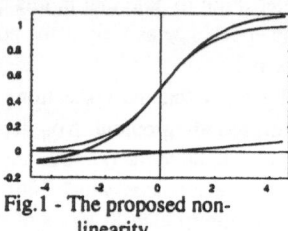

Fig.1 - The proposed non-linearity.

fig.2) and then $\Delta\omega_{ji}\neq 0$ when the conditions above considered hold. It seems rather more correct that in the case of large error it is the error itself to drive the correction process. f_o can be in any case used for the output layer. The introduction of such a function *in lieu* of a sigmoid has to be carefully evaluated for hidden layers. This is why sigmoids introduce at the upper and lower limits a saturation which constrain the output smoothly within fixed limits. Indeed, the representation theorems are valid only for *bounded* functions. All of the other required properties are verified by the proposed activation function. However, to stabilize the processing we act both on the learning rates and α , by properly reducing them. Furthermore, the use of stabilizers can reduce the possibility of fast growing weights (see point 5 above).

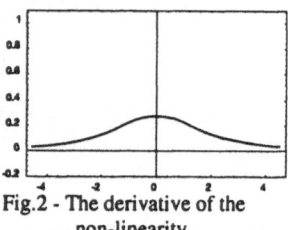

Fig.2 - The derivative of the non-linearity.

3. Results e conclusions.

We have applied the above described method has been to a number of different problems in electromagnetics. The results are always good in the first phase of trainings. In some cases, if appropriate actions are not taken the RMS error curve shows large oscillations. This can be explained by noting that the negative effect which we wish to contrast is significantly present in the initial phase of the training. So, we have to reduce the learning rate during the training·to avoid the oscillatory behavior. Fig.3 illustrates a typical case of convergence in a function approximation problem, for ANNs with sigmoids and modified sigmoids.

Fig.3 - RMS error curves.

Acknowledgment

The author is grateful to M. Campolo and F. Cirianni for technical assistance.

References

Fahlmann, S.E. (1988).An empirical study of learning speed in backpropagation networks. CMU Technical Report, **88-162**.

Hornik, K., Stinchcombe, M. and White, H. (1989). Multilayer feedfoward networks are universal approximators. *Neural Networks*, **2**, 359-366.

Rumelhart, D.E. and McClelland, J.L. (1986). Parallel distributed processing: exploration in the microstructure of cognition. Foundation, **1**, MIT Press.

Jacobs, R.A. (1988), Increased rates of convergence through learning rate adaptation. Neural Networks, **1**, 295-307.

Input Contribution Analysis in a Double Input Layered Neural Network

Z.Shen, M.Clarke, R.W.Jones

Department of Electrical Engineering, Brunel University
Uxbridge, Middlesex, U.K.

1. Introduction

An unique structure of Multi-layered Perceptron (MLP) with double input layers is proposed. By applying a second input layer to a traditional MLP, we obtain a set of input weights between the two single connected input layers, which we use to determine the contribution of each input to the decision making. We also find that the learning process is accelerated when the additional layer is used. In this paper, we present results to demonstrate these features and compare our network with a traditional MLP. Weight analysis allows us to determine the relative contribution of each input and explain the decision made by the network, so far one of major disadvantages of neural networks. It also allows reduction of input dimension of the network based on the contribution which can give improved network performance.

2. Development of the idea

The MLP with back-propagation learning algorithm (McClelland and Rumelhart, 1986) has been gaining popularity in medical applications. However, they have also often been criticised for having little or no capability to explain the decision made by the network (Caudill, 1991). This makes them less attractive than expert systems to clinicians who often desire a reasoning for the diagnosis. The problem lies in the architecture of a multi-layered network, with its multiplicity of connections between each layer, the interaction between the layers and the consequent difficulty of giving a simple explanation of any one weight. It is not appropriate or possible to determine the relative importance of a particular input by simply looking at the strength of the weight between input and output. In a traditional MLP trained through back-propagation, weights between each layer are adapted according to the error term δ back propagated from its upper layer (refer to figure 1). In a three layered neural network, the δ_j in the output layer are propagated backwards to the hidden layer and used to adapt the weights between the nodes in the hidden and output layer (w_{hj}); the δ_h from the hidden layer are propagated to the input layer and used to adapt the weights between the nodes of the input and hidden layer. Since input values are normally passed directly through the input layer, no weight adaptation is required, and the backpropagation is terminated at this stage and the term δ_i to input layer is neglected. We believe that this error term is significant and have investigated a new network structure. This has a second input layer so that the weights normally incorporated in the weights between the input and hidden layer are separated out into the first input layer. This makes them visible and thus can be analyzed for their significance and allow some explanation of the decision making process. These weights may also be used to produce a subset of inputs restricted to only those with significance. We show that networks trained on such subsets exhibit significant improvements in performance.

3. The double input layered MLP

Figure 1 shows the structure of the double input layered MLP. It consists of a traditional three layered fully connected feedforward network, and an additional input layer, the number of nodes in this layer being the same as the number of inputs. During training the extra input layer has its weights adapted according to the error term δ_i from the layer immediate above it. During testing this layer only multiplies its input by a constant, the weight. However we have separated an input weight which can be used to determine the relative significance of each input. The transfer functions in the output and hidden layers are sigmoid and in the input layer are linear.

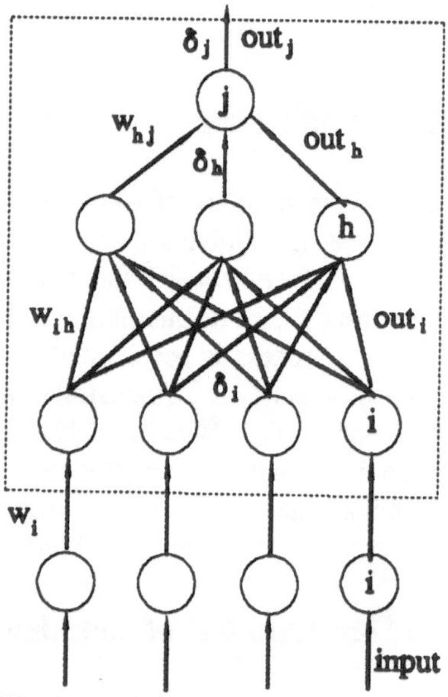

Figure 1 Structure of the double input layered MLP.

4. Results

We use a medical application: the early diagnosis of coronary heart disease (CHD) as a case study. The data comes from a self-applied questionnaire. 22 questions are selected from the questionnaire and 32 patterns are used for training and test the networks (Shen et al., 1992). The performance of the network was evaluated by ROC analysis (Campbell and Machin, 1990) which shows the false positive values against the true positives. The more area covered by a ROC curve, the better is the system. A perfect diagnosis system has a ROC covering unit area.

4.1 Reduce 22 inputs to 16 inputs based on the rank order of input weights

Twenty two inputs are initially selected from the questionnaire using a statistical data preprocessing method based on a root mean square metric which selects those questions with a reasonable significance of being related to CHD (Shen et al., 1992). Our MLP network is retained on these 22 inputs, and those with an input weight greater than 1 are used for further training. In our case, a network with 16 inputs was found to perform better than the original one with 22 inputs, based on ROC analysis (figure 2).

4.2 Comparison with "Contribution Analysis"

"Contribution analysis" or "sensitivity analysis" (Utans and Moody, 1991) defines the contribution of each input by calculating the sensitivity of the output to any individual input. This is determined by calculating the partial derivative of the output with respect to each input using the chain-rule for differentiation. It has also been applied to the early diagnosis of heart attack by neural networks by others (Harrison et al., 1991). When we compare the rank order of the weights, w_i, of our method and the rank order found by contribution analysis for the 22 and

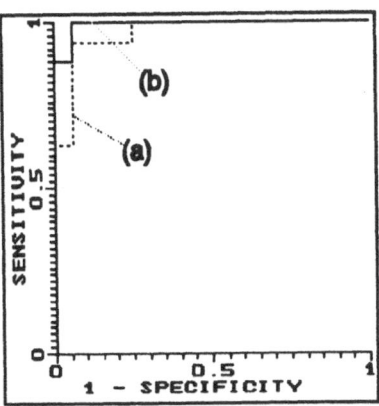

Figure 2 ROC curves (a) 22 input network (b) 16 input network

16 input networks, we find they are almost identical (a straight line if identical) particularly for the inputs at the top of the rank order (figure 3 and 4).

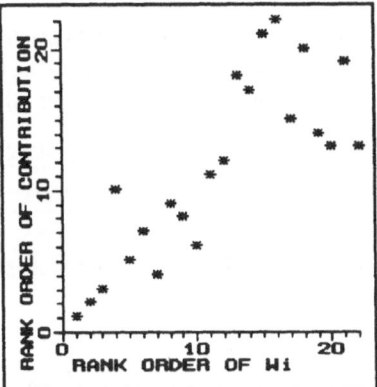

Figure 3 The rank orders of 22 inputs listed by w_i and contribution analysis.

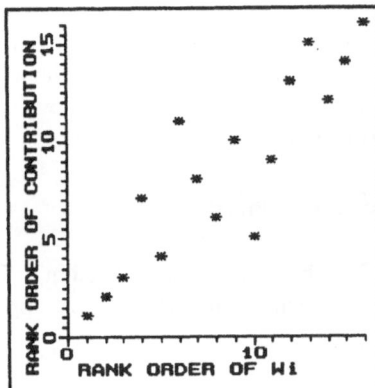

Figure 4 The rank orders of 16 inputs listed by w_i and contribution analysis.

4.3 The benefit for learning

When linear functions are used as the transfer function in the first two layers, the learning speed of the double input layered network is increased compared with a traditional MLP network. However, note that if the sigmoid function is used for the input layer, training is unstable, and it would appear no solution will be found (figure 5). The networks use the 22 general inputs mentioned in the last chapter. 32 patterns are used for training and training stop after 100 iterations.

5. Discussion and conclusion

544

The double input layer structure of MLP gives us a set of input weights which directly reflect the contribution of each input. There is close agreement of the rank order of the inputs obtained by weight analysis and "contribution analysis" which gives us confidence in our method and to further develop our technique. We have used the technique to remove the inputs giving least contribution and shown this to result in better network performance.

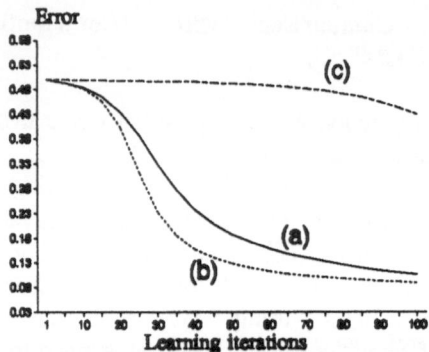

Figure 5 The error curves during training. (a) traditional MLP (b) double-input-layered MLP with linear function and (c) double-input-layered MLP with sigmoid function.

"Contribution analysis", by calculating the sensitivity of output to each input, has the disadvantage of suppressing larger outputs (Lisboa et al., 1993). The weight study we have proposed overcomes this shortcoming since the rank order of the weights from a trained network indicates the contribution of each input directly, no further calculation is needed.

Note that the input weights in our system are all positive and indicate the magnitude of the contribution of each input but not the sense. Further weight study can be carried out to determine the positive or negative contribution. A paper regarding this subject is in preparation and is to be published elsewhere.

6. References

Campbell, M.J. and Machin, D. (1990) Medical Statistics: A Commonsense Approach. John Wiley & Sons, pp36-38.

Caudill, M. (1991) Expert Networks. Byte, Oct.1991, pp108-116.

Harrison, R.S., Marshall, S.J. and kennedy, R. (1991) The Early Diagnosis of Heart Attacks: A Neurocomputational Approach. in Proc. Int. Joint Conf. on Neural Networks, (Seattle, USA), July 1992, vol.I, pp1-5.

Lisboa, P.J.G., Mehridehnavi, A.R. and Martin, P.A. (1993) The Interpretation of supervised Neural Networks. Workshop on neural Networks, Liverpool University, Sept. 1993.

McClelland, J.L. and Rumelhart, D.E. (1986) Parallel Distributed Processing: Explorations in the Microstructures of Cognition. Cambridge: MIT Press.

Shen, Z., Clarke, M., Jones, R. and Alberti, T. (1992) A Neural Network for Detecting Coronary Heart Disease. in Proc. Int. Joint Conf. on Neural Networks. (Beijing, China), Nov.1992, vol.I, pp104-108.

Utans, J. and Moody, J. (1991) Selecting neural network architectures via the prediction risk: application to corporate bond rating prediction. In Proc. of the Int. Conf. on Artif. Intel. Applic., IEEE Computer Society Press, Los Alamitos, CA.

A Unified Approach to Derive Gradient Algorithms for Arbitrary Neural Network Structures

Françoise Beaufays†, Eric A. Wan‡

† Dept. of Electrical Engineering, Stanford University
Stanford, CA 94305-4055 USA
‡ Dept. of Elec. Eng. and Applied Physics, Oregon Graduate Institute
P.O.box 91000, Portland, OR 97291 USA

1 Introduction

Deriving backpropagation algorithms for time-dependent neural network structures typically requires numerous chain rule expansions, diligent bookkeeping, and careful manipulation of terms. In this paper, we present a unified approach to derive such algorithms via a set of simple block diagram manipulation rules.

2 Adaptation Algorithms, Error Gradient Propagation

Adapting a feedforward multilayer neural network amounts to finding the set of variable weights W that minimizes the cost function

$$J = \sum_{k=1}^{K} \mathbf{e}(k)^T \mathbf{e}(k), \tag{1}$$

where the sum is taken over K samples in a training sequence, and $\mathbf{e}(k)$ is the error vector. According to gradient descent, the contribution to the weight update at each time step is

$$\Delta W(k) = -\mu \, \frac{\partial J}{\partial W(k)}, \tag{2}$$

where μ controls the learning rate.

At the architectural level, a variable weight w_{ij} may be isolated between two points $a_i(k)$ and $a_j(k)$ in the network (i.e. $a_j(k) = w_{ij}\, a_i(k)$). Using the chain rule, we get

$$\frac{\partial J}{\partial w_{ij}(k)} = \frac{\partial J}{\partial a_j(k)} \frac{\partial a_j(k)}{\partial w_{ij}(k)} = \frac{\partial J}{\partial a_j(k)} \, a_i(k), \tag{3}$$

and the weight update becomes

$$\Delta w_{ij}(k) = -\mu \, \delta_j(k) \, a_i(k), \tag{4}$$

where we define the error gradient

$$\delta_j(k) \triangleq \frac{\partial J}{\partial a_j(k)}. \tag{5}$$

The error gradient $\delta_j(k)$ depends on the entire topology of the network. Reported methods for deriving the delta terms rest on chain rule expansions that must be carried out analytically for the specific network topology.

3 A New Method for Error Gradient Evaluation

An alternative method consists of viewing an arbitrary neural network as a block diagram whose building blocks are: summing junctions, branching points, univariate functions, multivariate functions, and delay operators. Using the set of simple rules listed below and reversing the flow direction in the original network, we construct a *reciprocal network*. If we feed this network with $-2e(k)$, the signals propagating in the network are precisely the delta terms (wan 1993, wan *et al.* 1994). The weight update follows immediately (Eq. 4), without any extra algebraic derivation.

Transformation Rules:

- *Summing junctions are replaced with branching points.*
- *Branching points are replaced with summing junctions.*
- *Univariate functions are replaced with their derivatives.*
- *Multivariate functions are replaced with their Jacobians.*
- *Delay operators are replaced with advance operators.*

4 An example: Cascaded Neural Networks

As an example, let us consider the cascaded neural networks illustrated in Figure 1. The inputs to the first network are samples from a time sequence $x(k)$. Delayed outputs of the first network are fed to the second network. Typically, the last network represents the model of some physical system, and the first network is used to prewarp or equalize the driving signal.

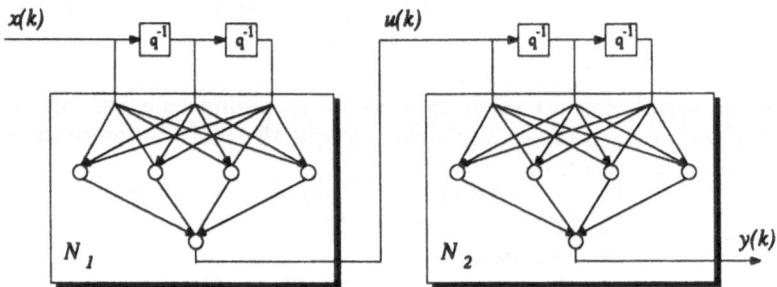

Figure 1: Cascaded neural network filters (q^{-1} represents a unit delay).

The cascaded networks are defined as

$$u(k) = \mathcal{N}_1(W_1, x(k), x(k-1), x(k-2)), \qquad (6)$$
$$y(k) = \mathcal{N}_2(W_2, u(k), u(k-1), u(k-2)), \qquad (7)$$

where W_1 and W_2 represent the weights parameterizing the networks, $x(k)$ is the input, $y(k)$ the output, and $u(k)$ the intermediate signal. Given a desired response for the output y of the second network, it is a straightforward procedure to use backpropagation (rumelhart *et al.* 1986) for adapting the second network. It is not obvious, however, what the effective error should be for the first network. In this case, the chain rule is simple enough to apply directly to find the instantaneous error gradient:

$$\frac{\partial e^2(k)}{\partial W_1} = -2e(k)\frac{\partial y(k)}{\partial W_1} \tag{8}$$

$$= -2e(k)\left[\frac{\partial y(k)}{\partial u(k)}\frac{\partial u(k)}{\partial W_1} + \frac{\partial y(k)}{\partial u(k-1)}\frac{\partial u(k-1)}{\partial W_1} + \frac{\partial y(k)}{\partial u(k-2)}\frac{\partial u(k-2)}{\partial W_1}\right]$$

$$= \delta_1(k)\frac{\partial u(k)}{\partial W_1} + \delta_2(k)\frac{\partial u(k-1)}{\partial W_1} + \delta_3(k)\frac{\partial u(k-2)}{\partial W_1}, \tag{9}$$

where we define

$$\delta_i(k) \stackrel{\triangle}{=} -2e(k)\frac{\partial y(k)}{\partial u(k-i)} \qquad i = 1,2,3. \tag{10}$$

The δ_i terms are found simultaneously by a single backpropagation of the error through the second network. Each product $\delta_{i+1}(k)\,\partial u(k-i)/\partial W_1$ is then found by applying backpropagation to the first network with $\delta_{i+1}(k)$ acting as an error. However, since the derivatives used in backpropagation are time-dependent, *separate* backpropagations are necessary for each $\delta_{i+1}(k)$. These equations, in fact, imply backpropagation through an *unfolded* structure as illustrated in Figure 2. In situations where there may be hundreds of taps in the second network, this approach leads to a very inefficient adaptation algorithm.

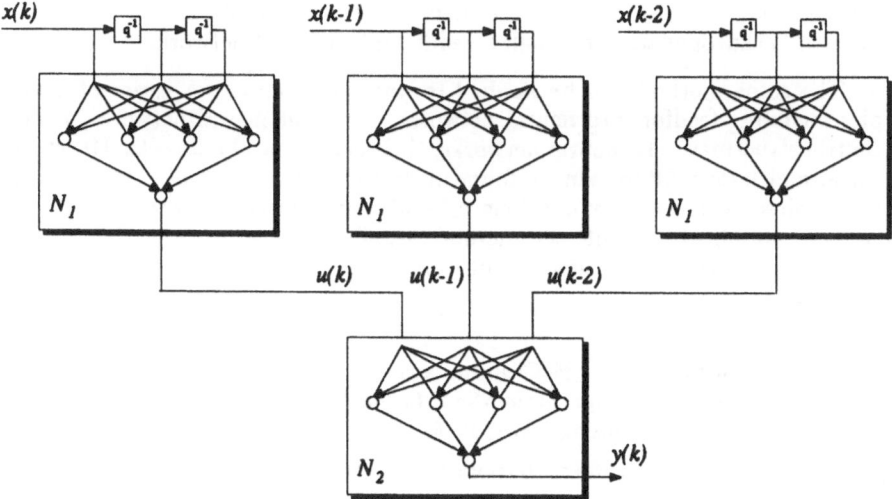

Figure 2: Cascaded neural network filters unfolded-in-time.

A more efficient algorithm for finding the delta terms may be arrived at by using our new method. The original cascaded networks are transformed into the reciprocal structure shown in Figure 3. Simply by labeling signals, gradient relations may be written down directly:

$$\delta_u(k) = \delta_1(k) + \delta_2(k+1) + \delta_3(k+2), \tag{11}$$

548

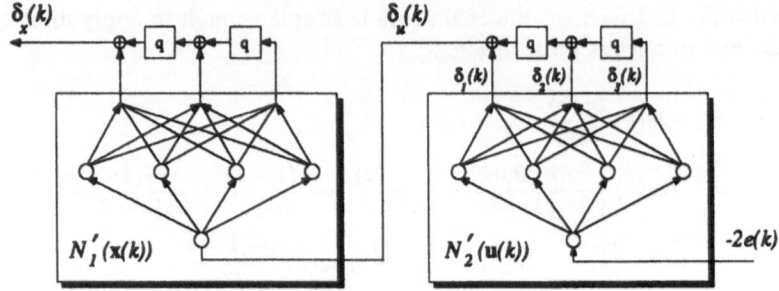

Figure 3: Reciprocal network for cascaded neural filters.

with

$$[\delta_1(k)\ \delta_2(k)\ \delta_3(k)] = -2e(k)\mathcal{N}_2'(\mathbf{u}(k)), \tag{12}$$

i.e., each $\delta_i(k)$ is found by backpropagation through the output network, and the δ_i's are summed together. The weight update is given by

$$\Delta W_1(k) = -\mu\,\delta_u(k)\,\frac{\partial u(k)}{\partial W_1(k)}, \tag{13}$$

in which the product term is found by a *single* backpropagation with $\delta_u(k)$ acting as the error to the first network. Equations can be made causal by simply delaying the weight update for a few time steps. Generalization to an arbitrary number of taps is straightforward. This new algorithm is far more efficient than the earlier direct gradient calculation method: we completely avoided backpropagation through a redundant unfolded network.

The same method can be applied to *any* architecture. A series of examples including feedforward neural networks, recurrent neural networks, neural control structures, FIR neural networks, time-delay neural networks, IIR structures, and lattice filters can be found in (wan 1993, wan *et al.* 1994). In all cases, direct chain rule expansions or equivalent unfolded structures are extremely complicated, while the method described above provides a quick and easy way to derive the desired adaptation algorithm.

Bibliography

1. D.E. Rumelhart, J.L. McClelland, and the PDP Research Group. *Parallel Distributed Processing: Explorations in the Microstructure of Cognition.* Vol. 1. MIT Press, Cambridge, MA, 1986.

2. E. Wan and F. Beaufays. Network Reciprocity: A Unified Approach to Derive Gradient Algorithms for Arbitrary Neural Network Structures. Submitted to *Neural Computation.*

3. E. Wan. *Finite Impulse Response Neural Networks with Applications in Time Series Prediction.* Ph.D. dissertation. Stanford University, Nov. 1993.

This work was sponsored by EPRI under contract RP8010-13 and NSF under grant NSF IRI 91-12531.

Interpretation of BP-trained net outputs

S. Gómez, Ll. Garrido

Dept. d'Estructura i Constituents de la Matèria,
Facultat de Física, Universitat de Barcelona,
Diagonal 647, 08028 Barcelona, Spain,
and
Institut de Física d'Altes Energies,
Universitat Autònoma de Barcelona,
08193 Bellaterra (Barcelona), Spain.

1   Introduction

During the last few years neural networks have become increasingly popular. Their ability to classify and predict from examples is what has made them most interesting. Among the different types of networks the ones which have found more applications are those whose training is based in the minimization of a squared error criteria. Although this method has proven to work well in many situations, the only problem which has received theoretical justification is that of multiclass recognition using unary desired-ouput representation. Our purpose in this article is to show that it is possible to achieve a simple interpretation of the output of any net trained to minimize the usual quadratic error, regardless the desired-output representation is discrete or continuous. As a consequence, the importance in the way the representation of the desired-output patterns is chosen will be discussed.

It must be stressed that our results will be derived with the only assumption that global minima are always possible to be calculated, without any reference to the intrinsic difficulty of this problem nor to its dependence on the shape of the net; in fact, it need not be a neural network. Thus, the word 'net' should be understood as a short for 'big enough family of functions', and could be applied indistinctly to a multilayer neural network, to a recurrent neural network and to any user-defined parametric set of functions.

2   Analytical interpretation of the net output

Let $\omega \in \mathcal{Z}$ denote the output corresponding to a certain $\xi \in \mathcal{X}$ input pattern. Since the sets $\mathcal{X}$ and $\mathcal{Z}$ are arbitrary, it is convenient to represent each pattern by a real vector in such a way that there is a one-to-one correspondence between vectors and feature patterns. Of course, we take for granted that such representations exist. We will make use of the vectors $x \in \mathbb{R}^n$ for the input patterns and $z(x) \in \mathbb{R}^m$ for the output ones.

If $\{(x^\mu, z^\mu), \mu = 1, \ldots, N\}$ is a representative random sample of pairs input-output, our goal is to find the net $o : x \in \mathbb{R}^n \longmapsto o(x) \in \mathbb{R}^m$ which closely resembles the unknown correspondence process. The least squares estimate is that which produces the lowest mean squared error $E[o]$, where

$$E[o] \equiv \frac{1}{2N} \sum_{\mu=1}^{N} \sum_{i=1}^{m} \left(o_i(\boldsymbol{x}^\mu) - z_i(\boldsymbol{x}^\mu)\right)^2 . \tag{1}$$

It is easy to realize that for large N, and due to the Strong Law of Large Numbers, the limiting value of $E[o]$ is given by

$$\begin{aligned}
E[o] &= \frac{1}{2} \sum_{i=1}^{m} \int_{\mathbb{R}^n} d\boldsymbol{x}'\, p(\boldsymbol{x}') \int_{-\infty}^{\infty} dz_i\, p(z_i|\boldsymbol{x}')\, [o_i(\boldsymbol{x}') - z_i]^2 \\
&= \frac{1}{2} \sum_{i=1}^{m} \int_{\mathbb{R}^n} d\boldsymbol{x}'\, p(\boldsymbol{x}') \int_{\mathbb{R}^m} d\boldsymbol{z}\, p(\boldsymbol{z}|\boldsymbol{x}')\, [o_i(\boldsymbol{x}') - z_i]^2
\end{aligned} \tag{2}$$

where $p(\boldsymbol{x})$ stands for the probability density function of the random variable $\boldsymbol{x}$ in the sample, and $p(\boldsymbol{z}|\boldsymbol{x})$ is the conditional probability density of $\boldsymbol{z}$ knowing that the former random variable has taken on the value $\boldsymbol{x}$.

Assuming no constraint in the functional form of $o(\boldsymbol{x})$, the minimum $o^*(\boldsymbol{x})$ of E is easily found by annulling the first functional derivative:

$$\begin{aligned}
\frac{\delta E[o]}{\delta o_j(\boldsymbol{x})} &= \sum_{i=1}^{m} \int_{\mathbb{R}^n} d\boldsymbol{x}'\, p(\boldsymbol{x}') \int_{\mathbb{R}^m} d\boldsymbol{z}\, p(\boldsymbol{z}|\boldsymbol{x}')\, [o_i(\boldsymbol{x}') - z_i]\, \delta_{ij}\, \delta(\boldsymbol{x} - \boldsymbol{x}') \\
&= p(\boldsymbol{x}) \int_{\mathbb{R}^m} d\boldsymbol{z}\, p(\boldsymbol{z}|\boldsymbol{x})\, [o_j(\boldsymbol{x}) - z_j] = p(\boldsymbol{x})\, [o_j(\boldsymbol{x}) - \langle z_j \rangle_{\boldsymbol{x}}] = 0
\end{aligned} \tag{3}$$

implies that

$$o^*(\boldsymbol{x}) = \langle \boldsymbol{z} \rangle_{\boldsymbol{x}}, \ \forall \boldsymbol{x} \in \mathbb{R}^n \text{ such that } p(\boldsymbol{x}) \neq 0 , \tag{4}$$

where

$$\langle z_j \rangle_{\boldsymbol{x}} = \int_{\mathbb{R}^m} d\boldsymbol{z}\, p(\boldsymbol{z}|\boldsymbol{x})\, z_j , \ j = 1, \ldots, m \tag{5}$$

is the mean of the output vectors in the training sample for each input pattern represented by $\boldsymbol{x} \in \mathbb{R}^n$. This is the key expression from which we will derive the possible interpretations of the net output.

As a particular case, if the output representation is chosen to be discrete, say $\boldsymbol{z}(\boldsymbol{x}) \in \{\boldsymbol{z}^{(1)}, \boldsymbol{z}^{(2)}, \ldots, \boldsymbol{z}^{(a)}, \ldots\}$, then equation (5) reads

$$\langle z_j \rangle_{\boldsymbol{x}} = \sum_{a} P(\boldsymbol{z}^{(a)}|\boldsymbol{x})\, z_j^{(a)} , \ j = 1, \ldots, m \tag{6}$$

where $P(\boldsymbol{z}^{(a)}|\boldsymbol{x})$ is the probability of $\boldsymbol{z}^{(a)}$ conditioned to the knowledge of the value of the input vector $\boldsymbol{x}$.

From a practical point of view unconstrained nets do not exist, which means that the achievable minimum $\tilde{o}(\boldsymbol{x})$ is in general different to the desired $o^*(\boldsymbol{x})$. The mean squared error between them is written as

$$\varepsilon[\tilde{o}] \equiv \frac{1}{2} \sum_{i=1}^{m} \int_{\mathbb{R}^n} d\boldsymbol{x}\, p(\boldsymbol{x}) \int_{\mathbb{R}^m} d\boldsymbol{z}\, p(\boldsymbol{z}|\boldsymbol{x})\, [\tilde{o}_i(\boldsymbol{x}) - \langle z_i \rangle_{\boldsymbol{x}}]^2 . \tag{7}$$

However, it is straightforward to show that

$$E[o] = \varepsilon[o] + \frac{1}{2} \sum_{i=1}^{m} \int_{\mathbb{R}^n} d\boldsymbol{x}\, p(\boldsymbol{x}) \int_{\mathbb{R}^m} d\boldsymbol{z}\, p(\boldsymbol{z}|\boldsymbol{x})\, [z_i - \langle z_i \rangle_{\boldsymbol{x}}]^2 , \qquad (8)$$

and, since the second term of the sum is a constant —it does not depend on the net—, the minimizations of both $E[o]$ and $\varepsilon[o]$ are equivalent. Therefore, $\tilde{o}(\boldsymbol{x})$ is a minimum squared-error approximation to the unconstrained minimum $o^*(\boldsymbol{x}) = \langle \boldsymbol{z} \rangle_{\boldsymbol{x}}$.

3 Discrete representations and bayesian decision rule

It is well known that nets trained to minimize eq. (1) are good approximations to bayesian classifiers, provided a unary representation is taken for the output patterns [2, 3, 4]. That is, suppose the input patterns have to be separated in C different classes $\mathcal{X}_a$, $a = 1, \ldots, C$, and let $\boldsymbol{z}^{(a)} \equiv (\overbrace{0}^{1}, \ldots, \overbrace{0}^{a-1}, \overbrace{1}^{a}$, $\overbrace{0}^{a+1}, \ldots, \overbrace{0}^{m})$ be the desired output of any input pattern $\boldsymbol{x} \in \mathcal{X}_a$. This assignment specializes each output component to recognize a distinct class ($m = C$). According to our eqs. (4) and (6),

$$o_a^*(\boldsymbol{x}) = \sum_b P(\boldsymbol{z}^{(b)}|\boldsymbol{x}) z_a^{(b)} = P(\boldsymbol{z}^{(a)}|\boldsymbol{x}) , \qquad (9)$$

i.e. the a-th component of the net output turns out to be a minimum squared approximation to the conditional probability that pattern $\boldsymbol{x}$ belong to class $\mathcal{X}_a$. Therefore we recover the familiar bayesian decision: $\boldsymbol{x}$ is most likely a member of class $\mathcal{X}_b$, where $\tilde{o}_b(\boldsymbol{x}) = \max\{\tilde{o}_1(\boldsymbol{x}), \ldots, \tilde{o}_C(\boldsymbol{x})\}$.

The applicability of eq. (9) goes beyond classifications. For example, suppose that you have a certain Markov chain $\{s_t, t \in \mathbb{N}\}$ of discrete states with constant transition probabilities, and you train a net to learn s_t as a function of $s_{t-1}, \ldots, s_{t-\tau}$. Hence, the output of the net will tend to give these transition probabilities $P(s_t|s_{t-1}, \ldots, s_{t-\tau})$, which by hypothesis do not depend on t.

Moreover, eqs. (4) and (6) permit a simple generalization to perform classification tasks. In the discrete and finite case, it is always possible to make an approximated bayesian decision provided the representation $\{\boldsymbol{z}^{(1)}, \ldots, \boldsymbol{z}^{(C)}\}$ is chosen such that the linear system

$$\begin{cases} \displaystyle\sum_{b=1}^{C} P(\boldsymbol{z}^{(b)}|\boldsymbol{x})\, z_a^{(b)} = o_a^*(\boldsymbol{x}), \ a = 1, \ldots, d, \ d \in \{C - 1, C\} \\ \displaystyle\sum_{b=1}^{C} P(\boldsymbol{z}^{(b)}|\boldsymbol{x}) = 1 \text{ needed if } d = C - 1 \end{cases} \qquad (10)$$

has a non null determinant, in order to be able to find the conditional probabilities as a function of the net outputs. Examples of these representations could be the 'thermometer-like' ones [1], where now is clear that to obtain a bayesian decision rule, the previous system has to be solved. Of course, any other representation that does not fulfill these conditions (e.g. 'binary' representations) cannot be used to obtain a bayesian classifier.

4 Continuous output representations

Discrete and finite output representations arise quite naturally in the treatment of classification problems. On the other hand, prediction and interpolation tasks usually amount to finding the 'best' value of several continuous variables for each given input. One possible but unsatisfactory solution is the discretization of these variables, which has to be made carefully in order to skip various problems. If the number of parts is to big, the number of training patterns should be very large. Otherwise, if the size of the parts is relatively big, the partitioning may fail to distinguish relevant differences, specially if the output is not uniformly distributed. Therefore, it may be stated that a good discretization needs a fair understanding of the unknown output distribution!

Fortunately, neural nets have proven to work well even when the output representation is left to be continuous, without any discretization. For instance, feed-forward networks have been applied to time series prediction of continuous variables, outperforming standard methods. The explanation to this success lies precisely in eq. (4), which reveals the tendency of nets to learn, for each input, the mean of its corresponding outputs in the training set. Thus, the net is automatically doing what everyone would do in the absence of more information.

5 Conclusions

The election of the statistical parameter which best fits a distribution has a high dependence on its shape and on the nature of the problem. The most repeatedly used are the 'mean' and the 'most probable value'. We have shown that both can be achieved using nets trained to minimize the standard quadratic error. However, the latter may present serious difficulties if the output variables are continuous.

References

[1] Gallant, S. (1990) IEEE Trans. Neural Networks, Vol. 1, No. 2, 179–191.

[2] Garrido, Ll and Gaitan, V (1991) Int. J. of Neural Systems, Vol. 2, No. 3, 221–228.

[3] Ruck, D.W. Rogers, S.K., Kabrisky, M., Oxley, M.E. and Suter, B.W. (1990) IEEE Trans. Neural Networks, Vol. 1, No. 4, 296–298.

[4] Wan, E.A. (1990) IEEE Trans. Neural Networks, Vol. 1, No. 4, 303–305.

Fluctuated-Threshold Effect in Multilayered Neural Network

Ken'ichi Iwami, Nobuyuki Matsui and Toshiharu Araki
Department of Computer Engineering
Faculty of Engineering, Himeji Institute of Technology
2167 Shosha, Himejishi, Hyogo, 671-22, Japan

1 Introduction

There has been significant progress, in recent years, in an attempt to achieve brain-like performances in computing systems. Almost of these performances are described as the collective behavior of highly interconnected networks of "formal neurons" which are simple computational elements with two possible states changing state according to an input-output transfer function with a time-invariant threshold. Real neurons in biological nervous systems, however, have a time-variant threshold owing to fluctuations such as occur in membrane potential (Hayashi et al., 1987). From both practical and theoretical points, therefore, it is important to incorporate such fluctuated neurons into brain-like computer systems by a computable and simple method and to investigate the fluctuated-threshold effect. We, here, define "fluctuated neurons" as neurons which have the input-output transfer function with the time-variant threshold fluctuating periodically.

The approach taken in this paper, from such a view point, aims to estimate learning capabilities of the multilayered neural network that includes "fluctuated neurons" in the hidden layer and to show that this type of modeling constitutes a simple and promising alternative in the study of introducing real neuron-like function into neural networks.

2 A method of incorporating "fluctuated neurons" into neural network

We begin with constructing a three-layered feedforward neural network that is able to solve the XOR problem. The architecture of this network is shown in Fig.1. To incorporate fluctuated neurons into this network, we consider as follows: a threshold value of a neuron, $h(x)$, is the sum of the

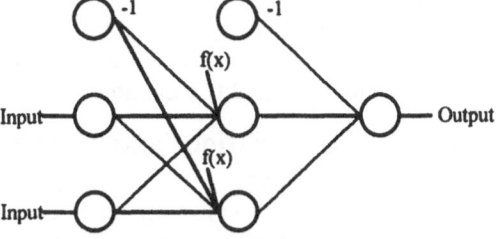

Fig.1 Three-layered feedforward network

weighted input $-w$ from the biased neuron of -1 in the preceding layer and the fluctuation function $f(x)$ that fluctuates periodically as the learning in this network progresses, namely,

$$h(x) = -w + f(x), \tag{1}$$

where x is the number of learning times for this network. If $f(x)=0$, then this network reduces to conventional one with "formal neurons".

We, in this article, fluctuate thresholds of neurons in the hidden layer only, and do not fluctuate thresholds of neurons in the output layer. We now examine two types

554

of fluctuation function $f(x)$, a sine wave and a sine wave added small sine wave as a ripple shown in Fig.2, described as follows:

$$f(x) = \alpha \sin(\beta \frac{\pi}{180} x), \qquad (2a)$$

and

$$f(x) = \alpha \{ \sin(\beta \frac{\pi}{180} x) + \gamma \sin(\delta \frac{\pi}{180} x) \}, \quad (2b)$$

where α and β are parameters that represent the amplitude and the period of the threshold fluctuation respectively, and γ and δ are similar parameters for the ripple, respectively.

3 Computer experiment

In order to estimate the learning capabilities of our neural network, we make this network solve the XOR problem and the 4-bit parity check problem.

We assume here that our network:

•Consist of three layers of interconnected units ("neurons").

•Each unit calculates the weighted sum of its inputs (plus a threshold) to form a net activation ("net"), and emits some function of net, g(net).

•This transfer function g(net) is described by following equation:

$$g(net) = \frac{\exp(net) - 1}{\exp(net) + 1} \qquad (3)$$

•Every unit in a hidden layer and an output layer has the sigmoidal function of Eq.(3), and every unit in an input layer has the liner function.

•The number of hidden units is 2 for the XOR problem, and 4 or 5 for the 4-bit parity check problem.

We adopt the back propagation algorithm (Rumelhart et al., 1986) as our learning algorithm of this multilayered neural network, but do not use the momentum method in order to simplify conditions.

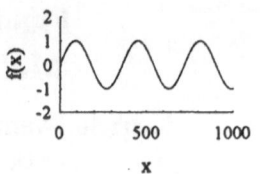

(a) The case of Eq.(2a)
(α=1.0, β=1.0)

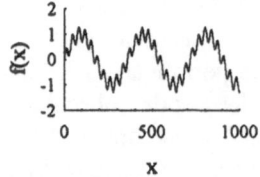

(b) The case of Eq.(2b)
(α=1.0, β=1.0, γ=0.3, δ=10.0)

Fig.2 Fluctuations of threshold

4 Results of simulations and discussions

4.1 The XOR problem

The success rate and the average number of learning times for 100 trials on the XOR problem are shown in table 1. For comparison, the case of no fluctuation and the case of adding white noise as the threshold fluctuation are also shown in table 1. Parameters α, β, γ and δ used in these simulations are optimum values discussed later. Table 1 shows that the learning results of our network are better than those of conventional one. The success rate of our network is 100(%), and the average number of learning times is about half as much as that of conventional one. The learning results for the fluctuation of Eq.(2b) are a little better than those of Eq.(2a). Although it has been showed that the learning results are improved by giving white noise to the network (Ohguchi, 1991), we further obtain better results by the network with periodic fluctuations.

Table 1 Learning results of the XOR problem

	(i)	(ii)	(iii)	(iv)
Success Rate(%)	100	100	88	100
Learning Times	135	129	318	282

(i)   The case of Eq.(2a): $\alpha=3.0$, $\beta=7.0$
(ii) The case of Eq.(2b): $\alpha=1.0$, $\beta=7.0$, $\gamma=1.0$, $\delta=10\beta$
(iii) The case of no fluctuation
(iv) The case of white noise

4.2 The 4-bit parity check problem

Table 2 shows similar results on the 4-bit parity check problem for the case of the network using 5 hidden units. In this case, we also find that our network improves learning results. The learning results for the fluctuation of Eq.(2b) are fairly better than those of Eq.(2a). We, hence, see that the fluctuation with the ripple affect the learning of networks effectively.

In table 2 (iv)* and (v)*, the learning results of the case that fluctuations with different amplitude are added to each hidden unit are also shown. Each amplitude of fluctuations is appropriately assumed, for example α, 2α, 4α, 8α and 16α. Compared (i), (ii) with (iv)*, (v)* in table 2, we find that the learning results can be improved by adding fluctuations with different amplitude to each hidden unit.

For the case of the network using 4 hidden units, similar results on the 4-bit parity check problem are shown in table 3. The success rate of the network with no fluctuation is very low (25%), but that of the network with fluctuation is much higher (89%). Thus, these observations suggest that the fluctuated-threshold has the equivalent effect that networks avoid falling into local minima.

Table 2 Learning results of the 4-bit parity check problem using 4-5-1 network

	(i)	(ii)	(iii)	(iv)*	(v)*
Success Rate(%)	100	100	96	100	100
Learning Times	833	672	3643	616	534

(i)   The case of Eq.(2a): $\alpha=7.0$, $\beta=0.5$
(ii) The case of Eq.(2b): $\alpha=5.0$, $\beta=0.5$, $\gamma=0.3$, $\delta=10\beta$
(iii) The case of no fluctuation
(iv) The case of Eq.(2a): $\alpha=0.7$, $\beta=0.7$
(v)   The case of Eq.(2b): $\alpha=0.7$, $\beta=0.5$, $\gamma=0.1$, $\delta=10\beta$
*   The case of adding fluctuations with different amplitude to each hidden unit

Table 3 Learning results of the 4-bit parity check problem using 4-4-1 network

	(i)	(ii)	(iii)
Success Rate(%)	89	89	25
Learning Times	6244	4735	6051

(i)   The case of Eq.(2a): $\alpha=30.0$, $\beta=0.05$
(ii) The case of Eq.(2b): $\alpha=10.0$, $\beta=0.07$, $\gamma=0.3$, $\delta=10\beta$
(iii) The case of no fluctuation

4.3 Consideration for parameter α and β

Figures 3(a) and 3(b) show the α-dependence of learning capabilities using the fluctuation Eq.(2a) on the XOR problem and the 4-bit parity check problem (5 hidden units), respectively. We observe that these figures have similar variations under the appropriate β value. In both cases, the bigger parameter α grows, the better these results are gradually, because the effects of fluctuation become noticeable. But if α is over a certain value, learning results become bad, because the fluctuation interferes with normal learning processes. So, we have the best results at $\alpha=3.0$ for the XOR problem, and at $\alpha=7.0$ for the 4-bit parity check problem. The bigger value α for the 4-bit parity check problem is caused by the bigger total-input into a hidden unit.

On the same problems, figures 4(a) and 4(b) show the β-dependence of learning capabilities at the above optimum α values. We observe that this β-dependence is insensitive in the easy problem such as the XOR problem, while in the 4-bit parity

556

check problem the learning results rather depend on β.

As mentioned above, we see that the optimum values (α, β) exist and the learning characteristics are fairly improved by our simple method.

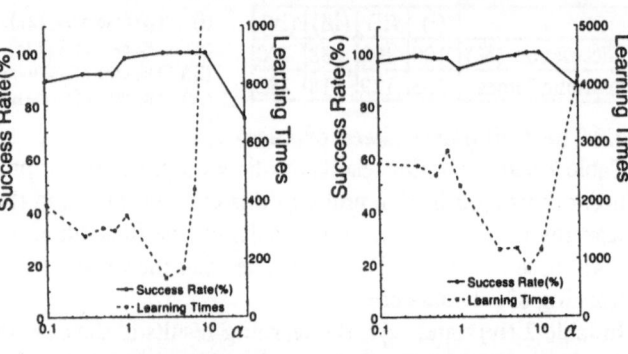

(a) The XOR problem (β=7.0)

(b) The 4-bit parity check problem (β=0.5)

Fig.3 The α-dependence of learning capabilities using the fluctuation Eq.(2a)

5. Conclusions

We have shown that the learning capabilities of networks improve noticeably by fluctuating threshold functions of neurons. We, therefore, know that fluctuations in neurons play an important role in information processing in the brain. Although we adopt a very simple method in this study, we consider this method is very promising in neural network architecture. In future, we have to clarify their capabili-

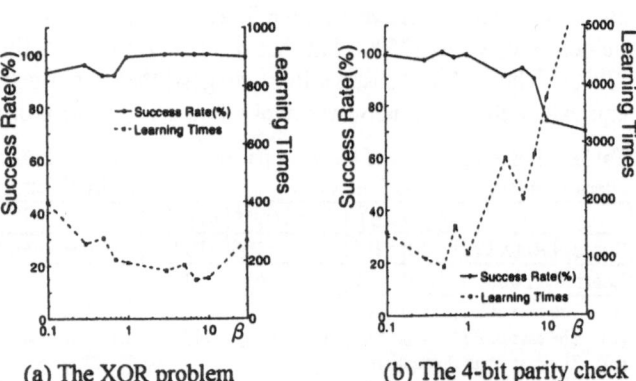

(a) The XOR problem (α=3.0)

(b) The 4-bit parity check problem (α=7.0)

Fig.4 The β-dependence of learning capabilities using the fluctuation Eq.(2a)

ties for information processing by applying this method to some applications.

References

Hayashi,H. and Ishizuka,S. (1991) Chaos in molluscan neuron. Chaos in biological systems (H. Degn, A.V. Holden and L.F. Olsen eds.), NATO ASI series, 138, 157-166. Plenum Press.

Ohguchi, Y. (1991) High Speed Back-Propagation Learning. IEICE Technical Report, NC91-47, 127-131. (In Japanese)

Rumelhart, D.E., McCleland, J.L., and the PDP Research Group (1986) Parallel Distributed Processing, MIT press.

On the Properties of Error Functions that Affect the Speed of Backpropagation Learning

R. J. Gaynier, T. Downs
Electrical and Computer Engineering, University of Queensland
Brisbane, Australia

1 Introduction

In the last few years several alternatives to the mean-squared error function have been proposed as a means of improving the speed of backpropagation learning. Perhaps best-known among these are the functions considered in (Fahlman 1988) and the cross-entropy function discussed in (Solla et al., 1988). In this paper we investigate properties of error functions that influence the rate of learning when backpropagation (BP) is employed.

In BP learning it frequently happens that a substantial proportion of the training set is learnt very quickly, but the remaining patterns take much longer to be learnt, and in some cases may not be learnt at all. This behavior can occur for a number of reasons, two of them being:

(i) *static exceptions* in the training data, which can be due to errors in the training data (e.g. misclassified or noise-affected patterns) or to training data that are abnormal or contradictory in some way;

(ii) *dynamic exceptions*, which can arise when the training data contain no static exceptions; they occur when a network's weight values make it difficult to accommodate those patterns not yet learnt, causing them to appear to the network as exceptions.

A recent paper (Lister et al., 1993) identified three particular properties of error functions that tend to cause dynamic exceptions to occur and in this paper we propose a new error function that allows these properties to be investigated rather more closely than was attempted in (Lister et al., 1993). These investigations allow us to draw conclusions about the desirable shape of the error function used in BP learning and we show that, when our error function is appropriately shaped, it provides very substantial increases in learning speed over existing functions for problems with dynamic exceptions. We then go on to show how the shape of our error function can be made variable during learning in order to achieve good performance on problems containing static exceptions.

2 Error Signal Properties Affecting Learning

In BP learning, an error signal is propagated backwards from the output units and is used to update network weights. The error signal is equal to the product of the derivative of the error function and the derivative of the activation function of the units in the network. In the process of learning a set of patterns,

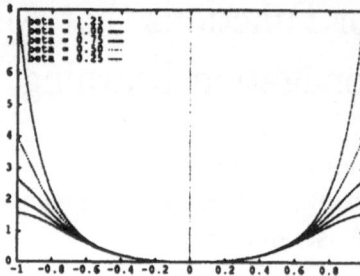 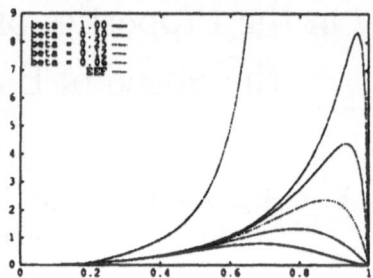

Figure 1: (a) TEF for various β with $\alpha = 1$; (b) error signals of the TEF (the error signal of the EEF, which goes rapidly to infinity, is also shown).

the error signal for the full set is equal to the sum of the individual contributions produced by each pattern. And clearly, if the contribution made to the error signal by poorly-learned patterns is not sufficiently large, its effect will be overridden by the combined effect on the error signal of the better-learned patterns. In (Lister et al., 1993), three different properties of error signals that can cause this to occur were identified:

(i) The error signal may approach zero too slowly for small error values so that the contributions to the error signal from well-learned patterns will be large.

(ii) The error signal from poorly-learned patterns may never reach a large enough value to have any effect on the learning process (i.e. the maximum value of the error signal may be too small).

(iii) Large errors drive activations toward saturation values in the sigmoidal characteristic causing the derivative of the activation function, and hence the error signal, to tend to zero. Thus very poorly-learned patterns contribute little to the overall error signal.

Having identified these three properties, the authors of (Lister et al., 1993) then proposed a new error function which they argued should perform well on problems involving dynamic exceptions. Using this error function, which they referred to as the Exception Error Function (EEF), they demonstrated good learning performance on N-2-N encoder problems (which are known to be prone to dynamic exceptions when standard BP is applied.

We have investigated at some length the influence on learning speed of each of the above properties, and this was achieved by means of the error function

$$E = \frac{\alpha(x^2 + x^4)}{1 - x^2 + \beta x^4} \tag{1}$$

where x is the error at the network output and α, β are parameters that allow the shape of E (and, more particularly, the shape of the error signal) to be varied. We refer to E as the Test Error Function (TEF) and its shape for $\alpha = 1$ and a range of values of β is shown in fig 1(a). Note that although α is merely a scale factor in E, it can be used to vary the value of the maximum of the error signal and so has a useful role to play. The error signals corresponding

14-2-14 Epochs				19-2-19 Epochs			
Func.	Least	Median	Most	Conv	Least	Median	Most	Conv
TEF	575	1606	2021	10/10	8178	12414	23524	10/10
EEF	2234	4397	5175	10/10	59039	97274	161484	8/10
X-Ent	126696	169554	405960	10/10	-	-	-	0/10
M Sq	221102	441074	735487	10/10	-	-	-	0/10

Table 1: N-2-N encoder for the TEF, EEF, Cross Entropy and Mean Squared error functions. Results for the latter three from (Lister et al., 1993)

Training Errors				Test Errors		
Func.	Epochs	Least	Median	Most	Least	Median	Most
VEF	500	6	10	12	0	1.5	4
X-Ent	500	4	7	8	2	3	6
EEF	500	7	14	31	4	5	19
TEF	500	15	20	57	6	13	28
Mean Sq	500	386	386	638	386	386	638

Table 2: Results for the Contiguity problem with 10 static exceptions.

to the error functions in fig 1(a) are shown in fig 1(b), where the error signal for the EEF is included for comparison purposes [1].

Our investigations of the effects of different shapes of error signal on learning speed led us to conclude that an additional property of the error signal (not considered in (Lister et al., 1993)) appears to have the greatest influence. This property is the location of the maximum of the error signal - note that in fig 1(b) the maximum shifts to the right with decreasing β. Our investigations also indicate that if the maximum of the error signal is too large it has a detrimental effect on learning. The overall effects of the shape of the error signal are discussed below.

3 Results and Discussion

Table 1 presents our results for the 14-2-14 and 19-2-19 encoder problems using what we believe is an approximately optimal shape for the error signal of the TEF. These results indicate a clear superiority of the TEF over the other three error functions for N-2-N encoder problems. The reasons for the superiority of the TEF over the EEF are, in our view, evident from fig.1(b). The fact that the error signal for the EEF tends to infinity for large error and that it is also large for "moderate" error indicates to us that the EEF places too much emphasis on exceptional patterns to the detriment of learned, or nearly learned, patterns. In contrast, the error signal for the TEF is small for low and moderate error values and then reaches a *finite* maximum near maximum error.

Given our success with a problem that involves dynamic exceptions, we decided to investigate the possibility of using a similar approach to improve learning performance for problems involving static exceptions. To this end, we

---

[1]The activation function employed was $1/(1 + e^{-x})$.

chose to consider the contiguity (2 or more clumps) problem which was used in (Solla et al., 1988) as a benchmark problem for the cross-entropy function. Static exceptions can be introduced into this problem by simply misclassifying some of the patterns.

In the case of the N-2-N encoder, our prime concern was learning speed, but in a case where the training data contains misclassified patterns, our main interest is in how these patterns are dealt with by the learning process. And, in the latter case, it is generally desirable that a network should fail to learn most (preferably all) of the misclassified patterns because otherwise its generalisation capabilities are likely to be impaired. For this to be achieved, we require that the large error at the output due to a misclassified pattern lead to only a small error signal. Thus, in order to deal with static exceptions in this way, an error signal like that of the EEF in fig.1(b) is out of the question. The other error signals in fig.1(b) do, however, offer the possibility of suitable treatment of static exceptions.

But in attempting to deal satisfactorily with static exceptions, we must not forget that dynamic exceptions have to be accommodated also. Our approach is based on the assumption that dynamic exceptions will gradually be accommodated so long as the error is large for large error in the earlier stages of training. We commence with an error signal like the one with the maximum peak ($\beta = .06$) in fig.1(b) and, over time, gradually increase β so that the peak decreases and shifts to the left. In this way we seek to accommodate the dynamic exceptions (whose error should gradually reduce) and exclude the static exceptions (whose error will remain large). We refer to an error function that changes in this way during the learning process as a VEF (Variable Error Function).

Table 2 shows the results we obtained on a ten-input, ten-hidden unit, single-output network given the task of learning the full set of 1024 patterns in the 10-bit contiguity problem. In this table, the data entitled "Training Errors" relate to the training of the network with 10 patterns misclassified; the data entitled "Test Errors" relate to the testing of the trained network on the complete contiguity problem with no misclassifications. It is interesting to note that the VEF tends to reject the 10 misclassified patterns during training and, as a consequence, performs best on the test set. The concept of a variable error function for dealing with static errors appears worthy of further investigation.

References

[1] Fahlman, S. (1988) "Faster Learning Variations on Backpropagation: An Empirical Study" Proceedings of the 1988 Connectionist Models Summer School, pp38-51.

[2] Lister, R., Bakker, P., Wiles J. (1993) "Error Signals, Exceptions and Back Propagation", Proceedings of the International Joint Conference on Neural Networks, Nagoya, Japan, pp573-576.

[3] Solla, S. A., Levin, E., Fleisher, M. (1988) "Accelerated learning in Layered Neural Networks" Complex Systems 2, pp625-639.

Neural Network Optimization for Good Generalization Performance *

Jieyu Zhao and John Shawe-Taylor

Department of Computer Science

Royal Holloway, University of London

Egham, Surrey, TW20 0EX, U.K.

1 Introduction

Generalization is one of the most important abilities of artificial neural networks. In order to improve the generalization performance, we have to choose a network which will not overfit the input data. There are however very few known theoretical ways of making this choice. We usually choose a network with an excess parameters and then try to simplify the network during the training.

A common way of simplifying neural networks is to use a cost function including a weight decay term that will penalize complexity. Since the cost function becomes more complicated, convergence may take a longer time when a gradient descent algorithm is used. The choice of an appropriate decay constant is also a very difficult task. Another approach that has been proposed to simplify the network is soft weight-sharing (Nowlan and Hinton 1992), which can lead to better generalization than simpler forms of weight decay. But it comes at the cost of greater complexity.

In this paper a new approach for neural network optimization is proposed. The training process of neural networks is divided into two phases, learning and optimizing. Regularization is carried out during the optimizing phase. The feasibility of this method is discussed. Simulations on several problems demonstrate this approach is very effective and ensures that the neural network will generalize well.

2 Generalization and Training Strategies

The generalization ability in neural networks is measured by the probability that the networks trained on t examples correctly predict a novel example. It is too difficult to calculate the exact generalization error since the relation between the error, the architecture of the network and the training samples is very complicated. The generalization error also depends on the training process which decides the final learning precision.

If the cost function is composed of the sum of squared errors and a weight decay term, the generalization performance(expected test set error) can be approximately estimated as follows (J.E.Moody, 1992),

$$< \varepsilon_{test}(\lambda) >_{\xi\xi'} \approx < \varepsilon_{train}(\lambda) >_{\xi} + 2\sigma_{eff}^2 \frac{p_{eff}(\lambda)}{n}$$

*The research was funded by the British Council and the Natural Science Foundation Council of China.

where n is the size of the training sample ξ, σ_{eff}^2 is the effective noise variance, λ is a weight decay parameter, and $p_{eff}(\lambda)$ is the effective number of parameters in the nonlinear model.

In order to improve the generalization performance, the network structure needs to be optimized and a low training error to be kept at the same time.

In practice, the training strategy is to allocate limited resources, such as network size and structure, training time, and available examples, to reach good generalization performance. It always seems difficult to find a strategy which can train a network to converge on a compact structure and keep a low learning error at the same time. Weight decay during learning can efficiently reduce the number of free parameters in the network, but it will probably cause an increase in learning error which significantly affects the generalization performance and results in a long time to converge.

An alternative training strategy for good generalization is to divide the training process into two phases, learning and optimizing. The former focuses on convergence with a larger network and the latter searches for a compact structure while the training error is still kept at a low level. Interestingly in our experiments we have found that there always appears to exist a continuous path in weight space from a global minimum with large structure to another global minimum with compact structure. At any point on this continuous path, the output error of the network remains at a very low level. In other words, the solutions of a slightly larger network in weight space are usually not separated points but continuous lines or even regions. This reveals the possibility of optimizing the network structure after convergence while not raising the output error at the same time. An additional benefit of this training strategy is that using networks with a few more parameters than the most constrained ones may reduce the learning epochs if a gradient descent algorithm is used.

During the learning phase, various gradient descent methods can be used to search for a global minimum. As soon as the algorithm converges (the output error becomes less than a certain small value), the algorithm enters into the optimizing phase in which weight regularization (weight decay and weight tying) and learning are carried out alternately. If the output error surpasses the given boundary because of the regularization, the learning algorithm is then called again; otherwise, the regularization continues. The length of the optimizing phase depends on the complexity of the problem and the structure of the network being used.

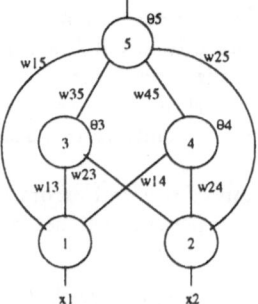

Figure 1: A redundant cascade architecture for XOR problem.

Whether or not such optimizing routes exist in general is still an open question. But as a simple example, the XOR problem with a two-layered cascade network (Figure 1) using a sigmoid type activation function f(x), can be demonstrated to possess such optimizing routes in weight space. After finding a solution with two hidden units, if $[f(w_{14}+\theta_4)+f(w_{24}+\theta_4)-f(\theta_4)-f(w_{14}+w_{24}+\theta_4)] \neq 0$, $w_{14} \neq 0$ and $w_{24} \neq 0$, the whole structure can always be changed smoothly into a more compact one (with only one hidden unit) by making w_{45} tend to 0, while the output error is kept unchanged.

3 Simulation Studies

The above ideas are illustrated on following examples. All experiments employ simple multi-layer or cascade structures and the generalization performance is compared based on the same training errors. The simulator has been developed based on Aspirin/MIGRAINES 6.0 in which an improved algorithm with structure optimization has been developed.

3.1 The Parity Problem

One of the classical problems which has received much attention is the parity problem, in which the output required is 1 if the input pattern contains an odd number of 1's and 0 otherwise. When a two-layered feed-forward network with a logical threshold activation function is used, it requires at least N hidden units to solve parity with patterns of length N. In practice, it is not so easy to train a network with N hidden units to solve N-bit parity problem when $N \geq 5$. Generally a slightly larger network is used.

Using the training strategy with learning and optimizing phases, the following results have been obtained.

n-bit	the most compact structure using BP	the optimized structure
3	3 hidden units	3 hidden units
4	4 hidden units	4 hidden units
5	6 hidden units	5 hidden units
6	10 hidden units	7 hidden units
7	16 hidden units	8 hidden units
8	24 hidden units	9 hidden units

It has been found that the optimized structures are equal to the most compact ones or have only one more hidden unit.

The recognition rates of the 10-bit parity problem are as follows: (with 150 learning examples and 256 test examples)

BP(small initial weights): 76.3%

After regularization: 85.5%

3.2 A Real Problem

Another task chosen to evaluate the effectiveness of regularization was handwritten digit recognition. The database consists of a clean set of 500 samples, 400 for training and 100 for testing. The input is on a 16x16 pixel array with 0 for the background and 1 for the foreground. Several samples are shown in Figure 2.

The network used 256 input units, 12 hidden units, 10 output units, and was fully connected.

According to the results of M.D.Richard and R.P.Lippmann, the outputs of neural networks classifiers estimate Bayesian a posteriori probabilities. Therefore, the real-valued output nodes were used. The final results were detected with a threshold of 0.15, which is the minimum acceptable Euclidean distance between the output vector and target vector.

	Train % correct	Test % correct
BP(small initial weights)	100%	81%
After regularization	100%	89%

The results for the same problem with threshold outputs are as follows

	Train % correct	Test % correct
BP(small initial weights)	100%	86%
After regularization	100%	93%

Figure 2. Several training samples.

4 Conclusion

An optimal network structure is crucially important for the generalization performance. Regularization after the learning phase is not only possible but also necessary to improves the generalization significantly. The structural optimization also makes the continuous learning of new samples much easier. However, the more theoretical investigation into the existence of the optimizing routes in general is needed.

References

[1] J.E.Moody. The Effective Number of Parameters: An Analysis of Generalization and Regularization. *Advances in Neural Information Processing Systems. 1992.*

[2] E.B.Baum and D.Haussler. What Size Net Gives Valid Generalization? *Neural Computation 1, 1989.*

[3] S.Amari and N.Fujita. Four Types of Learning Curves. *Neural Computation 4, 1992.*

[4] H.Drucker and Y.L.Cun. Improving Generalization Performance Using Double Backpropagation. *IEEE Trans. on Neural Networks, Vol. 3, No. 6, 1992.*

[5] E. Levin, N.Tishby and S.A.Solla. A Statistical Approach to Learning and Generalization in Layered Neural Networks. *1990.*

[6] S.J.Nowlan and G.E.Hinton. Simplifying Neural Networks by Soft Weight-Sharing. *Neural Computation 4, 1992.*

[7] M.D.Richard and R.P.Lippmann. Neural Network Classifiers Estimate Bayesian a posteriori Probabilities. *Neural Computation 3, 1991.*

[8] H.Schwarze and J.Hertz. Generalization in Fully Connected Committee Machines. *1992.*

[9] V.Tresp and J.Hollatz. Network Structuring and Training Using Rule-based Knowledge. to appear in *Advances in Neural Information Processing Systems. 1993.*

[10] J.Utans and J.Moody. Selecting Neural Network Architectures via the Prediction Risk: Application to Corporate Bond Rating Prediction. *The First International Conference on Artificial Intelligence Application on Wall Street. 1991.*

Block-Recursive Least Squares Technique for Training Multilayer Perceptrons

R. Parisi, E.D. Di Claudio, G. Orlandi

INFOCOM Dpt. - University of Rome "La Sapienza"
Via Eudossiana 18, 00184 Roma - Italy

1 Introduction

The multilayer perceptron is one of the most commonly used types of feed-forward neural networks and it is used in a large number of applications. Its strength resides in its capacity of mapping arbitrarily complex non-linear functions by a convenient number of layers of sigmoidal non-linearities (Rumelhart et al., 1986). The Backpropagation algorithm is still the most used learning algorithm; it consists in the minimization of the Mean-Squared Error (MSE) at the network output performed by means of a gradient descent on the error surface in the space of weights.

The backpropagation algorithm suffers from a number of shortcomings; above all the relatively slow rate of convergence and the final misadjustment that can not guarantee the success of the training procedure in real applications. The choice of a different learning algorithm, based on a different minimization criterion, can help to overcome these drawbacks (see (Azimi-Sadjadi et al., 1992) and (Scalero et al., 1992) for some examples of LS-based fast learning algorithms). The BRLS training algorithm next described allows to obtain considerable improvements from the point of view of both the numerical accuracy and the speed of convergence.

2 Algorithm description

The presence of the non-linearity makes it difficult to apply to multilayer perceptrons a number of techniques so popular in the field of adaptive filtering (Haykin, 1991). A kind of linearization is needed in order to make available to the problem of learning a large number of experimented and efficient algorithms.

The algorithm herein presented is based on the idea of separating each layer of the network in a linear part (the multiplication by the weights) and a non-linear one (the activation functions). Defining the error immediately before the non-linearity allows to use the method of QR Recursive Least Squares (Golub, 1989, and Haykin, 1991) to update the weights of each layer.

In backpropagation algorithm the output error at step n is defined as:

$$E(n) = \sum_p E_p(n) \tag{1}$$

where E_p is the output squared error for the p-th pattern.

The weights are updated by computing the derivatives of E according to the formula:

$$\Delta w_{ij}^{(k)}(n) = - \eta \frac{\partial E(n)}{\partial w_{ij}^{(k)}(n)} \tag{2}$$

where $w_{ij}(k)$ is the weight from the i-th neuron in layer (k-1) to the j-th neuron in layer (k) and η is the learning rate.

The learning rule thus derived is :

$$e_{pj}^{(L)}(n) = f'[y_{pj}^{(L)}(n)] [d_{pj}(n) - x_{pj}^{(L+1)}(n)] \qquad (3)$$

where $e_{pj}(k)$ is the *error signal* for the j-th unit in layer (k) and $x_{pi}(k-1)$ is the output of the i-th unit in layer (k-1), relatively to the p-th input pattern. The *error signal* is computed as

$$e_{pj}^{(L)}(n) = f'[y_{pj}^{(L)}(n)] [d_{pj}(n) - x_{pj}^{(L+1)}(n)] \qquad (4)$$

for the output layer, and as

$$e_{pi}^{(k)}(n) = f'[y_{pi}^{(k)}(n)] \sum_j e_{pj}^{(k+1)}(n) \, w_{ij}^{(k+1)}(n) \qquad (5)$$

for all the other layers. In these formulas d_{pj} is the j-th desired target output for the p-th pattern, $y_{pi}(k)$ is the input to the generic non linearity, $f[]$ is the non-linear activation function (typically the sigmoidal one) and $f'[]$ is its first derivative.

The error signals above defined can be used to form a linear system for each layer of the network; after each presentation of a learning epoch, this system can be solved in the LS sense yielding the optimal weights.

The new algorithm can be formulated in matrix notation in the following way.

Let P be the length of a generic epoch. For each layer we define first the following matrices :

$$\mathbf{X}(n) = \begin{pmatrix} \mathbf{x}_1^T(n) \\ .. \\ \mathbf{x}_P^T(n) \end{pmatrix}; \mathbf{Y}(n) = \begin{pmatrix} \mathbf{y}_1^T(n) \\ .. \\ \mathbf{y}_P^T(n) \end{pmatrix}; \mathbf{E}(n) = \begin{pmatrix} \mathbf{e}_1^T(n) \\ .. \\ \mathbf{e}_P^T(n) \end{pmatrix} \qquad (6)$$

where the layer index has been omitted and n indicates the generic iteration. In these expressions $\mathbf{x}_i^T$, $\mathbf{y}_i^T$ and $\mathbf{e}_i^T$ are the input, output and error row vectors relative to the linear part of the generic layer, for the i-th learning pattern (T indicates the matrix transposition operation). Moreover we indicate with $\mathbf{Q}(n)$ and $\mathbf{R}(n)$ the matrices deriving from the QR decomposition of the system coefficient matrix (Golub, 1989).

The BRLS algorithm consists of the following steps:

1- the weights are randomly initialized;

2- the triangular matrix $\mathbf{R}$ is initialized to $\mathbf{R}(0)=\text{diag}(\varepsilon)$, where ε is a properly chosen small value;

3- each pattern of the current epoch is presented to the network and forward propagated through it; during this phase the matrices $\mathbf{X}(n)$ and $\mathbf{Y}(n)$ for each layer are formed;

4- for each pattern, the output of the network is compared to the desired output; the error signals (4) and (5) at the output of the linear part of each layer are computed. For each layer the perturbation matrix $\mathbf{E}(n)$ is formed ;

5- after the presentation of an entire training epoch, for each layer the following linear system is formed:

$$\begin{pmatrix} \lambda^{1/2} \ \mathbf{R}(n-1) \\ (1-\lambda)^{1/2} \ \mathbf{X}(n) \end{pmatrix} \ \mathbf{W}(n) = \begin{pmatrix} \lambda^{1/2} \ \mathbf{C}(n) \\ (1-\lambda)^{1/2} \ (\ \mathbf{Y}(n) + \eta \ \mathbf{E}(n) \ ) \end{pmatrix} \quad (7)$$

where η is the learning rate (measuring the entity of the perturbation on matrix $\mathbf{Y}$) and λ is the forgetting factor. This system is solved for $n>0$ by performing first a QR decomposition of the coefficient matrix, yielding the matrices $\mathbf{Q}(n)$ and $\mathbf{R}(n)$. In (7) $\mathbf{C}(1)=\mathbf{0}$ while for $n>1$ $\mathbf{C}(n)$ is computed from the relation:

$$\mathbf{Q}^T(n-1) \begin{pmatrix} \lambda^{1/2} \ \mathbf{C}(n-1) \\ (1-\lambda)^{1/2} \ (\mathbf{Y}(n-1) + \eta \ \mathbf{E}(n-1) \ ) \end{pmatrix} = \begin{pmatrix} \mathbf{C}(n) \\ \mathbf{D}(n) \end{pmatrix} \quad (8)$$

Then a procedure of backsubstitution on matrix $\mathbf{R}(n)$ yields the optimal set of weights, in the sense of the minimal 2-norm of the weight solution matrix $\mathbf{W}(n)$;
6- if the global output error with the new weights is still higher than a specified threshold the procedure is repeated from point 3 by appending a new epoch (e.g. the QR decomposition is recursively performed as new data come); otherwise the training has terminated successfully.
The QR decomposition (performed with either the Householder transformation or the Givens rotations) gives to the algorithm stability and robustness from a numerical point of view. In some cases it can be replaced by a Singular Value Decomposition (SVD), yielding a complete control over the internal structure of matrix $\mathbf{X}$ and the regularity of the weight matrix $\mathbf{W}$, at the expenses of a higher computational cost.

3 Experimental results

The performance of the BRLS algorithm have been evaluated in several problems: parity (2,3 and 4 bits), generalized XOR, pattern recognition (circle in a square and character recognition). In all cases comparison with backpropagation has been made on the basis of a proper number of trials with different configurations of initial weights and different values of learning parameters. Main results of this analysis are much faster rate of convergence and higher accuracy of the new algorithm. For instance Fig. 1 reports the MSE as a function of the number of iterations in a typical case for the XOR problem; both the rapidity of convergence (about 30 iterations to get MSE<0.01) and its depth (MSE~10^{-4} after 100 iterations) can be verified.
The algorithm has shown also the ability of forming sharper transition regions, as shown in Fig. 2 referring to circle in a square problem.
The possibility of varying the length P of the training epoch (differently from the algorithms described in (Azimi-Sadjadi et al., 1992) and (Scalero et al., 1992)) is a peculiar feature of BRLS algorithm. Its efficacy is proved by the good performance of the algorithm in problems where the training patterns are totally randomly selected during the learning phase. Moreover, with respect to previous approaches, the numerical stability is enhanced by the fact that the proposed procedure works only on raw data matrices, without forming any correlation matrix.

568

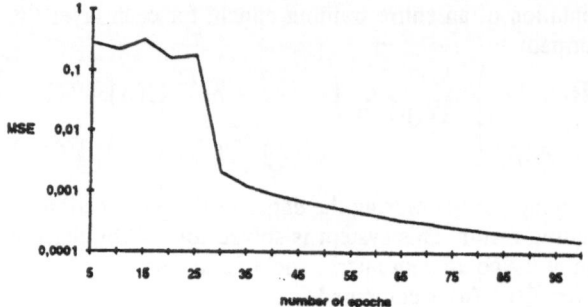

Fig. 1 : MSE versus number of epochs for XOR problem

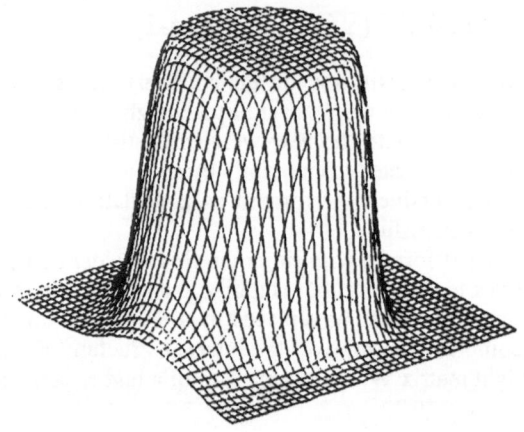

Fig. 2 : 3D output for circle in a square problem

References

Azimi-Sadjadi, M.R., and Liou, R.J., (1992) Fast Learning Process of Multilayer Neural Networks Using Recursive Least Squares Method, IEEE Transactions on signal processing, Vol.40, No. 2, February 1992.

Golub, G.H., and Van Loan, C.F., (1989) Matrix computations, John Hopkins Universiy Press, Second edition.

Haykin, S., (1991) Adaptive filter theory, Prentice Hall.

Rumelhart, D.E., Hinton, G.E., and Williams, R.G., (1986) Learning internal representations by error propagation in Parallel Distributed Processing: Exploration in the Microstructure of Cognition, D.E Rumelhart-J.L. McLelland (Eds.), Vol. 1, MIT Press/Bradford books, Cambridge, MA.

Scalero, R.S., and Tepedelenlioglu, N., (1992) A Fast New Algorithm for Training Feedforward Neural Networks, IEEE Transactions on signal processing, Vol.40, No. 1, January 1992.

Neural Networks for Iterative Computation of Inverse Functions

S. Anoulova

Department of Informatics, University of Dortmund
Dortmund Germany

1   Introduction

We have analyzed mathematically a curious effect, observed by engineers while applying neural networks for robot eye-hand configurations. Namely, Hirzinger et al. in [2] have trained the network to approximate the stabilization control motion, which is the inverse function of the mapping "coordinates" → "image". When used in the iterative routine, the network demonstrated convergence properties , similar to that of standard numerical procedures such as Newton method, for example. At the first glance it seems paradoxal: the iterations schould cause fluctuations about the stabilization point, the amplitude of the fluctuations being bounded by the accuracy of the approximation. We have found the reason of the convergence and propose a general method of training neural networks for iterative computation of inverse functions.

2   Stabilizing a Robot End-Effector: Neural Net in the Iterative Routine

2.1   The description of the problem

The sensomotorik kinematik problem can be essentially stated so: basing on what is "seen", derive correct control commands to move the end-effector to the desired position and orientation. Principally the problem consist in the following. To every state of the system are assigned "coordinates" (for example, euclidean coordinates and rotation parameters of the end-effector) and "visual data" (say, a vector of features extracted from camera image). We observe "visual data" (say, a vector of features extracted from camera image) and must retrieve "coordinates", that is, express "coordinates" as a function of "visual data". After that we can modify any control law, developed for "coordinate" description, for using in "visual data" terms. The direct approach is to calculate the mapping "visual data" → "coordinates" and substitute it into the given control law. This rather involved procedure can be replaced by an empirical approximation of the composition mapping "visual data" → "control movement" using neural networks [2]. Hirzinger et al. [2] have constructed their network for a stabilization problem in an iterative set-up. Their approximation :

- is accurate in a small neighborhood of a stabilization point

- moves the points outside the neighborhood into it

The control process is realized adaptively. The system is in a starting position, the "visual data" is given as input to the network, the output of the network indicates the control movement. After performing the latter, the system is in a new state. The cycle is iterated.

Let c and B denote the vectors of "coordinates" and "visual data" respectively. The control law is obvious: being in a state c move about $-c$ and you are at the point $c = 0$. To implement it, calculate c as a function of B: $c = c(B)$ and set the control law (denoted by M) in terms of B:

$$M = M(B) = -c = -c(B).$$

To approximate M take a set of training data and train a neural network N on this data (with conventional back-propagation algorithm).

Now perform the iterative stabilization procedure. Let the initial point be c^0 and the correspondent image vector $B^0 = B(c^0)$. Define

Procedure

$$
\begin{aligned}
c^1 &= c^0 - N(B^0) \\
B^1 &= B(c^1)
\end{aligned}
$$

$$\ldots$$

$$
\begin{aligned}
c^{n+1} &= c^n - N(B^n) \\
B^{n+1} &= B(c^{n+1}), \quad n = 1, 2, \ldots
\end{aligned}
$$

Question: must $c^n \to 0$ with $n \to \infty$?

The results of [2] give the following

Answer: no; but a rapid convergence (3 iterations suffice) $c^n \to c^\infty$ with c^∞ lying close to 0 was observed.

The following items need consideration:

- What causes the convergence?

- How must be corrected the procedure in order to stabilize finally at $c = 0$?

They will be treated in the following section

2.2 Convergence of the procedure

The phenomenon of convergence can be explained in the following way.

Suppose that the function $c(B)$ is "good", that is, within reasonable areas of B it is almost linear. Then the back-propagation algorithm produces an approximating neural network N which imitates these linear functions. That is, N approximates c together with its derivatives:

$$|c - N|, \qquad \left| \frac{\partial c}{\partial B} - \frac{\partial N}{\partial B} \right|$$

are small[1]. The sequence $c^0, c^1, \ldots$, defined in the Introduction, is in fact obtained by iterations of the following mapping $R: R(c) = c - N(B(C))$:

$$c^n = R^n(c^0), \quad n = 1, 2, \ldots,$$

---

[1] The principal possibility of such approximation is proven in [3]. Naturally, we cannot apply the relevant theorem rigorously, but it makes our reasoning plausible.

where R^n is n-times superposition of R. We "prove" below that R is a contraction, consequently, it has a fixed point c^∞ for which holds $R^n(c^0) \to c^\infty$, or $c^n \to c^\infty$ with $n \to \infty$. For the fixed point $R(c^\infty) = c^\infty$, which implies $N(B^\infty) = 0$ for $B^\infty \triangleq B(c^\infty)$.

Now the "proof". Take two arbitrary values e and d of the coordinate vector c. Then

$$R(e) - R(d) = e - d - \frac{\partial(N \circ B)}{\partial c} \big|_{c=e} \times (e - d) + o(|e - d|).$$

Assumed $\left| \frac{\partial c}{\partial B} - \frac{\partial N}{\partial B} \right|$ is small,

$$\frac{\partial(N \circ B)}{\partial c} \big|_{c=e} = \frac{\partial N}{\partial B}(B(e)) \cdot \frac{\partial B}{\partial c}(e) = I + \varepsilon,$$

where I is the identity matrixd and ε is a "small" matrix. Under reasonable assumptions it implies

$$|R(e) - R(d)| \le \rho|e - d|$$

with $\rho < 1$ independent of e, d, that is, R is a contraction.

To stabilize the system finally in $c = 0$, Hirzinger et al. modify the network N so that $c^\infty = 0$: take a new network $N' = N - N(0)$ and perform the same iterations as earlier with N' instead of N. As $N'(0) = 0$, 0 is a fixed point.

In fact we have to do with Newton method with a constant matrix:

$$\begin{aligned}
c^{n+1} &= c^n - N'(B^n) \\
&= c^n - (N'(B^n) - N'(B(0))) \\
&\approx c^n - \frac{\partial N'}{\partial B}(0)(B^n - B(0)) \\
&= c^n - \frac{\partial N}{\partial B}(0)(B^n - B(0)).
\end{aligned}$$

The matrix $\frac{\partial N}{\partial B}$ arises by itself while training the network. We can improve convergence choosing instead intentiously the matrix $\frac{\partial c}{\partial B} \big|_{B=B(0)}$ (or a certain pseudoinvers, if necessary). That is, take c^0, make 2 iterations c^1, c^2 with the network N and then run the procedure

$$\begin{aligned}
c^{n+1} &= c^n - \frac{\partial c}{\partial B} \big|_{B=B(0)}(B^n - B(0)), \\
B^{n+1} &= B(c^{n+1}), n = 2, 3, \ldots
\end{aligned}$$

Such combination of neural networks and iterative numerical methods is used, for example, in [1, p.443]. In the next section we explain how to build the numerical method into the neural network in order to unify the whole procedure.

3 Heuristic algorithm for training a neural network

We are going to compute the inverse of a one-to-one mapping $B : c \to B(c)$ from a region in Euclidean space into Euclidean space (of equal dimension, but this assumption is not principal). We have to choose heuristically the following values:

- r_{lin} so that $|b - \frac{\partial B}{\partial c}c|$ is small for $|c| \leq r_{lin}$

- r_{exact} so that with the available size of the constructed network N we are able to achieve $|c - N(B)| \leq r_{lin}$ for $|c| \leq r_{exact}$

- a large factor μ to force $|\frac{\partial N}{\partial B} - \frac{\partial c}{\partial B}(B(0))|$ to be small for $|c| \leq r_{lin}$

Generate examples $(c, B(c))$ so that:

- they suffice to calculate $\frac{\partial B}{\partial c}(0)$ (and consequently $\frac{\partial B}{\partial c}(B(0))$

- they fill tightly the region $\{|c| \leq r_{exact}\}$

- they fill sparsely the region $\{|c| > r_{exact}\}$

Set the cost function for training the network:

$$\mu \sum_{|c| \leq r_{lin}} (\frac{\partial c}{\partial B}B - N(B))^2 + \sum_{r_{lin} < |c| \leq r_{exact}} (c - N(B))^2 +$$

$$\sum_{|c| > r_{exact}} (c - N(B))^2 I(|c - N(B)| > r_{exact}).$$

After training make a correction shift: $N \Rightarrow N - N(B(0))$.

Given B^0, we have to calculate the correspondent value c^0 such that $B^0 = B(c^0)$. Perform procedure of section 2.1. According to subsection 2.2 $c^n \to 0$ with $n \to \infty$. As $c^{n+1} = c^0 - \sum_{i=1}^{i=n} N(B^i)$, $n = 0, 1, \ldots$, $c^0 = \sum_{i=1}^{\infty} N(B^i)$.

References

[1] S.Berkovitch et al., Vector Quantization Algorithm for Time Series Prediction and Visuo-Motor Control of Robots, in Verteilte Künstliche Intelligenz und kooperatives Arbeiten, 4. International GI-Kongress Wissenbasierte Systeme, München, October 1991, Proceedings, Springer, 1991.

[2] G.-Q.Wei, G.Hirzinger, Proceedings of the 11 International Conference on Pattern Recognition (IAPR), 1992, v.I, p.189.

[3] H.White et al., Artificial Neural Networks, Blackwell, 1992.

Cascade Correlation convergence theorem

Gian Paolo Drago
Istituto per i Circuiti Elettronici - CNR
via Opera Pia, 11 - 16145 Genova, Italy

Sandro Ridella
DIBE - Universita' di Genova
via Opera Pia, 11A - 16145 Genova, Italy

abstract- In this paper we prove a theorem which guarantees that the Cascade Correlation algorithm has at least a speed of convergence of order $O(1/n_h)$, where n_h is the number of hidden neurons, when approximating a function which is series of sigmoids with a finite number of terms.

1   Introduction

Cascade Correlation (CC) [4] is an incremental algorithm for training MLP: its advantages as an incremental tool for function approximation has been described in [2]. In this paper we prove a theorem which guarantees a speed of convergence of order $O(1/n_h)$, where n_h is the number of hidden neurons, when approximating a function which is series of sigmoids with a finite number of terms. This result is similar to the one obtained by Barron [1] for smooth functions.

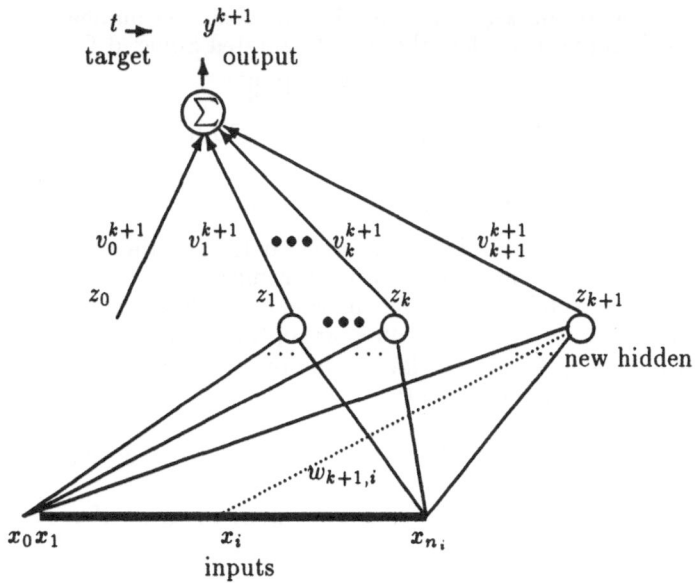

Figure 1: The cascade correlation architecture considered in this paper.

2 The cascade Correlation algorithm: definitions

Let us define $\mathbf{x} = (x_0, x_1, ..., x_{n_i})^T$ the input vector of the neural network and $\mathbf{z}^k = (z_0, z_1, ..., z_k)^T$ the output vector of the neurons of the hidden layer, as shown in Fig.1. Both x_0 and z_0 are equal to 1: they are used for giving a bias to the hidden neurons and the output neuron respectively.

The vector $\mathbf{v}^k = (v_0^k, v_1^k, ..., v_k^k)^T$ represents the weights connecting the hidden neurons to the output neuron. The components of $\mathbf{v}^k$ change when a new hidden neuron is added to the network: the superscript k indicates that the vector $\mathbf{v}^k$ has been modified after the insertion of the k_{th} hidden neuron. The matrix $\mathbf{W}^k = \{w_{ji} : j = 1, k; \ i = 0, n_i\}$ contains the weights connecting the inputs to the hidden neurons. Every hidden neuron is characterized by

$$z_j = \tanh(\sum_{i=0}^{n_i} w_{ji} x_i)$$

while the output neuron is characterized by

$$y^k = \sum_{j=0}^{k} v^k{}_j z_j = \mathbf{v}^{k^T} \mathbf{z}^k$$

The quantity

$$\delta^k = t - y^k \tag{1}$$

is the error at the k_{th} iteration, that is when the number of active hidden neurons is equal to k: then the ISE (Integrated Squared Error) is defined as

$$C^k = \|\delta^k\|^2 \tag{2}$$

Let us define, finally, the covariance relative to the $(k+1)_{th}$ hidden neuron

$$r^{k+1} = \langle \delta^k - \bar{\delta}^k, z_{k+1} - \bar{z}_{k+1} \rangle \quad where: \quad \bar{\delta}^k = \langle \delta^k, 1 \rangle \ ; \quad \bar{z}_{k+1} = \langle z_{k+1}, 1 \rangle \tag{3}$$

Training in CC is performed in two steps: in the first the weights at the inputs of the new hidden unit are found (by maximizing the magnitude of the covariance), while in the second step the weights connecting both the hidden units and the network inputs to the output units are optimized (by minimizing the ISE).

At beginning we have to find v_0^0 in order to minimize C^0. This is obtained by $v_0^0 = \langle t, 1 \rangle$ and, consequently,

$$C^0 = \|t\|^2 - \langle t, 1 \rangle^2 \quad and \quad \bar{\delta}^0 = 0 \tag{4}$$

When C^k is less than a predefined quantity, the algorithm stops to add new hidden neurons, since the value of the ISE is considered satisfactory.

3 The Cascade Correlation convergence theorem

THEOREM: Consider, in a space $\Re^{n_i}$, a function $t \in \Re$ given by

$$t = \sum_{j=0}^{n^*} v^*{}_j z^*{}_j$$

$$z^*{}_j = \tanh(\sum_{i=0}^{n_i} w^*{}_{ji} x_i); \quad ; \quad 1 \le j \le n^* \quad ; \quad z^*{}_0 = 1$$

Than the Cascade Correlation algorithm is able to approximate the function t, with a speed of convergence of order $O(1/n_h)$.

We will give now a sketch of the proof: detailed steps may be found in [3].

The first step of CC algorithm consists of obtaining

$$|r^{k+1}| = \max_{\{w_{k+1,i}: \, i=0,n_i\}} |\langle \delta^k - \bar{\delta}^k, z_{k+1} - \bar{z}_{k+1}\rangle| \tag{5}$$

The second step of CC algorithm consists of obtaining

$$C^{k+1} = \min_{\Delta \mathbf{v}^{k+1}, v_{k+1}^{k+1}} \|\delta^{k+1}\|^2 \tag{6}$$

We start analyzing eq.6 in order to understand how the CC algorithm works: later we will take into account properties of equation eq.5.

The minimum in eq.6 can be found by solving the linear system

$$\langle \delta^{k+1}, \mathbf{z}^k \rangle = 0 \quad , \quad \langle \delta^{k+1}, z_{k+1} \rangle = 0 \tag{7}$$

From equation 7 one can show that:

$$C^{k+1} = C^k - \frac{(r^{k+1})^2}{\|z_{k+1}\|^2 - q^{k+1}} \le C^k - \frac{(r^{k+1})^2}{\|z_{k+1}\|^2} \tag{8}$$

and, since $\|z_{k+1}\|^2 \le 1$,

$$C^{k+1} \le C^k - (r^{k+1})^2 \tag{9}$$

where

$$q^{k+1} = \langle z_{k+1}, \mathbf{z}^{k^T} \rangle (\mathbf{R}^k)^{-1} \langle \mathbf{z}^k, z_{k+1} \rangle > 0$$

having defined

$$\mathbf{R}^k = \langle \mathbf{z}^k, \mathbf{z}^{k^T} \rangle$$

Eq.9 gives a very simple relation among C^{k+1}, C^k and $(r^{k+1})^2$: in order to have $C^{k+1} < C^k$, in eq.9, we must show that, from eq.5, $(r^{k+1})^2$ is not equal to zero. We stress that this property is independent from eq.6.

Following [5] and using the hypothesis of the theorem and eq.1, we consider

$$J = \sum_{j=0}^{n^*} v^*_j \langle \delta^k, z^*_j \rangle = \langle \delta^k, \sum_{j=0}^{n^*} v^*_j z^*_j \rangle = \langle \delta^k, t \rangle = \langle \delta^k, \delta^k + y^k \rangle = \langle \delta^k, \delta^k \rangle + \langle \delta^k, y^k \rangle$$

from which

$$J = C^k \tag{10}$$

since $\langle \delta^k, y^k \rangle = 0$ because y^k is a linear combination of z^k and from eq.7. Then following [5], one obtains the necessary condition:

$$\max_{j=1,n^*} v^*_j \langle \delta^k, z^*_j \rangle \geq \frac{C^k}{n^*} > 0$$

This shows that we are always able to find a $(k+1)_{th}$ hidden unit such that

$$(r^{k+1})^2 > \frac{C^{k^2}}{b_f^{\,2}} \tag{11}$$

where

$$b_f = n^{*2} \left(\max_{j=1,n^*} |v^*_j| \right)^2 \tag{12}$$

We stress that $b_f^{\,2}$ does depend on t only and it is independent from the iteration number k.
From eq.4, 9 and 11

$$C^{k+1} < \frac{b_f^{\,2} C^0}{(k+1)C^0 + b_f^{\,2}} \tag{13}$$

which shows the convergence of order $O(1/k)$ as stated in the theorem.

This result guarantees a very efficient convergence of CC. Computer simulation shows the capability of the implemented CC algorithm to obtain this convergence speed [3].

References

[1] Barron A.R., *IEEE Trans. I.T.*, vol.39, no.3, May 1993, pp.930-945.

[2] Drago G.P. and Ridella S., Proc. of the ICANN '93, pp.750-754.

[3] Drago G.P. and Ridella S., submitted

[4] Fahlman S.E. and Lebiere C., *CMU-CS-90-100*, February 14, 1990.

[5] Jones L.K., *Annals of Statistics*, vol.20, no.1, 1992, pp.608-613.

Optimal weight initialization for neural networks

R. Rojas *

International Computer Science Institute
Berkeley, California 94704

1   Introduction

Apart from the learning algorithm, the most basic point to consider before training a neural network is *where* to start the iterative learning process. This has lead to an analysis of the best possible weight initialization strategies and their effect on the convergence speed of learning [Nguyen, Widrow 89]. We show in this paper that there is a relation between the width of the probability distribution for the weights and the optimal size of the learning step in a neural network. A criterion is provided which allows to calculate the range of weights most appropriate for starting the learning process. It is also shown that different interval sizes are needed for the different layers of weights in a network.

A well known initialization heuristic for a feed-forward network with sigmoidal units is to select its weights with uniform probability from an interval $[-\alpha, \alpha]$. But very small values of α paralyze learning, whereas very large values can lead to saturation of the nodes in the network and to flat zones of the error function in which, again, learning is very slow. Learning stops then at a suboptimal local minimum [Pfister, Rojas 93]. Therefore it is natural to ask what is the best range of values for α in terms of the learning speed of the network.

Some authors have conducted empirical comparisons of the best values for α and have found a range in which convergence is best [Wessels, Barnard 92]. But these results were obtained from a limited set of examples and the relation between learning step and weight initialization was not considered. Others, like [Drago, Ridella 92], have studied the percentage of nodes in a network which become paralyzed during training and have sought to minimize it with the "best" α. Their empirical results show, nevertheless, that there is not a single α which works best, but a very broad range of values with basically the same convergence efficiency. This paper provides a theoretical explanation for these results.

2   Maximizing the derivatives at the nodes

Let us consider first the case of an output node. If n different edges with associated weights $w_1, w_2, \ldots, w_n$ point to this node, then after selecting weights with uniform probability from the interval $[-\alpha, \alpha]$, the expected total input to the node is

$$E(\sum_{i=1}^{n} w_i x_i) = 0$$

---

*on leave from Freie Universität Berlin

where $x_1, x_2, \ldots, x_n$ are the input values transported through each edge. We have assumed that these inputs and the initial weights are not correlated. By the law of large numbers we can also assume that the total input to the node has a gaussian distribution. In this case the expected value $E(s')$ of the sigmoid's derivative becomes

$$E(s') = \frac{1}{\sqrt{\pi}\sigma} \int_{-\infty}^{\infty} \frac{e^{(-x)}}{(1+e^{-x})^2} e^{-x^2/\sigma} dx.$$

The above equation is just the convolution of the derivative of the sigmoid for each possible total input x with the probability distribution of these input values, where σ stands for their variance. Numerical integration shows that the expected value of the derivative is a decreasing function of σ. But the expected value falls slowly with an increase of the variance. For $\sigma = 0$ the expected value is 0.25 and for $\sigma = 4$ it is still 0.12, that is almost half as big.

The variance of the total input to a node is

$$\sigma^2 = E((\sum_{i=1}^{n} w_i x_i)^2) - E((\sum_{i=1}^{n} w_i x_i))^2 = \sum_{i=1}^{n} E(w_i^2) E(x_i^2),$$

since inputs and weights are uncorrelated. For binary vectors we have $E(x_i^2) = 1/3$ and the above equation simplifies to $\sigma = \sqrt{n}\alpha/3$. If $n = 100$, selecting weights randomly from the interval [-1.2, 1.2] leads to a variance of 4 at the input of a node with 100 connections and to an expected value of the derivative equal to 0.12.

Our first conclusion is thus: in small networks, in which the maximal input to each node comes from less than 100 edges, the expected value of the derivative is not very sensitive to the width α of the random interval, when α is small enough. For an interval of width $[-0.5, 0.5]$ the expected value of the derivative is at least 0.195, that is, almost 80% of its maximum possible value.

3 Maximizing the backpropagated error

In order to make corrections to the weights in the first block of weights (weights between input and hidden layer) easier, the backpropagated error should be as large as possible. Very small weights between hidden and output nodes lead to a very small backpropagated error, and this in turn to insufficient corrections to the weights. In a network with m nodes in the hidden layer and k nodes in the output layer, each hidden node h receives a backpropagated input δ_h from the k output nodes, equal to

$$\delta_h = \sum_{i=1}^{k} w_{hi} s_i' \delta_i^0,$$

where the weights $w_{hi}, i = 1, \ldots, k$, are the ones associated with the edges from hidden node h to output node i, s_i' is the sigmoid's derivative at the output node i, and δ_i^0 is the difference between output and target also at this node. After initialization of the network's weights the expected value of δ_h is zero. But in the first phase of learning we are interested in *breaking the symmetry* of the hidden nodes. They should specialize in the recognition of different features of the input. By making the variance of the backpropagated error larger, each hidden node gets a greater chance of pulling apart from its neighbors. The above equation shows, that by making the initialization interval $[-\alpha, \alpha]$ wider, two contradictory forces come into play. On the one hand, the variance of the weights becomes larger, but in the

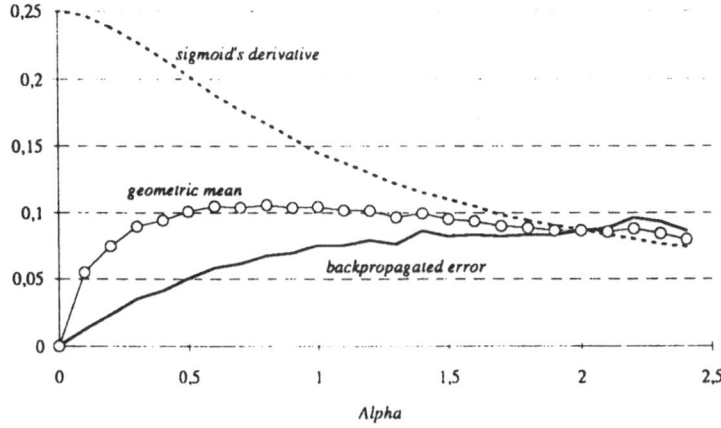

Figure 1: Expected values of the backpropagated error and the sigmoid's derivative

other hand, the expected value of the derivative s'_k becomes lower, as we saw in the last section. We would like to make δ_h as large as possible, but without making s'_i too low, since weight corrections in the second block of weights are proportional to s'_i. Figure 1 shows the expected values of the derivative at the output nodes, the expected value of the backpropagated error for the hidden nodes as a function of α, and the geometric mean of both values. The data in the figure was obtained from Monte Carlo trials, assuming a constant expected value of δ_i^0. Forty hidden and output units were used.

The figure shows, again, that the expected value of the sigmoid's derivative falls slowly with an increasing α, but the value of the backpropagated error is sensitive to small values of α. In the case shown in the figure, any possible choice of α between 0.5 and 1.5 should lead to virtually the same performance. This explains the flat region of possible α values found by [Drago, Ridella 92] and [Wessels, Barnard 92] in their respective experiments.

Our second conclusion is: the best values for α depend on the exact number of input, hidden and output units but the learning algorithm should not be very sensitive to the exact α chosen from a certain range of values.

4 Relation with the learning step

An additional difficulty makes the selection of the initialization interval a hard problem. Gradient descent learning normally employs a single global learning step γ and weight corrections are made according to $\Delta w_i = -\gamma \partial E / \partial w_i$, where E is the error function. We saw from Figure 1 that the backpropagated error can be optimized at around 0.1, whereas the sigmoid's derivative can be kept at around the same value for some choices of α. But since corrections to the weights between input and hidden layer are proportional to the product of the sigmoid's derivative at the hidden node with the backpropagated error, this means that in the best case the backpropagated error up to a weight in the first block is around 0.01. This is one order of magnitude smaller that the backpropagated error used for the corrections of the weights between hidden and output layer. In this case the same learning step γ produces much smaller corrections in one block of weights as in the other. The risky path is then increasing γ in order to *shake* the network faster. But this can lead to excessive corrections at the second block of weights and to oscillations in weight space.

A better alternative to alleviate this problem could be to use different initialization intervals for the two blocks of weights in the network. A larger α in the first layer of weights would make swifter adjustments of these weights easier, together with adequately sized adjustments in the second block of weights. Another possibility that should be studied empirically is to initialize the weights in the second block with a zero mean, but in such a way that the weights coming from each hidden node have a mean value different from zero. This would help to increase the variance of the backpropagated error without decreasing the expected value of the sigmoid's derivative.

Our third conclusion is thus: weight initialization is not independent of the learning strategy used afterwards. Authors who have addressed the problem of optimal weight initialization have failed to show the connection between their initialization strategies and learning methods.

5   Conclusions

We have shown in this paper that the empirical results of some authors regarding the existence of a range of widths for the initialization interval for which learning works better can be explained by considering the expected magnitude of the sigmoid's derivative at the nodes and the magnitude of the backpropagated error. This kind of criterion shows that the initialization interval normally used $[-0.5, 0.5]$ or $[-1, 1]$ performs adequately for small networks with less than 100 connections per node. The value of α scales inversely proportional to $\sqrt{n}$. We have also shown that the optimal initialization can not be separated from the kind of learning algorithm used. Since one of the main problems at the beginning of the learning phase is the discrepancy between the magnitudes of the backpropagated errors for each block of weights, this provides further evidence that algorithms using a different learning rate for each weight should perform better than algorithms which use a global learning rate. This has been observed in practice [Schiffmann et al. 93]. We have suggested also a differential initialization strategy in which different sets of weights are initialized with different variances. In the case of hidden nodes, the mean value of the weights associated with the edges coming out of each node could be made different from zero, without affecting the mean zero value of the weights connected to each output node. This would improve symmetry breaking when learning starts.

References

[1] G. P. Drago and S. Ridella, "Statistically Controlled Activation Weight Initialization (SCAWI)", *IEEE Transactions on Neural Networks*, vol. 3, no. 4, 1992, pp. 627-631.

[2] D. Nguyen and B. Widrow, "Improving the learning speed of 2-layer neural networks by choosing initial values of adaptive weights", *IJCNN*, 1989, pp. III-21-26.

[3] M. Pfister, R. Rojas, "Speeding-up Backpropagation – A Comparison of Orthogonal Techniques", *IJCNN*, Nagoya, 1993.

[4] W. Schiffmann, M. Joost and R. Werner, "Comparison of Optimized Backpropagation Algorithms", in: M. Verleysen (ed.), *European Symposium on Artificial Neural Networks*, Brussels, 1993, pp. 97-104.

[5] L. Wessels and E. Barnard, "Avoiding False Local Minima by Proper Initialization of Connections", *IEEE Transactions on Neural Networks*, vol. 3, no. 6, 1992, pp. 899-905.

Neural Nets with Superlinear VC-Dimension

(Extended Abstract)

Wolfgang Maass

Institute for Theoretical Computer Science, Technische Universitaet Graz
Klosterwiesgasse 32/2, A-8010 Graz, Austria; e-mail: maass@igi.tu-graz.ac.at

The Vapnik-Chervonenkis dimension VC-dimension($\mathcal{N}$) of a neural net $\mathcal{N}$ with n input nodes is defined as the size of the largest set $S \subseteq \mathbf{R}^n$ which is "shattered" by $\mathcal{N}$ in the sense that every function $F : S \to \{0,1\}$ can be computed by $\mathcal{N}$ with some assignment of real numbers to its weights.

The VC-dimension of a neural net $\mathcal{N}$ is an important measure for the expressiveness for $\mathcal{N}$, i.e. for the variety of functions that can be computed by $\mathcal{N}$ with different choices for its weights. In particular it has been shown in Blumer et al. (1987) and Ehrenfeucht et al. (1989) that the VC-dimension of $\mathcal{N}$ essentially determines the number of training examples that are needed to train $\mathcal{N}$ in Valiant's model (Valiant, 1984) for probably approximately correct learning ("PAC-learning").

It has been known for quite a while that the VC-dimension of a neural net with linear threshold gates and w edges (respectively w weights) is at most $O(w \cdot \log w)$. This result, which holds for arbitrary real valued input patterns, was first shown by Cover (1964 and 1968) and later by Baum and Haussler (1989). It has frequently been conjectured that the "true" upper bound is $O(w)$. This conjecture is quite plausible, since a single linear threshold gate with w edges has VC-dimension $w + 1$. Furthermore it is hard to imagine that the VC-dimension of a network of linear threshold gates can be larger than the sum of the VC-dimensions of the individual linear threshold gates in the network.

We disprove this popular conjecture by showing that for any depth $d \geq 3$ quite a number of neural nets $\mathcal{N}$ of depth d have a VC-dimension that is superlinear in the number w of edges in $\mathcal{N}$. In particular, we exhibit for arbitrarily large $w \in \mathbf{N}$ neural nets $\mathcal{N}$ of depth 3 (i.e. with 2 hidden layers) with w weights that have VC-dimension $\Omega(w \cdot \log w)$. This shows that the quoted upper bound of $O(w \log w)$ is in fact asymptotically optimal. Our lower bound also shows that the well-known upper bound $2w \log(eN)$ for the VC-dimension of a neural net with w weights and N computation nodes (due to Baum and Haussler, 1989) is asymptotically optimal.

The result of this paper may also be viewed as mathematical evidence for a certain type of "connectionism thesis": that a network of neuron-like elements is more than just the sum of its elements. We show that in a large neural net a single edge may add more than a constant to the VC-dimension of the neural net: its contribution may increase with the logarithm of the total size of the neural net. The proof of this result relies on a new method that allows us to encode more "program-bits" in the weights of a neural net than previously thought possible. The same result can also be derived for neural nets with other activation functions such as $\sigma(y) = \frac{1}{1+e^{-y}}$.

The proof of our result employs classical circuit construction methods due to Neciporuk (1964) and Lupanov (1972). We refer to Maass (1993) for details of the proofs. Bartlett (1993) has independently derived lower bounds for the VC-dimensions of various neural nets of depth 2 and 3 that are *linear* in the number w of weights.

The *neural nets* which are considered in this paper are feedforward neural nets with linear threshold gates (or simpler: threshold gates), i.e. gates which apply the heaviside activation function to the weighted sum $\sum_{i=1}^{m} \alpha_i y_i + \alpha_0$ of their inputs $y_1, \ldots, y_m$. The parameters $\alpha_1, \ldots, \alpha_m$ and α_0 are the *weights* of such gate. We will consider in this paper only neural nets with boolean inputs and one boolean output.

The *depth* of a neural net is the length of the longest path from an input node to the output node (= output gate). The depth of a gate in a neural net is the length of the longest path from an input node to that gate. We refer to all gates of depth d as "level d" of a neural net.

We will focus our attention on neural nets that are *layered* in the sense that only gates on successive levels are connected by an edge, and input nodes are connected by an edge only with gates on level 1. This is not a serious restriction, since for $i + 1 < j$ one may replace an edge between nodes on levels i and j by a path of length $j - i$ (by introducing $j - i - 1$ "dummy" gates on the intermediate levels). It is obvious that a layered neural net of depth d has exactly $d - 1$ *hidden layers*. One calls a layered neural net *fully connected* if any two nodes on successive levels (as well as any input nodes and any gates on level 1) are connected by an edge.

We use the standard notation $f = \Theta(g)$ for arbitrary functions $f, g : \mathbf{N} \to \mathbf{N}$ to indicate that both $f = O(g)$ and $f = \Omega(g)$.

Theorem 1: *Assume that $(\mathcal{N}_n)_{n \in N}$ is any sequence of fully connected layered neural nets of depth $d \geq 3$. Furthermore assume that $\mathcal{N}_n$ has n input nodes and $\Theta(n)$ gates, of which $\Omega(n)$ gates are on the first hidden layer, and at least $4 \log n$ gates are on the second hidden layer of $\mathcal{N}_n$.*
Then $\mathcal{N}_n$ has $\Theta(n^2)$ edges and VC-dimension$(\mathcal{N}_n) = \Theta(n^2 \log n)$.

The *proof* of Theorem 1 proceeds by "embedding" into the given neural nets $\mathcal{N}_n$ of Theorem 1 for suitable $\tilde{n} \leq n$ the special neural nets $\mathcal{M}_{\tilde{n}}$ that are constructed in the following Theorem 2.

Theorem 2: *Assume that n is some arbitrary power of 2. Then one can construct a neural net $\mathcal{M}_n$ of depth 3 with n input nodes and at most $17n^2$ edges such that VC-dimension$(\mathcal{M}_n) \geq n^2 \cdot \log n$.*

Sketch of the Proof of Theorem 2: We assume that n is of the form $2^{n'}$ for some non-zero $n' \in \mathbf{N}$. This implies that n is even and that $\log n \in \mathbf{N}$. We construct a neural net $\mathcal{M}_n$ with $2n + \log n$ binary inputs and $O(n^2)$ weights that shatters the following set $S \subseteq \{0,1\}^{2n+\log n}$ of size $n^2 \cdot \log n$:

$$S := \{\underline{e}_p \underline{e}_q \tilde{\underline{e}}_m : p, q \in \{1, \ldots, n\} \text{ and } m \in \{1, \ldots, \log n\}\},$$

where $\underline{e}_r \in \{0,1\}^n$ denotes the r-th unit vector of length n $(r = 1, \ldots, n)$, and $\tilde{\underline{e}}_m \in \{0,1\}^{\log n}$ denotes the m-th unit vector of length $\log n$ $(m = 1, \ldots, \log n)$. Thus each $\underline{u} \in S$ contains exactly three 1's.

Fix any map $F : S \rightarrow \{0,1\}$. One can encode F by a function $g : \{\underline{e}_1, \ldots, \underline{e}_n\}^2 \rightarrow \{0,1\}^{\log n}$ where the m-th output bit of $g\left(\underline{e}_p, \underline{e}_q\right)$ equals 1 if and only if $F\left(\underline{e}_p \underline{e}_q \tilde{\underline{e}}_m\right) = 1$. One can show that for any function $g : \{\underline{e}_1, \ldots, \underline{e}_n\}^2 \rightarrow \{0,1\}^{\log n}$ there exist for $k = 1, \ldots, 4$ functions $g_k : \{\underline{e}_1, \ldots, \underline{e}_n\}^2 \rightarrow \{0,1\}^{\log n}$ such that $g_k\left(\cdot, \underline{e}_q\right)$ is $1 - 1$ for every fixed $q \in \{1, \ldots, n\}$, and such that for all $p, q \in \{1, \ldots, n\}$:

$$g\left(\underline{e}_p, \underline{e}_q\right) = \begin{cases} g_1\left(\underline{e}_p, \underline{e}_q\right) \oplus g_2\left(\underline{e}_p, \underline{e}_q\right), & \text{if } p \le \frac{n}{2} \\ g_3\left(\underline{e}_p, \underline{e}_q\right) \oplus g_4\left(\underline{e}_p, \underline{e}_q\right), & \text{if } p > \frac{n}{2}. \end{cases}$$

The symbol $\oplus$ denotes here the bit-wise exclusive OR (i.e. parity) on bit-strings of length $\log n$. In the following we write $(\underline{x})_j$ for the j-th coordinate of some vector $\underline{x}$, and $\mathrm{bin}(i)$ for the binary string that represents the number $i - 1$.

The neural net $\mathcal{M}_n$ computes F in the following way. The output gate on level 3 is an OR of $4 \log n$ threshold gates. These threshold gates consist of $\log n$ blocks of 4 threshold gates, such that for any $b \in \{1, \ldots, \log n\}$ some threshold gate in the b-th block outputs 1 for network input $\underline{e}_p \underline{e}_q \tilde{\underline{e}}_m$ if and only if $m = b$ and $\left(g\left(\underline{e}_p, \underline{e}_q\right)\right)_b = 1$ (i.e. $F\left(\underline{e}_p \underline{e}_q \tilde{\underline{e}}_m\right) = 1$). More precisely, the a-th threshold gate in block b outputs 1 if and only if the a-th one of the following 4 conditions is satisfied:

$$
\begin{array}{llllll}
(1) & m = b & \wedge & p \le \frac{n}{2} & \wedge & \left(g_1\left(\underline{e}_p, \underline{e}_q\right)\right)_b = 1 & \wedge & \left(g_2\left(\underline{e}_p, \underline{e}_q\right)\right)_b = 0 \\
(2) & m = b & \wedge & p \le \frac{n}{2} & \wedge & \left(g_1\left(\underline{e}_p, \underline{e}_q\right)\right)_b = 0 & \wedge & \left(g_2\left(\underline{e}_p, \underline{e}_q\right)\right)_b = 1 \\
(3) & m = b & \wedge & p > \frac{n}{2} & \wedge & \left(g_3\left(\underline{e}_p, \underline{e}_q\right)\right)_b = 1 & \wedge & \left(g_4\left(\underline{e}_p, \underline{e}_q\right)\right)_b = 0 \\
(4) & m = b & \wedge & p > \frac{n}{2} & \wedge & \left(g_3\left(\underline{e}_p, \underline{e}_q\right)\right)_b = 0 & \wedge & \left(g_4\left(\underline{e}_p, \underline{e}_q\right)\right)_b = 1.
\end{array}
$$

The subconditions involving $g_1, \ldots, g_4$ are tested with the help of $8n$ threshold gates on level 1 that use weights $w_{k,i,j} \in \{1, \ldots, n\}$, which are defined by the condition $w_{k,i,j} = r \Leftrightarrow g_k\left(\underline{e}_r, \underline{e}_j\right) = \mathrm{bin}(i)$. There are $4n$ threshold gates $G_{k,i}^+\left(\underline{e}_p, \underline{e}_q\right)$ on level 1 that output 1 if and only if $\sum_{r=1}^n r \cdot \left(\underline{e}_p\right)_r \ge \sum_{j=1}^n w_{k,i,j} \cdot \left(\underline{e}_q\right)_j$ (i.e. $p \ge w_{k,i,q}$), and $4n$ threshold gates $G_{k,i}^-\left(\underline{e}_p, \underline{e}_q\right)$ on level 1 that output 1 if and only if $\sum_{r=1}^n r \cdot \left(\underline{e}_p\right)_r \le \sum_{j=1}^n w_{k,i,j} \cdot \left(\underline{e}_q\right)_j$ (i.e. $p \le w_{k,i,q}$); for $k = 1, \ldots, 4$ and $i = 1, \ldots, n$. These are the only weights in the neural net $\mathcal{M}_n$ which depend on the function $F : S \rightarrow \{0,1\}$. By definition one has that for each k, i at least one of the two gates $G_{k,i}^+\left(\underline{e}_p, \underline{e}_q\right)$, $G_{k,i}^-\left(\underline{e}_p, \underline{e}_q\right)$ outputs 1. Furthermore for any $k \in \{1, \ldots, 4\}$ and any $\underline{e}_p \underline{e}_q \tilde{\underline{e}}_m \in S$ there is exactly one $i \in \{1, \ldots, n\}$ such that both of these gates output 1. This index i is characterized by the equality $g_k\left(\underline{e}_p, \underline{e}_q\right) = \mathrm{bin}(i)$. Hence one can check whether $\left(g_k\left(\underline{e}_p, \underline{e}_q\right)\right)_b = 1$ by testing

584

whether $\sum_{\substack{i=1,\ldots,n \\ \text{with } (\text{bin}(i))_b=1}} G_{k,i}^+ \left(\underline{e}_p, \underline{e}_q \right) + G_{k,i}^- \left(\underline{e}_p, \underline{e}_q \right) \geq \frac{n}{2} + 1$, and one can check

whether $\left(g_k \left(\underline{e}_p, \underline{e}_q \right) \right)_b = 0$ by testing whether $\sum_{\substack{i=1,\ldots,n \\ \text{with } (\text{bin}(i))_b=0}} G_{k,i}^+ \left(\underline{e}_p, \underline{e}_q \right) +$

$G_{k,i}^- \left(\underline{e}_p, \underline{e}_q \right) \geq \frac{n}{2} + 1$. Furthermore the sums on the left hand side of both inequalities can only assume the values $\frac{n}{2}$ or $\frac{n}{2} + 1$. Therefore one can test the AND of two subconditions of this type and of the subconditions "$m = b$" and "$p \leq \frac{n}{2}$" ("$p > \frac{n}{2}$") by a *single* threshold gate on level 2 of $\mathcal{M}_n$. Hence one can test each of the conditions $(1), \ldots, (4)$ by a separate threshold gate in the b-th block on level 2.

Altogether the constructed layered neural net $\mathcal{M}_n$ consists of $2n + \log n$ input nodes, $8n + 8\log n + 3$ computation nodes, and $16n^2 + (8\log n + 2)n + 16\log n$ edges. Obviously the nodes and edges of $\mathcal{M}_n$ are independent of the given function $F : S \to \{0,1\}$, whereas the weights on $8n^2$ of the edges depend on F (these weights range over $\{1, \ldots, n\}$).

Since the function $F : S \to \{0,1\}$ that is computed by $\mathcal{M}_n$ was choosen arbitrarily, the construction implies that the set S is shattered by $\mathcal{M}_n$. Hence VC-dimension$(\mathcal{M}_n) \geq |S| = n^2 \cdot \log n$. We refer to (Maass, 1993) for details of this proof. ∎

References

Bartlett, P. L. (1993). *Lower bounds on the Vapnik-Chervonenkis dimension of multi-layer threshold networks*, Proc. of the 6th Annual ACM Conference on Computational Learning Theory, 144 - 150

Baum, E. B., Haussler, D. (1989). *What size net gives valid generalization?*, Neural Computation, **1**, 151 - 160

Blumer, A., Ehrenfeucht, A., Haussler, D., Warmuth, M. K. (1989). *Learnability and the Vapnik-Chervonenkis dimension*, J. of the ACM, **36**(4), 929 - 965

Cover, T. M. (1964). *Geometrical and statistical properties of linear threshold devices*, Stanford PH. D. Thesis 1964, Stanford SEL Technical Report No. 6107-1, May 1964

Cover, T. M. (1968). *Capacity problems for linear machines*, in: Pattern Recognition, L. Kanal ed., Thompson Book Co., 283 - 289

Ehrenfeucht, A., Haussler, D., Kearns, M., Valiant, L. (1989). *A general lower bound on the number of examples needed for learning*, Information and Computation, **82**, 247 - 261

Lupanov, O. B. (1972). *On circuits of threshold elements*, Dokl. Akad. Nauk SSSR, **202**, 1288 - 1291; engl. transl. in: Sov. Phys. Dokl., **17**, 91 - 93

Maass W. (1993). *Neural nets with superlinear VC-dimension*, IIG-Report 366 of the Technische Universität Graz, June 1993

Neciporuk, E. I. (1964). *The synthesis of networks from threshold elements*, Probl. Kibern. No. 11, 49 - 62; engl. transl. in: Autom. Expr., **7**(1), 35 - 39

Valiant, L. G. (1984). *A theory of the learnable*, Comm. of the ACM, **27**, 1134 - 1142

A Randomised Distributed Primer for the Updating Control of Anonymous ANNs

Antonio Calabrese[†], Felipe M. G. França[‡,§]

†Istituto di Cibernetica CNR, I-80072 Arco Felice, Italy
‡Neural Systems Engineering, Imperial College, London SW7 2BT, England

1. Introduction

The computational potentiality of ultra-large numbers of processing units working in parallel may be nullified if proper co-ordination mechanisms do not exist. Surprisingly, most simulations and implementations of ANN models under discrete time (digital-based hardware and software in conventional and parallel computers) do not necessarily have built-in distributed forms of controlling the updating of their neuronal units. In general, these simulation/implementation strategies are tailored to particular classes or topologies of neural paradigms and do not present explicit scalability, i.e. whether or not the strategy is prepared to cope with leaps in orders of magnitude in the number of neural units. In this sense, the implications of associating a distinct identification to each processing unit are definitely very strong if a massive number of these units are considered. Differentiated by this critical aspect, *anonymous* distributed mechanisms have also an intrinsic biological plausibility appeal in the context of ultra-large ANNs.

Barbosa's *scheduling by edge reversal* — SER — (Barbosa and Gafni, 89) (Barbosa, 93) is a powerful and fully distributed algorithm that has been applied to the simulation of ANN models in multiprocessor computers (Barbosa and Lima, 90) (Barbosa, 93) (França, 93). SER can work on anonymous networks (as well as on named networks) as a distributed synchronizer.

SER's only requirement for proper operation is the definition of an initial direct acyclic orientation (DAO) on the oriented graph $G(N, E)$, or simply G, representing neural units $n_i \in N$ and its connections $e_i \in E$. A straightforward analogy between SER and dynamic attractors (limit cycles) can be made. Interestingly, the problem of finding any initial acyclic orientation on G is much more complex than SER itself. In fact, to find any acyclic orientation which gives a SER period of optimal concurrency was proved to be a NP-complete problem (Barbosa and Gafni, 89). This work introduces an alternative probabilistic distributed algorithm which works as a SER *primer*, i.e. puts any anonymous network into a random SER dynamics. Its complexity in time is $O(N)$ and its communication complexity is $O(E)$.

2. SER

Let G be a connected unidirected graph representing a general discrete system composed of a finite number of processing units (PUs) where each node in G corresponds to one PU. A node is linked to another by only one oriented edge whenever both PUs share an atomic resource (only one PU can possess it at a time). These two nodes are called *neighbours*. PUs are always looking for all resources shared with neighbours to *operate*. Starting from any acyclic orientation ω in G

§ On leave from COPPE-Sistemas, Universidade Federal do Rio de Janeiro. Supported by grant n° 1400/88-5 from CAPES, Ministry of Education, Brazil.

there is at least one node that has all its edges directed to itself, called *sink*. Only sinks are allowed to operate while other nodes remain *idle*. After operation a sink reverses the orientation of its edges by sending a message to all its neighbours, becoming a *source*. Again another acyclic orientation is formed and new sinks can operate. A cyclic sequence of period length p, $p \geq 2$ is established. During one such period each node becomes a sink m times, $1 \leq m \leq p/2$, guaranteeing that neither starvation nor deadlock will ever occur (Barbosa and Gafni, 89). Figures 1(a) and 1(b) illustrate two possible SER dynamics for a single network topology.

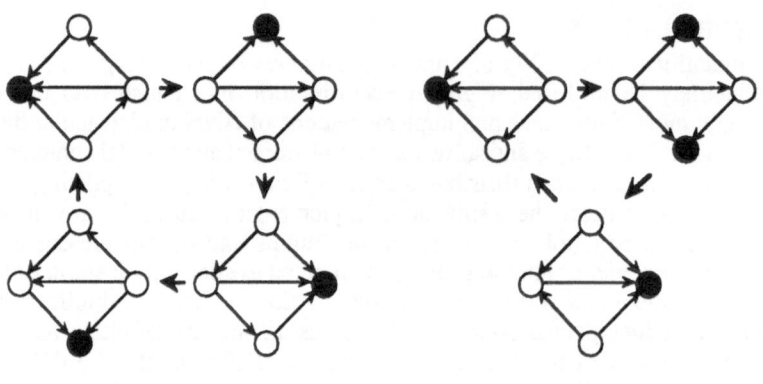

Figure 1. (a) $p = 4, m = 1$.   (b) $p = 3, m = 1$.

The amount of concurrency over synchronous time is given by $\gamma_0(\omega) = m/p$. Notice that m and p are highly dependent upon the initial ω (compare figures 1(a) and 1(b)) and on G's connectivity; sparse graphs will tend to provide greater concurrence than dense ones. However, the problem of finding $\gamma^*(G)$, i.e., to find an initial ω that maximises concurrence, is a *NP*-complete problem. Formal proofs for all SER properties can be found in (Barbosa, 93).

3. Finding a Randomised Acyclic Orientation

The scheduling mechanism proposed by Barbosa and Gafni requires an initial acyclic orientation onto the network to work properly. Such a network can be viewed as (or modelled with) an undirect graph U, in which nodes represent the processing units (PUs) and arcs represents the communication lines. PUs are all supposed to execute the same identical algorithm (i.e. PUs are "symmetric" during the computation) and no topological information ("sense of direction", "topological awareness") is available to nodes. The definition of an acyclic orientation onto such a graph implies on breaking the initial symmetry of the network. Actually, any acyclic orientation leads to a substantial diversification of the nodes (sinks, sources, intermediate nodes) and of the arcs (a well-defined sense of direction).

In the case of networks composed of processing units each having a distinct name, an initial acyclic orientation can be easily obtained by a deterministic distributed algorithm (Barbosa and Lima, 90). However, the initial symmetry of anonymous networks cannot be "broken" by any deterministic algorithm hence, any solution must rely on probabilistic mechanisms (Angluin, 80). Figure 2 shows what is, to our knowledge, the first distributed algorithm for imposing a DAO onto anonymous networks of arbitrary topology which has a well-defined proof of convergence (Calabrese, 93). Although it has been experimentally demonstrated that the *self-organising oscillatory network* (SOON) model (França, 91) is able to obtain solutions inside bounded classes (by a predefined minimum amount of concurrency), it does not have yet a formal proof of convergence.

```
Algorithm  SER_Primer:        (* at each node n_i *)
    Variables:
        active_i , oriented_i , SER_deterministic_i : boolean;
        neighbours_i : [active_j , oriented_j , SER_deterministic_j ];
        All_Neighbours_i : array [1..n] of neighbours_i ;        (* n_i hasn
    Initial  Input:                                               neighbours*)
        SER_deterministic_i := false;
        With All_Neighbours_i do oriented_j := false;
    Action at synchronous pulse:
        If SER_deterministic_i then
        begin
            If with All_Neighbours_i all oriented_j := true then
                                (* n_i  is a sink *)
            begin
                operate;
                reverse_edges;
            end
        else  begin
            active_i: = local_probability_i;
            If (active_i and (in All_Neighbours_i every not
                    SER_deterministic_j is not active_j ) then
            begin
                SER_deterministic_j : = true;
(* n_i  is a sink *)  With All_Neighbours_i do all oriented_j := false;
            end
            Else (not active_i and (in All_Neighbours_i every not
                    SER_deterministic_j is active_j ) then
            begin
                SER_deterministic_j : = true;
                With All_Neighbours_i do all oriented_j := true;   (* n_i is
            end                                        a source *)
        end.
```

Figure 2. The probabilistic distributed primer.

The algorithm shown in Figure 2 is a synchronous implementation of the distributed meta-algorithm for the creation of a DAO on a general graph proposed in (Calabrese, 93). Its convergence mechanism is based upon the principle of each PU reaching, randomly, a differentiated state from all its neighbours (e.g. a single PU *active* whilst all of its neighbours *inactive*). At this point such PU come out of the probabilistic process putting itself in the sink or source state and start to operate according to the deterministic SER dynamics. By the progressive auto-elimination from the random process of differentiation, a network configuration operating according to the SER dynamics is finally reached.

The algorithm is fully distributed and relies only on information locally available to each of the system's PUs, i.e. each PU can determine its state exclusively by using locally defined operations and messages exchanged with its neighbours. Notice also that the locality of the algorithm allows the operation according to the SER dynamics to start before the final acyclic orientation have been achieved. Furthermore, the algorithm can accommodate dynamic changes to the network topology , i.e. addition or deletion of nodes and edges, because a coherent acyclic orientation can still be maintained whenever a sink or a source (together with its associated edges) is added or deleted to the original directed acyclic graph.

In assessing the computational complexity, a relevant factor is which probability distribution is assigned to define the activation scheme of each processor. If the probability distribution proposed in (Braca and Calabrese, 93) is adopted, both the time and the message complexity of the algorithm are linear with the size of the problem (number of nodes or of edges in the network). It must be noted that in such a case no hypothesis about the dynamic capabilities of the algorithm can be made since the given distribution takes into account the initial size of the network in determining the probability of activation of each node. In order to fully exploit the dynamic capabilities of the algorithm, different probability distributions are currently under investigation, which can better exploit the locality property it possesses.

4. Conclusions

As massive parallel and distributed processing is a major appeal of neurocomputing, the investigation of updating strategies for the effective use of multiple quasi identical units working in parallel constitutes a very attractive subject. In this sense, fault-tolerance is also a strong appeal. SER is a potentially optimal distributed updating mechanism in terms of concurrency and localised control. This work introduced a new fully distributed algorithm: a SER primer for anonymous networks which, working together with SER itself, is not only able to control the updating of ultra-large numbers of artificial neural units, but also to accommodate the loss or addition of new nodes and connections to the system.

References

Angluin D. (1980). Local and global properties in networks of processors. *Proceedings of 12th ACM Symposium on Theory of Computing*. Los Angeles.

Barbosa V.C. (1993) *Massively Parallel Models of Computation: Distributed Parallel Processing in Artificial Intelligence and Optimization*. Ellis Horwood.

Barbosa V.C. and Gafni E. (1989). Concurrency in heavily loaded neighborhood-constrained systems. *ACM Trans. on Prog. Lang. and Systems* 11(4), 562-584.

Barbosa V.C. and Lima P.M.V. (1990). On the distributed parallel simulation of Hopfield's neural networks. *Software, Practice and Experience*, 20(10), 967-983.

França F.M.G. (1991). A self-organising updating network. In T. Kohonen, K. Mäkisara, O. Simula, and J. Kangas (Eds.), *Artificial Neural Networks* (vol. 2, pp. 1349-1352). North-Holland: Elsevier Science.

França, F.M.G. (1993). Scheduling weightless systems with self-timed Boolean networks. *Workshop on Weightless Neural Networks*. York, pp. 87-92.

Braca A. and Calabrese A.(1992). Leader election on a synchronous anonymous network with a known number of arcs. In S. Tzafestas, P. Borne and L. Grandinetti (Eds), *Parallel and Distributed Computing in Engineering Systems* (pp 425-428). North-Holland: Elsevier Science. IMACS.

Calabrese A. (1993). Un algoritmo distribuito per l'orientazione aciclica di reti anonime. IC-RI 101/93. Istituto di Cibernetica-CNR.

CATASTROPHIC INTERFERENCE IN LEARNING PROCESSES BY NEURAL NETWORKS

E.Pessa†, M.P.Penna†
†Dipartimento di Psicologia
Università di Roma "La Sapienza"
Roma, Italy

1 Introduction

Starting from the Eighties , there has been a fluorishing of models of memorization and learning processes , based on neural networks . As shown by McCloskey e Cohen (1989) , and Ratcliff (1990) , the ones characterized by a multilayer feedforward architecture and the supervised back-propagation learning rule are plagued by the so-called *catastrophic interference* problem .This latter arises when , after a network learned a certain number of items belonging to a suitable learning set , the same network is submitted to a second learning process with a new learning set . The simulations show that this latter learning destroys the previous one , i.e. the performance relative to the items belonging to the first learning set becomes very low (and even null). This cirumstance is particularly dangerous for the development of connectionist paradigm , first all because , also if interference phenomena appear in information recall processes by human beings , they cannot be defined as catastophic , and , secondarily , because , when we want to add new knowledge to a network which yet learned a previous one , we are forced to repeat again all learning process , relative to the entire knowledge base , the old and the new one. This makes impratical and very heavy the process of adding new information to a neural network through the backpropagation rule.

When comparing the behaviour of feedforward networks with the human one , one sees that the catastrophic interference is due to the fact that , in neural networks , we have interference processes both in the storage and in the recall phase , whereas the experimental data on human subjects suggest that in human beings interference takes place only in the recall phase . This can be inferred (see, e.g., Murnane and Shiffrin, 1991) from the existence of a positive lenght effect (i.e. the performance becomes worse with increasing lenght of the item list to be recalled) and from the absence of a strenght effect (i.e. recall performance is nearly the same both for "strong" items and for "weak" items). These circumstances suggest that a competitive learning network , with bottom-up and top-down connections, should be a better candidate than a multilayer perceptron for modeling human recall performance and for avoiding catastrophic interference problem. In this paper we will present some results about performance of a particular model network belonging to this category, i.e. Carpenter and Grossberg's ART 2 network. We obtained that this model is free from catastrophic interference and , at the same time, its performance bears some resemblance with the human one . It appears , thus , suited both for modeling human memory behaviour and for building artificial learning systems.

2. ART 2 model

The model ART 2 has been built by Carpenter and Grossberg (1988) and is described in a detailed way in literature. Here we remember briefly its basic equations : the model contains an input network , a categorization layer and a comparison network. The input network is made by three different layers : bottom, middle, and top .The dynamics of the model can be explicited as follows :

Bottom layer

$$w_i = I_i + a u_i \; , \; x_i = w_i/(e + |w|) \qquad (i = 1 ,...., M) \tag{1}$$

$$|w| = (S_i \; w_i^2)^{1/2}$$

Middle layer

$$v_i = f(x_i) + b f(q_i) \; , \; u_i = v_i/(e + |v|) \tag{2}$$

$$f(x) = 0 \; \text{if} \; 0 \le x \le \theta \; \text{and} \; f(x) = x \; \text{if} \; x \ge \theta$$

Top layer

$$p_i = u_i + \Sigma_j g(y_j) z_{ji} \tag{3}$$

$g(y_j) = d$ if the unit j of the categorization layer is the winner one
$g(y_j)=0$ otherwise

categorization layer
It obeys a WTA law , where the winner unit , whose output is 1 , is the one associated to the greater activation potential.

Comparison network

$$r_i = (u_i + c p_i)/(e + |u| + |c p|) \tag{4}$$

If it happens that

$$\rho /(e + |r|) > 1 \; , \; \rho = \text{vigilance parameter} \tag{5}$$

then the winner unit is reset .Otherwise learning law applies to bottom-up and top-down connections.

Bottom-up Learning Law

$$z_{ij} (t + 1) = z_{ij}(t) + g(y_j)[p_i - z_{ij}(t)] \tag{6}$$

$$z_{ji}(t+1) = z_{ji}(t) + g(y_i)[p_i - z_{ji}(t)] \qquad (7)$$

3 The simulation

We simulated the following experimental situation : a subject learns a first list of items , the study list , and then it is confronted with a new list of items , the test list . The task is to determine , for every item of the test list , if it was contained or not also in the study list. The "strenght" of every item of the study list is given by the number of times it appears in this list .If there is catastrophic interference ,i.e. interference in the storage phase , it would appear trough interference of every item of the study list on the others of the same list.

In the simulation the items were random binary vectors each one with 10 components . A first simulation with 10 items per list , by using normal backpropagation , with a succession of trainings relative to the single items , gave immediately catastrophic interference , yet after the training of the second item . We then used an ART 2 network whose behaviour was subdivided in two phases : the first one of ordinary training according to (1)-(7), corresponding to the learning of the study list, and the second in which reset operation triggered by comparison network was forbidden, corresponding to the recognition task relative to the test list . We interpreted , in this phase , the situation in which (5) was not satisfied as equivalent to the recognition of the input pattern as belonging also to the study list . The parameters used have been a=10 , b=10 , c=0.1 , d=0.9 , e=0.1 , θ=0.2 , ρ=0.98 .We varied the lenght of both lists , from 5 to 30 items , by doing 10 simulations for every lenght and obtaining a clear indication of a positive lenght effect (the correlation coefficient between the mean percent of correct recognitions and list lenght was -0.67).

To reveal an eventual strenght effect we built 4 types of study lists, each one with 20 items : a pure weak one (every item presented for 6 times) , a pure strong one (every item presented for 30 times), an ordered mixed list (the first 10 items presented for 6 times , and the others for 30 times), a random mixed list (as in the ordered mixed , but with the patterns presented in random order). For every study list, we built 5 different types of test lists, characterized by a different old items/new items ratio. Because we had 10 different versions of every type of test list , we did a total of 200 simulations. The network recognition performance with the various items has been measured through the so-called discriminability d, obtained from raw data through the Receiver Operating Characteristics (ROC) Curves .

If there is a catastrophic interference effect, then the performance on "weak" items should be better in the pure weak (d_{pw}) than in mixed lists (d_{mw}), whereas the performance on "strong" ones should be better in mixed lists (d_{ms}) than in pure strong ones (d_{ps}). As a consequence the value of the ratio :

$$R = (d_{ms}/d_{ps})(d_{pw}/d_{mw}) \qquad (8)$$

should be very greater than 1 .The numerical value obtained from our simulations has been R= 0.31 when dealing with random mixed lists , and R_0 = 0.99 when dealing with ordered mixed lists. We must conclude that there is no evidence of a catastrophic interference effect in ART 2 network .As a comparison the value obtained from experiments on human subjects is R=1.21 (Ratcliff, Sheu and Gronlund, 1992) .It is , thus , evident that ART 2 network is a very good candidate for modeling the behaviour of human memory .

References

Carpenter, G.A., and Grossberg, S. (1987) ART 2 : Self-organization of stable category recognition codes for analog input patterns . Applied Optics , **26** , 4919-4930 .

Mc Closkey , M., and Cohen , N.J. (1990) . Catastrophic interference in connectionist networks : The sequential learning problem .In G. Bower (Ed.) , The psychology of learning and motivation (vol.24 , pp. 109-165) . S. Diego , CA : Academic Press.

Murnane,K. and Shiffrin R.M. (1991) . Interference and the representation of events in memory. Journal of Experimental Psychology : Learning , Memory , and Cognition ,**17** , 855-874 .

Ratcliff , R. (1990) .Connectionist models of recognition memory : Constraints imposed by learning and forgetting functions. Psychological Review , **97**, 285-308 .

Ratcliff , R. , Sheu , C. and Gronlund , S.D. (1992) . Testing global memory models using ROC curves. Psychological Review, **99**, 518-535 .

Systematicity in IH-analysis

Dan Lundh

Department of Computer Science, University of Skoevde
Skoevde, Sweden

1 Introduction

AI researchers can agree that any non-trivial cognitive system must be able
to represent complex structured items. A complex structured item is typically
seen to be built up systematically from parts drawn from a determinate set.
The process by which such parts are combined to form complex structured
items is called the mode of composition.

The aim of this paper is to describe research focused on the equal treatment
of parts from the determinate set in a complex structure, i.e. systematicity
(Fodor and Pylyshyn, 1988). The paper is organized in the following way. First,
a hyperplane analysis (input to hidden unit weight analysis, IH) is presented
mathematically. Secondly, a refinement to the hyperplane analysis is presented
to show the systematicity. Finaly, some results are presented according to the
proposed analysis.

2 Hyperplane Analysis

A hyperplane (or decision line) is a line that partitions the hidden layer units
(by means of a threshold) such that the hidden layer function enables the output
[1]. That is, a mapping where the input x $\epsilon\{0,1\}^n$ maps to an output o $\epsilon\{0,1\}^m$
(where n and m are the number of input and output units, repectively), o =
f(x).

If the network consists of one hidden layer, the function (f) is a composition
of the functions $g_1(x) = h$ and $g_2(h) = o$, where g_1 corresponds to the input
to hidden layer function and g_2 the hidden layer to output function. It can be
showed that f(x) = $g_2(g_1(x))$ = o.

The partition that is done in a input to hidden weight analysis separates the
outcome of $g_1(x) = h$ (i.e. it separates the hidden values h) by means of a
threshold for each hidden unit j. The formula that is used for the threshold is

$$\theta_j = \sum_{i=1}^{n} w_{ij} + w_{bj} \tag{1}$$

where θ_j is the threshold for hidden unit j and w_{bj} is the weight from the bias
unit. This threshold (or decision line) is simply the point where the activation
"decides" to go towards a minimum (negative) or a maximum (positive) activation. This threshold is compared to the activation of the hidden layer units,

---

[1]That is, if the training has been successful

594

which is equal to the outcome of function $g_1(x)$ [2]

$$h_j = \sum_i (w_{ij} \cdot x_i) + w_{bj} \cdot x_b \qquad (2)$$

It can be showed that equation (2) is a linear combination [3] of all possible inputs (i.e. the set $\{0,1\}^n$). Notice that only some are used as input representations.

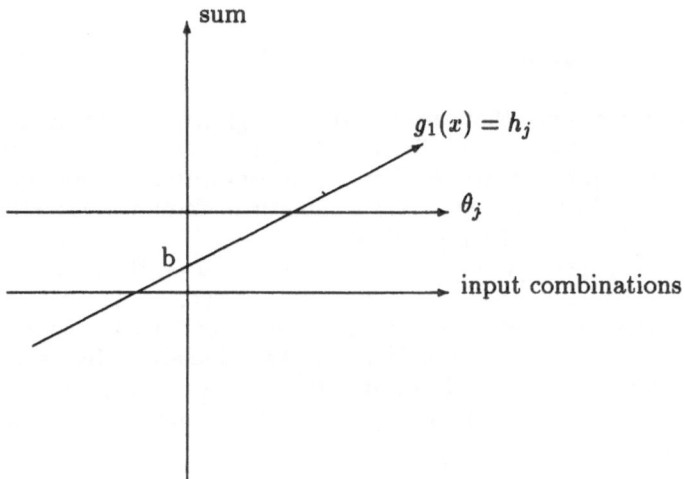

In the above situation b denotes the crossing with the Y axis which corresponds to the bias weight and θ_j denotes the descision line for the jth unit. Notice that the above situation is only used to demonstrate one type of relationships that can occur. In fact, the decision line could in reality be placed on anywhere on the Y axis. In the above picture, the input dimension lie on the X axis.

2.1 Limitations of Hyperplane Analysis (IH)

The first limitation of the hyperplane analysis (IH) is that it does not say anything about the network output. This is since the decision line can lie any where on the Y axis.

The second is if the training set contain feature vectors with non used representation bits (i.e. zeros). The problem lies in that the decision line is affected by weights connected to those representation bits while the input activation is not affected. The first limitation can be avoided by including Euclidian distance and the second can be avoided by having a good representation.

2.2 Systematicity

Systematicity is the term used by Fodor and Pylyshyn (1988) to denote the systematic treatment of different parts in complex structured items [4]. That is,

---

[2] The transfer function, e.g. a sigmoidal function is not accounted for.
[3] The sum is linearly independent of $g_1(x)$.
[4] This is not explicitly said in their paper, but it is the interpretation made by the author.

if a system is able to handle one type of sentence then the system also must be able to handle similar sentences in a equal mannor (e.g. Bo Relation-1 Object-1 then the system must handle Bo Relation-2 Object-2 in a similar mannor) [5].

In the hyperplane analysis (IH) this would be evident if a certain part of a complex structured item was partitioned equally for all inputs containing the part.

3 Hyperplane Analysis - revisited

If the hyperplane analysis is used on a correctly trained network, a clustering can be found among the input combinations (i.e. input combinations that have equally sums lie closer to each other on the X axis). That is, input combinations that have similar features are clustered together. The clustering that occurs among the input combinations are of course varying in the different hidden units (since this activation must enable the output in the mapping adopted). This means that the partitioning that occur within some (one or more) hidden units coincide with the input semantic (i.e. the semantics subscribed to each feature vector). This coincidence of semantics and partitioning is also named group thesis (Lundh, 1993).

The different partitionings that coincide with the semantics are creating hyperplane groups (thereby the name group thesis). To calculate the solidity of the hyperplane group(s), the variety formula has the means for providing this measure. That is, the variety is minimized if the input combinations used are tightly clustered. The formula used for this variety measure is

$$W(\varphi) = (\sum_i d_i^2)^{\frac{1}{2}} \tag{3}$$

$$d_i^2 = \frac{N \sum_j (sc_j)^2 - (\sum_j (sc_j))^2}{N(N-1)} \tag{4}$$

where d_i denotes hidden unit variety, sc (semantic contribution) denots the distance between the activation and the decision line (perpendicular) and φ is the group variety.

3.1 Systematicity - revisited

The systematicity can be found in some of the hidden unit partitionings (group of hidden units). But, since the partitioning in itself does not provide a measure of similarity, the need for the variety measure is obvious. That is, a variety measure close to zero has the input combinations close to each other in several hidden units.

4 Results

A feedforward network, consisting of 15-8-15 units, was trained with back-propagation to encode some sentences of the form Agent Act Object. In the

---

[5]It is the ability of the system that is of issue here, not whether thoughts or expressions are true.

representation there was no unique representation within a feature (the input region has several overlapping areas). The representation was as follows:

Types	Representation	
Agents:	Bill	1 0 0 0 1
	Sue	0 0 1 0 1
	John	1 0 0 1 0
	Mary	0 0 1 1 0
Acts:	hit	1 1 0 0
	move	0 0 1 1
	push	0 1 0 1
	throw	1 0 1 0
Objects:	chair	0 1 0 1 0 0
	glass	1 0 0 1 0 0
	plate	1 0 0 0 1 0
	table	0 1 0 0 1 0

Table 1: The Representation used.

This representation created the following hyperplane groups. The index of χ_j^p denotes hidden unit number, p+ denotes that the patterns are above the threshold (and p− denotes that the patterns are below the threshold).

"Semantics"	Group vs Variety	
Bill Y Z	$\chi_3^{p+} \cap \chi_5^{p-}$	2.642969
John Y Z	$\chi_3^{p-} \cap \chi_5^{p-}$	2.642963
Mary Y Z	$\chi_3^{p-} \cap \chi_5^{p+}$	2.642971
Sue Y Z	$\chi_3^{p+} \cap \chi_5^{p+}$	2.642967
X hit Z	$\chi_1^{p-} \cap \chi_4^{p-}$	2.194551
X move Z	$\chi_1^{p+} \cap \chi_4^{p+}$	2.194544
X push Z	$\chi_1^{p-} \cap \chi_4^{p+}$	2.194548
X throw Z	$\chi_1^{p+} \cap \chi_4^{p-}$	2.194534
X Y chair	$\chi_6^{p+} \cap \chi_7^{p-}$	6.371460
X Y glass	χ_8^{p+}	0.835898
X Y plate	$\chi_6^{p-} \cap \chi_7^{p+}$	6.371460
X Y table	χ_8^{p-}	0.835898

Table 2: The results of the hyperplane groups

In the above tavle it is shown that the intersection of some hidden units provide a small variety. This also allows other relationships among internal representation to emerge (e.g. x move table) without recourse to decomposition (the variety for this supergroup is $W(\varphi) = 16.833862$).

5 References

Fodor, J.A., Pylyshyn, Z.W. (1988) Connectionism and Cognitive Architecture: A Critical Analysis, Cognition 28.

Lundh, D. (1993) Pushing the Limits: An Analysis of Variable Binding with Hyperplane Groups, Master Dissertation MD-93-4, University of Skoevde.

Integrating Distance Measure and Inner Product Neurons

F. Mana, D. Albesano, R. Gemello
CSELT - Centro Studi E Laboratori Telecomunicazioni
Via G. Reiss Romoli, 274 - 10148 TORINO (ITALY)
Tel: 39-11-2286258; Fax: 39-11-2286207; Email: mana@cselt.stet.it

1. Introduction

Neural networks are systems based on simple computational elements (units) massively connected by weighted links. The unit performs a measure of similarity between its input vector $x=(x_1,..,x_n)$ and its weight vector $w_l=(w_{i1},..,w_{in})$. In literature there exist two main ways to compute this similarity: the former is the inner product (IP) of the vector x and the vector w_l (related to the projection of x on w_l) while the latter is based on some distance measure (DM) between the x and w_l vectors. This work continues the activity reported in [Gemello,91] and [Gemello,92] devoted to introduce hypersphere units, a kind of units based on DM similarity, and to describe their capability to improve the rejection of extraneous patterns. The focus of this paper is to extend the previous DM unit in order to create an hyperelliptical shaped decision region and to perform feature selection. The architecture we propose, the Closed Decision Region MLP net (CDRMLP), integrates both IP and DM unit, that in the following we call hyperellipse unit (HE), into the same architecture. We show how a gradient descent approach can be successfully used to train this network. The kind of unit we propose is not new: there already exist in the literature some connessionist approaches based on distance metric such as Learning Vector Quantization [Kohonen,92], Restriced Coulomb Energy [Cooper,87], Probabilistic Neural Networks [Specht,92], Radial Basic Functions [Musavi,92]. The differences rely on the learning strategy (gradient descent), the kind of computation, and the feature selection capability. The interest about this kind of unit is increased by the fact that other research fields, such as Hidden Markov Models [Singer,92] and Fuzzy Systems [Dickerson,93], use similar computational elements.

2. The Hyperellipse Definition

The activity o_i of the hyperellipse unit is ruled by the following equations:

$$net_i = \sum_j \left(w_{ij} - x_j \right)^2 \Big/ \sigma_{ij}^2 + bias_i \tag{2.1}$$

$$a_i = f(net_i, T_i) = 1 - \frac{1}{1 + e^{-\left(net_i / T_i + \alpha_i T_i \right)}} \tag{2.2}$$

$$o_i = a_i * range + (0.5 - range/2) \tag{2.3}$$

The decision region generated by this unit in a bi-dimensional pattern space is shown in Fig 2.1. As you can see, it is a closed decision region with an hyperelliptical shape. The position of the hyperellipse depends on the weights vector w_l, its area is controlled by the parameters $bias_i$ and T_i, while the variances vector $\sigma_l = (\sigma_{i1},..., \sigma_{in})$ determine its orientation along the x axes. At the first glance, it is clear how this unit exhibits a cluster-oriented way of working: it fires 1 only when an input pattern falls inside the hyperelliptical region. Looking at the HE unit definition in more detail, the net input (2.1) is the weighted Euclidean distance between the weights vector w_l and the input pattern $x = (x_1, .. ,x_n)$ - each component x_j is weighted by σ_{ij}. The net value is passed to the reversed sigmoid function (2.2). The factor $\alpha_i T_i$ shifts the reversed sigmoid in order to outputs about 1.0 (0.98) when the input pattern x agrees with the weights vector w_i, that is $net_i=0$. The α_i parameter is determined by the expression:

$$\alpha_i = -(1/T_i)\ln\left(1/(1-o_i)-1\right).$$

As much as the net value increases the activation a_i decreases depending on the parameter T_i. Bigger values of T_i cause a slowly decrease whereas lower values can approximate the step activation function. The bias value, controlling the magnitude of the net value, is able to shift the activation function along the net-axis. In the following we are going to call receptive field the region where the unit activation is about 1.0, and influence field the region where a_i ranges between 0.0 and 1.0. The output function (2.3), determines the value that the unit sends to the other units connected to it and it depends on the range value ($0<range\le1$). If range takes value 1.0 then $o_i=a_i$, otherwise the range represents the interval of values, centered on 0.5, that o_i can assume. The output function makes possible mitigate the influence of the position of the weights vector w_i in the pattern space: if the pattern under analisys is very away from w_i the unit can outputs a value near 0.5, for example, despite the activation a_i is 0.0.

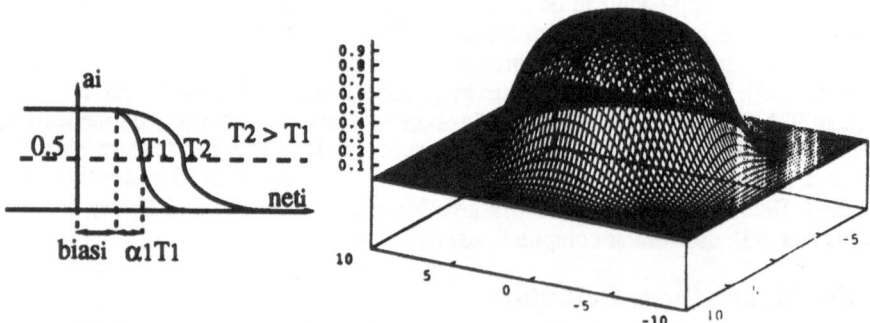

Fig 2.1 - The output function of the HE unit and its decision region.

3. The CDRMLP Architecture

The HE unit can be seen as a simple computational element and used together other ones of the same or different kind to build a network which connect all of them in same architecture. It can be placed everywhere in the network, however in the following we are going to focus toward a network made up by three or more layers: an input layer, a first hidden layer made up by HE units only, and one or more layers of IP units. Using hyperellipses on the first hidden layer instead of hyperplanes, there is a different approach with which the decision regions are built. Whereas hyperplanes are arranged in order to fit decision boundaries among classes, hyperellipses must be arranged in order to cover the pattern space of the classes. The next IP layers are used to combine simple decision regions to accomplish a more complex decision region by means of and-or operators. The final decision regions are naturally enforced to be closed: when an input pattern do not belong to any class on which the network was trained, it will fall outside any receptive fields and the output layer outputs zero. For this reason, we call it Closed Decision Regions Multy Layer Perceptron (CDRMLP).

4. The Learning Rules

When a HE unit is involved into the network, training means find the most proper value for the position of the hyperellipse, its receptive and influence fields amplitude, and the orientation of the hyperellipse in order to solve a particular task. It is possible to use a descent gradient approach [Rumelhart, 86] to train a HE unit involved at any layer into the architecture. This result is attractive because it is possible to train a network which integrates both kind of neurons using a well known tecnique. Before to focus our attention about the derivation of the learning rules for a HE unit, we face the problem present in training multy layer networks called credit assignment problem: how much hidden units partecipate in producing the errors? A solution to this problem is to realize a back propagation of the output error through the network. In literature this blame is addressed as δ_i and formally is computed by the derivative of the output error E respect each unit input net_i. The computation of δ_i depends on the kind of the unit u_i, if it is an output or hidden unit, and from the kind of unit u_k is connected to. In the next, for sake of semplicity, we give the δ_i for the CDRMLP architectures; however, the derivability of all functions involved in the forward computation assure that for any architecture can be easly computed the right δ_i expression.

Thus for the output units we have:

$$u_i \equiv IP \Rightarrow \delta_i = f'\left(net_i\right)\left(t_i - o_i\right) \tag{4.1}$$

and for hidden units connect to IP units:

$$u_i \equiv IP \Rightarrow \delta_i = f'\left(net_i\right)\left(t_i - o_i\right) \tag{4.2}$$

$$u_i \equiv HE \Rightarrow \delta_i = range \; f'\left(net_i\right)\sum_k \delta_k w_{ki} \tag{4.3}$$

where in all the equations $f()$ is the activation function of the unit u_i.

Faced the problem of determining the blame of each unit involved in the network, let us now move back our attention toward the original problem: which learning rules must be used to train a HE unit in order to reduce its blame? Our solution uses a gradient descent approach to solve it. In accordance with this framework, we have to minimize the output error function E with respect the hyperellipse position, orientation, receptive and influence fields amplitude, and so we have the following learning rules:

$$\Delta w_{ij} = -\eta \frac{\partial E}{\partial w_{ij}} = \eta \delta_i \frac{2}{\sigma_{ij}^2}\left(w_{ij} - x_j\right) \tag{4.4}$$

$$\Delta \sigma_i = -\eta \frac{\partial E}{\partial \sigma_i} = -\eta \delta_i \frac{2}{\sigma_{ij}^3}\left(w_{ij} - x_j\right)^2 \tag{4.5}$$

$$\Delta bias_i = -\eta \frac{\partial E}{\partial bias_i} = \eta \delta_i \tag{4.6}$$

$$\Delta T_i = -\eta \frac{\partial E}{\partial T_i} = \eta \left(-\delta_i \frac{net_i}{T_i^2} + \delta_i \alpha_i T_i\right) \tag{4.7}$$

5. The Feature Selection Problem

Looking at the net definition (2.1), it is possible to see that does not exist any value of the weight which assures that the associated feature does not influence the net value, whereas a null weight for a IP unit causes no contribution to the net input. In order to give the feature selection capability to HE unit, we are going to extend its definition. The new definition we propose uses the previous ones (2.2)(2.3) jointly to the following new equation for the net computation:

$$net_{ij} = \left(w_{ij} - x_j\right)^2 \Big/ \sigma_{ij}^2 \tag{5.1}$$

$$net_i = \sum_j r_j * net_{ij} + bias_i \tag{5.2}$$

The net input is still based on the weighted Euclidean measure between the weight vector w_i and the input one x, but now at each feature is assigned a relevance r_j in its contribution in the net input value (5.2). In such a way, the influence of the feature x_j can be omitted by a null value of its relevance r_j. The net definition (5.1) (5.2) is similar to that proposed in [Franzini,92]. During the training phase, it is possible learn automatically the magnitude of each parameter r_j obtaining the feature selection capability we are looking for. Once again, using a gradient descent approach:

$$\Delta r_j = -\partial E / \partial r_j = \delta_i net_{ij} \tag{5.4}$$

where δ_i is the backpropagated error discussed in the section 4. Since the neuron definition is a little bit different from the previous one introduced in the section 2, the learning rules of the HE with feature selection must be recomputed.

REFERENCES

[Cooper,87] - L.N. Cooper et al., "Learning System Architectursa Composed of Multiple Learning Modules". IEEE First International Conference on Neural Networks - June 1987.

[Dickerson, 93] - J.A. Dickerson and B. Kosko, "Fuzzy Function Approximation with Supervised Ellipsoidal Learning". Proc. WCNN 93 Portland OR - July 1993 - pp. II9-II17.

[Franzini,92] - M. Franzini, "The TARGET Architecture: A Feature-Oriented Approach to Connectionist Word Spotting". Proc. IJCNN Baltimora 92 - June 1992 - pp. I761-I768.

[Gemello, 91] - R. Gemello and F. Mana, "An Enhancement to MLP Model to Enforce Closed Decision Regions". Proc. IJCNN Singapore, November 1991, pp. 729-733.

[Gemello, 92] - R. Gemello and F. Mana, "Improving Rejection in Neural Network Classifiers: An Application to Isolated Word Recognition". Proc. ICANN 92 Brighton, September 1992, pp. 775-778.

[Kohonen,92] - T. Kohonen et ali, "LVQ_PAK: A Program Package for the Correct Application of Learning Vector Quantization Algorithms". Proc. IJCNN 92 Baltimora - June 1992 - pp. 725-730.

[Musavi,92] - M.T. Musavi et ali, "On the Training of Radial Basic Function Classifiers". Neural Networks, Vol. 5, 1992, pp. 595-603.

[Rumelhart, 86] - D.E. Rumelhart and R.J. McClelland, "Learning Internal Representations by Error Propagations". In Parallel Distributed Processing, D.E. Rumelhart and R.J. McClelland, editors, Vol 1 Fundutions, pp 318-362, MIT press, 1986

[Singer, 92] - E. Singer and P. Lippmann, "A Speech Recognizer Using Radial Basic Function Neural Networks in a HMM Framework". Proc. ICASSP-92 San Francisco - March 1992 - Volume I - Pages 629-632.

[Specht,92] - D. F. Specht, "Enhancements to Probabilistic Neual Networks". Proc. IJCNN Baltimora 92 - June 1992 - pp. I761-I768.

Teaching by Showing in Kendama Based on Optimization Principle

Mitsuo Kawato[†], Francesca Gandolfo[‡],
Hiroaki Gomi[†], Yasuhiro Wada[*]

† ATR Human Information Processing Res. Lab., Kyoto Japan
‡ Department of Brain and Cognitive Sciences, MIT, Boston USA
* Systems Laboratory, Kawasaki Steel Corporation, Chiba Japan

1 Representations for Task-Level Learning

Much progress has been made in the past decade regarding computational understanding of motor learning both in neuroscience and robotics. Our recent interests have been drawn to higher-level task learning rather than simple trajectory following. Learning algorithms such as reinforcement learning and genetic algorithms can be efficiently used for task-level learning if adequate representations of the task are selected. However, Schaal, Atkeson and Botros (1992) pointed out that this selection is the most difficult and critical part of motor learning and if one assumes the pre-existence of proper representations, this amounts to abandoning plans to tackle a major part of the problem. In this paper, based on the dynamic optimization theory for trajectory formation, we propose a general computational theory that derives representations for a wide variety of motor behaviors.

When humans learn novel motor skills, they often do so through teaching by showing. People watch professional athletes, sisters, brothers or their parents who demonstrate new motor behaviors. They perceive these movement patterns, extract essential features, and try to perform the movements by themselves. In many cases, it is useless or at least not a meaningful ultimate goal for the learner to accurately imitate the position and/or force trajectory demonstrated by the teacher. Indiscriminate imitation could be an efficient strategy in an ideal world where the following conditions are all satisfied. First, the teacher and the learner have exactly the same kinematic and dynamic properties in their bodies. Second, all the position and force data are measured with infinite precision with an infinitely high sampling rate. Third, the measured data are transferred from the teacher to the learner without any delay with an infinite bandwidth. Fourth, the environment of the teacher and the learner is exactly the same. Finally, the learner can perfectly control the body according to the transferred data. However, in almost all realistic situations in human life, robotics, and teleoperation, all or part of the above conditions are violated. In this case, a more abstract understanding of the teacher's motor behaviors at a higher level is essential. In another extreme ideal case, if the learner had an infinite power of intelligence, he would be able to understand the will or motor intention of the teacher by movement perception, and would be able to translate this information into a stream of actual motor commands by taking account of laws of physics describing the mechanics of bodies and environments of the teacher and learner. Traditional AI-style robotics has long been aiming at this goal with little success. The main reason for this difficulty has been an astounding gap between natural or computer languages describing the task and actual, physical motor commands which must be issued.

Thus, we believe that representations for motor primitives or languages describing motor behaviors and tasks should neither be AI-type natural/computer languages nor mere descriptions of position and force trajectory. The representations should take account of the dynamics of the controlled object and the external world as well as computational principles adopted by the CNS in motor control. The via-points extracted from movement trajectories based on a dynamic optimization principle are attractive candidates for this representation. To examine their potential power, we control a robot arm to execute Japanese Kendama using via-points extracted from a human demonstration.

2 Motor Theory of Movement Perception based on Dynamic Optimization

One invariant feature of human multi-joint arm movements is that the hand paths between two points are roughly straight, and the hand-speed profiles are bell-shaped (Morasso, 1981). Uno, Kawato and Suzuki (1989) extended the purely kinematic minimum-jerk model and proposed a dynamic optimization model (minimum-torque-change model) to account for a wider range of human behaviors, including the above invariant features and also to introduce versatility which is frequently exhibited in dynamic interactions between the body and external world. Recently, Wada and Kawato (1993) developed FIRM (the Forward-Inverse Relaxation Model) for optimal trajectory generation and control. FIRM can generate an arm trajectory within a few iterations. It contains both the forward dynamics model and the inverse dynamics model of the controlled object.

In reaching movements, the location of the start and end points and the specified movement duration provide the two-point boundary conditions for the optimization problem where the performance index is the time integral of the sum of the squares of the rate of change of torques. The nonlinear dynamics of the arm governs the relationship between the control variable (joint torques) and the trajectory. The subject can pass through specified via-points at the optimal times. According to Marr's three-level understandings of the brain function (Marr, 1982) we can summarize our study of reaching movements as follows: (1) The computational theory of reaching is the minimum-torque-change model. (2) Representation of the task is given as the start, via- and end points and the movement duration. (3) Algorithm and hardware are FIRM.

We hypothesized that this computational framework should be valid for other classes of voluntary movements such as handwriting or speech. For this extension of the theory, the above first and third levels are easily transferred, but the representation level needs careful consideration. For speech, we believe that each phoneme determines the via-point target location. For handwriting, we developed an algorithm to extract the minimum number of via-points from a given trajectory X_{data} with some level of error threshold θ. If a fixed number of via-points $S = \{P_1, P_2, \cdots, P_N\}$ are given and the arm dynamics is known, we can calculate the optimal trajectory $X_{opt}(S)$ passing through these via points. The above problem is to find the minimum value of N giving a trajectory which satisfies $\|X_{data} - X_{opt}(\mathcal{S})\| < \theta$. Note that this via-point extraction problem is again a nonlinear optimization problem. Our via-point extraction algorithm uses FIRM again and this suggests a duality between movement pattern formation and movement pattern perception (see Wada et al., 1994 for details).

We succeeded in reconstructing a cursive handwriting trajectory quite accurately from about 10 via-points for each character. The extracted via-points included not only kinematically definable feature points with maximum curvature and lowest velocity but also other points not easily extracted by any purely kinematic method which does not take account of the dynamics of the arm or the dynamic optimization principle. A simple word recognition system from cursive connected handwritings was constructed based on this via-point representation and it worked without a word dictionary. When the same algorithm was applied to speech articulator motion during natural speech, the extracted via-points corresponded fairly well to phonemes that were determined from a simultaneously recorded acoustic signal. Natural speech movement was reconstructed well from those phoneme-like via-points (Wada et al., 1994).

In the motor-theory of speech perception (Liberman et al., 1967), a neural network for motor control is supposed to play an essential role in the perception of speech. Our algorithm gives one specific computational realization of this psychological theory. Human movement data, either visual (e.g. handwriting, biological motion) or auditory (e.g. speech) is very severely constrained by the dynamics of the controlled objects and interactions with the external world as well as the motor control strategy adopted by the CNS. Thus, any efficient motor-pattern perception scheme must either implicitly or explicitly take account of these physical and physiological constraints. Here, we advocate a rather radical stand-point by stating that the movement-pattern generation network actively participates in movement-pattern perception in a dualistic way. The proposed theory of movement pattern perception based on dynamic optimization can be used for motion capture, movement pattern recognition (handwriting or speech), telecommunication and teleoperation, and teaching by showing in robotics.

3 Computational Study of Kendama

A Japanese Kendama consists of two parts connected by a thin string: a ball with a hole and a cross-like object with three cups of different sizes and a stick opposite the handle. In the initial condition, a player holds the handle with the ball connected to the string hanging down. The player makes an upward handle movement yanking the ball to fly over the handle. After a 0.5 sec flight the ball is hopefully caught in one of the cups or by the stick. Thus, there are several different ways to play Japanese Kendama. We studied the easiest one (the largest cup and ball) and the second hardest one (the stick and ball) in behavioral studies and only the former for robot control.

We measured the hand, elbow, shoulder and sometimes the ball position using the OPTOTRAK system. The effects of different-size Kendamas, external loads and individual differences were studied (Gandolfo et al., in preparation). Figure 1 compares human trajectories (thin curves) and fairly accurately reconstructed trajectories (thick curves) from 9 via-points (*). The via-points

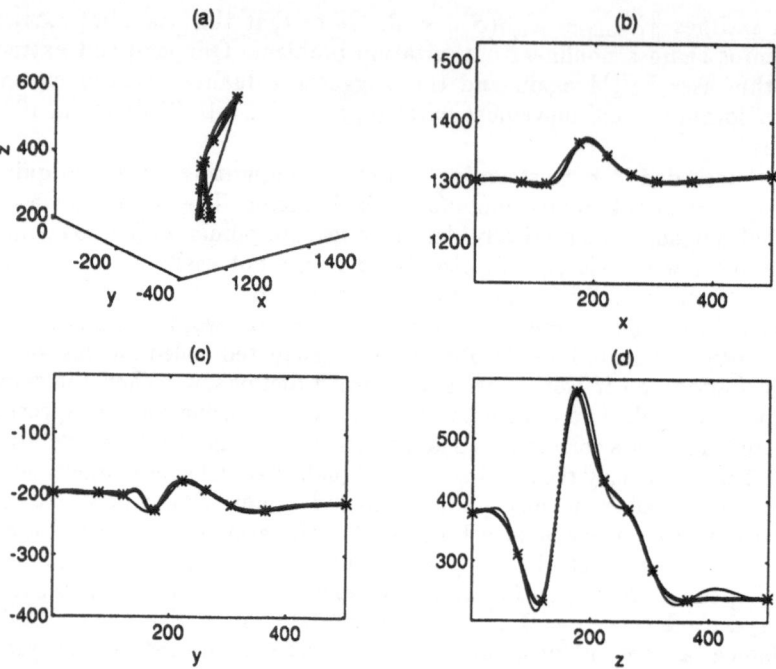

Figure 1: Measured hand trajectories and reconstruction from via-points. Ordinate is in mm. Abscissa is sampling number at 250 Hz. (a) Three-dimensional plot. (b) x (front and rear). (c) y (right and left). (d) z (up and down).

and the reconstructed trajectories were calculated by our method described in the previous section. Here, the minimum-jerk model as a first approximation to the minimum-torque-change model was used.

Figure 2 shows simultaneous measurement of ball and hand positions. First, around the beginning of the upward motion of the hand, the string is slightly lengthened by the tension between the handle and the ball (slight increase of ball-hand distance around time 0.2 to 0.3 sec in the uppermost row). Second, the time of maximum velocity of the hand (circle in the 6th row) almost coincides with the time of maximum velocity of the ball (circle in the third row) and also the beginning of free-fall of the ball (ball acceleration is equal to -9.8 m/s^2 slightly after the circle in the fourth row). Note that if the string were not extensible, the maximum velocity (zero acceleration) of the hand would lead to zero exerted force to the ball that would be then freed. Third, probably because of the elastic property of the string, the maximum velocity and acceleration of the ball are about 20 to 30 % larger than those of the hand. Thus, we can conclude that the hand trajectory (position, velocity and acceleration) from the beginning of z-axis acceleration to the maximum velocity determines the position and maximum velocity of the ball at the beginning of its free fall, and thus determines its upward free-fall (rise) trajectory until the catch. It must be noted that the force exerted on the ball is controlled during the above phase although we can not directly measure it. This is the violation to the second condition for indiscriminate imitation listed in section 1.

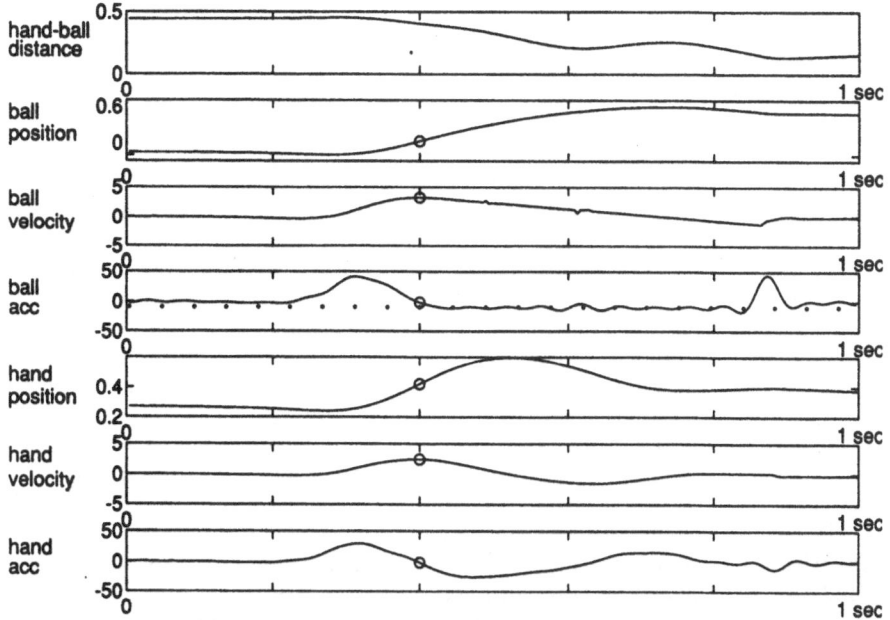

Figure 2: Hand and ball movement data. All scales are in MKS. The dots in the ball acceleration trace (4th row) show acceleration of gravity.

4   Control of Robot using Via-Points

The controlled object is the slave arm of the dexterous arm teleoperation system of the Sarcos Research Company. It has almost the same kinematic structure (7 degrees-of-freedom) as the human arm and 3 degrees-of-freedom at the hand. We do not use a visual sensing system in this study, to allow Kendama to be done in a purely feedforward manner. The kinematic parameters of the robot arm are different from the subject's arm's. The robot dynamic parameters such as the mass are very different from the human arm's; much heavier (violation of condition 1 for indiscriminate imitation). Because the robot arm is much stiffer than a human arm, even if the robot hand is able to follow exactly the same trajectory in Cartesian space as the human arm, it can not exert the same force to the ball at the catch and therefore can not succeed. In general, we want to control the force trajectory in most of interesting tasks. But, usually we can just perceive position trajectory. Indiscriminate imitation of the position trajectory even in the task-oriented coordinates gives no guarantee for the same force trajectory if the teacher and the learner have different dynamic and kinematic properties. Moreover, there are two further difficulties. First, in the kinematic transformation from the human trajectory to the robot trajectory, we can not simply require the same joint angle because the size of the robot arm is different from that of the human arm. If we ask for the same trajectory of the hand in Cartesian coordinates, the joint angles can not uniquely be determined because of redundancy in the robot arm. Furthermore, the combination of feedforward control using an inverse dynamics model and feedback control still induces some error (violation of condition 5 for indiscriminate imitation).

Via-points are the key in resolving these difficulties. Only the via-points are transformed from the subject coordinates to the robot coordinates. The Cartesian coordinates of the via-points are given as the hard constraint which must be strictly satisfied. The soft constraint necessary for resolving the redundancy is that the robot joint angles have to be as close as possible to those of the subject. Thus, we obtained 7 joint angles from all of the via-points. From this, a spline trajectory in the joint angle space passing through these via points was generated and given to the fixed feedforward/feedback controller as the desired trajectory. We know that the two via-points just before and after the peak hand velocity are critical in determining the ball's free-fall. Several via-points including these two are treated as independent control variables. A number of learning schemes, such as feedback-error-learning, direct inverse modeling, forward-inverse modeling, genetic algorithm, and reinforcement learning, can be used once there are a small number of appropriate representations for the task; via-points are selected here. Let us assume that a functional relationship $T = F(S)$ between the via-points S and the task target T is acquired by learning from examples while using a fixed trajectory generation and control scheme. Then, a simple learning scheme is $S_{n+1} = S_n + \epsilon \partial F / \partial S_n (T_{desired} - T_n)$. This learning scheme can be regarded as an extension of the task-level learning algorithm proposed by Aboaf, Atkeson and Reinkensmeyer (1988), in the sense that the control variables are not necessarily the task target but are as abstract as the task target. In this study, however, as a starting point, we manually adjusted the positions of the via-points. After a few hard working nights, F.G. and H.G. finally succeeded in making a robot play Kendama.

In a special case of Japanese Kendama, we examined the potential power of the following proposed strategy for teaching by showing. (1) Human demonstration of the task is the teaching information. (2) Perceive the demonstrated movement trajectory based on dynamics of the controlled objects and the dynamic optimization principle. (3) Extract via-points which can reconstruct trajectories using FIRM. (4) Learn the functional relationship between the via-points and the task evaluation from examples. (5) Treat via-points as abstract control variables while using a fixed trajectory generation and control scheme. (6) Adaptively modify locations and timing of via-points so that task is executed while using the above learned relationship.

Acknowledgements We express our sincere thanks to Drs. Chris Atkeson and Stefan Schaal for their discussions and also their efforts to establish an efficient environment to run robot experiments in our lab. We also thank Prof. Emilio Bizzi and Dr. Sandro Mussa-Ivaldi to allow F.G. to spend two months in ATR where most of the work was done. We thank Dr. Yoh'ichi Toh'kura for his continuous encouragement.

References

[1] Aboaf, E.W., Atkeson, C.G. & Reinkensmeyer, D.J. (1988) Task-level robot learning. *Proc IEEE Int Conf Robot Auto*, April 24-29, Philadelphia.

[2] Schaal, S., Atkeson, C.G., & Botros, S. (1992). What should be learned? *Proc Seventh Yale Workshop on Adaptive and Learning Systems*, 199-204.

[3] Wada, Y., & Kawato, M. (1993). A neural network model for arm trajectory formation using forward and inverse dynamics models. *Neural Networks*, 6:919-932.

[4] Wada, Y., Koike, Y., V-Bateson, E., & Kawato, M. (1994). A computational model for cursive handwriting based on the minimization principle. In *Adv Neural Inf Process Syst 6*. Morgan Kaufmann, in press.

From Coarse to Fine : a Novel Way to Train Neural Networks*

Li-Qun Xu & Trevor Hall

Department of Physics, King's College London

London WC2R 2LS

1 Motivation & Background

In supervised learning of neural networks, an overall cost function, $f(\mathbf{W}, \mathbf{D})$, ($\mathbf{W} = w_{ij}$, and $\mathbf{D} = \{t_m|x_m\}$, $m = 1, 2, \cdots, M$) is normally defined to measure the *mismatch* between the actual and desired output activation patterns, given a training data set $\mathbf{D}$ and a proper learning algorithm to adjust connection weights $\mathbf{W}$. The cost function $f(\mathbf{W}, \mathbf{D})$ can be of various reasonable forms [4, 8], though certain local minima are unavoidable in practical application domains, in some cases the search for the global minimum proved to be quite difficult and inefficient by classic gradient-descent-based learning algorithms which calculate the weight updates for each iteration according to

$$\Delta \mathbf{W}(t) = -\eta \nabla_w f(\mathbf{W}, \mathbf{D}) \tag{1}$$

where η is a constant learning rate, and $\nabla_w f(\mathbf{W}, \mathbf{D})$ is the gradient calculation.

Various ways exist to combat the deficiency of Eq: 1, including the design of adaptive learning rate $\eta_{ij}(t)$ [5] or the "search-and-then-converge" η_t [2], though this kind of measures leave the gradient term of r.h.s. of Eq: 1 wholly untouched; alternatively, the introduction of some stochastic perturbations into different bunches of connection weights at a time [3] or into the whole connection weights at the same time [1] to approximate the gradients $\hat{\nabla}_w f(\mathbf{W}, \mathbf{D})$. In these cases, however, the learning rate η is only kept a very small constant.

In this paper we describe a new strategy, originally studied by [10, 11, 12] in the context of function minimisation, which has potential to tackle the learning problems by dealing with the parameters η and $\nabla_w f(\mathbf{W}, \mathbf{D})$ simultaneously, it turns out that the two kinds of methods described above are specific examples of this new perspective [15].

The central idea of this research for an algorithm to circumvent the problem is instead to modify the cost function $f(\mathbf{W}, \mathbf{D})$ in a way to generate a sequence of its versions, $\{\tilde{f}_{\beta_t}(\mathbf{W}, \mathbf{D})\}_{t=1}^{T} \rightarrow f(\mathbf{W}, \mathbf{D})$, the function $\tilde{f}_{\beta_t}(\mathbf{W}, \mathbf{D})$ at stage (time) t, bears different *details* – steepness, roughness and ravine – of the original function, depending on the parameter β_t, the learning of connection weights or the search for the global minimum of original cost function is actually

---

*This research is partly supported by a UK DTI/SERC LINK project under contract number 3209.

$\mathbf{P}_k$ is a random perturbation vector generated by $G(\mathbf{P})$.

3 Learning Rate

This section is to deal with the adaptive learning rate η_t and compute the weight updates, this is often, however, related to the concept of a search *direction set* constructed from gradient information, like, especially, the example of conjugate gradient algorithm where the learning rate, or step size '$xmin$', [9] is obtained through the construction of a *direction set* and a sequence of line minimisation.

In this case, the gradient estimate is very crude (the number of samples K in Eqs: 3 and 4 only adopts 1 or 2 for the sake of efficiency,), in order to incorporate the gradient information into the stochastic approximation algorithms [6], the filtering of gradient information along time-scale is necessary. Therefore, we employ a method similar to the one suggested by [11], and the equations for weights update are given below :

$$\mathbf{d}(t) = \rho_t \cdot \hat{\nabla}_w \tilde{f}(\mathbf{W}(t), \beta) + i_t \cdot (1 - \rho_t) \cdot \mathbf{d}(t-1), \tag{6}$$

$$\Delta\mathbf{W}(t) = -\eta_t \cdot \mathbf{d}(t) \tag{7}$$

where $\mathbf{d}(t)$ is the *direction set*, η_t the adaptive step size (or learning rate), ρ_t the weighting factor (or aggregation factor), and i_t the gating factor adopting 0 or 1. The on-line changes, at time t, of local parameters η_t, ρ_t and i_t, are controlled by consecutive weights variations $|\Delta\mathbf{W}(t)|$, $|\Delta\mathbf{W}(t-1)|$. Eqs: 6 and 7 should be compared with Eq: 1.

4 Experiments

Experiments have been conducted on various neural network simulation problems, both the gradient estimator of Eq: 3 and 4 were used, and the "batch" learning mode adopted. The important facts explored include the learning ability, generalisation performance, scalibility. The results were compared with those of other algorithms and techniques, in the case of generalisation, for example, with the results of [8]. More details can be found in our recent work [13, 14, 15]. Figure 1 shows the learning curves of a typical "encoder/decoder" neural network. The gradient estimator of Eq: 4 was used.

5 Discussions

Several interesting issues are worthy of more attention. Firstly, the number of stages, T, and the discrete smoothing schedule $\{\beta_t\}_{t=1}^{T}$, were chosen by hand, we have actually been able to modify β_t continuously in a way similar to the so called "search-and-then-converge" idea for learning rate update discussed in [2]. The function is of the form $\beta_t = \beta_0/(1 + \frac{t}{T_d})$, where t is the number of iterations experienced so far, T_d is a time constant whose value is problem-dependent. This choice has been used in our experiments and proved quite successful. Further research to adaptively modify β_t is underway. Secondly, in performed on this sequence of functions. As a result of the undesired details (or idiosyncrasy) of its landscape are largely removed at the beginning, a search for the global minimum $\tilde{\mathbf{W}}_1$ of the function $\tilde{f}_{\beta_1}(\mathbf{W}, \mathbf{D})$ is obviously easy and efficient. Moreover, when the learning proceeds with $\tilde{f}_{\beta_t}(\mathbf{W}, \mathbf{D})$ and its global minimum $\tilde{\mathbf{W}}_t$ is obtained, it provides a sound starting point for the search at the following time $t+1$ when more details of original cost function are added back, leading to the finding of a minimum $\tilde{\mathbf{W}}_{t+1}$. Finally, the solution sequence, $\{ \tilde{\mathbf{W}}_t, \}_{t=1}^T \rightarrow \hat{\mathbf{W}}$ leads to $\hat{\mathbf{W}}$ - the desired global minimum of $f(\mathbf{W}, \mathbf{D})$.

Similar ideas can be found in the simulated annealing algorithm and other intuitive realisations including the one suggested by [7].

2 Gradient Estimation

Above objective can be effectively achieved if we convolve the cost function $f(\mathbf{W}, \mathbf{D})$ with a properly chosen kernel function, $\hat{G}(\mathbf{P}, \beta)$,

$$\tilde{f}(\mathbf{W}, \beta) = \int_{R^N} \hat{G}(\mathbf{P}, \beta)[f(\mathbf{W} + \mathbf{P}) + f(\mathbf{W} - \mathbf{P})]d\mathbf{P} \qquad (2)$$

where the parameter β actually describes the *broadness* of the function $\hat{G}(\mathbf{P}, \beta)$, hence determining the smoothness of the obtained function $\tilde{f}(\mathbf{W}, \mathbf{D}, \beta)$, and $\mathbf{P}$ is a perturbation vector. With $\beta \rightarrow 0.0$, $\hat{G}(\mathbf{P}, \beta)$ approaches $\delta(\mathbf{P})$ – the Dirac's delta functional. Note that for clarity we have dropped the training data note $\mathbf{D}$ in all functions $f(\cdot)$ without causing confusion.

The integral $\tilde{f}(\mathbf{W}, \beta)$ does not normally have an analytic solution, though a numerical solution employing Monte Carlo methods is not difficult to obtain. What we are interested in here is the gradients of the smoothed functional, towards this direction, $\hat{G}(\mathbf{P}, \beta)$ is chosen as a *probabilistic density function* satisfying the condition, $\hat{G}(\mathbf{P}, \beta) = \frac{1}{\beta^N} G(\frac{\mathbf{P}}{\beta})$.

From Eq: 2, it is easy to derive two forms of smoothed gradients expressions [10] and their unbiased gradient estimators are, respectively :

$$\hat{\nabla}_w \tilde{f}(\mathbf{W}, \beta) = \frac{1}{K} \sum_{k=1}^K [\nabla_w f(\mathbf{W} + \beta \mathbf{P}_k) + \nabla_w f(\mathbf{W} - \beta \mathbf{P}_k)] \qquad (3)$$

when the *original* cost function is differentiable (in the case of sigmoid units, the gradients $\nabla_w f(\cdot)$ can be obtained by error back-propagation algorithm), and

$$\hat{\nabla}_w \tilde{f}(\mathbf{W}, \beta) = \frac{(N+1)}{K} \frac{1}{\beta} \sum_{k=1}^K \frac{\mathbf{P}_k}{1 + |\mathbf{P}_k|^2} [f(\mathbf{W} + \beta \mathbf{P}_k) - f(\mathbf{W} - \beta \mathbf{P}_k)] \qquad (4)$$

when only the function value $f(\cdot)$ is available (in the case of discretised weights or hard thresholding units [13]).

where $G(\mathbf{P})$ adopted a normalised multi-dimensional Cauchy p.d.f.,

$$G(\mathbf{P}) = \frac{1}{\pi^{(N+1)/2}} \frac{\Gamma(\frac{N+1}{2})}{(1 + \sum_{i=1}^N (\frac{p^{(i)}}{\beta})^2)^{(N+1)/2}} \qquad (5)$$

610

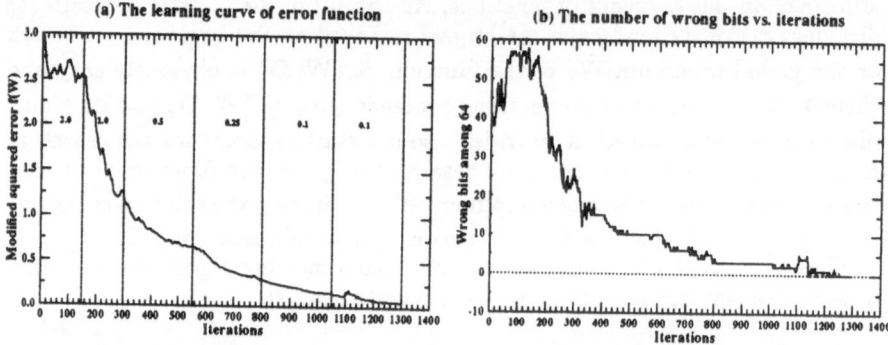

Fig.1 The learning profiles for an 8 × 3 × 8 fully connected neural network.

previous discussions, β is normally required to approach 0.0 at the final stage of the search procedure in order that the local minimum $\tilde{W}_T$ approximates the global minimum $\tilde{W}$. In our practical tests, however, we found that in many cases the global minimum can be achieved when the β is still quite large in its value, say 0.1, which accounts for 10% perturbation in estimating the gradient. In neural network applications, this solution provided a better generalisation performance than the solution that is obtained when $\beta \to 0.0$.

References

[1] Cauwenberghs, Gert (1993), *NIPS-5 proceedings*, 244-251.

[2] Darken, C., J. Chang, and J. Moody (1992), *Proc. of IEEE workshop on neural networks for signal processing 2*, IEEE Press, 445 Hoes Lane, Piscataway, NJ 08854.

[3] Flower, Barry & Jabri, Marwan (1993), *NIPS-5 proceedings*, 212-219.

[4] Hertz, J, A. Krogh, and R.G. Palmer (1991), Addison-Wesley Publishing Co. 1991.

[5] Jacobs, R. (1988), *Neural Networks*, 1, 295-307.

[6] Kushner, H.J. and D.S. Clark, New York: Springer-Verlag, 1978.

[7] Makram-Ebeid, S., J.A. Sirat, and J.R. Viala (1989), *Proc. of IJCNN*, II, 373-380, (Washington 1989).

[8] Nowlan, S.J. & Geoffrey E. Hinton (1992), *Proc. of NIPS-4*.

[9] Press, W.H., S.A. Teukolsky, W.T. Vetterling, and B.P. Flannery (1992), 2nd Edition, Cambridge University Press, 1992.

[10] Rubinstein, R.Y. (1981), Chapter 7. New York : John Wiley.

[11] Ruszczynski, A. & W. Syski (1984), *9th world conference of the IFAC*, Vol. VII, 230-234, Budapest, Hungary, July 2-6.

[12] Styblinski, M.A. and T.S. Tang (1990), *Neural Networks*, 3, 467-483, 1990.

[13] Xu, Li-Qun & T.J. Hall (1993), *Technical Report, Physics Department, King's College London*, Strand London WC2R 2LS, June 1993.

[14] Xu, Li-Qun & T.J. Hall (1993), *Proc. of 2nd Int. Conf on Fuzzy Theory & Technology*, Durhum, NC, USA, October 1993.

[15] Xu, Li-Qun & T.J. Hall (1993), *in preparation*. November, 1993.

Learning the Activation Function for the Neurons in Neural Networks

G.P. Fletcher & C.J.Hinde
Department of Computer Studies
Loughborough University
Loughborough

1 Introduction

Ever since the sigmoid replaced the threshold as the main activation function used in artificial neural networks, the properties of the activation function have been largely ignored. Most research aimed at improving the quality of the hypothesis or the speed at which it is obtained has concentrated on the topology or enhancing the learning algorithm (normally back propagation).

Even though the sigmoid function is intended to be an approximation to a real neurons response, artificial systems do not necessarily need to copy. Real neural networks model very complex information by using huge numbers of neurons. By developing a set of more complex activation functions it should be possible to produce better models of complex information resulting in smaller and faster networks and the development of more powerful application areas for artificial neural networks.

Other recent work has presented the idea of parametrisation of the sigmoid function (Horejs et al. 1993). By training these parameters using back propagation it is possible to alter the sigmoid's characteristics; i.e. such things as the maximum and mid values of the function. The idea was developed as a method of trying to remove the input weights as a training parameter. Instead of attempting to produce smaller networks with more powerful nodes the other work was aimed at allowing massive nets of very low complexity to be trained. The results show that although these nets will converge they require many more iterations. This paper shows that, by taking the converse view and improving the power of each neuron, the number of training iterations is significantly reduced.

2 Sine as an activation function for back propagation

Back propagation relies on the use of a differentiable activation function. The sigmoid function was originally chosen because of its similarity to the then popular threshold activation function of the simple perceptron model (Rosenblatt 1962). There is no reason why back propagation needs to use the sigmoid, replacing it with another differentiable function such as sine does not affect the convergent properties but does affect the types of hypotheses that can be represented.

If a two input neuron uses a sigmoid then it is perfectly able to model transfer functions such as "Boolean AND" and "Boolean OR", in fact it is difficult to imagine a better activation function for these two problems.

The "Boolean X-OR" or two input parity is the simplest problem that cannot be modelled in a single neuron using a sigmoid activation function. Figure 1 shows how the sine and weights interact to model the X-OR function in a single neuron.

612

Case 'A' shows that the threshold with two inputs set to false give 'Sum' a value of $-\pi/2$, and therefore an output value of -1.0 or false.

Case 'B' shows that if one of the inputs is true and the other false then 'Sum" is $\pi/2$, and the output value is 1.0 or true.

Case 'C' shows that with both inputs set to true then sum is $3\pi/2$ and output is -1.0 or false.

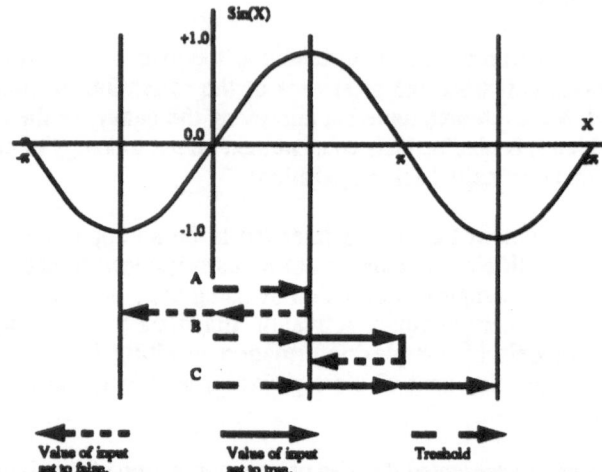

Figure 1 Demonstration of the sine activation function, and the combination of weights and bias to produce the X-OR function.

However, the sine activation function does not effectively model "Boolean OR". Even using such a simple network, it is apparent that by choosing the correct activation function for the problem, modelling becomes much easier.

3 The general activation function

Given a large network it is not feasible for the designer to allocate the correct activation function to every neuron. It therefore becomes necessary to develop a general parametrised activation function with enough flexibility to be able to model many different types of transfer function. If the parameters of the general activation function are trained together with the input weights the neuron will develop the activation function that most easily models the data.

An example activation function is constructed using both the traditional sigmoid and the sin activation functions. Let the activation function be $f(\beta)$ where β is the product of the weight and input vectors. i.e.

$$\beta = bias + weight_1 * input_1 + weight_2 * input_2 \dots$$
$$f(\beta) = a \cdot sigmoid(\beta) + (1 - a) \cdot sin(\beta)$$

Each element of the activation function has a -1 to +1 range. The output of the neuron has been maintained in the traditional -1 to +1 range by making the sum of the activation function weights equal 1. By back propagating the errors and using them to modify the parameter 'a' in each neuron the individual neurons will develop different characteristics. If a is high, i.e. 0.9, then the activation function becomes very close to the traditional sigmoid. This is very good when attempting to model problems like the "Boolean AND". Dropping the value of a to 0.1 produces an activation function easily capable of representing the "Boolean X-OR" problem.

4 Training with the general activation function with two dimensional problems

All two dimension problems can be represented using just a single neuron with a general activation function. To demonstrate the neurons ability to derive the correct activation function the solutions for "Two input AND" and "Two input X-OR" are presented.

TWO INPUT AND

Input 1 Weight : +1.7142
Input 2 Weight : +1.7159
Bias : -1.7126
Activation Function

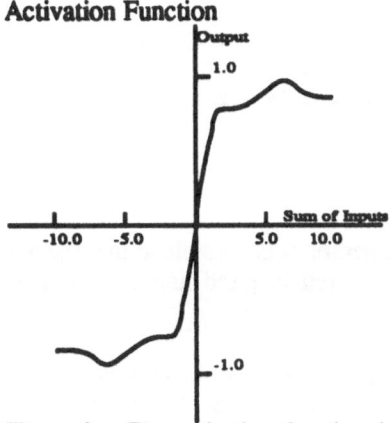

Figure 2a: The activation function in response to the "AND" training set.

TWO INPUT X-OR

Input 1 Weight : +1.5881
Input 2 Weight : +1.5776
Bias : +1.5718
Activation Function

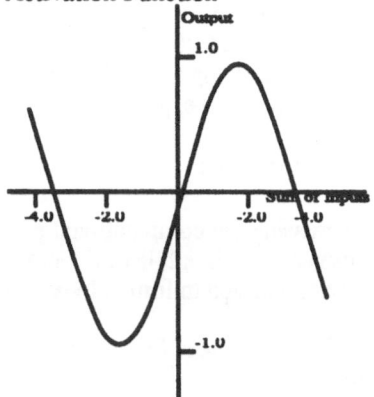

Figure 2b: The activation function in response to the "X-OR" training set.

Both of the neurons have developed activation functions capable of representing the problems being demonstrated. One having produced a threshold type function and the other a cyclic function, ideal for the two different problems.

5 Neurons with general activation function as a method of reducing the number of training iterations required.

Networks constructed of neurons that use the general activation function are more flexible in the representations they can adopt. If it is easier to represent the information within the network, the time required to find an acceptable representation will be reduced.

A four input, three midnode one output network forms the basis for the illustrative experiment, which is shown using standard and general activation functions. The graphs show the error over the training cycle for the two different types of neurons.

614

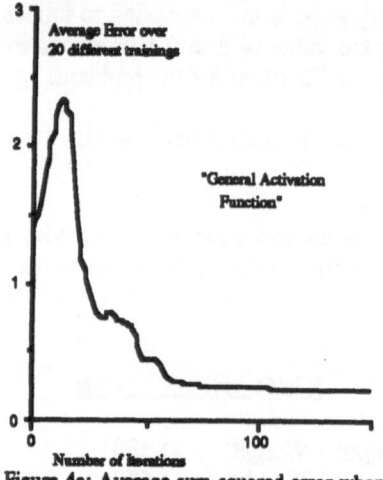

Figure 4a: Average sum squared error when using a general activation function.

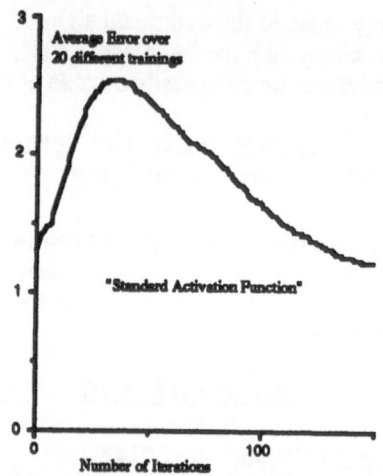

Figure 4b: Average sum squared error when using a standard activation function.

These results show the potential for very large savings in training time. The results are an average of 20 different training cycles. Each one uses a randomly chosen, consistent, training set and start point. The same training sets and starting points were used for both experiments.

6 Conclusion

By increasing the computational power of each neuron it is possible to increase the representational flexibility of neural networks while reducing the number of training iterations required to form a hypothesis.

By reducing the need for very large networks implementation becomes much easier. With the smaller networks it is tractable to produce a clear representation of their derived hypotheses (Fletcher et al. 1993). The developer will then be able to verify that the design has correctly represented the required hypothesis. Greater confidence in neural networks will enable them to be used on a much wider basis.

7 References

Fletcher, G.P., Hinde, C.J., West, A.A. & Williams, D.J., (1994) Neural networks as a paradigm for knowledge elicitation, ibid.

Horejs, J. & Kufudaki, O., (1993) Neural networks with local distributed parameters, Neurocomputing, Vol 5 NΩ 4. pp 211 - 219.

Rosenblatt, F., (1962) Principles of neurodynamics, Spartan books.

Projection Learning and Graceful Degradation

K. Weigl† and M. Berthod†

† INRIA Sophia Antipolis
F-06902 Sophia Antipolis Cedex France

1 Neural Networks as Bases in Function Space

We have presented in (Weigl et al. 1992 - 1993 b) a paradigm in which we consider neural networks such as Multi-layer Perceptrons as bases in a function space; the basis functions are the functions computed by the hidden layer neurons, and the function approximated by the network is the projection of the function to be approximated onto the manifold spanned by these basis functions. We have presented a learning algorithm based on that paradigm, which consists in shifting the manifold spanned by that base in function space in such a way that the distance to the function to be approximated is minimal.

Using that algorithm, we compare in the present paper two types of networks as far as their capacity for graceful degradation is concerned:
a) Networks with a hidden layer of perceptrons;
b) Networks with a hidden layer of multivariate gaussians;
Graceful degradation, in our context, is the ability to recover initial performance as far and fast as possible by a dynamical process. The dynamical process is in our case the original learning algorithm, called *projection learning*. The remainder of the paper is organized as follows: Aftern the next section, which introduces briefly the algorithm, we present the problem and compare the reactions of the systems. We shall furthermore interpret the results, discuss an intuitive measure which can be applied to any such network to judge its capacity for graceful degradation, and outline further research.

2 The Algorithm

We use the classical cost function:

$$E = \sum_{k=1}^{M} (F(x_k) - A(x_k))^2 \tag{1}$$

$F(x_k)$ being the function to be approximated, $A(x_k)$ the approximating function, x_k the set of input values, $k \in \{1, .., n\}$, $g_i, i \in \{1, .., N\}$ the set of arbitrary differentiable functions computed by the N hidden-layer neurons, and $g_i(x_k), k \in \{1, .., M\}, i \in \{1, .., N\}$ the output values computed by the hidden-layer neurons for given inputs x_k. We shall assume a linear output neuron [1].

---

[1] We have shown in (Weigl et al. 1993 b) the extension to a non-linear output neuron with invertible activation function; extension to multiple output neurons is trivial

The difference to backpropagation is that we are computing the weights from hidden-layer to output layer *directly* at each step for given input- to hidden-layer weights via a projection operator computed from the base functions, and then differentiate the weights from input- to hidden layer based upon these weights as in standard backpropagation. We have proven that this approach, of which an iterative variant is similar to an approach taken by Poggio (Poggio et al. 1990) for Hyperbasis functions, is mathematically exact (Weigl et al. 1993 b). The general expression for gradient descent, assuming for simplicity only one output neuron, is thus:

$$\frac{dparams_j}{dt} = -\frac{dE}{dparams_j} = -\sum_x 2(F(x_k) - \sum_i A^i g_i(x_k))(A^j \frac{dg_j(x_k)}{dparams_j})$$
(2)

where $\sum_i A^i g_i(x_k) = A(x_k)$ is the approximation, A^j is the weight from hidden-layer neuron j to the output neuron, and *params_j* are the weights from input to that neuron.

3 The Problem

We took a toy problem to solve, namely the XOR; we represented the function via eight examples. Fig 1 top left shows the function. Initial training was via the algorithm above, taking 170 msecs for the MLP, and 400 msecs for the network of multivariate gaussians, both on Sparc 10. We then simulated the "death" of a neuron of the hidden layer via setting its output to zero whatever the input. Afterwards, reconvergence of the network to the optimal approximation was monitored.

4 Results and Discussion

At each iteration, our algorithm computes the projection $\sum_i A^i g_i(x_k) = A(x_k)$ of the function to approximate onto the manifold of basis functions computed by the hidden-layer neurons; it does this by recomputing the weights A^i from hidden layer to output. If now one of the neurons of that layer dies, the manifold spanned by the remaining ones will be lower-dimensional than before, assuming initially linear independence. Immediate recomputation of the weights A^i amounts then to an automatic compensation for the loss of the neuron by the other neurons, as far as that is possible.

The effectiveness of that recomputation depends crucially on the *overlap* between neurons in sample space: In the case of XOR, this overlap can be verified after initial convergence. The requirement is simply the following:

For each input example for which we want a non-zero output, at least two neurons of the hidden layer must have a non-zero output. Fig. 1 illustrates that criterion: Here, the network learned the XOR, i.e. among others to map the input [1, 0] to the output [1]. However, only one neuron of the hidden layer had a non-zero output for the input [1, 0]. When that neuron dies, as can be seen in the top row of fig. 1, third from the left, none of the other neurons can compensate immediately for the loss: They have to be repositioned in order to represent again the function optimally. Fig. 2 shows what happens if the condition is fulfilled: Here, more than one neuron has a non-zero output for the input $[1, 0]$, so that after one iteration, the function is again represented nearly optimally, and only slight readjustment of the neurons of the hidden layer is necessary for the representation to be again optimal.

5   Conclusion and Outlook

Graceful degradation is a property ascribed to neural networks which is rarely fulfilled, and which could usually not be verified beforehand. It is obvious that such a property is of high importance to time-critical systems for which the risk of sudden breakdown cannot be tolerated, such as airspace control, medical equipment, et al. We have shown that projection learning allows a system of neurons to compensate rapidly for partial losses, if certain conditions are met, which are simple for examples such as XOR.

Obviously, in the case of more complex approximation problems, the requirements will be less obvious than in the case of the example above: That requirement is only a necessary, but not a sufficient condition: We shall study in the future a more elaborate approach, consisting in the verification of the density of overlap over all input examples, e.g. by the metric tensor of the base: All off-diagonal elements should be roughly equal among themselves, as well as all diagonal elements.

Furthermore, we shall compare the verification of robustness by study of the metric tensor with explicit verification, which consists in "killing" in turn one out of all the neurons of the hidden layer, and verifying the quality of the approximation by the remaining neurons, using the projection operator recomputed again in one iteration.

In case of an overcomplete base, i.e. where not all of the hidden-layer neurons are linearly independent, the robustness could be even greater (Weigl et al. 1993 c).

References

- Poggio, T., and Girosi, F., Networks for Approximation and Learning (1990) Proc. IEEE, vol. 78. No. 9, 1488-ff

- Rumelhart, D.E., McClelland, J.L., et al., (1986) Parallel Distributed Processing, Vol. 1, MIT-press.

- Weigl, K., and Berthod, M. (1992) Metric Tensors and Dynamical Non-Orthogonal Bases: An Application to Function Approximation. Proc. WOPPLOT 1992, Workshop on Parallel Processing: Logic, Organization and Technology, Springer Lecture Notes in Computer Sciences, to be published.

- Weigl, K., and Berthod, M., (1993 a) Non-orthogonal Bases and Metric Tensors: An Application to Artificial Neural Networks. New Trends in Neural Computation, Proc. IWANN'93, International Workshop on Artificial Neural Networks, Springer Lecture Notes in Computer Sciences, vol. 686, 173-178.

- Weigl, K., and Berthod, M., (1993 b) Neural Networks as Dynamical Bases in Function Space, Research Report INRIA no. 2124, 1-40.

- Weigl, K., and Berthod, M., (1993 c) Some Remarks about Boundary Creation by Multi-layer Perceptrons, submitted to WCNN 94

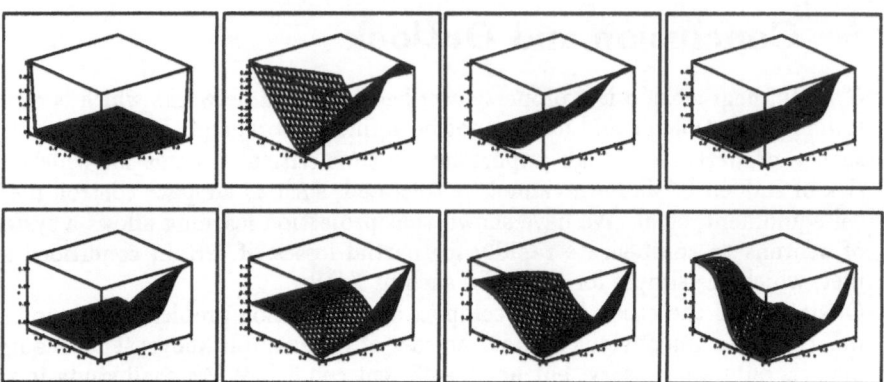

Figure 1: XOR-function to be approximated shown on top left; next to the right approximation learned by network, originally eight samples, three filters; system with one neuron dead, leaves two, shown next; immediate recomputation of the projection operator cannot correct here because of lack of overlap, as we can see in the image immediately to the right; remaining images, from left to right and top to bottom, show the recovery; data: 17 iterations, total time 160 msecs on a Sparc 10

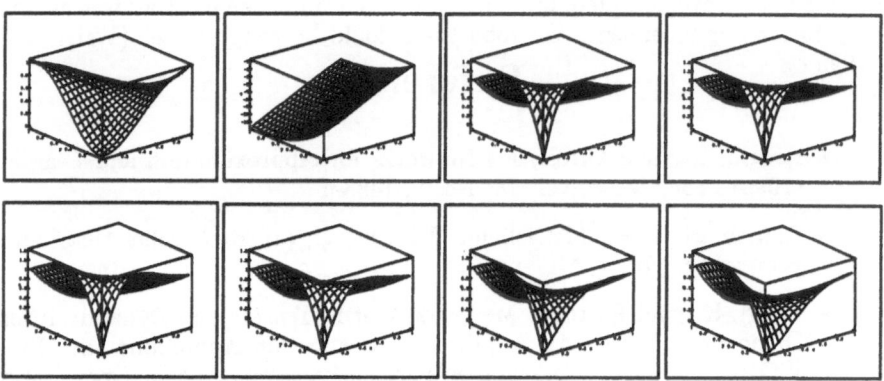

Figure 2: As above, but originally eight samples, five gaussian filters, shown on top left; system with one neuron dead, leaves four, shown next to it; immediate recomputation of projection operator results in almost perfect correction, shown next to the right; remaining images, from left to right and top to bottom, show the recovery with only minimal further corrections necessary; data: 12 iterations, total time 220 msecs on a Sparc 10

Learning with zero error in Feedforward Neural Networks.

M. L. Lo Cascio, G. Pesamosca

Dip. Metodi e Modelli Matematici Univ. 'La Sapienza', 00162 ROMA

1 Introduction.

This paper investigates the number of 'Training Patterns' which can be submitted to a Feedforward Neural Network (FFNN), in order to obtain a quadratic Learning Error equal to zero. The sygmoid function is supposed differentiable and invertible, and only three layer FFNN are up to now considered.

2 The learning problem.

In the following a three layer network will be considered, with n_1, n_2, n_3 Processing Elements respectively in the first, second and third layer. Moreover, two elements of indices n_1+1 and n_2+1 (bias) are added respectively to the first and second layer. Such a network will be shortly denoted as 'network n_1, n_2, n_3 '. According to (Rumelhart et al, 1988) we introduce the following notations:

$\vec{O}^{p,i} = \left[o_1^{p,i}, o_2^{p,i}, \ldots, o_{n_i+1}^{p,i} \right]^T$ - Output vector of the layer i, for the training pattern p

(i=1,2,3; p=1,2,...,M). $\vec{O}^{p,1}$ is the network input. The bias components are $o_{n_1+1}^{p,1}$

$= o_{n_2+1}^{p,2} = 1$. The vector $\vec{O}^{p,3}$ has n_3 components only.

$\vec{T}^p = \left[t_1^p, \ldots, t_{n_3}^p \right]^T$ - Target vector at the third layer. $\vec{T}^p$ can assume real values, compatible with the sigmoid function.

$W^i = \left[w_{jk}^i \right]$ (i=1,2; j=1,2,...,n_{i+1} ; k=1,2,...,n_i+1) - Weight matrices of the network.

More precisely, w_{jk}^i is the weight associated with the edge connecting the kth element of the layer i with the jth element of the layer i+1.

$\vec{W}$ - Vector whose components are the w_{jk}^i 's, disposed row-wise.

$\vec{W}_j^i$ - j^{th} row of the matrix W^i, considered as a column vector.

The $\vec{O}^{p,i}$ (i=2,3) are computed by means of the recursive formula

$$\vec{H}^{p,i} = \left[h_1^{p,i}, \ldots, h_{n_i}^{p,i} \right]^T = W^{i-1} \vec{O}^{p,i-1} \qquad (i=2,3; \quad p=1,\ldots M) \qquad (1)$$

$$\vec{O}^{p,2} = \left[f(\vec{H}^{p,2}),1 \right]^{T}, \qquad \vec{O}^{p,3} = f(\vec{H}^{p,3}) \tag{1'}$$

The notations in (1') mean that the sigmoid function f is applied to every component:

$$o_j^{p,i} = f(h_j^{p,i}) \qquad (i=2,3; \; j=1,...,n_i, \; p=1,...M)$$

Examples of sygmoid functions are $f(x) = b^{-1}tgh(bx)$, $f(x) = b(1+e^{-bx})^{-1}$.

In the learning problem, M pairs of vectors $(\vec{O}^{p,1},\vec{T}^{p})$ (p=1,2,...M) are
submitted to the network, and the weights $w_{j,k}^{i}$ are determined so as to minimize a
suitable error function $E(\vec{W})$. In this paper the following $E(\vec{W})$ is considered:

$$E(\vec{W}) = \frac{1}{2} \sum_{p=1}^{M} \sum_{j=1}^{n_3} (o_j^{p,3} - t_j^p)^2 = \frac{1}{2} \sum_{p=1}^{M} \sum_{j=1}^{n_3} E_{pj}^2 \tag{2}$$

that is, taking (1), (1') into account

$$E(\vec{W}) = \frac{1}{2} \sum_{p=1}^{M} \sum_{j=1}^{n_3} \left\{ f\left[\sum_{i=1}^{n_2} w_{ji}^2 f\left(\sum_{k=1}^{n_1} o_k^{p,1} w_{i,k}^1 + w_{i,n_1+1}^1 \right) + w_{j,n_2+1}^2 \right] - t_j^p \right\}^2 \tag{3}$$

Our problem leads to study the system

$$E_{pj}(\vec{W}) = 0 \quad (p=1,2,...M; \; j=1,2,...n_3) \tag{4}$$

since any solution of (4) assures an error $E(\vec{W})$ and a gradient $G(\vec{W})$ equal to
zero. Let f^{-1} be the inverse function of f and $t_j^{*p} = f^{-1}(t_j^p)$. Then the system
(4), taking (3) into account, can be written

$$\sum_{i=1}^{n_2} w_{ji}^2 o_i^{p,2} + w_{j,n_2+1}^2 = t_j^{*p} \quad (p=1,..,M; \; j=1,...,n_3) \tag{5}$$

where

$$o_i^{p,2} = f\left(\sum_{k=1}^{n_1} o_k^{p,1} w_{ik}^1 + w_{i,n_1+1}^1 \right) \quad (p=1,...,M; \; i=1,...,n_2) \tag{6}$$

Putting $\vec{T}_j = \left[t_j^{*1}, t_j^{*2},...,t_j^{*M} \right]^T$ $(j=1,...,n_3)$ and grouping the equations with
the same index j, (5) can be formally divided into n_3 systems of M equations:

$$B\vec{W}_j^2 = \vec{T}_j \quad (j=1,2,...,n_3) \tag{7}$$

having the same coefficient matrix B=[b_{pi}] with

$$b_{pi} = o_i^{p,2}, \; b_{p,n_2+1} = 1 \quad (p=1,...,M; \; i=1,...,n_2) \tag{8}$$

3 Learning with zero error

The following theorems show that for small values of M there are infinite minimum points $\vec{W}*$, with $E(\vec{W}*) = 0$. Moreover the learning problem can be solved by means of linear systems. For the proofs, see (Lo Cascio et al. 1994).

<u>Theorem 1.</u> If $M \le n_2 + 1$ the minimum value $E(\vec{W}*)$ is zero. The infinite solutions $\vec{W}*$ can be evaluated by choosing arbitrary values for the weights W^1, and then solving the systems (7) with respect to W^2.

<u>Theorem 2.</u> If $M \le n_1 + 1$ and $n_3 \le n_2$ the minimum value $E(\vec{W}*)$ is zero. The infinite solutions $\vec{W}*$ can be found by means of the following steps:

1. Choose arbitrarily the weights W^2.

2. Solve with respect to the $o_1^{p,2}, ..., o_{n_2}^{p,2}$ the systems

$$W^2 \vec{B}_p = \vec{T}^p \qquad , \qquad (\quad \vec{T}^p = \left[t_1^{*p}, ..., t_{n_3}^p \right]^T, \vec{B}_p = \left[o_1^{p,2}, ..., o_{n_2}^{p,2}, 1 \right]^T)$$

(p=1,...,M)

3. Set $*o_i^{p,2} = f^{-1}(o_i^{p,2})$, and solve the system

$$\sum_{k=1}^{n_1} o_k^{p,1} w_{ik}^1 + w_{i,n_1+1}^1 = *o_i^{p,2} \quad (p=1,2,...,M).$$

If M exceeds the values indicated in Theorems 1 and 2, the occurrence of $E(\vec{W}*) = 0$ depends on both the network structure and the input/target data.

Let us firstly consider the networks n_1, n_2, 1. By shortly denoting as $\vec{W}^2$ and $\vec{T}$ the vectors $\vec{W}_1^2$ and $\vec{T}_1$, the group of systems (7) reduces to the system

$$B\vec{W}^2 = \vec{T} \qquad (9)$$

in the $n_2 + 1$ unknowns $\vec{W}^2$. The complete matrix is

$$D = \begin{bmatrix} o_1^{1,2} & ....o_{n_2}^{1,2} & 1 & t_1^* \\ o_1^{2,2} & ....o_{n_2}^{2,2} & 1 & t_2^* \\ .... & .... & ... & ... \\ o_1^{M,2} & ....o_{n_2}^{M,2} & 1 & t_M^* \end{bmatrix} \qquad (10)$$

From the Rouchè theorem, the compatibility of (9) requires the vanishing of the $\mu = = M-(n_2 + 1)$ maximun order minors $D_1(W^1), ..., D_\mu(W^1)$ of D, where $D_j(W^1)$ is obtained by the first $n_2 + 1$ rows plus the $(n_2 + 1 + j)^{th}$ row of D. Hence, the weights $W^1 = \left[w_{11}^1, ..., w_{1,n_1+1}^1, w_{21}^1, ..., w_{n_2,n_1+1}^1 \right]$ must satisfy the following non-linear system of μ equations in $n_2(n_1 + 1)$ unknowns:

$$D_1(W^1) = 0, D_2(W^1) = 0, ..., D_\mu(W^1) = 0 \tag{11}$$

It follows from (10) and (6) that (11) admits the n_2 solutions

$$w^1_{i,1} = w^1_{i,2} = ... = w^1_{i,n_1} = 0 \quad (i=1,...n_2), \quad \text{all the others } w^1_{r,k} \text{ arbitrary} \tag{12}$$

These solutions lower the rank of B and do not allow the computation of the W^2. So it is necessary to consider other solutions. Let $M^* = n_2(n_1+2)+1$ be the value of M for which the number of equations (11) equals the number of unknowns W^1.

- If $M=M^*$, the system (11) may admit solutions $\hat{W}^1$ different from (12) for suitable, infinite sets of input/target data. The example 1 in (Lo Cascio et al., 1994) illustrates this occurrence. If such solutions exist, they are continuously varying with the data.

- If $M>M^*$ there are more equations than unknowns, and hence solutions different from (12) can exist only for particular, isolated data.

- If $M<M^*$ there are in (11) $M-M^*$ free variables $w^1_{r,k}$, , and hence the possibility to find solutions different from (12) increases with respect to the case 1.

The analysis of the case $n_3 >1$ is analogous to the case $n_3 =1$. With the positions

$$\tilde{W}^2 = \left[\vec{W}^2_1,...,\vec{W}^2_{n_3}\right]^T, \quad \tilde{T} = \left[\vec{T}_1,...,\vec{T}_{n_3}\right]^T, \quad \tilde{B} = diag(B,B,...,B)$$

the group of systems (7) can be written as a single linear system $\tilde{B}\tilde{W}^2 = \tilde{T}$. The compatibility requires again the vanishing of the Rouchè theorem minors $D_1(W^1),..., D_\mu(W^1)$ [$\mu=n_3$ ($M-n_2$ -1)] of the matrix

$$D = \begin{bmatrix} B & 0 & ...0 & \vec{T}_1 \\ 0 & B & ...0 & \vec{T}_2 \\ .. & .. & ..... & .. \\ 0 & 0 & ...B & \vec{T}_{n_3} \end{bmatrix}$$

The maximum integer M for which the number of equations (11) is less or equal to the number of unknowns W^1 is now $M^* = n_2 +1+ \left\lfloor \dfrac{n_2}{n_3}(n_1 +1) \right\rfloor$

References.

Lo Cascio M.L., Pesamosca G. - Learning con errore nullo nelle reti neuronali di tipo Feedforward. Preprint Dip. MeMoMat, Roma 1994.

Rumelhart D.E., Hinton G.E., Williams R.J. Learning Internal Repres. by Error Propag. - Parallel Distrib. Process. Vol. I, Bradford Books, Cambridge MA 1988.

Robustness of Hebbian and anti-Hebbian Learning

T. Fomin and A. Lőrincz

Institute of Isotopes, The Hungarian Academy of Sciences

Budapest, Hungary

1   Introduction

Fault tolerance of artificial neural networks (ANN) has been studied mostly for passive systems, that does not react in any special way to compensate for the effect of internal failures (Protzel, Palumbo, and Arras, 1993). Systems with active fault-tolerance reorganize their resources to counteract the fault effects. Studied examples describe adaptation or retraining after internal faults (Anderson, 1983 and Sejnowski and Rosenberg, 1986). Other examples suggest prewired self-repair mechanisms (Petsche and Dickinson, 1990).

Here the the fault tolerance of a self-organizing Hebbian and anti-Hebbian (HAH) network was studied. In case of self-organized learning the question of 'performance' arises, since the network always does something. Our starting point was that HAH networks perform soft competition and thus HAH neurons search and compete for high order correlations and divide the 'world' between themselves. IN this sense the network should provide a 'quasi-orthogonal representation' and network performance may be judged by considering orthogonality of neural filter vectors. Different learning algorithms will perform in a different fashion, since orthogonality of receptive fields depend strongly on e.g. the postsynaptic or presynaptic nature of learning. A geometrical problem - forming spatial filters - is studied here, since it offers easy judgement. In addition, we restricted our studies to cases where the networks were started from 'scratch'. Neural metwork parameters, such as learning rates, neural activities, sharpness of nonlinearities were considered different for different neurons.

2   Hebbian and anti-Hebbian learning

We have performed our computations on a HAH network consisting of 18 neural output units. In a HAH network each neuron receives inputs from an input vector x via Hebbian weights. Let us denote the weight connecting the ith neuron with the jth component of the input vector by q_{ij} ($i=1,...,18$; $j=1,...270$). The input vector x_j ($j=1,...,270$) represents a 15 by 18 pixel sized grid. The neurons are connected with each other through the anti-Hebbian connections w_{ij} ($i,j=1,...,18$). When an input is connected to the system the outputs of the neural units develop through a relaxation process governed by one of the following non-linear differential equations:

$$\dot{y}_i = -y_i + f\left(\sum_j q_{ij}x_j - \sum_{i\neq k} w_{ik}y_k - \theta_i\right) \qquad (1)$$

where y_i and f denote the output of the ith neural unit and the sigmoid function,

$$\dot{y}_i = -y_i + f\left(\sum_j S(q_{ij})x_j - \delta\sum_k S(w_{ik})y_k - t_i\right) \tag{2}$$

respectively. Function S denotes a similar nonlinearity to f. During the training process q and w are both modified according to the following equations:

$$\epsilon\dot{q}_{ij} = -q_{ij} + \alpha x_j y_i \tag{3}$$

$$\epsilon\dot{w}_{ik} = -w_{ik} + \beta y_i y_k \tag{4}$$

ϵ^{-1} determines the learning rate as compared to the settling time of the network. The equation for the threshold may be given as:

$$\epsilon\dot{t}_i = y_i - p \tag{5}$$

Neurons lower their threshold if their outputs are low, and are trying to be more selective when their outputs are high. The network used here is similar to the one introduced in (Foldiak).

3   Examples with spatial filters

The case of formation of spatial local filters is suitable for the study of different network parameters, since neurons play equal role and should be similar from every respect if neural parameters are identical for individual neurons. The network was trained on different inputs of the size of 6 by 6 pixel. It was found that the network is extremely tolerant against deviations in learning parameters. The most sensitive parameter was the saturation value (SV) of neural activity function, f of Eqs. (1) and (2). The sensitivity to the other parameters, including λ, the sharpness of the sigmoid function was much smaller. Spatial filters for 0%, 30%, 50% and 80% SV deviations are shown in Fig. 1. 50% SV deviation means that the SV value was randomly selected between 0.5 and 1.5 as opposed to 1.0 for the perfect system. As it may be seen from the figures filters at these very high values become distorted.

HAH networks work by searching and competing for correlations. Neurons with very different receptive fields do not compete with each other. Strong anti-Hebbian connections develop between neurons of neighbouring (overlapping) receptive fields. The connection structure reflects this porperty. Figures 2 and 3 show our tests, how well the network represents neighbouring relations. Results for perfect neurons are shown in Fig. 2. This figure depicts the overlap between receptive fields of two neurons as a function of the anti-Hebbian weight between the same neurons. The overlap vs. anti-Hebbian weight 'function' is monotone, has a small noise. It does not start from zero as could have been expected. Larger inputs may bring that feature. There are two clusters in the figure: (i) neuron pairs with negligible overlap and small anti-Hebbian weight and (ii) neuron pairs with considerable overlap and large anti-Hebbian weight. The latter correspond to neighboring neurons. The distribution is large since filters can overlap by edges or by corners. Figure 3 shows the results for 30% ('+' signs), 50% (squares) and 80% ('x' signs) SV deviations. (Note the change in the scales of the axis.) There is very little change for 30% SV deviation. The network can tolerate such high values of neural saturation deviations: it can range between 0.7 and 1.3. For higher SV deviations the curve continues to higher values and the relation becomes more 'noisy'. It is, however, interesting, that even for the case of 80% SV deviation (saturation value of f ranges between 0.2 and 1.8) there are nine neurons with receptive fields of similar shape that covers the whole region. In this respect it is important to recall that the object size and the image size were 5x6 and 15x18, respectively, that is our 12 neurons give a redundant representation and 9 neurons are enough for developing spatial filters.

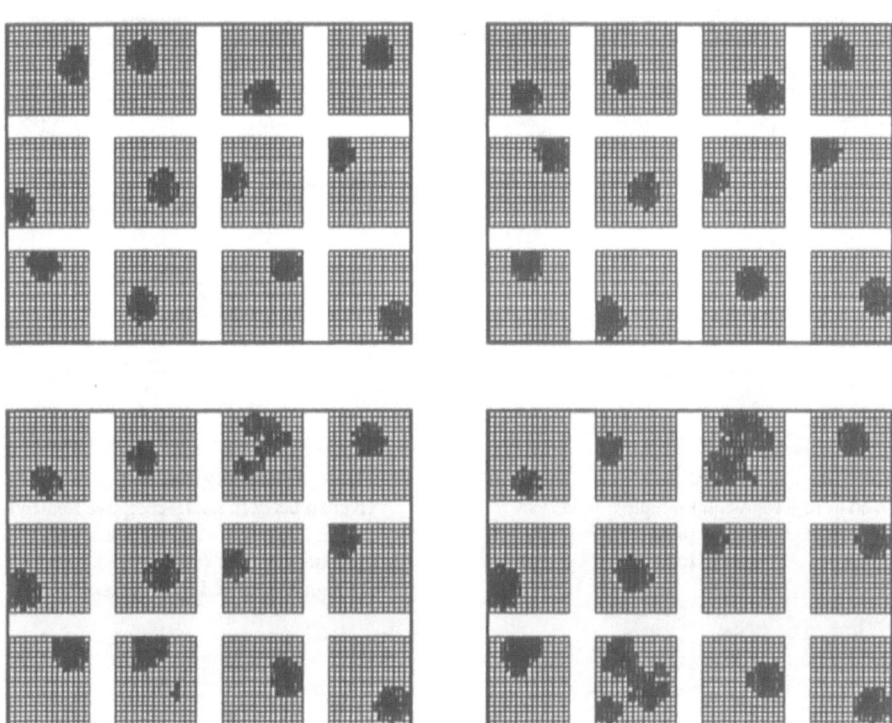

Figure 1:
Receptive fields formed by 12 neurons with 270 dimension input vector after receiving inputs at different positions. Deviations of neuron activity saturation value (SV) for the four sets from the top to the bottom are, respectively: 0%, 30%, 50% and 80%. 50% deviation means, for example, that SV ranges between 0.5 and 1.5.

Learning is somewhat slower, but curves are somewhat better for learning equation (3). Learning under noisy conditions (with noise amplitudes up to 40% of signal amplitudes were tested under similar conditions. Noise degraded the receptive fields in a minor fashion: (i) the network learnt the presence of noise, receptive fields became a noisy, and (ii) that was compensated by the inhibitory connections between far away neurons distorting the overlap vs. anti-Hebbian weight 'function', but thus improving performance.

4 Conclusions

Self-organized Hebbian and anti-Hebbian learning is capable to perform 'quasi-orthogonalization' with a very broad distribution of neural parameters within the network. It is not only the different learning rates, the neural activity values, the and the sharpness of the sigmoid function that may differ from the neuron, but the very form of neural equation can be very different and the network still performs the same task. The network was tested against tolerance to input noise, both during training when evaluated.

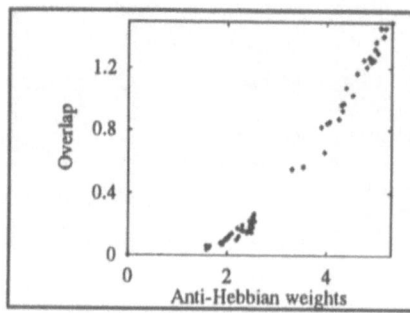
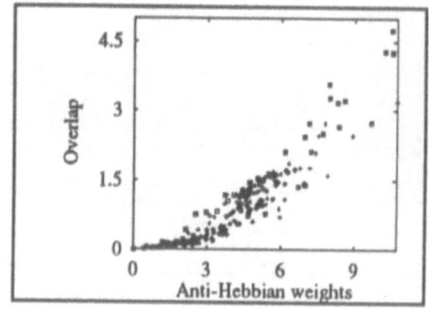

Figure 2:

Overlap between neural receptive fields vs. anti-Hebbian weights for the 0% case of Fig. 1.

Figure 3:

Overlap between neural receptive fields vs. anti-Hebbian weights for the 30%, (diamonds), 50% (squares) and 80% ('x' signs) cases of Fig. 1, respectively.

References

Földiák, P., (1991). *Biological Cybernetics*, **64**, 165-170.

Petsche, T. and Dickinson, B.W., (1990). *IEEE Trans. on Neural Networks*, **2**, 154-166.

Protzel, P.W., Palumbo, D.L., and Arras, M.K. (1993). *IEEE Trans. on Neural Networks*, **4**, 600-614.

Sejnowski, T.J., and Rosenberg, C.R., (1986). Tech. Rep. JHU/EECS-86/01, *John Hopkins University*.

Computational Experiences of New Direct Methods for the On-line Training of MLP-Networks with Binary Outputs

M. Di Martino†, S. Fanelli†, M. Protasi†

†Dipartimento di Matematica, Università di Roma - Tor Vergata
Roma (Italy)

1 Introduction

In this paper the authors consider some recent "direct methods" for the training of a three-layered feedforward neural network with thresholds: the Algorithm II and III (Bärmann et al., 1992) , here named $FBFBK$, the Least Squares Backpropagation (LSB) Algorithm (Bärmann et al., 1993), and their innovative method called Iterative Conjugate Gradient Singular Value Decomposition ($ICGSVD$) Algorithm (Di Martino et al., 1993).

These methods are based on the direct determination of the matrices of weights V and W (respectively the matrix of weights from the input layer to the hidden layer and the matrix of weights from the hidden layer to the output layer), by solving the general equation on the weights matrices, derived by the nonlinear condition of perfect learning.

With a suitable choice of the activation function (f.i. invertible) the latter condition can be reduced to a system of linear equations in the unknown matrix W. These equations may be solved in the classical or in the least-squares sense by means of a numerical direct method or a numerical iterative method.

A typical disadvantage of classical "direct methods", like f.i. the simple SVD (Franzoi et al., 1991), depends upon the fact that a large number of neurons in the hidden layer is in general a necessary condition to obtain satisfying results. The methods illustrated in this paper overcome the latter difficulty.

Numerical experiences show that the ICGSVD algorithm is particularly efficient even with a low number of units in the hidden layer and is the most competitive in general.

2 Direct Methods for MLP-networks

Let us consider a three-layered feedforward neural network having m input units, n hidden units and r output units. Let T be the matrix of the p targets, f be an invertible activation function, and $Q = (q_{ik})$ $i = 1, \ldots, p$ $k = 0, 1, \ldots, m$ be the matrix of patterns, being $q_{i0} = 0.5$ for $i = 1, \ldots, p$. Furthermore, let $V = (v_{kj})$ $k = 0, 1, \ldots, m$ $j = 1, \ldots, n$ denote the matrix of weights from the input layer to the hidden layer and $W = (w_{jl})$ $j = 0, 1, \ldots, n$ $l = 1, \ldots, r$ denote the matrix of weights from the hidden layer to the output layer.

The methods we are dealing with are based on the direct evaluation of the matrices of weights V and W, by solving in the classical or in the least-squares sense a set of systems of linear equations, derived by the condition of perfect learning.

A general scheme for the corresponding algorithms can be summarized by the following steps:
Choose V and W randomly.

Step 1. Compute $A = [0.5 \, f(QV)]$, i.e. the output matrix of the hidden layer. Evaluate V properly, such that $\| AW - f^{-1}(T) \|_2$ is minimized.

Step 2. Recompute A. Evaluate W by solving the system of linear equations $AW = f^{-1}(T)$.

Step 3. If $O := f(AW) = $ the output matrix of the output layer is close enough to T Stop, otherwise continue at Step 1.

In the FBFBK algorithm (Bärmann et al., 1992) the matrices of weights are computed by an iterative procedure. More precisely, in the Step 1 the columns of V are determined as follows:
For $j = 1, \ldots, n$:

- Compute $D^j = f^{-1}(T) - F^j W$, being $\mathbf{F}^j_h = \begin{cases} 0 & \text{for } h = j \\ \mathbf{a}_h & \text{otherwise} \end{cases}$

  $h = 0, 1, \ldots, n$.
  Define a *targetvector* $\mathbf{D}^j_{.\lambda}$, chosen by means of suitable criteria (Bärmann et al., 1992).

- Solve $max_{\mathbf{v}_j} Z(\mathbf{v}_j) = <\mathbf{a}_j, \mathbf{D}^j_{.\lambda}>^2 / \| \mathbf{a}_j \|^2$, using a step-size-controlled gradient method.

- Compute $w_{jl} = <\mathbf{a}_j, \mathbf{D}^j_{.l}> / \| \mathbf{a}_j \|^2 \quad l = 1, \ldots, r$.

In the Step 2 W is evaluated with another cycle of the following iterations:
For $j = 0, 1, \ldots, n$:

- Compute D^j. Set $w_{jl} = <\mathbf{a}_j, \mathbf{D}^j_{.l}> / \| \mathbf{a}_j \|^2 \quad l = 1, \ldots, r$.

The ICGSVD algorithm (Di Martino et al., 1993) determines the matrix V by the same procedure of the Step 1 of FBFBK, improved by the use of the Fletcher and Reeves version of the Conjugate Gradient Method. On the other hand, Step 2 is performed by the following direct numerical method:

- Compute A^+, the Moore-Penrose pseudoinverse matrix of A, using the Singular Value Decomposition (SVD). Set $W = A^+ f^{-1}(T)$.

In the *LSB* algorithm (Bärmann et al., 1993), the matrix V is determined by solving a linear system of equations in the following way:

- Define $R := A + \Delta R$ i.e. the desired output matrix of the hidden layer. Solve $(\Delta R)W = f^{-1}(T) - AW$ in the following way:

  – Compute W^+. Set $\Delta R = (f^{-1}(T) - AW)W^+$.

- Evaluate R. Let S be the matrix containing all the columns of R except the first one. Solve $QV = f^{-1}(S)$, i.e. the linear equation derived by the condition of perfect learning for the hidden layer, as follows:

  – Compute Q^+. Set $V = Q^+ f^{-1}(S)$.

In the Step 2 the matrix W is determined just as in the ICGSVD algorithm, i.e.:

- Compute A^+. Set $W = A^+ f^{-1}(T)$.

However, in the LSB algorithm the QR decomposition and the Householder transformations are used to determine all the pseudoinverse matrices W^+, Q^+, A^+.

3 Numerical Results

We have compared the performances of FBFBK, LSB and ICGSVD. All the experiments were run on a VAX 6000-410. In the tables n indicates the number of hidden neurons, h the number of the iterations in the procedures, and $E = \sum_{i=1}^{p} \sum_{l=1}^{r} (t_{il} - o_{il})^2 / (p * r)$ the value of the global error function of the network.

The indepth analysis carried out in a forthcoming paper (Di Martino et al., to appear) has shown that a crucial aspect to estimate the performances of FBFBK, LSB and ICGSVD is the relationship among the cardinalities of m (input units), n (hidden units) and p (patterns). More precisely, it is useful to point out the following general considerations, clearly shown by the results of tables 1-2-3:

1. ICGSVD is the most competitive algorithm, particularly if $m < (p - 1)$ and $n < (p - 1)$
2. if $m \geq (p - 1)$ LSB has the same reliability of ICGSVD
3. FBFBK is in general less effective (Di Martino et al., 1993)

On the other hand, it is necessary to observe that if $m >> p$ the use of the QR decomposition in the LSB method is quite unsatisfactory since the latter algorithm is based on the evaluation of the pseudoinverse of a "strongly rectangular" matrix.

Table 1. The T-C problem (Di Martino et al., 1993).
In this example the SVD method was chosen to determine all the pseudoinverse matrices required by LSB since $m >> p$.
Table 2. The peak, valley and ridge problem (PVR-problem).
A set of fifteen patterns in $[-1,1]$, describing the characteristics of a surface, is presented to a network with five input units and three output units. Any output represents a type of surface.
Table 3. The PVR-problem.
The same problem using a network with thirty input units.

TABLE 1 - THE T-C PROBLEM (400 inputs, 1 output, 8 patterns)

	FBFBK			LSB		
n	h	E	CPU	h	E	CPU
2	100	8.361328×10^{-2}	37:59.68	1	8.468748×10^{-2}	0:07.05
4	17	1.296955×10^{-17}	13:50.56	2	7.792236×10^{-2}	0:07.76
7	10	1.462432×10^{-17}	15:41.37	1	1.745483×10^{-31}	0:06.99

	ICGSVD		
n	h	E	CPU
2	1	3.423875×10^{-34}	0:01.25
4	1	8.559689×10^{-35}	0:01.23
7	1	7.703720×10^{-34}	0:01.39

TABLE 2 - PVR-PROBLEM (5 inputs, 3 outputs, 15 patterns)

	FBFBK			LSB		
n	h	E	CPU	h	E	CPU
3	100	2.836594×10^{-3}	8.10	1	3.355680×10^{-2}	1.37
	100	2.520542×10^{-3}	7.94	1	3.494360×10^{-2}	1.18
5	100	1.364242×10^{-3}	12.50	2	5.202524×10^{-3}	1.45
	100	2.970451×10^{-4}	13.07	2	2.095847×10^{-2}	1.51
10	100	3.287814×10^{-6}	26.75	2	2.277521×10^{-3}	1.58
	100	1.642942×10^{-5}	28.20	2	2.766738×10^{-3}	1.67
14	100	2.038733×10^{-6}	41.90	1	4.718214×10^{-27}	1.32
	100	1.194361×10^{-5}	40.37	1	9.575108×10^{-25}	1.36

n	h	ICGSVD E	CPU
3	1	5.797037×10^{-17}	1.44
	1	2.116126×10^{-16}	1.46
5	1	2.913713×10^{-16}	1.78
	1	1.730792×10^{-17}	1.56
10	1	1.234135×10^{-17}	2.97
	1	1.140278×10^{-16}	2.85
14	1	1.174866×10^{-31}	3.99
	1	8.330894×10^{-18}	3.22

TABLE 3 - PVR-PROBLEM (30 inputs, 3 outputs, 15 patterns)

n	h	FBFBK E	CPU	h	LSB E	CPU
3	100	3.601286×10^{-8}	0:44.05	9	8.823223×10^{-17}	0:02.68
	200	4.479864×10^{-7}	1:29.03	7	9.339960×10^{-18}	0:02.37
5	200	1.121476×10^{-10}	2:32.98	7	1.150134×10^{-17}	0:02.69
	200	3.329009×10^{-8}	2:30.98	9	2.672660×10^{-17}	0:03.10
10	200	7.229494×10^{-16}	5:04.13	6	9.076152×10^{-8}	0:03.74
	200	3.436129×10^{-9}	5:14.99	6	1.167667×10^{-10}	0:03.46
14	200	2.155892×10^{-13}	7:24.22	9	3.167915×10^{-17}	0:05.52
	100	8.978214×10^{-10}	3:34.73	5	9.160435×10^{-17}	0:04.06

n	h	ICGSVD E	CPU
3	1	6.091646×10^{-17}	0:03.01
	2	4.762375×10^{-17}	0:04.29
5	1	1.583670×10^{-17}	0:03.47
	1	3.471988×10^{-18}	0:03.41
10	1	4.903112×10^{-18}	0:03.03
	1	1.197774×10^{-18}	0:03.27
14	1	1.756031×10^{-19}	0:03.96
	1	1.405982×10^{-20}	0:04.36

References

Bärmann, F., and Biegler-König, F. (1992) *On a class of efficient learning algorithms for neural networks.* Neural Networks, 5 , 139-144.

Bärmann, F., and Biegler-König, F. (1993) *A learning algorithm for multilayered neural networks based on linear least squares problems.* Neural Networks, 6 , 127-131.

Di Martino, M., Fanelli, S., and Protasi, M. (1993) *An efficient algorithm for the binary classification of patterns using MLP-networks.* IEEE International Conference on Neural Networks, San Francisco, II, 936-943.

Di Martino, M., Fanelli, S., and Protasi, M. (1993) *A new improved on-line algorithm for multi-decisional problems based on MLP-networks using a limited amount of information.* IJCNN'93 - Nagoya, I, 617-620.

Di Martino, M., Fanelli, S., and Protasi, M. *Exploring and comparing the best direct methods for the efficient training of MLP-networks with binary outputs.* - to appear.

Fletcher, R., and Reeves, C.H. (1964) *Function minimization by conjugate gradients.* Comput.J., 7.

Franzoi, P., Johnson, O.G., and Pieroni, G.G. (1991) *Comparing the performance of the generalized delta rule with singular value decomposition based global algorithms for simple vision tasks.* IV Italian Workshop: Parallel Architectures and Neural Nets, Vietri sul Mare 181-190.

Optimising Local Hebbian Learning: use the δ-rule

J.W.M. van Dam[*] B.J.A. Kröse, F.C.A. Groen

Faculty of Mathematics and Computer Science

University of Amsterdam

Kruislaan 403

NL - 1098 SJ Amsterdam

1 Introduction

We consider problems where the *correlation* between signals $\mathbf{x} = (x_1, \ldots, x_n)^T$ and $\mathbf{y}^* = (y_1^*, \ldots, y_m^*)^T$ is to be learned supervisedly[1]. This is useful in, e.g., sensor fusion applications, to correlate different representations of the same scene (see, e.g., [Gielen et al., 91]). Traditionally, *local Hebbian learning* is suggested to approach these problems. Here, it is proven, that if continuous-valued learning samples are available, the δ-rule gives better performance. For binary-valued learning samples, it is shown that the Hebb rule does not converge to the optimal weight values. Therefore, a modification to the Hebb rule is introduced and it is shown that the δ-rule is to be modified similarly.

Problems of the kind mentioned above can be approached with a single layer feed-forward neural network with input units $x_1, \ldots, x_n$ and output units $y_1, \ldots, y_m$. Each unit y_i is connected to all input units x_j with weights w_{ij}. The optimal weight values w_{ij}^* are given by the correlations:

$$w_{ij}^* = E[y_i^* x_j] = \int \int y_i^* x_j f(y_i^*, x_j) dy_i^* dx_j \tag{1}$$

In this article, we compare two different learning algorithms to train such a network: the Hebb rule and the δ-rule.

2 Continuous-valued learning samples

We assume learning samples $(\mathbf{y}^*, \mathbf{x})$ are available where the inputs $\mathbf{x}$ are uncorrelated, $E[x_j x_k] = 0$ for $k \neq j$, with zero mean, $E[x_j] = 0$, and with variance $\alpha_j = \int x_j^2 f(x_j) dx_j$. The outputs $\mathbf{y}^*$ are characterized by:

$$E[\mathbf{y}^*] = \mathbf{H}\mathbf{x}$$

[*]The investigations were supported by the Foundation for Computer Science in the Netherlands (SION) with financial support from the Netherlands Organisation for Scientific Research (NWO).

[1]We write $\mathbf{y}^*$ for the "optimal" or "true" signal, whereas $\mathbf{y}$ denotes an estimation of this signal.

where $\mathbf{H} = \mathbf{R}_{yx} \cdot (\mathbf{R}_{xx})^{-1}$, $\mathbf{R}_{yx} = E[\mathbf{y}^*\mathbf{x}^T]$, the correlation matrix, and $\mathbf{R}_{xx} = \alpha\mathbf{I}$, the autocorrelation matrix.

The Hebb rule initializes all weights to 0 and updates according to:

$$\Delta w_{ij} = \frac{1}{N}y_i^* x_j \qquad (2)$$

with N the number of learning samples. Therefore, the network converges to

$$\lim_{N \to \infty} w_{ij} = \mathcal{E}[y_i^* x_j] = \int \int y_i^* x_j f(\mathbf{y}^*, \mathbf{x})dy^* dx \qquad (3)$$

The value $\mathcal{E}[y_i^* x_j]$ is an *estimation* of the true correlation $E[y_i^* x_j]$ and this is not necessarily a correct estimation since the Hebbian algorithm estimates all $m \times n$ correlations in parallel. We have

$$\mathcal{E}[y_i^* x_j] = \int \int y_i^* x_j f(\mathbf{y}^*, \mathbf{x})dy^* dx = \ldots = \int x_j f(x_j)\mathcal{E}[y_i^* \mid \mathbf{x}]dx_j$$

with

$$\mathcal{E}[y_i^* \mid \mathbf{x}] = \mathcal{E}[\frac{1}{\alpha_j}\sum_j w_{ij}^* x_j \mid \mathbf{x}] = \frac{1}{\alpha_j}w_{ij}^* x_j + \mathcal{E}[\sum_{k \neq j}\frac{1}{\alpha_k}w_{ik}^* x_k \mid \mathbf{x}] = \frac{1}{\alpha_j}w_{ij}^* x_j$$

which gives $\mathcal{E}[y_i^* x_j] = w_{ij}^*$; the algorithm converges to the optimal value.

Similarly, we can also calculate the *variance* $var[y_i^* x_j]$. This value gives a measure of how close to our optimum we may expect our algorithm to converge. Naturally, this measure can also be calculated if instead of the Hebb-rule, the δ-rule is applied. If linear activation values are used, the δ-rule takes

$$w_{ij}^{\text{new}} = w_{ij} + (y_i^* - y_i)x_j$$

and the variance in the δ-rule is given by (see [Hammersley, Handscomb (1965)]):

$$var[w_{ij} + y_i^* x_j - y_i x_j] = var[y_i^* x_j] + var[y_i x_j] - 2cov[y_i^* x_j, y_i x_j] \qquad (4)$$

So, if we can prove that $2cov[y_i^* x_j, y_i x_j] > var[y_i x_j]$, we may expect the δ-rule to converge closer to the optimal weight values than the Hebbian learning rule. We take for the activation function $y_i = \sum_j \frac{w_{ij}}{\alpha_j}x_j$. The calculation of the terms in (4) is straightforward, and differs little from the calculation of $\mathcal{E}[y_i^* x_j]$. The results are, that if $\forall i, j \; 2|w_{ij}^*| > |w_{ij}|$, we have $2cov[y_i^* x_j, y_i x_j] > var[y_i x_j]$. And thus, if we are careful with the initialisation of the weights w_{ij}, the δ-rule can be expected to give better performance than the Hebb-rule.

3 Binary-valued Learning samples

In some cases, we may only have binary-valued, non-linear learning samples (see [Van Dam et al., (1993)] for an example). In this case, the correlation between y_i^* and x_j is interpreted as:

$$w_{ij}^* = E[y_i x_j] = P(y_i = 1 \mid x_j = 1)$$

We assume that all possible patterns are equally distributed, i.e., $P(x_j = 1) = \frac{1}{2}$.

For the Hebb rule, we take $\forall i, j\ y_i^*, x_j \in \{-1, +1\}$. The algorithm converges to:

$$\mathcal{E}[y_i^* x_j] = \sum_{a=+1,-1} \sum_{b=+1,-1} a \cdot b \cdot P(x_j = a\ \&\ y_i^* = b)$$

Using elementary probability theory, we obtain:

$$\mathcal{E}[y_i^* x_j] = w_{ij}^* u_{ij}^*$$

where

$$u_{ij}^* = P(y_i^* \text{ not set to } +1 \text{ by any } x_k, k \neq j) = \prod_{k \neq j}(1 - w_{ik}^* \tilde{x}_k)$$

with $\tilde{x}_k = \frac{x_k+1}{2}$ such that $\tilde{x}_k \in \{0, 1\}$. This follows directly from the definition of the learning samples.

In other words, the Hebb rule does not converge to the optimal weight value. We can, however, formulate a modified Hebbian Learning rule, which is applicable to binary-valued learning samples, by simply modulating (2) with $1/u_{ij}^*$. We can calculate estimations of u_{ij}^* by using the current weight values w_{ij}. However, the Hebb rule is very sensitive to bad estimations.

If we want to train the network with the δ-rule, we take $\forall i, j\ y_i^*, x_j \in \{0, 1\}$. The network activation function must be chosen to suit the definition of the learning samples:

$$y_i = P(y_i^* \text{ set to } +1 \text{ by } at\ least\ one\ x_j)$$

From combinatorial probability theory this probability is given by:

$$y_i = \sum_{k=1}^{n}(-1)^{k-1} \sum_{0 \leq j_1 < \ldots < j_k \leq n} w_{ij_1} x_{j_1} \cdot \ldots \cdot w_{ij_k} x_{j_k} = 1 - \prod_j (1 - w_{ij} x_j) \quad (5)$$

Now that we have a different activation function, the δ-rule gives:

$$\Delta w_{ij} = -\gamma \frac{\partial}{\partial w_{ij}}(y_i^* - y_i)^2 = 2\gamma(y_i^* - y_i)x_j \prod_{k \neq j}(1 - w_{ik} x_k) \quad (6)$$

Not surprisingly, the modified δ-rule introduces a modification which is similar to the proposed modification of the Hebb-rule.

We have compared the δ-rule to the Hebb rule in a similar way to section 2, but a fair comparison is difficult to make (see [Van Dam et al., (1994)]). Nevertheless, the δ-rule can be expected to give better performance since it is less sensitive to bad estimations of the modification term.

4   Implementation and results

We tested the algorithms presented in the previous sections on two grids $\mathbf{y}^*$ and $\mathbf{x}$ of 16×16 square cells each, which are rotated and translated with respect

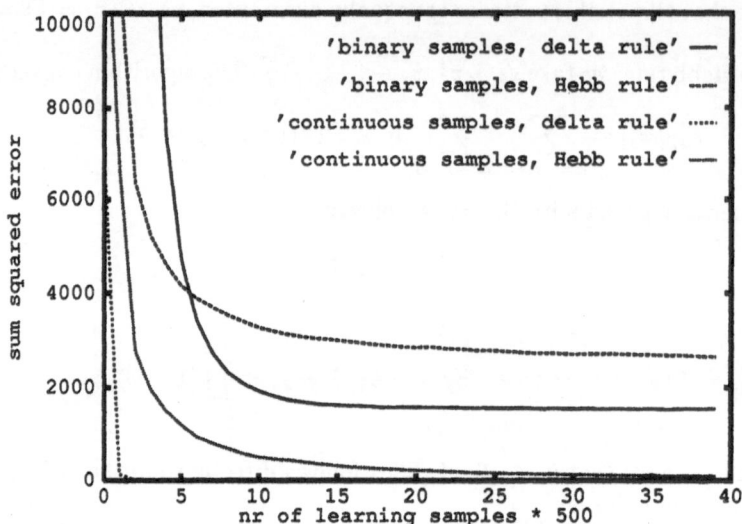

Figure 1: Comparing the δ-rule to the Hebbian learning rule. Plotted are the squared errors summed over all 256 output units and over 1000 continuous-valued test samples against the number of learning samples presented.

to each other. We define the correlation w_{ij}^* as the *intersection* of cell y_i^* with cell x_j. Continuous valued learning samples are constructed by taking random values for each x_j and by taking $y_i^* = \sum_j w_{ij}^* x_j$, where weight values w_{ij}^* are given by calculating the intersections of cells based on a-priori knowledge. To create binary patterns, we take a number of random points in the grid. We set $x_j = 1$ if a point is present somewhere in its cell. The same is done for values y_i^*. All values not set to 1 are set to -1 for the modified Hebb-rule and to 0 for the modified δ-rule . For the Hebb rule, we calculated u_{ij}^* using w_{ij}^* (which makes the Hebb rule look good). Note that to create binary patterns, we don't need a-priori knowledge on the values of w_{ij}^*. Results for all four learning rules are given in figure 1. Both if continuous valued learning samples are available and if only binary samples are available, the δ-rule gives better performance.

References

[Gielen et al., 91] C.Gielen, K.Krommenhoek, J. van Gisbergen (1991), "A procedure for self-organised sensor-fusion in topologically ordered maps". Proceedings of the Second International Conference on Autonomous systems, 417-423.

[Hammersley, Handscomb (1965)] J.M. Hammersley, D.C. Handscomb (1965), "Monte Carlo Methods", 50-76.

[Van Dam et al., (1993)] J.W.M. van Dam. B.J.A. Kröse, F.C.A. Groen (1993), "Transforming Occupancy Grids Under Robot Motion", ICANN93 Amsterdam, the Netherlands, 318.

[Van Dam et al., (1994)] J.W.M. van Dam. B.J.A. Kröse, F.C.A. Groen (1994), "On Local Hebbian Learning", internal report.

Efficient Neural Net α-β-Evaluators

Alois P. Heinz[*]

Institut für Informatik, Universität Freiburg

D-79104 Freiburg, Germany

1 Introduction

We describe a new artificial neural network for the conceptual realization of α-β-evaluators which are extensively used as the central part of many game playing programs. Then we present a very efficient implementation of this network that provides a speed-up of recall time in the order of magnitudes on conventional computers as compared to sequential implementations of traditional neural network architectures.

The α-β-evaluator is a function of three arguments, a game position p and the boundaries α and β of a real interval $[\alpha, \beta]$, also called the α-β-*window*. It returns a real value $v(p, \alpha, \beta) \in [\alpha, \beta]$ that may be considered as a measure of the real strength of the position adjusted to the nearest point of the α-β-window. During the evaluation inquiries on a set of *features* of p can be made for the cost of additional computation time. A strong demand on the shape of any α-β-evaluation function is that it should be smooth in all parameters to prevent the so-called *blemish effect* which deteriorates the quality of the overall decision making process heavily (Berliner, 1980).

2 Setting up the network

Conceptually the α-β-evaluator is realized by a feedforward artificial neural network. Finding the right network architecture is an example of a *supervised learning* problem. The set of training feature vectors $X_c = \{x_{c,1}, \ldots, x_{c,m}\}^t$ with assignments of values v_c may be generated by an automatic procedure (Heinz et al., 1993). The topology of the network and the settings of its parameters are developed in a sequence of four steps.

The first step consists of generating from the training set a binary decision tree that is able to approximate the desired function. Each inner node k of the tree is labeled with an index k_i and a constant k_c and each leaf L is labeled with a value $v(L)$. Informally, when a new vector X is to be evaluated by the tree, a path from the root to a leaf is traced. At an inner node k the path is extended to the left, if $x_{k_i} - k_c \leq 0$, and to the right in the other case. At the uniquely defined leaf L the value $v(L)$ is returned. Any of the acknowledged tree building procedures (Mingers, 1989a, 1989b) that are based on methods from information theory and statistics may be used in this step.

In the second step the tree is transformed into a network N. The input layer of N consists of one neuron for each of the feature values and for the values of α and β. The *decision layer* consists of the decision neurons k, which correspond to the inner nodes of the tree. Each neuron k receives input only from the

---

[*]This research was partly supported by the Academy of Finland.

feature neuron for x_{k_i} and is equipped with a threshold k_c. It computes the *sharp decision function*

$$d_k(X) := \begin{cases} 0 & \text{if } x_{k_i} - k_c \leq 0, \text{ and} \\ 1 & \text{else.} \end{cases}$$

The neurons of the *AND layer* correspond to the leaf nodes L of the tree. They compute the product of the input values which they receive from all neurons in the decision layer that correspond to ancestor nodes of the corresponding leaf of the tree. Any connection between these two layers computes for a pair (k, L) the *decision contribution* $D_{k,L}(X)$ that is defined to be $d_k(X)$ if L belongs to the right subtree of k and $(1 - d_k(X))$ if L belongs to the left subtree. Each neuron L of the AND layer feeds its output via a connection with weight $v(L)$ into a OR neuron that computes the sum of all its inputs. The maximum of its output value and the value of α is computed by a MAX neuron. Then a MIN neuron computes the minimum of this value and the value of β as the output of the network.

In the third step the network that up to now only mimics the behavior of the decision tree is changed by replacing the sharp decision functions by softer versions. These are based on a real function $f(x)$, that is constant 0 for $x \in (-\infty, -1)$, constant 1 for $x \in (1, \infty)$ and $f(x) = 1/2 + x - x|x|/2$ for $x \in [-1, 1]$. The *soft decision function* of neuron k is defined to be

$$d_k^*(X) := f\left(\frac{x_{k_i} - k_c}{R_k}\right).$$

The strictly positive parameter R_k controls the *radius of softness* of neuron k. Larger values of R_k cause softer decisions. For $R_k = 4$ the decision function of k approximates very closely to the logistic function $t(x) = \frac{1}{1+e^{-x}}$ with a maximal aberration of less than 3 %. The evaluation of the network N for an input vector X and α and β can be described by the formula

$$v_N(X, \alpha, \beta) = \min\left(\beta, \max\left(\alpha, \sum_{L \in \text{AND-layer}(N)} v(L) \prod_{k \in \text{pred}(L)} D_{k,L}^*(X)\right)\right),$$

where the set $\text{pred}(L)$ contains all predecessor decision neurons of neuron L. The functions $D_{k,L}^*(X)$ are defined according to $D_{k,L}(X)$ but with the soft decision functions instead of the sharp ones.

The network's topology is now fully determined and with some choice of small R_k values its evaluations are close to the already good results of the tree. But in the fourth step the net is retrained on its free parameters, namely the c_k, R_k, and $v(L)$ values to improve the evaluation. Only a few steps of incremental adaption are needed as compared to randomly initialized nets. As training method the quickprop algorithm (Fahlman, 1988) is used because it shows faster convergence than back-propagation (Rumelhart et al., 1986) and does not have problems with the derivative of function f that is non-zero only in the interval $(-1, 1)$.

3 Implementing the network

The fastest implementation of our network would make use of extensive parallel hardware. But this is not always a feasible way, for several reasons. We therefore present a sequential implementation of the α-β-evaluation network on a binary decision tree. We only give the details of the data structure and the evaluation algorithm. From this description it should become clear that the incremental learning algorithm can be carried out on the tree structure as well.

Using the same correspondence between network and tree as in the last section we only need to describe what information from the net is stored where in the tree and then state the evaluation algorithm. Each inner node k is labeled with the index k_i of the feature that is used in the decision, with the threshold k_c, and with the radius of softness R_k. And k has pointers to its left and right descendants k_l and k_r. Each leaf L of the tree is labeled with its value $v(L)$. Additionally, any node q is labeled with two values, q_α and q_β which are the minimum and maximum of all $v(L)$ values in the subtree of the tree with root q, respectively. The complete recursive evaluation algorithm is given as follows:

```
function evaluate (p: position; α, β: real; k: node): real;
begin
    if kβ ≤ α then return (α);
    if kα ≥ β then return (β);
    if leaf (k) then return (v(k));
    if unknown [kᵢ] then begin
        x[kᵢ] := get_feature (p, kᵢ);
        unknown [kᵢ] := false
    end;
    d* := f ((x[kᵢ] − kc)/Rk);
    if d* = 0 then return (evaluate (p, α, β, kₗ));
    if d* = 1 then return (evaluate (p, α, β, kᵣ));
    αₗ := max ((α − k_rβ · d*)/(1 − d*), −1);
    βₗ := min ((β − k_rα · d*)/(1 − d*), +1);
    vₗ := evaluate (p, αₗ, βₗ, kₗ) · (1 − d*);
    αᵣ := max ((α − vₗ)/d*, −1);
    βᵣ := min ((β − vₗ)/d*, +1);
    vᵣ := evaluate (p, αᵣ, βᵣ, kᵣ) · d*;
    return (vₗ + vᵣ)
end;
```

Initially this function is called with the root of the tree as last argument and the global vector of unknown-flags for the features completely set to true. This is to assure that each feature will be computed only if and only the first time it is needed. After the first reference it is stored in the vector x for repeated later reuse.

One idea of the algorithm is to use the fact that the $D^*_{k,L}(X)$ terms are hierarchically ordered. It's therefore sufficient to trace only those paths from the root to the leafs that proceed along branches with a non-zero decision contribution. The other idea exploits the available knowledge about the range of values that are contained in a given subtree with root k. If the intersection of this range $[k_\alpha, k_\beta]$ with the α-β-window is empty then further computation is needless and the nearest value of this window is returned. If in fact k is a leaf, its stored value $v(k)$ can be retuned. Otherwise the soft decision function $d^*_k(X)$ is computed. If it is 0 then the decision contribution of k and the leafs of the right subtree is 0. So only the evaluation of the left subtree has to be returned. The case of $d^*_k(X) = 1$ is handled symmetrically.

638

The interesting case is when $d_k^*(X)$ is somewhere between 0 and 1. Then both branches may contribute to the evaluation, the left one with a factor of $(1 - d_k^*(X))$ and the right one with a factor of $d_k^*(X)$. The α-β-windows of the two branches have to be adapted according to these scaling factors and because some pre-knowledge is available. In this case the sum of the evaluations of left and right subtrees is returned.

To estimate the algorithm's average case runtime we count the average number of branches $\ell_{ac}(n,p)$ traversed during an evaluation without α-β-cut-offs occurring as a function of the number of leafs n and p, which is assumed to be the probability that an inner node operates within its radius of softness. By imposing a balancing constraint during the tree building procedure it can be guaranteed that the height of the tree is limited by $c \log_2 n$ for some $c \geq 1$. Thence $\ell_{ac}(n,p)$ can be derived as

$$\ell_{ac}(n,p) = (1+p)\left\{ c \log_2 n + \binom{c \log_2 n}{2} p + \binom{c \log_2 n}{3} p^2 + \cdots \right\},$$

which grows slowly with n for small p. At any rate, $\ell_{ac}(n,p)$ is much smaller than the $O(n^2)$ steps required by a sequential implementation of the fully connected network with n neurons in the AND layer and the $O(nc \log_2 n)$ steps required for the sparsely connected network derived from a decision tree. A further speed-up is caused by α-β-cut-offs that even may occur at the root.

4 Conclusion

We have presented a new artificial neural network with a very efficient sequential implementation. The efficiency is achieved by mapping the network onto a tree and its algorithms that exploit the sparse connection structure, hierarchical connection ordering, vanishing decision contributions in the exterior of the radius of softness, and α-β-cut-offs. This strategy results in a system that combines the advantages of the neural net, namely smoothness, adaptability, and simplicity with the profits of the tree, namely efficiency.

References

Berliner, H. J. (1980). Backgammon computer program beats world champion — performance note. *Artificial Intelligence*, **14**, 205–220.

Fahlman, S. E. (1988). *An empirical study of learning speed in back-propagation networks* (CMU-CS-88-162). Pittsburgh, PA: Carnegie-Mellon University.

Heinz, A. P., & Hense, C. (1993). Bootstrap learning of α-β-evaluation functions. *Proceedings of the Fifth International Conference on Computing and Information (ICCI'93)*, Sudbury, Canada, 1993.

Mingers, J. (1989a). An empirical comparison of selection measures for decision-tree induction. *Machine Learning*, **3**, 319–342.

Mingers, J. (1989b). An empirical comparison of pruning methods for decision-tree induction. *Machine Learning*, **4**, 221–243.

Rumelhart, D. E., Hinton, G. E., & Williams, R. J. (1986). Learning internal representations by error propagation. In D. E. Rumelhart, J. L. McClelland and the PDP Research Group (Eds.), *Parallel distributed processing: Explorations in the microstructure of cognition, Vol. 1* (pp. 318-362). Cambridge, MA: MIT Press.

A Parallel Algorithm for a Dynamic Eta/Alpha Estimation in Backpropagation Learning

M. Raus, W. Ameling

Rogowski Institute at the Aachen University of Technology

Aachen, Germany

1 Introduction

Since Rumelhart published his backpropagation method, many variations about various aspects of the algorithm have been developed. Most of them are trying to improve the generalization ability and the learning speed, but comparable efforts were made to reduce the degree of freedom in designing a neural network for special applications. The standard backpropagation algorithm with momentum term offers the following influence quantities:

- number of hidden layers h,

- number of hidden neurons $n(1) \ldots n(h)$,

- learning rate η,

- momentum rate α.

In general the dimensionalities of the input and output layer are determined by the application demands and/or the data representations.

The *optimal* setting of these quantities represents a hard problem, every domain or neural system engineer has to cope with. Even finding a *suboptimal* but sufficient solution requires a huge amount of experience. Heuristic methods, trial–and–error, and the use of analogies are the today's most commonly used strategies.

2 Dynamic estimation of learning and momentum rate

On the way to a reduced set of parameters and therefore to a more efficient neural network design, the presented algorithm allows the estimation of a *proper* (η, α) combination. During the training phase a search strategy moves the network through the η/α parameter space (see fig.1). This training procedure supplies a pair $(\tilde{\eta}, \tilde{\alpha})$ that can be used to train the network to a succesful state. But furthermore, the resulting network of the estimation procedure converges in a stable state that generally turns out to be very near to the state after a training with the optimal set of *constant* parameters. In this state the network shows a good generalization behaviour.

This work has been supported by the Deutsche Forschungsgemeinschaft, Bonn, Germany.

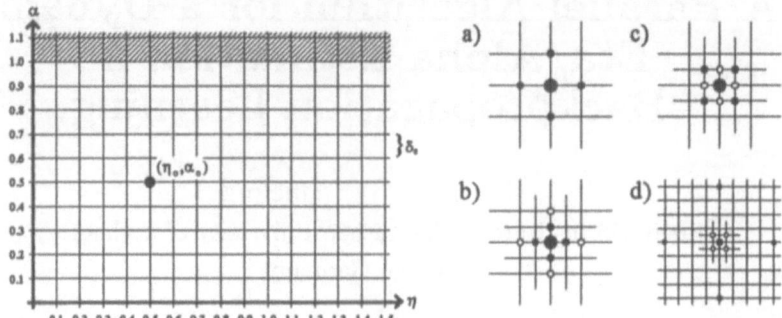

Figure 1: The η/α parameter space with the starting point (η_0, α_0) of the iterative procedure. The grid represents the reachable combinations with a distance value $\delta = 0.1$. The momentum rate is limited to values $\alpha < 1.0$. The detail drawings a)–d) show the main strategies used by the algorithm, drawing d) is presented in a reduced scale. O = former values, ● = new values

The iterative process runs through several processing stages. During the *initialization* phase, the grid width δ is set to the standard value δ_0 and the *rate vectors*

$$\mathbf{r}_i = (\eta_i, \alpha_i) \quad \text{with} \quad i = 0, \dots, 4$$

of the *pivot network* and its four orthogonal neighbouring nets (see figure 1.a) are initialized as follows:

$$
\begin{aligned}
\eta_0 &= \eta(n) & \alpha_0 &= \alpha(n) \\
\eta_1 &= \eta(n) & \alpha_1 &= \alpha(n) + \delta_0 \\
\eta_2 &= \eta(n) - \delta_0 & \alpha_2 &= \alpha(n) \\
\eta_3 &= \eta(n) & \alpha_3 &= \alpha(n) - \delta_0 \\
\eta_4 &= \eta(n) + \delta_0 & \alpha_4 &= \alpha(n)
\end{aligned}
\tag{1}
$$

with $\eta(n = 0)$ and $\alpha(n = 0)$ set to reasonable values, e.g. in the center of the parameter space.

In the following *training* phase these five networks are trained with a fixed number of patterns or during a fixed number of epochs. Each of the networks starts from the same point in the weight space, i.e. with the same random initialization of the weight matrices. At the end of the training phase the *fittest* network is chosen as the pivot net of the next iteration $n + 1$, i.e.

$$\eta(n + 1) = \eta_\nu \quad \text{and} \quad \alpha(n + 1) = \alpha_\nu \tag{2}$$

if ν is the index of the fittest network. The criterion of fitness is the total error over a complete training set recall.

In figure 1 the detail drawings a)–d) show the main strategies used by the algorithm in the subsequent *alteration* phase:

- in the *standard* mode (a) the orthogonal neighbours in a distance of δ_{std} are trained,

- for the *reduced* mode (b) the distance δ_{std} is reduced to δ_{red},

- the *diagonal mode* (c) uses the diagonal neighbours in a distance of $\delta_{dia} = \sqrt{2} \cdot \delta_{red}$,

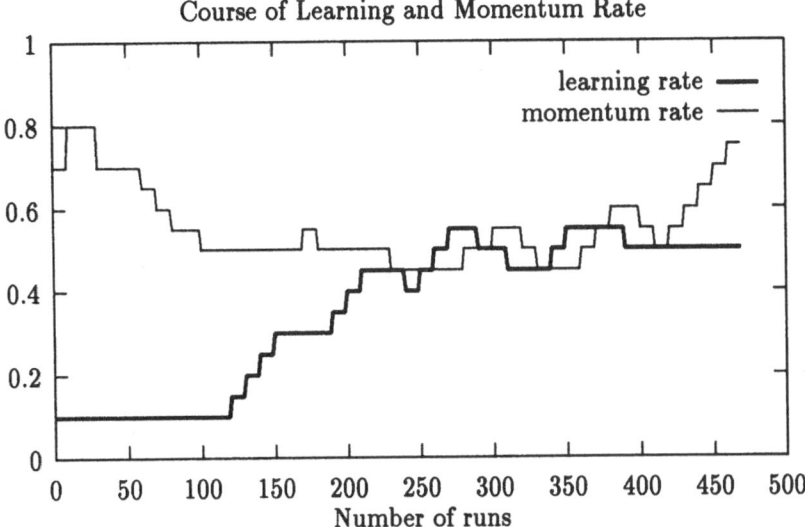

Figure 2: The diagram shows the development of the learning and momentum rate during the training of a 45–300–300–100 network, trained to recognize the 3D rotation of a simple object by generalizing from 160 *characteristic* out of 23 040 *possible* views.

- and for the *escape* mode (d) again the orthogonal neighbours but now in a extremely enlarged distance δ_{esc} define the training parameters.

The selection of the adequate modification strategy is primarily based on the number of iterations without changing at least one of the parameters. So, if

$$\eta(n+1) = \eta(n) \quad \text{and} \quad \alpha(n+1) = \alpha(n)$$

the *no changes counter* (nc) is incremented by one. Two iterations without parameter changes are tolerated $(nc \leq 1)$ before the first modification strategy becomes active:

$$\delta(n+1) = \frac{\delta(n)}{2} \tag{3}$$

as long as δ exceeds a predefined threshold value (see figure 1.b). If the actual grid width $\delta(n)$ falls below this value, it remains unchanged $(\delta(n+1) = \delta(n))$ and the diagonal neighbours in the η/α space are used instead (see figure 1.c). Therefore the rate vector components for the next iteration $n+1$ are set to:

$$
\begin{aligned}
\eta_0 &= \eta(n+1) & \alpha_0 &= \alpha(n+1) \\
\eta_1 &= \eta(n+1) - \delta(n+1) & \alpha_1 &= \alpha(n+1) + \delta_0(n+1) \\
\eta_2 &= \eta(n+1) - \delta(n+1) & \alpha_2 &= \alpha(n+1) - \delta_0(n+1) \\
\eta_3 &= \eta(n+1) + \delta(n+1) & \alpha_3 &= \alpha(n+1) - \delta_0(n+1) \\
\eta_4 &= \eta(n+1) + \delta(n+1) & \alpha_4 &= \alpha(n+1) + \delta_0(n+1)
\end{aligned}
\tag{4}
$$

The diagonal mode is only used once, i.e. if the parameters still remain constant after the next iteration, the algorithm attempts to *escape* from this point in the parameter plane by enlarging the grid width (see figure 1.d). Only networks with legal parameter combinations take part in the training phase. Whenever the pivot network moves inside the η/α space, the nc counter is set to 0.

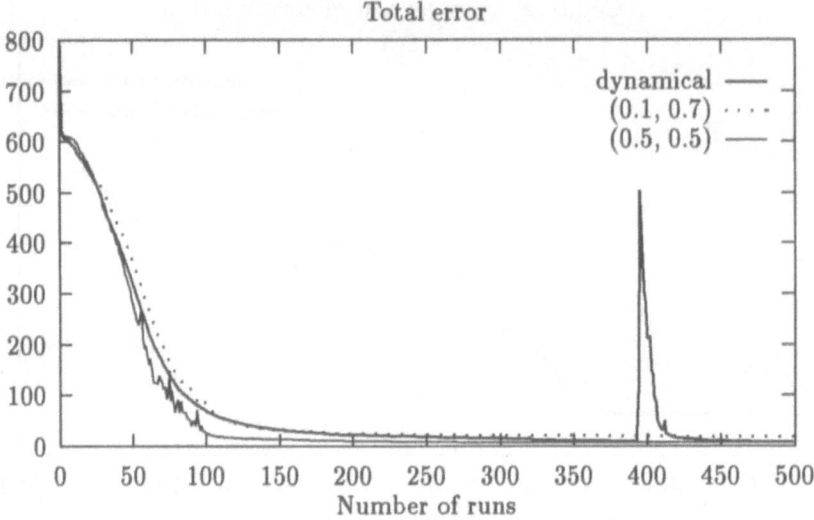

Figure 3: The diagram shows three different total error curves from the training example described in figure 2. One error curve represents the results of the dynamic algorithm and the other two are gained from experiments with constant parameter sets, the initialization set and the finally resulting set (see fig. 2).

After the alteration phase the new pivot network and its neighbour nets start again from the same weight space location, that is defined by the final state of the fittest network after the last iteration. This procedure is continued until one of the networks fulfils a defined breaking condition.

3 Experimental results

This estimation algorithm was used for many applications in different domains like image processing, welding control, or the manufacturing of plastic pieces. In all of these applications the procedure showed a very robust behaviour, i.e. the networks converged in a slightly increased number of epochs with nearly optimal generalization results. Figure 2 shows a typical course of the learning and momentum rates during a training phase. They start from an initialization $r_{start} = (\eta_0 = 0.1, \alpha_0 = 0.7)$ and iterate towards a state of *dynamic convergence*. In figure 3 the learning curve of the dynamic algorithm shows an error peak in the last part of the diagram, i.e. this dynamic convergence helps the network to escape even from deep local minima. Furthermore fig. 3 presents the training curve of a network that uses the initialization rates as constant values and reaches a worse minimum while the resulting network with constant rates determined by the algorithm converges very quickly into an optimal state.

It generally turned out that the single training of five networks with this dynamic algorithm is much more economic than repeated training runs with different parameter settings without a profitable strategy. Especially the possibility of a parallel implementation of the described algorithm remarkably increases the efficiency of a network design.

Dynamic Pattern Selection: Effectively training Backpropagation Neural Networks

A. Röbel

Department of Applied Computer Science, Technical University of Berlin

10587 Berlin, Germany

1 Introduction

It is well known that the generalization properties of neural networks used in function approximation are strongly affected by the size and distribution of the training set. However, the problem of selecting the optimal training set has not yet been solved. For a good generalization the training set has to contain enough information to fix the network function f_n not only at the training patterns, but on the domain $\mathbb{X}$ of the target function f_t.

To achieve this, we want to adapt the training set during training and employ the net function to decide which pattern should be chosen. Following this strategy the *active pattern selection* algorithm, which has been demonstrated to select surprisingly small training sets has been developed (Plutowski *et al.*, 1993). In contrast to this algorithm, the *dynamic pattern selection* proposed here, is constructed to effectively select the training sets by continually validating the generalization properties of the net (Röbel, 1993b).

2 Dynamic Pattern Selection

Assume that the task to solve consist of approximating a smooth target function f_t on its domain $\mathbb{X}$. For training a backpropagation network there are only a number of examples available which are in the set of available patterns $\mathbb{D}_a$.

As the distribution of the data in $\mathbb{D}_a$ is not adapted for neural network training, using all the data in $\mathbb{D}_a$ will, in general, not lead to optimal training results (Röbel, 1993b). We would like to select the training set $\mathbb{D}_t \subset \mathbb{D}_a$ such that minimizing the net error $N(\mathbb{D}_t)$ forces the network function f_n to regularly converge to f_t over the complete domain $\mathbb{X}$. The two questions to solve are: When should the new pattern be selected and which one should be chosen.

To answer the first question, we state that the training set shall be increased, whenever the generalization properties of the net becomes poor. As is known from cross validation, we divide $\mathbb{D}_a$ into the training and validation repertoire $\mathbb{D}_T$ and $\mathbb{D}_V$. The training data is selected from $\mathbb{D}_T$, while the validation set $\mathbb{D}_v$ is randomly chosen from $\mathbb{D}_V$ where $|\mathbb{D}_v| = |\mathbb{D}_t|$[1]. Based on theoretical investigations of the regular generalization of training sets, we define the generalization

---

[1] $|\mathbb{A}|$ denotes the cardinality of $\mathbb{A}$

training set size	training error $E(\mathbb{D}_s)$	generalization error $E(\mathbb{D}_u)$	generalization factor ρ_u	for/backward propagations
15	4.29e-3 ± 2.0e-3	2.50e-2 ± 1.7e-2	5.83	0.54e+6
25	3.77e-3 ± 1.8e-3	4.80e-3 ± 2.0e-3	1.27	0.90e+6
35	2.46e-3 ± 1.2e-3	2.90e-3 ± 1.4e-3	1.18	1.26e+6
50	3.02e-3 ± 1.6e-3	3.46e-3 ± 1.8e-3	1.15	1.80e+6
75	2.99e-3 ± 1.5e-3	3.14e-3 ± 1.7e-3	1.05	2.70e+6
100	2.90e-3 ± 1.6e-3	3.21e-3 ± 1.7e-3	1.10	3.60e+6
150	4.53e-3 ± 2.1e-3	5.08e-3 ± 2.4e-3	1.12	5.40e+6
69 ± 10	3.69e-3 ± 1.7e-3	3.06e-3 ± 1.1e-3	0.83	0.74e+6

Tab. 1. Predicting the Henon model. Comparison of the average training and generalization rms error using a 2-7-1 neural net and different training sets. The patterns in the fixed sets are approximately equally spaced on the henon attractor.

factor

$$\rho = \frac{N(\mathbb{D}_v)}{N(\mathbb{D}_s)}, \tag{1}$$

which is a normalization of the generalization error measured by means of the backpropagation error function $N()$. The generalization factor indicates the error we make in optimizing on $\mathbb{D}_s$ instead of $\mathbb{X}$. For regular generalization we demand that

$$\rho \leq 1.0. \tag{2}$$

As the selection overhead should be as small as possible, it is reasonable to select the training patterns based on an easily calculable criterion. We select the maximum error pattern out of $\mathbb{D}_T$ which is not in $\mathbb{D}_t$ such that the resulting training set resembles a worst case collection of patterns.

The training set may now be controlled to achieve a faithful generalization. The straightforward strategy, namely to enlarge $\mathbb{D}_t$ whenever the generalization factor is greater than one, results in very small training sets. For high precision training, however, the selection process turns out to be too slow. Estimating the tendency of the generalization factor and inserting the new training pattern whenever the generalization factor increases, fixes the problems (Röbel, 1993b).

3 Experimental Results

The properties of the *dynamic pattern selection* will be demonstrated by solving two nonlinear signal prediction tasks. For training, a batch mode backpropagation algorithm with adapted learning rate and momentum is used (Salomon, 1991).

In the first experiment the chaotic time series of the henon model (Grassberger and Procaccia, 1983)

$$\begin{pmatrix} x_{n+1} \\ y_{n+1} \end{pmatrix} = \begin{pmatrix} y_n + 1 - a \cdot x_n^2 \\ b \cdot x_n \end{pmatrix} \quad \text{with} \quad (a = 1.4, b = 0.3) \tag{3}$$

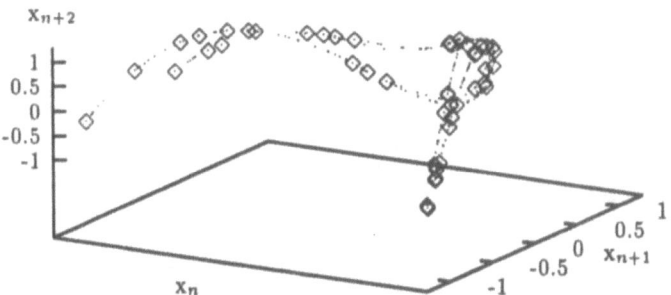

Available data
Selected Patterns ◇

x_{n+2}

1
0.5
0
-0.5
-1

1
0.5
0
-0.5 x_{n+1}
-1

x_n

Fig. 1. A typical distribution of training patterns on the prediction function of the henon model. Depicted is the distribution of training patterns subsequent to 1000 training cycles.

is predicted. In table (1) the results for a number of fixed training sets are compared with the dynamic pattern selection. The normalized rms error $E()$ for the training set and an independent validation set $\mathbb{D}_u$ is given. The cost is estimated by the number of forward/backward propagations through the net. The fixed training sets with 35-100 patterns give the best results. In the case of dynamic training sets a mean of 69 patterns is selected. Due to the initially small training sets the cost for the dynamic selection training is considerably lower than for all appropriate fixed training sets.

Further investigation reveals that only the fixed training sets with more than 50 patterns achieve a stable generalization factor. Compared to this, the dynamically selected mean of 69 patterns is quite reasonable.

Along with the size of the training set, the distribution of the training patterns has a considerable effect on the generalization results. A typical distribution obtained by the dynamic pattern selection is shown in figure (1). The distribution is not uniform, but reflects the error distribution of the net function.

The second example is the prediction of the Mackey-Glass model

$$\dot{x}(t) = \frac{a \cdot x(t-\tau)}{(1 + x(t-\tau)^{10})} - b \cdot x(t) \quad \text{with} \quad (a = 0.2, b = 0.1). \qquad (4)$$

Lapedes and Farber (1987) demonstrated the prediction of the Mackey-Glass time series ($\tau = 30$) using a neural net with six input units, two hidden layers with ten units each and a linear output unit. These settings are used here, too. Lapedes and Farber used a fixed training set with 500 examples. Results for this training set and the dynamic pattern selection are shown in table (2). Achieving the same average generalization precision, the dynamically selected training sets contain a mean of 207 training patterns. The cost is lower by a factor of three. Due to the suboptimal distribution of the fixed training set, the generalization factor increases steadily.

training set size	training error $E(\mathbb{D}_s)$	generalization error $E(\mathbb{D}_u)$	generalization factor ρ_u	for/backward propagations
500	2.47e-2 $\pm$ 3.6e-3	2.73e-2 $\pm$ 3.4e-3	1.11	60e+6
207 $\pm$ 11	4.62e-2 $\pm$ 3.9e-3	2.74e-2 $\pm$ 1.4e-3	0.59	18e+6

Tab. 2. Predicting the Mackey-Glass model. Comparison of the average training and generalization rms error using a 6-10-10-1 neural net and different training sets.

4 Real world data and online training

The dynamic pattern selection algorithm has been successfully applied to optimize predictors of high quality musical signals (Röbel, 1993a). The experiments revealed that the dynamic pattern selection is not disturbed by noise, as long as the noise level is well below the prediction error.

It is widely accepted that, in the case of redundant data, the online mode training will yield superior results to training in batch mode. We therefore compared the computational expenses for online training with *Search-Then-Converge* learning rate schedule (Darken and Moody, 1991) and the dynamic pattern selection algorithm. The neural nets have been trained to predict a piano signal given 15000 training patterns.

Despite the preceding optimization for the online training algorithm and the fact that, in the case of dynamic training sets, only 50 out of the 15000 patterns have been selected (resulting in a considerable overhead for the selection procedure), the total expense for the dynamic pattern selection has been lower by a factor of three.

References

Christian Darken and John Moody (1991) Note on learning rate schedules for stochastic optimization. *Neural Information Processing Systems*, 3, 832–838.

P. Grassberger and I. Procaccia (1983) Estimation of the Kolmogerov entropy from chaotic signal. *Physical Review A*, 28(4),2591–2593.

A. Lapedes and R. Farber(1987) Nonlinear signal processing using neural networks: Prediction and system modelling. Technical Report LA-UR–87-2662, Los Alamos National Laboratory.

Mark Plutowski, Garrison Cottrell, and Halbert White (1993) Learning Mackey-Glass from 25 examples, plus or minus 2. *Preprint*.

Axel Röbel (1993a) Neuronal models of nonlinear dynamic systems and the application to musical signals. PhD thesis, Technical University of Berlin, (In German).

Axel Röbel (1993b) The dynamic pattern selection algorithm: Effective training and controlled generalization of backpropagation neural networks. Technical Report 93-23, Technical University of Berlin.

R. Salomon (1991) Improving connectionsts learning, based on gradient descend. PhD thesis, Technical University of Berlin, (In German).

A Learning Rule which implicitly Stores Training History in Weights *

Ferdinand PEPER†, Hideki NODA†

† Communications Research Laboratory (CRL),

Japan Ministry of Posts and Telecommunications,

588-2, Iwaoka, Iwaoka-cho, Nishi-ku, Kobe 651-24, Japan.

1 Introduction

In unsupervised learning Neural Networks one often encounters learning equations involving a learning rate who's value decreases as training of the network progresses. The learning rate is usually implemented as a monotonically non-increasing function of a counter, which is increased by 1 for every training step. The problem of such an approach is that the counter is not an intrinsic part of the network, but just an external resource, which is unlikely to find its counterpart in biological Neural Networks.

Weight vectors are internal resources of Neural Networks, and usually encode *directional* and *length* information. In many Neural Network models the lengths of weight vectors (and input vectors) are not considered important information, rather than the direction of the vectors. Examples are the ART-models (Carpenter & Grossberg, 1987a; Carpenter & Grossberg, 1987b; Carpenter et al., 1991), which heavily depend on normalization of their input-signals, and mainly record directional information in their weight-vectors. The storage capacity left open to record vector length information in the weight vectors of such networks can be used to record otherwise external resources.

The Self-Organizing Feature Map employs an external counter k to control its learning rate, which is usually a function decreasing linearly from 1 to 0 as the counter increases. A learning rate of the form $\beta(k) = 1/k$ is also sometimes used (Kohonen, 1989 (p.133); Sanger, 1989). In (Meyering & Ritter, 1992) the learning rates decay exponentially from 0.9 to a value close to 0. In (Xu et al., 1992) the learning rate decreases linearly from 0.1 to 0.001 at the first 500 learning steps, and then remains constant at 0.001. In all these cases β is a function which is monotonically non-increasing during the training process.

This paper presents a learning rule which incorporates a mechanism to implicitly keep track of the learning rate by expressing training experience by the lengths of weight vectors. The learning rule increases the length of a weight vector each time the vector is trained. The resulting simultaneous increases of the weights in each weight vector are independent from the changes of the

---

*This work was financed by the Japan Ministry of Posts and Telecommunications as part of their Frontier Research Project in Telecommunications.

individual weights caused by learning. By employing the growth of the length of weight vectors, we obtain not only a biologically inspired implementation of the Hebb-rule on individual synapses, but also obtain such a rule for a group of synapses. That is, the more a group of related pre-postsynaptic cells is activated, the more the corresponding synapses are strengthened simultaneously, independently of individual synaptic strengthening.

The structure of the paper is as follows. The learning rule is described in the next section. After proving its convergence properties in section 3, we prove in section 4 that a certain function K of the vector length behaves like a counter, and thus can be employed as a substitute of an external counter k. We finish with some simulation results and a short discussion in section 5.

2 The Learning Rule

The weight vector z obtained after k learning trials, denoted as $z^{(k)}$, is determined by the following equations:

$$z^{(k+1)} = \frac{(a^2+1)\|z^{(k)}\|+a}{a\|z^{(k)}\|+1} \cdot N(\, z^{(k)} + \beta(K(\|z^{(k)}\|)\,)\,(\mathbf{I}^{(k)}-z^{(k)})) \quad (1)$$

$$z^{(0)} = a.N(\mathbf{I}^{(0)}) \quad\quad (2)$$

where $\mathbf{I}^{(k)}$ is the input vector in trial k, a is a positive constant, N is a vector normalizer, and β expresses the learning rate as a function of the number of times vector z has been trained. The function K is defined by

$$K(x) = -\alpha.\log(z_\infty - x),$$

where the constant $\alpha\epsilon\mathbb{R}^+$ and $z_\infty = \lim_{k\to\infty}\|z^{(k)}\|$. The expression $K(\|z^{(k)}\|)$ increases approximately linearly with k and thus behaves like a counter, as will be shown in section 4.

Equation (1) consists of two parts: a vector length encoding part, constituted by the quotient in (1), and a vector direction encoding part, constituted by the expression normalized. It is easily seen that the vector length encoding part develops completely independently from the directional part. The opposite is not true though, since the vector sum constituting the direction encoding part depends on the length of z. In the next section we will examine the convergence properties of $\|z^{(k)}\|$.

3 Convergence Properties of $\|z^{(k)}\|$

Consider the recurrence equation

$$z_{k+1} = a + \frac{1}{a+\frac{1}{z_k}}. \quad\quad (3)$$

It is easily derived from (1) and (3) that $z_k = \|z^{(k)}\|$. The series $z_0, z_1, z_2, ...$ constitute the even-indexed convergents of the Infinite Continued Fraction (e.g.

see (Burton, 1980))

$$a + \cfrac{1}{a + \cfrac{1}{a + \cfrac{1}{a + \cdots}}} \qquad \left(= \frac{a + \sqrt{a^2 + 4}}{2} \right).$$

The theory of Continued Fractions tells us that z_∞ is equal to this, and $z_0 < z_1 < z_2 < ... < z_\infty$. Consequently $\|z\|$ converges to z_∞ in accordance with the series $z_0, z_1, z_2, ...$, and thus $\|z\|$ increases monotonically as training progresses.

4   Behavior of the Counter Function K

In order to learn more about the behavior of K, we analyze the expression $|z_\infty - z_k|$. It can be proved (e.g. see (Burton, 1980)) that

$$|z_\infty - z_k| < \frac{1}{q_{2k}^2}, \tag{4}$$

where q_k is determined by the recurrence equation

$$q_k = a.q_{k-1} + q_{k-2} \tag{5}$$

with boundary conditions $q_0 = 1$ and $q_1 = a$. In a similar way as for (4) it can be deduced that

$$|z_\infty - z_k| > \frac{1}{2q_{2k+1}^2}. \tag{6}$$

Equation (5) can be solved by standard techniques (Liu, 1968). Substitution of the solution of (5) in (4) and (6) and some algebraic manipulations give

$$\frac{1}{(a+1)^{4k+4}} < |z_\infty - z_k| < \frac{a^2 + 4}{(a^2 + 1)^{2k+1}}. \tag{7}$$

Consequently $-\log|z_\infty - z_k| = O(k)$, so $K(z_k) = O(k)$. To make K behave like a counter, we set constant α such that $K(\|z^{(k)}\|) \approx k$ for large k.

For the learning rule to be practically applicable, the length of $z^{(k)}$ should converge slowly to z_∞. For, otherwise the value of $|z_\infty - z_k|$ would converge to 0 quickly and, because of the finite word-length of computers, actually would become 0. To obtain slow convergence of z_k towards z_∞, we set the value of a close to 0. Then, according to (7) the expression bounding $|z_\infty - z_k|$ from below will converge only slowly towards 0. A good choice for a would be $a = 0.001$. It provides values for K of over 10000 using 64 bit floating point arithmetic.

5   Simulation and Discussion

The proposed learning rule has been tested using a learning scheme as in (Peper & Noda, 1993) with the expression $I \cdot z_j / (\|I\|.\|z_j\|)$ used as the similarity measure between input vector I and weight vector z_j. This measure is also used in ART-models. The classification behavior of this scheme was compared to the behavior of the same scheme employing a conventional learning rule and an external counter k, i.e.

$$z_j^{(k+1)} = z_j^{(k)} + \beta(k)(I^{(k)} - z_j^{(k)}).$$ (8)

For both schemes function β was defined by $\beta(x) = 1/(x+1)$. The two schemes were compared on the task of classifying the 50 patterns used in (Carpenter & Grossberg, 1987b; Carpenter et al., 1991; Peper et al., 1993), setting the number of classes to 8. The scheme with learning rule (1) and parameter settings $a = 0.001, \alpha = 1604.6$ divided the patterns into 7 classes (1 class remained unused), while the scheme with rule (8) divided the patterns into 5 classes. Though the number of used classes was different for the two schemes, 76% of the patterns was classified identically by them. The observed differences were mainly due to the fact that, unlike the weight vectors in (1), the weight vectors in (8) were of about the same length during the whole training process, which influences the outcome of the vector sums in (1) and (8). We conclude that a scheme employing the proposed learning rule (1) is able to classify patterns without relying on an external counter to control its learning rate.

In this paper we used weight vector lengths to control the learning rate in Neural Networks. Weight vector lengths can also be used to control other network functionalities evolving during the training process, such as the vigilance of an ART-like network. Research into this direction is in progress.

References

[Burton, D.M. (1980)] *Elementary Number Theory*. Boston: Allyn and Bacon Inc.

[Carpenter, G.A. & Grossberg, S. (1987a)] A Massively Parallel Architecture for a Self-Organizing Neural Pattern Recognition Machine. *Computer Vision, Graphics, and Image Processing*, 37, 54-115.

[Carpenter, G.A. & Grossberg, S. (1987b)] ART2: Self-organization of stable category recognition codes for analog input patterns. *Applied Optics*, 26, 4919-4930.

[Carpenter, G.A., Grossberg, S., & Rosen, D.B. (1991)] ART 2-A: An Adaptive Resonance Algorithm for Rapid Category Learning and Recognition. *Neural Networks*, 4, 493-504.

[Kohonen, T. (1989)] *Self-Organization and Associative Memory*. Springer-Verlag.

[Liu, C.L. (1968)] *Introduction to Combinatorial Mathematics*. McGraw-Hill.

[Meyering, A. & Ritter, H. (1992)] Learning 3D-Shape Perception with Local Linear Maps. *Proceedings of IEEE International Conference on Neural Networks*, Baltimore, IV.432-437.

[Peper, F. & Noda, H. (1993)] A Generalized Unsupervised Competitive Learning Scheme. *IEICE Transactions on Fundamentals of Electronics, Communications, and Computer Science*, E76-A, pp. 834-841.

[Peper, F., Shirasi, M.N., & Noda, H. (1993)] A Noise Suppressing Distance Measure for Competitive Learning Neural Networks. *IEEE Transactions on Neural Networks*, 4, pp. 151-153.

[Sanger, T.D. (1989)] Optimal Unsupervised Learning in a Single-Layer Linear Feedforward Neural Network. *Neural Networks*, 2, 459-473.

[Xu, L., Oja, E., & Suen, C.Y. (1992)] Modified Hebbian Learning for Curve and Surface Fitting. *Neural Networks*, 5, 441-457.

A Comparison Study of Unbounded and Real-valued Reinforcement Associative Reward-Penalty Algorithms

R. Neville & T.J. Stonham

Dept Elec Eng, Brunel University, Uxbridge, Middx

Abstract

A comparison study was carried out between two Associative Reward-Penalty, or A_{R-P}, algorithms. The regimes solve nonlinear supervised learning tasks utilising multi-layer feed-forward networks. We introduce a variant of the A_{R-P} algorithm, called the 'Unbounded' reinforcement A_{R-P} algorithm. The 'Unbounded' reinforcement A_{R-P} is compared with the real-valued reinforcement A_{R-P} algorithm. The 'Unbounded' reinforcement method utilises a quantised real-valued reinforcement, which is a payoff metric optimised by an Associated Critic Net.

1 Introduction

The research presented by Barto [1] puts forward two classes of Associative Reward-Penalty A_{R-P} algorithm. In the first, the reinforcement signal, $r \in [0, 1]$, probabilistically takes on one of two values

$$r = 1 \; with \; probability \; 1 - e_o \; or \; r = 0 \; with \; probability \; e_o \qquad (1)$$

where the error e_o is the mean-square output error of the net.

The second more informative reinforcement signal was a real-valued signal $0.0 \leq r^{\Re} \leq 1.0$, defined by

$$r^{\Re} = 1 - e_o \qquad (2)$$

The two regimes deduce their reinforcement values solely as a function of the output error and the present input stimuli, which we term *primary* information, obtained from an external training environment (R) that provides input stimuli to the network and monitors the output action of the net. These algorithms do not utilise *secondary* information, such as past data obtained from the environment R. We contrast Bartos' real-valued reinforcement regime with an extension of the scalar reinforcement A_{R-P} which uses *secondary* information to derive its reward, based on traces of the frequency of 'stimuli' occurrence.

2 Reinforcement Training

2.1 The Sigma-pi Neuron Model

The neuron model we utilise has previously been termed a Sigma-pi unit [2] , these units are similar to pRAM units [3], PLN units [4] and GNU units [5]. We use the stochastic model Direct Output Node (DON). The activation of the DON, for the Analogue case, is defined as:

$$a = \frac{1}{2^n} \sum_{\mu} \sigma(S_\mu) \prod_{i=1}^{i=n} (1 + \bar{\mu}_i z_i) \qquad (3)$$

where z_i defines a set of probability distributions for an input address formed from a set of Boolean variables $\{X_i\}$, given $P_{\underline{\mu}}(\underline{x}_i) = \frac{1}{2}(1 + \underline{\mu}_i z_i)$, given $x \in \{x_1, x_2, ...x_i\}$ is a binary input vector which may be represented as a set of bits in positions x_1 to x_i. The site address $\mu \in \{\mu_1, \mu_2, ...\mu_i\}$ is represented by a set of bits in positions μ_1 to μ_i. The site value S_μ is addressed by the binary string μ. The site value S_μ stores a value $S_\mu \in \{-S_m, S_m\}$. Then for the stochastic model

$$a = \sum_\mu \sigma(S_\mu) P(\mu) = < \sigma(S_\mu) > \tag{4}$$

The output y of the DON is defined as equal to the activation a and

$$P(\underline{y} = 1|\eta) = \sigma(S_\mu) = \frac{1}{1 + e^{\frac{-S_\mu}{\rho}}} \tag{5}$$

Then the output $y = < \sigma(S_\mu) >$.

2.2 Training Artificial Neural Networks by Error minimisation

The goal of the learning regime is to minimise a mean-square output error term:

$$\forall_{I_v} \quad e_o = \frac{1}{N_V} \sum_{I_V} [y_t^j - \sigma(S_\mu^j)]^2 \tag{6}$$

where $[.]^2$ is the square error per input stimuli, defined on the output. This is summed over all N_V output units or visible units. The sum is over the set I_v of these visible units. The error is the difference between the target response $y_t^j \in [0, 1]$ of output j for a given input/output pattern pair, and the sigmoidal value of the site $\sigma(S_\mu^j)$, where μ specifies a site address.

2.3 Real-valued Reinforcement $A_{R-P}^{\Re}$ Training

In the case of Barto's real-valued regime, the reinforcement $r^{\Re}$ is deduced utilising (2). This means that large values of $r^{\Re}$ correspond to better matches between the network's output pattern and the desired pattern. The units are then updated using this reward signal. Each node, j, given site address μ, updates its site value according to the following equation:

$$\Delta S_\mu^j = \alpha[y^j - \sigma(S_\mu^j)]r_{(t)} + \alpha\lambda[1 - y^j - \sigma(S_\mu^j)](1 - r_{(t)}) \tag{7}$$

where $r_{(t)} = r_{(t)}^{\Re}$ for Real-valued Reinforcement $A_{R-P}^{\Re}$ Training.

2.4 Unbounded Reinforcement A_{R-P} Training

The external reward utilised by the unbounded A_{R-P} has been previously defined in (1), where $r_{(t)} \in [0, 1]$ is a binary scalar value. This external reinforcement is then scaled:

$$r'_{(t)} = (2 * r_{(t)}) - 1 \tag{8}$$

where $r'_{(t)} \in [-1, +1]$ is of the form utilised by Barto, Sutton and Anderson [6]. The scaled reward signal is then used to derive an improved or internal reinforcement signal, given by:

$$\hat{r}_{(t)} = r'_{(t)} + \gamma P_{v(t)} - P_{v(t-1)} \tag{9}$$

where $P_{r(t)}$ is the present prediction and $P_{v(t-1)}$ the past prediction. It should be noted that this is not the same as Barto's [6] original work, where he uses the prediction values $P_{(t)}$ and $P_{(t-1)}$. We use the present and previous prediction values for the given site address r. The coefficient $0.0 < \gamma \leq 1.0$ has previously been termed the "discount factor" by Barto et. al. [6]. In our case we store the internal reinforcement signal as a quantised real-valued number. The prediction value is updated by:

$$\Delta P_{v(t+1)} = \beta \hat{r}_{(t)} \bar{x}_{v(t)} \tag{10}$$

where $0 < \beta < 1$ is a positive constant determining the rate of change of $P_v(.)$. In our case we store the prediction values in n-tuples. Where $P_{(.)} \in [0, +P_n]$, giving $D = P_n + 1$ discrete levels,' which are stored as q bit numbers. Hence if $P_n = 8$ then $P_{(.)} = \{0.125n \mid n = 0.1,....N\}$ where $N = 1.0/0.125 = P_n$.

All the input eligibility traces are updated using:

$$\forall_u \quad \bar{x}_{u(t+1)} = \lambda \bar{x}_{u(t)} + (1 - \lambda)x_v \tag{11}$$

In our case we store the eligibility values in an n-tuple. Hence for all input addresses $0 \leq u \leq \eta$. where η is the maximum input address (i.e. for an 8-tuple $\eta = [(2^8) - 1]$ or 255 decimal or FF hexidecimal). and where lambda $0 < \lambda < 1$ determines the eligibility traces decay rates. The binary value x_v is a trigger for the eligibility trace, and when the site v is addressed $x_v = 1$ and all other non-addressed traces are updated with $x_{\bar{v}} = 0$. The input eligibility $\bar{x}_{v(t)}$ is interpreted as: given an 'i' bit input vector $\{x_1, x_2,....x_i\}$, which addresses location 'v' in an eligibility n-tuple. giving an eligibility $\bar{x}_{v(t)} \in [0, +\bar{x}_n]$, that is specified as a 'q' bit number. having $D = \bar{x}_n + 1$ discrete levels. Hence if $\bar{x}_n = 8$ then $\bar{x}v_{(.)} = \{0.125n \mid n = 0, 1,...N\}$ where $N = 1.0/0.125 = \bar{x}_n$. The internal reinforcement $r_{(t)}$ is then re-scaled

$$r^*_{(t)} = \frac{1}{2}(\dot{r} + 1.0) \tag{12}$$

which denotes a quantised real-valued reinforcement $-1.0 \leq r(t)* \leq +2.0$, that is defined as the unbounded internal reinforcement $r^*_{(t)}$. this permits penalisation even when $\lambda = 0$. We define the unbounded reward signal as $r^*_{(.)} \in [-r_n, +2r_n]$ which is a q bit number with $D = 3r_n + 1$ discrete levels. Then the unbounded reward utilises in (7) would be defined as $r^*_{(.)} = \{0.125n \mid n = -N,...0.1,......2N\}$. given $N = r_n$, given the unbounded internal reinforcement the site values are updated utilising (7). where $r_{(t)} = r^*_{(t)}$.

3 Simulation Results

3.1 The 8-3-8 encoder

We utilise the 838-encoder of Hinton et. al. [7], which they used for their research into the Boltzmann machine. In all our simulations we begin by setting all the site values at the start of the training to $S_\mu = 0$. then $\sigma(S_\mu) = 0.5$, giving $P(Y = 1 \mid \mu) = 0.5$, i.e. 50% probability of the output Y obtaining a value "1", i.e. no prior information has been bestowed on the network. Hence finding a solution to such a problem requires that the two visible groups (v) come to agree upon the meaning of a set of codes without any prior conventions for communicating through the network's hidden units (h).

3.2 Experimental Delimitations

The results presented show a graph of the error $\bar{e}_o$ where

$$\bar{e}_o = \frac{1}{N_{th}} \sum_{n=1}^{n=N_{th}} \frac{1}{N_p} \sum_{k=1}^{k=N_p} e_o \tag{13}$$

is the average error over $N_{tn} = 100$ trained networks, after 6000 training cycles have elapsed, over all N_p training patterns. where e_o is the mean-squared output error (6) of each training vector. The training vector set used was hexidecimal numbers $\{F0, 78, 3C, 1E, 0F, 87, C3, E1\}$, hence $N_p = 8$. The training set was randomly ordered for each sample and a different seed was given to the stochastic operator of the net at the start of each training session. The training vectors each have four adjacent set-bits. This means that there are 192 valid codes, which represent 0.0011% of all possible code solutions. For all the experiments $\rho = 0.3$, $\lambda = 0.0$, $S_m \in [-10, +10]$ and $P_n = \bar{x}_n = r_n = 8$. The plots of log error $\bar{e}_o$ against log α are shown in Figure 1. The learning rates used were $\alpha = 0.1, 0.25, 0.5, 1.0, 2.0, 5.0, 10.0$ & 20.0. The graphs show that unbounded A_{R-P} reduces the average percentage error over all eight learning rates, when compared with standard A_{R-P}, by 10% and the real-valued reward A^R_{R-P} by 9%.

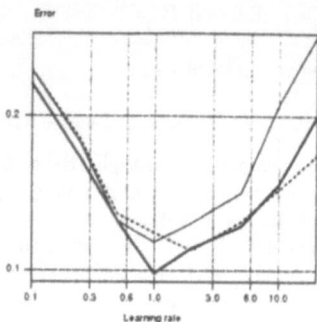

Figure 1 Average log error $\bar{e}_o$ against log α. (The lighter solid line shows the error for the Standard A_{R-P}, the dotted line shows the error for the Real-valued Reinforcement A_{R-P} and the heavy solid line shows the error for the unbounded Reinforcement A_{R-P}.)

4 Concluding Remarks

The unbounded Reinforcement A_{R-P} gives increased efficiency of training when compared to the standard A_{R-P}. The unbounded Reinforcement A_{R-P} performs as well as, if not better than, the real-valued Reinforcement A_{R-P}. This we hypothesize is due to the fact that the unbounded reward signal is able to reward/penalise the net to a higher degree. If, for example, the external reward is a penalty signal and the temporal difference between the predictions is a negative quantity (i.e. $P_{v(t)} < P_{v(t-1)}$), then the internal reinforcement is reduced, and the net is then penalised to a greater degree. The converse is also true, as the internal reinforcement would be increased if the external reinforcement signal is a reward and the temporal difference between the predictions is positive (i.e. $P_{v(t)} > P_{v(t-1)}$). It is of interest to note that the unbounded Associative Reward Penalty training methodology permits penalisation of the net even when the penalty coefficient is set to zero, as the internal reward signal may be negative, normally the net is only penalised if the penalty coefficient λ is non-zero.

References

[1] A.G. Barto and M.I. Jordan. Gradient following without back-propagation in layered networks. In *Proceedings 1st IEEE Conference on Neural Networks*, pages II.629–II.636. IEEE, 1987.

[2] K.N. Gurney. Training nets of hardware realizable sigma-pi units. *Neural Networks*, 5:289–303, 1992.

[3] D. Gorse and J.G. Taylor. A continuous input ram-based stochastic neural model. *Neural Networks*, 4:657–665, 1991.

[4] I. Aleksander. Canonical neural nets based on logic nodes. In *1st IEE International Conference on Artificial Neural Networks*, pages 110–114, 1989.

[5] I. Aleksander. Weightless neural tools: Towards cognitive macrostructures. In *CAIP Neural Network Workshop*, New Jersey, 1990. Rutgers University.

[6] R.S. Sutton A.G. Barto and C.W. Anderson. Neuronlike adaptive elements that can solve difficult learning problems. *IEEE Transactions on systems, man, and cybernetics*, SMC-13(5):834–846, September/October 1983.

[7] T.J. Sejnowski G.E. Hinton and D.H. Ackley. Boltzmann machines: Constraint satisfaction networks that learn. Technical Report CMU-CS-84-119. Carnegie Mellon University, Pittsburgh, PA, 1984.

To swing up an inverted Pendulum using stochastic real–valued Reinforcement Learning[1]

A. Standfuss, R. Eckmiller

Department of Computer Science VI (Neuroinformatics)

University of Bonn

D – 53117 Bonn, F.R. Germany

1 Introduction

This paper deals with the problem of learning to swing up an inverted pendulum, which belongs to the class of highly nonlinear, non–minimum phase control problems without a general control methodology. It is thus a challenge for reinforcement learning over time (Sutton, 1988).

An inverted pendulum consists of a cart with a pole mounted on it. The cart can be moved along a track through the application of a horizontal force. Thus the state of such a system consists of four components:

1. cart position x_1,

2. pole angle x_2,

3. cart velocity x_3 and

4. pole velocity x_4.

Fig. 1 schema of an inverted pendulum

The problem to swing up an inverted pendulum means to move the pole from hanging upside down ($180°$) to a position in which a controller can stabilize it in the upright position ($0°$).

We present first results gained in experiments using the simulation of a real system being described by the following set of nonlinear coupled differential equations (amira GmbH, 1992):

$$\dot{x}_1 = x_3 \qquad\qquad \dot{x}_2 = x_4$$
$$\dot{x}_3 = \beta \left(a_{32} \sin x_2 \cos x_2 + a_{33}x_3 + a_{34} \cos x_2 x_4 + a_{35} \sin x_2 x_4^2 + b_3 F \right)$$
$$\dot{x}_4 = \beta \left(a_{42} \sin x_2 + a_{43} \cos x_2 x_3 + a_{44}x_4 + a_{45} \cos x_2 \sin x_2 x_4^2 + b_4 \cos x_2 F \right)$$

$$\beta = \frac{1}{1 + \frac{N^2}{N_{01}^2} \sin^2 x_2}; \quad N = M_1 l_S; \quad N_{01}^2 = \Theta M - N^2; \quad \Theta = \Theta_S + M_1 l_S^2$$

a_{32}	a_{33}	a_{34}	a_{35}	b_3	a_{42}	a_{43}	a_{44}	a_{45}	b_4
$-\frac{N^2 g}{N_{01}^2}$	$-\frac{\Theta F_r}{N_{01}^2}$	$\frac{CN}{N_{01}^2}$	$\frac{\Theta N}{N_{01}^2}$	$\frac{\Theta}{N_{01}^2}$	$\frac{MNg}{N_{01}^2}$	$\frac{NF_r}{N_{01}^2}$	$-\frac{MC}{N_{01}^2}$	$-\frac{N^2}{N_{01}^2}$	$-\frac{N}{N_{01}^2}$

[1] Supported by Ministry for Research and Technology (BMFT–Project SENROB), under grant 01 IN 105 G/6

$M = M_0 + M_1$; M_1: pole mass; l_S: pole length up to center of mass; Θ_S: pole's moment of inertia; C pole's friction coefficient; M_0: cart mass; F_r cart's friction coefficient proportional to velocity; F force applied to cart. Values for these quantities were measured for the available real system (amira GmbH, 1992).

To balance the pole we used a conventional PD–controller (amira GmbH, 1992), which later on will be replaced by a neural–controller.

2   Methods

The applied method is stochastic real–valued reinforcement learning (Gulla-palli, 1990; Williams, 1992) combined with temporal difference (TD)–learning applied to backpropagation (Sutton, 1987; 1988).

In stochastic real–valued reinforcement learning the action a is determined by a *Gaussian unit* which at first calculates the mean μ and variance σ, based on its weights and input, and then choses an output with respect to the resulting normal distribution, $g(a, \mu, \sigma) = \frac{1}{\sqrt{2\pi}\,\sigma}\, e^{-(a-\mu)^2/2\sigma^2}$. A learning algorithm for μ and σ is given by:

$$\Delta\mu = \alpha_\mu\, r_s\, \frac{a - \mu}{\sigma^2} \qquad\qquad \Delta\sigma = \alpha_\sigma\, r_s\, \frac{(a-\mu)^2 - \sigma^2}{\sigma^3}$$

where α_μ and α_σ are learning rates and r_s is the secondary reinforcement. μ and σ are approximated via a multi–layer perceptron, with the state of the inverted pendulum as input.

As secondary reinforcement r_s we used the TD–error computed with a multi-layer perceptron trained via TD–learning, using the state of the inverted pendulum as input:

$$w_{ij}^{t+1} = w_{ij}^t + \alpha\,(y^{t+1} - y^t)\,e_{ij}^t\,; \qquad\qquad e_{ij}^t = \sum_{n=1}^{t} \lambda^{t-n}\, \frac{\partial y^n}{\partial w_{ij}}\,;$$

where e_{ij}^t is recursively computed as $e_{ij}^{t+1} = \lambda e_{ij}^t + \delta_j^{t+1}\, y_i^{t+1}$ and δ_j^{t+1} as $\delta_j^{t+1} = \begin{cases} y_j^t\,(1 - y_j^t), & \text{if unit j output unit} \\ \sum_k \delta_k^t\, w_{jk}^t\, y_j^t\,(1 - y_j^t) & \text{otherwise.} \end{cases}$

The TD–error is then computed as $r_s^t = r^t + \gamma\, y^t - y^{t-1}$, where r^t is the primary reinforcement and y^t the activation of the output unit at time t.

3   Results

At first we tried to solve the problem as a whole using the absolute angular velocity as primary reinforcement. It was found that the net was not able generate an action sequence moving the pole far above the horizontal (see Fig. 2). We proposed the hypothesis that the net learned a strategy, which had to be inverted above the horizontal. We tested this through changing the sign of the action, that means the direction of the force applied to the cart, as soon as the pole moved over the horizontal. The hypothesis proved to be true, since the pole could be swung up to angles less than 2°, so that the given controller could stabilize it in the upright position (see Fig. 3).

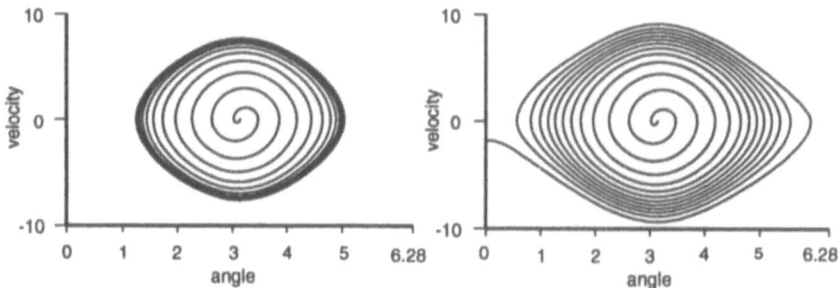

Fig. 2 phase plane of pole using one net   Fig. 3 phase plane of pole using one net with inversion above horizontal

We therefore divided the problem into two subproblems, 1. to swing up below the horizontal and 2. to swing up above the horizontal. We used separate neural nets for each of them (in the following called $N1$ for 1. and $N2$ for 2.), both having the same architecture. While for $N1$ we used the same primary reinforcement as before, we used $\frac{1}{2}\left(\frac{|x_4|}{mv}\frac{|x_2|}{\pi/2} + \frac{\pi/2-|x_2|}{\pi/2}\right)$ for $N2$, mv being the maximum angular velocity. After training $N1$, we began to train $N2$, while using $N1$ in the pole angle range below the horizontal, without stopping its learning process. In combination, the two nets were able to learn to swing up the inverted pendulum, as can be seen in Fig. 4.

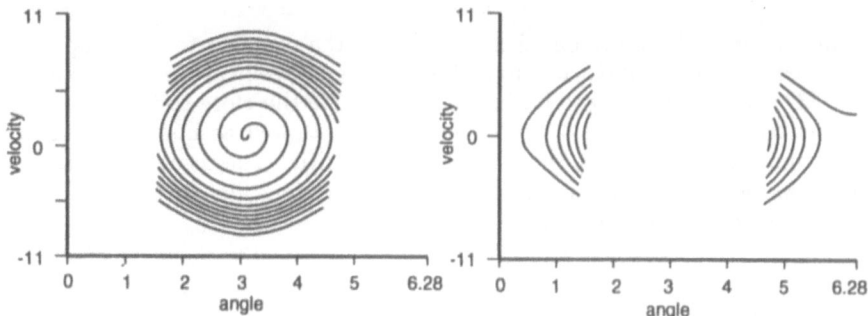

Fig. 4 phase plane of pole using two interactive nets (the diagrams show the alternating contributions of $N1$ (left) versus $N2$ (right))

It was found that both nets learned a bang–bang control strategy using the maximum force allowed for each direction. $N1$ learned to pull the cart right if the pole moves clockwise and to pull the cart left when it moved counter-clockwise. $N2$ learned the inverse strategy, pulling the cart left when the pole moved clockwise and right when it moved counterclockwise (see Fig. 5).

For the learning process, the variance σ is a crucial parameter. It determines the exploratory behaviour of the Gaussian unit and should converge to zero. In that case the choice of action is deterministically relying on the learned policy, given through the mean μ. While this was the case for $N1$, it did not hold for

$N2$. Nevertheless the policy converged to a valid solution of the posed problem. The learning process was stopped once such a policy was found.

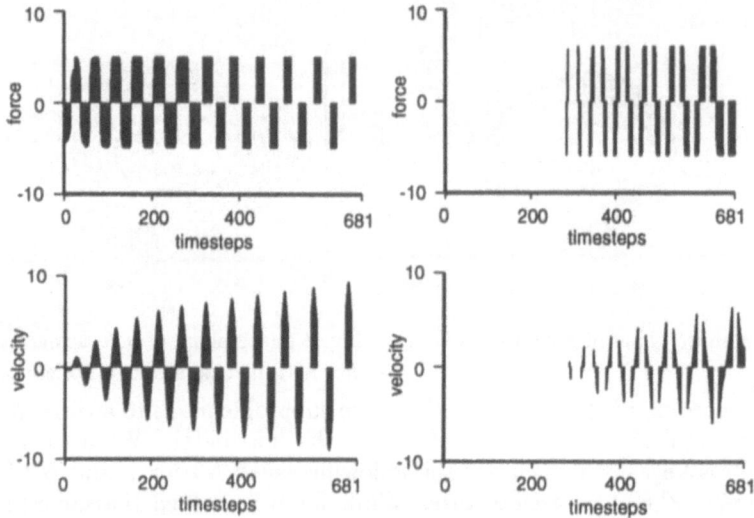

Fig. 5 The upper two diagrams show the cart force generated by the two nets (the left one for $N1$ and the right one for $N2$) and the lower two diagrams show the corresponding angular velocity of the pole.

4 Conclusions

In this paper we have presented a solution for the problem to swing up an inverted pendulum using stochastic real–valued reinforcement learning combined with TD–learning applied to backpropagation. We therefore could use the continuous state– and action space of the system as well as nonlinear function approximators to represent the policy and the critic. The results are based on the simulation of an available real inverted pendulum; first experimental data from the real system will be presented.

References

amira GmbH, (1992), LIP100 Invertiertes Pendel, *Dokumentation*.

Gullapalli, V. , (1990), A stochastic reinforcement learning algorithm for learning real–valued functions, In: *Neural Networks*, vol. 3, pp. 671 - 692.

Sutton, R. S. , (1987), Implementation details of the TD(λ) procedure for the case of vector predictions and backpropagation, *GTE Laboratories Technical Note TN87-509.1*.

Sutton, R. S. , (1988), Learning to predict by the methods of temporal differences, In: *Machine Learning*, vol. 3, pp. 9 - 44.

Williams, R. J. , (1990), Simple statistical gradient–following algorithms for connectionist reinforcement learning, In: *Machine Learning*, vol. 8, pp. 229 - 256.

Efficient Reinforcement Learning Strategies for the Pole Balancing Problem

D. Kontoravdis, A. Likas and A. Stafylopatis
National Technical University of Athens
Department of Electrical and Computer Engineering
Computer Science Division
157 73 Zographou, Athens, Greece

1 Introduction

According to the general framework of delayed reinforcement learning, a system accepts input from the environment, selects and executes sequences of actions and, at the end, receives a reinforcement signal which is usually a scalar value rewarding or penalizing the sequence of selected actions []. The objective of learning is the discovery of an optimal policy, which for each state x maximizes the expected reinforcement $e(x)$.

The main approach to deal with delayed reinforcement learning problems corresponding to relatively large state spaces, is based on the use of two networks [1, 2]. The *evaluation network* provides as output a prediction $e(x)$ of the evaluation of state x. Learning of $e(x)$ is achieved through the use of the method of temporal differences, i.e., we try to minimize the error $\delta = e(x) - (r + \gamma e(y))$, where y is the new state, r the received reinforcement and γ is a discount factor. The second network is the *action selection network* (or policy network), that provides the action to be executed and is trained through backpropagation, with δ being the error corresponding to the selected output, while the error of the remaining outputs is considered to be zero. We will refer to this algorithm as $AHCON$ [4].

The performance of the learning algorithm can be improved in case an *action model* is available. Such a model provides for each state-action pair (x, α) the next state y and the corresponding reinforcement r. Using an action model, it is possible to perform a kind of planning called *look-ahead planning* that leads to improved learning times. In [4], a reinforcement algorithm (called $AHCON - M$) has been proposed that is derived from $AHCON$ by incorporating a planning phase before action selection.

In the following sections several techniques will be described that improve the performance of $AHCON$ and $AHCON - M$ algorithms.

2 Switching Between Stochastic and Deterministic Action Selection

In order for reinforcement learning algorithms to discover the optimal policy an exploration mechanism is necessary, that is implemented by introducing stochasticity in the action selection procedure. Such a way of operation is not always effective especially in the case of delayed reinforcement problems.

This is due to the fact that the AHCON algorithm estimates the quantities $e(x)$ according to the current policy, by selecting the actions suggested by the current policy (policy actions). The existence of stochasticity in the action selection procedure may lead to choices that are not suggested by the current policy. Consequently, the evaluation of some states may be underestimated causing problem to the backpropagation of the corresponding reinforcement signal.

In order for such effects to be avoided, when the performance of the learning system reaches a satisfactory level, it is necessary for the action selection network to operate in a detrministic way, by selecting the action with the highest probability. Since the way in which the network weights are adapted does not change, only step 2 of the $AHCON$ algorithm is modified. This necessecity for swithiching to deterministic action selection as learning proceeds, has also been observed in [5] in the case of an immediate reinforcement problem.

When the action selection network operates in a deterministic way, it is possible that the learning system has discovered a suboptimal policy that is not satisfactory enough. For this reason, we have allowed the possibility of *switching back* to stochastic operation, in case an unsuccessful deterministic cycle has been encountered. The switching between the two ways of operation may be repeated many times until the system converges to a satisfactory policy. We shall call this algorithm $AHCON - D$ (AHCON-Deterministic).

Futhermore, in order to decrease the effect of interference, during the deterministic phase the weight updates of the action selection network do not take place at each step. Instead, the updates are stored and a *cumulative* (batch) update takes place only in the case where the action network switches back to stochastic operation. This modified version will be called $AHCON - DB$ (AHCON-Deterministic Batch).

3 Model-Based Techniques

The idea of switching between stochastic and deterministic operation can also be easily applied in the case of the $AHCON - M$ algorithm. Of course in this case the model will not be used in the deterministic phase. We shall refer to this strategy as $AHCON - DM$ and we shall consider that during the deterministic phase, batch update of the weights of the action selection network is performed.

In the $AHCON - M$ algorithm, the model is used only for learning the evaluation function. However, under certain conditions, it is possible for the action with maximum probability to be deterministically selected for execution. Let α_k denote the action with maximium selection probability after look-ahead planning, and let y_k and r_k the new state and the reinforcement signal as provided by the system model. If the quantities $e(x)$ and $r_k + \gamma e(y_k)$ are very close, then the action α_k can be selected for execution.

This criterion is very strict. The fact that the above quantities are very close means that action α_k has the best prediction according to the planning algorithm. In addition, since training using the backpropagation algorithm is not possible to be achieved in one pass, the fact the quanities $e(x)$ and $r_k + \gamma e(y_k)$ are very close implies that action α_k has been predicted many times as the best action (in the current state) and, additionally, the evaluation function has converged to relatively constant values.

The incorporation in the $AHCON - M$ algorithm of the above criterion for deterministic action selection leads to a new algorithm which will be called $AHCON - M^*$. Moreover, the $AHCON - M^*$ algorithm can be used in conjuction with the switching technique of the previous section. In the resulting algorithm (called $AHCON - DM^*$), we will again consider that batch update of the weights in the deterministic phase is performed.

4 Application to the Pole Balancing Problem

We have examined the proposed reinforcement strategies on the Pole Balancing problem, which constitutes a benchmark problem for testing the efficiency of learning algorithms for delayed reinforcement tasks. Details concerning the motion equations and the values of the parameters of the problem can be found in [3, 1]. The architecture of the action selection network and the evaluation network as well as the parameter values of the learning algorithms were the same with the ones described by Anderson [1], which has tested the $AHCON$ algorithm on the same problem. The input to both networks is a normalized state vector of the four state variables [3, 1].

In order to implement an action model of the controlled system, two feedforward networks were considered that were trained on-line using the backpropagation algorithm. The *state predictor* which predicts the next system state and the *reinforcement predictor* which predicts the reinforcement signal that will be received. Both networks have one hidden layer (with 8 and 4 hidden units) and direct connections between the input and the output layer. The input to both networks consists of a state-action pair, where the two possible actions are encoded as 10 and 01. Moreover, in order to facilitate learning, we have maintained a list (of size 100) containing the training examples (x, α, y, r) as they were created by the operation of the system. Each time the list was full, an off-line batch update of these examples was performed and then the list became empty. Care was taken (using a similarity measure) to maintain the necessary variation in the examples of the list.

5 Results and Conclusions

The following algorithms have been tested: $AHCON$, $AHCON-D$, $AHCON-DB$, $AHCON - DM$ and $AHCON - DM^*$. For each of the proposed strategies, 30 runs were performed with each run consisting of a number of trials. A trial ends either with a failure signal (unsuccessful trial) or when the balancing succeeds, i.e., no failure signal is received for 100000 consequtive steps (successful trial). A run was unsuccessful, if no balancing was achieved in 10000 trials. In all cases, the switching from stochastic to deterministic operation takes place when a trial ends with a failure signal and the system has operated for more than 250 steps during that trial.

Table 1 presents the average number of trials (over 30 runs) until balancing for the $AHCON$, $AHCON - D$ and $AHCON - DB$ algorithms that do not consider a model of the system. Moreover, the number of unsuccessful runs is also presented. It is obvious that the algorithms based on switching between stochastic and deterministic action selection, perform substantially

Algorithm	Steps	Failures
AHCON	7645	6/30
AHCON-D	4850	0/30
AHCON-DB	4310	0/30

Table 1: Comparative results of algorithms that do not use an action model

Algorithm	Perfect Model		Learned Model	
	Steps	Failures	Steps	Failures
AHCON-DM	2230	0/30	2910	1/30
AHCON-DM*	1390	0/30	2620	1/30

Table 2: Comparative results of model-based algorithms

better and require almost half number of trials. Table 2 displays results for the $AHCON - DM$ and $AHCON - DM^*$ algorithms that exploit an action model. Experiments have been performed both for the case where the model is known in advance (perfect model) and the case where the model is learned on-line. As in the previous case, it is obvious that the incorporation of steps where deterministic selection takes place, results in improved learning times.

References

[1] Anderson, C.W., *Strategy Learning with Multilayer Connectionist Representations*, Technical Report TR87-509.3, GTE Laboratories, Waltham, MA, 1987.

[2] Anderson, C.W., *Learning to Control an Inverted Pendulum Using Neural Networks*, IEEE Control Systems Magazine, Vol. 9, No. 3, April 1989.

[3] Barto, A.G., Sutton, R.S. and Anderson, C.W., *Neuronlike Elements that Can Solve Difficult Learning Control Problems*, IEEE Transactions on Systems, Man, and Cybernetics, 13, 835-846, 1983.

[4] Lin, L., *Self-Improving Reactive Agents Based on Reinforcement Learning, Planning and Teaching*, Machine Learning, vol. 8, pp. 293-321, 1992.

[5] Millan, J. R. and Torras, C., *A Reinforcement Connectionist Approach to Robot Path Finding in Non-Maze-Like Enviroments*, Machine Learning, vol. 8, pp. 363-395, 1992.

Reinforcement Learning in Kohonen Feature Maps

N.R. Ball
Engineering Design Centre
Department of Engineering
University of Cambridge
Trumpington Street
Cambridge U.K.

1 Introduction

The effective application of Kohonen Feature Maps (KFM) to the control of artificial organisms has been demonstrated using a single map to provide an associative memory of the task environment (Ball & Warwick, 1993). The nature of the control problem demands the concurrent exploration and exploitation of the domain by the organism. This can result in the generation of a series of bad training examples during the early stages of the exploration resulting in serial correlation of the data presented to the map. This paper presents recent research into mechanisms that maintain plasticity in the map until useful training examples have been uncovered and rehearsed by the organism.

2 Architecture

A schematic of the modified KFM, known as the Feature Correlation Network (FCN), is shown in figure 1. The key modifications are the additon of a Reinforcement Layer (which is used to generate rehearsal examples back onto the input layer) and a Correlation Layer (to map preset goals into achievable targets).

Training examples are extracted from the domain state by the organism's sensory array. Each example consists of a vector of normalised feature elements $\{E\}$. Elements can be of two types - external feature $\{EF\}$ or internal feature $\{IF\}$. External features can be controlled directly by the organism's classifiers e.g. change X/Y location by applying movement classifiers. Internal features are dependent on external features and are derived from them e.g. take up food if present at X/Y location. Thus the job of the FCN is to provide accurate predictions of external feature targets given preset internal feature goals.

3 Mechanism

The objective of the organism is to satisfy internal feature goals. Each goal is mapped onto the correlation layer and applied to the output layer to produce a best match specified by -

$$\min_{ij} (((G_{IF} - o_{IF}^{ij}) + (G_{IF} - r_{IF}^{ij}) * (1.0 - f^{ij}))^2$$

where IF = internal feature

$\quad G_{IF}$ = current goal for internal feature IF

$\quad o_{IF}^{ij}$ = element of output node ij encoding feature IF

$\quad r_{IF}^{ij}$ = element of reinforcement node ij encoding feature IF

$\quad f^{ij}$ = number of training examples captured by node ij (Ahalt et al., 1992)

This produces a set of output nodes $\underline{O}$ that predict the domain states required to satisfy the internal feature goals. These are then ranked in order of

$\max_{IF} | o_{IF}{}^{ij} - S_{IF} |$

where S_{IF} = organism's internal feature state IF.

The external elements of the winning node $o_{EF}{}^{ij}$ are then passed to the classifier feature networks (Ball &Warwick, 1994) as external targets for classifier selection and activation.

Classifier activation changes the organism's state and produces a new set of training examples for the CFN. Each example is used in two ways at each time step -

(1) as a training example applied to the output layer with standard Kohonen selection (Kohonen, 1984) and local adaptation (Whittington & Spraken, 1990) defined by -

    (a) for winning node c -

        state match $s^c = | S_{EF} - o_{EF}{}^c |$

        $f^{ij} := f^{ij} + 1$

        where S_{EF} = organism's external feature state EF

    (b) for each node ij in output layer -

        $o_F{}^{ij} := o_F{}^{ij} + Hf * (A^{123}*s^c)* g^{ij} * (S_F - o_F{}^{ij})$

        where Hf = local habituation factor

                      A^{123} = fixed adaptation rate associated with the three zones of the Mexican Hat profile

                      g^{ij} = goal match of node ij (see (2))

(2) as reinforcement feedback onto the reinforcement layer defined by -

    (a) for each IF on winning node c (see figure 1) -

        if $|$ actual-error$_{IF} | < |$ predicted-error$_{IF} |$ then $r_F{}^c = S_F$

    (b) for each node ij in output layer -

        goal match $g^{ij} = \text{sum}_{IF}(0.5* | G_{IF} - r_{IF}{}^{ij} |+ | o_{IF}{}^{ij} - r_{IF}{}^{ij} |)^2$

        $o_F{}^{ij} := o_F{}^{ij} + Rf * A^1 * (1.0 - g^{ij}) * (r_F{}^{ij} - o_F{}^{ij})$

        where Rf = reinforcement factor

                A^1 = adaptation rate of primary zone of Mexican hat profile.

Parameter Hf governs the behaviour of the local adaptation which in turn effects the plasticity of the network. Parameter Rf governs the feedback mechanism which effects the accuracy of predictions generated at the output layer.

4 Simulation

The simulation has focussed on the effectiveness of the reinforcement mechanism by varying the Hf and Rf parameters. A 2D test environment (figure 2) is provided for the organism (+) with single food(F) and water(W) sources (*) surrounded by auras that can be sensed (----). Four features are configured - two internal(F and W) and two external (X and Y). Goals are preset on F and W (high) with goals for X andY being undefined. Typical behaviour is shown in figure 3(a) with an initial exploration period of ~ 300 steps followed by stable oscillation between food and water sources during the exploitation phase.

Increasing Hf with Rf constant increases the plasticity of the network making it more suspectible to sequences of poor training examples during the early exploratory period (figure 3(b)). In this case the early feedback on the food signal is swamped by later sensory data around the water source with the result that the network produces grossly inaccurate predictions for the food source.

Increasing *Rf* with *Hf* constant decreases the reinforcement feedback degrading the exploitation behaviour (figure 3(c)). In this case the exploration phase is completed by step 250 but the predicted location for the food source is unstable (under-rehearsed) and the organism oscillates around the source before eventually switching to water.

5 Discussion

Combining exploration with exploitation is a difficult learning task that highlights the plasticity / stability dilemma inherent in any self-organising system. In the Feature Correlation Network, stability has been achieved through the introduction of an additonal reinforcement layer that captures optimal training examples and allows them to be repeatedly rehearsed. Plasticity is maintained throughout the exploitation phase by applying a local adaptation mechanism which permits opportunistic discovery of new goal states at any time during this phase.

Current work on this system is testing the scaleability of the approach in domains with larger feature vectors and conflicting internal goals.

References

Ahalt S., Krishnamurthy A., Chen P., Mellon D. (1992). Competitive Learning Algorithms for Vector Quantization. Neural Networks, 3, 277 - 290. Pergamon Press.

Ball N. & Warwick K. (1993). Using Self-Organising Feature Maps for the control of artificial organisms. Proc. IEE, Part D, 140, 176-180.

Ball N. & Warwick K. (1994). Applying Self Organising Feature Maps to the control of Artificial Organisms in maze running tasks. Submitted to IEEE Trans, Systems Man and Cybernetics.

Kohonen T.(1984). Self Organization and Associative Memory. Springer Verlag.

Whittington G., Spraken T. (1990). The Application of Neural Networks to Industrial spectral Analysis, Identification and Classification. Proc IEEE Workshop on Genetic Algorithms, Simulated Annealing and Neural Nets. University of Glasgow.

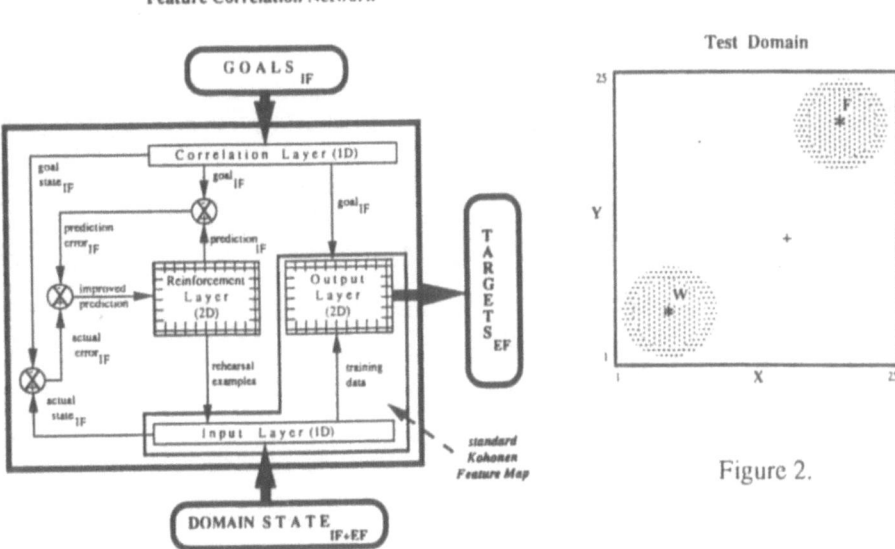

Figure 1.

Figure 2.

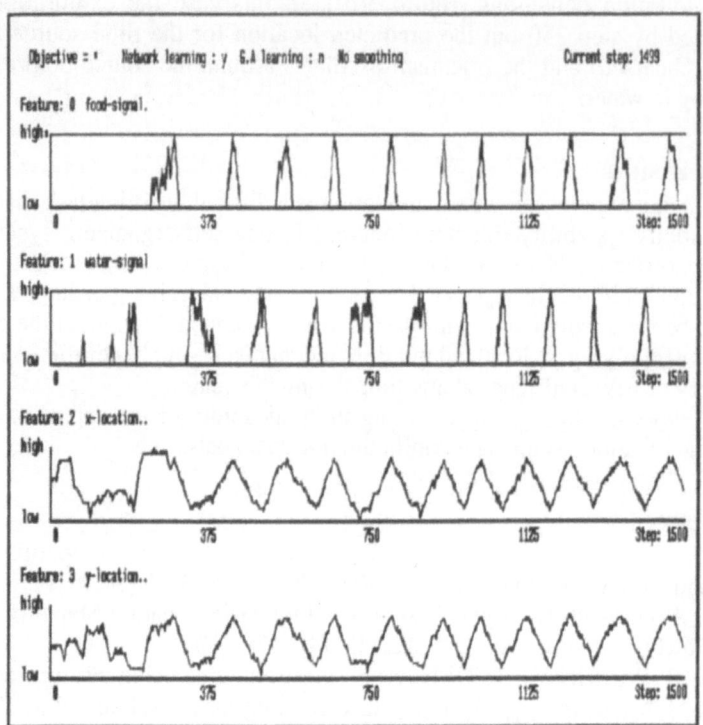

Figure 3(a).

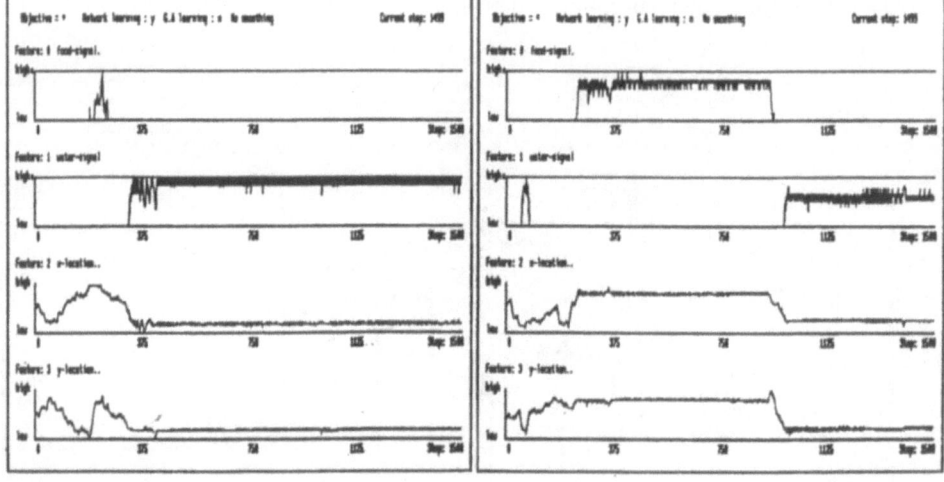

Figure 3(b).   Figure 3(c).

CMAC Manipulator Control Using a Reinforcement Learned Trajectory Planner

Darin P.W. Graham and Gabriele M.T. D'Eleuterio
University of Toronto, Institute for Aerospace Studies
Downsview, Ontario, Canada

1   Introduction

In studies performed to date that use the CMAC (Albus, 1981) for manipulator control (for example, (Miller *et al.*, 1987) and (Graham & D'Eleuterio, 1991)) most of the effort has concentrated on the neural network control module that learns the inverse-dynamics of the system. One area that has yet to receive its due attention is the trajectory planning module. The planner estimates a parameter that is used as an input by the CMAC controller and therefore directly impacts the performance of the system. The controller will have difficulty performing the desired task without a good planner.

The trajectory planner that is implemented in the CMAC-based manipulator control architecture is a simple mathematical equation. Given the current joint position (θ) and velocity ($\dot{\theta}$), the equation predicts a joint acceleration ($\ddot{\theta}_{estimate}$) that will bring the joint from its current actual trajectory to the desired trajectory ($\theta_{desired}$).

If the actual and desired trajectories are similar, this heuristic will calculate predicted accelerations that are acceptable for recalling the appropriate torques needed to control the manipulator. Quite often, such as during the learning phases, this is not the case and the estimated accelerations tend to hinder the control of the robot. Accelerations are calculated that are too large in magnitude. The farther apart the actual and desired paths, the larger the acceleration. If the actual joint acceleration is sufficiently different from the calculated acceleration, the input vector used for learning ($\theta, \dot{\theta}, \ddot{\theta}$) will be significantly different from the recall input vector ($\theta, \dot{\theta}, \ddot{\theta}_{estimate}$). As a result, recalled information from the CMAC memory, to be used for control, is from a different region of the input domain than where learning had occurred. In addition, the heuristic is fixed *a priori* and does not have the ability to adapt and learn as the rest of the control structure learns. As such, this type of trajectory planner is often inappropriate as it exists now and merits improvement.

2   Reinforcement ANN for Trajectory Planning

The correct predicted accelerations are not known and appropriate error values for learning cannot be constructed. As such, a supervised learning approach is not appropriate for the trajectory planner. The planner is however an ideal candidate for a reinforcement learning scheme that uses a reinforcement signal based upon tracking performance. CMAC-based ANNs are architectures well suited to this type of input domain problem. For these reasons, MARCH (Graham, 1994), a reinforcement learning paradigm with a CMAC decoder, will be used.

MARCH is an ASE/ACE-based ANN system (Barto *et al.*, 1983) that uses reinforcement learning. This system does not require a specific heuristic to generate a continuous reinforcement signal, but rather only needs an occasional failure signal. It is also computationally fast as it only updates the few active cells every control cycle, not all the cells in the system as in a typical reinforcement technique.

The ASE/ACE class of reinforcement learning schemes has been successful in applications that implement binary decisions, such as the pole balancing problem using "bang-bang" control. For these applications, weight saturation is an indication of how strongly the ANN thinks the given control action should be taken. The use of this type of reinforcement scheme, in its current form, is not appropriate for control schemes that require real-valued outputs. This section describes a new ANN architecture, MAP, that uses the strengths of these schemes to address the trajectory planner problems.

A single MARCH module as used in the pole-balancing control example has two possible output decisions: left or right. If it were possible to have N modules work in conjunction with each other, the system can provide 2^N discrete output decisions. In this approach, each MARCH module represents one bit in a base-two representation (2^i). Consider a system of N ordered modules, numbered 0 to $N-1$, with each capable of producing a binary decision, y_i, of 0 or 1. Module i is in the ith position in the bit string: $y_{N-1} \cdots y_i \cdots y_2 y_1 y_0$. In general, the output, z_i, of the ith module is:

$$z_i = 2^i y_i \tag{1}$$

The total output representation, Z, of a system of N modules is the sum of the individual output representations:

$$Z = \sum_{i=0}^{N-1} z_i = \sum_{i=0}^{N-1} 2^i y_i \tag{2}$$

If $y_i = 0 \, \forall \, i$, then Z is zero. If $y_i = 1 \, \forall \, i$, then $Z = 2^N - 1$. Therefore, the system is able to determine 2^N integer output values ranging from 0 to $2^N - 1$. The integral output values can be used to produce discrete real-values through scaling and by using an offset.

As described earlier, the heuristic planner has many problems, but it can provide an estimate of the acceleration that is often suitable. The new ANN trajectory planner can be used to modulate the equation estimates in an effort to obtain this goal. Undesirable estimates can be adjusted so that they are within a range of values that would improve trajectory tracking. As such, adaptability has been provided to the heuristic planner and results in continuous real-valued estimates for each of the 2^N binary levels.

Since the values as produced by the heuristic planner are larger than required, scaling of the ANN trajectory planner output from zero to one would modulate the equation planner values appropriately. Thus, the modulated heuristic acceleration estimate would become the new planner estimate:

$$\ddot{\theta}_{estimate,MAP}(t) = \frac{Z}{2^N} \ddot{\theta}_{estimate,equation}(t) \tag{3}$$

The input to the ANN trajectory planning modules will consist of actual past and desired future trajectory information: $\theta(t - 3\kappa\Delta t), \theta(t - 2\kappa\Delta t), \theta(t - \kappa\Delta t), \theta(t), \theta_{desired}(t + \kappa\Delta t), \theta_{desired}(t + 2\kappa\Delta t), \theta_{desired}(t + 3\kappa\Delta t)$. The value κ is an integer that determines the sampling frequency of the data to be used.

When implementing the MARCH paradigm, failure for the manipulator is assumed to occur if any of the joint angles deviate from the desired joint trajectory by a predetermined error limit. To find better joint acceleration predictions, the joint error limit is gradually reduced after each successful trial. The process continues until a final error limit is reached, at which time the network has achieved the desired accuracy and learning stops.

3 Simulation Results

Results of using the new reinforcement ANN trajectory planner with several simulation examples are described in this section. These examples include cases that did not learn using a standard heuristic trajectory planner. A 2-DOF manipulator simulation has been used for these tests. Details about the CMAC architecture that is in the neural network controller to capture the inverse-dynamics, as well as MARCH, are found in (Graham, 1994).

For the first test of the new trajectory planner, a sample trajectory is taken from (Miller *et al.*, 1987). In this type of trajectory, large accelerations occur, and the heuristic trajectory planner gives poor acceleration estimates. When ten different runs (each with a different CMAC hashing structure) were made with this trajectory, only six runs learned successfully (i.e., converged to the desired trajectory). When MAP was included, not only did all ten runs converge, but the final average RMS error (about 0.0025 radians) was almost half that of the six runs that managed to converge without MAP. Figure 1 compares average trajectory RMS errors at the end of each trial with and without the new trajectory planner.

The second simulation test was used to investigate whether the same CMAC controller and MAP could learn to track more than one trajectory. The test was conducted by alternately learning two trajectories until the failure limit of 0.01 radians was reached. That is, Path 1 was presented for learning over several trials until the limit was reached, the memories of the neural networks were maintained, and then Path 2 was learned. A second set of learning trials of Path 1 was then done, followed by another learning set for Path 2. Trajectory RMS error learning trends for this series of learning sets is shown in Figure 2. Final trajectory following RMS errors at the end of each learning set were reduced to approximately 0.0016 radians. Low errors achieved during the second set of learning trials show that information in both the CMAC controller and MAP had been kept. The second learning trials were over two times faster than the first.

The sensitivity of the planner to important learning and structural parameters is investigated in (Graham, 1994). It has been shown that MAP learns for a wide range of parameter values.

4 Conclusions

The objective of this paper was to investigate the trajectory planner module of the manipulator control architecture in order to improve upon the equation-based (heuristic) trajectory planner currently used. A reinforcement learning algorithm, such as the MARCH paradigm that uses a CMAC-like architecture, was identified as suitable for the trajectory planner.

The new CMAC-based trajectory planner, MAP, was tested by placing it in the manipulator control architecture. Results from several simulations show that MAP can assist the CMAC controller module to control the manipulator where the controller could not do so with only the heuristic trajectory planner. Trials included trajectories that required large acceleration estimations and switching between two paths.

670

The technique has applications to problems that require a real-valued output where reinforcement learning is necessary, and will take advantage of the strengths of the CMAC architecture. One such application is the use of MAP for the generation of trajectories that avoid objects.

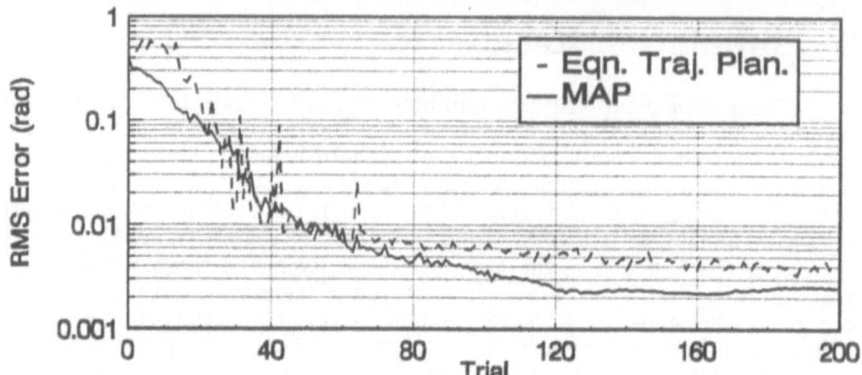

Figure 1: Comparison of RMS Errors for different Trajectory Planners.

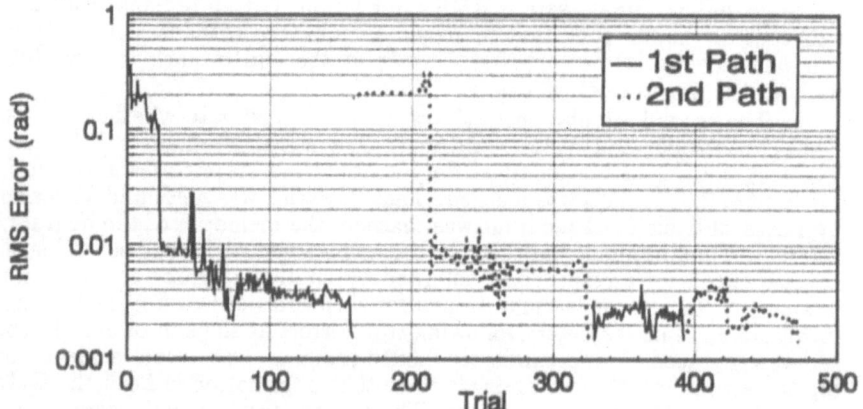

Figure 2: RMS Errors using MAP when Switching Between Paths.

Albus, J.S., *Brains, Behavior, & Robotics*, BYTE Publications Inc., 1981.

Barto, A.G., Sutton, R.S., and Anderson (1983), C.W., "Neuronlike Adaptive Elements that can Solve Difficult Learning Control Problems", *IEEE Transactions on Systems, Man and Cybernetics*, Vol. SMC-13, No. 13, pp834-846.

Graham, D.P.W. (1994), *Manipulator Operations using Value Encoding*, Ph.D. Thesis, University of Toronto.

Graham, D.P.W., and D'Eleuterio, G.M.T. (1991), "Robotic Control Using a Modular Architecture of Cooperative Artificial Neural Networks", *Artificial Neural Networks*, Kohonen, T., Ed., North Holland, Vol. 1, pp365-370.

Miller III, W.T., Glanz, F.H., and Kraft III, L.G. (1987), "Application of a General Learning Algorithm to the Control of Robotic Manipulators", *The International Journal of Robotics Research*, Vol. 6, No. 2, pp84-98, Summer.

A Fast Reinforcement Learning Paradigm With Application to CMAC Control Systems

Darin P.W. Graham and Gabriele M.T. D'Eleuterio
University of Toronto, Institute for Aerospace Studies
Downsview, Ontario, Canada

1 Introduction

The ASE/ACE (Associative Search Element / Adaptive Critic Element of (Barto *et al.*, 1983) is a common architecture that uses reinforcement learning. The major shortfalls of reinforcement learning are that the algorithms are very slow to learn and are computationally intensive. An improved algorithm is needed to reduce the computations required when updating weights every cycle as well as to reduce the total number of trials necessary for learning.

The problem domains that have been addressed using the ASE/ACE like architectures and learning schemes have used only a small number of cells in their input domain space. For example, the broom-balancing problem of (Barto *et al.*, 1983) used only 162 cells. A practical robotic manipulator control problem using neural networks may require on the order of tens of thousands of cells. With this domain size, updating of all cells every control cycle, as required by the ASE/ACE, is computationally intensive. For small domain spaces this is not an issue, but for control problems of the size and complexity of a multidegree-of-freedom manipulator, the standard ASE/ACE approach is computationally unmanageable. In this paper, a technique called *"Catch-up Reinforcement Learning* is developed, based upon the ASE/ACE architecture to perform reinforcement learning with reduced computational requirements.

2 Development of the Catch-up Paradigm

The equations used in the ASE/ACE network are of the form such that given the values of the network parameters at the current learning step the updated values for the next step can be calculated. The use of the equation in this form requires extensive calculations at each cycle in a learning trial as all cells that have become active some time in the past must be updated every cycle. The work presented in this section describes the development of a reinforcement learning paradigm that only requires cell weights to be updated as they are used, and not at every control cycle. Such an approach has potential for reducing the computational overhead of reinforcement learning as the calculations required at each control cycle remain constant regardless of the size of the input domain space. The method maintains the approach where only an occasional reinforcement is provided and does not need a heuristic to determine a continuous reinforcement signal at each time step.

Since the equations for the ASE structure also apply to the ACE, we will first examine the ASE and then use our findings with the ACE. The equations that represent the full development of the ASE and ACE are found in (Barto *et al.*, 1983). The first step in developing a faster paradigm is to reformulate the learning equations in a series formulation. Consider the following scenario. Cell i is used at t_0 and not used again until t_n, and that failure does not occur during these times. Updating ASE weights between uses can be written as:

$$w_i(t_n) = w_i(t_1) + \alpha e_i(t_1)R_\delta^n \quad \text{where} \quad R_\delta^n \equiv \sum_{j=1}^{n-1} \delta^{j-1}\hat{r}(t_j) \tag{1}$$

The reinforcement is directly related to the prediction values and can be used to calculate an *n-step prediction* as follows:

$$R_\delta^n = \begin{cases} [\gamma p(t_1) - p(t_0)] & \text{for } n = 2 \\ [\gamma \delta^{n-2} p(t_{n-1}) - p(t_0)] + (\gamma - \delta)P_\delta^n & \text{for } n \geq 3 \end{cases} \tag{2}$$

$$\text{where} \quad P_\delta^n = \sum_{j=1}^{n-2} \delta^{j-1} p(t_j) \tag{3}$$

The n-step prediction is a weighted sum of the ACE reinforcement predictions through time. In keeping with the objective of not performing calculations unless a cell is active, the calculations done for cell i during times between t_0 and t_n would not be done until the start of time step t_n, and the cell would have to "catch up" on the missed calculations. If these missed calculations were done iteratively no computational saving would occur, and the process would require large amounts of memory for retaining the reinforcement signals. To limit the number of calculations and obtain an estimate of P_δ^n, an arbitrarily fixed number of previous values could be stored. This would require each cell to maintain its own history. If the choice of how many values to keep is too small, the arbitrary truncation of values could lead to important predictions being lost. Accordingly, such techniques will not be used.

For this study, it is assumed that the desired history of predictions can be approximated by an average value over the time segment between a cell's activations. Thus, the sum in Equation (3) can be approximated by:

$$P_\delta^n(t_n) \approx \sum_{j=1}^{n-2} \delta^{j-1} \bar{p}_i(t_1, t_{n-2}) \tag{4}$$

where $\bar{p}_i(t_1, t_{n-2})$ is the average value of the prediction signal for cell i between times t_1 and t_{n-2}, inclusive. Each cell will have its own average since the internal reinforcement signals will be different for each cell between successive activations. This will require only one more value to be stored per cell.

Since $\bar{p}_i(t_1, t_{n-2})$ is a constant, it can be taken out of the sum, and substituting for the sum of exponential terms, Equation (4) becomes:

$$P_\delta^n(t_{n-2}) = \bar{p}_i(t_1, t_{n-2}) \frac{1 - \delta^{n-2}}{1 - \delta} \tag{5}$$

The equations for the ACE are of identical form as those above with learning rate β instead of α and decay rate λ instead of δ. A failed control cycle uses the Catch-up P^n Method to update all cells and then applies the standard ASE/ACE calculations using the failure signal. Full details of the Catch-up P^n Method can be found in (Graham, 1994).

The application of the CMAC to the ASE/ACE learning technique is described in (Lin & Kim, 1991). It was shown that far fewer training trials were required by the CMAC in order to learn to the same level of control as the nonoverlapping cells. To improve the computational performance, the Catch-up scheme is now applied to the CMAC. This combination of architecture and learning will be called MARCH.

Since responses and learning are a result of several cells within the CMAC being activated, credit is assigned equally to the active cells. If there are K cells activated by the CMAC decoder the weight updating equations use learning gains of $\alpha' = \alpha/K$ and $\beta' = \beta/K$ for the ASE and ACE. The remainder of the Catch-up P^n equations remain unchanged.

3   Simulation Results

Historically, reinforcement algorithms have been tested and compared using the *broom-balancing* control problem. This will also be used here. The input discretization by the decoder uses 162 cells as in (Barto *et al.*, 1983). The parameter values as determined by (Barto *et al.*, 1983) will also be used.

In addition to using the simple discretization described above for the ASE/ACE, MARCH will be used for a larger input domain. Each of the four domain variables will be divided into seven equal discretizations with the following limits: $x \in [-2.4, 2.4]$ m, $\theta \in [-12, 12]°$, $\dot{x} \in [-1.5, 1.5]$ m/s, $\dot{\theta} \in [-115, 115]°$/s. With this discretization scheme, a larger domain ASE/ACE decoder will have 2401 cells. For this study, three coarse grids were chosen for the CMAC decoder, resulting in a total of 243 coarse cells.

Each method was run 10 times, each with a different seed for the random noise generator, and the resulting data from the different runs were averaged. If during a given trial the system was able to perform 500,000 iterations without failure, it was assumed to have learned completely.

Figure 1 shows the results for the various learning algorithms and architectures. MARCH performed as well as the ASE/ACE learning CMAC decoder. As expected, both CMAC-based methods learned quickly, and better than the small domain 162-cell ASE/ACE. The ASE/ACE algorithm had difficulty learning to control the system when the domain space was increased.

The advantage of the new method can be seen from the computational requirements of the simulation summarized in Table 1. For the same sized domain problem, MARCH is about 4.4 times faster than the CMAC decoder (with an ASE/ACE learning scheme) and over 6.2 times faster than an equivalent 2401-cell ASE/ACE. Thus, MARCH is computationally faster and requires the same, if not fewer, learning trials to control the system to a comparable degree than any of the other techniques. Comparing MARCH with a 162-cell ASE/ACE, it requires about half the computational time per iteration, but the input domain is nearly 15 times larger.

Method	Seconds/Iteration
CMAC (with ASE/ACE)	0.0074
MARCH	0.0017
2401-cell ASE/ACE	0.0106
162-cell ASE/ACE	0.0031

Table 1: Computational requirements of the different reinforcement methods.

674

4 Conclusions

In summary, the following observations can be made regarding the new reinforcement learning paradigm:

- The ASE/ACE/CMAC and the Catch-up P^n with CMAC require about the same number of trials to learn to control the example system.

- The Catch-up P^n method requires about one-sixth of the computational time as the ASE/ACE for the small input space of the broom-balancing problem.

- The computational requirement for the new paradigm is constant for successful learning iterations, regardless of the input domain size.

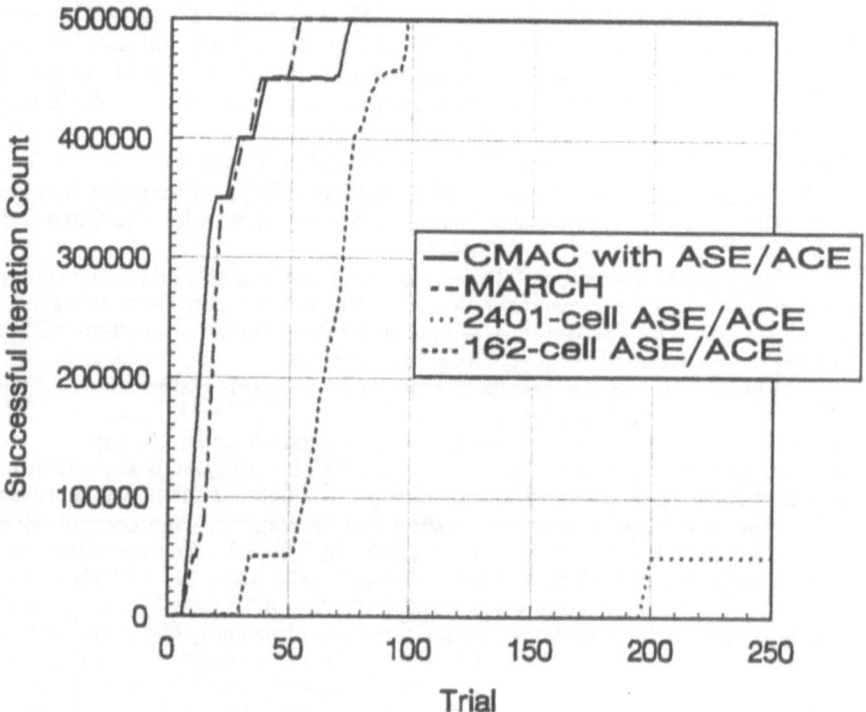

Figure 1: Simulation Results for the Broom-Balancing Control Problem.

Barto, A.G., Sutton, R.S., and Anderson (1983), C.W., "Neuronlike Adaptive Elements that can Solve Difficult Learning Control Problems", *IEEE Transactions on Systems, Man and Cybernetics*, Vol. SMC-13, No. 13, pp834-846.

Graham, D.P.W. (1994), *Manipulator Operations using Value Encoding*, Ph.D. Thesis, University of Toronto.

Lin, C.-S., and Kim, H. (1991), "Use of CMAC Neural Networks in Reinforcement Self-learning Control", *Artificial Neural Networks* (Proc. of ICANN-91), Kohonen, T., Ed., Elsevier Science Publishers, Vol. 2, pp1285-1288.

Information Geometry and the EM Algorithm

Shun-ichi Amari

Department of Mathematical Engineering and Information Physics
Faculty of Engineering, University of Tokyo
Bunkyo-ku, Tokyo 113, JAPAN

Abstract

Information geometry is a mathematical method of studying intrinsic properties of a manifold of probability distributions. It provides a geometrical method for studying stochastic neural networks including hidden variables. The present paper elucidates the geometrical structure underlying the *EM* algorithm, which is used for a number of interesting neural networks models. A learning algorithm is easily derived from this approach.

1 Introduction — Stochastic multilayer perceptron

Let us consider a stochastic multilayer perceptron with one hidden layer as a typical example. Let $\mathbf{x}$ be an input signal, and let $\mathbf{z}$ be output from the hidden layer. Let y be the final output. The hidden output $\mathbf{z}$ is determined stochastically depending on the input $\mathbf{x}$, so that the behavior of the hidden units is described by the conditional probability $p(\mathbf{z}|\mathbf{x})$. We consider the stochastic neuron (Amari [1991]) such that the conditional distribution is given by

$$p(z_i|\mathbf{x}) = \frac{\exp\{z_i\mathbf{w}_i \cdot \mathbf{x}\}}{1 + \exp\{\mathbf{w}_i \cdot \mathbf{x}\}},$$

where $\mathbf{z} = (z_1, \cdots, z_m)$ and z_i, the output of the ith hidden unit, takes on 0 and 1, and $\mathbf{w}_i$ is the connection weight vector of the ith hidden unit.

The output unit emits the final output y taking values on 0 and 1 by receiving $\mathbf{z}$. Its probability is

$$p(y|\mathbf{z}) = \frac{\exp\{y\mathbf{v} \cdot \mathbf{z}\}}{1 + \exp\{\mathbf{v} \cdot \mathbf{z}\}}.$$

Hence, when $\mathbf{x}$ is input, the conditional joint probability is written as

$$p(y, \mathbf{z}|\mathbf{x}; \mathbf{u}) = \prod p(z_i|\mathbf{x})p(y|\mathbf{z}), \tag{1}$$

where $\mathbf{u} = (\mathbf{v}, \mathbf{w}_1, \cdots, \mathbf{w}_m)$ is a vector summarizing all the network parameters. The conditional probability of y itself is written as

$$p(y|\mathbf{x}; \mathbf{u}) = \sum_{\mathbf{z}} p(y, \mathbf{z}|\mathbf{x}; \mathbf{u}). \tag{2}$$

Let $(\mathbf{x}_1, y_1), \cdots, (\mathbf{x}_T, y_T)$ be T examples of an input-output relation which we would like to realize by a stochastic perceptron by modifying the network parameter $\mathbf{u}$. Here, the network includes a hidden variable $\mathbf{z}_t$, the output of the hidden units at time t, which is not observed nor specified from the outside. However, if we can estimate or assign the values of the hidden variables adequately, the probability structure (1) is simpler than (2), so that the estimation of $\mathbf{u}$ becomes easier and its learning algorithm can be implemented also in a simple form. This is the problem of hidden variables and may be regarded as a version of the credit assignment problem.

Many stochastic neural networks include hidden variables. They are, for example, the Boltzmann machines with hidden units (Amari et al. [1992], Byrne [1992]), the hierarchical mixture of experts (Jacobs et al. [1991], Jordan and Jacobs [1994], Jordan and Xu [1993]), the normal mixtures (Neal and Hinton [1994]), and many others.

There are two methods for estimating the hidden variables $\hat{\mathbf{z}}_t$ iteratively. One is the *EM* algorithm known in statistics. This estimates $\hat{\mathbf{z}}_t$ by the conditional expectation of $\mathbf{z}_t$ with respect to the probability distribution given by the current estimated $\hat{\mathbf{u}}$,

$$\hat{\mathbf{z}}_t = E[\mathbf{z}_t|y_t; \hat{\mathbf{u}}]. \tag{3}$$

By using the estimated data $(\mathbf{x}_t, \hat{\mathbf{z}}_t, y_t)$, $t = 1, \cdots, T$, $\hat{\mathbf{u}}$ is further modified by maximizing the likelihood.

Information geometry (Amari [1985], Amari [1987], Amari and Han [1989]) gives another method. It has already been applied to neural networks (Amari [1991], Amari et al. [1992].) Let S be the manifold consisting of all the conditional probability distributions $p(y, \mathbf{z}|\mathbf{x})$,

$$S = \{p(y, \mathbf{z}|\mathbf{x})\}$$

which is assumed to form an the exponential type distribution. When y and $\mathbf{z}$ are discrete variables, this assumption always holds. Most models of neural networks admit the exponential type distribution. A conditional distribution which is realizable by a stochastic neural net has a form of $p(y, \mathbf{z}|\mathbf{x}; \mathbf{u})$ specified by $\mathbf{u}$. The set of all the realizable distributions forms a submanifold M in S,

$$M = \{p(y, \mathbf{z}|\mathbf{x}; \mathbf{u})\},$$

where $\mathbf{u}$ plays the role of a coordinate system of M. When a set of complete data $(\mathbf{x}_t, \mathbf{z}_t, y_t)$, $t = 1, \cdots, T$ is given, the sufficient statistics calculated therefrom define a candidate distribution $\hat{q}$ in S. The maximum likelihood estimator $\hat{\mathbf{u}}$ is the one that minimizes the Kullback-Leibler divergence $K(\hat{q} \| p)$, $p \in M$.

When observation or specification is incomplete because $\mathbf{z}_t$ are missing, we can assign any values to $\mathbf{z}_t$. In this case, the candidate point $\hat{q}$ is not uniquely determined. Instead, by assigning arbitrary values to the hidden $\mathbf{z}_t$, we have a candidate submanifold D which any candidate points belong to. When D and M are given, we can choose the two points $\hat{p} \in M$ and $\hat{q} \in D$ such that they jointly minimize $K(q \parallel p)$, $q \in D$, $p \in M$. . This determines the hidden $\hat{\mathbf{z}}_t$. Such an algorithm was proposed in studying the Boltzmann machines (Amari et al. [1992], Byrne [1992], see also Amari [1991]).

We give a unified framework of information geometry (Amari, 1985) to elucidate the above two algorithms. A more detailed paper will appear in Amari [1994].

2 Information geometry of neural networks

Since $p(y, \mathbf{z}|\mathbf{x})$ belongs to an exponential family, it is written as

$$p(y, \mathbf{z}|\mathbf{x}) = \exp\{\boldsymbol{\theta}(\mathbf{x}) \cdot \mathbf{r}(y, \mathbf{z}) - \psi\} \tag{4}$$

where $\boldsymbol{\theta}(\mathbf{x})$ is the canonical parameter to specify the distribution, it depends on $\mathbf{x}$ and $\mathbf{r}(y, \mathbf{z})$ is a sufficient statistic. The expectation of $\mathbf{r}(y, \mathbf{z})$ is written as

$$\boldsymbol{\eta}(\mathbf{x}) = E[\mathbf{r}(y, \mathbf{z})] \tag{5}$$

where the expectation is taken with respect to the distribution specified by $\boldsymbol{\theta}(\mathbf{x})$. When observation is repeated, we have

$$\begin{aligned}
&p\{(y_1, \mathbf{z}_1), \cdots, (y_T, \mathbf{z}_T) \mid (\mathbf{x}_1, \cdots, \mathbf{x}_T)\} \\
&= \prod_t p(y_t, \mathbf{z}_t \mid \mathbf{x}_t),
\end{aligned} \tag{6}$$

and this belongs to the direct product of the exponential families $S_t = \{p(y_t, \mathbf{z}_t|\mathbf{x}_t)\}$,

$$S_T^* = S_1 \times \cdots \times S_T. \tag{7}$$

In the case of the stochastic perceptron, we have sufficient statistics

$$\bar{\mathbf{r}} = \frac{1}{T} \sum_{t=1}^{T} \mathbf{r}(y_t, \mathbf{z}_t, \mathbf{x}_t), \tag{8}$$

such that the product distribution is summarized as

$$p\{(y_1, \mathbf{z}_1)\cdots, \{(y_1, \mathbf{z}_1), \cdots, (y_T, \mathbf{z}_T) \mid (\mathbf{x}_1, \cdots, \mathbf{x}_T)\} = \exp\{T(\boldsymbol{\theta} \cdot \bar{\mathbf{r}} - \psi)\}, \tag{9}$$

where $\boldsymbol{\theta}$ does not depend on $\mathbf{x}$. We show the case of the stochastic perceptron: Let $\mathbf{k}$ be a vector whose components are 0 or 1, $\delta_{\mathbf{k}}(\mathbf{z})$ is the delta function and $\mathbf{w}_{\mathbf{k}} = \sum k_i \mathbf{w}_i$. Then, $\mathbf{r}$ is divided into two components whose indices are $\mathbf{k}$,

$$r_{1,\mathbf{k}} = y\delta_{\mathbf{k}}(\mathbf{z}), \qquad r_{2,\mathbf{k}} = \mathbf{x}\delta_{\mathbf{k}}(\mathbf{z}) \tag{10}$$

and (8), (9) hold.

In the distributions (9), not all the distributions are realizable by perceptrons. The manifold M of distributions realizable by neural networks is specified by

$$\boldsymbol{\theta} = \boldsymbol{\theta}(\mathbf{u}) \tag{11}$$

in the coordinate system $\boldsymbol{\theta}$ of S. The data manifold D_T which is specified by the partial observation is given in the coordinate system $\boldsymbol{\eta}$ of S as

$$D_T = \{\boldsymbol{\eta} | \boldsymbol{\eta} = \bar{\mathbf{r}}\}, \tag{12}$$

which includes $\mathbf{z}_t$ as free parameters.

Let Q be a point in D, and P be a point in M, whose distributions are $p(y, \mathbf{z}|\mathbf{x}; Q)$ and $p(y, \mathbf{z}|\mathbf{x}; P)$, respectively. The KL-divergence is defined by

$$K(Q \parallel P) = E_{Q,X} \left[\log \frac{p(y, \mathbf{z}|\mathbf{x}; Q)}{p(y, \mathbf{z}|\mathbf{x}; P)} \right], \tag{13}$$

where $E_{Q,X}$ denotes the expectation with respect to Q and $p(\mathbf{x})$, the distribution of input $\mathbf{x}$.

Given P, let $\hat{Q}$ be a point in D that minimizes $K(Q, P)$, $Q \in D$. The information geometry (Amari, 1985), see also Amari [1991], Amari et al. [1992] proves that $\hat{Q}$ is given by the e-geodesic projection from P to D. In order to define the geodesic projection, we need the concept of the orthogonality and the concept of the geodesic. The orthogonality is given by the Riemannian metric which is provided by the Fisher information matrix. In an exponential family, an e-geodesic is a linear curve with respect to the $\boldsymbol{\theta}$-coordinate system. (In the general case, we need the notion of the e-affine connection.)

On the other hand, given Q, the poitn $\hat{P} \in M$ that minimizes $K(Q \parallel P)$, $P \in M$, is given by the m-geodesic projection of Q to M. Here, an m-geodesic is a linear curve in the $\boldsymbol{\eta}$-coordinate system. In general, the e- and m-geodesics, or the underlying e- and m-affine connections, are dually coupled with respect to the Fisher information Riemannian metric. This is a new structure of differential geometry originated from information science.

The information-geometric algorithm, which we call the em-algorithm, is formulated as follows.

1. Choose an intitial guess $\hat{\mathbf{u}}_0$ or $\hat{P}_0 \in M$.

2. Repeat the following for $i = 0, 1, 2, \cdots$.

   (a) e-step : e-project $\hat{P}_i$ to D, obtaining $\hat{Q}_i$.

   (b) m-step : m-project $\hat{Q}_i$ to M, obtaining $\hat{P}_{i+1}$.

The procedure converges to the pair $(\hat{Q}, \hat{P})$ that minimizes

$$K(Q \parallel P), \quad Q \in D, \ P \in M$$

in the dual manner.

3 The equivalence of the EM and em algorithms

We have shown the two algorithms. The EM algorithm replaces the e-projection of $2(a)$ by the conditional expectation to obtain $\hat{Q}'_i$. That is, by assigning the conditional expectation to the hidden $\mathbf{z}_t$, we have a new candidate distribution $\hat{Q}'_i \in D$. An important question is if the two algorithms are equivalent, that is, if $\hat{Q}_i = \hat{Q}'_i$. They are not in general. However, they are equivalent for most neural networks. We show

Theorem. Let y_{obs} be the value of the observable variable and let

$$\mathbf{a}(y) = E[\mathbf{z}|y; \hat{\mathbf{u}}].$$

The two algorithms are equivalent, if and only if

$$E[\mathbf{a}(y)] = \mathbf{a}(y_{\text{obs}}),$$

where expectation is taken at $\hat{Q} \in D$. When random variables are discrete, the condition holds.

Once the equivalence of the two algorithms are established, we can analyze the EM algorithm by the information-geometric framework (Amari [1994], giving new fruitful results including learning.

4 Conclusions

We have formulated the information-geometrical procedure for the problem of hidden variables in stochastic neural networks. Once we prove the equivalence of the EM- and em-algorithms, we can easily study the convergence property of the algorithms, their differential forms and learning rules. Information geometry proves its usefulness in the field of neural networks.

5 References

- Amari, S. (1985) *Differential Geometrical Methods in Statistics*, Springer Lecture Note in Statistics, **28**.

- Amari, S. (1987) Differential Geometry of a Parametric Family of Invertible Linear Systems — Riemannian Metric, Dual Affine Connections and Divergence, *Mathematical Systems Theory*, **20**, pp.53–82.

- Amari, S. (1991) Dualistic Geometry of the Manifold of Higher-Order Neurons, *Neural Networks*, **4**, pp.443–451.

- Amari, S. (1994) Information Geometry of the EM algorithm for neural networks, to appear in *Neural Networks*.

- Amari, S. and Han, T. S. (1989) Statistical Inference under Multi-Terminal Rate Restrictions — A Differential Geometrical Approach, *IEEE Trans. on Information Theory*, **IT-35**, pp.217–227.

- Amari, S. Kurata, K. and Nagaoka, H. (1992) Information Geometry of Boltzmann Machines, *IEEE Trans. Neural Networks*, **Vol. 3**, No.2, pp.260 271.

- Byrne, W. (1992) Alternating Minimization and Boltzmann Machine Learning, *IEEE Trans. Neural Networks*, **3**, pp612–620.

- Jacobs, R. A., Jordan, M. I., Nolwan, S. J. and Hinton, G. E. (1991) Adaptive Mixtures of Local Experts, *Neural Computation*, **3**, pp.79–87.

- Jordan, M. I. and Jacobs, R. A. (1994) Higherarchical Mixtures of Experts and the *EM*-Algorithm, *Neural Computation*, to appear.

- Jordan, M. I. and Xu, L. (1993) Convergence Properties of the EM Approach to Learning in Mixture-of-Experts Architectures, *MIT Comp. Cog. Sci.*, **TR9302**.

- Neal, R. N. and Hinton, G. E. (1993) A New of the EM Algorithm that Justifies Incremental and Other Variants, to appear.

SSM : A Statistical Stepwise Method for Weight Elimination

M. Cottrell†, B. Girard†, Y. Girard†, M. Mangeas†, C. Muller‡

† Samos, Université Paris 1

90, rue de Tolbiac, 75634 Paris Cedex 13, France

‡ DER-EDF

1, Avenue du Général de Gaule, 92141 Clamart, France

1 Introduction

The multilayer feedforward network[1] has a now well-known approximation capability, which can be used to achieve any *association task* with a very good accuracy. See Leshno et al. [8] for the most recent results and references. However, the choice of the architecture of the network, and especifically of the number of hidden units is not easy, because greater is this number, better is the accuracy ! Much work is currently directed towards this problem, see e.g. [6], [7]. It is clear for all the researchers that it would be worthwhile to have at their disposal a good method to eliminate the superfluous weights, in order to increase the *robustness* of the network, that is to increase the *so-called generalization property*, instead of learning *by heart*.

In this study, following the direction initialized in previous works in [3], [4], we throw a statistical light on this problem and provide a *Statistical Stepwise Method* (SSM) *to prune the superfluous weights*, based on the testing of hypotheses.

The paper is organized as follows : in Section 2, we present the statistical formulation of the problem and the theoretical results. In Section 3, we present some examples in function approximation, non linear regression, and time series analysis. Section 4 is devoted to discussion and comparison with the OBD scheme [7].

2 The Model and the Main Results

Let consider a multilayer perceptron, with n input-units, one hidden layer and one output-unit. Only the hidden units are non linear and their common activation function is denoted ϕ. In the following, $W = (W_l)$, with $l = 1, \ldots, M$ represents the vector of the weights and the thresholds, M is the number of parameters. The function implemented by the network is f_W.

Let us assume that the relation between the target output Y and the input vector X can be represented by the model

$$Y = f(X) + \eta$$

---

[1]This research is partially supported by EDF, Direction des Etudes et des Recherches, Clamart, France.

where η is a noise (gaussian or not) with zero mean and finite variance, and f a real-valued function defined on R^n.

From the approximation property of the feedforward networks [8], we can emulate the function f by a function f_W implemented by such a network. The model becomes $Y = f_W(X) + \delta + \eta$, where δ is the error term due to the approximation.

So, denoting $\delta + \eta = \epsilon$, we consider the resulting model, called Model (W)

$$Y = f_W(X) + \epsilon$$

where we assume that the global error term ϵ is a noise with zero mean and finite variance σ^2.

Then we are concerned with inference about the parameter W for a given k, and with the choice of the best k, given T independent observations $(X_t, Y_t), t = 1, \ldots, T$, of the pair (X, Y). For a given k, as usually, the least squares estimator $\widehat{W}$ is defined to be the value of W that minimizes the sum of squared residuals $S(W) = \sum_{t=1}^{T}(Y_t - f_W(X_t))^2$. The computation of $\widehat{W}$ is *the training of the network*. The minimization can be carried out with any minimization method, and we know that using a second order method (like BFGS or Levenberg-Marquardt methods [9]) can be more efficient and reliable to avoid local minima than the simple gradient-descent method like back-propagation.

To consider the *deletion of a weight* w_l is equivalent to consider the test of the null hypothesis "$w_l = 0$" against the alternative hypothesis "$w_l \neq 0$", i.e. *to test the Model* (W_l) *against the Model* (W), where parameters W and W_l are equal, except in w_l which is zero in W_l and not zero in W. We say that W_l is a *sub-model* of W.

To consider *successive deletions of weights* $w_{l_1}, w_{l_2}, \ldots w_{l_L}$ is equivalent to test iteratively a sequence of embedded Models $W_{l_1}, W_{l_1 l_2}, \ldots, W_{l_1 l_2 \ldots l_L}$, where successively $w_{l_1} = 0$, $w_{l_1} = w_{l_2} = 0$, etc...

The problems to solve are then
(1) how to achieve a *test* of "W_l" against "W"
(2) how to determine the best *sequence* $l_1, l_2, \ldots, l_L$
(3) how to choose a *stop criterion*

It is well known that the least squares estimators satisfy *asymptotic normality* in the linear case (when function f is linear). The very interesting result is that under weak conditions (regularity conditions of function f_W, for what it can be valuable to choose the sigmoid function as activation function ϕ), the asymptotic properties of the least squares estimators remain valid in much more general cases (linear or not, autoregressive, etc...)[2].

So for a very general class of models, one has when $T \to \infty$

$$\sqrt{T}(\widehat{W} - W) \xrightarrow{D} \mathcal{N}_M(0, \sigma^2 \Sigma^{-1})$$

where σ^2 is the residual variance, estimated by $\hat{\sigma}^2 = \frac{S(\widehat{W})}{T}$, and Σ is the information matrix estimated by $\frac{1}{2T} \nabla^2 S(\widehat{W})$ and approximated by

$$\frac{1}{T} \sum_t (\nabla f_{\widehat{W}}(X_t))(\nabla f_{\widehat{W}}(X_t))'.$$

So the procedure to answer to the three questions is :

a)−Choose a comfortable architecture W, and train the network.

b)−Compute all the quotients $Q(l) = \hat{w}_l/\hat{\sigma}(\hat{w}_l$ where $\hat{\sigma}(\hat{w}_l) = \frac{\hat{\sigma}}{\sqrt{T}}\sqrt{(\hat{\Sigma}^{-1})_{l,l}}$ is the estimated standard deviation of $\hat{w}_l$.

c)−Define l_1 corresponding to the minimum value of these quotients.

d)−Test the Model (W_{l_1}) (null hypothesis) against the Model (W) (alternative hypothesis) : accept the elimination of w_{l_1} if and only if $Q(l_1)$ is less than a typical value. For a level 5% of signification, the typical value is 1.96.

e)−In case of reject, stop the elimination process, and keep the previous Model. In case of acceptation, retrain the network corresponding to Model (W_{l_1}), and repeat step b), from Model (W_{l_1}), to look for the indice l_2, and so on.

The *stop criterion* is very natural : one stops the elimination process when no more weight can be eliminated. That gives an objective criterion, well-founded, resting on the statistical properties of the weights estimators. We are actually able to decide what is a "small" weight, that we can eliminate. Note that a weight equal to 10^{-5} can be "great" if its estimated standard deviation is 10^{-6}.

The proposed algorithm SSM belongs to the family of *stepwise backward algorithms* widely used in *regression analysis* [5].

3 Examples

Ex 1 : A *deterministic function* (plus noise) : $Y = (\sin(X_1 + X_2) + \sin(X_2 + X_3) + \sin(X_3 + X_1))/3 + \epsilon$, with $\sigma^2 = 0.1$, $T = 343$. From $M_{initial} = 51$, the method leads to $M_{final} = 22$ and $\hat{\sigma}^2 = 0.101$.

Ex 2 : A *regression model* : The Y variable is the number of telephones, as a function of the population, the number of employees, the area, the income per capita, the total amount of salaries, in each district of California, $T = 58$. From $M_{initial} = 61$, the method leads to $M_{final} = 19$ and $\hat{\sigma}^2 = 1.39$. The model is strictly better that the corresponding linear one.

Ex 3 : An artificial simulated *time series* : $Y_t = 1 + \phi(0.7T_{t-1} - 1.3) + \phi(0.9Y_{t-2} - 0.4) + \epsilon$, with $\sigma^2 = 0.25$, $T = 1000$. From $M_{initial} = 16$, the method leads to $M_{final} = 7$, $\widehat{W}$ very close to W, and $\hat{\sigma}^2 = 0.247$.

Ex 4 : Example of the *prevision of the complete half-hourly load curve of electrical energy for the following day*, with $T = 2738$. From $M_{initial} = 300$, the method leads to $M_{final} = 150$. The model has better performances than the currently used model, but it is for the moment more heavy to use.

See more examples (simulated ones, sunspot series, daily electrical consumption data) in [3] or [4].

4 Discussion

The *stepwise backward method* is very classical. However, some difficulties can be encountered. In particular, the final model can depend on the initial model W. So to compute a criterion of the quality of the resulting model and to compare two models which are not embedded, one uses the *Akaike information criterion* [1] defined as the $BIC(W) = \ln \frac{S(W)}{T} + \frac{M \ln T}{T}$ where M is the total number of parameters of the model W. This criterion contains a penalty term, which gives a cost to any weight. The Akaike information criterion is more valuable than the *generalization* one : in fact, when the network works well on the *test set*, one can only be sure that the *test set* and the *learning set* are governed by the same model.

At last, let us discuss the relation between the SSM and the OBD defined by Le Cun et al [7], and based on the *saliency* that they use to decide which weight is candidate to elimination. They define $s_l = \frac{1}{2T}(S(\widehat{W_l}) - S(\widehat{W}))$, with the previous notations. One can verify that $Q(l)^2 = (\hat{w}_l/\hat{\sigma}(\hat{w}_l))^2 = \frac{2Ts_l}{S(\widehat{W})/(T-M)}$ so $Q(l)^2$ and s_l are proportional because the denominator $S(W)/(T - M)$ can be considered as a constant when T is large. So the comparison of the $Q(l)$ statistics is equivalent to the comparison of the saliencies. However OBD does not take into account the relative value of the saliency. Overmore, as the complete Hessian matrix is computed, it is possible to test simultaneous deletion in an exact way, without any diagonal approximation as in [7], which would be correct if the estimators were independent.

Currently, we are applying the SSM to numerous data problems, resulting of our collaboration with Electricity of France.

References

[1] H. Akaike, "Fitting autoregressive models for prediction", Ann. Inst. Stat. Math., 21, 243-247, 1969

[2] T. Amemiya, *Advanced Econometrics*, Basil Blackwell, 1986

[3] M. Cottrell, B. Girard, Y. Girard and M. Mangeas, "Times Series and Neural Network : a Statistical Method for Weight Elimination", Proc. of ESANN'93, 157-165, M. Verleysen Ed, Quorum, Bruxelles, 1993.

[4] M. Cottrell, B. Girard, Y. Girard, M. Mangeas and C.Muller, " Neural Modeling for Time Series : a Statistical Stepwise Method for Weight Elimination", submitted, Tech. Rep.# 20, SAMOS, Université Paris 1, 1993

[5] N. Draper and H. Smith, *Applied Regression Analysis*, 2d Edition, New York, John Wiley & Sons, 1981

[6] J. Gorodkin, L.K. Hanses, A. Krogh, C. Svarer and O. Winther "A Quantitative Study of Pruning by Optimal Brain Damage", Int. J. of Neural Systems, Vol. 4, No. 2, 159-169, 1993

[7] Y. Le Cun, J.S. Denker and S.A. Solla, "Optimal Brain Damage", in Advances in Neural Information Processing Sust. II, ed. D.S. Touretzsky, Morgan Kaufman, 598-605, 1990

[8] M. Leshno, V. Ya Lin, A. Pinkus and S. Schocken "Multilayer Feedforward Networks with a Nonpolynomial Activation Function can Approximate any Function", Neural Networks, Vol. 6, 861-867, 1993

[9] D. Luenberger, *"Introduction to linear and nonlinear programming"*, Addison-Wesley, 1973

Computing the Probability Density in Connectionist Regression

Ashok N. Srivastava† and Andreas S. Weigend‡

†Department of Electrical and Computer Engineering
University of Colorado at Boulder
Boulder, CO 80309-0529 USA
srivasan@sebastian.colorado.edu

‡Department of Computer Science and Institute of Cognitive Science
University of Colorado at Boulder
Boulder, CO 80309-0430 USA

We introduce a non-parametric method for determining the degree of uncertainty in prediction and show its use in a regression problem.

1 Problem Statement and Network Architecture

Often, it is not difficult to train a network to generate a number– but what confidence can we have in that number? We would like to compute a probability density whose mean is the target value, and the variance is the confidence.

The problem can be formally expressed in the following manner: given a function $f(\underline{x}): R^n \rightarrow R^1$, we construct a neural network which performs a nonlinear regression on f with a vector of inputs $\underline{x} \in R^n$ whose outputs $\underline{y} \in R^m$ are an estimate of the continuous probability density function $p(\underline{y}|\underline{x})$.

The key idea is to represent a continuum output as a *soft histogram*. The continuos distribution is approximated with m classes. In order to create a training set, we sort the data in ascending order and binned them into m bins with an equal number of points per bin. This approach minimizes quantization errors and shows a method for recasting the function approximation problem as a classification problem by partitioning the range-space of the target function.

The mean and variance of the output distribution can be computed directly via

$$\mu(\underline{x}) = \sum_{i=1}^{m} y_i \zeta_i \tag{1}$$

$$\sigma^2(\underline{x}) = \sum_{i=1}^{m} y_i \zeta_i^2 - [\mu(\underline{x})]^2, \tag{2}$$

where ζ_i is the center of bin i and y_i is the activation of unit i. Figure 1 shows the network and an example of the soft-histogram coding technique (target of 0.24).

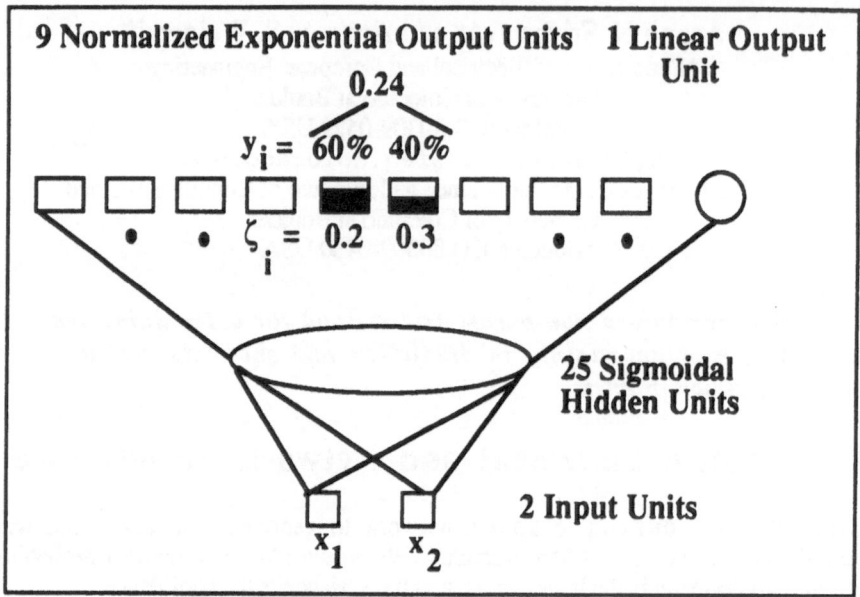

Fig. 1. Network architecture for probability density estimation

The normalized exponential units (also called soft-max) have the functional form

$$p_j = e^{h_j} \Big/ \sum_i e^{h_i} . \tag{3}$$

h_j is the total weighted input received by output unit j, $1...m$ (Rumelhart et al.).

In our experience, learning can be facilitated by having the network perform an additional, easier, task. We added a linear output unit which performs a direct prediction of the mean $\mu(\underline{x})$. This additional output focuses the hidden units to learn the mean. The network has a single hidden layer composed of 25 sigmoidal units, and is trained with backpropagation minimizing the sum-squared error. All units have biases associated with them.

2 An Example

The network's task is to perform a nonlinear regression on a function of the form

$$f(\underline{x}) = g(\underline{x}) + n(\underline{x}). \tag{3}$$

This function is composed of the target function $g(\underline{x})$ corrupted with additive noise $n(\underline{x})$. The network reconstructs $g(\underline{x})$ and estimates $n(\underline{x})$ from a finite number of samples of $\underline{x}, f(\underline{x})$. The noise estimate can be interpreted as the uncertainty, while the estimate of $g(\underline{x})$ can be thought of as the desired output.

In order to be able to visualize an example, we chose to approximate a mapping $f(\underline{x}): R^2 \rightarrow R^1$. The target function, $g(\underline{x})$, which we chose is shown in Fig. 2 and is composed of Gaussians weighted with polynomials (see Matlab User's Guide, 1992, pp. 2-98).

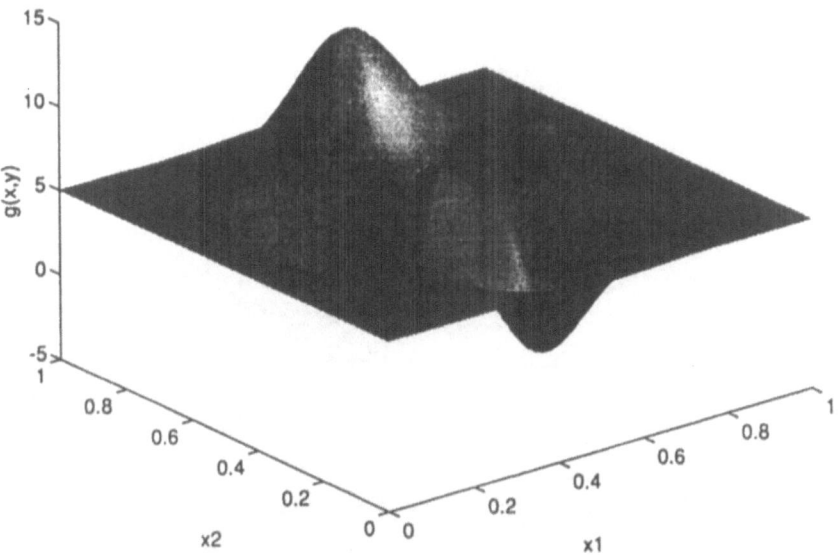

Fig. 2. The target function.

The noise model $n(\underline{x})$ is a Gaussian random variable

$$n(\underline{x}) \sim N(0, \frac{1}{2}(\sqrt{x_1} + x_2)) \tag{6}.$$

The variance is bounded between 0 and 1 for values of (x_1, x_2) in the unit square. The training set is composed of 5000 random samples of (x_1, x_2) from the unit square and the corresponding values of f.

We trained the neural network with a learning rate of 10^{-5}, which is sufficiently small to avoid artifacts. We used no momentum and trained the network for 850 epochs. The direct-prediction output greatly reduced the training time, and it estimated the mean as expected. The output units are weighed equally in the sum-squared error model since the range of the output units is comparable.

Figure 3 shows a surface plot of the mean of the distribution along with error bars (variance) which indicate the uncertainty in the prediction for evenly distributed pairs of (x_1, x_2) in the unit square. The error bars are symmetric, so only the positive

688

portion of the bars are shown. A contour plot of the uncertainty surface is also shown. It indicates that the uncertainty in the estimate is low near the edges of the surface, but increases toward the center. The uncertainty is highest where the estimated surface is in greatest deviation from the function $f(\underline{x})$. Thus, we see that the uncertainty surface appropriately reflects the regions in which the approximation is in error. Apart from the added uncertainty $n(\underline{x})$, there are other errors reflected in the surface, such as sampling error due to the finite sample size.

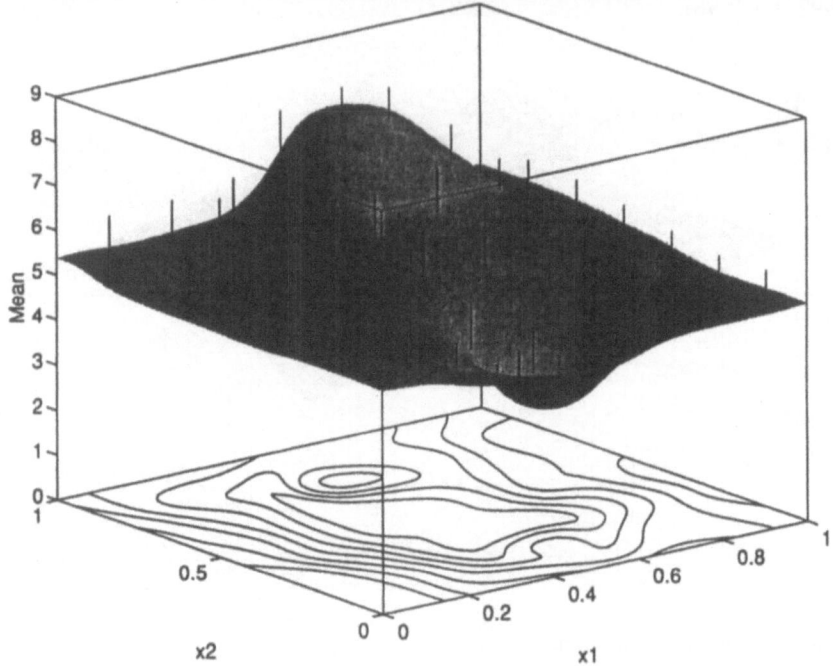

Fig. 3. A surface plot of the mean, a contour plot of the uncertainty surface and error bars.

4 Summary

We designed and tested a neural network which performs a prediction and gives a measure of the precision of the prediction. Outputs consist of a set of normalized exponential units which produce an estimate of the continuous probability density function $p(y|\underline{x})$. The variance of this distribution can be interpreted as the uncertainty in the classification. The distribution can also be used for calculations of percentiles and higher order moments.

5 Reference

Rumelhart, D. E., R. Durbin, R. Golden, and Y. Chauvin (still in press), "Backpropagation: the Basic Theory," in Backpropagation: Theory, Architectures, and Applications, edited by Y. Chauvin and D. E. Rumelhart. Lawrence Erlbaum

Estimation of Conditional Densities: A Comparison of Neural Network Approaches

R.Neuneier‡, F.Hergert‡, W.Finnoff*, D.Ormoneit†

‡ Siemens AG, Corporate Research and Development
Otto-Hahn-Ring 6, D 81730 Muenchen, Germany
Email: Ralph.Neuneier@zfe.siemens.de
† Dept. of Computer Science, TUM, D-80290 Munich
*Prediction Company, 234 Griffin St., Santa Fe, NM 8750

1 Introduction and the Task

In recent years, neural networks have been successfully used to attack a wide variety of difficult nonlinear regression and classification tasks and their effectiveness, particularly when the dimension of the problem measured in the number of variables involved, has been widely documented (Finnoff 1993).

To adequately address certain issues, coming from the field of stochastic control or portfolio optimization, more complete information about the distribution of the data is required than can be provided by a simple regression function. This information can only be obtained by an estimator of the complete probability density of the data. Although there is a wealth of literature from the statistics community dealing with this subject, close inspection reveals that most is either of purely theoretical nature or only considers very low dimensional problems or tasks with a relatively simple probablistic structure.

This area has recently attracted growing interest in the neural network community (White, 1992, Mackey, 1991 and Nowlan, 1991), still many issues remain to be addressed as to the effectiveness of competing techniques when applied to difficult problems.

The task is to estimate the conditional densities $p(y|x)$ given empirical data points (y, x) produced by a time series, for example, we wish to know the probability for a (future) value y having observed the value x. In this paper we will compare several architectures and learning algorithms using both an artificially generated and a real world data set.

The first example is a Monte Carlo simulation with known probability density which can be viewed as a bounded Brownian process. A particle is allowed to move freely in a double well potential (Ormoneit, 1993) with momentum and a friction term, perturbed by small normally distributed random shocks, presenting the price dynamics of a market with two long term equilibria perturbed by small information "shocks" in which some of the market participants use trend following strategies. The task is to predict the position of the particle 25 time steps ahead.

The second example is a real world time series consisting of US$-SFR exchange rates sampled every 10 minutes. Using information of the past as input the task is to produce the conditional probability distribution of the exchange rate 30 minutes into the future. Estimates of that kind will be used in a system under development for making sell or buy decisions.

2 Conditional Density Estimation with NN

To estimate probability densities we restrict ourselves to the function class of linear mixtures of probability densities $p(z) = \sum_{i=1}^{c} P(i)p(z|i)$ where the component densities $p(z|i)$ are multi-dimensional Gaussians with weighting factors $P(i)$ satisfying $P(i) \geq 0$, $\sum_i P(i) = 1$. From the perspective of a data generating process the weighting factors $P(i)$ can be interpreted as a priori probabilities for a data point belonging to the component distribution i, and $p(z|i)$ is a measure of belief that data point z was produced by component i.

There are two basic approaches to find the conditional density. Either one can estimate the conditional density directly, or, first estimate the joint density and then compute the conditional density using the relation $p(y|x) = \frac{p(y,x)}{p(x)}$.

Gaussian Mixture Network and EM-Algorithm: For the Gaussian Mixture Network (GMN) we follow the latter approach, estimating the joint density. Linear mixtures of Gaussians are easily translated to GMNs, by letting each hidden unit i represent a component density $p(z|i)$ and summing up the activation values of these units in a linear output neuron with hidden-to-output weightings given by the $P(i)$s. Denoting by $z_k = (y_k, x_k)$ the set of training patterns the GMN will try to maximize the joint likelihood $L = \prod_k p(z_k)$. This is equivalent to minimizing the logarithm of the inverse joint likelihood, yielding the error function $E := -\log L = -\sum_k \log p(z_k)$, which is well suited for use in conjunction with the backpropagation algorithm (Tresp, 1993).

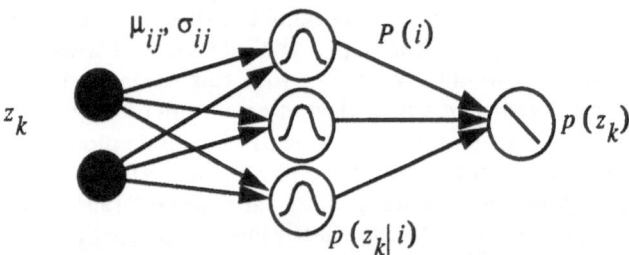

Instead of using the very general approach of gradient descent the Expectation Maximization (EM) Algorithm (Redner, 1984 and Duda, 1973) tries to exploit more specifically the properties of the likelihood function. This algorithm can be modified to become an on-line, incremental learning scheme (Ormoneit, 1993) which allows to integrate the algorithm easily into a usual neural network setting, in particular one can use 'standard' backpropagation to initialize the parameters and then continue fast learning with the EM-algorithm.

Probabilistic Neural Network: The Probabilistic Neural Network (PNN) provides a conceptually very simple way to estimate a probability density (Parzen, 1962 and Specht, 1990). The data points define the centers of the Gaussian component densities. Since all a priori probabilities are assumed to be equal the only parameters that remain to be adjusted are the variances of the Gaussians (exhaustive search!). In our examples the quality of the probability density estimate was very sensitive to the appropriate choice of the variance.

Conditional Density Estimation Network: The last two networks considered here estimate the conditional densities $p(y|x)$ directly. The networks try to maximize the conditional density likelihood $L^c = \prod_k p(y_k|x_k)$ for the empirical data points $z_k = (y_k, x_k)$, or equivalently minimize

$E^c := \sum_k E^c_k = -\sum_k \log p(y_k|x_k) = -\sum_k \log p(y_k, x_k)/ \int p(y, x_k)dy.$

For a Conditional Density Estimation Network (CDEN) one assumes that the parametric conditional density $p(y|x)$ can be defined as $p(y|\theta(x))$, where $\theta(x) = \{q_1(x), \ldots, q_r(x)\}$ denotes a set of parameters functionally depending on x. The function $\theta(x)$ is realized as a network whose outputs are the parameters for the parameterized conditional density estimator.

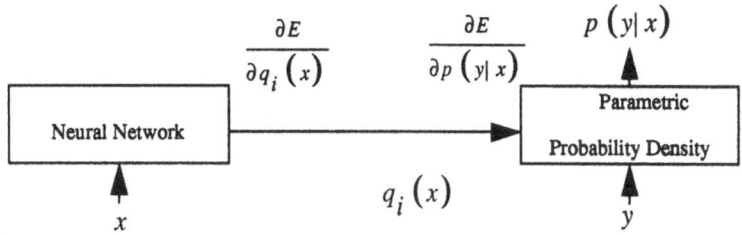

During forward pass the network outputs are used as density parameters, while during error backpropagation the derivatives of the error function E^c_k with respect to the parameters are used to initialize the external error of the parameter estimation network (see schematic CDEN in figure above).

Distorted Probability Mixture Network: Finally we investigated the Distorted Probability Mixture Network (DPMN), which is another possibility to estimate the conditional density directly. The network is based on the assumption that we can model the densities with elementwise independent component densities (diagonal matrices for the variances of the Gaussian units). This allows to rewrite $p(y, x|i)$ as $p(y|i)p(x|i)$ (Tresp, 1993) yielding the following expression for the conditional density which can be directly implemented into a network topology (Ormoneit, 1993) as shown in figure below

$$p(y|x) = \frac{p(y,x)}{p(x)} = \frac{\sum_i p(y,x|i)P(i)}{\sum_l p(x|l)P(l)} = \sum_i p(y|i)\frac{p(x|i)P(i)}{\sum_l p(x|l)P(l)} = \sum_i p(y|i)p(i|x).$$

3 Results

The results for the Brownian process and the exchange rate are summerized in the table below. Using a validation set to detect overfitting and to stop the training process the performance of the resulting network was evaluated on a generalization set (for more details on this stopped training technique and related concepts see Finnoff, 1993). The first column of the table gives the network used. Several network topologies were explored for each experiment. The reported performance always refers to the best network found. The performance given in columns Example 1 resp. 2, represents the negative average logarithm of the conditional likelihood of the data points of the generalization set – smaller values therefore denote better estimates. Note that values on different data sets are not directly comparable.

Network	Algorithm	Brownian	Exch. Rate
GM	Bp.	0.525	-0.04
GM	EM Bp.	0.413	-0.07
PNN	random search	0.556	—
CDE	Bp.	0.372	-0.32
DPM	Bp.	0.334	> 1

In the case of the Brownian motion the training set consisted of 3,000 data points, drawn from the 12,000 data points produced by a Monte Carlo simulation of the process. The 13-dimensional input vector encoded the increments during the last ten steps, the current position itself, and exponentially smoothed averages of the previous increments and positions. The target is the position of the process 25 time steps ahead.

In the second example the network used 20 inputs computed from the last six values of the exchange rate plus the value of the future exchange rate itself. Training, validation, and generalization set contained 3691, 1230, and 1610 data points, respectively. Due to the large size of the data set and the inferior performance of the PNN in example 1, no trials with that network were undertaken on the second task.

From our tests we conclude that the direct approaches to conditional density estimation (CDEN, DPMN) appears superior to the indirect methods (GMN, PNN). On the other hand the failure of the DPM-network on the second task indicates that the performance of alternative algorithms can be very domain dependent especially for problems with high dimensional and noisy data.

Acknowledgements: The authors are grateful to the influential ideas of Volker Tresp and the extensive discussions with him.

References:
• Duda R. O. and Hart P. E. (1973), Pattern Classification and Scene Analysis
• Finnoff W., Hergert F. and Zimmermann H. G. (1993), Improving Model Selection by Nonconvergent Methods, Neural Networks Vol. 6, pages 771-783
• MacKay D. J. C. (1991), Bayesean Modelling and Neural Networks, PhD-Thesis at California Institute of Technology, Pasadena
• Nowlan S. J. (1991), Soft Competitive Adaptation: Neural Network Learning Algorithms based on Fitting Statistical Mixtures, PhD-Thesis at School of Computer Science, Carnegie Mellon University, Pittsburgh
• Ormoneit D. (1993),Estimation of Probability Densities using Neural Networks, Master-Thesis: Dept. of Computer Science, TU Munich
• Parzen E. (1962), On Estimation of a Probability Density Function and Mode, Annals of Mathematical Statistics 33
• Redner R. A. and Walker H. F. (1984), Mixture Densities, Maximum Likelihood and the EM Algorithm, SIAM Review, 26
• Specht D. F. (1990), Probabilistic Neural Networks, Neural Networks 3
• Tresp V., Hollatz J. and Ahmad S. (1993), Network Structuring and Training Using Rule-Based Knowledge, Advances in NIPS 5
• White H. (1992), Parametrical Statistical Estimation with Artificial Neural Networks, Techreport University of California, San Diego

Regularizing Stochastic Pott Neural Networks by Penalizing Mutual Information

G. Deco and T. Martinetz
Siemens AG, Corporate Research and Development, ZFE ST SN 41
Otto-Hahn-Ring 6, 81739 Munich, Germany

Abstract

In this paper we present a method for eliminating overtraining during learning on small and noisy data sets. The key idea is to reduce the complexity of the neural network by increasing the stochasticity of the information transmission from the input layer to the hidden-layer. The architecture of the network is a stochastic multilayer perceptron the hidden layer of which behaves like a Pott-Spin. The stochasticity is increased by penalizing the mutual information between the input and its internal representation in the hidden layer. Theoretical and empirical studies validate the usefulness of this novel approach to the problem of overtraining.

1.0 Introduction

Two principal complications appear when a neural network is trained for extracting the underlying structure in the data of a real world problem. Firstly, the data is noisy, and secondly, there is usually only a finite amount of training patterns. These two properties of the training data set make it difficult for a neural network to learn only the useful structure and ignore the particularities of each training data, e.g., the noise. Learning particularities like noise leads to overtraining and bad generalization capabilities. Overtraining can be avoided by employing regularization techniques. In learning with neural networks this is usually done by controlling the complexity of the network based on penalty terms (Hinton 1986; Weigend et al., 1991). These penalty terms reduce the effective number of parameters, which, however, still causes a number of problems (Deco et al, 1993; Nowlan and Hinton, 1991).

The present work introduces an alternative approach which regulates the complexity of the network not by reducing the effective number of parameters but by increasing the stochasticity of the data representation. Nowlan and Hinton use a similar but different approach. They insert noise on each weight of the network and control the amount of noise in order to regularize the model (Nowlan and Hinton, 1991). In the present work we use a stochastic network and control the stochasticity of the network by reducing the mutual information between the input and its internal representation. The motivation of this approach is the fact that most of the information describes the noise and should not be transmitted from the input to the hidden layer.

2.0 The Neural Network Architecture

The first layer is just to represent the input data ξ_i^a, with ξ_i^a denoting pattern a of dimension n. The second layer is a layer of m probabilistic Boolean hidden units S_j; i.e., each hidden unit can have the discrete output value 1 with probability P_j and the discrete output value 0 with probability $(1 - P_j)$. We choose P_j such that the hidden layer represents a Pott spin, i.e.,

$$P_j = \frac{e^{(\omega_j \cdot \xi^a)}}{\sum_k e^{(\omega_k \cdot \xi^a)}} \tag{2.1}$$

(Peterson and Soederberg, 1989). The output layer is given by a set of T neurons with linear activation functions. The mean output values of the network are then given by

$$O_t^a = \sum_j W_{tj} P_j^a. \tag{2.2}$$

The mean output values O_t^a are used to establish a continuous input-output mapping $\xi^a \to O_t^a$. In order to learn this input-output mapping, we train the stochastic network such that the squared error between the desired outputs and the mean output values O_t^a is minimized.

To reduce the complexity of the network we increase the stochasticity in the internal representation of the input patterns. This is achieved by reducing the amount of information conveyed from the input layer to the hidden layer, i.e., by reducing the mutual information between the input and the internal representation. Shannon defined the mutual information as the amount of information transmitted in a stochastic channel. In our case the stochastic channel lies between the input layer and the hidden layer of our network and is defined by the Pott probability function (2.1). The mutual information M between input layer and hidden layer is given by

$$M = \sum_a p(a) \sum_j P_j^a \log \left(\frac{P_j^a}{\sum_a p(a) P_j^a} \right) \tag{2.3}$$

We add the mutual information M as penalty term to the quadratic cost function, obtaining

$$E = \sum_a \sum_t (Y_t^a - O_t^a)^2 + \lambda M \tag{2.4}$$

with λ as a Lagrange multiplier and Y_t^a as the desired outputs. The network learns the training data, and at the same time the penalty term avoids the excessive transmission of information, i.e., information which might describe the noise. Note that by minimizing the mutual information we are increasing the stochasticity of the network. The gradient descent learning rule which corresponds to the quadratic cost function (2.4) can easily be derived. After some algebra we obtain

$$\Delta W_{tj} = \eta \sum_a (Y_t^a - O_t^a) P_j^a \tag{2.5}$$

$$\Delta \omega_{ji} = \eta \sum_a \xi_i^a \sum_{t,k} (Y_t^a - O_t^a) W_{tk} (P_j^a \delta_{kj} - P_j^a P_k^a) \tag{2.6}$$

$$-\lambda \eta \sum_a p(a) \sum_k (P_j^a (\delta_{kj} - P_k^a) \xi_i^a) \log \left(\frac{P_k^a}{\sum_a p(a) P_k^a} \right)$$

with η as the learning step size.

3.0 Simulations

In this section we present the result we obtained by applying our model to a synthetic data set. For demonstrating the performance of our approach we choose a common benchmark from the statistic community. The benchmark, which was introduced by Friedman (1991), is a function of ten variables and is given by

$$f(x_1, ..., x_{10}) = 0.1 e^{4x_1} + \frac{4}{(1 + e^{-20(x_2 - 0.5)})} + 3x_3 + 2x_4 + x_5 \qquad (3.1)$$

This function has a nonlinear additive dependence on the first two variables, a linear dependence on the next three, and is independent of the last five variables (pure noise). Random values within the unit hypercube were chosen for the ten variables x_i. Then the corresponding response values were calculated according to

$$Y_t = f(\underline{x}_t) + v_t \qquad , \qquad 1 \le t \le N, \qquad (3.2)$$

with v randomly generated from a standard normal. The signal to noise ratio is 3.28 so that the true underlying function accounts for 91% of the variance of the response. Two data sets, one for training and one for testing, were generated using equation (3.1) and (3.2), with 100 and 300 data points, respectively. The network architecture consisted of 10 inputs, 15 hidden units and one output. The learning step size was $\eta = 0.01$, and the Lagrange multiplier had a value of $\lambda = 1$.

Without penalty term the neural network learns the noise and the spurious dependence on the last five variables, which leads to overtraining and a very bad generalization. Figure 1 shows the evolution of learning with and without the mutual information regularizer. Without the mutual information regularizer the typical overtraining occurs. After adding the mutual information regularizer to the cost function the overtraining disappears and the error on the test set remains asymptotically constant. This indicates that the deterioration of the generalization due to the learning of noise (real or semantic) is now avoided by limiting the amount of information transmitted from the input layer to hidden layer.

Table 1 shows the average relative variance (arv) calculated on the test set after training with and without mutual information penalty term. We see that the mutual information penalty term leads to a significant reduction of the generalization error.

4.0 Discussion

In the neural network architecture two regularization effects are incorporated: First, the mutual information penalty term stops excessive decorrelation between the hidden units, which usually takes place in multilayer perceptrons trained with backpropagation learning rules. Backpropagation activates all the resources of the network for learning the training set, which leads to a strong decorrelation between the hidden units. However, by decorrelating the hidden units also the noise is learnt. The mutual information penalty term acts as an entropy that tries to uniform the activations in the hidden layer, i.e., tries to stop decorrelation. The second regularization effect is provided by the stochasticity of the network. An increase of the stochasticity of the network leads to a reduction of its complexity. The stochasticity is regulated by regulating the transmission of entropy between the input and the hidden layer through the Lagrange multiplier in the cost function.

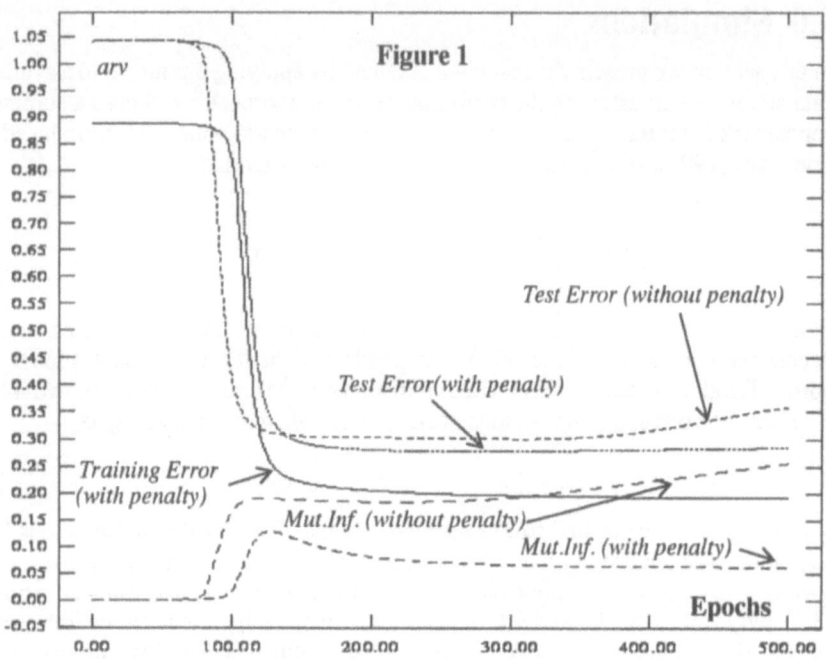

Figure 1

TABLE 1.

Model	arv (test set)
BP-Pott-Net	0.36
BP-Pott-Net with mut. inf.	0.28

References:

- Deco G, Finnoff W. and Zimmermann H.G., 1993, "Elimination of Overtraining by a Mutual Information Network", ICANN'93, Amsterdam, Proc. p. 744-749.

- Le Cun Y., Denker J. and Solla S., 1990, "Optimal Brain Damage", in Proceedings of the Neural Information Processing Systems, Denver, 598-605.

- Nowlan S. and Hinton G., 1991, "Adaptive Soft Weight Tying using Gaussian Mixtures", Neural Information Proccesing Systems, Vol. 4, 847-854, San Mateo, C.A. Morgan Kaufmann.

- Peterson C. and Soederberg B., 1989, "A new method for mapping optimization problems onto neural networks", Int. J. Neural Syst., 1, 68.

- Weigend A., Rumelhart D. and Huberman B., 1991, "Generalization by weight elimination with application to forecasting", in Advances in Neural Information Proccesing, III, Ed. R. P. Lippman and J. Moody, Morgan Kaufman, 1991.

- Friedman J.H., "Multivariate adaptive regression splines", 1991, Annals of Statistics, 19, 1-141.

Least Mean Squares Learning Algorithm in Self Referential Linear Stochastic Models

E. Barucci†, L. Landi‡

† DIMADEFAS, University of Florence

‡ Dipartimento di Sistemi ed Informatica, University of Florence

Florence, Italy

1 Introduction

In this paper, we address a problem very important in economic theory: the interaction between the evolution of an economic system described by a set of stochastic linear difference equation, i.e., self-referential linear stochastic model, and a bounded rationality learning process for the economic agents based on a *perceived* economic model. In the economic life, agents are able to affect the evolution of a system by means of their expectations. In this context, if they know the *actual* model then the so called *Rational Expectation Equilibrium*, i.e., REE, is obtained where agents take their expectations consistently with the economic model. If they do not know the *actual* economic model then the REE can be reached only if agents, believing in a mispecified model, learn it by comparing their expectations on the mispecified model with the real economic data. In such models, a topic of interest is the agent learning mechanism. By assuming that the models are linear, economists have considered a learning process based on *Recursive Ordinary Least Squares*, i.e., ROLS, algorithm (Marcet and Sargent, 1989), and have studied its convergence to the REE. In this paper, we study the same problem by assuming that agents learn by means of the *Least Mean Squares*, i.e., LMS, algorithm (Widrow, 1971). In the literature, it has been shown that ROLS algorithm is not able to converge to the REE in many economic models, see (Marcet and Sargent, 1989), (Bray and Savin, 1986). In our study, we analyze a learning process based on LMS algorithm and prove that it is able to converge to REE in a wider set of self-referential linear stochastic models than ROLS algorithm does.

2 Convergence in self-referential models

A self-referential linear stochastic model is described, at time t, by a n-dimensional vector of random variables, i.e., $z_t \in \Re^n$. Self-referential means that the model evolution is fully determined by itself, by an exogenous component, i.e., random variables, and also by expectations taken by the economic agents on some variables. The self-referential feature requires that expectations are taken with a signal extraction activity on some of the n variables. In order to represent this class of models, we denote with two subvectors of z_t, not necessarily disjoint, the set of economic variables that the economic agents are interested to predict, i.e., $z_{1t} \in \Re^{n_1}$, and the set of economic variables, i.e.,

$z_{2t} \in \Re^{n_2}$, that the agents think are relevant to predict the first subvector of variables. The vector z_t can be written, without loss of generality, as:

$$z_t = \begin{pmatrix} z_{1t} \\ z_{1t}^c \end{pmatrix} = \begin{pmatrix} z_{2t}^c \\ z_{2t} \end{pmatrix}$$

where the superscript c expresses the complement with respect to z_t. As in (Marcet and Sargent, 1989), it is necessary to define the model by which agents perceive the law of motion for z_{1t}, at time t, using observations up to time $(t-1)$. If an *adaptive linear combiner* is used as a model for agent believes (Widrow and Stearns, 1985), the variables in z_{1t} are estimated as

$$z_{1t}^e = \mathcal{B}_t^T z_{2(t-1)} \tag{1}$$

where $\mathcal{B}_t \in \Re^{n_2 \times n_1}$ is the weight matrix representing the *perceived* law of motion of z_{1t}. The agents believe (1) causes the *actual* law of motion for the entire vector z_t to be given by:

$$z_t = \begin{bmatrix} z_{1t} \\ z_{1t}^c \end{bmatrix} = \begin{bmatrix} 0 & T(\mathcal{B}_t)^T \\ & A(\mathcal{B}_t)^T \end{bmatrix} \cdot \begin{bmatrix} z_{2(t-1)}^c \\ z_{2(t-1)} \end{bmatrix} + \begin{bmatrix} V(\mathcal{B}_t)^T \\ B(\mathcal{B}_t)^T \end{bmatrix} \cdot u_t \tag{2}$$

where $u_t \in \Re^n$ is a stationary white noise, $T(\cdot)$ is the application which, given $\mathcal{B}_t$, describes the *actual* law of motion for z_{1t} at time t, i.e., $T : \mathcal{B}_t \to T(\mathcal{B}_t) \in \Re^{n_2 \times n_1}$ and

$$A : \mathcal{B}_t \to A(\mathcal{B}_t) \in \Re^{n \times n_2}, V : \mathcal{B}_t \to V(\mathcal{B}_t) \in \Re^{n \times n_1}, B : \mathcal{B}_t \to B(\mathcal{B}_t) \in \Re^{n \times n_2}.$$

Let us remark that the agents' model in (1) defines, together with the economic features of the model, the *actual* law of motion. Moreover, note that the data generating process in (2) does not imply that z_{2t} is a stationary process. Although the LMS algorithm has been proved to converge also in non-stationary environments (Widrow and Stearns, 1985), we restrict our study to the stationary case. Let us define the set D_s where the model defined by (2) is stationary.

$$D_s = \{\mathcal{B} \in \Re^{n_2 \times n_1} \,|\text{the eigenvalues, i.e., } \gamma_i, \text{ of}$$
$$\begin{bmatrix} 0 & T(\mathcal{B})^T \\ & A(\mathcal{B})^T \end{bmatrix}$$

are less than unity in absolute value, i.e., $|\gamma_i| < 1\}$

If the LMS algorithm is used to estimate the law of motion of the self-referential model described in (2), the weight matrix $\mathcal{B}$ is estimated to minimize the *mean square error*, i.e., MSE, given by the mean value of the square of the *instantaneous* error, i.e., $\varepsilon_t = z_{1t} - z_{1t}^e$. Let us denote the MSE with ξ so that its i-th component is $\xi_i = E\{\varepsilon_{t_i}^2\} = E\{(z_{1t_i} - z_{1t_i}^e)^2\}$, where it has been assumed that $E\{\varepsilon_{t_i} \varepsilon_{t_j}\} = 0$ for $i \neq j$. Assuming $\mathcal{B}_t$ fixed over time, i.e., $\mathcal{B}_t = \mathcal{B}$, ξ_i is given as:

$$\xi_i = E\{z_{1t_i}^2\} + \beta_i^T E\{z_{2(t-1)} z_{2(t-1)}^T\} \beta_i - 2E\{z_{1t_i} z_{2(t-1)}^T\} \beta_i$$

where $\beta_i \in \Re^{n_2}$ is the i-th column of the matrix $\mathcal{B}$. Let us express the functions $T(\cdot)$ and $V(\cdot)$ when evaluated for β_i, for $i = 1, \ldots, n_1$: $T : \beta_i \to T(\beta_i) \in \Re^{n_2}$ is the i-th column vector of the matrix $T(\mathcal{B})$ and $V : \beta_i \to V(\beta_i) \in \Re^n$ is the

i-th column vector of the matrix $V(\mathcal{B})$. At time instant t, let us denote the matrices $R = E\{z_{2(t-1)}z_{2(t-1)}^T\}$ with $R \in \Re^{n_2 \times n_2}$ and $R = R^T$, $\Sigma = E\{u_t u_t^T\}$ with $\Sigma \in \Re^{n \times n}$ and $\Sigma = \Sigma^T$, and $C = E\{u_t z_{2(t-1)}^T\}$ with $C \in \Re^{n \times n_2}$. In economics, the assumption that $C = 0$ is done because the observations z_{2t} are independent of u_t. Then, the i-th MSE surface can be expressed as:

$$\xi_i(\beta_i) \;\; = (T(\beta_i) - \beta_i)^T R (T(\beta_i) - \beta_i) + V(\beta_i)^T \Sigma V(\beta_i)$$

The central question to be answered is whether the learning mechanism based on the LMS algorithm for the mispecified model described above, is able to converge to a REE point β_i^*, see (Marcet and Sargent, 1989), that is a fixed point of $T(\beta_i)$, i.e., $T(\beta_i^*) = \beta_i^*$, for $i = 1, \dots, n_1$.

Proposition 2.1 *Given a REE point β_i^* such that $\beta_i^* \in D_s$ for $i = 1, \dots, n_1$. Assume that the correlation matrix C is the null matrix, i.e., $C = 0$, $V(\beta_i) = K_i$ for $i = 1, \dots, n_1$, where $K_i \in \Re^n$ is a constant vector, $T(\beta_i)$ is linear, i.e., $T(\beta_i) = K_{1i}\beta_i + K_{2i}$ where $K_{1i} \in \Re^{n_2 \times n_2}$, $K_{2i} \in \Re^{n_2}$ for $i = 1, \dots, n_1$. Then, the LMS learning algorithm converges on the average to the REE, i.e., $\beta_i \to \beta_i^*$, given that the constant η satisfies the condition $0 < \eta < \Omega(i)^{-1}$ for $i = 1, \dots, n_1$ where $\Omega(i)$ is the matrix with the eigenvalues of $(K_{1i} - I)^T R (K_{1i} - I)$.*

Proof. The LMS algorithm updates the i-th weight vector at time t, i.e., β_{t_i}, according to the following rule, for $i = 1, \dots, n_1$ (Widrow, 1971):

$$\beta_{t+1_i} = \beta_{t_i} - 2\eta(z_{1t_i} - z_{1t_i}^e)\frac{\partial \varepsilon_{t_i}}{\partial \beta_i} \; .$$

At time t, the i-th *instantaneous* error can be computed as follows:

$$\varepsilon_{t_i} = (T(\beta_{t_i}) - \beta_{t_i})^T z_{2(t-1)} + V(\beta_{t_i})^T u_t \; .$$

Differentiating ε_{t_i}, the updating rule becomes:

$$\beta_{t+1_i} \;\; = \beta_{t_i} - 2\eta \left[\frac{\partial T(\beta_{t_i})}{\partial \beta_{t_i}}^T z_{2(t-1)} z_{2(t-1)}^T (T(\beta_{t_i}) - \beta_{t_i}) \right.$$
$$\left. - z_{2(t-1)} z_{2(t-1)}^T (T(\beta_{t_i}) - \beta_{t_i}) + \left(\frac{\partial T(\beta_{t_i})}{\partial \beta_{t_i}} - I \right)^T z_{2(t-1)} u_t^T V(\beta_{t_i}) \right] .$$

Since $T(\beta_i)$ is a linear function, then $\partial T(\beta_{t_i})/\partial \beta_{t_i} = K_{1i}$. Moreover, let us assume that the weight vectors β_{t_i} are independent of both z_{2t} and u_t. Let us take the mean values over time t of both sides of the last equation:

$$E\{\beta_{t+1_i}\} = E\{\beta_{t_i}\} - 2\eta (K_{1i} - I)^T R E \{T(\beta_{t_i}) - \beta_{t_i}\} \; .$$

Let us change the coordinate system from β_i to $v_i = T(\beta_i) - \beta_i$, so that adding and subtracting from both sides $T(\beta_{t+1_i})$, the LMS updating rule becomes:

$$E\{v_{t+1_i}\} = \left[I - 2\eta (K_{1i} - I)^T R (K_{1i} - I) \right] E\{v_{t_i}\} \; . \tag{3}$$

Let us rotate the coordinate system to v_{t_i}' such that $v_{t_i}' = Q^T v_{t_i}$, where Q is the matrix with the eigenvectors of $(K_{1i} - I)^T R (K_{1i} - I)$. If $\Omega(i)$ is the diagonal matrix with the associated eigenvalues, let us define $\Omega(i)$:

$$(K_{1i} - I)^T R (K_{1i} - I) = Q\Omega(i)Q^{-1}.$$

Note that, being R symmetric and positive definite, then also the matrix $(K_{1i} - I)^T R (K_{1i} - I)$ is positive definite and symmetric. The general updating rule in (3) becomes:

$$E\{v'_{t+1_i}\} = [I - 2\eta\Omega(i)] E\{v'_{t_i}\}.$$

The updating rule can also be expressed with respect to the initial condition v'_{0_i} as follows:

$$E\{v'_{t_i}\} = [I - 2\eta\Omega(i)]^t E\{v'_{0_i}\}.$$

A necessary and sufficient condition for the expression above to converge to zero is $|I - 2\eta\Omega(i)| < I$, for all i that is $0 < \eta < \Omega(i)^{-1}$. If such condition is satisfied, then:

$$\lim_{t \to \infty} E\{v'_{t_i}\} = 0, \quad \forall i.$$

Therefore, $(T(\beta_{t_i}) - \beta_{t_i}) \to 0$ and since there is a unique REE, it follows that $\beta \to \beta^*$. $\square$

3 Evaluations and Conclusions

We have proved the convergence of the LMS learning algorithm to the REE of a class of self-referential linear stochastic models. This result is accomplished for a wider set of parameters of the function $T(\beta)$ in (2) than that allowed by the ROLS algorithm. For example, let us consider the self-referential model in (Bray and Savin, 1986) where $T(\beta) = a\beta + m \in \Re$ and the REE point is given by $\beta^* = \frac{m}{1-a}$. If the ROLS learning algorithm is used, convergence is achieved only for $a < 1$ while the LMS algorithm converges for $a \neq 1$, see (Barucci and Landi, 1993).

References

Barucci, E. and Landi, L. (1993). Convergence of least mean squares error learning algorithm in self referential linear stochastic models. Technical report. University of Florence, Florence, Italy.

Bray, M. M. and Savin, N. E. (1986). Rational expectations equilibria, learning, and model specification. *Econometrica*, 54:1129–1160.

Marcet, A. and Sargent, T. J. (1989). Convergence of least squares learning mechanisms is self-referential linear stochastic models. *Journal of Economic Theory*, 48:337–368.

Widrow, B. (1971). Adaptive filters. In *Aspects of Network and System Theory*, pages 563–587, New York: Holt, Rinehart and Winston. R. Kalman and N. DeClaris.

Widrow, B. and Stearns, S. (1985). *Adaptive Signal Processing*. Prentice-Hall.

An Approximation Network
with maximal Transinformation

Rüdiger W. Brause

J.W. Goethe-University, Frankfurt, Germany
(brause@informatik.uni-frankfurt.de)

1 Introduction

One of the most important applications of artificial neural networks is the approximation of an unknown function. It is well known (e.g. Hornik et al., 1989) that a two layer feedforward neural network can approximate each function arbitrarily well when a sufficient number of neural units are provided, see fig. 1.

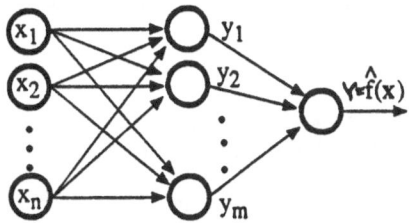

Fig.1 The two layer feedforward approximation network

Very often, units with sigmoidal activation functions and, as learning algorithm, the popular backpropagation algorithm are used, minimizing as objective function the mean squared error and using a gradient descend for the two layers. This approach has some flaws, especially the problems of getting trapped in suboptimal local minima and learning the training samples too specific (*overfitting*).

This paper proposes another approach which avoids those problems. Let us see the approximation as the task to find the parameters of splines interpolating sampled function values f(x). For a linear interpolation (*linear spline*) figure 2 illustrates the situation.

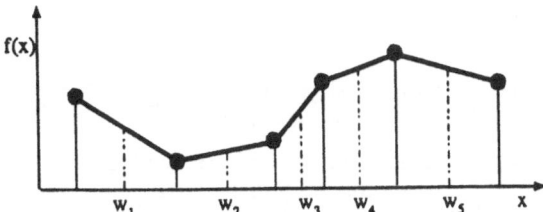

Fig. 2 A piecewise-linear approximation

Here, the approximation task is broken up into two parts. The first part performs the division of the input space {x} into disjoint sets (intervals in fig.2) of events (*classes*) and assigns a typical input value w_i to each class i, called a *class prototype*. This is essentially a vector quantization and can be implemented e.g. by a topology-approximating mapping. The second part, the approximation of the bias $f(w_i)$ and the slope of the linear spline, is performed locally by a neuron for each class. For constant-valued splines (*bars*) it is well known that such a mapping has the ability to approximate every arbitrary function sufficiently well if the necessary number of neurons (classes) is provided (Blum et. al. 1991). We implement each part as a separate, independent network layer.

In fact, this kind of two layer neural networks has been already used for robot control by Ritter, et al. (1989). In this paper, we propose new learning rules for the two layers which relay on the objective function of *maximal transinformation*.

2 The non-linear topology-approximating mapping

A topology-approximating mapping version using the least squared error, made by (Kohonen, 1982), became quite popular by its broad range of applications.

This paper makes a new, synthetic approach by using a performance measure of *transinformation* or *mutual information* between the input events and the response of the neurons, see (Linsker, 1988). Following (Shannon, 1949) we preserve the maximal amount of information in the data compression stages and model the information processing process as a pipeline of actions by optimal layers.

Let us consider a clustering or mapping (vector quantization) of an input pattern x to a class c as it is defined by

$$|x-w_c| = \min_k |x-w_k| \qquad\qquad x, w_k \in \Re^n \qquad (2.1)$$

Knowing the input pattern x, the Shannon average transmitted information H_{trans} for all inputs and outputs is with the expectation operation $\langle\,\rangle$ for N output points (class prototypes) w_i

$$H_{trans} = \langle I_{trans} \rangle_{w_i, x} = \langle I_{out} \rangle_{w_i, x} - \langle I_{out/inp} \rangle_{w_i, x} \qquad (2.2)$$
$$= -\Sigma_i P(w_i) \log[P(w_i)] - \Sigma_x P(x) \Sigma_i P(w_i/x) \log[P(w_i/x)]$$

The average transmitted information H_{trans} is maximized when

$$\langle I_{out} \rangle_{w_i, x} \overset{!}{=} \max \quad (2.3) \qquad\qquad \text{and} \qquad \langle I_{out/inp} \rangle_{w_i, x} \overset{!}{=} \min \quad (2.4)$$

It can be easily shown that (2.3) is satisfied when $P(w_i) = P(w_j) = 1/N \quad \forall i,j$. Additionally, for deterministic mappings (see Brause, 1993b) we have $\langle I_{out/inp} \rangle = 0$ which is the minimal achievable value.

Let us use directly a stochastic approximation algorithm using the information as an error criterion for a gradient search

$$w_k(t+1) = w_k(t) + \gamma(t+1) \, \partial H_{Trans} / \partial w_k \qquad (2.5)$$

for the maximum of the transinformation H_{Trans}. Evaluating the gradient $\partial H_{Trans}/\partial w_k$ we get (see Brause, 1993b)

$$w_k(t+1) = w_k(t) - \gamma(t+1) \sum_{i \neq k} \frac{(w_i - w_k)}{|w_i - w_k|^2} (P_k \log P_k - P_i \log P_i + P_k - P_i) \qquad (2.6)$$

As we can see, for equally probable classes with $P_i = P_k$ no change takes place in equation (2.6); the class prototypes remain fixed.

To test this algorithm, we can use the random test which was conceived by (Ritter and Schulten, 1986) with an 1-dim increasing distribution p(x)=2x. For the Kohonen map, they found that the point density M(x) of the class prototypes we have $M(x) \sim p(x)^{2/3}$. It was shown (Brause, 1992) that for the topology approximating mapping which preserves the maximum of information, M(x) must approximate the probability distribution of the input patterns directly proportional $M(x) \sim p(x)$. For the random test, the new algorithm showed good results (Brause, 1993b).

3 The local, linear interpolation

Let us consider an input x which is disturbed by an independent, additive error deviation η. After the linear layer the error still exists and is linearly transformed to the output

$$F(x) = \bar{y} = W(x + \eta) = y + y_\eta \quad \text{with} \qquad y_\eta = \bar{y} - Wx \qquad (3.1)$$

How should we choose the coefficients of W (weights of the neurons) to maximize the transinformation for the approximation of the function values $\bar{y}$? Here, the basic assumption lies in the nature of the random process as a physically generated, normally, spacial distributed error deviation.
Thus, we know

$$p(y_\eta) = p(y_\eta|x) = A \exp(-y_\eta^T y_\eta / 2\sigma_\eta^2) \qquad (3.2)$$

The transinformation becomes

$$H_{trans} = H(\bar{y}) - H(\bar{y}|x) = H(x) + H(\eta) - H(y_\eta|x)$$

Since the random variables x and η and therefore $H(x)$ and $H(\eta)$ are given, we can only *maximize* the transinformation by *minimizing* $H(y_\eta|x)$. This means

$$H(y_\eta|x) = \langle -\log p(y_\eta|x)\rangle = \langle -\ln A\rangle + \langle(\bar{y}-Wx)^2/2\sigma_y^2\rangle = min \qquad (3.3)$$

which in turn is minimal when the squared error becomes minimal. Certainly, this is not true if the patterns are *not* normally distributed as it is often the case for classification purposes. Here, the general criterion of maximal information yields better results than the ordinary least squared error, see for instance (Bridle 1990).

Now let us evaluate the conditions for the best fitting of the data points. In each class region, the unknown vector-valued function $f(x)$ is approximated in each component by the a linear spline, i.e. in the multivariate case by a hyperplane. Let us regard this more closely. For each vector component, equation (3.3) means

$$\langle(\bar{y}-w^T x)^2\rangle = min \qquad (3.4)$$

By the notation $w \rightarrow w = (w_1,..,w_n,-1)^T$, $x \rightarrow x = (x_1,..,x_n,\bar{y})$ and the introduction of a bias s this becomes

$$R(x,w,c) = \langle(w^T x - s)^2\rangle = min \qquad (3.5)$$

The objective function $R(x,s)$ takes its minimum for s at $s = w^T\langle x\rangle$. The objective function $R(w)$ easily becomes zero if we reduce the length of w to zero - but this is only the trivial solution. To get the best base vectors of our linear transformation, let us assume a non-zero, constant transformation with $det(W)=const$.
Therefore, (3.5) becomes

$$R(w,s) = \langle(w^T x - s)^2\rangle = \langle(w^T(x-\langle x\rangle))^2\rangle = w^T C w \overset{!}{=} min \qquad (3.6)$$

with the covariance matrix $C=\langle(x-\langle x\rangle)(x-\langle x\rangle)^T\rangle$. By the method of Lagrange multiplier it can be shown (see Brause 1992b) that $R(w)$ has only one minimum and takes it at w being the normalized eigenvector of the covariance matrix C with the smallest eigenvalue. Thus, we are looking for the direction of the smallest diameter of the data cloud {x} and project all samples in the remaining subspace orthogonal to the direction. This method is known as *eigenvector fitting* (Duda et al., 1973) and has some approximation advantages over the simple least squared error method in the sections of a fast changing f(x), see e.g. (Xu et al.,1992). An neural implementation of this kind of eigenvector search was proposed by (Brause 1992a,b, 1993a) who used an anti-Hebb rule for learning.

For each neuron selected, the input $x = (x_1,..,x_n, x_{n+1})$ with $x_{n+1}= f(x_1,..,x_n)$ is centered by the preprocessing stage

$$x(t) \rightarrow x(t)-b_i(t) \quad b_i(t) = \langle x\rangle_i$$

If the random variable x_i is stationary, i.e. the intervall is stable, this can be done iteratively, otherwise, when the first layer is still learning, the iterative average should be replaced by a floating one.

The *training* is accomplished by an Anti-Hebbian learning rule and an normalization for the weights (see Brause 1992a,b)

$$\tilde{w}(t) = w(t-1) - \gamma(t) x(t) y \qquad y = w^T x = \sum_i^{n+1} w_i x_i \qquad (3.7)$$
$$w(t) = \tilde{w}(t) / |\tilde{w}(t)|$$

In the *activity phase*, the approximation $F(x_1,..,x_n)$ can be obtained directly by the input $(x_1,..,x_n,0)$. We know that for all points of the hyperplane we have no deviation, i.e. $g(x)=w^T x -s =0$ with $x = (x_1,..,x_n, x_{n+1})$ and $x_{n+1}= F(x_1,..,x_n)$. So, we get

$$F(x_1,..,x_n) = b_{n+1} - \frac{1}{w_{n+1}} \sum_{i=1}^{n} w_i(x_i - b_i) \qquad (3.8)$$

This can be seen as using the neural network as an *continuous autoassociative memory*: for the input data $x_1,..,x_n$ and zero $f(x_1,..,x_n)$ the neuron computes an output which is the (linearly transformed) function value $F(x_1,..,x_n)=b_{n+1} - y/w_{n+1}$.

4 Convergence analysis

As the main result of this section the convergence of the whole network to one unique stable state is shown. All proofs are given in (Brause, 1993b).

Lemma 1 The weights of the quantization network will converge to a stable state which corresponds to the optimal configuration of equal probable classes.

Lemma 2 Given a stable state of the quantization network, each weight of the interpolation network converge to a unique, stable fixpoint.

Theorem Given a stationary distribution of the input samples, the approximation network will converge to a unique, stable fixpoint.

References

E. Blum, L. Li (1991) Approximation Theory and Feedforward Networks; Neural Networks, 4, pp.511-515,

R.Brause (1992) Optimal Information Distribution and Performance in Neighbourhood-conserving Maps for Robot Control; Int. Journal for Computers and Artif. Intelligence, Vol. 11, No.2, pp. 173-199,

R. Brause (1992a) The minimum entropy neuron- A new building block for clustering transformations; in: I. Aleksander, J.Taylor (eds.), Artificial Neural Networks 2, North Holland Publ, Amsterdam , pp. 1095-1098,

R. Brause(1992b) The minimum entropy network; Proc. IEEE Int. Conf. on Tools with Artificial Intelligence TAI-92, pp. 85-92

R. Brause (1993a) A VLSI-Design of the Minimum Entropy Neuron; in: J.Delgado-Frias, W. Moore (Eds.): VLSI for Artificial Intelligence and Neural Networks, Plenum Publ. Corp.,

R. Brause (1993b) An Information-based Approximation Network; (preprint) submitted to IEEE Transactions on Neural Networks

J. Bridle (1990) Probabilistic Interpretation of Feedforward Classification Networks Outputs, with Relationship to Statistical Pattern Recognition; in: F. Fogelman Soulié, J.Hérault (eds.), Neuro-computing, NATO ASI Series, Vol. F68, Springer Verlag, pp. 227-236

R. Duda, P. Hart (1973) Pattern Pattern classification and Scene Analysis; John Wiley&Sons, New York

K. Hornik, M. Stinchcombe, H. White (1989) Multilayer Feedforward Networks are Universal Approximators; Neural Networks, Vol.2, pp.359-366

T. Kohonen (1982) Self-organized Formation of Topologically Correct Feature Maps; Biological Cybernetics, , Vol 43, pp 59-69

R. Linsker (1988) Towards an Organizing Principle for a Layered Perceptual Network; in : D. Anderson (ed), Neural Information Processing Systems, Amer. Inst. of Physics (NY)

H. Ritter, K. Schulten (1986) On the Stationary State of Kohonen's Self-Organizing Sensory Mapping; Biolog. Cybernetics Vol 54, pp. 99-106

C.E. Shannon, W.Weaver (1949) The Mathematical Theory of Information; University of Illinois Press, Urbana

L. Xu, E. Oja, C. Suen (1992) Modified Hebbian Learning for Curve and Surface Fitting; Neural Networks, Vol.5, pp.441-457

EXTENDED FUNCTIONALITY FOR PROBABILISTIC RAM NEURONS

D Gorse
Department of Computer Science, University College London

J G Taylor
Department of Mathematics, King's College London
T G Clarkson
Department of Electrical and Electronic Engineering, King's College London

A feedback circuit is described which when attached to the output bit stream of a probabilistic RAM (pRAM) neuron results in a sigmoid-like squashing transformation. This transformation does not disturb the flow of bits from inputs to final output, and so does not disrupt pulse-based stochastic learning procedures. This output transform circuit can itself be made adaptive, so that the degree of 'squashing' required can be learned by the system. It is demonstrated how the use of such transform circuits can greatly reduce the number of free parameters needed for successful function approximation by pRAM nets.

The pRAM Model

An N-input pRAM [1] has 2^N memory locations, indexed by the N-bit binary vector $u = (u_1, u_2, ...u_N)$. A binary signal $i = (i_1, i_2, ...i_N)$ on the input lines will access just one of these locations, that one for which $u = i$. The variable stored at this location, α_i, gives the probability that a spike value $a=1$ will be produced on the output line, given an input i. Thus in this binary input case the mean value of the output is given by $y = \alpha_i$. If there is a real-valued signal $x = (x_1, x_2, ...x_N)$ on the input lines each x_j is represented by a stream of pulses i_j in which $Prob(i_j = 1) = x_j$. Over $r=1..R$ time steps a spread of locations (as in general each i(r) will be different) will be accessed and contribute stochastically a spike value $a=1$ or $a=0$ to the output stream. In this case the mean value of the output is

$$y = \sum_u \alpha_u \prod_{j=1}^N [u_j x_j + (1-u_j)(1-x_j)] \equiv \sum_u \alpha_u X_u(x) \tag{1}$$

where the variable X_u is the probability location u is accessed.

When hardware is used [2] the output of a pRAM is not computed from (1) directly, but as an approximation by averaging over the bits in the output stream:

$$\hat{y}(R) = \frac{1}{R} \sum_{r=1}^R a(r) \tag{2}$$

As $R \to \infty$, $\hat{y}(R) \to y$, the mean value of the output. In practice a value of $R = R_{max}$ in the order of hundreds of spikes gives a suitably close approximation to the mean.

The pRAM is not a 'weightless' model; the 2^N variables α_u are the weights of the system. It is not a look-up table since these variables represent probabilities continuous in the range [0,1] and a pRAM output in general contains spikes that have been contributed by a wide range of locations. The output function (1) can be seen to be a polynomial of O(N); in biological terms pRAMs can describe synapse-synapse interactions, in connectionist terminology they are sigma-pi nodes.

Adding Non-linear Output Transforms

Under certain circumstances, in particular when multi-pRAM networks with hidden units are used, it may be desirable to apply a non-linear 'squashing' transformation to the pRAM output function. This greatly extends the functionality of pRAM nets, and leads to more efficient function approximation. By applying a sigmoid-like output transformation a given function can usually be approximated by a more compact network which also uses pRAMs of lower input dimensionality. This reduces the number of free parameters, thereby improving generalisation.

How should such an output transform be applied? It is of no use to apply a squashing transformation $s(\alpha_u)$ to the contents of individual memory locations, since this simply maps the weight parameter set $\{\alpha_u\}$ into another set $\{\alpha_u^{new} = s(\alpha_u)\}$ and does not affect the polynomial character of the output function. The output transform must be applied to the bit stream of (2), integration of which gives an approximation to the polynomial function (1), and applied in such a way as not to disrupt the free flow of pulses from input to final output (performance assessment), on which hardware-realisable learning rules [2,3] depend.

The usual sigmoid output transform $\sigma(y) = \dfrac{1}{(1 + e^{-y})}$ is designed for models with output activities $y = \sum_j w_j x_j \in (-\infty, \infty)$, not for the current case, where the activity y, as given by (1), is restricted to the interval [0,1]. The transform function

$$s(n, y) = \frac{y^n}{y^n + (1-y)^n} \tag{3}$$

illustrated for various values of n in Figure 1, preserves all the essential features of the sigmoid, being steepest (with a gradient equal to n) and closest to linear in its midrange, and with a gradient of zero at the limits of its argument ($y = 0, 1$ in this case). As well as being a more natural choice than the sigmoid for arguments restricted to [0,1], it has the very useful feature that it can be implemented by a simple pulse-based feedback circuit. The output bit stream (mean value y) is tapped and n+1 independently generated bits diverted to a buffer $a_0 a_1 \cdots a_n$. a_0 is input to the interpolation unit to be described below; the n spikes $a_1 \cdots a_n$ are input to a recurrent (n+1)-input pRAM with fixed deterministic memory contents

$$\alpha_u = (1-u_0) \prod_{j=1}^{n} u_j + u_0 \left(1 - \prod_{j=1}^{n} (1-u_j) \right)$$

(the 0th input is the self-feedback line). The output of this recurrent pRAM has the mean value (3).

By applying an output transform like (3) we are able to effectively increasing the range of the probabilistic weights (the α_u), which are of necessity restricted to the interval [0,1]. However it is commonly observed in the training of conventional neural networks that not all problems require the development of large weights (and the corresponding step-like output function). So the transform we are looking for should be one which can be applied adaptively. It is clear that the parameter n in (3) (the number of external inputs to the recurrent 'squashing' pRAM) is not a good choice for adaptation, since it must necessarily change in discrete (integer) steps, and the effect of such a step is large (see Figure 1). A much better choice is the interpolation parameter β in the new output function

$$f(\beta, y) = (1 - \beta) y + \beta s(n, y) \tag{4}$$

as this function can be accomplished by a 2-input pRAM with memory contents $(0, \beta, 1-\beta, 1)$ and inputs (i) from the untransformed bit stream with mean value y and (ii) from the spike output of the recurrent pRAM whose mean value is given by (3). The variable β in (4) plays a similar role to the 'inverse temperature' parameter in the Boltzmann model. When it is smallest (highest 'temperature', $\beta = 0$) the output function is flattest, when it is largest (lowest 'temperature', $\beta = 1$) the output is most strongly compressed into a step-like form. Of course there will be a limit to the amount of squashing ultimately available which is determined by n. A suitable value for n must therefore be decided by simulation before any hardware is constructed. It appears from work done so far that n does not need to be large; a value between 2 and 4 was adequate for all applications we have investigated.

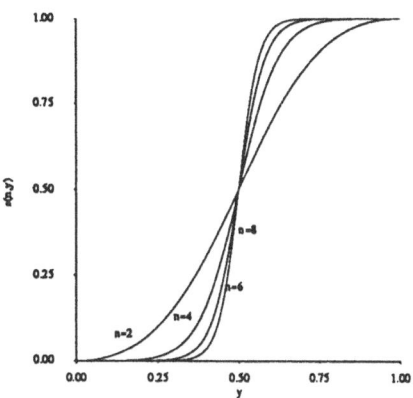

Figure 1

Since the parameter β (with its complement, $1-\beta$) is itself contained in a pRAM it can be learned using any techniques appropriate to the modification of pRAM memory contents, in particular the new continuous output, pulse-based stochastic reinforcement rule first described in [3] (with a small amount of additional circuitry to ensure that the sum of the second and third locations in the interpolation pRAM always sum to 1, and that the contents of the first and fourth locations are not adapted). Thus the learning procedure is uniform over all modules; there is only one kind of adaptive circuit required to change both the parameters involved in polynomial fitting (the α_u) and the parameters which allow a non-polynomial 'tuning' of the output behaviour (the βs).

The task chosen to illustrate the use of non-linear output transforms was the learning of the chaotic logistic series $x(t+1) = 3.97x(t)(1 - x(t))$, from a 100-pattern training sequence beginning with $x(0)=0.5$. The architecture used had six 2-input pRAMs in the hidden layer and one 6-input pRAM in the output layer (there was no attempt to optimise the architecture and it is likely that similar results could have been obtained with a somewhat smaller network). Adaptive output transform circuits (with n=2) were attached to all the pRAM neurons. The learning rule of [3] was used to adapt both the 88 α_u and the 7 βs. When the βs are initially set to zero, as in this case, the network begins by trying to fit a polynomial approximation to the data ; if this is unsatisfactory the βs then begin to increase, otherwise these parameters remain negligible. Figure 2a shows the decrease of mean absolute error with epoch for this example. The final error produced by this multilayer net is significantly less than that produced by a single layer net with no output transforms and 256 weights α_u, more than twice the number used here. Figure 2b shows the evolution of the seven βs in the network during training. The top curve is the transform parameter for the output unit, the cluster of six curves below it the βs associated with the hidden units. Notice that all seven βs initially

708

show a rapid increase - indicating that the network has discovered that pure polynomial approximation in a network of this architecture is not adequate - but thereafter level off. None of the βs achieve a value close to the theoretical maximum of 1, indicating that a transform circuit with n=2 is easily sufficient in this example. However, although the βs remain quite small, their presence is crucial to get a result of this quality - with all the βs fixed at zero (no output transforms) this network converges to a final error about twice that achieved in Figure 2a.

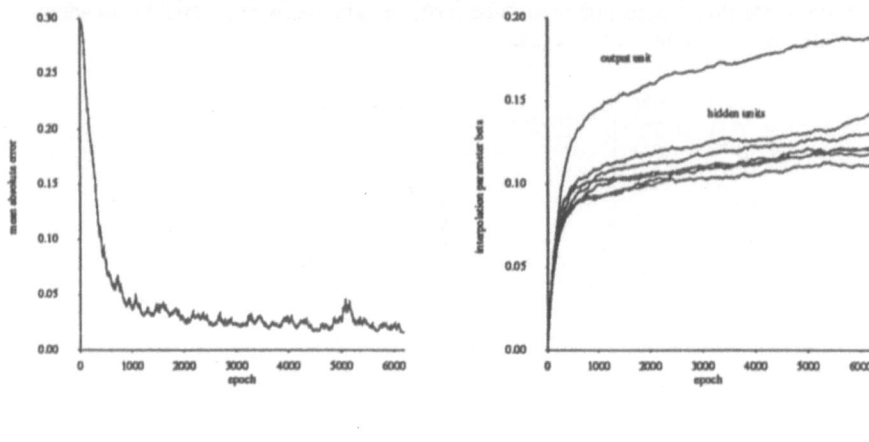

Figure 2a   Figure 2b

Discussion

A construction has been presented which allows probabilistic RAM (pRAM) nets to perform function approximation more compactly, with fewer adjustable parameters. A small adaptive circuit can learn by how much to pass the normally pure polynomial output of a pRAM neuron through an additional sigmoid-like transform; control over the slope of the resulting output function allows the range of the weights of the pRAM neuron to be effectively extended.

References

[1]   D Gorse and J G Taylor, "An analysis of noisy RAM and neural nets", *Physica* , D34, 90-114 (1989); T G Clarkson, D Gorse and J G Taylor, "From wetware to hardware: reverse engineering using probabilistic RAMs", *Journal of Intelligent Systems* , 2, 11-30 (1992).

[2]   T G Clarkson, C K Ng, D Gorse and J G Taylor, "Learning probabilistic RAM nets using VLSI structures" , *IEEE Transactions on Computers* , 41(12), 1552-1561 (1992).

[3]   D Gorse, J G Taylor and T G Clarkson, "Learning real-valued functions using a stochastic reinforcement algorithm", in: *Proceedings of WCNN '93* , III-301 - III-304 (1993).

Statistical Biases in Backpropagation Learning

Chris Thornton. Cognitive and Computing Sciences, University of Sussex
Email: Chris.Thornton@cogs.susx.ac.uk

1 Introduction: learning by the capture of statistical effects

The process of learning is conveniently conceptualized in terms of the acquisition of a target input/output mapping. To have any chance of success the learner requires some source of feedback regarding the mapping to be acquired. In the much studied supervised learning scenario. this feedback takes the form of a set of examples taken from the target mapping. [1] The learner's aim is to arrive at the point at which it is able to map any input taken from the mapping onto its associated output. In more general terms. the learner's aim is to be able to give a high probability to the correct output for an arbitrary input taken from the mapping.

If the learner is to have any chance of achieving this goal. the feedback it receives must contain information which justifies the assigning of particular probabilities to particular outputs. Thus we see that supervised learning is essentially the process of discovering and exploiting such justifications. To understand the nature of the process we need to analyze the ways in which supervisory feedback can provide justifications for assignments of particular probabilities to particular outputs.

There are two main cases to consider. Supervisory feedback (i.e., sets of examples) can justify probability assignments either *directly* or *indirectly*. Direct justification is provided if the probabilities in question are observed directly within the training examples. as statistical effects. They are justified indirectly if they cannot be observed directly but can be systematically derived from those examples; or — which amounts to the same thing — if they can be observed in data which are derived from the original data. [2].

In this paper I will be concerned only with the direct (i.e., statistical) form of justification. and with the ways it which it is exploited by the backpropagation learning algorithm (but see [3]). The nature of this form of justification can be illustrated with an example. Consider the following training set. This is based on two

input variables (x1 and x2) and one output variable (y1). There are six training examples in all. They are laid out with one example per line. An arrow separates the input part of the example from the output part. The values of the two input variables appear on the left of the arrow. The value of the output variable appears on the right.

```
x1  x2      y1

1   3   -->  1
2   1   -->  0
3   2   -->  1
3   1   -->  0
1   2   -->  1
3   1   -->  0
```

A wide variety of probabilities can be observed directly in these training examples. some of which relate specifically to the output variable (and are thus relevant for a supervised learner). To begin with we have the probabilities for first-order cases (i.e., instantiations of a single variable). These are shown in Table 1. The 'C' column shows the case in question and the 'P(C)' column shows the observed probability of that case.

C	P(C)
	1
x1=3	0.5
x2=1	0.5
y1=1	0.5
y1=0	0.5
x2=2	0.33
x1=1	0.33
x2=3	0.17
x1=2	0.17

Table 1:

We can also observe the probabilities of many second-order cases — cases involving the instantiation of two variables. These are shown in Table 2.

C	P(C)
x2=1, y1=0	0.5
x1=3, x2=1	0.33
x2=2, y1=1	0.33
x1=1, y1=1	0.33
x1=3, y1=0	0.33
x1=1, x2=3	0.17
x1=3, x2=2	0.17
x2=3, y1=1	0.17
x1=3, y1=1	0.17
x1=1, x2=2	0.17
x1=2, x2=1	0.17

Table 2:

The probabilities for the third-order cases (i.e., the cases that specify values for all three variables) are, of course, degenerate. Assuming there is no duplication in the training data, each third-order case occurs exactly once. Thus its probability is necessarily $1/n$ where n is the size of the training set.

The probabilities introduced so far are all unconditional. A variety of conditional probabilities can also be observed. For example, we can observe the conditional probability of observing a particular instantiation of the output variable for given first-order cases of the input variables. These probabilities are shown in Table 3. By the argument used previously, the second-order conditional probabilities here are degenerate since there is necessarily exactly one occurrence of each second-order case of the constrained variables.

C	P(C)	P(y1=0|C)	P(y1=1|C)
	1	0.5	0.5
x1=3	0.5	0.67	0.33
x2=1	0.5	1.0	0.0
x2=2	0.33	0.0	1.0
x1=1	0.33	0.0	1.0
x2=3	0.17	0.0	1.0
x1=2	0.17	1.0	0.0

Table 3:

As mentioned above, in the supervised learning scenario, it is the probabilities affecting the output variable(s) which are of interest. Thus, Table 3 in conjunction with the table listing the first-order unconditional probabilities for the output variable, provide an exhaustive enumeration of all directly observed, probability-assignment justifications (henceforth just 'statistical effects'). A quick perusal of the two tables shows that the probabilities for possible values of y1, conditional on

values of x2 and x1 are extreme. These might form the underlying justification for the summary rule: y1=1 if x2=1 or x2=3; otherwise y1=0.

2 Statistically-oriented learning methods

By definition, directly-observed justifications for probability assignments can be discovered more readily than indirectly-observed justifications. It is to expected therefore that general-purpose learning algorithms such as ID3 [4] and Backpropagation [5] will be able to exploit them most effectively. However, it is also to be expected that such learning algorithms will exhibit some form of bias or predisposition to deal more effectively with statistical effects of particular types. For example, we might find that a particular learning algorithm exploits first-order effects very easily but is effectively insensitive to higher-order effects.

3 The study

The study involved training a standard backpropagation implementation (the PDP package of [6] was used) on a variety of artificial learning problems. The solution of each artificial learning problem was based on a statistical effect of a particular order. The results showed clearly that the generalization performance of backpropagation varies monotonically with the order of the underlying statistical effect. Best generalization performance was consistently produced on nth-order problems, where n was the number of input units used and thus the maximum order of statistical effect that could be represented.

All the artifical learning problems were classification problems. They involved learning to correctly allocate an input to one of five classes (to produce an output correctly classifying the input as a member of one of five different input classes). The classification rule — the solution to the learning problem — was, in all cases, prototype-based. That is to say, the five input classes were defined in terms of five input prototypes. To obtain solutions based on low-order effects, I used prototypes defined over *subspaces* of the input space. Thus, for a solution based on an mth-order effect, I used a prototype defined over an m-dimensional subspace of the input space.

Note that, in the present context, a 'prototype is just a set of input-variable instantiations. Low-order prototypes are instantiations for a small subset of input variables. High-order prototypes are instantiations for a large subset of input variables. An example of a prototype, i.e., an instance of the input class defined by the prototype, is derived simply by generating an input vector which 'matches' the prototype. An input vector forms a good match to a prototype if the values it shows for the prototype variables are close to (within 1 standard deviation of) the instantiations specified by the prototype. Thus, to generate an example of a prototype we construct a vector which features small random variations of the prototypical-variable instantiations and purely random values elsewhere. To generate a complete training set we repeatedly cycle through our N prototypes generating prototype examples and appending, as target output, the appropriate class 'label'.

All the data shown is based on experiments with 'minimal' backpropagation architectures. These were strictly layered, feed-forward, networks with one layer of hidden containing just enough units to guarantee perfect acquisition of the *training cases*.[1] All the networks had eight input units and one output unit. Thus all the problems presented involved producing an 'output classification' expressed as a single activation value. In practice, the artificial problems involved making 5-way classifications and this effectively necessitated using output values in the set $\{0.0, 0.25, 0.5, 0.75, 1.0\}$. Having networks of eight input units allowed statistical effects from first-order up to eight-order to be tested.

To minimize interference effects (i.e., ambiguities) between prototype definitions, the same input variables were always used for the specification of all five, nth-order prototypes. Thus all first-order prototypes were defined in terms of a single instantiation of a single input variable. All second-order prototypes were defined in terms of a instantiations of two particular input variables, and so on.

4 The results

The main results of the study are summarised by the graph shown in Figure 1. This shows the mean generalization performance (expressed in terms of mean-difference error on the testing set) for learning problems based on statistical effects of different orders. Note how the generalization performance improves monotonically with the order of the relevant effect. The general implication is that backpropagation is biased towards higher order statistical effects.

In some of the runs performed, overtraining was observed but in general this did not seem to be a major

---
[1] The precise number of units in each case was determined by trial and error.

Mean generalization error

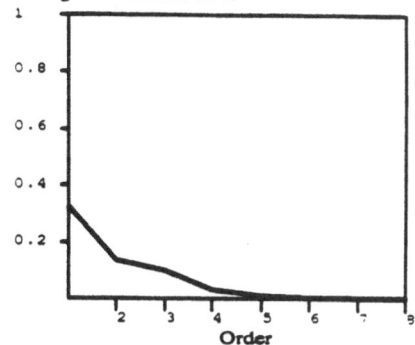

Figure 1:

feature of the learning. A typical error curve for a run on a low-order training problem is shown in Figure 2. Note the rapid descent of the error curve for the training set and the much more gradual descent of the error curve for the testing cases.

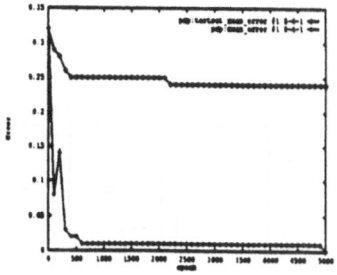

Figure 2:

5 Explaining the results

At first sight, the results of the study seem somewhat paradoxical. In some sense, high-order statistical effects are more 'complex' than low order effects since they involve the specification for larger numbers of values. Therefore, one might expect a learning algorithm

712

to deal more readily with low-order effects. On the other hand, exploiting low-order statistical effects involves 'factoring out' all those input variables which are not involved in the statistical effect, and whose values are therefore pure noise.

As it turns out, it is this 'factoring out' aspect of the low-order exploitation task which appears to be backpropagation's weakness. The problem generator used in the study instantiates prototypes working upwards through the input dimensions. Thus all the first-order prototypes are instantiated over input variable 1. This input variable corresponds to the leftmost, lowest numbered input unit in the input layer of the network. In a first-order problem, the instantiations for all input variables bar the first one are effectively pure noise. Thus, in a backpropagation solution for such a problem, we would hope to see all other variables factored out of the internal representation. In other words, we would hope to see zero weights on all the connections from all input units except input unit number 1. However, when we come to examine the Hinton diagram [7] for non-input nodes in the the network (see Figure 3), we certainly do not see this effect. In fact what we see is that the learning has produced quite pronounced positive and negative weightings for connections from higher-numbered input units.

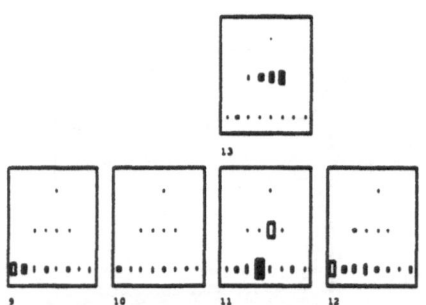

In a sense, this is only to be expected. Backpropagation is relying on the assumption that no part of the input space will contain pure noise. For the low-order problems, this assumption is invalid. And thus we see a quite clear deterioration in generalization performance for problems of this type.

6 Summary

The paper has discussed the purely statistical effects which form the most readily-obtained justifications for solutions to supervised learning problems. It has observed that such effects can be classified according to their conditionality and their order. It seems plausible that learning algorithms able to exploit statistical effects will have some form of bias towards particular types of effect. In the case of backpropagation it appears that this bias takes the form of a strong predisposition towards higher-order effects. This, somewhat counter-intuitive result may be explicable in terms of backpropagation's inability to properly discount input variables whose instantiations are randomly acquired and thus form pure noise.

References

[1] Duda, R. and Hart, P. (1973). *Pattern Classification and Scene Analysis*. New York: Wiley.

[2] Thornton, C. (1993). Supervised learning of conditional approach: a case study. CSRP 308. Cognitive and Computing Sciences, University of Sussex, UK.

[3] Clark, A. and Thornton, C. (1993). Trading spaces: computation, representation and the limits of learning. Cognitive Science Research Paper 291, Brighton BN1 9QH: University of Sussex (Price:1.50).

[4] Quinlan, J. (1986). Induction of decision trees. *Machine Learning, 1* (pp. 81-106).

[5] Rumelhart, D., Hinton, G. and Williams, R. (1986). Learning representations by back-propagating errors. *Nature, 323* (pp. 533-6).

[6] McClelland, J. and Rumelhart, D. (1988). *Explorations in Parallel Distributed Processing: A handbook of Models, Programs, and Exercises* Cambridge, Mass.: MIT Press.

[7] Hinton, G. and Sejnowski, T. (1986). Learning and relearning in boltzmann machines. In D. Rumelhart, J. McClelland and the PDP Research Group (Eds.), *Parallel Distributed Processing: Explorations in the Microstructures of Cognition. Vol I and II* (pp. 282-317). Cambridge, Mass.: MIT Press.

AN APPROXIMATION OF NONLINEAR CANONICAL CORRELATION ANALYSIS BY MULTILAYER PERCEPTRONS

Hideki ASOH†, Osamu TAKECHI‡

† Electrotechnical Laboratory
Tsukuba-shi Ibaraki 305 Japan
‡ Kochi Prefectural Industrial Technology Center
Kochi-shi Kochi 781-51 Japan

1   Introduction

Canonical Correlation Analysis (CCA) is one of the most known multivariate data analysis methods for analysing and summarizing the correlation structure between two multi-dimensional information sources (see e.g. Johnson and Wichern 1988). In ordinal CCA, linear transformations are constructed and applied to each datum from both of two information sources. These transformations are dimension reduction mappings which make the correlation matrix C_{XY} in the reduced dimensional spaces be diagonal and maximize $J = tr(C_{XY})$.

Asoh, Kurita and Otsu showed that CCA can be extended to nonlinear one if we can know or estimate probabilistic structure of the information sources (Asoh et al. 1987). Nonlinear Canonical Correlation Analysis (N-CCA) constructs **nonlinear** dimension reduction mappings. Using N-CCA it is expected to be able to reveal and summarize the deeper probabilistic correlation structures between the two information sources than ordinal linear CCA.

In this paper, we present a method to approximate the nonlinear canonical correlation mappings with multilayer perceptrons. Multilayer Perceptrons (MLP) have successfully applied to many nonlinear function approximation problems. So far some authors have applied MLP to approximate nonlinear data analysis methods such as Regression Analysis and Discriminant Analysis (Asoh and Otsu 1989, 1990, Webb and Lowe 1990, Ruch et al. 1990, Lowe and Webb 1991, Richard and Lippmann 1991, Kurita, Asoh et al. 1992). However, according to Canonical Correlation Analysis, no result has not been obtained previously.

2   Nonlinear Canonical Correlation Analysis

Let $\mathbf{u} \in R^m$, $\mathbf{v} \in R^n$ be continuous random variables, and $p(\mathbf{u}, \mathbf{v})$ denotes the probability density function. Then, the problem of N-CCA is formalized as :

FIND such mappings $\Phi : R^m \to R^L$; $\mathbf{x} = \Phi(\mathbf{u})$ and
$\Psi : R^n \to R^L$; $\mathbf{y} = \Psi(\mathbf{v})$ that satisfy the following.

1) $\bar{\mathbf{x}} = \bar{\mathbf{y}} = 0$ (zero means)
2) $\Sigma_X = \Sigma_Y = I_L$.   (covariance matrices to be unit matrix)

3) $tr(\Sigma_{XY}) \to$ maximized.

Solving this problem turns out to be reduced to solving the following simultaneous integral equation system for Φ and Ψ;

$$\Lambda\Phi(u) = \int \Psi(v)\,(p(v|u) - p(v))\,dv \tag{1}$$

$$\Lambda\Psi(v) = \int \Phi(u)\,(p(u|v) - p(u))\,du, \tag{2}$$

where Λ is a diagonal matrix. Eliminating Φ from the above equations, we obtain the following integral eigen equation for Ψ.

$$\Lambda^2\,\Psi(v) = \int \Psi(v')(s(v'\,|\,v) - p(v'))\,dv', \tag{3}$$

where $s(v'\,|\,v)$ is defined as

$$s(v'\,|\,v) = \int p(v|u)p(u|v')\,du,$$

and called the intersection coefficients (Otsu 1975).

3 Approximating N-CCA with MLP

If we have all information about $p(v, u)$ and can solve the equation (3), we will get the optimal nonlinear canonical correlation mapping $\Psi(v)$. And from $\Psi(v)$ we can obtain $\Phi(u)$ using the equation (1). However, this procedure is not expected to be feasible in usual situations. Here, instead of solving (3)and(1) directly, we try to approximate the optimal nonlinear canonical correlation mapping Ψ and Φ using multi-layer perceptrons.

3.1 Architecture of the Network

Hence N-CCA regards two information sources u and v, for the approximation of N-CCA we need to combine two neural networks as shown in Fig.1. Transformed outputs x and y are obtained from the output layer of each network.

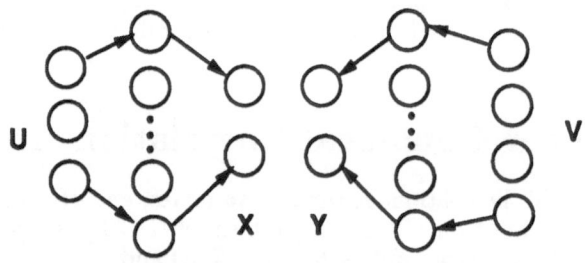

Network A   **Network B**

Figure 1: Architecture of the Network

3.2 Learning Algorithm

We train our networks along the following learning procedure. It makes the outputs of the networks be satisfy the normalization conditions mentioned in the section 2.

1. Propagate input **u** through the network A (see Fig.1).

2. Normalize the output of the network A (make $\Sigma_X = I$).

3. Train the network B several cycles using the normalized output of A as teacher signals.

4. Propagate input **v** through the network B.

5. Normalize the output of the network B (make $\Sigma_Y = I$).

6. Train the network A several cycles using the normalized output of B as teacher signals.

7. Goto 1 till the learning process converges.

4 Experimental Results

Here we show a result of a simple experiment. We apply CCA and N-CCA to an example dataset from a textbook of multivariate data analysis. Original dimensions are three for **u** and four for **v**. Data are taken from salesmen in a company. **u** represents the sales capability of each salesman and **v** represents the result of some intelligence tests. We use two 3-layer networks with 10 hidden units for reducing the dimension of the data to two and plot the results. That is, for the mapping $\Phi(\mathbf{u})$ we use a network with 3-10-2 structure (network A in Fig.1) and for the mapping $\Psi(\mathbf{v})$ a network with 4-10-2 structure (network B). In this experiment, each network is trained 100 cycles in the step 3 and 6 of the learning procedure. Fig.2 and 3 show the results of linear CCA and N-CCA respectively.

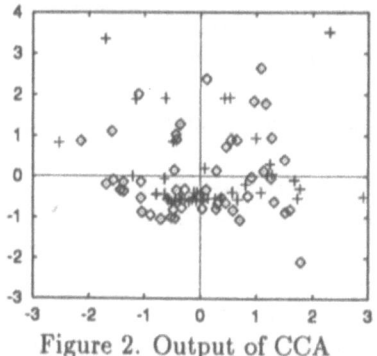

Figure 2. Output of CCA

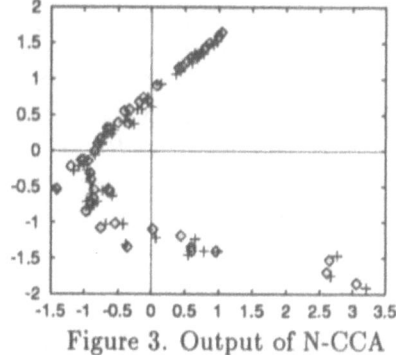

Figure 3. Output of N-CCA

In both plots, "◇" represents $\mathbf{x} = \Phi(\mathbf{u})$ and "+" represents $\mathbf{y} = \Psi(\mathbf{v})$.

5   Discussions and Conclusions

Apparently approximated N-CCA outperforms linear CCA in the above experiment. The maximum value of $tr(C_{XY})$ attained by out networks is 1.99 (After 250 learning cycles) while the value attained by linear CCA is 1.65. Moreover, the plot of the output from approximated N-CCA reveals the one dimensional simple structure of the dataset. Each data point corresponding to each salesman is on one dimensional curve, and the order of the points represents the "capability" of each salesman. We executed several runs of the learning process in some different conditions and found that this structure is stable.

In this paper we proposed a method to approximate Nonlinear Canonical Correlation Analysis with two combined neural networks. Experimental results show that the proposed method works well and can reveal deeper correlation structure between two information sources than usual linear CCA.

Recently several researchers proposed to use CCA for building the interface of image data base (e.g. Kurita, Kato et al. 1992). There the CCA is used to extract good internal expressions of images or key words. Our method will be useful for improving such internal expressions.

This research was supported by the Real-World Computing Program and Kochi Prefecture.

References

[1] Asoh,H., Kurita,T., and Otsu,N. (1987) An interpretation of canonical correlation analysis and discriminant analysis by nonlinear extensions, (in Japanese). *The Japanese Journal of Behaviormetrics*, 14, 1-9.

[2] Asoh,H. and Otsu,N. (1989) Nonlinear data analysis and multilayer perceptrons. *Proc. of IJCNN'89*, II, 411-415.

[3] Asoh,H. and Otsu,N. (1990) An approximation of nonlinear discriminant analysis by multilayer neural networks. *Proc. of IJCNN'90*, III, 211-216.

[4] Kurita,T., Asoh,H., and Otsu,N. (1992) Nonlinear discriminant features constructed by using outputs of multilayer perceptron, submitting to ISSIPNN'94.

[5] Kurita,T., Kato,T., Fukuda,I. and Sakakura,A.(1992) Sense Retrieval on a Image Database of Full Color Paintings (in Japanese), *Trans. of IPSJ*, 33.

[6] Lowe,D. and Webb,A.R. (1991) Optimized feature extraction and the Bayes decision in feed-forward classifier networks. *IEEE Trans. on Pattern Analysis and Machine Intelligence*, 13, 355-364.

[7] Otsu,N. (1975) Nonlinear discriminant analysis as a natural extension of the linear case. *Behavior Metrica*, 2, 45-59.

[8] Richard,M.D. and Lippmann,R.P. (1991) Neural network classifiers estimate Bayesian *a posteriori* probabilities. *Neural Computation*, 3, 461-483.

[9] Ruch,D.W., Rogers,S.K., Kaabrisky,M., Oxley,M.E., and Suter,B.W. (1990) The multilayer Perceptron as an approximation to a Bayes optimal discriminat function. *IEEE Trans. on Neural Networks*, 1, 296-298.

[10] Webb,A.R. and Lowe,D. (1990) The optimised internal representaion of multilayer classifier networks performs nonlinear discriminant analysis. *Neural Networks*, 3, 367-375.

Information Minimization to Improve Generalization Performance

Ryotaro Kamimura†, Shohachiro Nakanishi‡

† Information Science Laboratory,

‡ Department of Electrical Engineering, Tokai University

1117 Kitakaname Hiratsuka Kanagawa 259-12, Japan

1 Introduction

Many techniques have been developed for the improvement of the generalization performance. One of the most successful methods consists in the minimization of complexity in network architectures, for example, addition of weight decay or weight pruning. If network architectures are too complex, they can store everything including noises in addition to the necessary part of input patterns. If the complexity is too simple, it is impossible to learn training patterns. Suppose that the complexity represents a kind of information capacity of networks. If the information regarding training input patterns is excessively stored, meaning that networks store every details of training patterns, the generalization performance is not improved. Thus, the information is appropriately stored in network architectures so as to improve the generalization performance.

We suppose that the information to be stored in network architectures is as small as possible, under the condition that networks can generate targets with appropriate accuracy. For demonstrating this hypothesis, let us define the information, stored in network architecture. The Information can be defined by the difference between maximum entropy and observed entropy:

$$I = H_{max} - H \tag{1}$$

where H^{max} is a maximum entropy and H is an observed entropy. Thus, using this information, our objective is to show that the generalization performance can be improved by minimizing the information content. Finally, one of the difficult problems is to define the information. We have defined an entropy for the internal representation. Thus, to reduce the information corresponds to the minimization of the information, defined for the internal representation.

2 Theory and Computational Methods

Let us formulate an entropy function for the internal representation. Suppose that a network is composed of three layers: input, hidden and output layers. Hidden unit activities are denoted by v_i and input terminals by ξ_j. Then, input-hidden connections are denoted by w_{ij} and hidden-output connections are denoted by W_{ij}. A hidden unit produces an output

$$v_i^k = f(u_i^k),$$

718

where f is a logistic function defined by

$$f(u_i^k) = \frac{1}{1 + e^{-u_i^k}}$$

and where u_i^k is a net input to ith hidden unit and defined by

$$u_i^k = \sum_{j=1}^{L} w_{ij}\xi_j^k.$$

where ξ_i^k is ith element of an input pattern and L is the number of elements in the pattern. An entropy for kth input pattern is defined by

$$H_k = -\sum_i p_i^k \log p_i^k, \tag{2}$$

where

$$p_i^k = \frac{v_i^k}{\sum_m v_m^k},$$

where the summation is over all the hidden units. Averaging over all the input patterns, we have

$$H = -\frac{1}{N}\sum_k^N \sum_i^M p_i^k \log p_i^k, \tag{3}$$

where M is the number of hidden units and N is the number of input patterns. The information is defined by the difference between the maximum entropy and the observed entropy. That is,

$$R = H_{max} - H \tag{4}$$

$$= \log M + \frac{1}{N}\sum_k^N \sum_i^M p_i^k \log p_i^k.$$

Following the derivation by Deco (Deco et al. 1993), we have the following update rule:

$$\Delta w_{ij} = \sum_k [\alpha\phi_i^k + \beta\delta_i^k]\xi_j^k - \gamma w_{ij}, \tag{5}$$

where

$$\phi_i = p_i^k(1 - v_i^k)(-\log p_i^k + \sum_m p_m^k \log p_m^k),$$

and δ is an ordinary delta for the back-propagation, and the final term should be added to prevent weights to grow overwhelmingly. Intuitively, this equation can easily be understood. That is, weights are updated so as to minimize the difference between individual information $-\log p_i^k$ and average information or entropy $-\sum_m p_m^k \log p_m^k$.

3 Results

In experiments, we trained networks to produce correct past tense forms, given various verb stems of artificial languages, close to English. All the artificial languages were composed of strings: CVC, CCV, and VCC, where V is a vowel, and C is a consonant. Each string was represented in a phonological representation with eight bits, used by Plunkett et al. (Plunkett et al., 1990). The number of training patterns was 100 for the regular verbs. The number of validation patterns was 500 and testing patterns was also 500 patterns for all the experiments. The number of input, hidden, and output units was 18, 20, 20 respectively.

Now, let us see how the generalization errors can be improved by using the method of information minimization. Figure 1 shows generalization errors as a function of the number of epochs. A figure at the top was computed with training data, and a figure at the bottom was with testing data. Let us see an upper figure for the training data. As you can see from the figure, errors are decreased gradually and finally zero both for the weight decay method and information minimization method. Little difference can be seen in the generalization errors. However, for the testing data, a big difference between two methods can be seen. A lower figure shows generalization errors for the testing data. As the learning is advanced, the difference between two method becomes more significant. We can clearly see that the generalization is much improved by using the information minimization. This result shows that the method of information minimization can be effective to improve the generalization performance. In other words, the information content must be as small as possible to improve the generalization performance.

4 Conclusion

In the present paper, we have shown that for the good generalization performance, the information stored in network architecture must be as small as possible, under the condition that networks have the sufficient capacity to learn the input patterns with appropriate accuracy. By applying the method of information minimization to the problem of language acquisition, we have demonstrated that the generalization is really improved by minimizing the information content, stored in the internal representation.

References

G. Deco, W. Finnof and H. G. Zimmermann (1993) "Elimination of overtraining by a mutual information network," in *Proceeding of the International Conference on Artificial Neural Networks*, Springer-Verlag, pp.744-749.

K. Plunkett, V. Marchman, and S. L. Knudsen (1990) "From Rote Learning to System Building: Acquiring Verb Morphology in Children and Connectionist Nets," in *Connectionist Models: Proceedings of the 1990 Summer School*, D. S. Touretzky, J. L. Elman and G. E. Hinton, (Eds), Morgan Kaufmann Publishers, Inc, San Mateo: California, pp.201-219.

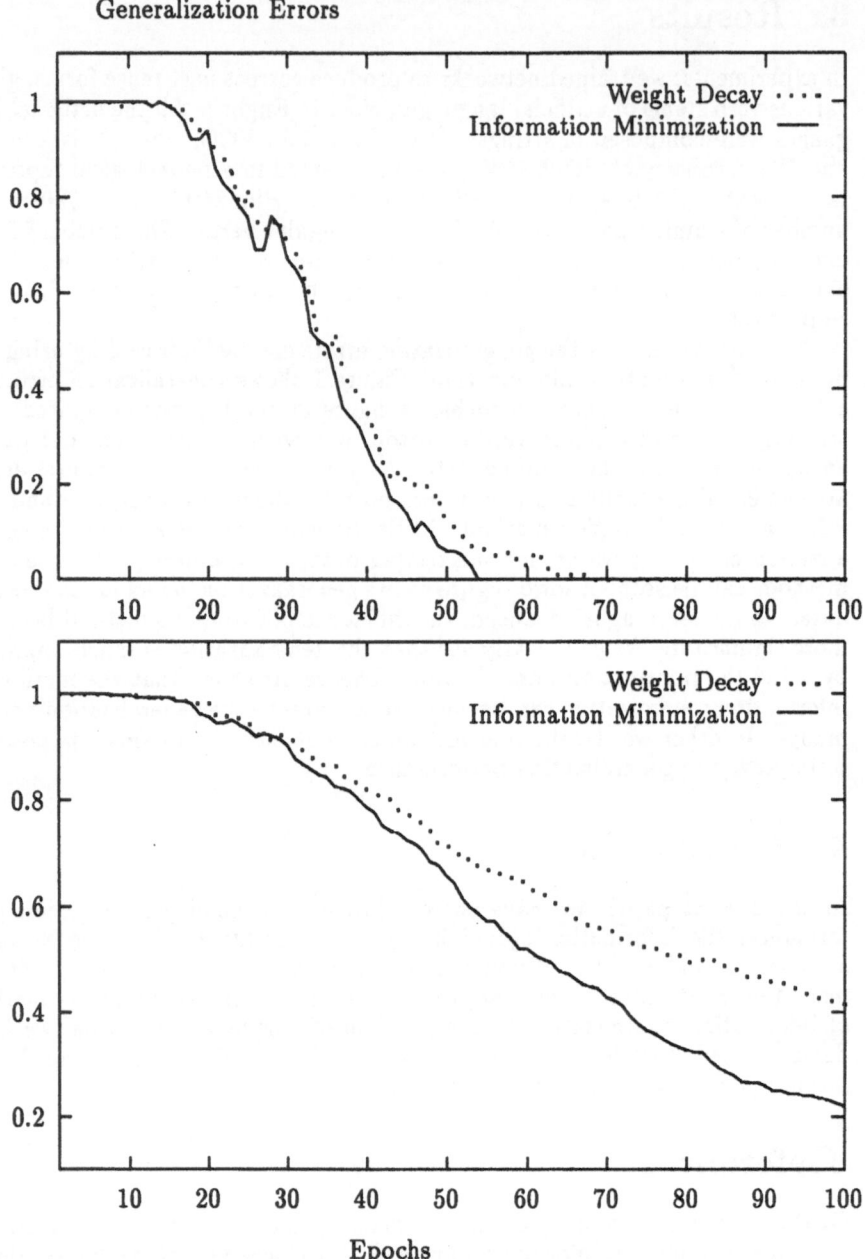

FigureOB 1: Generalization errors as a function of the number of epochs, computed by training data (upper figure) and by testing data (lower figure). The parameters, α, β, γ were set to 0.0001, 0.05 and 0.001 respectively.

Learning and Interpretation of Weights in Neural Networks

C.C.A.M. Gielen

Dept. of Medical Physics and Biophysics, University of Nijmegen,
Nijmegen, The Netherlands

1   Introduction

Contrary to traditional approaches in which an explicit algorithm is constructed
to solve a particular problem, Neural Networks are trained by presenting examples and a learning rule ensures that after some period of training the desired
functionality in terms of input/output relations is achieved.

It is well known that when the architecture of the neural network has been
chosen properly in the sense that the network can learn the desired input/output
relation, the optimal performance of the network is obtained by gradually decreasing the learning parameter to zero, since this reduces the fluctuations in
performance due to random selection of training samples (see e.g. Wasan, 1969).
This causes a conflict with learning in a situation, in which the stimulus ensemble may change and in which a learning parameter different from zero is essential
to store new information. This is a realistic problem both in biological systems
and in artificial systems.

In this chapter, we will review some of the literature on how to chose the
optimal learning parameter in a changing environment and we will develop some
theoretical tools to gain insight in the performance of a trained neural network
from the weights. This is important for neurophysiological studies, in which the
functional and neurophysiological properties of neuronal structures have to be
derived from correlations between neurones (see e.g. Eggermont, 1990) as well
as for artificial neural networks if one wants to understand and explain *how* the
neural network does the job.

2   Learning processes and their average behavior

Let the adaptive elements of a neural network, such as synapses and thresh-
olds, be given by a weight vector $\mathbf{w} = (w_1, \ldots, w_N)^T \in \mathbb{R}^N$. At distinct
iteration times $\mathbf{w}$ is changed due to the presentation of a training pattern
$\vec{x} = (x_1, \ldots, x_n)^T \in \mathbb{R}^n$, which is drawn at random according to a probability distribution $\rho(\vec{x})$. The new weight vector $\mathbf{w}' = \mathbf{w} + \Delta\mathbf{w}$ depends on the

old weight vector and on the training pattern:

$$\Delta \mathbf{w} = \eta \, \mathbf{f}(\mathbf{w}, \vec{x}) . \tag{1}$$

where function $\mathbf{f}$ is a general representation of the learning rule which captures all familiar learning rules and where η is the learning parameter.

Because of the *random* pattern presentation, the learning process is a stochastic process. We have to talk in terms of probabilities, averages, and fluctuations. It can be shown (see Heskes and Kappen, 1993) that with the transition probability $T(\mathbf{w}'|\mathbf{w})$ to "walk" in one learning step from state $\mathbf{w}$ to state $\mathbf{w}'$ given by

$$T(\mathbf{w}'|\mathbf{w}) = \int d^n x \, \rho(\vec{x}) \, \delta^N(\mathbf{w}' - \mathbf{w} - \eta \mathbf{f}(\mathbf{w}, \vec{x})), \tag{2}$$

the probability $P(\mathbf{w}, t)$, that a network is in state $\mathbf{w}$ at *time t* obeys the master equation reads

$$\frac{\partial P(\mathbf{w}', t)}{\partial t} = \int d^N w \, [W(\mathbf{w}'|\mathbf{w})P(\mathbf{w}, t) - W(\mathbf{w}|\mathbf{w}')P(\mathbf{w}', t)], \tag{3}$$

with the transition probability per unit time

$$W(\mathbf{w}'|\mathbf{w}) = \frac{1}{\tau} T(\mathbf{w}'|\mathbf{w}). \tag{4}$$

Through τ, which is related to the interval between subsequent stimulus presentations, we have introduced a physical time scale. In this way a discrete time random walk equation has been transformed into a continuous time master equation. For the rest of this chapter we will choose $\tau = 1$, i.e., the average time between two learning steps is our unit of time.

For future use we define the drift vector, which is just the average learning rule

$$\mathbf{f}(\mathbf{w}) \stackrel{\text{def}}{=} \langle \mathbf{f}(\mathbf{w}, \vec{x}) \rangle_\Omega ,$$

and the diffusion matrix

$$D \stackrel{\text{def}}{=} \langle \mathbf{f}(\mathbf{w}, \vec{x}) \, \mathbf{f}^T(\mathbf{w}, \vec{x}) \rangle_{\Xi(t)} , \tag{5}$$

containing the fluctuations in the learning rule. Here $\Xi(t)$ represents the ensemble of learning networks. Furthermore, we define the Hessian matrix $H(\mathbf{w})$ with components

$$H_{ij}(\mathbf{w}) = -\frac{\partial f_i(\mathbf{w})}{\partial w_j} . \tag{6}$$

If and only if the Hessian matrix is symmetric, an energy function or error potential $E(\mathbf{w})$ can be defined such that the learning rule performs a (stochastic) gradient descent on this error:

$$\mathbf{f}(\mathbf{w}) = -\nabla_\mathbf{w} E(\mathbf{w}) . \tag{7}$$

The Hessian matrix gives the curvature of the error potential in the different directions. This approach is the "small fluctuations expansion" and is valid in regions of weight space with positive definite Hessian $H(\mathbf{w})$. These regions will be called attraction regions.

The attractive fixed point solutions of this "average learning dynamics" will be denoted by $\mathbf{w}^*$. If there exists an error potential, then these fixed points are simply the (local) minima. At a fixed point $\mathbf{w}^*$ we have no drift, i.e., $\mathbf{f}(\mathbf{w}^*) = 0$ and a positive definite Hessian $H(\mathbf{w}^*)$. The typical local relaxation time towards these fixed points is

$$\eta_{\text{local}} = \frac{1}{\eta \lambda_{\min}(\mathbf{w}^*)} , \qquad (8)$$

with $\lambda_{\min}(\mathbf{w}^*)$ the smallest eigenvalue of the Hessian $H(\mathbf{w}^*)$. To study the asymptotic convergence, we can make an expansion around the minimum $\mathbf{w}^*$. In Heskes and Kappen (1991) and van Kampen (1981) it is shown that for sufficiently small η the evolution equations for the weightvector $\mathbf{w}$ and for the fluctuations of the weightvector around $\mathbf{w}^*$ are given by

$$\begin{aligned}
\frac{1}{\eta}\frac{d\mathbf{m}(t)}{dt} &= -H\mathbf{m}(t) \\
\frac{1}{\eta}\frac{d\Sigma^2(t)}{dt} &= -H\Sigma^2(t) - \Sigma^2(t)H^T + \eta D ,
\end{aligned} \qquad (9)$$

where the Hessian and the diffusion matrix are both evaluated at the fixed point $\mathbf{w}^*$ and with definitions for the bias $\mathbf{m}(t)$ and covariance matrix $\Sigma^2(t)$

$$\mathbf{m}(t) \stackrel{\text{def}}{=} \langle \mathbf{w} \rangle_{\Xi(t)} - \mathbf{w}^*, \quad \Sigma^2(t) \stackrel{\text{def}}{=} \left\langle \left[\mathbf{w} - \langle \mathbf{w} \rangle_{\Xi(t)} \right] \left[\mathbf{w} - \langle \mathbf{w} \rangle_{\Xi(t)} \right]^T \right\rangle_{\Xi(t)} . \quad (10)$$

These results demonstrate that the asymptotic probability distribution for small learning parameters η is a simple Gaussian, with its average at the fixed point $\mathbf{w}^*$ and a covariance matrix Σ^2 obeying

$$H\Sigma^2 + \Sigma^2 H = \eta D . \qquad (11)$$

So, there are persistent fluctuations of order η that will only disappear in the limit $\eta \to 0$.

3 A conflict in a changing environment

Equation (11) states that we must decrease the learning parameter to zero in order to prevent asymptotic fluctuations in the network state. However, a truly adaptive system can always adapt itself to changes in the environment. This kind of adaptability is also desirable for artificial neural networks, e.g., for neural networks for the modeling of economic processes.

Mathematically speaking, a changing environment corresponds to a time-dependent input probability $\rho(\vec{x}, t)$. The probability density of network states $\mathbf{w}$ now follows a continuous-time master equation with a time-dependent transition probability $T_t(\mathbf{w}'|\mathbf{w})$:

$$T_t(\mathbf{w}'|\mathbf{w}) = \int d^n x \rho(\vec{x}, t) \delta^N(\mathbf{w}' - \mathbf{w} - \eta \mathbf{f}(\mathbf{w}, \vec{x})) \stackrel{\text{def}}{=} \langle \delta^N(\mathbf{w}' - \mathbf{w} - \eta \mathbf{f}(\mathbf{w}, \vec{x})) \rangle_{\Omega(t)},$$

where $\Omega(t)$ stands for the set of training patterns, the "environment", at time t. The fixed points $\mathbf{w}^*(t)$ of the deterministic equation

$$\frac{1}{\eta} \frac{d\mathbf{w}(s)}{ds} = \langle \mathbf{f}(\mathbf{w}(s), \vec{x}) \rangle_{\Omega(t)}, \tag{12}$$

may depend on time. We define the "misadjustment" $\mathcal{E}$ as the average squared Euclidian distance with respect to this fixed point $\mathbf{w}^*(t)$:

$$\mathcal{E} \stackrel{\text{def}}{=} \frac{1}{T} \int_0^T dt \, \langle |\mathbf{w} - \mathbf{w}^*(t)|^2 \rangle_{\Xi(t)} = \frac{1}{T} \int_0^T dt \, |\mathbf{m}(t)|^2 + \text{Tr}\left[\Sigma^2(t)\right]. \tag{13}$$

The bias $\mathbf{m}(t)$ is a measure of how well the ensemble of learning networks follows the environmental change *on the average*. It gives the typical delay between what the average network state is, $\langle \mathbf{w} \rangle_{\Xi(t)}$, and what it should be, $\mathbf{w}^*(t)$. The covariance matrix $\Sigma^2(t)$ gives the width of the distribution and thus a measure of "confidence". T is a time window which should be chosen such that it is long enough to allow a large number of stimuli to be presented to the network, but not longer than the time constant which characterizes the changes in $\Omega(t)$.

In order to calculate a learning parameter that yields a good compromise between fast adaptability (a small bias) and high confidence (a small variance), we will make the approximation that the learning parameter is so small that it is allowed to make the usual small fluctuations expansion, and that the rate of change $\mathbf{v} \stackrel{\text{def}}{=} \dot{\mathbf{w}}^*$ is much smaller than the typical weight change ηf. Then the evolution of the bias $\mathbf{m}(t)$ and the covariance $\Sigma^2(t)$ is governed by

$$\frac{d\mathbf{m}(t)}{dt} = -\eta H(t)\mathbf{m}(t) - \mathbf{v}(t),$$

$$\frac{d\Sigma^2(t)}{dt} = -\eta H(t)\Sigma^2(t) - \eta \Sigma^2(t) H^T(t) + \eta^2 D(t), \tag{14}$$

with notation $H(t) \stackrel{\text{def}}{=} H(\mathbf{w}^*(t))$, and so on. With the assumption that the changes in the environment are slow relative to the local relaxation time η_{local} [see equation (8)], the bias and covariance matrix tend to stationary values. The stationary bias is inversely proportional to the learning parameter η and proportional to the speed v, whereas the variance is proportional to the learning parameter and more or less independent of the speed. So, for nonlinear learning rules a misadjustment is obtained (see (Heskes and Kappen, 1991) of the form

$$\mathcal{E} \approx \frac{\alpha v^2}{\eta^2} + \beta\eta \, ,$$

with α and β constants that depend on the diffusion D and the curvature H at the fixed point. Here the time window T must be larger than the local relaxation time η_{local} and smaller than the time in which at least one of the quantities $\mathbf{v}$, D, or H, changes substantially. Differentiation of $\mathcal{E}$ with respect to η gives that the optimal learning parameter is proportional to $v^{2/3}$.

Since one could argue that the neural network may not have explicit information about the environment (the diffusion and the curvature at the fixed point) other parameters can be used to estimate the adjustment of the learning parameter (see Heskes and Kappen, 1992).

4 Interpretation of performance from weights in neural network.

Although research into opening of the "black box" of a trained neural network is still beginning, a few methods are available. For Kohonen-type of networks, this is a more or less trivial activity. For all other neural network types with continuous neurons or with neurons for which the probability for an output is a continuous function of it's input (such as multi-layer perceptrons with continuous or Boolean units) one could study the analytic expression for the gradient of the output of the network relative to the input stimulus in order to find extrema. This approach is practically useful only when the number of extrema is small.

Seung and Sompolinsky (1993) studied the conditional probability $p(\vec{z}|\vec{x})$ for the output $\vec{z}$ given a stimulus $\vec{x}$ and investigated the extrema of this function to find the optimal value for $\vec{x}$. A better procedure is to investigate the conditional probability $p(\vec{x}|\vec{z})$ for the stimulus x given the output $\vec{z}$ (see Gielen et al., 1998), which follows from Bayes equality when $p(\vec{z}|\vec{x})$, $p(x)$ and $p(\vec{z})$ are known. When the stimuli are randomly chosen from a flat probability distribution and when all outputs $\vec{z}$ are equally probable, both methods are equivalent. Otherwise, the most plausible stimulus is found based on differentiation of the conditional probability $p(\vec{x}|\vec{z})$. This is a straightforward procedure since the conditional probability $p(\vec{z}|\vec{x})$ (which for a single neuron is related to it's receptive field) can be easily derived from the weights in the network.

The procedure outlined in this section can be used without explicit information on the probability $p(x)$. Most learning rules (such as back propagation) give the set of weights which minimize a cost function averaged over the input/output pairs:

$$< (\vec{z}_{desired} - \vec{z}(\vec{x}))^2 >_{\Omega(t)} = \int_{\Omega(t)} d\vec{x} p(\vec{x})(\vec{z}_{desired} - \vec{z}(\vec{x}))^2. \qquad (15)$$

This already includes an averaging over the stimuli which have been presented to the network in the past, such that explicit information does not have to be stored in the network. Since $p(\vec{z})$ does not explicitly depend on the stimulus $\vec{x}$, this probability function is not relevant for finding the extrema of $p(\vec{x}|\vec{z})$.

5 Acknowledgment

This work was supported by the Dutch Foundation for Neural Networks and the Japanese Real World Computing Program.

6 References

1. Gielen C.C.A.M., Hesselmans G.H.F.M., Johannesma P.I.M. (1988) Sensory interpretaion of neural activity. Math. Biosci. 88, 15-35. Math. Biosci.

2. Heskes, T.M. and Kappen B. (1993) On-line learning processes in artificial neural networks. In: Mathematical Approaches to Neural Networks. J.G. Taylor (Ed.) Elsevier Science Publishers, pp. 199-233.

3. Heskes, T., and Kappen B. (1991) Learning processes in neural networks. Phys. Rev. A 44, 2718-2726.

4. Heskes, T. and Kappen B. (1992) Learning-parameter adjustment in neural networks. Phys. Rev. A 45, 8885-8893.

5. Seung, H.S., Sompolinsky H. (1993) Simple models for reading neuronal population codes. Proc. Natl. Acad. Sci. USA 90, 10749-10753.

6. van Kampen N. (1981) Stochastic processes in physics and chemistry. North-Holland, Amsterdam.

7. Wasan, T. (1969) Stochastic Approximation. Cambridge University Press.

Variable Selection with Optimal Cell Damage

Tautvydas Cibas *, Françoise Fogelman Soulié **, Patrick Gallinari °, Sarunas Raudys[#]

* LRI, bât. 490, Université de Paris-Sud, F-91405, Orsay, (France).
** SLIGOS,Tour Anjou, 33 quai de Dion Bouton, F- 92814 Puteaux cedex (France)
° LAFORIA-IBP, Univ. Paris 6, 4 place Jussieu, F-75232 Paris cedex 05, (France).
Dep. Data Analysis, Inst.Math.Informatics, Akademijos 4,Vilnius 2600, (Lithuania).

1. Introduction

Neural Networks -NN- have been used in a large variety of real-world applications. In those, one could measure a potentially large number N of variables X_i; probably not all X_i are equally informative: if one could select n << N "best" variables X_i, then one could reduce the amount of data to gather and process; hence reduce costs. *Variable selection* is thus an important issue in Pattern Recognition and Regression. It is also a complex problem; one needs a criterion to measure the value of a subset of variables and that value will of course depend on the predictor or classifier further used. Conventional variable selection techniques are based upon statistical or heuristics tools [Fukunaga, 90]: the major difficulty comes from the intrinsic combinatorics of the problem. In this paper we show how to use NNs for variable selection with a criterion based upon the evaluation of a variable usefulness. Various methods have been proposed to assess the value of a weight (e.g. *saliency* [Le Cun *et al.* 90] in the *Optimal Brain-Damage* -OBD- procedure): along similar ideas, we derive a method, called *Optimal Cell Damage* -OCD-, which evaluates the usefulness of input variables in a Multi-Layer Network and prunes the least useful. Variable selection is thus achieved during training of the classifier, ensuring that the selected set of variables matches the classifier complexity. Variable selection is thus viewed here as an extension of weight pruning. One can also use a regularization approach to variable selection, which we will discuss elsewhere [Cibas *et al.*, 94]. We illustrate our method on two relatively small problems: prediction of a synthetic time series and classification of waveforms [Breiman *et al.*, 84], representative of relatively hard problems.

The paper is organized as follows: section 2 describes notations and results from the literature; section 3 the problems used for testing our methods and section 4 results of variable selection by OCD.

2. Variable Selection

2.1. Definitions

Let a random variable pair $(X, Y) \in \mathfrak{R}^N x \mathfrak{R}^P$ be given, with probability distribution P. Based on a sample $D_m = \{(x^1,y^1)...(x^m,y^m)\}$, drawn from (X, Y), we train a NN α, and produce an estimator F, which depends upon α and D_m. Different estimators, $\alpha_1, \alpha_2,..., \alpha_M$ may have different errors: here we will select models depending upon the comparison of their respective empirical errors. In this paper, we chiefly compare estimators α_n differing only in their input dimension: some components of the original input vector $X \in \mathfrak{R}^N$ are eliminated to produce a vector $x \in \mathfrak{R}^n$, with $n \leq N$. The problem in variable selection is to extract the best set of features, and to possibly determine, for a given criterion, the optimal number n* of features.

2.2. Neural Network methods for sensitivity analysis

One major issue with NNs is to ensure that the trained network correctly generalizes, i.e. ensure that the discrepancy between empirical and theoretical errors is small. When the amount of data is restricted, one has to control the network complexity: network *pruning* is one possible technique for that. Let N be a NN picked from a family of multi-layer networks with n inputs. We use the Mean Squared Error -MSE- defined as:

$$C(W, D_m) = \frac{1}{2m} \sum_{k=1}^{m} \| F(x^k) - y^k \|^2 \tag{2-1}$$

where $F(x^k)$ is the computed output for input x^k, y^k the desired output.

Various authors have proposed methods to *prune* least useful connections [Chauvin, 90], [Weigend *et al.*, 91]. We focus here on the *Optimal Brain Damage* technique [Le Cun *et al.*, 90]: a weight *saliency* is defined as the cost variation resulting from the weight suppression. Le Cun *et al.* have shown that, if the Hessian can be approximated by a diagonal matrix H, and the cost is locally quadratic, then weight i *saliency* is:

$$s_i = \frac{1}{2} H_{ii} W_i^2 = \frac{1}{2} \frac{\partial^2 C}{\partial w_i^2} W_i^2 \tag{2-2}$$

which can be computed by an additional pass to the usual Back Propagation algorithm [Bishop, 92]. The weight of smallest saliency is the weight whose pruning will least increase the cost. Notice that OBD explicitly requires that the network has reached convergence (W_0 is a local optimum). Exact methods could be used to compute the Hessian, instead of assuming diagonality [Hassibi *et al.*, 93].

3. Test problems

In this section, we describe the version of Gradient Back Propagation -GBP- and testbed problems we used.

3.1. Training procedure

Our training procedure is an on-line GBP implemented as follows: the gradient step ε is decreased as soon as the relative average cost variation becomes too small.

GBP	- run on-line GBP with learning step ε as long as $\varepsilon \geq \varepsilon_1$				
-	if	$r(W, D_m^v, t, T_1, 1, T_1) \leq \theta_1$			(3-1)
then	$\varepsilon(t+1) = \alpha \cdot \varepsilon(t)$	else	$\varepsilon(t+1) = \varepsilon(t)$		

$$[C(W,D)]_t^{T_1} = \frac{1}{T_1} \sum_{\tau=0}^{T_1-1} C(W(t-\tau), D), \quad r(W,D,t,T_1,T_2,T_3) = 1 - \frac{[C(W,D)]_{t+T_2}^{T_3}}{[C(W,D)]_t^{T_1}} \tag{3-2}$$

3.2. Time Series

We have used an artificial time series y_t, where ε_t is a white noise:

$$y_t = 0.3\, x_{t-6} - 0.6\, x_{t-4} + 0.5\, x_{t-1} + 0.3\, x_{t-6}^2 - 0.2\, x_{t-4}^2 + \varepsilon_t \tag{3-3}$$

We used a linear network 12-1 (with all 10 input variables x_t plus variables x_{t-6}^2 and x_{t-4}^2): variable selection should pick the "true" inputs $x_{t-6}, x_{t-4}, x_{t-1}, x_{t-6}^2, x_{t-4}^2$ (y_t is linear in these variables). We also used a network 10-7-1 (with all 10 variables x_t as inputs, 7 non linear hidden units and linear output). Now, selection should pick input variables ($x_{t-6}, x_{t-4}, x_{t-1}$). Three independent samples of size m=1000 each were drawn from series y_t: D_m^l for learning, D_m^v validation and D_m^t test. Results are shown in figure 1 for net 10-7-1 ($\varepsilon=0.0001$, $\theta_1=0.0075$)and net 12-1 ($T_1 = \alpha = 1$). We use the MSE on D_m^v to choose the best parameter values $T_1 = 3$ and $\alpha = 5/6$.

3.3. Waveforms

We have extended a problem introduced in [Breiman *et al.*, 84]: 3 vectors or *waveforms* in 21 dimensions, H^i, i=1,...,3, are defined. Noisy convex combinations of 2 of these vectors are formed [de Bollivier *et al.*, 91] and the problem is to classify in one of the 3 classes corresponding to the 3 pairs (m,n):

$$X_i = \frac{1}{5}[u\,H_i^m + (1-u)H_i^n] + \varepsilon_i, \quad 0 \le i \le 20; \qquad X_i = \varepsilon_i, \qquad 21 \le i \le 40 \qquad (3\text{-}4)$$

D_m^l has 300 elements, D_m^v 1000, D_m^t 5000. Simulations were run with $\varepsilon_1 = 0.005$, $\theta_1 = 0.0075$, on a network 40-30-20-3 (2 hidden layers of 30 and 20 neurons). Results shown in figure 2 allow us to set parameter values at: $T_1 = 3$ and $\alpha = 5/6$.

T_1	3			5			7		
α\Set	D_m^l	D_m^v	D_m^t	D_m^l	D_m^v	D_m^t	D_m^l	D_m^v	D_m^t
5/6	4491	**5159**	4657	4479	5190	4679	4479	5214	4697
7/8	4471	5184	4674	4460	5188	4660	4447	5209	4681
9/10	4461	5211	4701	4447	5222	4712	4440	5269	4779
net 12	4461	4629	4209						
3/4	96.0	81.86	81.9	96.33	83.0	81.46	96.66	82.1	81.86
5/6	96.66	**83.1**	82.0	97.66	82.3	81.56	96.66	81.0	81.86
7/8	97.33	82.4	81.54	97.66	83.0	82.48	96.33	81.9	82.26
9/10	97.66	82.9	81.96	97.33	81.9	82.34	97.0	82.3	81.8

Figure 1: final 10^5 MSE for time series. and NN 10-7-1 and 12-1 (top lines)
final classification performance for Breiman's noisy waveforms problem (bottom)

4. Optimal Cell Damage
4.1. The formalism
We now describe our variable selection procedure, based on an extension of OBD. Discarding variable X_i is implemented by setting to 0 all weights W_j in Fan_Out(i), the Fan-Out of input neuron i, and the resulting cost is just the sum of the costs associated to the suppression of the various weights. The *saliency* of variable i is thus defined as:

$$\zeta_i = \sum_{j \in \text{Fan\_Out}(i)} s_j \qquad (4\text{-}1)$$

```
OCD   if      r(W, Dᵥₘ, t, T2, T2, T2)          ≤     θ2      and

              r(W, Dᵥₘ, t+T2, T2, T2, T2)       ≤     θ2                    (4-2)

then compute ζi at t+T2 and order them ζi1 ≥ ζi2 ≥ ... ≥ ζiN

                        n          N
choose n such that:    Σ ζil ≥ q  Σ ζil , eliminate variables in+1,...,iN
                       l=1        l=1

continue OCD
else continue GBP
```

Our algorithm for OCD runs as follows: if the relative average cost variation in 2 successive periods is too small, then prune a fraction q of inputs. Pruning starts early if θ_2 is large, late otherwise. Parameters θ_2 and q are set by cross-validation.

4.2. Results
We ran our OCD algorithm with $T_1 = 3$ and $\alpha = 5/6$ as determined in 3.2-3.3, with $T_2 = 3$ and cross-validation to choose parameters θ_2 and q. Results are shown in figure 2. In the time series problem, for the optimal values q = 0.99 and $\theta_2 = 0.05$, our variable selection procedure selected the appropriate variables (variables 5, 7 and 10 in figure 2). in the waveforms case, for the optimal values q = 0.985 and $\theta_2 = 0.05$, our variable

selection procedure selected all variables 0 to 20, except variables 0, 1, 2, 7, 19 and 20; and rejected all variables 21 to 39 (white noise!), except variables 21, 22, 29, 31, 33, 34, 38 and 39. Comparison with conventional techniques is on-going to assess the validity of this selection: notice that after pruning of these input variables, performances are increased from 82.0 (fig.2) to 84.16 on test set D_m^t.

q	0.985			0.99			0.995		
θ_2	0.03	0.05	0.06	0.03	0.05	0.06	0.03	0.05	0.06
D_m^l	6886	6613	6613	4789	4638	4638	4804	4725	4725
D_m^v	7476	7004	7004	5237	**4991**	**4991**	5422	5123	5123
D_m^t	7049	6590	6590	4884	4579	4579	5041	4738	4738
Set	5,7	5,7	5,7	5,7, 10	5,7, 10	5,7, 10	1,2,5 7,10	5,7, 10	5,7, 10
θ_2	0.03	0.05	0.07	0.03	0.05	0.07	0.03	0.05	0.07
D_m^l	95.33	95.33	95.66	96.33	95.66	96.00	97.33	97.33	97.00
D_m^v	83.4	**83.9**	83.0	82.0	82.5	82.4	82.9	83.2	82.5
D_m^t	84.06	84.16	83.74	83.12	83.68	83.52	82.48	82.4	82.28

Figure 2: 10^5 MSE and selected variables for quadratic time series., after OCD (top) performance after OCD on Breiman's noisy waveforms (bottom)

5. Conclusion

We have given a method to evaluate the usefulness of a variable in a NN: this method, derived from Optimal Brain Damage, has proved efficient on two artificial relatively complex problems. However, our method suffers from two drawbacks: there are various parameters which much be set by cross validation, and that is costly (although results seem to indicate that the results are rather robust with respect to the exact values of these parameters); parameter-free algorithms such as conjugate gradients could be used too. We have also approximated the Hessian by a diagonal matrix, which is known to be untrue in many applications. Further work should indicate whether this hypotheses hinder the variable selection process.

6. References

BISHOP C.: Exact calculation of the Hessian matrix for the multilayer perceptron. Neural Comp. vol. 4, 494-501, (1992).

BOLLIVIER de M., GALLINARI P., THIRIA S.: Cooperation of neural nets and task decomposition. IJCNN'91, vol. II, 573-576, (1991).

BREIMAN L., FREIDMAN J., OLSHEN R., STONE C.: Classification and regression trees. Wadsworth Int. Group. (1984).

CHAUVIN Y.: Dynamic behavior of constrained back-propagation networks. In NIPS'89, D. Touretzky ed., Morgan Kaufmann, vol. 2, 643-649. (1990)

CIBAS T., FOGELMAN SOULIE F., GALLINARI P., RAUDYS S.: Variable selection with neural networks (in preparation).

FUKUNAGA K.: Statistical Pattern Recognition. Academic Press, 2nd edition. (1990).

HASSIBI B., STORK D.G.: Second order drivatives for network pruning: Optimal brain Surgeon. In NIPS'92, S.J. Hanson, J.D. Cowan, C. L. Giles eds., Morgan Kaumann, vol. 5, 164-171, (1993).

LE CUN Y., DENKER J.S. , S.A. SOLLA: Optimal brain damage. In "Neural Information Processing Systems", NIPS'89, D. Touretzky ed., Morgan Kaufmann, vol. 2, 598- 605, (1990).

WEIGEND A.S., RUMELHART D.E., HUBERMAN B.A.: Generalization by weight elimination with application to forecasting. In NIPS'90, R.P. Lippmann, J.E. Moody, D. S. Touretzky eds., Morgan Kaufmann, vol. 3, 875-882, (1991).

Comparison of Constructive Algorithms for Neural Networks

F.M. Frattale Mascioli, G. Martinelli, and G. Lazzaro
Dip. INFO-COM, Università di Roma "La Sapienza"
Via Eudossiana 18, Rome, Italy

1 Introduction

The determination of an optimal architecture of a supervised neural network is still an open problem. A promising approach is represented by the constructive algorithms. The basic idea is to start from minimal network structures and properly add further elements. The strategies to be pursued in this growth can be very different. Several algorithms have been proposed with different performances in terms of resulting architectural complexity, generalisation capability and resistance to overfitting, computational cost required for extracting the architecture from the training examples, time delay for processing the input signal, and so on. In the present paper, we will discuss some preliminary results regarding a comparison among four constructive algorithms carried out on a classical benchmark problem: the two-spirals classification problem (Lang *et al.*, 1988). This problem, for its extreme nonlinearity, is out of the possibility of the multilayer perceptron trained by the classical back-propagation (BP) learning algorithm. Instead, the considered constructive algorithms are always able to arrive to a solution.

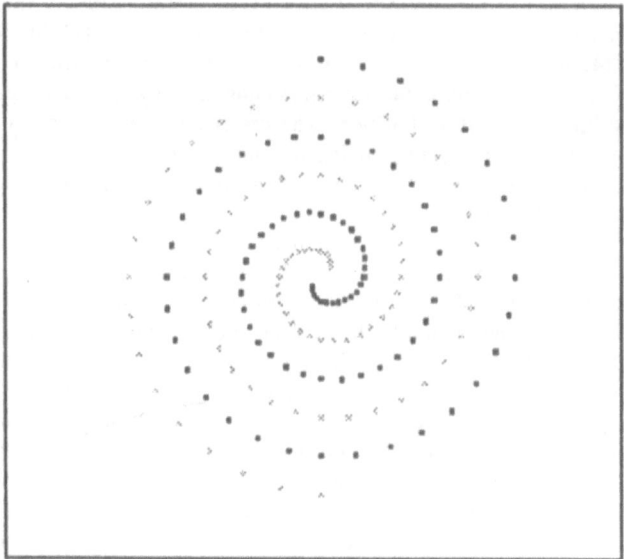

Fig. 1: The two-spirals problem training set.

2 Two-spirals problem

The input to the net is represented by the continuous coordinates of the plane where two spirals are drawn. The output is the classification result which is either +1 or -1 depending on which set the input belongs. The training set consists of 194 X-Y real values, half of which are to produce a +1 output and half a -1 output. These training points are arranged in two interlacing spirals that go around the origin three times, as shown in fig. 1. The spiral with black dots corresponds to class +1. After the net is constructed by the algorithm under consideration, it is tested by presenting 1000 points uniformly distributed in the square embedding the spirals. The result of the test is a square where a black dot denotes a +1 result. The examination of this square enables us to judge the generalisation capability of the constructed nets.

3 Constructive algorithms under comparison

The algorithms taken into consideration will be shortly described in the following:
UPSTART ALGORITHM (UA): the algorithm constructs the net during training by adding perceptron-like units as they are needed in order to eliminate errors. The growth of the net is leaded by a recursive rule. The characteristic resulting architecture is a pyramid with the the output at the top. Each neuron is connected to the input and to two neurons following a feedforward arrangement (Frean, 1990).
CASCADE CORRELATION ALGORITHM (CCA): also in this case a perceptron-like unit is added at each step of the algorithm. However, both the error-correction strategy and the architecture of the net are different. The former consists in choosing from a pool of candidates the unit having the maximum correlation with the network error and in discarding the rest. The latter is a cascade where each neuron is connected to the input and to the two neighbouring units (Fahlman et al., 1990). The alternative use of the same error-correction strategy of UA is possible with good results (Burgess et al., 1992).
LINEAR-PROGRAMMING PERCEPTRON ALGORITHM (LPPA): neural networks having pyramidal or cascade architecture can be also constructed by the algorithms proposed in (Martinelli et al., 1990; Martinelli et al., 1992; Martinelli et al., 1993). These algorithms are based on a linear-programming strategy with perceptron-like units, having a hard-limiter activation function. The cascade architecture can be transformed into a two-layer perceptron by rearranging its units in the form of the first layer of a multilayer perceptron. The second layer is constituted by only one unit, which properly shapes the desired decision region. The resulting two-layer perceptron is hence obtained by a rearrangement, based on linear-programming strategy, of the algorithms proposed in (Martinelli et al., 1990; Martinelli et al., 1992; Martinelli et al., 1993).
OILSPOT ALGORITHM (OSA): it operates with binary data directly in the input space. Namely, it determines the separating hyperplanes of perceptron-like units by means of topological properties of the vertices of the hypercube, corresponding to the training set . The constructive algorithm presently used requires to transform the real input data into binary form by a suitable codification, which must preserve the neighbourhood of data (i.e., two points which are contiguous in the real input space must be coded into two contiguous vertices of the multidimensional binary hypercube). The resulting two-layer perceptron is constituted by a hidden layer with neurons determined as in (Frattale Mascioli et al., 1993), and a second layer constituted by only one output unit performing an OR operation.

4 Results and comments

The result of the comparison is shown in fig. 2 and commented in the following:
GENERALISATION: the decision region of LPPA and OSA are better than those of UA and CCA. It is also interesting to compare them with that obtained in (Lang et al., 1988) with a multilayer perceptron having an ad-hoc structure (fig. 3). In spite of a very

complex architecture, an extremely large amount of processing time and the necessity for several preliminary attempts, the resulting decision region is in this case worst than those obtained by the considered constructive algorithms.

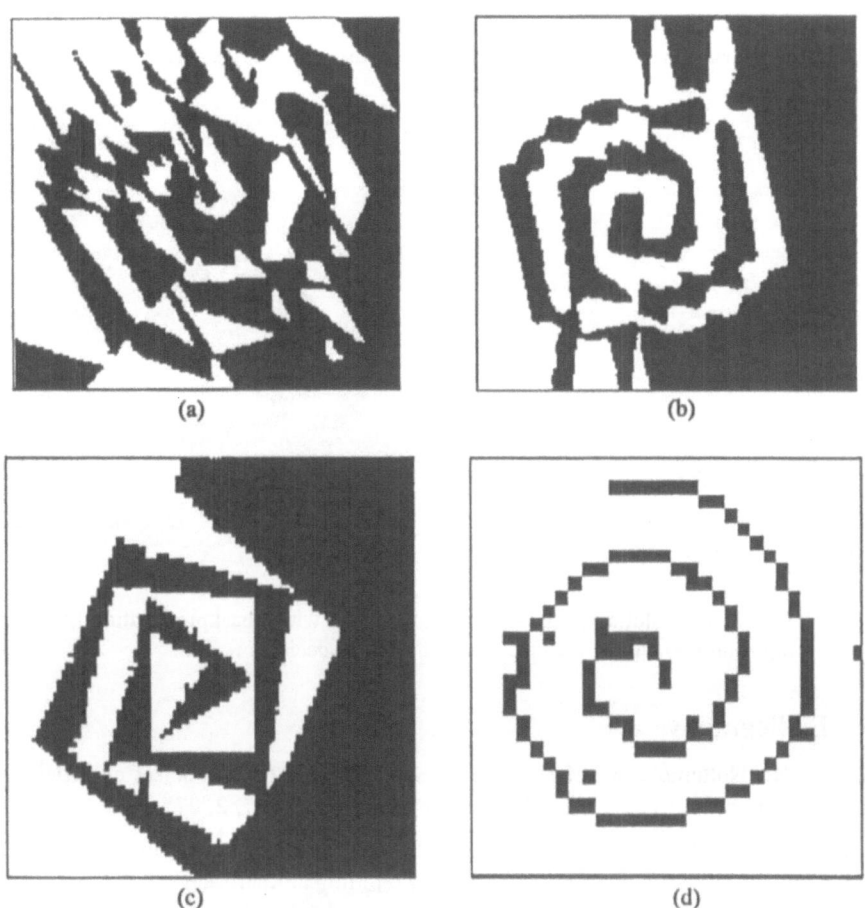

Fig. 2: The resulting decision regions of the considered constructive algorithms:
a) Upstart; b) Cascade-Correlation; c) Linear-Programming Perceptron; d) OilSpot.

TRAINING COST: the time required for yielding the net, with reference to a 486-based computer, has the following order of magnitude: UA = less than one hour; CCA = less than one hour; LPPA = less than five minutes; OSA = less than five seconds; modified backpropagation = more than five hours. The previous values are already a comment regarding this performance. Moreover, it is important to note that LPPA and OSA do not need of specific parameters to set or stopping criterion to use during training. Finally, we note that the time delay for processing the input data is better in the case of the two-layer architecture of LPPA and OSA.

OVERFITTING: LPPA and OSA are both inherently self-pruning. In fact, they are able to measure the importance of the hyperplane introduced in the input space by a neuron with respect to the shaping of the decision region. Consequently, it is easy to introduce in

734

their case a criterion for discarding a neuron on the basis of its importance, in order to avoid a dangerous overfitting during training.

STRUCTURAL COMPLEXITY: it is measured by the number of units present in the constructed nets. They are: UA = 18; CCA = 18; LPPA = 21 and OSA = 45. Finally, we remark that, in the case of OSA, the higher number of units is justified by the augmented dimensionality (from 2-D to 8-D) of the input space, due to the adopted codification.

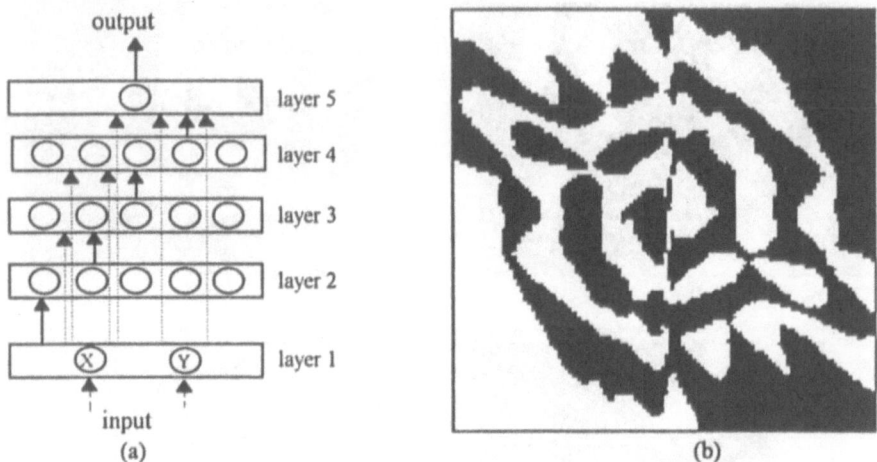

Fig. 3: The solution of the two-spirals problem with a backpropagation-like algorithm: a) ad-hoc architecture of the net; b) decision region.

5 Bibliography

Burgess N., Notturno Granieri M., and Patarnello S. (1992). 3-D object classification: application of a constructive algorithm. *Int. Journ. of Neur. Sys.*, 2, 275—282.

Fahlman S.E., Lebiere C. (1990). The cascade-correlation learning architecture. *Adv. in Neur. Inf. Proc. Sys.*, 2, D. Touretzky, 524—532, Morgan Kaufmann.

Frattale Mascioli F.M., Martinelli G. (1993). A constructive algorithm for binary mapping. *Proc. of ICANN '93*, 776.

Frean M. (1990). The upstart algorithm: a method for constructing and training feedforward neural networks. *Neural Computation*, 2, 198—209.

Lang K.J., Witbrock M.J. (1988). Learning to tell two spirals apart. *Proc. Connectionist Models Summer School*, Morgan Kaufmann.

Martinelli G., Prina Ricotti L., Ragazzini S., and F. Mascioli F.M. (1990). A pyramidal delayed perceptron. *IEEE Trans. on CAS*, 7, 1176—1181.

Martinelli G., F. Mascioli F.M. (1992). Cascade Perceptron. *IEE Electronics Letters*, 28, 947—949.

Martinelli G., F. Mascioli F.M., and G. Bei (1993). Cascade neural network for binary mapping. *IEEE Trans. on NN*, 1(4), 148—150.

Task Decomposition and Correlations in Growing Artificial Neural Networks*

Jan Matti Lange†, Hans-Michael Voigt†‡ & Dietrich Wolf†

† Center for Applied Computer Science (GFaI)

‡ Bionics & Evolution Techniques, Technical University Berlin

Berlin, Germany

1 Introduction

To reduce the engineering efforts for the design of neural network architectures a data driven algorithm is desirable which constructs a network during the learning process. For structure adaptation different approaches with evolutionary algorithms (Voigt et. al., 1993), growth algorithms (Fahlmann et. al., 1990), and others are used.

To solve large problems successfully it is necessary to divide the problem into subproblems and to solve them separately. This is a fundamental principle of nature. There are different approaches (Jordan et. al., 1992), (Smieja et. al., 1992) following this principle, but they yield fixed network structures. The proposed architecture includes both sides of the medal – structure adaptation and **TA**sk decomposition – by **CO**rrelation Measures (TACOMA). For task decomposition it is necessary that the subnetworks do not mutually disturb each other. This is achieved by locally relevant neurons with a new kind of activation function and a connection routing algorithm which connects only cooperative units. The new activation function combines the local characteristics of radial basis function units (Lee et. al., 1991) with sigmoid units. Structure adaptation is done by a data driven cascading of hidden layers inspired by CASCOR (Fahlmann et. al., 1990).

2 The TACOMA Learning Architecture

The growing process works bottom up starting with only output units and inserts hidden layers one by one. It stops if the error at the output units is smaller than a threshold. The output units are trained cyclically after each insertion of a new hidden layer to minimize the squared error at the output of the network. After each cycle the residual errors are computed. The input sites of the output units are connected with the network input vector $\vec{x}$ and the outputs of the units of all hidden layers. At each cycle the number of inputs of an output unit is incremented by the number of units in the new inserted hidden layer.

If a new hidden layer is to be inserted than the number of units in this layer has to be calculated. A hidden layer unit works locally restricted in the input space. Its output is weighted by a window function in the input space. For the

*This work is supported by the Bundesminister für Forschung und Technologie (BMFT) as part of the project *REFLEX*.

computation of the number of hidden layer units a mapping procedure is used which gives the maxima of the mapping of the output units residual error to the input space. The number of maxima which are above a threshold corresponds to the number of units to be inserted. Inserting locally restricted units at the points of these maxima in the input space yields a significant reduction of the network error with high probability, i.e. the window function of a unit is centered at a mapping maximum in the input space. Because the number of maxima is different from layer to layer the number of units is different too. The range of a window function is adjusted such that the patterns which caused the output units residual error are grasped by this window.

The connection routing algorithm is used to estimate the connections between previously inserted units and a new unit. Each hidden layer unit gets the network input vector $\vec{x}$. Additionally, it is connected with the output of those previously inserted hidden layer units which are influenced by the same subspace of the input space, i.e. if the window functions of hidden layer units have a significant overlap than the units will be connected.

The hidden layer units are trained to maximize the correlation between its output and the residual error of the output units and to maximize the anticorrelation between all hidden layer units. These two optimization goals are aggregated to a compound functional. Because this functional has a lot of local extrema the above described initialization of the window function is of essential importance for the success of the algorithm.

Training of output units: The output units are simple sigmoid units. They are trained using the Quickprop algorithm.

Mapping of the residual error to input space: The mapping procedure works similar to a Kohonen Net without neighborhood relations. In the input space a map with K units ($K = 8$ for example) is randomly initialized, each unit has a reference Vector $\vec{v}_k$ with the same dimension as the input space and an output g_k. The mapping works as follows. For a number of epochs compute for each pattern $\vec{x}$ the $\vec{v}_*$ for which $\|\vec{v}_* - \vec{x}\| < \|\vec{v}_k - \vec{x}\|$ holds for all $* \neq k$ and update $\vec{v}_*$ by

$$\vec{v}_{*,t+1} = \vec{v}_{*,t} + \alpha(t) \sum_{o=1}^{O} |E_o^r| \ (\vec{x} - \vec{v}_{*,t}) \qquad , \alpha \text{ decrease with time.}$$

This yields that the reference vectors $\vec{v}_k$ are located at the maxima of the mapping of the residual error in input space. Now the outputs g_k will be computed. $g_k = \overline{\sum_{o=1}^{O} |E_{o,k}^r|}$ with $E_{o,k}^r$ the residual errors for which the corresponding $\vec{x}$ activates the unit k. If $g_k > \overline{g}$ a unit will be inserted in the new hidden layer. The hidden units have sigmoid activation functions weighted with a gaussian window function.

$$y = e^{-u} \frac{1}{1 + e^{-s}} \qquad \text{with} \qquad u = \sum_{i=1}^{I_{net}} \left(\frac{x_i - m_i}{r_i} \right)^2 \qquad s = \sum_{i=1}^{I_{hunit}} x_i w_i$$

I_{net} is the dimension of the network input space and I_{hunit} is the dimension of the hidden unit input vector. To initialize the ranges r_i of a hidden unit the means $\overline{d_{k,i}} = 1/N_k \sum_{j=1}^{N_k} |x_{i,j} - m_{k,i}|$ for each dimension i of the network input space for all patterns x_j activating unit k will be computed. Than the r_i of a unit are initialized such that the value of the window function for $\overline{d_{k,i}}$ is equal to β (for example $\beta = 0.55$).

Connection routing: If the window functions of a new hidden layer unit and any previously inserted hidden layer unit significantly overlap than the units will be connected. To check the overlap the correlation $R_{l,m}$

$$R_{l,m} = \frac{\sum_{i=1}^{N} [w_l(\vec{x}_i)w_m(\vec{x}_i)]}{\sum_{i=1}^{N} w_l^2 \sum_{i=1}^{N} w_m^2}$$

between the values of two window functions w_l and w_m about the pattern set $\{\vec{x}_i\}$, $i = 1 \cdots N$ is computed. If $R_{l,m} > \gamma$ (for example $\gamma = 0.001$) then the units will be connected.

Training of hidden layer units: The unit weights are trained to maximize the covariance S

$$S_o = \frac{\sum_p y_p E_{p,o}^r - N\overline{y}\overline{E_o^r})}{\sum_o \sum_p E_{p,o}^2} \qquad S = \sum_o |S_o| \quad \rightarrow \text{max}$$

between the unit output and the residual error of the output units for all output dimensions N_o and all patterns p by a gradient procedure using Quickprop. The m_i and r_i of the unit window function are trained to maximize the covariance S like previously mentioned and to maximize the anticorrelation between the unit output and the other hidden layer unit outputs. These goals are reflected by the aggregated functional F

$$F = \frac{\sum_{l=1}^{L} S_l}{\sum_{i=1}^{L-1} \sum_{j=i+1}^{L} |R_{i,j}| + \eta} \rightarrow \text{max} \qquad R_{i,j} = \frac{\sum_p y_{i,p} \, y_{j,p} - N\overline{y_i} \, \overline{y_j}}{\sqrt{\sum_p (y_{i,p} - \overline{y_i})^2 (y_{j,p} - \overline{y_j})^2}}.$$

L is the number of units in the hidden layer. The maximization of F is done by a simple gradient procedure with respect to the window function parameters r_i and m_i. The update of weights and window function parameters is done at the same time. The training of a hidden layer stops if F is stagnant or after a maximal number of epochs.

Task decomposition: The three components of local hidden units, maximation of anticorrelation between hidden layer units and the connection routing leads to the network property of task decomposition. At the hidden layer level the units are trained to reduce the residual error for different areas of the input space, i.e. to solve different tasks. At the network level hidden layer units which are relevant in same areas of the input space, which try to solve the same task, are cooperating by connections. At the end of training substructures of the networks can be specified solving different tasks. Because of the connection routing there is a soft transition between different substructures. That seems to be more plausible in the sense of parallel distributed systems.

3   Results

The performance of the proposed algorithm has been evaluated for difficult problems. Here we present the results for the Two Spirals Problem of Kevin

Lang and Michael Witbrock, which was used as a benchmark for different algorithms, and the Two Twin Spirals Problem, which shows especially the limitations of the Cascade–Correlation Learning Architecture. For the CASCOR experiments the program of (Fahlmann et. al., 1990) was used.

Figure 1: Two-Spirals (upper row) and Two-Twin-Spirals (lower row), typical solutions, left CASCOR, right TACOMA

In figure 1 the decision regions are shown. In all cases the networks are trained to a recognition rate of 100 %. The better results of the TACOMA architecture are quite obvious. For the second example it is shown that CASCOR can't solve the task because of the mutual disturbances of the hidden units. The network solution of the TACOMA approach distinguishes well between the two intertwined spirals. The network structure reflects the two domains of expertise.

References

[S.E. Fahlman & C. Lebiere, 1990] "The Cascade-Correlation Learning Architecture", in Touretzky (ed.) *Advances in Neural Information Processing Systems 2*, Morgan-Kaufmann

[M. Jordan & R. A. Jacobs, 1992] "Hierarchies of adaptive experts", in *Proceedings of 1992 Neural Information Processing Systems Conference*, vol. 4, pp. 985-992

[S. Lee & R.M. Kil, 1991] "A Gaussian Potential Function Network With Hierarchically Self-Organizing Learning", in *Neural Networks*, vol. 4, pp. 207-224

[E. Littman & H. Ritter, 1993] "Generalization Abilities of Cascade Network Architecture", in *Artificial Neural Networks 5*

[F. Smieja & H. Mühlenbein, 1992] *Reflective modular neural network systems*, GMD Technical Report 633, Sankt Augustin, Germany

[H.-M. Voigt, J. Born & I.Santibanez-Koref, 1993] *Evolutionary Structuring of Artificial Neural Networks*, Technical University Berlin, Bionics and Evolution Techniques Lab, Technical Report TR-93-002

XNeuroGene: A system for evolving artificial neural networks

C. Jacob, J. Rehder, J. Siemandel, A. Friedmann

Lehrstuhl für Programmiersprachen, Universität Erlangen-Nürnberg

Postfach 3429, D-91022 Erlangen, Germany

Email: *jacob@informatik.uni-erlangen.de*

1 Hierarchical design of artificial nervous systems

NeuroGene [2] is a hierarchical evolutionary system for artificial neural networks. Individual genotypes representing partial descriptions of components of artificial nervous systems are evolved on three hierarchical levels: net topology, neuron functionality and weight settings (fig. 1) representing coarse and fine tuning levels, respectively.

The network model used consists of a predefined, problemdependent set of input and output neurons and a set of cortex modules to be evolved. A cortex module is either a single neuron or another input-cortex-output network. Input neurons are connected in feedforward direction only, i.e. inputs can be passed to either hidden or output neurons (modules). Cortex modules are connected either to hidden or output neurons.

Functionality of the cortex neurons is modelled as a composition of the following three functions: An input function collects the neuron's incoming, weighted signals; this function might be a weighted summation of the inputs, a Sigma-Pi-function or another appropriate input processing function. The "summation value" is further processed by an activation function (linear, sigmoid, radial basis etc.) resulting in an internal neuron activity, which is taken by the output function (identity, linear, threshold etc.) to compute an externally visible output value that can be passed to other neurons.

The design process for a neural network then has to evolve a connectivity structure for the input, cortex and output neurons, a set of functional parameters defining the functionality of the hidden neurons, and a set of weight values for all existing connections (fig. 1). For each of these three, partly interdependent phases we use different string populations and codings.

2 The *XNeuroGene* user interface in detail

Experimenting with evolutionary algorithms is tedious when there is no proper user-interface available; the *XNeuroGene* graphical user interface (fig. 2) supports interactive control of the *NeuroGene* kernel system and external utility programs. Specialized scripts are available via the graphical interface to call

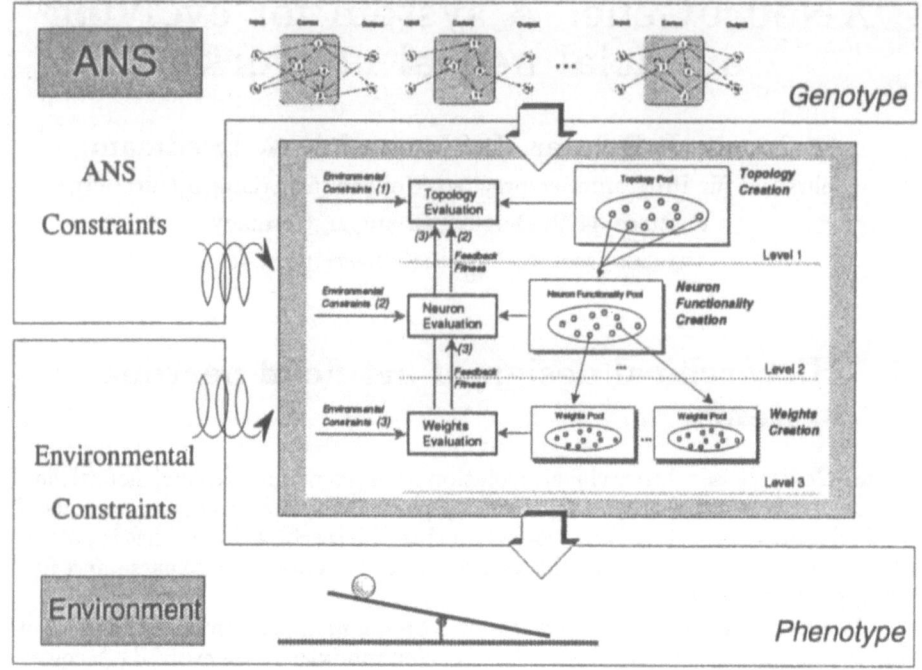

Figure 1: Design hierarchy for artificial nervous systems

filter programs which first extract selected data from protocol files produced by the NeuroGene kernel modules and then post-process, analyse and visualize these data with the help of (e.g.) Mathematica.[1]

A collection of evolution experiments (runs) for a certain problem domain is called a *project*. Each run might have slightly different control parameters (mutation rate, crossover rate, dump interval etc.).

For each project the following window classes are available:

- **Setup** windows control mode settings like coding (Gray or other), parameter representation (bit strings or floating point vectors), average performance statistics as well as parameter settings which are mostly floating point values out of a predefined interval (number of experiments, population size, report interval etc., see e.g. fig. 2, lower right).

- **Special setup** windows are used to edit problemdependent parameter settings like definitions of a test environment in which phenotypes controlled by the evolved networks compete against each other (fig. 3.a).

- **Control** windows interactively control simulation runs. Each run in the

---

[1] Mathematica is a registered trademark of Wolfram Research, Inc.

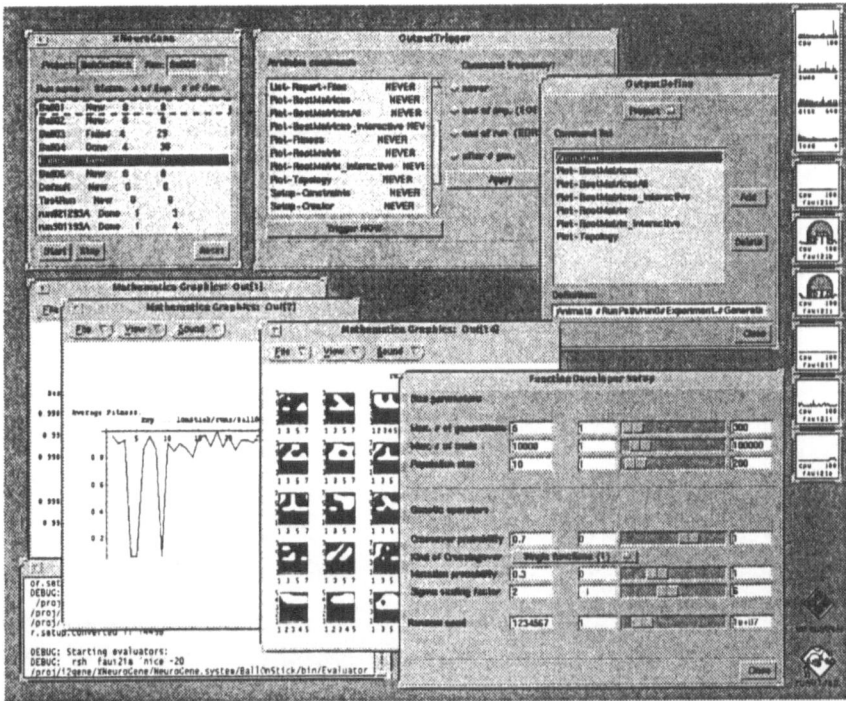

Figure 2: An example *XNeuroGene* session

project list can be started, interrupted and stopped independently; all other dialogs refer to the currently selected experiment from the project list (figure 2, upper left). For each control window there is a kind of console where system messages are directed to (see fig. 2, lower left).

- **Output control trigger** windows control which utility programs should be called at which stage of the simulation process (every nth generation, at the end of a run, after a series of runs etc.). How these external programs and scripts are called can be defined through **output define** windows (fig. 2, upper middle/right).

- **Graphics** windows serve as special visualization tools for different performance measures or developed genotypes and phenotypes. These windows range from simple functional plots of (average, best, worst) fitness values to more complex graphics as e.g. a collection of the best connection matrices evolved so far (fig. 2, lower left quarter). **Animation** windows provide realtime live visualization of evolved phenotypes in their problemspecific environment(fig. 3).

The developed *XNeuroGene* system [3] serves as our basic control interface

---

[3] A first version of this interface has been implemented for NeXTStep which provides an excellent software environment for the development of graphical user interfaces.For more

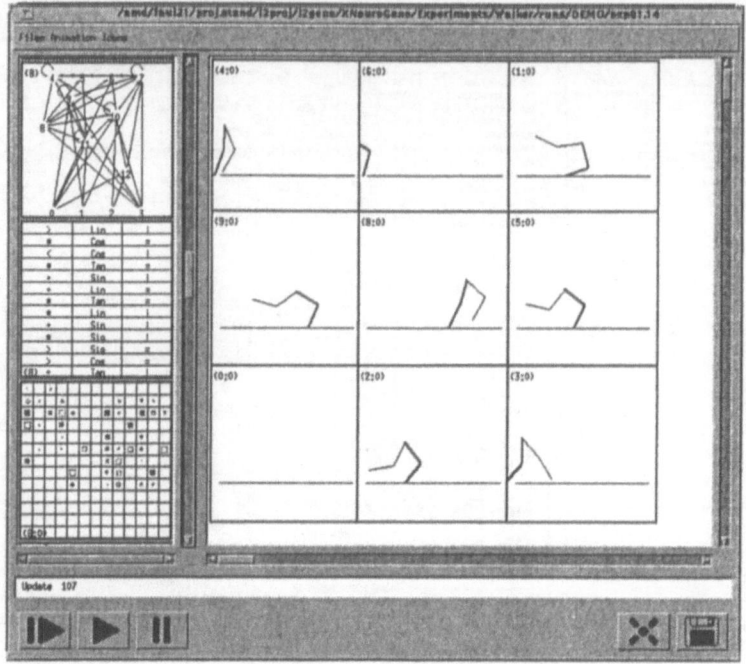

Figure 3: Animation of a population of 'stick walkers' controled by evolved neural nets. The topology, neuron functionality and connection matrix of a selected ANS are shown in the left scroll window.

for an evolution system of neural net topologies, connection weights and neuron functionality. Currently the system gets enhanced by special modules to support interactive evolution and genetic programming techniques [1], [4] and integration of parallel rewrite systems.

References

[1] de Garis, H. (1991), *Genetic Programming: Building Artificial Nervous Systems with Genetically Programmed Neural Network Modules*, in: Soucek, B. (ed.), *Neural and Intelligent Systems Integration*, New York.

[2] Jacob, C., and Rehder, J. (1993), *Evolution of neural net architectures by a hierarchical grammar-based genetic system*. ICNNGA, International Joint Conference on Neural Networks and Genetic Algorithms, Innsbruck, 1993, 72-79.

[3] Jacob, C., and Burghof, A. (1993), *NeXTGene: A graphical user-interface for GENESIS under NeXTStep*. ICNNGA, Innsbruck, 1993, 602-606.

[4] Koza, (1993) *Genetic Programming, On the Programming of Computers by Means of Natural Selection*, MIT Press, London.

information about the NeXT version using GENESIS as the kernel GA system can be found in [ICNNGA2].

INCREMENTAL TRAINING STRATEGIES

Ian Cloete   Jacques Ludik
ian@cs.sun.ac.za   jludik@cs.sun.ac.za
Department of Computer Science, University of Stellenbosch,
Stellenbosch 7600, SOUTH AFRICA

1   Introduction

The popular Back Propagation (BP) algorithm is probably the most widely used learning method for neural network projects. It is simple to implement and has been tested on very diverse applications. However, because it is a gradient descent method it can get stuck in a local minimum, where performance of the network is much worse than that obtained for a global minimum. This problem gets worse for recurrent networks, e.g. the Simple Recurrent Network (SRN) [Elman, 1990], and is aggravated further when the training set is very large.

To solve the problems of a large training set and slow (or no) learning we propose new training schedules called *incremental training strategies*. No attention, except for [Cottrell & Tsung, 1991], has been given to alternative methods for presenting training data to a neural network in order to speed up learning; in contrast, methods to set initial values and adjust parameters of the network during learning were investgiated, e.g. [Silva & Almeida, 1990]. These new incremental strategies improve learning time in simple recurrent networks by more that 50% compared to conventional training on a fixed training set, and almost attain the predicted lower bound for the least number of updates when following an incremental training schedule.

2   Previous Work

We previously proposed *increased complexity training* (*ict*) in which a fixed training set is partitioned into a sequence of disjoint subsets, each of which is "more complex" than the previous. Training occurs on the first subset until a termination criterion is met, then on the first two subsets (merged in random order), and so on. *ict* was compared with *cst* [Cottrell & Tsung, 1991] using both an Elman Recurrent Network and Temporal Autoassociation showing a vast improvement in learning time [Ludik & Cloete, 1993]. However, the complexity measure of the training data depended on the task to be learned (e.g. dividing training patterns into classes depending on the number of output neurons "switched on" by a particular training subset). Since that is a problem-dependent measure which is not suited to every task we propose a new class of training strategies which are independent of this complexity ordering and thus suitable for any BP trained network.

3 Incremental Training Strategies

An incremental training strategy trains a network with succesively enlarged subsets of examples from the training set, called the *fixed* set. The network is trained on an initial subset until an acceptable success criterion is met. Then this subset is augmented by randomly adding more examples. This training process is repeated until the final success criterion is reached on the fixed set. This schedule leaves room for much variation in the size of the initial subset, the increment in subset size, termination criteria for each subsequent subset, and whether training can be terminated earlier without learning on the entire fixed set. This last point is desirable since good generalization may be obtained on a subset of the fixed set. We therefore investigated the following incremental training schedule: As success criterion on a subset of examples the Root Mean Square (RMS) error value is used to take into account that subset sizes increase. The user specifies the desired final RMS error value which signifies successful training, as well as the increment in subset size, i.e. the number of patterns to be randomly added to the training subsets. If an independent test set is available, the user specifies the percentage of test examples which should be correctly given by the fully trained network in feedforward mode. This allows that training be terminated as soon as the network has generalized sufficiently. The initial subset size is the same as the subset increment. The training schedule then consists of two nested loops: The outer loop repeats until successful generalization is obtained on the test set, while the inner loop repeats training on the subset until the required RMS error value for that training subset is obtained. The RMS termination values for each training subset is decremented linearly from that obtained for the initial subset to the final user-specified value, thus gradually requiring stricter termination criteria. This prevents training to a too high accuracy on a small set of examples, which may cause an inferior solution to be selected instead of generalizing sufficiently. When the desired RMS value for a particular subset has been reached, the next subset is constructed, the new RMS value is computed and the process repeated. The number of incremental subsets (and number of RMS decrements) is $S = \left\lceil \dfrac{\text{Number of training examples in fixed set}}{\text{Subset increment}} \right\rceil$

The number of epochs (i.e. training passes over the training set) cannot be used as a measure for convergence "time" for these training strategies since the training sets differ in size; therefore the number of times weights are updated is suggested as a fair performance measure. Let I denote the subset increment. If one assumes that weights are updated after every example, that examples all have length 1, and that training is not repeated on any subset, then the lower bound for the number of weight updates on the entire training set is $IS(S + 1)/2$ since the incremental subset sizes are I, $2I$, $\ldots, SI$. The number of weight updates when an example is a sequence of single patterns is addressed shortly.

4 Experiments

The *addition* task [Cottrell & Tsung, 1991] for sequential processing with a SRN was used for experiments because conventional training on a fixed set for this task has been reported not to converge, and it requires a large number of weight updates. This task is very complex, because not only must the network learn a sequence of inputs, but for each input a different sequence of outputs is required.

The aim is to learn to sequentially add two base four numbers. Each base four number is given a two-digit binary representation. The network input is confined to one column of digits at a time. It has five inputs, four representing the one column of digits and one indicating the end of the input, 16 hidden and 16 state units, and six output units representing the sum (two units) of the one column of digits and the four possible actions (four units). Actions are to write the sum, to remember or output the carry, shift to the next column of digits, and indicate if done. Each single input pattern may generate a varying sequence of outputs. We call such an entire addition sequence a temporal pattern. The input and target patterns for each time step is called a single pattern.

The fixed training set for 3-column addition consisted of 320 temporal patterns of varying lengths. Of the possible 4096 temporal patterns 3700 were used as test set. All network initial conditions for every experiment were indentical. Weights were updated after every single pattern. The final required RMS error value was RMS $\leq$ 0.15, while the success percentage on the test set was 97%.

Five training strategies were compared – *fixed*: conventional training on the fixed set; *ict* and *cst*: previously developed schedules; and two new incremental training strategies denoted by *ist* and *iict*. Incremental Subset Training (*ist*) constructs subsequent incremental training subsets by randomly adding temporal patterns. Incremental Increased Complexity Training (*iict*) also randomly adds temporal patterns from the fixed set, except that additions of one column are exhausted first, then one and two columns, and finally up to three columns. For incremental schedules experiments were repeated with increments of 1, 2, 4, 8, 16 & 32 temporal patterns with corresponding linear RMS decrements.

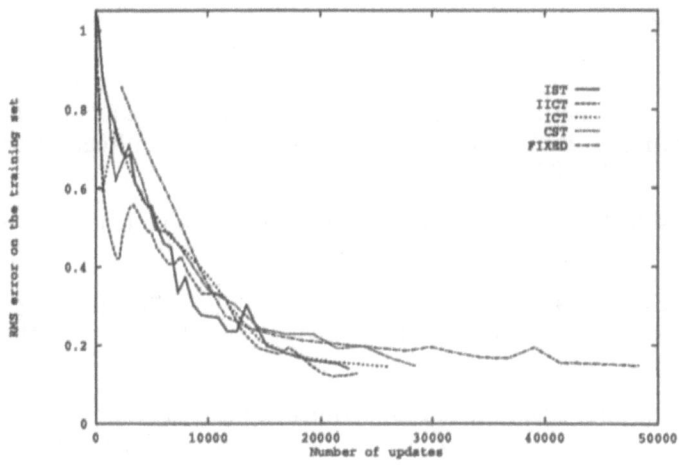

Figure 1: The performance of training schedules for *Addition*

5 Results

ist (with optimum increment of four temporal patterns) improved performance by 53% compared to the fixed set, by 21% compared to *cst*, and by 14% compared to *ict*. *iict* (again with optimum increment of four) improved performance by 52% compared to the fixed set, by 18% compared to *cst*, and 11% compared to *ict*. Figure 1 shows the performance of the different training schedules.

One can expect that the lower bound for the number of updates using *ist* with temporal patterns which vary in length is therefore $LIS(S+1)/2$, where L denotes the average length of the temporal patterns, and I the increment in temporal patterns. For increments 1, 2, 4 and 8 the number of updates required almost attained the optimum lower bound. The number of updates required for *ist* (with increment four) was 22582, whereas the calculated lower bound was 22464 for an average temporal pattern length of 7.2, and requiring only 156 of the 320 temporal patterns. When this set of 156 patterns were used for fixed set training, 45360 updates were necessary compared to the 22582 of *ist* to also reach 97% success on the test set. For increments 16 and 32 more updates were required due to the repetition of their final subsets to reach the desired RMS criteria and generalization.

6 Conclusion

The proposed training schedules improved the convergence rate compared to *cst* and fixed set training. *ist* and *iict* show that a good training set can be quite small to provide very good generalization. *ist*, for example, needed only 156 of the 320 patterns to double the performance of training on a fixed set. It also improved performance by 50% compared to fixed set training on the same subset of the training set.

References

[Cottrell & Tsung, 1991] Cottrell, G.W., and Tsung, F.S. (1991) "Learning Simple Arithmetic Procedures", *High-Level Connectionist Models*, eds. J.A. Barnden, J.B. Pollack, Advances in Connectionist and Neural Computation Theory, 1, 305–321.

[Elman, 1990] Elman, J.L. (1990) "Finding Structure in Time", *Cognitive Science*, 14, 179–211.

[Ludik & Cloete, 1993] Ludik, J., & Cloete, I. (1993) "Training Schedules for Improved Convergence", Int Joint Conf on Neural Networks, Nagoya, Japan.

[Silva & Almeida, 1990] Silva, F.M., Almeida, L.B. (1990) "Speeding up back-propagation", in *Advanced Neural Computers*, ed. Eckmiller, R., Elsevier Science Publishers, Amsterdam.

Modular Object-Oriented Neural Network Simulators and Topology Generalizations

G. Thimm, R. Grau, and E. Fiesler

IDIAP, Case postale 609, CH-1920 Martigny, Switzerland

1 Introduction

In current neural network research, simulation plays a crucial role. Although there is a wide range of neural network simulators available, it is impossible to keep up with the continuous surge of new neural networks and their variations. Consequently, the extensibility and modularity of neural network simulation software is an important issue. Implementation and modification of neural networks and their embedding into an simulation environment should be possible with minimal effort.

Object-oriented programming languages facilitate the fulfillment of these demands and allows the software developer to reuse software models and therefore reduce the overall implementation effort.

OpenSimulator[1] *3.1* and *Sesame*[2] *4.5*, two simulation packages written in *C++*, have the potential of fulfilling the need. Besides ready to use modules, these packages also provide a graphical user interface. Their flexibility is tested using higher order ontogenic neural networks as an example of non standard neural network topologies (see for example [Fiesler-94.1] and [Fiesler-94.2]).

2 Some Neural Network Topology Generalizations

Topologies of higher order ontogenic neural networks differ from standard neural networks topologies in two points. Firstly, the interconnections, which usually connect a source neuron with a sink neuron, are generalized by allowing a connection to combine several inputs by means of a so called *splicing function*, which maps one or more input values to a single output value.

The properties of the used connections characterize the network: the number of inputs of a connection define its order and the maximal order of a connection in a neural network determines the order of the network.

A second generalization allows a variable number of units (neurons or connections), where the number and kind of units to be added or removed is determined by a learning algorithm. This kind of network is called an *ontogenic* neural network.

3 Data Structures for High Order Ontogenic Networks

An intuitive approach for implementing higher order neural networks is to store the weights of connections of order ω in $\omega+1$-dimensional matrices, where each dimension of the matrix either refers to one of the inputs, or to one of the outputs of a connection. Unfortunately, this data structure has some drawbacks, since several matrices have to be reserved for the connections between two layers of neurons, one for each group of connections having the same order and splicing function. These matrices are often largely unused, since ontogenic neural networks are likely to be sparsely connected. Besides this, the numerous matrices complicate both the propagation of values through the network and the updating of weights.

An improved approach introduces only necessary connections, using a projection of the activation values of layer $l-1$ to a variable sized vector V_l. Each vector element stores the result of a splicing function applied to the activation values of layer $l-1$ (see figure). In the implementation, this projection is realized by instances of a class defining connections. Each connection instance has pointers to some elements of the output vector of layer $l-1$, as well as to an element of vector V_l. Whenever such a connection instance is called, it applies a splicing function to the values indicated as its inputs and the outcome is written to result vector V_l. This vector provides the inputs for a variable sized perceptron. Only if the ith connection (writing its output to the ith element of vector V_l) is expected to take part in the input for the jth neuron, the element in the ith column and the jth row of the weight matrix is allowed to be non-zero. If a connection is introduced, and another connection with

---

[1] *OpenSimulator* is copyrighted by the ETH Zürich.
[2] *Sesame* is copyrighted by the GMD Schloß Birlinghoven.

748

the same splicing function, order, and combination of inputs exists and writes to the
ith element of the vector V_l, the corresponding entry in the weight matrix is allowed to
become non-zero. Otherwise, the size of the weight matrix is increased by a column,
the size of the vector V_l by an element, and a new instance of the connection class is
initialized.

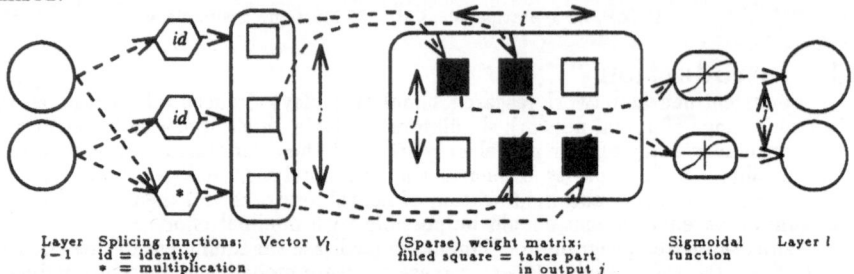

Layer   Splicing functions;   Vector V_l (Sparse) weight matrix;   Sigmoidal   Layer l
$l-1$ id = identity   filled square = takes part   function
      * = multiplication   in output j

The second approach is more efficient, both in memory and execution time. Furthermore, the vector V_l can be regarded as an input vector for a standard perceptron
layer, which implies only minor modifications to a Perceptron layer implementation.

4   The Neural Network Simulators

4.1   *OpenSimulator*

The *OpenSimulator* software package was developed by J.-F. Leber at the Institute
for Signal and Information Processing of the ETH Zürich [Leber-92] [Leber-93]. The
proliferation of the software is restricted. *OpenSimulator 3.1* requires *UNIX*[3], *XView*
and *X-Windows*[4]. The user can enter commands using a menu based interactive
graphical interface or a command interpreter, or restore a previously saved session,
since all data, such as the network topology, the patterns, and the weight matrices,
are saved in files. *OpenSimulator* allows a group of users to share a common version
of the simulator and to generate an extended version in the user's private workspace.

The functions of *OpenSimulator* are divided into two categories: functions of
the *simulation kernel* and functions of their *interactive counterparts*. Both have to be
created in order to introduce a new building block for a neural network. This creation
is supported by a menu, which allows the creation of templates for all necessary files,
to which the user is supposed to add code. In the source code, macros are used to
add variables, replace or add functions, add menus, and make plots. New macros
can easily be defined by the user and neural network topologies can either be defined
using the graphical interface or by a file containing calls to *C++* functions.

For the construction of neural networks the following tools are available:
- matrix libraries for bytes, shorts, integers, floats, doubles, and complex data types,
- graphics libraries, and
- a basic layer class for backpropagation training, a perceptron layer class, and a topology-
  preserving feature map class.

4.2   *Sesame*

Sesame was developed by A. Linden and Ch. Tietz at GMD Schloß Birlinghoven in
Germany [Linden-92] and is available as a freeware software package[5]. *Sesame*[6] 4.1
requires *UNIX*, *InterViews*[7], *X-Windows*, and several GNU tools and libraries. The
user interface is based on a command interpreter and the current state of a simulation
can be stored. *Sesame* is based on modules which can contain anything ranging from
a simple counter or a complex neural network to a graphical display. New modules
may be assembled from existing modules or designed in *C++*. In the latter case the
export of variables or command names to the command interpreter or the declaration
of *typed communication sites* is done by function calls and their handling is completely

---

[3] *UNIX* is a registered trademark of AT&T.

[4] *XView* and *X-Windows* are registered trademarks of the MIT.

[5] *Sesame* is at least available via ftp at the anonymous ftp site ftp.gmd.de.

[6] Meanwhile version 4.5 has been released. This version has some more features but also some
minor incompatibilities with version 4.1.

[7] *InterViews* is copyrighted by the Stanford University.

covered by their classes. This includes visualization of parameters in the user interface and saving in files for future experiments.

A simulation is defined by a sequence of interpreted commands (typed or read from files) which organize the static and dynamic data flow. The static data flow is described using connection commands, whereas the dynamic data flow is described by one or more procedures, which may contain calls to other procedures, calls to modules, or simple control structures. Assembly of these commands and procedures can be done in the command interpreter, and saved together with the current state of modules, in intelligible files.

The existing modules supply the following features:

- available neural network models: one and two hidden layer backpropagation, Kohonen's self organizing feature map, and Kanerva's sparse distributed memory,
- basic modules for neural networks, like neuron layers for backpropagation or radial basis function neural networks,
- modules for data handling like vector comparison, statistical analysis, and graphical displays,
- vector/matrix handling (only completely implemented for double floating point), and
- list, stack, pattern, and file handling modules.

4.3 Comparison of the Simulators

In the table below, some of the main differences between the two simulators are summarized, and marked as advantages ($\oplus$) or disadvantages ($\ominus$).

Sesame 4.5	OpenSimulator 3.1
$\oplus$ Provides on-line help functions, has good documentation, and a tutorial.	$\ominus$ Provides limited on-line help and user documentation (tutorial only in German).
$\ominus$ Has no interactive graphical user interface, therefore the interactive construction of experiments, must be done with the command interpreter.	$\oplus$ Has an interactive graphical surface and a command interpreter.
$\oplus$ Is easy to understand.	$\oplus$ Its simulation kernel and interactive counterpart are clearly distinct.
$\oplus$ Its source is easy to maintain and extend.	
$\oplus$ Is very flexible in combining basic modules to a complete simulation.	$\ominus$ Its graphical interface is mainly based on a vast hierarchy of menus. Often, one has to open a confusing amount of menus to get a particular piece of information or to set parameter.
$\ominus$ Its functionalities are highly divided up into small modules; this causes sometimes confusion in the conception of an experiment.	
	$\oplus$ Is based on shared libraries.

4.3.1 Remark: *OpenSimulator 4.0* has been announced

For December 1993, J.-F. Leber and A. Jarosch have announced version 4.0 of OpenSimulator, wherein the objects of a simulation can be distributed over a network containing different computer architectures, as the communication is quick and easy. Computationally expensive algorithms can therefore be executed efficiently on a super computer while the graphical interaction takes place on a workstation. Moreover, the programmer of a new module does not need to concern himself with any graphics, as they are generated on-line from an automatically generated ASCII description. The algorithms are thus ideally separated from the graphics. OpenSimulator 4.0 is more modular and flexible, simpler to use, and easier to maintain.

5 The Implementation

The first approach to implement higher order ontogenic neural networks was done using *OpenSimulator*. For the reasons given in section 3, the authors later switched to the second approach, which was implemented using *Sesame*.

5.1 Extending *OpenSimulator*

For a neural network implementation, *OpenSimulator* offers classes for neuron layers with input and output handling and optional forward, backward, and lateral feedback weights. It also provides different sigmoidal functions and functions controlling the training and testing phase.

For the implementation of higher dimensional matrices it is not appropriate to inherit from the available classes for two dimensional matrices, since these classes are highly specialized for other purposes. Also, the matrix class definition may cause some confusion, as they are distributed over a "kernel" file and other header files. This is done with the aim of realizing a kind of C++-template concept (which was not available when the library was written). The most specialized classes for the implementation of a neural network in *OpenSimulator* represent layers with optional weight matrices, which can be used as models for a first implementation approach to higher order ontogenic neural networks. Although these classes do have to be rewritten, the implementation does profit from existing classes through inheritance.

Changes to *OpenSimulator* are usually spread out over the software package: both the functions in the simulation kernel and the macros used in the simulation description files have to be changed (compare paragraph 4.3.1). In the interactive counterpart, facilities for handling the new features need to be created.

For an implementation of the second approach, the above mentioned modifications as well as the problems with using the existing matrix implementation are similar.

5.2 Extending *Sesame*

Due to the design of *Sesame*, a class representing a high order ontogenic neural network needed to be written, since no appropriate module exists. The new module reuses code of a backpropagation network class, which provides a good model. The variable size weight matrix and the vector V are realized by an oversized matrix and vector respectively. A more elaborate implementation requires the extension to variable sized subclasses, which is easy. A minor problem in the implementation is the access of array elements via addresses, which was not foreseen in the design.

All changes to the *Sesame* code are confined to the new class module (except for three lines to let *Sesame* know about the existence of the class). The concepts for communication, matrix handling, and other basic concepts are clear and simple to use, and the high order neural network class is therefore easily embedded.

Regarding the first approach, similar statements can be asserted: *Sesame* has no matrix class for higher dimensional matrices but offers a suitable basic matrix class.

6 Conclusions

Both the *OpenSimulator* and the *Sesame* software packages are very useful tools for the simulation of neural networks. Besides that both simulators have the (hopefully temporary) disadvantage of still being under development, they have other characteristics in common: the functionalities presented for the implementation of higher order and ontogenic neural networks are similar. As both software packages are written in *C++*, new code can often be easily generated using the inheritance mechanism.

The overall concept of *Sesame* is clearer and the method of modular and object oriented programming is better fulfilled, and therefore the implementation of a neural network and embedding in the simulator is simpler (a revised version of *OpenSimulator* has been announced, see paragraph 4.3.1). This can be seen from the fact that changes to *Sesame* are localized in the new module, whereas *OpenSimulator* requires extensions innon-neural networks parts. However, only *OpenSimulator* allows the manipulation of the neural networks through a graphical interface. It should also be noted that, *Sesame* and *OpenSimulator* can be considered generic simulator construction tools with special features for neural networks.

Acknowledgements

The authors would like to thank Jean-François Leber at the Institute for Signal and Information Processing of the ETH Zürich and Alexander Linden and Thomas Sudbrak at the GMD in Schloß Birlinghoven and their colleagues for their swift responses to questions, helpful discussions, and their overall support of the software.

References

[Fiesler-94.1] E. Fiesler (1994), **Neural Network Classification and Formalization**, accepted by *Computer Standards & Interfaces*, 16, special issue on Neural Network Standards, J. Fulcher (ed.), North-Holland/Elsevier, Amsterdam, The Netherlands.

[Fiesler-94.2] E. Fiesler (1994), **Comparative Bibliography of Ontogenic Neural Networks**, in these proceedings (ICANN'94).

[Leber-92] J.-F. Leber and G. S. Moschytz (1992), **An Acoustical Signal Recognizer Implemented on a Novel Interactive Object-Oriented Neural Network Simulator**, in I. Alexander and J. Taylor (ed.), *Artificial Neural Networks*, 2, pp. 1291–1294, North-Holland/Elsevier, Amsterdam, The Netherlands, Sep. 4–7.

[Leber-93] J.-F. Leber (1993), **The Recognition of Acoustical Signals Using Neural Networks and an Open Simulator**. Series in Microelectronics, 20, W. Fichtner, W. Guggenbühl, H. Melchior, and G. S. Moschytz (ed.), Hartung-Gorre Verlag, Konstanz, Germany.

[Linden-92] A. Linden and C. Tietz (1992). **SESAME - A Software Environment for Combining Multiple Neural Network Paradigms and Applications**, in I. Alexander and J. Taylor (ed.), *Artificial Neural Networks*, 2, pp. 1265–1268, North-Holland/Elsevier, Amsterdam, The Netherlands, Sep. 4–7.

Gradient-Based Adaptation of Network Structure

Bert de Vries

David Sarnoff Research Center and
the National Information Display Laboratory
CN 5300, Princeton, NJ 08543-5300
bdevries@sarnoff.com

Abstract

In this paper I introduce a technique which effectively performs like gradient-based optimization of the network dimension. This method directly applies to adaptation of the order of a tapped delay line.

1   Introduction

Effective application of neural networks not only requires adaptation of the network coefficients (weights) but also of the *number* of weights. The generalization performance of a neural net generally peaks for a fixed number of weights, that is, the performance degrades considerably for less or more than the optimal number of weights. Similar comments could be made for the number of nodes in a network.

There is an important difference between the adaptation of the weights and the adaptation of the network structure as implemented by typical constructive/pruning algorithms. In general, weight updates make use of the gradient ($\partial E / \partial w$) of the cost E with respect to the weights. Generally this gradient is directly computable from the network architecture (by a backpropagation algorithm for instance). However, such a gradient cannot be computed with respect to the dimensionality m of the weight vector (or the number of nodes), that is, $\partial E / \partial m$ does not exist. The reason that $\partial E / \partial m$ does not exist is that m is an integer.

Typical constructive/pruning algorithms perform *local search*: For a network of size m, we evaluate the costs $E(m-1)$, $E(m)$ and $E(m+1)$. If $E(m-1)$ or $E(m+1)$ performs better than $E(m)$, we switch to the better network (either $m \leftarrow m - 1$ or $m \leftarrow m + 1$) and continue this process until $E(m)$ is optimal (better than $E(m-1)$ and $E(m+1)$). In the following, I will label network structure altering algorithms of this type as *local search algorithms*.

Empirical results with many versions of local search in the network structure have been generally good. However a problem is that the criterion for deciding how and when to add or remove nodes/weights is heuristic and therefore very little can be said about the general applicability of any method. Moreover, training with local search generally requires the network to be trained until convergence before any evaluation regarding network dimension can be made. This in contrast again to gradient-based weight update procedures which adapt the weights without decision heuristic and continuously.

In this paper I make an attempt to extend the obvious advantages of (continuous space) gradient search over (discrete space) local search to network structure adaptation. The first problem that I choose concerns the order of a tapped delay line (that is, the number of taps). I am still several steps away from distilling the common ingredients of the algorithms for different applications in order to create a general applicable procedure.

2   Gradient-Based Adaptation of Delay-Line Order

The problem is to find the weights and order K which minimizes E for the following system of equations (an FIR filter):

$$x_0(t) = 1, \; x_1(t) = u(t), \; x_k(t) = x_{k-1}(t-1) \text{ for } k=2,\dots,K \qquad (1a)$$

$$y(t) = \sum_{k=0}^{K} w_k x_k(t), \; E = \sum_{t=1}^{T} (d(t) - y(t))^2 = \sum_{t=1}^{T} e^2(t). \qquad (1b)$$

For gradient-based adaptation, we need to compute $\partial E / \partial w$ and $\partial E / \partial K$. $\partial E / \partial w$ is easily computable and evaluates to $\partial E / \partial w_k = -2\sum_t e(t) x_k(t)$. $\partial E / \partial K$ however does not exist, because K is an integer. The strategy will be to introduce a real parameter β (beta) that effectively behaves as the order of the delay line, but as β is real, $\partial E / \partial \beta$ is computable. The order of the delay line is than effectively adapted by gradient descent with respect to β.

The algorithm proceeds as follows. Replace (1d) by

$$y(t) = \sum_{k=0}^{K} w_k x_k(t) s(k), \qquad (2)$$

where $s(k)$ is a saturation function of a *real* variable k which is parametrized by α and β. For small k ($k < \beta-\alpha$), $s(k)=1$ and thus the contribution of the kth tap to $y(t)$ is $w_k x_k(t)$. For large k ($k > \beta+\alpha$), $s(k)=0$, and effectively this tap does not contribute to $y(t)$. For $\beta-\alpha <= k <= \beta+\alpha$, $s(k)$ gradually falls from 1 to 0 and thus the contribution of $w_k x_k(t)$ for these taps are weighted. This scheme is shown in Figure 2. The effect is that the end of the delay line is "soft". If β increases the effective order of the delay line gets larger. Thus we can adapt the effective order by gradient-based adaptation of β. In the following, this algorithm will be referred to as SOFTORDER.

We could use a sigmoidal function for $s(.)$, but for computational simplicity I have used the following function:

$$s(k) = \begin{cases} 1 & \text{for } k < \beta - \alpha \\ (\beta - k)/(2\alpha) + 0.5 & \text{for } \beta - \alpha \leq k \leq \beta + \alpha \\ 0 & \text{for } k > \beta + \alpha \end{cases} \qquad (3)$$

which leads to $\quad \dfrac{\partial E}{\partial \beta} = -2\sum_t e(t) \sum_k w_k x_k(t) \dfrac{\partial s(k)}{\partial \beta},\quad$ where

$\dfrac{\partial s(k)}{\partial \beta} = \begin{cases} 1/(2\alpha) & \text{for } \beta - \alpha \leq k \leq \beta + \alpha \\ 0 & \text{else} \end{cases}$. Thus,

$$\frac{\partial E}{\partial \beta} = -\frac{1}{\alpha}\sum_{t}e(t)\sum_{k=ceil(\beta-\alpha)}^{floor(\beta+\alpha)} w_k x_k(t), \qquad (4)$$

where *floor(x)* and *ceil(x)* round to the nearest integer smaller and larger than x respectively. Equation (6) allows an easy interpretation if we assume gradient descent updating for β ($\Delta\beta \sim -\partial E/\partial\beta$). Assume that the contribution to $y(t)$ of the "soft end" of the delay line, $\sum_{k=ceil(\beta-\alpha)}^{floor(\beta+\alpha)} w_k x_k(t)$, is positive (>0). Also assume that $e(t)>0$. Then according to (6), $\partial E(t)/\partial\beta < 0$ and β will increase (by gradient descent). If β increases the signal contribution of the soft end of the delay line will also increase and therefore $y(t)$ will increase. This is exactly what we want because we assumed that $e(t)>0$ or $y(t)<d(t)$. Similar arguments hold for other sign combinations.

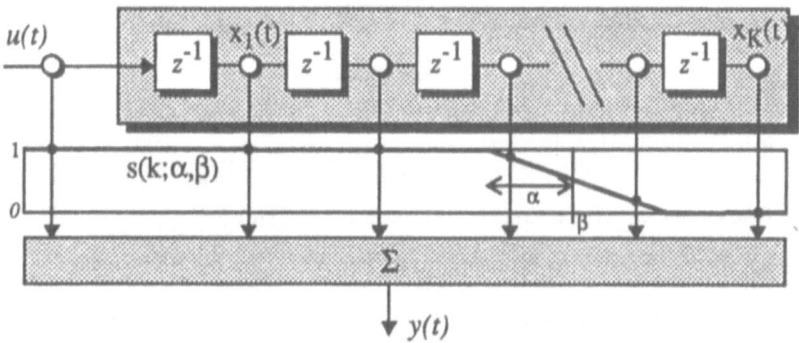

Figure 1. The "soft terminating" tapped delay line.

The physical interpretation of β is clearly the effective order (number of taps) of the delay line. α determines how soft the delay line terminates and establishes the equivalent of a search neighborhood in relation to local search algorithms.

3   Experiment

A simple system identification experiment was run to validate the algorithm described in the previous section. The input data set consists of 1000 samples from an N(0,1) distributed random number generator. The plant model is

$$y(t) = u(t) + u(t-1) - u(t-2) + u(t-3) + u(t-4) - u(t-5) + u(t-6).$$

The neural net model was also an FIR filter as described by equations (1a)--(1e).

The results are displayed in Figure 3. The cost converged to zero while β settled at 8. I used a search neighborhood $\alpha=1$, and hence the network converged to delay line of order $\beta-\alpha=7$, which is optimal. α determines the range of the search neighborhood. For $\alpha=0.5$, $\partial s/\partial\beta$ only has a value (unequal to zero) for $\beta - 0.5 \leq k \leq \beta + 0.5$. In this case, the "soft" end of the delay line consists of one tap only. The SOFTORDER algorithm reduces to a simple heuristic: if the last tap k (determined by $\beta - 0.5 \leq k \leq \beta + 0.5$) reduces the cost of the network, then increase β. If the last tap has a negative effect (cost increases), then decrease β.

When α gets larger than 0.5, more than one tap is involved in deciding how to

754

adapt β. This is interesting if we are modeling systems with non-uniform delay: for larger α it becomes possible to look beyond one tap. The possibility to determine a search range by setting an appropriate α is clearly an advantage of SOFTORDER over most constructive/pruning algorithms. Particular schemes of starting with large search neighborhood and reducing α over the course of training are conceivable.

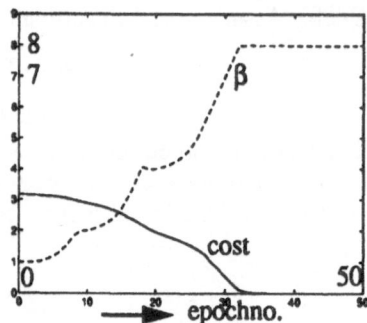

Figure 2. Results for the FIR system identification experiment with SOFTORDER. Experimental parameters: NDATA is 1000 samples, UpdateInterval = 200 samples, $w_{init} = N(0,.1)$, $\beta_{init} = 1.0$, $K_{max} = 10$, $\alpha=1.0$, the function s(k) as given in equation (6), cost $= (1/2N) \sum_{n=1}^{N} e^2(n)$. The cost and β are plotted as a function of epoch number. Note that β converges at 8 which means a delay line order of $\beta-\alpha = 7$, which is optimal.

4   Discussion

The goal of this research is to adapt complexity or effective degrees of freedom of a network along with the values of the weights. A large variety of constructive and pruning algorithms have been proposed and the sheer amount of the different algorithms reflects the volatile state of the art. Although several constructive and pruning algorithms have shown very encouraging results, the crucial step in a constructive algorithm (when to add or remove a node or weight) is typically based on heuristic grounds. This is the "bottleneck" step for any constructive algorithm and it will be hard to find a general applicable criterion.

Gradient-based adaptation does not require such a criterion. In a continuous space we do not have to make decisions regarding adding or removing one unit. Thus, I have tried to move gradient computation to the dimension of the network. The main idea is that we introduce real variables which effectively serve as the dimension of a network structure.

Acknowledgments

This research was supported by the Advanced Projects Agency of the Department of Defense and was monitored by the Air Force Office of Scientific Research under Contract No. F49620-92-C-0072. The United States Government is authorized to reproduce and distribute reprints for governmental purposes notwithstanding any copyright notation hereon.

A connectionist model using MULTIPLEXED OSCILLATIONS and SYNCHRONY to enable dynamic connections

J.-C. Martin

Laboratoire d'Informatique et de Mécanique pour les Sciences
de l'Ingénieur (LIMSI-CNRS), B.P. 133, 91403 Orsay Cedex, FRANCE
Ecole Nationale Supérieure des Télecommunications, Paris, FRANCE

1   Introduction

The final goal of our work is to build a system for multimodal man-machine communication. Processing multimodal data means taking into account a large number of possible combinations of features obtained through several modalities. In a connectionist system, a given combination can be represented by a pattern of connections converging on a single unit (Barlow, 1972).

Another possibility is to add a new coding dimension to those existing in connectionist systems and implement mechanisms enabling variable manipulation. Several solutions have already been proposed to solve the problem of variable manipulation in connectionist systems (Feldman, 1982 ; Touretzky, 1990). A possible solution consists in an evolving architecture in wich units and links can be added when necessary to memorize links between variables and values (Blanchet, 1992).

Instead of modifying the network structure, the internal signals of the network can be considered as propagating variables, provided that they can be distinguished from each other using their specific content (Béroule, 1990). For instance, the fine temporal structure of these signals has been proposed for dynamically bind units (von der Malsburg 1981). An example using this code is the model of propagation of role and value binding for inferences proposed in (Ajjanagade and Shastri, 1991).

In this paper, we present a model of variable and value binding through synchrony. In this model, the signals emitted by dynamically bound units are synchronized and all signals are multiplexed in a single unit. Complex assignment concerning several values or several variables can be achieved. The architecture uses a small number of units and links. The example chosen is the classical variable manipulation such as in symbolic programming languages.

2   An architecture dedicated to multiplexing

One of the interesting features of synchrony is the capacity to multiplex independant signals. In our model, each variable-unit is linked unidirectionally to a multiplexing unit (called MPX) and each value-unit is linked bidirectionally to MPX (Fig. 1). Thus, it uses only $nb\_var + 2 * nb\_val$ links, much less than

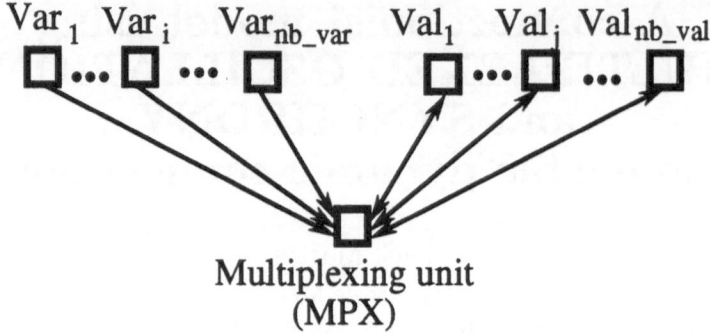

$$\text{Var}_1 \quad \text{Var}_i \quad \text{Var}_{\text{nb\_var}} \quad \text{Val}_1 \quad \text{Val}_j \quad \text{Val}_{\text{nb\_val}}$$

Multiplexing unit
(MPX)

Figure 1: *Architecture of the model of variable manipulation: signals sent by variables and values are multiplexed in a single MPX unit.*

a solution connecting each variable to each value wich needs $nb\_var * nb\_val$ links.

3 Distinct variables have distinct phases

A specific integer called "phase" is associated to each variable-unit. Each unit computes its input activity from the onset and offset of its input signals. As in the McCulloch & Pitts formal neuron, a unit sends a signal when its input activity reaches its threshold. A number called "phase" can be associated to each unit. This phase is used to compute the times at which this unit, if activated above its threshold, will send a signal. Distinct phases are associated to distinct variables, in a static way. Thus, information concerning distinct variables can be multiplexed without crosstalk (Fig. 2). Finally, MPX can dynamically associate a phase to values during a certain interval of time.

4 Algorithm

Three steps can be distinguished in a simple assignment such as $Var := Val$:

ASSIGNMENT: Var and Val must be simultaneously activated by external signals coming, for instance, from another network. Var has an associated phase and sends a signal at this phase to MPX. Val has no specific phase and thus sends a signal during a whole cycle (Fig. 2). MPX is activated by the time coincidence of both signals at the phase associated to Var. This phase is then associated to Val by MPX.

MEMORIZATION: the result of the assignment is a periodic signal sent by Val to MPX at the phase associated to Var.

CONSULTATION: Var is activated by an external signal and sends a signal to MPX. MPX is activated by the time coincidence of this signal and one of the signals sent by Val. Then MPX activates Val.

Since signals associated to distinct variables have distinct phases, several assignments such as $Var_1 := Val_4$ and $Var_2 := Val_5$ can be memorized in

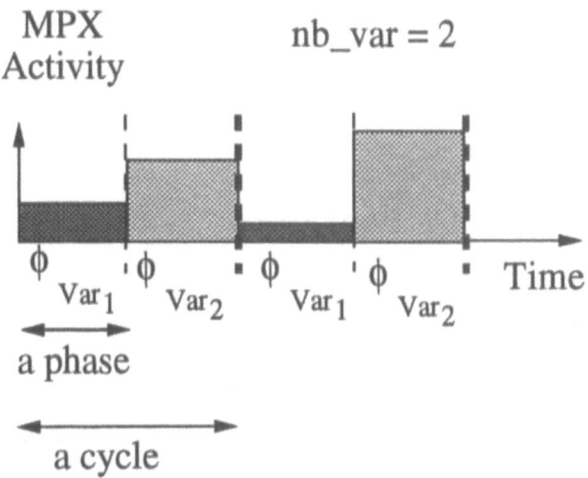

Figure 2: *Time is divided into periodic cycles. Each cycle contains nb_var phases. Signals concerning distinct variables are multiplexed in MPX without crosstalk.*

MPX without crosstalk: the signals memorizing the different assignments reach MPX at different times and do not overlap.

Several values can be assigned to a single variable (for instance, $Var_2 := \{Val_5, Val_7\}$). The result of the assignment is a periodic signal sent by each value at the phase associated to the variable.

A single value can be assigned to several variables (for instance, $Var_1 := Val_4$ and $Var_2 := Val_4$). The assignments cannot begin exactly at the same time, since the criterium used to bind a variable and a value is the time coincidence of the signals emitted by both units. If a value has been assigned to several variables, it will send signals at several phases without overlapping.

5 Conclusion

We have proposed a connectionnist system using multiplexed oscillations and synchrony to enable dynamic connections with the use of a small number of units and links. The problem of waiting a whole cycle (wich can be long if there are many variables) to get the value of a variable will have to be tackled. The integration of this model of variable manipulation is currently under way within a Guided Propagation Network for Man-Machine Communication purpose (Martin, Béroule, 1993a).

This model has been inspired by neurobiological hypotheses about the implementation of representation in the brain with the use of synchrony (Singer, 1993, Frégnac, 1991): entities could be represented in the brain by neural assemblies dynamically created by synchronization of several neurons (von der Malsburg, 1981), stimulus dependant synchronization of oscillations have been detected in several areas of the brain (Engel et al., 1992), synchrony could be achieved using feedback connections from convergence zones (Damasio, 1989), a neuron belonging to several networks may dynamically select one of them and oscillate with the selected network (Hooper and Moulin, 1989). This neurobiological inspiration is discussed in (Martin, 1993).

6 References

Ajjnagadde, V., Shastri, L. (1991) Rules and variables in neural nets. Neural computation 3, 121-134.

Barlow, H.B. (1972) Single units and sensation : A neuron doctrine for perceptual psychology ? Perception, 1, 371-394.

Béroule, D. (1990) Guided propagation : current state of theory and application. Neurocomputing, NATO ASI Series, Vol. F 68, 241-260. Edited by F. Fogelman Soulie and J. Herault, Springer Verlag Berlin Heidelberg.

Blanchet, P. (1992) Une architecture connexionniste pour l'apprentissage par l'expérience et la représentation des connaissances. Thèse de doctorat en Sciences de l'Université Paris XI, 8 Décembre.

Damasio A.R. (1989) The brain binds entities and events by multiregional activation from convergence zones. Neural Computation, 1, 123-132.

Engel, A.K., König, P., Kreiter, A.K., Schillen, T.B., Singer, W. (1992) Temporal coding in the visual cortex : new vistas on integration in the nervous system. TINS, vol15, n6.

Feldman, J.A. (1982) Dynamic connections in neural networks, Biol. Cybernet. 46, 27-39.

Frégnac, Y. (1991) How many cycles make an oscillation. In "Representations of Vision : Trends and Tacit Assumptions". Eds. Gorea, A., Frégnac, Y., Kapoula, Z., Finlay, J. Cambridge University Press, 97-109.

Hooper, S.L., Moulin, M. (1989) Switching of a neuron from one network to another by sensory-induced changes in membrane properties. Science, 244, 1587-1589.

Martin, J.-C., Béroule, D. (1993a). Trends in Human-Machine Multi-modal Interaction. In "Non-Visual Human-Computer Interactions", Eds D. Burger, J.-C. Sperandio. Colloque INSERM/John Libbey Eurotext Ltd., vol. 228, 145-166.

Martin, J.-C., Béroule, D. (1993b) Symbolic Variable Manipulations in a Connectionist Model Using Multiplexed Oscillations and Synchrony. "Temporal Coding in the Brain", IPSEN Foundation, Paris, France, 11 Oct.

Martin, J.-C. (1993) Interaction Homme-Machine Multimodale et Neurobiologie de la Perception. Colloque Interdisciplinaire du Comité National du C.N.R.S., "Images et Langages : Multimodalité et Modélisation Cognitive", Paris, 1er et 2 Avril.

Singer, W. (1993) Synchronization of cortical activity and its putative role in information processing and learning. In Annu. Rev. Physiol. 1993, 55:349-374.

Touretzki, D.S. (1990) BoltzCONS : Dynamic Symbol Structures in a Connectionist Network. Artificial Intelligence, 46 (1-2), 5-46.

von der Malsburg, C. (1981) The correlation Theory of Brain Function (Internal Report 81-2), Max-Planck-Institute for Biophysical Chemistry.

Some Results on Correlation Dimension of Time Series Generated by a Network of Phase Oscillators

R. Borisyuk†, A. Casaleggio‡, Y. Kazanovich†, G. Morgavi‡

† Inst. of Mathematical Problems of Biology, Russian Academy of Sciences, Pushchino, Moscow Region, 142292, Russia

‡ Istituto per i Circuiti Electronici, Consiglio Nazionale delle Ricerche, Via De Marini 6, 16149, Genova, Italy

1   Introduction

In the last years a number of investigations have been focused on applying chaotic time series analyses to investigate cortical activity recorded in electroencephalograms (EEG). It has been shown that some modes of brain activity can be characterized as low-dimensional chaos (Babloyantz, 1989). Several models of EEG rhythm generation have been proposed to reflect the main spatio-temporal characteristics of the EEG (Wright & Kydd, 1992). The important feature of this modeling is the capability to interpret the EEG parameters in terms of dynamical regimes of the network.

In most papers the modeling have been limited to spatially homogeneous networks. This contradicts to the concept which is now shared by some researchers of the brain that the activity of cortical columns may be governed by the central element whose role can be played by some subcortical structure. This direction of EEG modeling is represented by the paper (Destexhe & Babloyantz, 1991) where the authors investigate the influence of an oscillatory input from thalamic region on the synchronization of oscillations in a cortical tissue.

Our approach to this problem is also based on the network with the central element. It differs from the one of Destexhe and Babloyantz (1991) in the following points. First, we simulate the EEG by a network of phase oscillators in the regime of partial synchronization when only some part of cortical oscillators are synchronized by the central oscillator. This originates from the fact that our network have been developed as a model of attention in which the oscillators working synchronously with the central oscillator (the hippocampus) have been interpreted as *the attention focus* while other oscillators represent other stimuli (Kryukov, 1991). Second, the architecture of the network implies both forward and backward connections which radically changes the interaction between the cortex and subcortical central structure.

The main result of our studies is that the correlation dimension (D_2) of the simulated EEG can be used as an estimation of the number K of the groups of cortical oscillators working at different frequencies. This is an important feature of the attention system because it shows the diversity of stimuli which are out of *the attention focus*. We have also found that the network can show both quasyperiodic or chaotic behavior and that the chaotic behavior leads to a relatively small underestimation of K by D_2.

The results obtained can be considered as a first step in solving the problem of identification of a dynamical system underlying the EEG which will enable the detailed analysis of its dynamical behavior.

2 Model description and computation results

To model the cortico-hippocampal interplay, we use a network of phase oscillators which consists of the central oscillator (CO) and N peripheral oscillators (PO) and which is described by the following system of ODEs:

$$\begin{cases} \dot{\theta}_0 & = & \omega_0 + A \sum_{i=1}^{N} sin(\theta_i - \theta_0) \\ \dot{\theta}_i & = & \omega_i + Bsin(\theta_0 - \theta_i) \end{cases} \tag{1}$$

where θ_0 is the phase of the CO, θ_i are the phases of POs, ω_0, ω_i are the natural frequencies of the oscillators, A and B are coupling strengths, $A, B \geq 0$. For simplicity, no coupling is assumed between POs..

The EEG is simulated as the signal represented by a spatial average

$$U(t) = \frac{1}{N} \sum_{i=1}^{N} sin\theta_i.$$

After a time series U(t) is computed, we forget for a moment that we have the full description of the model. We consider U(t) as the only source of information which can be used to characterize the degree of temporal coherence of the network, thus imitating the conditions in which a researcher of the EEG have to work. At this stage one can apply to U(t) usual techniques of nonlinear time series analysis (Lin, 1989) such as phase portraits, Lyapunov exponents and correlation dimensions to extract dynamical parameters of the network. Then these parameters can be compared with those built in the model a priori or obtained in the direct analysis of its dynamical behavior.

There are two extreme cases in the dynamics of the network. If the connections between the CO and POs are eliminated (A=B=0), the POs oscillate independently with frequencies ω_i. If there are no multiple frequencies among ω_i, the dynamical regime of POs is quasyperiodic oscillations on the N-dimensional torus. Another extreme case is represented by periodic oscillations which correspond to the synchronization of the whole network (this happens when coupling is strong enough). In this case the differences $\theta_0 - \theta_i$ are constant for all POs.

The partial synchronization lies somewhere between these two extreme modes. In partial synchronization the differences $\theta_0 - \theta_i$ are limited for some POs (which are said to be partially synchronous with the CO) and are not limited for the others. During partial synchronization the network can show quasyperiodic or chaotic oscillations of POs of dimensions between 1 and N.

For detailed computer analysis of dynamical behavior, we have chosen a network of 8 oscillators whose natural frequencies are represented in Fig.1.

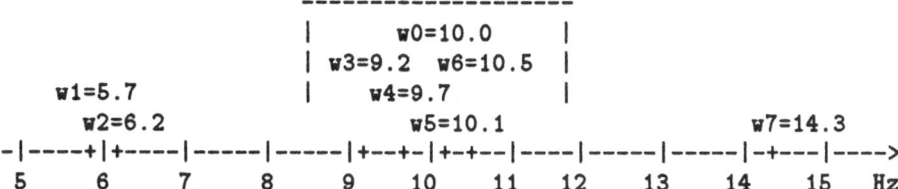

Figure 1: *Natural frequencies of the network of oscillators (w0 ≡ ω₀, wi ≡ ωᵢ).*
The window indicates the oscillators which run synchronously when B=1,2,3.

This choice of natural frequencies has not been conditioned by any biological reasons. We only wished to avoid multiple frequencies which could bring unnecessary complications in the results and to operate with not very small nor too big frequency values. For the values of coupling strengths which have been used in our experiments, the oscillators from 3 to 6 have been partially synchronous with the CO while other oscillators have been out of synchronization.

The computation of Lyapunov exponents has shown that for B=1,2,3 and $0 \leq A \leq 8$ the network demonstrates quasiperiodic oscillations. The chaotic behavior can be clearly seen when B=3 and $A \geq 16$. The border between these two modes is not very sharp. There is a region where a small variation of A can transfer the system from one mode to the other.

The correlation dimension D_2 of a time series is known to be strictly related to the number of degrees of freedom of the system which produced this series. Therefore, we expected that D_2 estimation of a time series U(t) must be about 4. Our intuition has been based on the simple case when the coupling of the CO with the oscillators 1,2 and 7 is eliminated. In this case we have 3 independent oscillators and one group of synchronous oscillators with the numbers from 3 to 6 (the coupling strengths B=1,2 and 3 are big enough to synchronize these oscillators with the CO). In all, 4 independent frequencies of oscillations would be represented in U(t) simultaneously, thus the attractor should lie on a four-dimensional torus.

In fact, the dynamics of the network is more complex than in the above case due to mutual interaction between non-synchronous POs and the CO, therefore it has been interesting to investigate the influence of different coupling strengths on the system dynamics and hence on the variation of D_2.

The results of estimation of correlation dimensions D_2 by using time series U(t) are shown in Table 1:

A B	0	0.5	1	2	4	8	16	24	32
1	3.4	3.5	4.0	3.9	3.9	4.0	4.0	3.6	3.7
2	3.5	3.5	3.7	4.1	4.0	4.0	4.0	3.8	3.7
3	3.2	3.3	4.0	3.9	2.9	2.8	3.2	3.0	2.9

Table 1: *Correlation dimension estimates obtained varying the coupling coefficients A and B.*

The method for the computation of the correlation dimension is explained in Corana et al. (1991). Referring to that paper, all the results reported in

762

the Table are obtained computing the correlation integral $C_m(r)$ for embedding dimension m varying from $m = 18$ to $m = 26$. Each time series is 9000 points long, and the best fitting of the slope of the $\log(C_m(r))$ vs. $\log(r)$ is always done in the range $-1.0 \le log(r) \le -0.5$ with the exception of two cases of parameter values B=2, A=0 and B=2, A=0.5 when the range was $-0.7 \le \log(r) \le -0.2$ (this change of the range was due to knees in the log-log plots). The numbers shown in the table are average values of D_2. In all cases the standard deviation was less than 0.2.

3 Discussion

One can see from the table that the correlation dimension gives the right estimation of the number K of the groups of POs with different current frequencies (in our example K=4) or slightly underestimate this value. In all cases the underestimation is less than 1.2, hence D_2 can be considered as a suitable estimation of K under the following precautions. The estimation can be lower than K for small values of coupling strength A. As we guess, this may be caused by small differences of current frequencies of POs which do not run synchronously with the CO (in our example, these are the oscillators 1 and 2). Also, when A is large this results in smaller values of D_2 especially when chaotic oscillations appear. This can be explained in the following manner. In this case the current frequency of the CO varies in a large range and the CO can phase-lock for short times those POs which are usually out of synchronization with it. This results in variation of K which sometimes becomes lower than 4, and then D_2 reflects the average value of K.

Acknowledgements: The work of R.B. and Y.K. was supported by James S. McDonnell Foundation, Grant 93-9; the work of A.C. and G.M. was partially supported by the Progetto Finalizzato *C.N.R. - S.I.C.P.*, Grant 9314132.

References

Babloyantz, A. (1989) Estimation of correlation dimension from single and multichannel recordings. In Brain Dynamics, E. Basar and T.H. Bullock (eds.), Springer-Verlag, Berlin - Heidelberg, 122-130.

Corana, A., Casaleggio, A., Rolando, C., and Ridella, S. (1991) Efficient computation of the correlation dimension from a time series on a LIW computer, Parallel Computing, 17: 809-820.

Kryukov, V. (1991) An attention model based on the principle of dominanta. In Neurocomputers and Attention I. Neurobiology, Synchronization and Chaos, A. Holden and V. Kryukov (eds.), Manchester University Press, UK, 319-352.

Lin, Hao Bai (1989) Elementary Symbolic Dynamics and Chaos in Dissipative Systems, World Scientific Pub.

Destexhe, A., and Babloyantz, A. (1991) Pacemaker-induced coherence in cortical networks, Neural Computation, 3: 143-154.

Wright, J. and Kydd, R. (1992) The electroencephalogram and cortical neural networks, Network, 3: 341-362.

Towards the Application of Networks with Synchronized Oscillatory Dynamics in Vision

H.-U. Bauer

Institut für Theoretische Physik and SFB "Nichtlineare Dynamik",
Universität Frankfurt, Robert-Mayer-Str. 8-10, 60054 Frankfurt, Germany

1   Introduction

The functional relevance of synchronized oscillatory responses observed in the visual cortex of various species (Gray and Singer, 1989; Eckhorn et al., 1988; Kreiter and Singer, 1992) is a hotly debated topic in vision. A widespread assumption holds that the synchronized responses code the results of reintegration processes in the visual system, i.e. of image segmentation, figure-ground segregation, feature binding etc. (Malsburg and Buhmann, 1992; Engel et al., 1992). Many numerical simulations showed that oscillations in general can occur and that spatial synchronization can be mediated in physiologically plausible models (see, e.g., Bush and Douglas, 1991). In addition, the models allowed to qualitatively recover the experimentally observed responses when analogous stimuli, i.e. long bars, interrupted bars etc. are applied (König and Schillen, 1991; Niebur et al., 1991).

However, the current models are rather simple compared to the functions envisaged for the oscillatory dynamics. It is therefore a fruitful enterprise to improve synchronization models with regard to application requirements. In this contribution we focus on two such requirements, the speed of the process which establishes synchronization and the synchronization/desynchronization dilemma.

2   Speed of Synchronization

First we consider the maximum time available when synchronization over extended regions of space (across a whole object) has to be established. Judging from the speed of shifts of the visual focus of attention ($\approx 30ms$, Saarinen and Julesz, 1991), or from the increase of search times in feature binding tasks (Treisman and Gelade, 1982) one can very roughly argue that the synchronization should be established within at most one oscillation period. This speed is very difficult to accomplish in models, where pure oscillators sucessively decrease their phase differences, and in this way synchronize. A solution for this problem arises when more then just the oscillatory aspect of the local dynamics is considered. Detailed analysis of the data reveals, that the oscillations as well as the synchronization between different spatial locations are not per-

764

sistent, but seem to switch between oscillatory and stochastic periods (Gray et al., 1992; Pawelzik et al., 1993). The switching of the local dynamics has been described in two models with bistable dynamics (Deppisch et al., 1993; Bauer and Pawelzik, 1993). Here, we extend these local models to a spatially extended model which can operate on synthetic or real world images. Simplifying the more realistic local models, the building blocks of the extended model are local Markov-elements which basically have a stochastic and an oscillatory state. The transition probabilities between these states depend on the amount of external and lateral excitation which is projected to each element. The oscillatory state is divided into several substates; a pulse-coupling mechanism in the spirit of the Mirollo-Strogatz-model (Mirollo and Strogatz, 1990) allows to synchronize responses between several Markov-elements which are in the oscillatory state. An adequate description of this network can be found in a forthcoming publication (Bauer and Geisel, 1994).

Getting back to the problem of synchronization speed, we can now evaluate the time necessary to synchronize purely oscillatory elements and bistable elements. Respective time series are shown in Fig. 1a,b. In brief, purely oscillatory elements need on the order of ten oscillation periods to establish a sufficient amount of synchronicity across an excited region. In contrast the bistable elements can establish synchronicity instantaneously due to simultaneous switching. In this way the biologically inspired bistable dynamics in the models helps to meet an application demand.

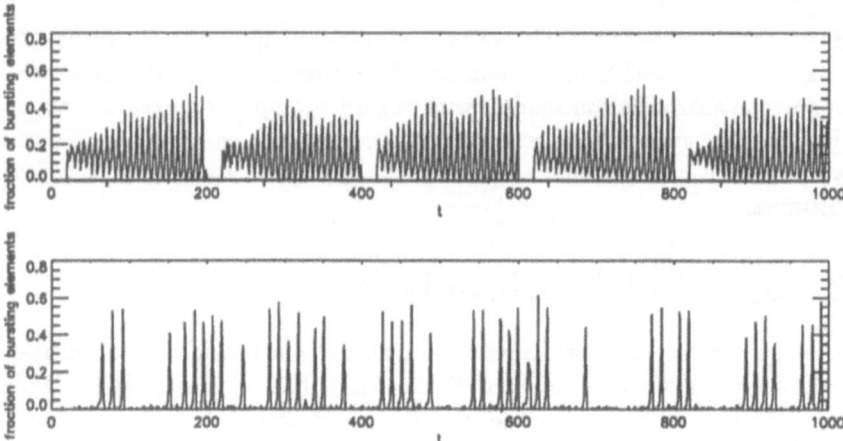

Fig. 1: Number of bursting elements in a layer of Markov elements. A square region with 12×12 elements is stimulated. Part a) shows the results with purely oscillatory elements (reinitialized to random phases at $t = 200, 400, \ldots$ Part b) shows the same network but with the bistable Markov elements (initialized to the stochastic state). Approximately at time $t = 60$, most elements switch simultaneously to the oscillatory states, and fire in synchrony until $t \approx 100$. Similar synchronized oscillatory periods occur later. In each case, the synchronization is established immediately due to simultaneous switching.

3 Synchronization vs. Desynchronization

If two objects are neighboring each other in a scene, the responses for each of them ought to be synchronized, but between them the responses ought to be desynchronized (This dilemma has already raised, e.g., in Schillen and König, 1991). The decision for two neighboring elements, whether to synchronize or desynchronize, could be made by comparing the difference of their local feature values to a threshold. However, suitably constructed synthetic example pictures as well as the results of physophysical experiments demonstrate that a fixed feature value difference is a too rigid criterion.

Instead, we here propose to include edge information in the decision. To this purpose, we include edge elements on our above described network. Local interactions are chosen such, that synchronous oscillations on both sides of an edge are prohibited. Simulations reveal that this mechanism suffices to separate segments which would otherwise wrongly be linked.

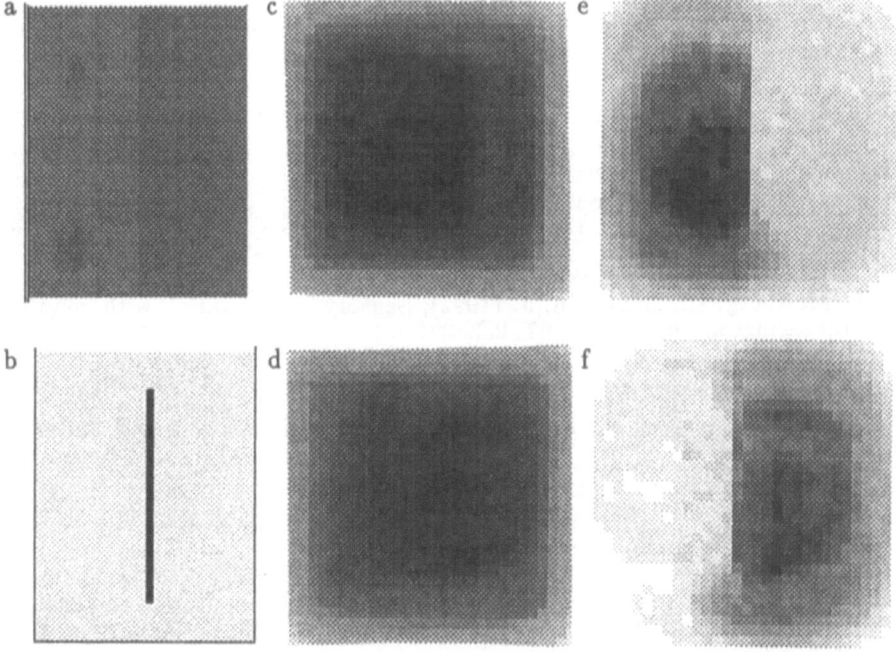

Fig. 2: a: Synthetic image with two segments. The feature on which the segmentation is based is gray value. The gray value difference is so small, that the segments would by linked if no edge interactions are present (indicated by the homogeneous level of black in the correlation images c,d). If the edge information from the image (b) is exploited, and the edge interaction in the network is operating, the two segments are separated (e,f).

So we can conclude for this chapter, that inclusion of edge information helps to solve the local synchronisation/desynchronisation problem.

4 Acknowledgements

Helpful discussions with Klàus Pawelzik are gratefully acknowledged. This work has been supported by the Deutsche Forschungsgemeinschaft (Sonderforschungsbereich 185 "Nichtlineare Dynamik", TP E3).

5 References

Bauer, H.-U., Pawelzik, K. (1993), Alternating oscillatory and stochastic dynamics in a model for a neuronal assembly, Physica D, in print (1993), and Proc. Icann 93, 136 (1993).

Bauer, H.-U. (1994) to be submitted to IEEE Trans. Neur. Netw.

Bush, P.C., Douglas, R.J. (1991), Synchronization of bursting action potential discharge in a model network of neocortical neurons, Neur. Comp. **3**, 19.

Eckhorn R., Bauer R., Jordan W., Brosch M., Kruse W., Munk M., and Reitboeck H.J. (1988), Biol. Cyb. **60**, 121-130.

Engel, A.K., König, P., Kreiter, A.K., Singer, W. (1992), Temporal coding in the visual cortex: New vistas on intergration in the nervous system, TINS **15**, 218.

Gray C.M., König P., Engel A.K., and Singer W. (1989), Nature **338**, 334-337.

Gray, C.M., Engel, A.K., König, P., Singer, W. (1992), Synch. of osc. neuronal responses in cat striate cortex: Temporal properties, Vis. Neurosci. **8**, 337-347.

König, P., Schillen T., (1991), Stimulus-dependent assembly formation of oscillatory responses: I. Synchronization, Neural Comp. **3**, 155-166.

Kreiter, A.K., Singer W. (1992), Eur. J. Neurosci. **4**, 369.

v.d. Malsburg, C., Buhmann, J. (1992), Sensory segmentation with coupled neural oscillators, Biol. Cyb. **67**, 233.

Mirollo, R.E., Strogatz, S.E. (1990), Synchronization of Pulse-Coupled Biological Oscillators, SIAM J. Appl. Math **50**, 1645.

Niebur, E., Kammen, D., Koch, C. (1991), Phase-Locking in 1-D and 2-D networks of oscillating neurons in: Nonlinear Dyn. and Neuronal Networks, ed. H.-G. Schuster, VCH Weinheim, 173.

Nothdurft, H.C. (1991), Texture Segmentation and Pop-Out from Orientation Contrast, Vis. Res. **31**, 1073.

Pawelzik, K., Bauer, H.-U., Geisel, T. (1993), Switching between predictable and unpred. states in data from cat visual cortex, in: Computation and Neural Systems (Proc. CNS 1992), eds. F. Eeckman, J. Bower, Kluwer Academic, 487.

Saarinen, J., Julesz, B., (1991), The speed of attentional shifts in the visual field, Proc. Nat. Acad. Sci. USA **88**, 1812.

Schillen, T., König, P. (1991), Stimulus-dependent assembly formation of oscillatory responses: II. Desynchronization, Neural Comp. **3**, 167-178.

Treisman, A., Gelade, G., (1980), A feature integration theory of attention, Cogn. Psych. **12**, 97.

New Impulse Neuron Circuit for Oscillatory Neural Networks

Jong-Han Shin

Research Department

Electronics and Telecommunication Research Institute

P.O. Box 8, Daeduk Science Town, Daejeon, Korea

(FAX)82-42-860-5033,(E-mail)jhshin@logos.etri.re.kr

1   Introduction

Recently oscillatory neural networks have received a great deal of attention because of their analogy to biological neural system which utilizes frequency and phase of nerve pulse trains for information processing[Gray et al., 1989; Jang et al., 1993]

In conventional oscillatory neuron circuits, the axon hillock circuit based on integration-fire pulse coding(IFPC) technique has been widely used to convert the analog voltage of the neuron body into a pulse sequence[Hamilton et al., 1992; Mead, 1989; and Meador et al., 1991].

In this paper, we describe a novel oscillatory neuron circuit based on noise feedback pulse coding(NFPC) and compare it with the conventional oscillatory neuron circuit with respect to the accuracy of the pulse encoding. The proposed oscillatory neuron circuit can be used as an elementary building block in the oscillatory neural networks.

2   Conventional Electronic Axon Hillock Circuit

Fig. 1(a) shows the conventional electronic axon hillock circuit[Meador et al., 1991], which uses IFPC. This axon hillock circuit has two discrete states-input integration and fire(nerve pulse generation), the schmitt trigger establishes two threshold potentials which determine when the axon hillock circuit moves between these states. Feedback to two FET's, one in series with and one shunting capacitor C_s, determines whether it is being charged by net input excitation during the integration state or being discharged during an nerve pulse output. The role of NMOS transistor M2 is to limit pulse duration by turning off I(t), when the output is turned on.

Consequently, the amplitude and duration of the output pulses are determined by the circuit characteristics, and the output pulse frequency of the axon hillock circuit is proportional to the current I(t).

The circuit of Fig.1(a) was built on a test protoboard, using discrete components. Fig. 2 shows four waveforms observed in the experiment; (i) shows the test input sinusoid of 400Hz and 1V~4V range, (ii) shows the voltage x_i at the integrating capacitor node, (iii) shows the output pulse sequence for the input signal. In order to see the validity of the encoded output signal of (iii), it was decoded via a postsynaptic neuron's body modeled by a simple RC low pass filter. The result is shown in curve (iv).

3 Novel Electronic Axon Hillock Circuit

Fig. 1(b) shows an electronic axon hillock circuit based on NFPC[Shin, 1993], in which the RC network acts as an integrator and the input current I(t) is integrated in capacitor C. The integrated signal is compared with the threshold voltage V_{th} of a comparator, and the output of the comparator is fed to a single pulse generator which generates a delayed single pulse. The single pulse output is applied to the reference voltage switch, which feeds back the reference voltage $-V_{ref}$ to the other input terminal of integrator. Then the capacitor C integrates the current difference $I(t)-I_r$.

As a result, the output pulse frequency is proportional to the amplitude of the input signal. Thus, the output pulses carry the information of the input signal amplitude.

The axon hillock circuit of Fig. 1(b) was built on a test protoboard, using discrete components.

Fig. 3 shows four waveforms observed in the experiment; waveform (i) shows the test input sinusoid of 400Hz and 1V~4V range, (ii) shows the waveform at the integrating capacitor node in the axon circuit, and (iii) shows the pulse sequence for the input signal. In order to see the validity of the encoded output signal of (iii), it was decoded via the simple RC low pass filter. The result is shown in curve (iv), which is very close to the waveform (i), compared with the one of Fig. 2.

The experimental result suggests that the NFPC axon hillock circuit has more computation accuracy than the IFPC axon hillock circuit of Fig. 1(a).

4 Oscillatory neuron circuit using the novel axon hillock circuit

An oscillatory neuron circuit using the novel axon hillock circuit is shown in Fig. 4, where the synapse is modeled by the combination of PMOS transistor M1 and NMOS transistors M2, M3 and M4. Transistors M1 and M4 properly biased output a current proportional to the weight voltage, W_{ij}, when M2 and M3 are on. In the synapse circuit, I_3 is controlled by V_{GS4} which represents the synaptic weight; in other words, the synapse circuit acts as an excitatory synapse $I_1 > I_2$ and as an inhibitory synapse when $I_1 < I_2$.

The resulting output current I_3 is integrated over an accumulation time in the neuron body (which is modeled by a current integrator) to represent the multiplication of W_{ij} by S_j. The current integrator representing the neuron body performs spaciotemporal summation of synaptic currents. If the operational amplifier has a high gain, then the output voltage x_i must satisfy the following equation:

$$\tau \frac{dx_i(t)}{dt} = -x_i(t) + \sum_j W_{ij} S_j(t), \quad \tau = R_f \cdot C_f. \tag{1}$$

In the oscillatory neuron circuit, the sigmoidal function is obtained by non-linear characteristics of the neuron body circuit and coding characteristics of the axon hillock circuit.

5 Conclusions

Two kinds of axon hillock circuits for oscillatory neural networks, which are based on IFPC and NFPC are described. Experimental results, tested in the electronic axon hillock circuits, show that the NFPC axon hillock circuit has more computation accuracy, compared with the IFPC axon hillock circuit. An oscillatory neuron circuit which consists of the synapse circuit, the current integrator and the NFPC axon hillock circuit is presented.

Acknowledgement

I would like to thank Dr. El Hang Lee for his continuing encouragement.

References

Gray, C. M. et al. (1989) Oscillatory responses in cat visual cortex exhibit intercolumnar synchronization which reflects global stimulus properties. Nature, vol. 338, pp. 334-337.

Hamilton, A. et al. (1992) Integrated Pulse Stream Neural Networks : Results, Issues, and Pointers. I.E.E.E. Transactions on Neural Networks, 3 , 385-393.

Jang, J. S. et al. (1993) Oscillatory neural network for integrated segmentation and recognition of patterns. Proceedings of World Congress on Neural Networks, vol. I, 1993, pp.33-35.

Mead, C. A. (1989) Analog VLSI and Neural Systems(America : Addison-Wesley).

Meador, C. A. (1991) Programmable impulse neural circuits. I.E.E.E trans. on Neural Networks, vol. 1, pp.101-109.

Shin, J. H. (1993) Novel neural circuits using stochastic pulse coding and noise feedback pulse coding. International Journal of Electronics, vol. I, pp.33-35.

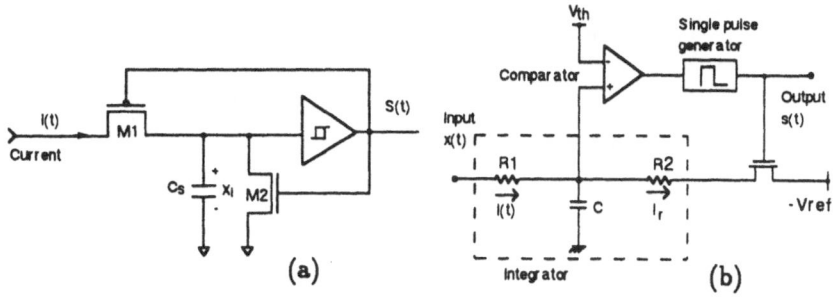

Fig. 1 (a) Conventional axon hillock circuit, (b) Novel axon hillock circuit.

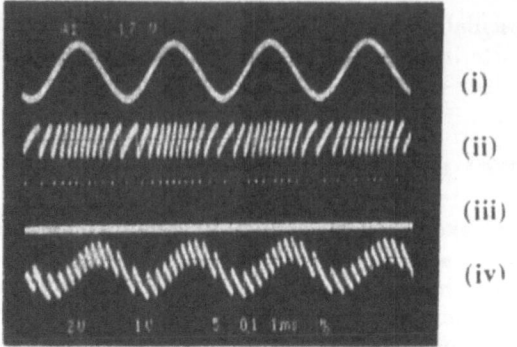

Fig. 2 Experimentally observed waveforms for Fig. 1(a).

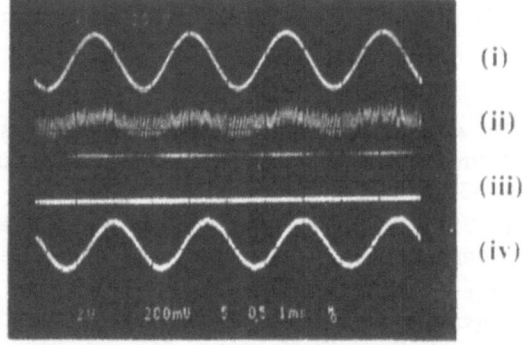

Fig. 3 Experimentally observed waveforms for Fig. 1(b).

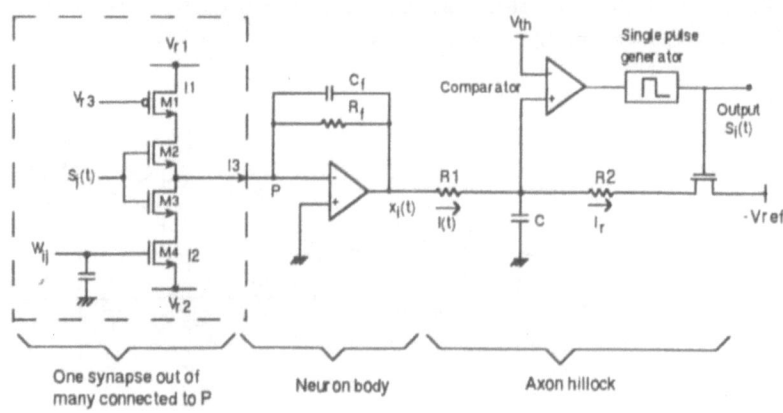

Fig. 4 Oscillatory neuron circuit.

Adaptive Topologically Distributed Encoding

Martin Eldracher*,   Hans Geiger
Technische Universität München, 80290 München, Germany

1   Introduction

Neural networks often suffer from long training time and local minima. To reduce these problems networks without hidden layers would be interesting.

One way to achieve this, is to use special encoding schemes (data preprocessing). Many proposed methods expand the original input representation into a higher dimensional representation in an unsupervised way. Afterwards training is performed using a simple supervised training method, e.g. a perceptron. Well known examples are radial basis functions (Moody & Darken, 1988), CMAC (Albus, 1981) and growing cell structures (Fritzke, 1991). Another approach is TDE, topologically distributed encoding (Geiger, 1990). The non-adaptive form of TDE is equivalent to one-dimensional radial basis functions, but TDE scales linearly rather than exponentially.

Working with fixed resolution in TDE, the resources can not be optimally used. This paper deals with a modified TDE, adaptive TDE, that takes less encoding units, and yields better performance in less training epochs.

Our example dataset will be Iris classification dataset, which is not linearly separable. Nevertheless, the combination of TDE and a single-layer perceptron suffices to solve the problem (Eldracher, 1992).

2   Topologically Distributed Encoding (TDE)

Especially for the classification of not clearly separated classes exact encoding of numerical values is frequently needed to find an area to place the classifier. Encoding numerical values either in single real-valued units or using a channel model, both show theoretical and practical disadvantages (Eldracher, 1992). Furthermore both encodings are biologically implausible.

TDE is biologically motivated. TDE achieves a high overall-accuracy even with comparatively inaccurate encoding in the single units (Kinder, 1990). For every variable input x a number of units ordered along a one-dimensional axis is used. To encode a value x_1 the corresponding units are activated according to their "receptive fields" consisting of overlapping Gaussian shaped curves. Since there are always several units active numerical information is encoded in topologically distributed fashion (Fig. 1).

Only two encoding units would be enough to get the accuracy to solve the Iris-dataset if it was linearly separable (Eldracher, 1992). But an increasing number of encoding units simplifies the separation of two different values. Independent

*Funded by German Ministry for Research and Technology (FKZ 01 IN 102 B/0).

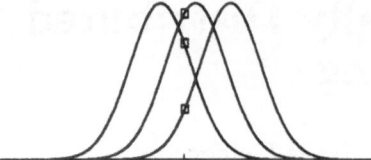

Fig. 1: *Topologically distributed encoding: X-axis: numerical value, y-axis: "receptive fields" of three units (Gaussian shaped curves) and their activation for a specific value (squares).*

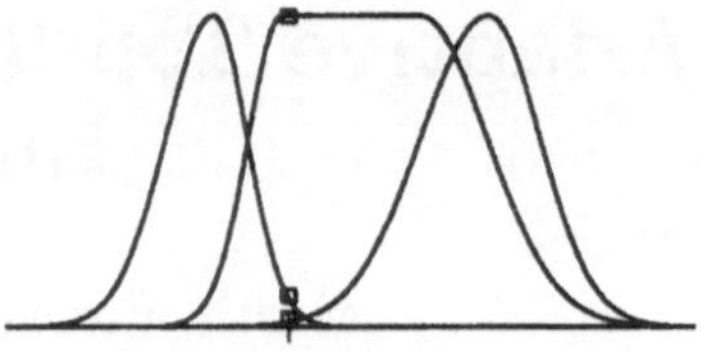

Fig. 2: *Adaptive filtering encoding: X-axis: numerical value, y-axis: "receptive fields" of three units (curves) and their activation for a specific value (squares).*

of the absolute values, slightly different widths σ of the "receptive fields" may allow or hinder the discrimination of two specific training patterns (Eldracher, 1992). Thus there is no optimal overall σ with respect to the classification ability. Nevertheless there can be calculated an optimal σ with respect to resolution capability (Kinder, 1990).

Though excellent training results can be obtained with fixed TDE (section 4) there remains the desire for a more efficient use of the resources.

3   Adaptive TDE

The aim of adaptive TDE is to use less encoding units in order to get quicker architectures with better classification ratio. This should be achieved via pushing the encoding units in these parts of the input space where they are really useful instead of distributing them uniformly over the entire possible input interval. In this way the resolution in all other regions is automatically reduced. This "pushing" could be performed by Kohonen-like variations of the centers of the "receptive fields". In particular, whenever an input pattern is presented to the network the center of the most activated encoding unit (i.e the unit that is placed nearest to the data point, the best matching unit bmu) is moved a little bit closer to the data point. The topological neighbors are also moved into the direction of the input sample. Afterwards the widths σ have to be adapted. In addition a usual Δ–rule learning step is performed. There are several slightly different possibilities to incorporate this basic scheme. For example the movement of the neighbors of the bmu can be scaled using a Gaussian shaped neighborship function, that narrows with ongoing training (Ritter, Martinetz, & Schulten, 1991). An other possibility is to use fixed parameters to move the bmu, next neighbors, second next neighbors, etc. (Fritzke, 1991).

It is clear, that the displacement of the receptive fields should not be done with simple Gaussian shaped encoding units as used before. In order to get the highest encoding precision in fixed topologically distributed encoding, σ should be chosen as $\frac{1}{\sqrt{2}}\Delta\mu$, where $\Delta\mu$ denotes the distance between the centers of each two adjacent encoding units (Eldracher, 1992). Having different distances to the left and right neighbor, but using only one σ for both sides, destroys several desired properties of the encoding:

1. Possibly not all numbers can be represented any more as intervals may occur that are not covered by any receptive field.

2. There is no more guarantee, that the activation value of the encoding units is a linearly decreasing function of the distance between the center of the receptive field and the data point.

3. At most one side can have an optimal value for σ.

These considerations lead to the choice of different left-side and a right-side σ, in order to get the best possible resolution and a well defined overlap of the receptive fields.

4 Experiments and Results

Several series of experiments were done with both, adaptive and non-adaptive, TDE using different numbers of coding-units (CU). Each series consists of experiments using an increasing number of training patterns (equally drawn from each class). The remaining (of 150) patterns are used as test patterns and the number of correct classifications on the whole dataset was measured (Tab. 1).

number of used training patterns	epochs to convergence						correctly classified patterns					
	non-adaptive			adaptive			non-adaptive			adaptive		
	number of CU			number of CU			number of CU			number of CU		
	10	16	20	10	16	20	10	16	20	10	16	20
60	6	8	7	6	5	4	143	145	143	146	145	143
90	69	53	39	15	12	10	146	146	147	147	145	145
120	800	800	257	18	16	15	147	146	150	149	148	146
150	800	800	226	14	19	31	146	148	150	150	150	150

Tab. 1: *Classification results of topolocical distributed encoding for Iris dataset.*

Trying to train more and more patterns results in quasi exponential increase of necessary episodes until the network's resolution ability is reached. Therefore learning was aborted after 800 epochs.

Using adaptive rather than fixed TDE always results in much fewer training epochs. Thus though single update steps are a little bit more complex, training is faster with adaptive TDE. The classification ratio remains approximately equal. Minor differences result of the fact, that for classifying test data the exact centers μ may be relevant. Furthermore the units' centers μ are adapted according to the density of the input distribution, and hence the classification difficulty in overlap regions (Fig. 3 and 4). It should be mentioned, that fixed TDE with less than 18 units can not learn the entire set. With adaptive TDE 8 units suffice to learn the set.

A further improvement to this encoding is not to use the input density but the size of classification error for the input units' adaption. Looking at the results a different encoding seems even more advantageous. In some areas, with patterns of only one class, one single responding encoding unit would suffice to respond rather than several unspecific units with large σ. Another kind of distributed encoding, working like an adaptive filter would help (Fig. 2). Adaptive filter encoding (AFE) allows wide receptive fields with steep borders. An AFE can be obtained, as the bounded combination of a sigmoid and a negated sigmoid unit form one adaptive filter. The steepness of each boarder can be varied using the temperature of the sigmoids. The width of the filter

774

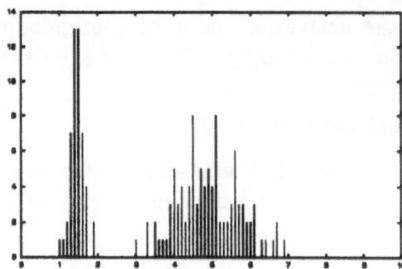 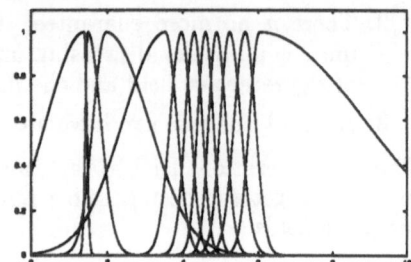

Fig. 3: *Density of the input distribu-* Fig. 4: *Receptive fields after training*
tion, third coordinate. *for third coordinate.*

can be adapted by changing the thresholds. Many such filters result in another
adaptive distributed encoding. Experiments with AFE are under progress.

5 Conclusion

Utilizing TDE, a sophisticated, biologically motivated input encoding of the
numerical values we solved a not linearly separable problem using a single-layer
perceptron. This allows us to train with Δ-rule yielding the double advantage
of efficient learning and the guarantee of finding an existing optimal solution.

Adaptive TDE uses the resources much more efficient than fixed TDE. Con-
centrating encoding units in areas, where high resolution is necessary to solve
classification problems, additionally lowers the number of training cycles until
convergence. This is due to the fact that if in an area many weights are present,
each of them must not be adapted very exact.

Reference

Albus, J. S. (Ed.). (1981). *Brains, Behaviour, and Robotics.* Byte Books, Subsidiary
of McGraw-Hill. ISBN 0-07-000975-9.

Eldracher, M. (1992). Classification of Non-Linear-Separable Real-World-Problems
using Δ-Rule, Perceptrons, and Topologically Distributed Encoding. In *Proceedings of SAC-92*, pp. 1098–1104. acm press.

Fritzke, B. (1991). Let It Grow – Self-Organizing Feature Maps with Problem Dependent Cell Structure. In *Proceedings ICANN-91 Helsinki, Finnland*, pp. 403–408.
Elsevier Science Publishers B.V., North Holland.

Geiger, H. (1990). Storing and Processing Information in Connectionist Systems. In
Eckmiller, R. (Ed.), *Advanced Neural Computers*, pp. 271–277. Elsevier Science
Publishers B.V. (North-Holland).

Kinder, M. (1990). Repräsentation mehrdimensionaler, funktionaler Abhängigkeiten
in konnektionistischen Systemen. Master's thesis, Technische Universität
München, Institut für Informatik.

Moody, J., & Darken, C. (1988). Learning with Localized Receptive Fields. Research
report YALEU/DCS/RR-649, Yale University, Dep of CS.

Ritter, H., Martinetz, T., & Schulten, K. (Eds.). (1991). *Neuronale Netze: Eine
Einführung in Neuroinformatik selbstorganisierender Netzwerke* (2 edition).
Addison Wesley.

On-line Learning with Momentum for Nonlinear Learning Rules

Wim Wiegerinck[1], Andrzej Komoda[1] and Tom Heskes[2]

[1] Dept. of Medical Physics and Biophysics, Univ. of Nijmegen,
Dutch Foundation for Neural Networks,
Geert Grooteplein Noord 21, 6525 EZ Nijmegen, The Netherlands.

[2] Beckman Inst. and Dept. of Physics, Univ. of Illinois at Urbana-Champaign,
405 N. Mathews Ave., Urbana, Illinois 61801, USA

1 Introduction

We study on-line learning with a momentum term. At each learning step i a training pattern x is drawn randomly from a training set Ω and is presented to the network. The network then changes its weights $w(i)$ according to

$$\Delta w(i) = \eta\, f(w(i), x) + \alpha\, \Delta w(i-1).\tag{1}$$

The function f is called the learning rule, η is the learning parameter and α the momentum parameter.

For a linear learning rule, i.e., if f is linear in w, the effect of the momentum term on the learning process has been recently studied by Tugay et al. [4]. They showed that for any $\alpha < 1$ and small η, the statistical properties of networks trained with parameters η and α are identical to those of networks trained without momentum, but with rescaled learning parameter $\tilde{\eta} = \eta/(1-\alpha)$.

However, the momentum term is often used in connection with *nonlinear* learning rules, such as backpropagation. In this paper, we will outline how the results for linear learning rules generalise to nonlinear learning rules. For notational convenience we will restrict ourselves to one-dimensional learning rules.

2 The Framework

By introducing the auxiliary variables $m(i) = \tilde{\eta}^{-1}\Delta w(i-1)$, (with the rescaled learning parameter $\tilde{\eta} \equiv \eta/(1-\alpha)$), the stochastic process (1) becomes a Markov process

$$\begin{aligned}
\Delta w(i) &= \tilde{\eta}\left((1-\alpha)\,f(w(i),x) + \alpha\,m(i)\right)\\
\Delta m(i) &= (1-\alpha)\left(f(w(i),x) - m(i)\right).
\end{aligned}$$

If the time intervals between subsequent learning steps are drawn from a Poisson distribution, the probability $\tilde{P}(w, m, t)$ for the network to be in state (w, m) at time t follows the master equation [1]

$$\partial_t \tilde{P}(w, m, t) = \int T(w, m|w', m')\tilde{P}(w', m', t)\, dw'\, dm' - \tilde{P}(w, m, t),\tag{2}$$

with the transition probability

$$T(w, m|w', m') = \langle \delta(w - w' - \tilde{\eta}((1 - \alpha)f(w', x) + \alpha m'))$$
$$\times \delta(m - ((1 - \alpha)f(w', x) + \alpha m')) \rangle_\Omega .$$

where $\langle . \rangle_\Omega$ denotes an average over the training set Ω.

The variable of interest is the weight w. The evolution of this variable is slow, with typical time scales of order $\tilde{\eta}^{-1}$. In the next section we will derive a series expansion for the evolution of $P(w, t) \equiv \int \tilde{P}(w, m, t) dm$ for large t.

3 Perturbation Theory

The moment vector $\vec{M}(w, t)$, with components

$$M_k(w, t) \equiv \int m^k \tilde{P}(w, m, t) \, dm , \qquad\qquad k = 0, 1, \ldots, \infty ,$$

is a different representation of $\tilde{P}(w, m, t)$. Note that $M_0(w, t) = P(w, t)$. The evolution equation for $\vec{M}(w, t)$ is obtained from the master equation (2),

$$\partial_t \vec{M} = H \vec{M} = \sum_{n=0}^{\infty} \tilde{\eta}^n H^{(n)} \vec{M} , \tag{3}$$

in which the elements of the matrices $H^{(n)}$ are the operators

$$H_{ij}^{(n)} = \left[\frac{(-1)^n}{n!} \frac{\partial^n}{\partial w^n} \sum_{l=0}^{n+i} \binom{n+i}{l} (1 - \alpha)^{n+i-l} \alpha^l f_{n+i-l} \delta_{lj} \right] - \delta_{n0} \delta_{ij} ,$$

for $i, j = 0, 1, 2, \ldots$, with $f_k(w) \equiv \langle (f(w, x))^k \rangle_\Omega$. The system (3) is studied using perturbation theory.

First we consider the unperturbed ($\tilde{\eta} = 0$) system

$$\partial_t \vec{M} = H^{(0)} \vec{M} . \tag{4}$$

From the triangular form of $H^{(0)}$, we can calculate its eigenvalues $\lambda_\gamma = \alpha^\gamma - 1$. So the solution of (4) can be written as

$$\vec{M}(w, t) = \sum_{\gamma=0}^{\infty} e^{-(1-\alpha^\gamma)t} \left[\mathcal{P}_\gamma^{(0)} \vec{M} \right] (w, 0) , \tag{5}$$

in which the operators $\mathcal{P}_\gamma^{(0)}$ project on various modes. The fast modes with $\gamma \neq 0$ will decay rapidly to 0. For large t, only the slow mode, the one with $\gamma = 0$, will remain.

The perturbed ($\tilde{\eta} \neq 0$) system is roughly treated as follows. The eigenvalues of the perturbed system are equal to the eigenvalues of the unperturbed system $1 - \alpha^\gamma$ plus terms of order $\tilde{\eta}$. So, if $\tilde{\eta} \ll 1 - \alpha$, we can still distinguish the fast modes from the slow mode. For large t, the fast modes will vanish and the dynamics of the system will be governed by the slow mode only, from which the time evolution of the distribution $P(w, t)$ can be calculated. A small $\tilde{\eta}$ expansion yields [6]

$$\partial_t P(w, t) = -\tilde{\eta}\, \partial_w f_1(w) P(w, t) + \frac{1}{2}\tilde{\eta}^2 \partial_w^2 f_2(w) P(w, t)$$

$$+ \frac{\tilde{\eta}^2 \alpha}{1 - \alpha} \left[\partial_w^2 f_1(w)^2 P(w, t) - \partial_w f_1(w) \partial_w f_1(w) P(w, t) \right] + \mathcal{O}(\tilde{\eta}^3). \quad (6)$$

4 Small fluctuations expansion

To study (6) in the limit $\tilde{\eta} \to 0$, we apply Van Kampen's small fluctuations expansion [5]. We make the following Ansatz:

$$w = \phi(\tau) + \sqrt{\tilde{\eta}}\xi, \quad (7)$$

with $\tau = \tilde{\eta}t$ and $\phi(\tau)$ a function to be determined. Equation (7) says that the stochastic variable w is given by a deterministic part $\phi(\tau)$ plus a term of order $\sqrt{\tilde{\eta}}$ containing the (small) fluctuations. A posteriori, this Ansatz should be verified. The rescaled time τ indicates that the evolution of the weights is very slow.

In the limit $\tilde{\eta} \to 0$, we obtain that $\phi(\tau)$ must satisfy the deterministic equation

$$\frac{d\phi(\tau)}{d\tau} = f_1(\phi(\tau)) \quad (8)$$

and that $\Pi(\xi, \tau)$, which is the probability P in terms of the new variable ξ, obeys the Fokker-Planck equation

$$\partial_\tau \Pi(\xi, \tau) = -f_1'(\phi(\tau))\partial_\xi \xi \Pi(\xi, \tau) + \frac{1}{2} f_2(\phi(\tau))\partial_\xi^2 \Pi(\xi, \tau). \quad (9)$$

The prime denotes differentiation with respect to the argument. It can be shown that the fluctuations ξ are bounded if $f_1'(\phi(\tau)) < 0$. In that case, the Ansatz (7) is a posteriori justified.

The small fluctuation expansion, straightforwardly applied to the learning process without a momentum term [2] also gives the results (8) and (9). We conclude that for learning parameters $\eta \ll (1 - \alpha)^2$ learning with momentum term is equivalent to learning without momentum term with rescaled learning parameter $\tilde{\eta}$.

5 Simulations

To illustrate the analytical results, we simulate the nonlinear learning rule of Oja [3] with momentum in two dimensions,

$$\Delta w(n) = \eta\, (x^T w(n)) \left[x - (x^T w(n))\, w(n) \right] + \alpha \Delta w(n - 1)$$

which searches for the principal component of the correlation matrix $\langle xx^T \rangle_\Omega$. We draw the inputs x at random from a rectangle centered at the origin with

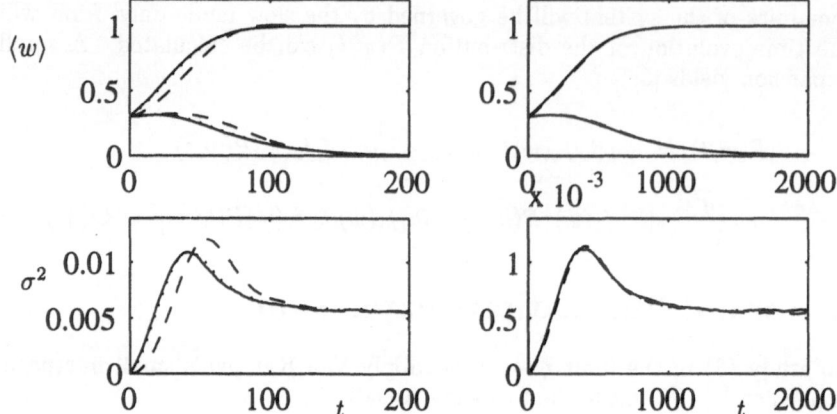

Figure 1: Oja learning with momentum. Means $\langle w_i \rangle$ and sum of variances σ^2 as a function of time. All 10000 networks in the ensemble started with $w = (0.3, 0.3)^T$. The momentum parameter is $\alpha = 0$ for the solid lines, $\alpha = 0.5$ for the dotted lines, and $\alpha = 0.9$ for the dashed lines. The rescaled learning parameter is $\tilde{\eta} = 0.1$ for the left column and $\tilde{\eta} = 0.01$ for the right column.

sides of length 2 and 1 along the x_1- and x_2-axis, respectively. Figure 1 shows the ensemble averages (denoted by $\langle . \rangle$) of the weights and the sum of variances $\sigma^2 \equiv \sum_{i=1}^2 \langle (w_i - \langle w_i \rangle)^2 \rangle$ as a function of time. The parameters η and α are such that the rescaled learning parameters $\tilde{\eta} \equiv \eta/(1 - \alpha)$ are the same for all curves in each of the graphs. Figure 1 shows that curves of fixed $\tilde{\eta}$ are almost overlapping. The curves get closer to each other when $\tilde{\eta}$ decreases (see left and right graphs) or as η becomes much smaller than $(1 - \alpha)^2$ (dotted curves).

References

[1] D. Bedeaux, K. Lakatos-Lindenberg, and K. Shuler (1971). On the relation between master equations and random walks and their solutions. *Journal of Mathematical Physics*, 12:2116–2123.

[2] T. Heskes and B. Kappen (1993). On-line learning processes in artificial neural networks. In J. Taylor, editor, *Mathematical Foundations of Neural Networks*. Elsevier, Amsterdam.

[3] E. Oja (1982). A simplified neuron model as a principal component analyzer. *Journal of Mathematical Biology*, 15:267–273.

[4] M. Tugay and Y. Tanik (1989). Properties of the momentum LMS algorithm. *Signal Processing*, 18:117–127.

[5] N. van Kampen (1981). *Stochastic processes in physics and chemistry*. North-Holland, Amsterdam.

[6] W. Wiegerinck, A. Komoda, and T. Heskes (1994). Stochastic dynamics of learning with momentum in neural networks. *Submitted for publication*.

Constructive Neural Network Algorithm for Approximation of Multivariable Function with Compact Support

Nicolay Magnitskii

Institute for Systems Studies Academy of Sciences,
9, Prospect 60-let Oktyabrya, 117312, RUSSIA

Let $f(x)$ be any square integrable function defined on n-dimensional Euclidean unit ball B^n. We show that the function $f(x)$ has an expansion in a series of orthogonal functions:

$$f(x) = \sum_{i=1}^{\infty} C_i F_i(x) = \sum_{i=1}^{\infty} C_i \int_{|y|=1} S_i(y) G_i(x*y) dy \tag{1}$$

where y is a unit vector, $x*y$ is a inner product of vectoors x and y, $S(y)$ is a real spherical harmonic, and $G_i(r)$ is known nonlinear function of one variable (Gegenbauer polynomial) For every $\epsilon > 0$ we may choose an integer $N > 0$ such that

$$\|f(x) - \sum_{i=1}^{N} C_i \int_{|y|=1} S_i(y) G_i(x*y) dy\|_{L_2} < \epsilon/2 \tag{2}$$

To relpace all integrals in (2) by the sums such that

$$\| \int_{|y|=1} S_i(y) G_i(x*y) dy - \sum_{k=1}^{M_i} a_{ik} S_i(y_k) G_i(x*y_k) \|_{L_2} < \epsilon^i (1-\epsilon)/2$$

we obtain a constructive parallel algorithm for approximation of the function $f(x)$ with arbitrary accuracy ϵ in L_2:

$$f(x) = f(x_1, \ldots, x_n) \simeq \sum_{i=1}^{N} C_i \sum_{k=1}^{M_i} a_{ik} S_i(y_k) G_i(\sum_{j=1}^{n} y_{ji} x_j) \tag{3}$$

We note that the expression (3) is a linear combination of N neural networks

$$W_i = \sum_{k=1}^{M_i} a_{ik} S_i(y_k) G_i(\sum_{j=1}^{n} y_{ji} x_j), \quad i = 1, \ldots, N$$

each of them has a single hidden layer with the activation function $G_i(y)$ and with known weights y_{ik} and $a_{ik} S_i(y_k)$. Thus, to learn the neural network (3) we must define only weights C_i. Such learning algorithm can be realized by minimization of the expression

$$min_{b_1,\ldots,b_N} \|f(x) - \sum_{i=1}^{N} b_i F_i(x)\|_{L_2} \to (C_1,\ldots,C_N$$

Our results is a constructive result unlike the results of the paper [1-3].

References

[1] G.Cybenko, Approximation by supperpositions of a sigmoidal functions, Univ. of Illinois (1988).

[2] S.M. Carrol, B.W. Dickinson, Construction of neural nets using the Radon transform, *Int. Joint Conf.Neural Networks*, New York, v.1 (1989),pp.607-611.

[3] M.Stinchombe, H.White, Universal approximation using feedforward networks with nonsigmoid hiddeen layer activation functions, *Int. Joint Conf.Neural Networks, New York*, v.1 (1989),pp.613-617.

A HEBB-LIKE learning rule
for CELL ASSEMBLIES formation

Francisco J. Vico[1], F. Sandoval[1] & J. Almaraz[2]
[1] Dpto. Tecnología Electrónica
[2] Dpto. Psicobiología
Universidad de Málaga
Málaga Spain

1   Introduction

The cell assembly was first described by Hebb (1949) as a functional unit composed of a group of strongly interconnected neurons. The activation of such groups were hypothesized to be internal representations of the outside information. Reverberatory activity (or after-activity) is another feature of cell assemblies, this implies the existence of direct or indirect excitatory feedback synapses among the neurons. Perceptual completion, selectivity, spike synchronization and noise suppression, are also relevant properties of hebbian cell assemblies.

The Hebb's theory found biological support in the studies on cortical physiology and anatomy (Braitenberg 1978, Palm 1982). But this biological evidence was not confirmed by the results of computer simulations. The Hebb's rule is a good candidate to form cell assemblies, but this learning rule predicts that a positive weight increases indefinitely, and it does not implement depression of synaptic weights. These limitations are not desirable in the design of a learning rule to form cell assemblies.

This work proposes a new Hebb-like rule that verifies cell assemblies formation in a neural network. The main feature of this model is that the weights grow until the assembly is formed, so the growing is bounded by the neural activity, not by maximum values. The assemblies formed in this way are flexible, and reorganization processes can be observed when new inputs are presented.

2   The Neuron Model

For simulation purposes we designed a neuron model that includes two biological parameters: membrane voltage decay and activation threshold. They determine the shape of the output function. The summation of the connecting neurons is given by:

$$a_j = \sum_i (w_{ij} x_i) \tag{1}$$

where w_{ij} is the weight from neuron i to neuron j, with a value in the range $[0,1]$ for excitatory synapses and in $[-1,0]$ for inhibitory synapses. x_i is the activity level of neuron i. This value is bounded in the range $[0,1]$ according to the function:

$$x_j = \begin{cases} 0 & , if(a_j < V_D V_T) \\ 1 + \dfrac{V_T - a_j}{V_T V_D - V_T} & , if(V_D V_T < a_j < V_T) \\ 1 & , if(a_j > V_T) \end{cases} \qquad (2)$$

where V_T represents the threshold of the neuron and V_D is the voltage decay of the membrane, both with values in the range $[0,1]$.

With this model the neuron presents an output function with a shape as in figure 1, where there is a range of analog functioning (the activity level has a value between 0 and 1), and a digital range, with the states of resting (null activity level) and maximum activation (activity level equal to 1). The width of the analog range will be a determinant factor in the neuronal functioning, as the learning rule will depend completely on the activation values in this range, considering that values 0 and 1 are stable values and nothing is learnt when these values are reached.

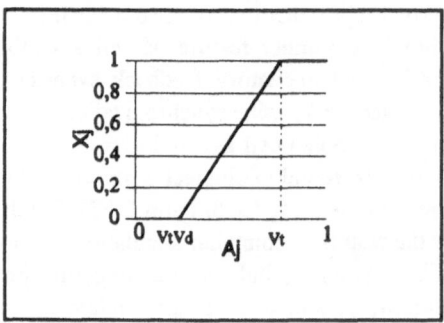

Figure 1. Activation function.

3   Learning rules and Cell assemblies formation

The process of cell assemblies formation involves competitive learning, as the neurons in the network can be taken by one assembly or another. If the patterns presented to the network are orthogonal then the assemblies should be completely independent, this is: no neuron could belong to more than one assembly. To implement this competition it is necessary a mechanism for synaptic depression. A form of heterosynaptic depression consists of decreasing connection strength when the post-synaptic site is active while the pre-synaptic site is inactive. Such a learning rule has the form:

$$\Delta w_{ij} = C(2x_i - 1)x_j \qquad (3)$$

being x_i and x_j activity levels between 0 and 1, and C is the learning rate. All the weights in the network were in the range $[0,1]$ and the sign was changed depending on the excitatory or inhibitory nature of the synapse.

In the refereed case (x_i equal to 0 and x_j equal to 1) the change is -1, so the activity induced by a neuron is not only increasing its own efficacy, but it is also decreasing the other connecting synapses that are resting. According to this competence mechanism, the neurons in the network form strong connections among themselves and weak connections with the rest of the network.

The problem of this rule is that the weights increase indefinitely. The weight can be bounded in the range [0,1] with a simple modification:

$$\Delta w_{ij} = C(2x_i - 1) x_j (1 - w_{ij}) \qquad (4)$$

so the increments are smaller as the weight reaches the value 1.

The formation of hebbian cell assemblies with this learning rule was verified in previous works (Hetherington, P. & Shapiro, M. 1993), and our simulations confirm these results. The problem of the resulting assemblies applying this rule is that they are rigid, in the sense that the weights always reach the limit values: 0 or 1 (for both excitatory and inhibitory synapses).

3.1 Model architecture and procedure

We used an array of fully connected neurons to verify the formation of cell assemblies presenting orthogonal patterns. All neurons were connected to the rest of the network with two weights (one excitatory and one inhibitory), that were randomly initialized.

After training the network with equation (4), cell assemblies raised in the network as sets of independent neurons. A first analysis of the weights after the training revealed that all of them reached a limit value, this is because training proceeded after the assemblies were formed. The final weights map showed that a neuron belonging to an assembly was connected with a weight of +1 to all the cells in the same assembly, and with a weight of -1 to all the cells in other assemblies. The cells that were not captured in any cell assembly were connected with a weight of -1 to all the cells belonging to a cell assembly, and with a weight of 0 to all the cells that did not belong to an assembly.

4 Simulation results

It was clear that the heterosynaptic depression was the key of the hebbian cell assemblies formation. So, we focused the design of the learning rule in an extension of the previous formulation.

Changes in the synaptic weights should occur until the assembly was formed, but not after this. This can be achieved introducing the post-synaptic activity in the learning rule to reduce the increment as the post-synaptic activity level goes up. In this way, the weights increase until the neurons are captured in the assembly, without reaching the value +1. The final value will depend only on the neuronal threshold (V_T); the lower it is, the smaller weights are needed to reach the maximum activity.

In this way we solve the problem of the excitatory weights saturation. About the inhibitory weights the reasoning is similar: competition makes sense among neurons with some activity, but when a pre-synaptic neuron is completely inactive then it is not involved in the formation of assemblies including that post-synaptic neuron. A neuron that captured another neuron in an assembly should compete only with other neurons with some pre-synaptic activity. In this way, an inhibitory weight in a synapse is augmented until the corresponding pre-synaptic neuron has a null activity level, and the inhibitory synapses does not reach the -1 value. The final value of inhibitory synapses will depend on the product of the membrane voltage decay and the neuronal threshold ($V_D V_T$), as this is the limit for a null activity level.

With these principles, the equation (4) was modified giving:

$$\Delta w_{ij} = C(2x_i - 1)x_j x_i (1-x_j)(1-w_{ij}) \tag{5}$$

where the term $(1-x_j)$ implements the bound for the growing of excitatory synapses, and the term x_i bounds the increase of inhibitory synapses.

Equation (5) is a Hebb-like rule as it introduces two terms in the original Hebb's formula.

The same simulations were made applying equation (5) as the learning rule, the result was that assemblies grew as before but the weights remained with small values; they just changed a little to obtain the assemblies.

After the formation of the assemblies, the network was fed with a bigger set of patterns including some patterns made as a combination of the initial set. The result was that the assemblies changed to represent all the training set.

5 Acknowledgments

This work has been partially supported by the Spanish Comisión Interministerial de Ciencia y Tecnología (CICYT), Project No. TIC91-0965.

6 References

Braitenberg, V. (1978). Cell assemblies in the cerebral cortes. In R. Heim & G. Palm (Eds.) *Theoretical Approaches to Complex Systems*, pp. 171-88, Springer-Verlag.

Hebb, D.O. (1949). *The organization of behabior*. Wiley, New York.

Hetherington, P.A. & Shapiro, M.L. (1993). Simulating Hebb cell assemblies: the necessity for partitioned dendritic trees and a post-not-pre LTD rule. *Network*, 4, pp. 135-153.

Palm, G. (1982). *Neural assemblies. An alternative approach to artificial intelligence*. Springer-Verlag, Berlin.

CARVE — A Constructive Algorithm for Real Valued Examples

Steven Young & Tom Downs
Department of Electrical and Computer Engineering
University of Queensland
Brisbane, Australia 4072

1 Introduction

We present a constructive neural network training algorithm. The algorithm builds a single hidden layer network and is guaranteed to implement any consistent training set of real-valued vectors classified into any number of classes. The algorithm extends the 'sequential' learning algorithm of (Marchand et al., 1990) from the binary input case to the real valued input case.

Most constructive neural network algorithms which build layered neural networks have been developed for binary input training problems only. There are two general reasons for the restriction to binary input problems: (1) The proofs of convergence for the algorithms sometimes rely on the input vectors being binary valued, eg. (Marchand et al., 1990; Mézard & Nadal, 1989); (2) For some, the computational scheme used by the algorithm requires binary inputs, eg. (Ruján & Marchand, 1989). Some algorithms have both restrictions, eg. (Biswas & Kumar, 1990). CARVE has neither of these restrictions. ((Barkema et al., 1993) is another constructive algorithm which builds layered networks that deal with real valued inputs, but the architecture has more than one hidden layer generally.)

For classification tasks on a point set, it is known that a single hidden layer is sufficient to implement any classification task (Huang & Huang, 1991), and so we know *a priori* that our algorithm can accommodate the problems we attempt.

The next section describes the algorithm. Section 3 presents some experimental results and compares performance with other constructive algorithms. Section 4 provides discussion and conclusions.

2 The CARVE Algorithm

CARVE constructs a threshold network with a single hidden layer of units. The algorithm adds units to the layer one at a time. Each unit implements a hyperplane in the input domain that separates a set of points of one class from the remaining training examples. The main work of the CARVE algorithm is similar to that of (Marchand et al., 1990) in that it seeks to find 'good' sets of points to separate or 'carve' from the remaining points. Points carved by the hidden unit from the remaining examples are removed from the training set, and the next hidden unit is sought, with the aim of finding a good carve set

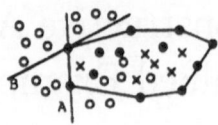

Figure 1: Searching for Hyperplanes around the boundary of the convex hull

in the remaining points. The hidden layer is complete when points of only one class remain in the training set. The implementation of the sequential learning scheme in (Marchand et al., 1990) works only for binary valued inputs. For the real valued vector case, to find hyperplane boundaries that separate points of one class from the remaining vectors, we construct the convex hull of the points of the other classes, then carve sets are chosen from the points outside the convex hull. We search the set of hyperplanes that touch the boundary of the convex hull to find the largest set of points. The construction of convex hulls is repeated for each class of points and the largest carve set for each class is found. The largest carve set is then chosen; a hidden unit is installed which implements the appropriate hyperplane boundary, and the carve set is removed from the training set. Thus, in the situation in fig. 1, a hidden unit implementing hyperplane B would be installed.

For a hidden layer constructed in this manner the internal representations (IRs) corresponding to points of each class are 'faithful' (Mézard & Nadal, 1989) and linearly separable. IRs are faithful if two points of different class always have different IRs. To see that CARVE produces faithful IRs, consider any two points of different class. When the hidden layer construction has finished, there must be at least one hyperplane that separates these points, otherwise the conditions of hidden layer construction (that points of one class only be carved and that the hidden layer is finished when points of a single class remain) would not be satisfied. A proof that the IRs corresponding to each class of points are linearly separable from the IRs of the points of all the other classes is given in (Marchand et al., 1990). The two conditions on the IRs (faithfulness and linear separability) ensure that the CARVE algorithm can construct a neural network that implements any classification task on a set of points.

Once the hidden layer has been constructed, the determination of the hidden to output unit weights can be done by the Perceptron algorithm, linear programming, or by using a convex hull method, where the convex hull of the hidden unit representations of each class is constructed and an appropriate hyperplane boundary which separates the disjoint convex hulls is determined. Our current implementation uses the Perceptron algorithm.

This completes the description of the algorithm. The complexity of the algorithm is dependent on the time complexity of the process of constructing the convex hulls. We used the Beneath-Beyond method described in (Edelsbrunner, 1987,ch 8.4). Beneath-Beyond runs in time $O(n^{\lfloor (d+1)/2 \rfloor})$, with $O(n^{\lfloor d/2 \rfloor})$ storage, where n is the training sample size and d is the input dimension. So CARVE has time and space complexity which is polynomial in the training sample size, but exponential in the input dimension. We are currently seeking methods which will have smaller computational complexity with respect to training size and input dimension (Seidel, 1986).

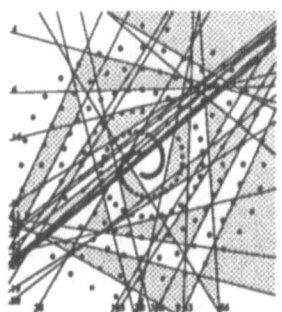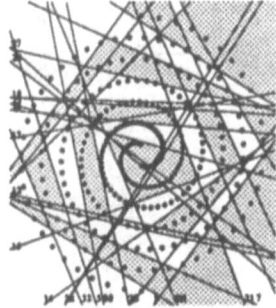

Figure 2: CARVE Solutions to two spirals benchmark (a) 194 point problem (b) 288 point problem

3   Experiments

In this section we present the results of applying CARVE to three training problems. The first two are real-valued training problems and the third is a binary input problem for the purpose of comparing CARVE to binary constructive schemes.

3.1   Two Spirals

Applying CARVE to the two spirals benchmark produces the solution given in figure 2(a). The only other solution to two spirals in a single hidden layer network, that we are aware of, has 50 hidden units (Baum & Lang, 1990). The solution constructed by CARVE has 35 hidden units. Figure 2(b) gives the CARVE solution for a more dense two spiral problem (288 training points instead of 194). The solution builds a network with 32 hidden units.

3.2   Iris data

The iris data is a classic pattern classification database (Fisher, 1936). The data consists of 150 patterns of 4 real-valued attributes which are classified into 3 classes corresponding to different iris flower types (50 of each class). We measured cross-validation error using the leave-one-out method. For each of the training runs, CARVE generated a network which correctly classified all 149 training examples. The average network size over the 150 training trials was 3.973 ± 0.053 and the cross-validation error was 0.046. For the offset algorithm (Martinez & Estève, 1992), which is a constructive scheme that builds then prunes a two hidden layer network, the average network size on the iris data was 3.41 ± 1.01. and cross-validation error was 0.033. Thus, the offset algorithm performs slightly better on this problem but it does, of course, employ pruning and also uses two hidden layers.

3.3   Random Boolean Mappings

Randomly generated Boolean functions have been widely used as a test problem for constructive algorithms. The average network size generated for random Boolean functions with 6 inputs are compared in Table 1 for a number of

N	CARVE < 100 >	Sequential < 100 >	Regular < 200 >	Tiling < 100 >	Upstart < 25 >
6	7.33±1.00	7.28±0.82	15.8±0.82	16.19 (3.75 layers)	8

Table 1: Network sizes generated by various constructive algorithms for random Boolean functions

constructive algorithms. The table contains reported results of four other algorithms. The value in '<>' brackets is the number of functions over which the network size was averaged. CARVE's performance is similar to that of sequential learning (which is to be expected) and better than the other algorithms in the table.

4 Conclusions

CARVE constructs networks which implement classification tasks on real-valued training data. Most existing constructive neural network methods are developed for binary input mappings only and so are not applicable to more interesting real-valued classification tasks. Currently the algorithm complexity is exponential in the input dimension. Methods for reducing the computational complexity of the algorithm appear possible, and are under investigation.

References

G. T. Barkema, H. M. A. Andree & A. Taal (1993), "The Patch algorithm: fast design of binary feedforward neural networks," *Network* 5, 393–407.

Eric B. Baum & Kevin J. Lang (1990), "Constructing Hidden Units using Examples and Queries," in *NIPS3*, 904–910.

N. N. Biswas & R. Kumar (25 June 1990), "A new algorithm for learning representations in Boolean neural networks," *Current Science* 59, 595–600.

Herbert Edelsbrunner (1987), *Algorithms in Combinatorial Geometry*, Springer Verlag, Berlin.

R. A. Fisher (1936), "The use of multiple measurements in taxonomic problems," *Annals of Eugenics* VII, 179–188.

Shih-Chi Huang & Yih-Fang Huang (Jan. 1991), "Bound on the Number of Hidden Neurons in Multilayer Perceptrons," *IEEE Transactions on Neural Networks* 2 , 47–55 .

M. Marchand, M. Golea & P. Ruján (1990), "A Convergence Theorem for Sequential Learning in Two-Layer Perceptrons," *Europhysics Letters* 11, 487–492.

D. Martinez & D. Estève (1992), "The Offset Algorithm: Building and Learning Method for Multilayer Neural Networks," *Europhysics Letters* 18, 95–100.

Marc Mézard & Jean-Pierre Nadal (1989), "Learning in feedforward layered networks: the tiling algorithm," *Journal of Physics A: Mathematical and General* 22, 2191–2203.

Pál Ruján & Mario Marchand (1989), "Learning by Minimizing Resources in Neural Networks," *Complex Systems* 3, 229–241.

Raimund Seidel (May 1986), "Constructing Higher-Dimensional Convex Hulls at Logarithmic Cost per Face," in *18th ACM Symposium on Theory of Computing*, Berkeley, CA, 404–413.

A Supervised Learning Rule for the Single Spike Model

K. Eder

Kratzer Automatisierung GmbH, Carl v. Linde Str. 38, D-85716 Unterschleißheim
eder@kratzer-automation.de

Abstract

A supervised learning rule for the single spike model, which a special implementation of an integrate and fire model, is presented. Because selforganisation mechanisms such as the modified Hebbian learning for temporal structure are not always applicable (e.g. classification tasks) there arises the necessitiy of a modified perceptron learning rule, which is able to incorporate the temporal information contained in the incoming spike trains. The learning rule proposed, fulfills the presumptions of the Perceptron convergence theorem by transforming the temporal information to a virtual quantitiy of input arriving simultaneously. The parallel classification of superimposed input patterns to hypotheses of classes shows an application of the learning rule.

Introduction

Latest results of neurophysiological research have shown synchronisation and desynchronisation mechanisms in laterally coupled layers of the cortex, e.g. [4]. Modelling the basic properties of such networks has been investigated in several works [3,5]. Thus it is possible to encode the relationship of superimposed patterns in the temporal structure of output activity of a synchronisation layer. This kind of encoding has been applied to several problems such as encoding parallel word hypotheses where one word is represented by a multidimensional vector of syntactic and semantic features [9] or figure ground separation [7]. When making the information about synchronisation of features available to following layers in a network the problem arises of how to process the output of the synchronisation layer. A very common problem is the classification of patterns in a feed forward network. Classification means the projection of a pattern to a class where it belongs to. When processing superimposed patterns, i. e. keeping parallel hypotheses of some input information, the task of classification consists of projecting patterns to hypotheses of classes. Because classification is a typical task for supervised learning, a learning rule has to be developed for supervised association of the temporal structure of hypotheses of parallel input to hypotheses of classes.

The model

In the following section a supervised learning rule for the single spike model is proposed. The single spike model consists of a part for modelling some properties of a neural cell membrane, a part for modelling the refractory and the threshold function and a part for producing spikes dependent on the firing probability (for further description see [2]). The membrane part of the model, which is the most interesting part for a supervised learning rule is described by the following equation

$$S_0[U_D(t)-U_0]+C_0\frac{dU_D(t)}{dt}+S_{exc}(t)[U_D(t)-U_{exc}]+S_{inh}(t)[U_D(t)-U_{inh}]=0$$

with the following definitions:

U_0:	ground potential of membrane	U_D:	membrane potential
U_{exc}:	excitatory subsynaptic potential	U_{inh}:	inhibitory subsynaptic potential
U_S:	threshold potential	U_R:	stochastic potential
U_{Ref}:	refractory potential	S_0:	membrane conductivity

S_{inh}: subsynaptic membrane conductivity caused by inhibitory input
S_{exc}: subsynaptic membrane conductivity caused by excitatory input
C_0: membrane capacity

Learning rule

Two kinds of target information have to be specified in order to perform learning of temporal information:

- class information, which means whether or not there has to be output activity at a specific output neuron at a given time
- temporal information, which means at which time the output activity has to occur.

Learning takes place when either an output spike or a target spike occurs at the currently considered time t_c. The learning rule changes locally the weight w_{lk} between the presynaptic neuron l and the postsynaptic neuron k. The learning rate depends on the time when the latest input occurred from neuron l and on the membrane depolarisation necessary for producing the proper output [8]:

t_c currently considered time, where output or target spike occurs
t_l time of last presynaptic spike at neuron l

Thus the complete learning rule can be written as follows:

$$\Delta w_{lk} = \eta \cdot \vartheta(t_c - t_l) \cdot \Delta U_k$$

where ϑ defines the contribution to the weight change dependent on the input spike train and η is some properly choosen learning coefficent. Figure 1 shows the shape of ϑ schematically:

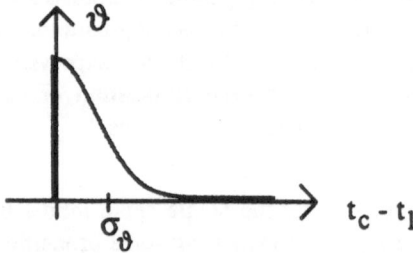

Figure 1: Weight change according to difference t_c-t_l

The function ϑ has to represent whether a spike at the presynaptic neuron can cause an output spike at time t_c. Thus the width σ_ϑ depends on the time constant in the integrating membrane part of the postsynaptic neuron which is $\tau_0 = \dfrac{C_0}{S_0}$

ΔU_k is defined as follows:

$$\Delta U_k(t_c) = \begin{cases} U_S - U_D(t_c), & \text{for active target and inactive output} \\ U_D(t_c) - U_0, & \text{for inactive target and active output} \\ 0, & \text{otherwise} \end{cases}$$

In contrast to the standard Perceptron learning rule and to Hebbian learning, the rule described above is able to take into account the integration facility of the single spike model. The synaptic strength between two neurons is rather changed according to simultaneous activity but to the temporal structure of incoming activity.

For the proof of Perceptron convergence theorem [6] a residual activity resulting from the incoming spikes of one presynaptic neuron in the learning history is defined. This residual activity depends on the parameters defined for the postsynaptic neuron and the times where the other incoming spikes arrived. The scalar product of the vector of residual activities with the weight vector is equal to the current membrane depolarization which defines - combined with a threshold function - whether or not an output spike occurs.

An application: classification of superimposed patterns

Next the proposed learning rule was applied to a classification problem. Therefore a simple feed forward network was combined with a laterally coupled layer for encoding the superimposed input patterns by a synchronization process (see fig. 2). The construction and training of the synchronization layer by a modified hebbian learning rule for temporal information is shown in [2].

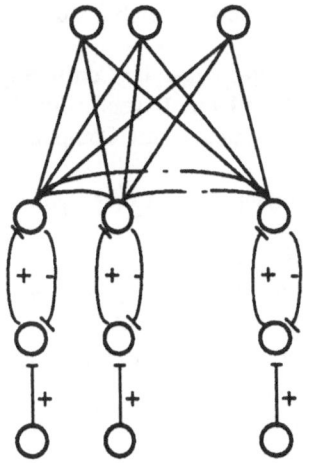

classification layer

classifying feed forward connections

synchronizing lateral connections
synchronization layer

processing layer

input layer

Figure 2: Synchronization structure connected to a feed forward network

The training of the classification layer was performed by applying multiple superimposed patterns combined with the class index of the patterns. This is one of the major advantages for the above learning rule, because the input patterns do not have to be separated for the training process. In another application [8] the learning rule was used for classification of word sequences to hypotheses of sentence frames.

Discussion

The proposed learning rule shows two major advantages. First, it fulfills the Perceptron convergence theorem. Thus learning rates and storage capacity of a network can be calculated. Second, the learning rule can be applied to the temporally structured output of a synchronisation layer without explicitly separating the superimposed patterns. In addition to the Perceptron learning rule it incorporates the temporal information containd in the arriving spike trains. Because the learning rule does not depend on mean firing rates, information such as synchronicity in the input is not lost during the classification process. Thus hypotheses about the input information are transformed into hypotheses of class information. Furter applications, which are currently implemented, are the processing of continuous speech signals [1], recognition of sequences of word hypotheses and associative recall of state sequences for process control.

References

[1]   R. Deffner, K. Eder, H. Geiger, M. Paping, H.-W. Strube: "Understanding continuous speech using a neural word model and an associative predictor network prestructured with linguistic knwoledge", submitted to ICANN '94.

[2]   K. Eder, H. Geiger, W. Brauer: "A Neurophysiologically Motivated Neural Network Model and its Application to the Superposition Problem". In ICANN '93, Proceedings, Springer-Verlag, 1993, pp 482-485.

[3]   W. Gerstner, J. L. v. Hemmen: "Spikes or Rates? - Stationary, Oscillatory, and Spatio-temporal States in an Associative Network of Spiking Neurons". In ICANN '93, Proceedings, Springer-Verlag, 1993, pp 633-638.

[4]   C. M. Gray, P. König, A. K. Engel, W. Singer: "Oscillatory responses in cat visual cortex exhibit inter-columnar syncronization which reflects global stimulus properties". Nature, Vol. 338, 1989, pp 334-337.

[5]   P. König, B. Schillen: "Segregation of Oscillatory Responses by Conflicting Stimuli - Desynchronizing Connections in Neural Oscillator Layers". In: R. Eckmiller et al. (eds.): Parallel Processing in Neural Systems and Computers, Elsevier North Holland, 1990, pp 117-120.

[6]   M. Minsky, S. Papert: "Perceptrons: An Introduction to Computational Geometry", MIT Press, Cambridge Massachusetts 1982.

[7]   A. Nischwitz, H. Glünder, P. Klausner: "Synchronization of Spikes in Populations of Laterally Coupled Model Neurons". In Kohonen et al. (eds.): Artificial Neural Networks, Elsevier North Holland, 1991, pp 1771-1774

[8]   A. Schütz: "Satzerkennung durch Klassifikation gelernter Merkmals-sequenzen mit dem Single-Spike-Modell", diploma thesis, TU-Munich, 1994.

[9]   M. Sturm: "Untersuchungen zur Rückgekoppelten Hypothesenbildung in Single Spike Netzen", diploma thesis, TU-Munich, 1993.

Comparative Bibliography of Ontogenic Neural Networks

E. Fiesler

IDIAP, Case postale 609, CH-1920 Martigny, Switzerland

One of the most powerful aspects of neural networks is their ability to adapt to problems by changing their interconnection strengths according to a predetermined *learning rule*. On the other hand, one of the main drawbacks of neural networks is the lack of knowledge for determining the topology of the network, that is, the number of layers and number of neurons per layer. A special class of neural networks tries to overcome this problem by letting the network also automatically adapt its topology to the problem. These are the so called ontogenic neural networks. Other potential advantages of ontogenic neural networks are improved generalization, implementation optimization (size and/or execution speed), and the avoidance of local minima.

Possible topology modifications are the addition and/or removal of layers, neurons, or connections. Note that modification of the number of layers usually implies modification of the number of neurons, and that modification of the number of neurons usually implies modification of the number of connections. This publication provides an extensive comparison of these ontogenic neural networks in a compact tabular form.

In this table, the following notations are used. "L" is the number of layers (including the input and output layer), $l+x$ means that the number of layers is initially l and can grow. The arrow-columns indicate whether the method increases ($\nearrow$) or decreases ($\searrow$) the network size by manipulating layers (L), neurons (N), or weights (W). (The symbol "N" is also used in the case of an initial two-layer network growing to become a three-layer network with the addition of the first neuron.) "*Lo.*" indicates whether the method is local, which means that the method is exclusively based on information available at the neuron in question, which includes information about the connections to its neighbors. "*I.*" denotes the initial conditions of the method which are encoded by a number, indicating the number of layers in the initial topology (0 for an empty topology), a first letter, indicating the initial connectivity ("F" for fully interlayer connected, "T" for a tree structure, "O" for other topologies), and a second letter, indicating the values of the initial weights ("R" when they are initialized with random values, "T" if the ontogenic method starts with an already trained network). And lastly, "*Pr.*" lists whether a convergence proof exists (M=mathematical proof, I=informal proof). In general, an "*X*" indicates that a condition is fulfilled, while a "-" indicates it is not. A "?" denotes that the information is not available in the quoted reference.

In order to sort the various kinds of ontogenic neural networks, they are classified firstly by learning rule (based on Perceptron, feature map, or other) and secondly by topology (layered or tree based), and thirdly by growing and/or shrinking.

The main goal for most **Perceptron based growing methods** is the determination of the (optimal) topology of a neural network, followed by the goal to avoid getting trapped in local minima, where new units are added in the hope that a higher dimensional error surface eliminates the local minima in question. It is interesting to note that all the tree based methods are growing methods.

Perceptron based pruning methods are usually targeted towards improving generalization, since a small network usually has a better generalization capability.

Most **Perceptron based methods** that both add and delete units are a mere combination of growing and pruning methods. Firstly, they increase the network size until it is large enough to handle the problem (or to eliminate possible local minima). This often yields an oversized topology. Next, redundant units are removed.

Reference	Model	Name	L	↗	↘	Lo.	Init.	Pr.
Perceptron based methods								
Ash-89.2]	BP[1]	Dyn. Node Creation	3	N	-	-	3FR	-
Bellido-91]	BP	-	3	N	-	X	2FR	-
Fahlman-90.2]	MLP[2]	Cascade Correlation	2+x	L	-	X	2FR	-
Fahlman-91.2]	MLP	Recurrent CasCor.	2+x	L	-	-	2OR	-
Hanson-90]	SDR[3]	Meiosis networks	3	N	-	X	2FR	-
Mezard-89]	MLP	Tiling	2+x	L	-	?	2FR	M
Nadal-89]	MLP	-	2+x	L	-	?	xOR	M
Rujan-89.2]	MLP	regular partitioning	3	N	-	-	2FR	M
Sirat-90]	MLP	-	3	N	-	X	1	I
Zollner-92]	MLP	-	3	N	-	-	2FR	I
Thodberg-91]	BP	Ockham's Razor	3	-	W	-	3FT	-
LeCun-90]	MLP	Opt. Brain Damage	3	-	W	-	3FR	-
Hassibi-93]	MLP	Opt. Brain Surgeon	3	-	W	-	3FR	-
Hagiwara-90.2]	BP	-	3	N	W	-	3FR	-
Hirose-91]	BP	-	3	N	N	-	?	-
Honavar-88] Honavar-89.2]	'cones' BP	units recruitment	L	N[4]	N W	X	x O	-
Deffuant-90]	BP	NEURAL[5]	tree	N	-	X	0	M
Golea-90]	BP	decision trees	tree	N	-	X	2	M
Frean-90]	BP	Upstart algorithm	tree	N	-	?	2OR	I
Self-Organizing Feature Map based methods								
Wen-92.1]	comb.[6]	SSGNN[7]	1+x	N	N	-	0	
Fritzke-91.1] Fritzke-91.2] Fritzke-91.3]	SOFM	-	2	N	N	X	1OR	-
Other ontogenic neural networks								
Lansner-87]	Hopfld.	-	2	N	-	X	2OR	-
Kadir.-92.1] Kadir.-92.2]	RBF[8]	-	3	N	-	-	2FR	-
Bonnlander-93]	RBF	latticed networks	2	X	-	-	2OR	-
Alpaydin-90.1]	-	Grow-and-Learn	3	N	N	-	1O	I
Diederich-88]	-	Neuron Recruitment	4	N	-	-	-	-
Reilly-82]	-	Category Learning	3	N	-	X	0	I
Ivakhnenko-68]	GMDH	-	1+x	L	N	X	1	I
Barron-75]	GMDH	McLaurin series	1+x	L	N	X	1	-
Tenorio-89]	GMDH	Simulated Annealing	1+x	L	N	X	1	-
Baum-91]	MLP	Query Learning	3	X	-	-	2FR	-
Fiesler-92.2]	HONN[9]	Superceptron	2	W	-	-	0	-
Ring-93.2]	HONN	-	2+x	L	-	-	2FR	-

[1] Backpropagation
[2] Multilayer Perceptron
[3] Stochastic Delta Rule
[4] transform = cluster of neurons
[5] NEural Units Recruitment ALgorithms
[6] combination of Self-Organizing Feature Map and Multilayer Perceptron
[7] Self-Generating Neural Networks
[8] Radial Basis Function
[9] High Order Neural Networks

The typically long training time for **Kohonen's Self-Organizing Feature Maps** can be shortened by starting with a small topology which grows according to some predefined criteria. Another reason for an ontogenic approach is to add neurons where the input space density is the greatest, in order to overcome the impossibility of dividing the input distribution into disjunct sub-structures. As is the case for Perceptron based neural networks, the generated network is often not optimal, and may be improved by removing neurons or otherwise reshaping the network.

The small capacity of **Hopfield Networks** can be improved with algebraic methods like the projection rule which reduce training set constraints.

The author acknowledges the support of J. Klotz, who participated in the creation of this document as well as its expanded and comprehensively annotated version, which is currently in preparation [Klotz-94].

References

[Alpaydin-90.1] E. Alpaydin. Grow-and-learn: An incremental method for category learning. In *Proc. INNC '90*, volume 2, pages 761–764, Dordrecht, The Netherlands, 1990. Kluwer.

[Ash-89.2] T. Ash. Dynamic node creation in backpropagation networks. *Connection Science*, 1(4):365–375, 1989.

[Barron-75] R. L. Barron. Learning networks improve computer-aided prediction and control. *Computer Design*, pages 65–70, August 1975.

[Baum-91] E. B. Baum and K. J. Lang. Constructing hidden units using examples and queries. In R. P. Lippmann et al., editor, *NIPS 3*. Morgan Kaufmann, 1991.

[Bellido-91] I. Bellido and G. Fernández. Backpropagation growing networks: Towards local minima elimination. In A. Prieto, editor, *Artificial Neural Networks; Proc. IWANN '91*, pages 130–135, Heidelberg, Germany, 1991. Springer.

[Bonnlander-93] B. Bonnlander and M. C. Mozer. Latticed rbf networks: An alternative to constructive methods. In *NIPS 5*, 1993.

[Deffuant-90] G. Deffuant. Neural units recruitment algorithm for generation of decision trees. In *Proc. IJCNN '90 - San Diego*, volume I, pages 637–642, Ann Arbor, MI, 1990. IEEE Neural Networks Council; Edward Brothers.

[Diederich-88] J. Diederich. Connectionist recruitment learning. In *Proc. of the 8th European Conference on A.I.*, pages 351–356, 1988.

[Fahlman-90.2] S. E. Fahlman and C. Lebiere. The cascade-correlation learning architecture. Technical Report CMU-CS-90-100, School of Computer Science, Carnegie Mellon University, Pittsburgh, PA, 1990.

[Fahlman-91.2] S. E. Fahlman. The recurrent cascade-correlation architecture. Technical Report CMU-CS-91-100, School of Computer Science, Carnegie Mellon University, Pittsburgh, PA, May 17 1991.

[Fiesler-92.2] E. Fiesler. Partially connected ontogenic high order neural networks. Technical Report 92-02, IDIAP, Martigny, Switzerland, August 1992.

[Frean-90] M. Frean. The upstart algorithm: A method for constructing and training feedforward neural networks. *Neural Computation*, 2(2):198–209, 1990.

[Fritzke-91.1] B. Fritzke. Let it grow - self-organizing feature maps with problem dependent cell structure. In T. Kohonen et al., editor, *Artificial Neural Networks; (ICANN-91)*, volume 1, pages 403–408, Amsterdam, The Netherlands, 1991. North-Holland; Elsevier Science Publishing Company B.V.

[Fritzke-91.2] B. Fritzke. Unsupervised clustering with growing cell structures. In *Proc. IJCNN '91 Seattle*, volume II, pages 531–536, 1991.

[Fritzke-91.3] B. Fritzke and P. Wilke. Flexmap - a neural network with linear time and space complexity for the traveling salesman problem. In *Proc. IJCNN-91*, 1991.

[Golea-90] M. Golea and M. Marchand. A growth algorithm for neural network decision trees. *EuroPhysics Letters*, 12(3):205–210, 1990.

[Hagiwara-90.2] M. Hagiwara. Novel back propagation algorithm for reduction of hidden units and acceleration of convergence using artificial selection. In *Proc. IJCNN '90 San Diego*, volume I, pages 625–630, Ann Arbor, MI, 1990. IEEE Neural Networks Council; Edward Brothers.

[Hanson-90]   S. J. Hanson. Meiosis networks. In D. S. Touretzky, editor, *NIPS 2*, pages 533–541, San Mateo, CA, 1990. Morgan Kaufmann.

[Hassibi-93]   B. Hassibi and D. G. Stork. Second order derivatives for network pruning: Optimal brain surgeon. In *NIPS 5*, 1993.

[Hirose-91]   Y. Hirose, K. Yamashita, and S. Hijaya. Back-propagation algorithm which varies the number of hidden units. *Neural Networks*, 4:61–66, 1991.

[Honavar-88]   V. Honavar and L. Uhr. A network of neuron-like units that learns to perceive by generation as well as reweighting of its links. In D. Touretzky et al., editor, *Proc. of the 1988 Connectionist Models Summer School*, pages 472–484, San Mateo, California, 1988. Morgan Kaufmann.

[Honavar-89.2]   V. Honavar and L. Uhr. Generative learning structures and processes for generalized connectionist networks. *Connection Science*, 1:139–159, 1989.

[Ivakhnenko-68] A. G. Ivakhnenko. The group method of data handling - a rival of stochastic approximation. *Soviet Automatic Control*, 1:43–55, 1968.

[Kadir.-92.1]   V. Kadirkamanathan and M. Niranjan. Application of an architecturally dynamic network for speech pattern classification. In *Proc. of the Institute of Acoustics*, volume 14, 1992.

[Kadir.-92.2]   V. Kadirkamanathan and M. Niranjan. A function estimation approach to sequential learning with neural networks. Technical Report CUED/F-INFENG/TR.111, Cambridge University Engineering Department, Cambridge, UK, September 13 1992.

[Klotz-94]   J. Klotz and E. Fiesler. Ontogenic neural networks. *Expert Systems*. (to be submitted).

[Lansner-87]   A. Lansner and Ö. Ekeberg. An associative network solving the "4-bit ADDER problem". In M. Caudill and C. Butler, editors, *Proc. ICNN*, volume II, pages 549–556, San Diego, CA, 1987. SOS Printing.

[LeCun-90]   Y. Le Cun, J. S. Denker, and S. A. Solla. Optimal brain damage. In D. S. Touretzky, editor, *NIPS 2*, pages 598–605, San Mateo, CA, 1990. Morgan Kaufmann.

[Mezard-89]   M. Mézard and J.-P. Nadal. Learning in feedforward layered networks: The tiling algorithm. *J. of Physics: A*, 22(12):2191–2203, 1989.

[Nadal-89]   J.-P. Nadal. Study of a growth algorithm for a feedforward neural network. *Int. J. of Neural Systems*, 1(1):55–59, 1989.

[Reilly-82]   D. L. Reilly, L. N. Cooper, and C. Elbaum. A neural model for category learning. *Biological Cybernetics*, 45:35–41, 1982.

[Ring-93.2]   M. Ring. Sequence learning with incremental high-order neural networks. CSE technical Report AI 93-193, Dept. of Computer Sc., University of Texas at Austin, January 1993.

[Rujan-89.2]   P. Ruján and M. Marchand. Learning by minimizing resources in neural networks. *Complex Systems*, 3:229–241, 1989.

[Sirat-90]   J. A. Sirat and J. P. Nadal. Neural trees: A new tool for classification. *Network; Computation in Neural Systems*, 1:423–438, 1990.

[Tenorio-89]   M. F. Tenorio and W.-T. Lee. Self organizing neural networks for the identification problem. In D. S. Touretzky, editor, *NIPS 1*, pages 57–64, San Mateo, CA, 1989. Morgan Kaufmann.

[Thodberg-91]   H. H. Thodberg. Improving generalization of neural networks through pruning. *Int. J. of Neural Systems*, 1(4):317–326, 1991.

[Wen-92.1]   W. X. Wen, H. Liu, and A. Jennings. Self-generating neural networks. In *Proc. IJCNN - Baltimore*, 1992.

[Zollner-92]   R. Zollner, H. J. Schmitz, F. Wünch, and U. Krey. Fast generating algorithm for a general three-layer perceptron. *Neural Networks*, 5(5):771–777, September–October 1992.

Controlled Growth of Cascade Correlation Nets*

Lars Kai Hansen and Morten With Pedersen

CONNECT Electronics Institute B349

Technical University of Denmark, DK-2800 Lyngby, DENMARK

1 Introduction

The optimal selection of neural network architecture is a problem of significant theoretical and practical importance. The functional capacity of the network architecture determines the generalization properties and the computational complexity of the algorithm. For supervised training of feed-forward networks, current generalization theories tell us to minimize the functional capacity, *i.e.*, the number of weights subject to the constraints imposed by the rule implicit in the training examples see *e.g.*, (Hertz et al., 91). Several schemes have been proposed for *pruning* of the network architecture. The *optimal brain damage* method for network pruning proposed in (LeCun et al., 90) has been successful for pruning of classifiers as well as for pruning of regression networks for time series processing (Svarer et al., 93). The *optimal brain surgeon* (OBS) advanced in (Hassibi and Stork, 92) provides a refined estimate of the weight *saliency* by incorporating the effects of quadratic retraining; however, at the expense of significant additional computation. Likewise, growth algorithms are *legio*. Among these, Fahlman's *cascade correlation* stands out for solving hard classification problems (Fahlman and Lebiere, 89). In the standard implementation cascade correlation adds hidden units that are fully connected to all input units and to all previously added hidden units, thereby rapidly increasing the number of weights in the network with the obvious danger of overfitting.

In this communication we propose a controlled growth version of cascade correlation in which each hidden unit is pruned before it is inserted in the network. Since the cascade correlation scheme adds hiddens in the form of simple perceptrons the OBS scheme is feasible. For general networks the surgeon is problematic because it involves the full Hessian of the network costfunction. As the cascaded hidden units of the cascade correlation algorithm are trained according to a correlation measure rather than the usual least squares error function, we are forced to modify the surgeon in two directions: We give the modifications needed for the optimal brain surgeon operating with regularized costfunctions *e.g. weight decay*. Furthermore, we present the OBS saliencies for the cascade correlation measure. The weight decay regularization term plays a dual role in this work. First, it bias the learning process in the direction of "simple explanations"; secondly, it prevents the weights of the cascaded hidden units from diverging to infinite values that would otherwise maximize the correlation measure.

---

*This work is supported by the Danish Natural Science and Technical Sciences Research Councils. We thank Jan Larsen for valuable comments.

To illustrate the operation of the controlled growth scheme we apply it to a hard discrimination problem the so-called "Wieland spirals". In particular we show that we can grow cascade correlation nets with about two-thirds of the weights that we obtain from cascade correlation without pruning.

2 Optimal brain surgeon with weight decay

Let the training cost function (*e.g.* the mean squared error on the training set) be denoted $E(\mathbf{w})$. The augmented cost function for regularized learning is

$$C(\mathbf{w}) = E(\mathbf{w}) + \frac{1}{2}\mathbf{w}^T\mathbf{D}\mathbf{w}, \tag{1}$$

where $\mathbf{D}$ is the regularization matrix. For a simple weight decay $\mathbf{D}$ is proportional to the unit matrix: $\mathbf{D}_{jj'} = \alpha\delta_{jj'}$. The OBS saliency of a weight is defined as the change in *training error* as the weight is eliminated and the remaining weights retrained to the new minimum of (1). The saliency is estimated through the following steps (Hassibi and Stork, 92)

- Expand the cost function to second order around an extremum.
- Find the change of the remaining weights resulting from elimination of the j'th weight, $\mathbf{w}_j$, and retraining of the remaining weights in the quadratic approximation.
- Compute the associated change in training error.

Let the extremum of the unconstrained cost be denoted $\mathbf{w}_0$. The expansion of the cost function around this point reads

$$C(\mathbf{w}) = C(\mathbf{w}_0) + \frac{1}{2}\delta\mathbf{w}^T(\mathbf{A} + \mathbf{D})\delta\mathbf{w}, \tag{2}$$

where $\mathbf{w} = \mathbf{w}_0 + \delta\mathbf{w}$. The first order term vanish by the assumed optimality of $\mathbf{w}_0$. $\mathbf{A}$ is the second derivative matrix of the training error; the *Hessian* of the cost function is $\mathbf{H} = \mathbf{A} + \mathbf{D}$. Secondly, we derive the constrained extremum. Elimination of the j'the weight can be expressed as $\delta\mathbf{w}^T\mathbf{e}_j = -\mathbf{w}_0^T\mathbf{e}_j$, where $\mathbf{e}_j$ is the j'th unit vector. The constrained extremum is found using a Lagrange multiplier:

$$\tilde{C}_j(\mathbf{w}) = C(\mathbf{w}) + \gamma(\delta\mathbf{w} + \mathbf{w}_0)^T\mathbf{e}_j \tag{3}$$

the extremum is given by $\delta\mathbf{w}_j = -\gamma_j\mathbf{H}^{-1}\mathbf{e}_j$, with

$$\gamma_j = \frac{\mathbf{w}_0^T\mathbf{e}_j}{\mathbf{e}_j^T\mathbf{H}^{-1}\mathbf{e}_j} \tag{4}$$

We are now ready to compute the saliency of the j'th weight

$$\delta E_j(\mathbf{w}) = \gamma_j\mathbf{w}_0^T\mathbf{D}\mathbf{H}^{-1}\mathbf{e}_j + \frac{1}{2}\gamma_j^2\mathbf{e}_j^T\mathbf{H}^{-1}\mathbf{A}\mathbf{H}^{-1}\mathbf{e}_j. \tag{5}$$

Note that this expression reduces to the one derived by (Hassibi and Stork, 92), if $\mathbf{D} = 0$. For further generalizations of OBS see also (Larsen, 93).

3 Cascade correlation

The basic idea of the cascade correlation algorithm is iteratively to reduce the error made by the output unit by inserting hidden units that correlate (or anti-correlate) well with the error signal. By freezing the network while optimizing the new hidden unit candidate the algorithm avoids the moving targets problem of standard backprop (Fahlman and Lebiere, 89). The current error (difference between network output and teacher signal) of example k is called E^k and the output of the candidate unit called V^k. Fahlman used the correlation measure: $S = \left| \sum_k (V^k - \overline{V})(E^k - \overline{E}) \right|$ to quantify the quality of a hidden unit candidate computing: $V^k = \tanh \left(\sum_{j=1}^{n_I + n_H} w_j x_j^k \right)$ and n_I, n_H being the number of input units and previously inserted hidden units respectively. Further $\overline{V}, \overline{E}$ are the training set averages of V^k and E^k. The Hessian of the correlation measure S is approximated by

$$\frac{\partial^2 S}{\partial w_j \partial w_{j'}} = \mathrm{sign}_S \sum_k (E^k - \overline{E}) \frac{\partial^2 V^k}{\partial w_j \partial w_{j'}}, \tag{6}$$

with $\mathrm{sign}_S = \mathrm{sign} \left(\sum_k (V^k - \overline{V})(E_k - \overline{E}) \right)$. This matrix plays the role of the second derivative of the training error $(\mathbf{A})$ in the OBS scheme, except for the trivial change of objective: here we *maximize* the magnitude of the correlation. The corresponding "correlation saliencies" read $(\mathbf{H} = \mathbf{A} - \alpha \mathbf{1})$:

$$\delta S_j(\mathbf{w}) = \frac{1}{2} \frac{(\mathbf{w}_0^T \mathbf{e}_j)^2}{\mathbf{e}_j^T \mathbf{H}^{-1} \mathbf{e}_j} - \alpha \left(\frac{\mathbf{w}_0^T \mathbf{e}_j (\mathbf{w}_0^T \mathbf{H}^{-1} \mathbf{e}_j)}{\mathbf{e}_j^T \mathbf{H}^{-1} \mathbf{e}_j} - \frac{1}{2} \frac{\mathbf{e}_j^T \mathbf{H}^{-2} \mathbf{e}_j}{(\mathbf{e}_j^T \mathbf{H}^{-1} \mathbf{e}_j)^2} \right). \tag{7}$$

4 Experiments and concluding remarks

The generic benchmark for cascade correlation networks is the spirals classification problem involving classification of 192 (x, y) points forming two intertwined spirals (Fahlman and Lebiere, 89). On this problem the algorithm outperforms standard algorithms like feed-forward nets with backprop. Our primary interest here is to show that the controlled growth algorithm can reduce the number of weights needed to implement the spirals. It turns out that the performance of cascade correlation is quite sensitive to the search algorithms used for optimization of the hidden units. We use a dual scheme: Initially we use a simple standard backprop optimizer for the hidden weights, when improvement stops we refine the solution by a few second order Newton steps. Our results are presented in Figure 1 accumulating 20 runs of standard cascade correlation and 20 runs of the controlled growth algorithm. We trained a pool of 10 candidate units at each level in the cascade; the candidate with optimal correlation was pruned by OBS. The weight decay parameter was set to: $\alpha = 10^{-4}$. Pruning was terminated when the magnitude of the correlation had dropped by 5%. For the fully connected cascades all twenty runs succeeded with a number of hidden units ranging from 11 to 17. Likewise, all pruning runs succeeded with a number of hiddens in the range 10-16. The average number of weights used by the fully connected nets is 146, while the average number of weights in the pruned nets is 103.

In conclusion we have derived OBS saliency expressions for weight-decay regularized training and for training with Fahlman's correlation measure. The proposed controlled growth version of cascade correlation produces nets with approximately two-thirds of the number of weights used by fully connected nets.

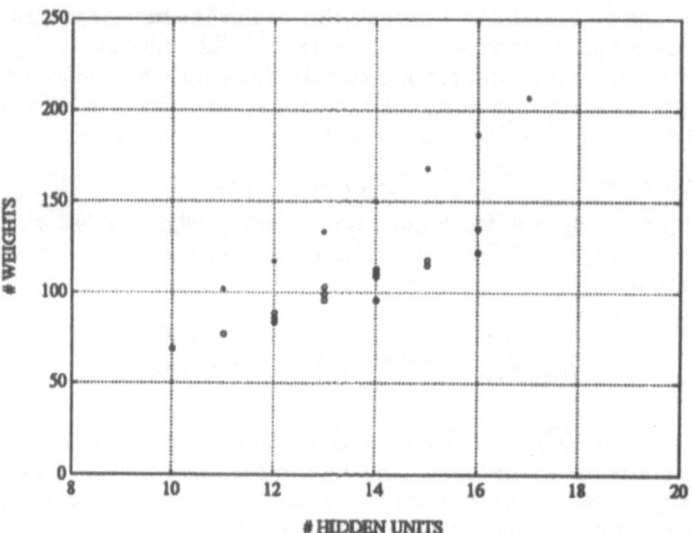

Figure 1: Number of weights versus number of cascaded hidden units for fully connected (*) and OBS-pruned cascade correlation networks(o). Twenty runs of both algorithms resulted in networks with varying number of hidden units.

References

Fahlman, S.E. and Lebiere, C.: *The Cascade-Correlation Learning Architecture.* In NIPS 2, Ed. D.S. Touretzsky, 524-532. San Mateo, Morgan Kaufmann (1990).

Hassibi, B. and Stork, D.G.: *Second Order Derivatives for Network Pruning: Optimal Brain Surgeon.* In NIPS 5, Eds. S.J. Hanson, et al., 164. San Mateo, Morgan Kaufmann (1993).

Hertz, J., Krogh, A., and Palmer, R.G.: *Introduction to the Theory of Neural Computation.* Addison Wesley, New York (1991).

Larsen, J.: *Design of Neural Networks Filters.* PhD thesis, Electronics Institute, Technical University of Denmark (1993).

Le Cun, Y., Denker, J.S., and Solla, S.A.: *Optimal Brain Damage.* In NIPS 2, Ed. D.S. Touretzsky, 598-605. San Mateo, Morgan Kaufmann (1990)

Svarer, C., Hansen, L.K., and Larsen, J.: *On Design and Evaluation of Tapped-Delay Neural Network Architectures.* The 1993 IEEE Int. Conference on Neural Networks, San Francisco. Eds. H.R. Berenji et al., 45-51, (1993).